Primate Evolution and Human Origins

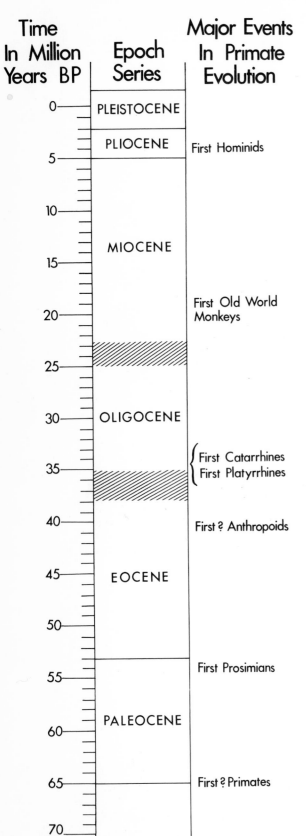

Time In Million Years BP

Epoch Series

Major Events In Primate Evolution

Time (Million Years BP)	Epoch Series	Major Events
0	PLEISTOCENE	
5	PLIOCENE	First Hominids
10 – 20	MIOCENE	First Old World Monkeys
25 – 35	OLIGOCENE	First Catarrhines / First Platyrrhines
40		First ? Anthropoids
45 – 50	EOCENE	
55		First Prosimians
60 – 65	PALEOCENE	First ? Primates
70		

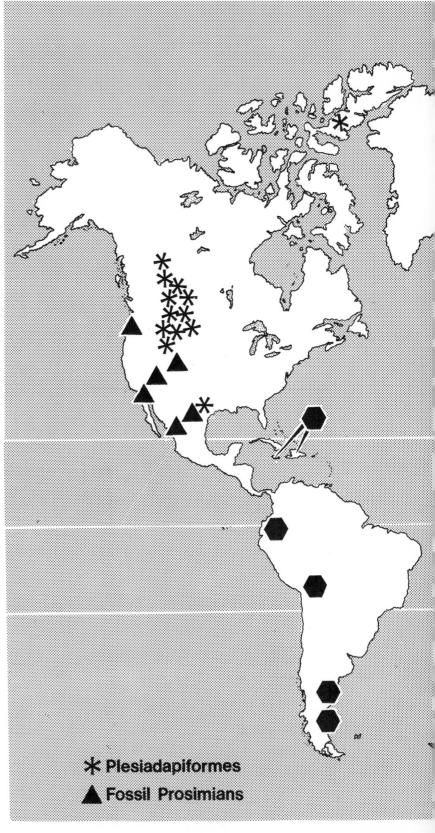

✳ Plesiadapiformes

▲ Fossil Prosimians

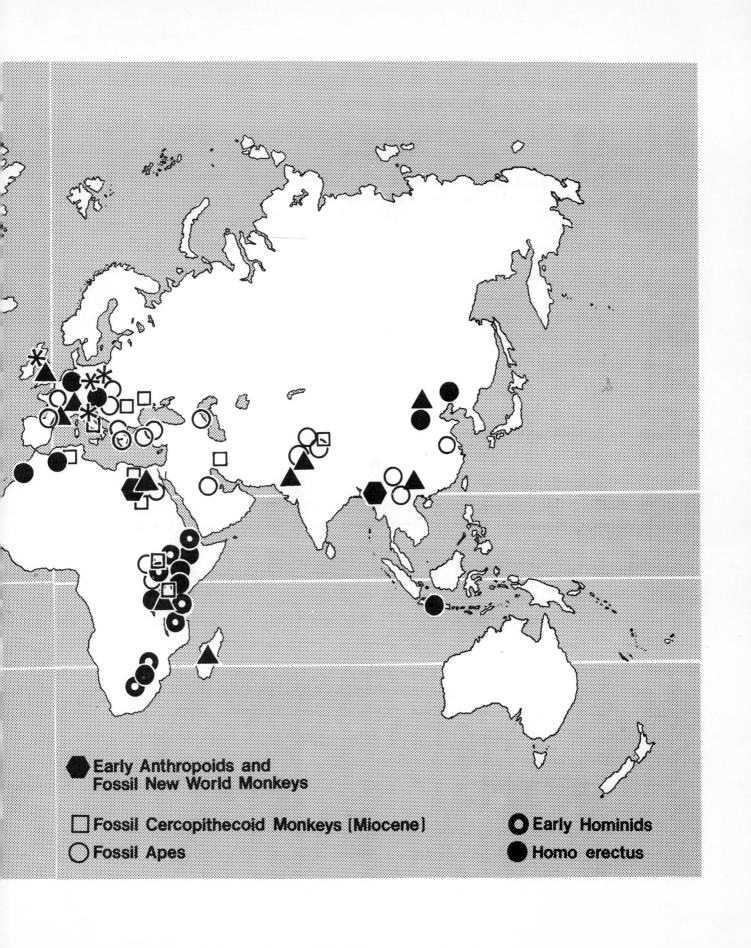

Early Anthropoids and Fossil New World Monkeys

☐ **Fossil Cercopithecoid Monkeys (Miocene)**

○ **Fossil Apes**

◉ **Early Hominids**

● **Homo erectus**

FOUNDATIONS OF HUMAN BEHAVIOR

An Aldine de Gruyter Series of Texts and Monographs

Edited by
Sarah Blaffer Hrdy, *University of California, Davis*
Melvin Konner, *Emory University*
Richard W. Wrangham, *University of Michigan*

Primate Evolution and Human Origins

RUSSELL L. CIOCHON

JOHN G. FLEAGLE

ALDINE DE GRUYTER
New York

ALDINE DE GRUYTER
(formerly Aldine Publishing Company)
A Division of Walter de Gruyter, Inc.
200 Saw Mill River Road
Hawthorne, New York 10532

Library of Congress Cataloging-in-Publication Data
Primate evolution and human origins.

   (Foundations of human behavior)
   Bibliography: p.
   1. Primates—Evolution. 2. Human evolution.
I. Ciochon, Russell L. II. Fleagle, John G. III. Series.
[QL737.P9P67249 1987]   599.8′ 0438   86-26570
ISBN 0-202-01175-5 (pbk.)

Printed in the United States of America
10 9 8 7 6 5 4 3 2 1

*Cover illustration by Stephen Nash, State University of
New York, Stony Brook.* Reconstruction of the early Miocene
paleoenvironment of Rusinga Island, Western Kenya, shows
the diversity of fossil primates.

Preface

The fossil evidence for the evolution of the primates has increased dramatically over the past two decades. This increase has engendered many new interpretations concerning primate ancestry. In some cases new discoveries and syntheses have clarified evolutionary histories; in others they have only heightened controversy. These collected readings bring together for the first time a selection of publications from the past twenty years to provide students, teachers, and researchers with material that has shaped current views of primate evolution and human origins.

We present these original articles in mostly unedited form so that students and teachers alike can judge directly the significance of the new discoveries and their interpretations. We have selected articles covering the entire scope of primate and human evolution with particular emphasis on the fossil record. Naturally, in a book of only 400 pages, it is impossible to include every major article of the last twenty years. It would be unfair not to acknowledge that our selection was in part determined by article length. Because it was important to cover the entire field of primate evolution within a predetermined page limit to make the book affordable to a wide audience, many important articles could not be included. However, most of these articles are listed in the bibliography at the end of the book. We urge students to make use of this section.

Primate Evolution and Human Origins is divided into seven parts that survey different aspects of the field. Each part includes articles that represent conflicting theories or interpretations of the fossil evidence for various aspects of primate and human evolution. We have ordered the articles chronologically to show how the debates and opinions on primate evolution have developed. Each part includes an introduction that focuses on the major issues in that section.

Part I, "Primate Origins," presents a variety of viewpoints dealing with the origin of the Order Primates, the problems associated with defining the Order, and the adaptive radiation of the earliest primates. Part II, "Evolution of Prosimians," discusses the radiation of prosimian primates during the Eocene epoch and also more recently and the possible relationships between extinct and living radiations. Part III, "Anthropoid Origins and New World Monkeys," reviews the conflicting theories concerning the origin of the higher primates from morphological, paleontological, and paleogeographical perspectives and focuses on the origin and diversification of South American primates. Part IV, "Evolution of Old World Monkeys and Apes," summarizes current knowledge concerning the implications of the Fayum primate radiation, the evolution of Old World monkeys, and the phyletic placement of the early Miocene apes. Part V, "*Ramapithecus* and Human Origins," catalogs the arguments for and against *Ramapithecus* as the earliest hominid and finishes with a viewpoint detailing its proposed orangutan affinities. Part VI, "Early Hominids," presents several different theories concerning the evolution and diversification of hominids in the Plio-Pleistocene. Finally, Part VII, "Diverse Approaches in Human Evolution," provides a widely varied collection of methodological and theoretical articles concerned with ape and human evolution at a variety of levels.

We gratefully acknowledge Dr. Ingrid Krohn for critical assistance and advice during the early planning stages of this book. Without her foresighted judgment this volume would not have been published. We also acknowledge Dr. F. Clark Howell for his support and encouragement. Joanna Viola and especially Stephanie Rippel contributed herculean efforts toward the completion of the bibliography, and Christa Sadler helped with typing and correction of the proofs. We thank artist Stephen Nash for drawing the endpaper illustration, section maps, and cover illustration. We also thank Ms. Joy Myers for her careful assistance with proof reading. For financial support during the preparation of this book we acknowledge funding from the L. S. B. Leakey Foundation and the Lokey Bequest (for R.L.C.) and the National Science Foundation and a John Simon Guggenheim Fellowship (for J.G.F.). Finally, we wish to note that the order of names on the title pages and preface of this work was determined alphabetically.

Russell L. Ciochon
John G. Fleagle

About the Authors

RUSSELL L. CIOCHON received his Ph.D. from the University of California at Berkeley. He has taught anthropology at the University of North Carolina at Charlotte and has held research appointments at the Institute of Human Origins in Berkeley and the State University of New York at Stony Brook. Dr. Ciochon is the author of numerous research papers and two edited books: *Evolutionary Biology of the New World Monkeys and Continental Drift* and *New Interpretations of Ape and Human Ancestry.* He has also organized paleoanthropological field research projects to Burma, India, China and Vietnam.

JOHN G. FLEAGLE is Professor of Anatomical Sciences at the State University of New York at Stony Brook. He has published numerous research papers on many aspects of primate evolution and comparative anatomy and has conducted primatological investigations in Asia, Africa and South America.

Contents

Contents

PLESIADAPIFORM PRIMATES

Part I

Primate Origins

The origin of the order Primates has been an area of considerable activity during the past two decades (e.g., Cartmill, 1982). Earlier theories regarding primate origins and early evolution have been questioned from a variety of standpoints, and new discoveries and methods of analysis have given us a much better understanding of the morphology and habits of the earliest fossil primates. The chapters in this part provide a sampling of the issues and controversies surrounding the origin and early evolution of primates.

In the first chapter, Farish Jenkins (1974) describes the locomotor habits of the living tree shrews based on both a survey of their natural history and his own radiographic studies of the locomotion in one species (*Tupaia glis*). Studies of the locomotor behavior of living primates in both natural and more technological settings have become important tools for interpreting the locomotor behavior of fossil primates (see also Tuttle, Chapter 42; Fleagle, 1979; Stern and Susman, 1983; Jenkins and Fleagle, 1975). Based on his analysis of tree shrew locomotion, Jenkins then focuses on the question of whether early mammals (and the ancestors of the earliest primates) were arboreal or terrestrial. He concludes that his debate is probably a meaningless one, because for small mammals such as tree shrews, both terrestrial and arboreal habits entail similar types of limb function.

The role of arboreal locomotion in primate evolution is also the focus of the second chapter, by Matt Cartmill (1974c). Whereas Jenkins questions the role of arboreality in early mammal evolution, Cartmill examines the role of arboreality in primate evolution. He reviews the arguments of Wood-Jones (1916) and Le Gros Clark (1934, 1959) that many distinctive primate features, such as stereoscopic vision, reduced olfaction, and grasping hands, are the evolutionary result of arboreal habits and finds that they are unconvincing and lack generality. In other groups of mammals, arboreal habits are not associated with such morphological adaptations. As an alternative, Cartmill suggests that primate stereoscopic vision and manual dexterity reflect an original adaptation for visual predation. Since the earliest fossil "primates" from the Paleocene Epoch lacked many of the cranial and locomotor features that characterize living primates, he argues that they should not be considered as part of the order.

Chapter 3, by F. S. Szalay (1975a), is primarily a response to the suggestions of Cartmill and others that the "primates" from the Paleocene Epoch are not really primates at all. Szalay argues that there was no absolute boundary between primates and early mammals, just as there can be no absolute boundary within any continuously evolving lineage; the boundary depends on the anatomical features chosen to define the order. In Szalay's view, the Paleocene species are primates because they resemble living primates in the structure of their auditory region (but see MacPhee et al., 1983) and dentition (Gingerich, 1974a). He also suggests that these archaic primates were probably herbivorous and arboreal, despite their differences from living species.

The fourth chapter, by Richard F. Kay and M. Cartmill (1977), compares the dental and cranial anatomy of one archaic primate, *Palaechthon nacimienti,* with that of a wide range of living primates and other mammals in order to reconstruct the behavior of this Paleocene species. The authors conclude that this species was probably nocturnal, insectivorous, and, on the basis of its size, terrestrial. In their reconstruction, *Palaechthon* and many other Paleocene primates foraged for insects among leaf litter on the forest floor (but see Szalay, 1981).

The final chapter, by K. D. Rose and J. G. Fleagle (1981), provides a brief overview of the archaic primates from North America and a summary of the controversies surrounding the group. Like Szalay, they note that the boundary between the earliest primates and insectivores is largely an arbitrary one, and they argue for a very broad definition of the order based on dental similarities.

Few of the topics discussed in this part have been resolved, and some have taken on new twists in recent years and months, even while this collection was in preparation. Most of the contributors to this part agree that tree shrews were not phyletically related to living primates and that similarities between these groups were almost certainly due to evolutionary convergence or retention of primitive mammalian features (see particularly Martin, 1968b; Cartmill, 1982). However, many researchers maintain that the tree shrews are a group of nonprimate mammals most closely related to our own order, suggesting that some of the primate-like features of the tree shrews may actually indicate a phyletic relationship at a supraordinal level. Others feel that the living orders of mammals are far too disparate to permit any conclusions about higher level relationships (see papers in Luckett, Ed. 1980; also Fleagle, 1981).

The primate affinities of the Plesiadapiforms, questioned by some of the authors in this part but supported by others, have come under renewed suspicion as a result of recent investigations of the pattern of vascular canals in the auditory region (see MacPhee, et al., 1983). The issue is clearly not resolved at present, and new fossil discoveries from the early epochs of the age of mammals will almost certainly lead to further revelations about this group's origins. This is especially true in the case of the earliest primate, *Purgatorius,* known from beds on the Cretaceous/Tertiary boundary in Montana. New discoveries will more precisely determine its phyletic affinities as well as help elucidate the features currently used to define the Plesiadapiforms as "primates."

1

Tree Shrew Locomotion and the Origins of Primate Arborealism

F. A. JENKINS, JR.

The family Tupaiidae (tree shrews) ranges from India and southern China throughout most of southeastern Asia. Tree shrews are found on Hainan, Sumatra, Java, Borneo, and Mindanao (Philippines) in addition to other smaller island groups, but they are not found southeast of Wallace's Line. Their typical habitat is montane and tropical rain forest, although few species are as strictly arboreal as their common name implies.

Squirrel-sized or smaller, tree shrews are represented by five genera (*Anathana, Dendrogale, Ptilocercus, Tupaia,* and *Urogale*) and about fourteen species. *Ptilocercus* is sufficiently distinct to warrant a separate subfamily (Ptilocercinae), the other genera being grouped in the Tupaiinae (Lyon, 1913). Although there is some disagreement on species taxonomy in *Tupaia* (e.g., contrast Martin, 1968a with Napier and Napier, 1967), the major taxonomic question concerns the phylogenetic relationship of tupaiids as a group. Early anatomical studies, particularly those of Carlsson (1922) and Le Gros Clark (e.g., 1924b, 1925, 1926, 1960), presented evidence that tupaiids were more closely related to prosimians than to insectivores. Recent reexamination of this evidence, in addition to histological and biochemical studies, has not confirmed this relationship. Thus the current taxonomic status of tree shrews is uncertain (for a review of literature on this question, see Luckett, 1969, and Sorenson, 1970). Some authors rank tupaiids in a separate order Tupaioidea (Martin, 1966), while others retain them among the Primates (Napier and Napier, 1967) or Insectivora (Hill, 1953b).

The uncertainty surrounding tupaiid phylogeny is a consequence of an inadequate fossil record. At present early Tertiary mammals are represented principally by teeth and fragmentary jaws. Dental similarity between tupaiids, marsupials, and insectivores makes it unlikely that an early Tertiary fossil tupaiid will be positively identified without the recovery of more of a skull than the dentition. A number of fossil genera have been proposed as tupaioid but these taxonomic assignments are doubtful (McKenna, 1966). Van Valen (1965) identifies *Adapisoriculus minimus* and *A.? germanicus* from the Paleocene as tupaiids, although other authors had variously identified the genus as insectivore or marsupial. Szalay (1968a), however, considers that these species "may or may not be tupaiids" and proposes that another Paleocene (Lutetian) genus, *Messelina,* is at least as similar to tupaiids as *Adapisoriculus* in lower dentition and more similar in upper dentition. Thus there is no consensus as to which fossil forms, if any, represent the tree shrew lineage.

Despite taxonomic ambiguities and the apparent remoteness of a tupaiid-primate relationship, there are several reasons why tree shrews merit inclusion among studies of primate locomotion. First, understanding anatomical and behavioral aspects of arboreal versus terrestrial activity in different tree shrew species may clarify similar phenomena among true primates. Second, tree shrews represent one possible "model" of a primitive primate or placental mammal. Gregory (1910, 1913) was among the first to allude to tupaiids in this sense; the increase in knowledge of early mammals, primates, and tree shrews since that time has not diminished the potential usefulness of this concept. Critics of the "model" approach to paleontology usually cite differences or specializations which purportedly invalidate comparison between the fossil species and the living "model." In this case, however, use of tree shrews as an experimental "model" does not imply equivalency with early Tertiary insectivores or primates, for differences of varying degree are well known; rather, there are sufficient similarities to support the expectation that a better understanding of form-function problems in tree shrews may contribute to understanding similar problems among early insectivores and primates. Simpson (1965a) summarized the point:

> ... The tupaioids arose, and still stand, somewhere between the earliest placental (nominally insectivore) stem and that of the Primates. Their reference to one group or the other is in part arbitrary or semantic. Use of them to represent the earliest primate or latest preprimate stage of evolution is as valid and useful, and subject to as much caution, as is any use of living animals to represent earlier phylogenetic stages.

As McKenna (1966) pointed out, one trend of current opinion is to regard "tupaiids as leptictid-like insectivores (*sensu lato*) with special similarity among primates to Malagasy lemurs, *Adapis* and *Notharctus*. Among living nonprimates the tupaiids are apparently the closest primate relatives, and these conclusions in no way lessen the value of tupaiids to primatology."

TREE SHREW LOCOMOTION: BEHAVIORAL OBSERVATIONS

From the very first zoological descriptions (Diard, 1820; Raffles, 1821), tupaiids have been associated with a squirrel-like habitus in name ("tree shrew"; the generic root "tupai" is the Malay word for squirrel) as well as description. At one time the tupaiid-squirrel comparison was satisfactory as a first approximation, but the association is vague and even misleading in view of the diversity of habitats among both squirrel and tree shrew species. However, there is some evidence that a mimicry relationship exists between certain species of *Tupaia* and squirrels of the genera *Sciurus* and *Funambulus* (Banks, 1931).

Discussions of tupaiid locomotion usually attempt to define their habitat and, in this context, have tended to focus on the question of an arboreal versus a terrestrial existence. As is well known in primatology, this distinction is seldom absolute; the question usually involves the relative percentage of time spent on the ground, in trees, or in some intermediate habitat (e.g., undergrowth). Little such data have been gathered for free-ranging tree shrew species, possibly because of their small size and shyness, and notwithstanding their apparent tendency to territorial restriction. Some reports of the preponderance of terrestrial over arboreal activity in tree shrews may be biased because of the animal's greater visibility at ground level; conversely, other reports may be biased in favor of an arboreal existence because of an arboreal escape reaction when confronted with an observer. Since most accounts of tree shrew activity are brief commentaries based on incidental rather than programmed observations, they should not be regarded invariably as definitive. However, it is certain that considerable locomotor differences exist between tree shrew genera, and even between species of *Tupaia*.

Observations on Tree Shrews in the Wild

Relatively few species are considered to be principally arboreal in the sense of seldom descending to the ground and being especially adept in arboreal locomotion. Kloss (1903, 1911) reported that *Tupaia glis longicauda* from East Perhentian Island (off the east coast of the Malay Peninsula) and *T. nicobarica* from the Nicobar Islands are truly arboreal. Bartels (1937) implied that *T. javanica* is arboreal but did not document his observation; Müller (quoted in Horsfield, 1851) claimed that this species nests some distance above the ground but feeds both on the ground and in trees. The only other certainly arboreal tupaiine is *T. minor* (Le Gros Clark, 1927). Davis (1962) reported that this species occupies small trees and vines of the lower middle story between 5 and 50 feet above the ground, and similarly Lim (1969) referred to *T. minor* as a canopy and undercanopy animal which ". . . is often seen running upside down along thin branches of tall trees." Although Lim (1969) claimed that this species seldom descends to the ground, Banks (1931) reported that they are often seen running on the ground and on fallen tree trunks. *Tupaia gracilis,* somewhat larger but otherwise similar to *T. minor,* is apparently also arboreal on the basis of a few field observations (Davis, 1962). The only other certainly arboreal tupaiid is *Ptilocercus lowii* (the feather- or pen-tailed tree shrew); its nocturnal habits are unique among tree shrews (Le Gros Clark, 1926; Davis, 1962; Muul and Lim, 1971; Lim, 1967, 1969).

Two tupaiines are predominantly terrestrial. The Philippine tree shrew, *Urogale everetti,* apparently nests in the ground or in cliffs and is active on the forest floor; nevertheless it is reputed to be an excellent climber (Wharton, 1950). *Tupaia (Lyonogale) tana,* the so-called terrestrial tree shrew, has similar habits (Banks, 1931; Davis, 1962). The extent to which tanas will climb is uncertain. Davis (1962) observed that they ". . . are rarely seen even in the lowest branches of the trees. When pressed they do not resort to climbing, but escape on the ground." However, observations on captive specimens (see below) are equivocal in regard to their climbing ability and propensity.

On the basis of relatively limited field observations, most other tree shrew species appear to be in some measure both arboreal and terrestrial. Most accounts favor the view that the majority of activity is on the forest floor or at least in low undergrowth. Banks (1931) observed that ". . . though they can and do run about in the trees most species spend the greater part of their time on the ground running over and under fallen tree trunks." Of *Tupaia (glis) ferruginea,* Robinson and Kloss (1909) write that ". . . it is quite exceptional to see one anywhere than on the ground, among the roots of trees and on low bushes." Similar descriptions were made by Banks (1931), Bartels (1937), Davis (1962), Lim (1969), and Ridley (1895), the latter noting that *T.*

glis, when alarmed, will partly ascend a tree but will usually descend if further pressed to escape. Pfeffer (1969) claims that the arboreal-terrestrial behavior of the species varies somewhat according to the forest type: "Le Tupaïa, qui en forêt dense ou en zone broussailleuse (forêts secondaires jeunes) est aussi bien terrestre qu'arboricole (sans monter toutefois dans les strates supérieures) est essentiellement terrestre et inféodé aux arbres morts et aux termitières en forêt claire."

Information is almost totally lacking for *Dendrogale murina* and *D. melanura* (the smooth-tailed tree shrews). The type specimen of *D. melanura* was procured in stunted montane jungle, and this species appears to be restricted to higher elevations. *Anathana wroughtoni*, the Madras tree shrew, is known to climb when pursued, and Verma (1965) suggested that certain anatomical features relate to an arboreal existence; definitive studies on these species' habits, however, remain to be made.

Observations on Captive Tree Shrews

Most studies on captive tupaiids confirm field observations as to terrestrial versus arboreal activity patterns. One exception is noteworthy, however. The terrestrial tree shrew (*Tupaia tana*) has been reported in the wild to be almost exclusively terrestrial (Banks, 1931; Davis, 1962). Sorenson and Conaway (1964) found that their captive tanas were clumsy climbers relative to other species, and that they often fell when descending. Yet in view of the fact that Sorenson and Conaway's tanas seldom ran and appeared to be rather lethargic—an unusual pattern for tree shrews generally and tanas in particular (Banks, 1931; Davis, 1962)—the observed pattern may have been abnormal. However, Sorenson and Conaway (1966) noted that tanas will engage in long chases that often result in a reversal of hierarchy. The captive tanas of Banks (1931) ". . . climbed well, were thoroughly at home on horizontal branches and slept at night in the top of their cage." Similarly, Schlott (1940) observed that "Die Tanas sind echte Baumtiere und in ihren Bewegungen unglaublich gewandt und jäh. Sie halten sich fast durchweg im Gezweige ihres Kletterbaumes auf . . . Auf den Erdboden kommen sie eigentlich nur, um zu trinken oder Nahrung zu holen." Steinbacher (1940) also reported that captive tanas are skillful climbers. Although the nature of the discrepancy in the foregoing accounts remains to be resolved, it appears nevertheless that *T. tana* has more ability in arboreal activity than is usually exhibited in the wild. The behavior of *Urogale everetti* is probably comparable. This species nests in the ground and is active on the forest floor but is also a good climber (Wharton, 1950). Captive specimens, although spending most of their time on the cage floor, climb frequently and skillfully (Polyak, 1957), yet to what extent free-ranging animals exhibit this behavior is uncertain.

In a comparative behavioral study of four species of *Tupaia* in captivity, Sorenson and Conaway (1964) confirmed field reports that *T. minor* is the most arboreal and *T. tana* the least. *T. gracilis* tends also to be arboreal, but less so than *T. minor*. *T. (glis) longipes* appears to be behaviorally intermediate; although it readily leaps up to 8 feet with excellent depth perception and sleeps in raised nest boxes, it often feeds and rests at ground level. Sorenson (1970), in a thorough behavioral study of eight tree shrew species, ranked each species according to its apparent affinity for arboreal activity. *Tupaia minor* and *T. gracilis* are the most arboreal (1), and *T. tana* is the most terrestrial (5). In comparison to *T. glis* and *T. montana,* which are intermediate (3), *T. longipes* is behaviorally more arboreal (2), and *T. chinensis* and *T. palawanensis* more terrestrial (4). It is interesting, however, that all species are capable of walking inverted along the underside of branches, although the arboreal forms exhibit this behavior most often. Sorenson observed other locomotor differences between species. At lower speeds, *T. gracilis* tends to hop whereas *T. longipes* uses a walking gait; *T. chinensis* moves more slowly and is less nervous than *T. longipes*. When descending vertically, *T. minor* and *T. gracilis* (as well as *T. glis*—personal observation) turn the feet around to a posteriorly directed position; in this posture, the claws are used as hooks to hang by (Fig. 6). *Tupaia chinensis* and *T. tana* do not reverse their feet in this situation as much as other species do, a trait that Sorenson believes may reflect a more terrestrial adaptation.

Other observers of captive *T. glis* also conclude that this species is both arboreal and terrestrial, but the precise balance of this dual behavior is not clear. Hofer's (1957) detailed study concluded that "Nach allen unseren Beobachtungen möchte ich *Tupaia glis* nur eine sehr beschränkte arboricole Lebensweise zubilligen." However, Kaufman's (1965b) *T. glis*, although spending most of their time on the ground, were frequently active in 15 feet tall trees in a large enclosure. Particularly interesting are Vandenbergh's (1963) data on five animals (Table I) which indicate that there may be considerable lability in arboreal-terrestrial behavioral patterns. Escape reactions apparently differ between males and females, the males ascending for display and vocalization, the females fleeing at ground level to secluded nesting sites (Sprankel, 1961; Vandenbergh, 1963). Resting on branches and preference for elevated nest boxes is not uncommon for captive animals.

Table I *Frequency of Arboreal and Terrestrial Activity of Tupaia glis*[a]

| Sex | Time spent | | Time in trees (%) | Mean changes per hour | Duration of observation (hour) |
	On ground (minutes)	In trees (minutes)			
♂	315.5	464.5[b]	59.6	27.8[c]	13
♂[d]	78.8	161.2[b]	76.2	6.3	4
♀	509.3[b]	270.7	34.8	16.2	13
♀[d]	401.2[b]	138.8	26.6	12.6	10
♀	768.1[b]	11.9	1.5	0.9	13

[a]From J. G. Vandenberg, 1963, *Folia Primatol.* 1, 199–207. Courtesy S. Karger and the author.
[b]Significantly more time at <0.001 level (X[2] test).
[c]Significantly more changes per hour than any other tupaia at <0.05 level (t test).
[d]Male and female died during the study.

In summary, the evidence for locomotor behavior in both captive and wild tree shrews indicates a moderate diversity of habitat preference. Clearly some tree shrew species are more arboreal than others; other species are more or less terrestrial. Perhaps the most significant fact is that all tree shrew species can climb, and at least occasionally, if not frequently, do so. Conversely, even the most arboreal species may be found on ground level. In tupaiids, the range of adaptive types is probably too subtle to be understood in terms of the gross categories of arborealism and terrestrialism, and as Sorenson (1970) has pointed out, there is need for further study of the behavior and ecology of wild tupaiids.

LOCOMOTOR PATTERNS OF *TUPAIA GLIS*

Rapid, scurrying movements punctuated by short pauses, often described as "nervous behavior," have been noted in almost every description of tree shrew activity. Sprankel (1961), Vandenbergh (1963), and Hofer (1957) described this behavior in detail, especially with regard to the agility and acrobatics of captive specimens.

The six *Tupaia glis* used in the present study were maintained in cages approximately 45 × 75 × 75 cm. In general, their locomo-

tor behavior conformed to that reported in previous studies; therefore, further detailed description based on visual impressions will be omitted. However, several characteristics of their locomotor behavior are worth emphasizing. Individuals of *Tupaia glis* are active climbers, moving about on branches and the cage mesh as frequently as they do on the cage floor. Commonly, they establish a repetitive locomotor pattern which involves some acrobatic maneuver such as springing off a vertical wall or turning about on a narrow branch. *Tupaia glis* is particularly adept at rapid locomotion in an environment which necessitates abrupt changes in direction or elevation, and is capable of upward leaps of approximately one meter.

Footfall Patterns and Speed

According to Hildebrand (1967), symmetrical gaits (footfalls of each pair of feet spaced evenly in time) are common in walking primates but are apparently rare in tree shrews. Hildebrand's conclusions, based on limited observations of terrestrial locomotion in *Tupaia (glis) belangeri* and *Urogale everetti*, may now be supplemented with data from over 600 feet of film of terrestrial and arboreal locomotion in six captive specimens of *T. glis*. Footfall patterns are not specific for arboreal or terrestrial surfaces and vary principally with speed. Movement during exploratory activity is interrupted by momentary pauses, and the pattern of footfalls is highly asymmetrical (Fig. 1A). The asymmetry is accentuated by the tendency for any given foot to make two half-steps in sequence (see LF, Fig. 1A). In a nonexploratory situation, such as a linear pathway along a narrow branch, the pattern may approach symmetry (Fig. 1B) and approximate Hildebrand's category of a fast, lateral-sequence, diagonal-couplets walk. With an increase in speed, the foot-substrate contact time decreases, but the footfall pattern remains essentially the same

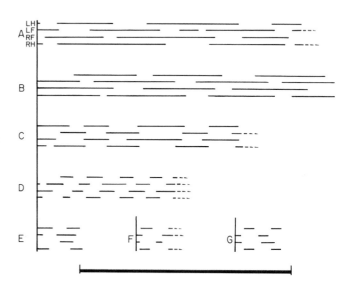

Figure 1 *Typical gait patterns of* Tupaia glis *at various speeds. Lines LH, LF, RF, and RH represent foot substrate contact of the left hind, left front, right front, and right hindfoot, respectively. Gaps in the lines represent periods when a foot is off the substrate. A 1-second interval is shown for scale. (A) Halting walking on a flat surface. (B) Slow walk on a horizontal branch 14 mm in diameter. (C) Fast walk on a flat surface. (D) Bounding run on a horizontal branch 17 mm in diameter. (E–G) Gait variations at top speed, exhibited on both terrestrial and arboreal surfaces that are more or less horizontal.*

(Fig. 1C) up to speeds of approximately 150–175 cm/second. With greater speed, *T. glis* begins a bounding hop with the hindfeet functioning more or less synchronously (Fig. 1D). At speeds greater than approximately 225 cm/second, a phase occurs in which no foot is in contact with the substrate (Fig. 1E–G); the contact pattern of the forefeet may vary from more or less synchronous (Fig. 1E) to disynchronous (Fig. 1F), but the hindfeet are almost always nearly synchronous. Two male *T. glis,* chased over a 7.5 meter linear course to provoke maximum speeds, averaged approximately 13.6 km/hour over thirteen trials; the slowest speed was 10.4 km/hour, the fastest 18.3 km/hour. This rate is comparable to that reported for the red squirrel (*Tamiasciurus hudsonicus,* 14.5 km/hour) but slower than the maximum speed of the grey squirrel (*Sciurus carolinensis*), 27.2 km/hour (Layne and Benton, 1954).

Postural Patterns in Locomotion

The following description of skeletal movement is based principally on cineradiography at speeds from 48 to 150 frames/second, but also upon black and white cine, single exposure X-ray and photographical films. For descriptive purposes, the locomotory stride of a limb is divided into four phases. The initial propulsive movement is phase I; the limb, ceasing forward movement and contacting the substrate, begins a propulsive thrust. At phase III the limb is maximally extended at the completion of propulsive thrust. Phase II represents the sequence intermediate between phase I and III in the propulsive thrust. During phase IV the limb is off the substrate and moves forward to begin another propulsive thrust.

Limb Movements   During exploratory activity (footfall pattern as in Fig. 1B and C), the forelimb in phase I is commonly positioned as in Fig. 2A and B (right side). The scapula is inclined dorsoposteriad, with the scapular spine at about 35° to the vertical (Fig. 2A, right side). The distal end of the humerus is abducted 30° in dorsoventral view (Fig. 2B) and depressed 45° in lateral view. The forearm intersects the horizontal at 30°; the manus initially contacts the substrate anterolateral to the glenoid. Variation from this common posture principally involves the degree of humeral depression and concomitantly the orientation of the forearm. The humerus may depress as much as 55° or may be essentially horizontal during phase I, with the forearm approaching the horizontal or vertical planes, respectively. During phase II the distal humerus is adducted and elevated, so that by phase III

the long axis is within about 15° of the parasagittal and the distal end is above the level of the glenoid (Fig. 2A and B, left side). The position of the manus in phase III is variable; seen in dorsoventral view, it may lie directly behind the elbow joint (Fig. 2B), posteromedial or posterolateral to the joint, or directly beneath the joint.

When resting, *Tupaia glis* adducts the humerus closer to parasagittal than is normal for locomotor postures (Fig. 3), but in other respects this stance resembles a phase II posture.

During phase I the mechanical axis of the femur (a line from head through condyles) typically is abducted from the parasagittal by 20 to 30° (Fig. 2B, left side) and is approximately horizontal (Fig. 2A). The crus, more or less parallel with the femur, is directed posteromedially and is inclined at 30° to the horizontal. The pes, in dorsoventral view, is usually slightly anterior to the hip joint (Fig. 2B, left side). Variations of this basic postural pattern principally involve the femoral orientation relative to horizontal; in some cases, the distal femur may be 10° or more above the proximal, in others, 25° below. The resulting variation in length of the stride is accompanied by variable foot positions which may be either medial or lateral to the femoral axis. During phase II the femur abducts and depresses relative to the hip joint. At phase III the distal femur may be 50° or more below the proximal end and, seen in dorsoventral view (Fig. 2B, right side), is commonly oriented from 40 to 50° to the parasagittal plane. The crus is almost invariably parasagittal and horizontal in orientation. The proximodistal axis of the pes shifts to a more abducted orientation during phase III (Fig. 2B), and the principal locus of extension (dorsiflexion) is at the metatarsophalangeal joints (Fig. 2A, right side).

During the bounding run used for maximum speed, limb excursions of *Tupaia glis* are somewhat modified. In the forelimb, the adduction-abduction pattern of the humerus in the parasagittal plane remains essentially the same as at slower speeds, but the distal humerus frequently moves through 90° (Fig. 4), resulting in greater protraction of the manus and hence longer bounds. Similarly, the femur has a greater anteroposterior excursion, often through more than 100° (Fig. 4A–C). Seen in dorsoventral view, the pattern of femoral abduction from phase I to the middle of phase II approximates the phase I—II—III pattern characteristic of slower speeds; in particular, the mechanical axis of the femur lies at 20 to 25° to the parasagittal in phase I, and by phase II has abducted an additional 20–25°. By phase III, however, the femur again is in an adducted position, lying at about 20° to the parasag-

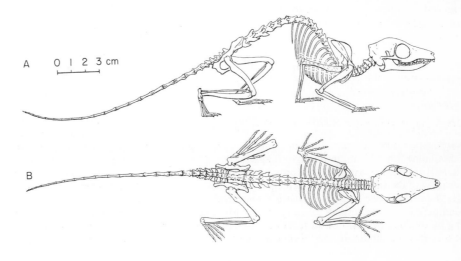

A   0 1 2 3 cm

B

Figure 2   *A typical skeletal posture of* Tupaia glis *in lateral (a) and dorsal (B) views. This figure is based on radiographical films of living specimens.*

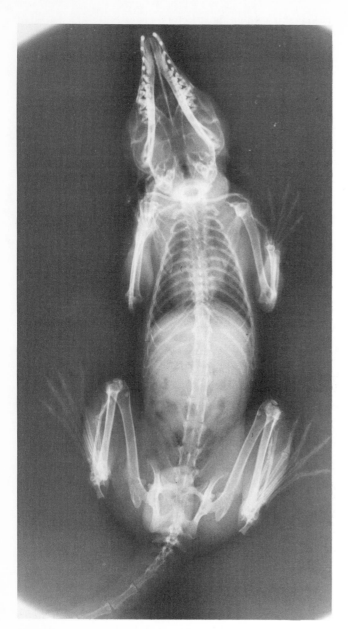

Figure 3   *Radiograph (dorsoventral projection) of* Tupaia glis *in a standing rest posture on a flat surface.*

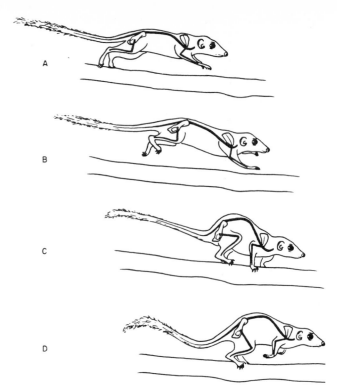

Figure 4   *(A–D) Sequential phases of the bounding run used by* Tupaia glis *at top speed. The sites of greatest curvature along the vertebral column (heavy dark line) are the lumbothoracic and cervicothoracic regions. Flexion and extension are pronounced only at the middorsal region (T11–T12, T12–T13, T13–L1 intervertebral articulations). Note also the orientations of the scapula and pelvis. As* T. glis *runs along a branch, each manus tends to be supinated (C) and each pes inverted (D), with the result that the branch is clasped between the feet. Based on radiographical and cine records.*

ittal. The biomechanical significance of this pattern is not yet known.

Gambarian and Oganesian (1970) experimentally measured the relative forces sustained by fore and hindlimbs in diverse small mammals. In galloping mammals the total muscle weight of fore and hindlimbs is approximately the same, and both pairs of limbs participate equally in propulsion. In mammals using a "primitive rebounding jump," hindlimb weight is appreciably greater than that of the forelimb; the hindlimbs provide most of the propulsive thrust with the forelimbs acting principally as shock absorbers. The locomotor patterns of *Tupaia glis* appear to conform to Gambarian and Oganesian's category of the "primitive rebounding jump," and on this basis a functional differentiation of the forelimbs from hindlimbs may be postulated.

Manus   The manus of *Tupaia glis* is best classified as convergent (see Napier, 1961), for as the digits flex the claws tend to converge on the palm (for a discussion of the bony anatomy of the *Tupaia* "Spreizhand," see Altner, 1971). Although several authors have suggested that the pollex (digit I) is in some measure opposable, no substantial documentation of this possibility has been made and pollical movement appears to be simple convergence. In a nonweight-bearing posture (either in locomotor phase IV or during stationary relaxation), the manus may be straight or slightly flexed at the carpus; the digits are usually not flexed. The longitudinal axis of digit III more or less coincides with that of the forearm. Digits II and IV symmetrically diverge from III at angles of less than 10°. Digit I, the pollex, is slightly more divergent (20°) than digit V (15°). In the weight-bearing manus on a flat surface, digital divergence may span an arc up to 150°. Digit III is usually directed exactly anteriad, and represents the central axis of the manus; however, slightly abducted and adducted manual positions are not uncommon, in which cases the central axis lies between digits II and III or III and IV, and digit V usually lies at 50 to 60°. The pollex is the most divergent and also the most variable in position, lying at least 60° and often as much as 85° from the axis of digit III. Thus, at maximum divergence, the axes of digits I and V are widely spread (as much as 150°); this finding modifies the report by Bishop (1964, Figs. 3 and 15d; see Fig. 15c, however) based on minimal divergence.

A characteristic digital posture, particularly in arboreal situations, is sharp flexion at the proximal interphalangeal joint; the distal (ungual) phalanx and the metacarpal remain relatively parallel to the substrate, and in this posture the holding effect of the claw is maintained. In an arboreal grip, the pollex frequently is abducted at the metacarpophalangeal joint (Fig. 5, top). This abduction, which in extreme cases is nearly 90°, appears to be passive and not the result of muscular control. In certain positions the body weight and the firm implantation of the claw apparently combine to produce forces acting perpendicularly to the digital axis. A similarly abducted posture is occasionally seen in other digits, notably IV and V and the hallux, but the degree of abduction is relatively small.

Tupaia glis uses the same manual posture when walking or running on relatively large branches (diameter greater than 2 cm) as on a flat surface. The palm is placed flatly on the substrate and at the end of phase III the palm and then the digits are lifted. However, the curvature of smaller branches (less than 2 cm diameter) forces the manus into varying degrees of supination which is accompanied by spreading and slight flexing of the digits. When running at maximum speed on small branches, *Tupaia glis* supinates each manus as much as 90°. As the contact of both hands is nearly synchronous, the effect is to grip the branch between the right and left manus (Fig. 4C), rather than to run on top of it.

Bishop (1964), studying prosimian hand orientation on slender, branch-like substrates, found the patterns used by *Tupaia glis* to be the most variable of all species studied (0.6 cm doweling was used for the tree shrews, larger doweling as appropriate for other species). Although by this criterion *T. glis* lacks a clearly defined manual grip pattern, Bishop cited the possible function of the large, protruding hypothenar pad for digital opposition. In Bishop's observations, the manus was almost always positioned across the branch in such a way that the branch lay between some of the digits and the hypothenar pad. The radiographical and anatomical data gathered in the present study give further support to Bishop's interpretation. The hypothenar pad in *T. glis*, far from being a passive cushion, is buttressed at its proximal end by the pisiform bone and is traversed by several muscles. Articulating proximally at the carpus with the ulna and scapholunate and attached distally to the carpal ligament, the rodlike pisiform projects well into the hypothenar pad (Fig. 5). The long axis of the pisiform is inclined somewhat proximally away from the wrist, and this position probably is maintained largely by activity of the flexor carpi ulnaris, for the pisiform is embedded in its tendon as a sesamoid. Undoubtedly the bone is capable of movement which would alter the conformation and pliancy of the hypothenar pad. In addition to the flexor carpi ulnaris, other muscles acting on the pisiform are the abductor digiti minimi (with lateral and distal force components), palmaris brevis (the few slips from the carpal ligament would engender a medial force), and possibly the flexor digiti minimi brevis (although not directly attached, this muscle has an origin from the carpal ligament and thereby might contribute to medially directed force). The anatomy appears to be essentially the same in *T. tana* and *Ptilocercus lowii* (Haines, 1955). In *T. glis*, at least, the tendon of the flexor carpi ulnaris envelops only the base (articular end) of the pisiform, leaving the remainder to protrude palmarward. For a sesamoid such morphology would be inexplicable except on the basis that the nontendinous portion is acting as a bony strut. Both the anatomy and functional employment of the hypothenar pad support the conclusion that it serves to oppose digital flexion, with the result that the manus is capable of adapting to, if not actively gripping, a cylindrical or uneven substrate.

Pes   The median, proximodistal axis of the pes of *Tupaia glis* passes through digit III to the heel. The foot is narrow and elongate, with linear plantar pads along the medial and lateral sides. Digital divergence is less than in the manus. When relaxed, digits II and IV diverge only 5° from III (and the proximodistal axis). Digit V diverges about 10° and the hallux about 20°. During movement on flat surfaces or large branches, divergence may increase up to 10° for digits II and IV, and 25° for V. The hallux, which has a certain degree of independent movement, is discussed in detail below. As in the manus, sharp flexion at the proximal interphalangeal joints is associated with use of the claws in securing a grip.

Locomotory movements of the pes involve considerable plantar flexion in an abducted posture. At phase I, the foot is placed almost flatly on the ground, only the heel remaining slightly elevated; the amount of pedal abduction, measured at the intersection of the proximodistal axis of the foot with a parasagittal plane, varies from 10 to 25°. The digits are not usually flexed. By phase III the plantar aspect of the foot beneath the tarsus and metatarsus is positioned more or less vertically. Approximately 90° of extension occurs at the metatarsophalangeal joints and the digits are flexed at the proximal interphalangeal joints. Most striking, however, is the fact that the proximodistal axis of the foot is commonly at 45 to 75° to a parasagittal plane, and thus the foot is in a more abducted posture in phase III than in phase I. Abduction is accompanied by slight inversion. These movements occur as the result of the simultaneous elevation and forward movement of the ankle which passes to the medial side of the digits instead of directly above them. Thus the foot "twists" in place, that is, assumes a more abducted posture relative to the direction of movement.

The abducted, inverted posture of the pes is characteristic of arboreal locomotion, particularly on vertical surfaces and on branches less than 2 cm in diameter. On branches, plantar contact is limited to the metatarsus and digits; the heel remains well elevated, even in phase I. *Tupaia glis* places its feet on the upper sides of the branch with approximately the same degree of abduction as on other surfaces. However, the curvature of the branch requires that the foot be inverted with the plantar surface facing medioventrally instead of ventrally (as on flat surfaces). Not only does inversion ensure maximal plantar contact, but it also promotes stability when the tree shrew is moving at top speed. In this instance, the hindfeet tend to move synchronously and by means of inversion are capable of grasping the sides of the branch which is thus held between two hindfeet (Fig. 4D). This posture, analogous to the manual supination used by the forelimb at top speeds, is usually accompanied by independent movements of the hallux to be discussed below.

Tupaia glis employs a hyperabducted foot posture when descending vertical or steep surfaces such as a tree trunk. The hindlimbs are usually sprawled out behind with the femora well abducted and the tibiae more or less parallel to the substrate. Normally, with the limbs in such a position the volar surface of the feet would face dorsad; however, through hyperabduction the volar surface is brought to bear against the substrate, and the claws are employed as hooks to control or prevent descent. Similar behavior is known in other arboreal mammals (see Cartmill, 1974a). Radiography of living *T. glis* demonstrates hyperabduction occurring principally by rotation of both the calcaneus and the remaining pes about the head of the astragalus (Fig. 6). The subtalar (astragalocalcaneal) joint system is the major articulation involved in this movement.

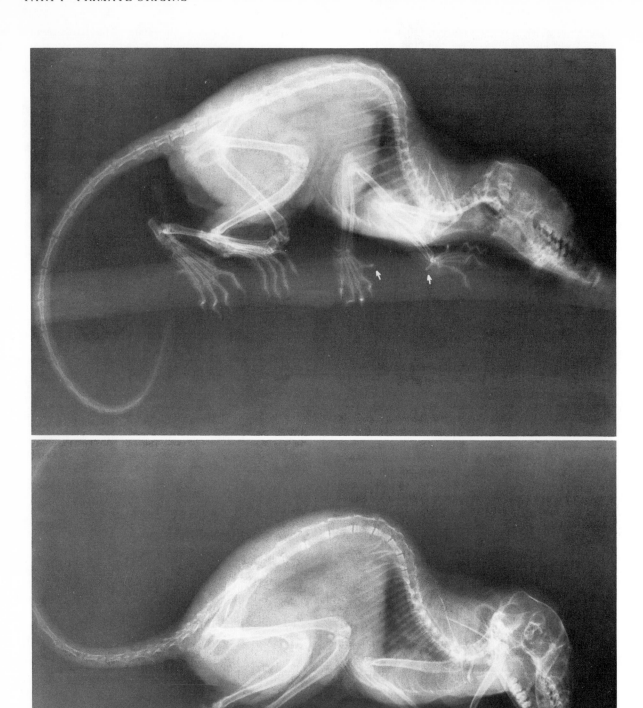

Figure 5   *Radiographs (lateral projection) of* Tupaia glis *on a horizontal branch 14 mm in diameter; in both cases the tree shrew had momentarily paused while walking along the branch. Note the sharp lumbothoracic and cervicothoracic curvatures. Arrows point to the large pisiform and hyperabducted pollex as described in text. About* **natural** *size.*

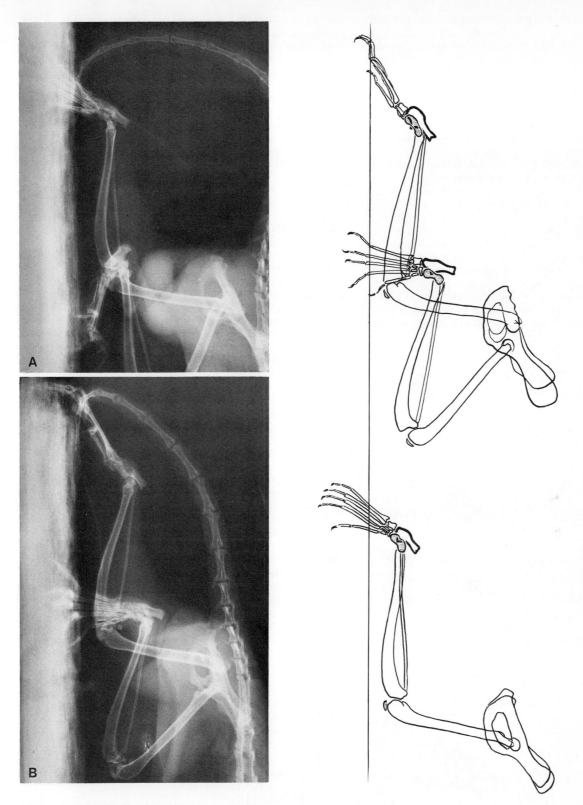

Figure 6   *Radiographs (lateral projection) of* Tupaia glis *making a vertical descent along a bark surface. Illustrated are the different degrees of pedal abduction used by the tree shrew to secure a grip. In the line diagram the calcaneus is indicated by a heavy line, and the astragalus by stippling. (A) Hyperabduction; the calcaneus and remaining pes begin to rotate about the head of the astragalus. (B) Extreme hyperabduction; note the change in calcaneal outline from (A) to (B), and especially the shift in the triangular sustentaculum tali (st).*

The most striking feature of pedal function in *Tupaia glis* is the independent action of the hallux on uneven or arboreal substrates. During late phase IV, the hallux is maximally extended and abducted (Fig. 7C). Thus the hallux is positioned to lie across the top of the branch (Fig. 7D and E). In this posture, the nail and phalanges prevent slippage off the side of the branch, and possibly the branch is actively gripped between the hallux and the remaining digits. The hallux maintains this posture throughout phases II and III, and can be clearly seen in its abducted position even when the foot is maximally inverted at the end of phase III (Fig. 7H). Although the tarsometatarsal joint is saddle-shaped and probably permits some flexion-extension and abduction-adduction, cinematical and radiographical evidence shows that these movements are limited. Metatarsal I is bound to the palmar fascia by a thick aponeurotic band which limits abduction but which also may help to stabilize the hallux when abducted and bearing weight. The principal site of movement appears to be the metatarsophalangeal joint. The distal articular surface of metatarsal I is an asymmetrical condyle; the lateral aspect, being much more parallel to the shaft than the medial, permits phalangeal abduction. Crossing this joint are the following muscles which control the independent movement of the hallux: a large abductor hallucis, one tendon each from the extensor digitorum communis and extensor hallucis longus (the latter crossing the dorsolateral aspect and thus

having an abducting component), the large tendon of the flexor hallucis longus, and the flexor hallucis brevis. The relative size and degree of morphological differentiation of the hallical muscles from those of the remaining digits confirms the present interpretation of the functional importance of the hallux.

Axial Skeleton   *Tupaia glis* typically has 7 cervical, 13 thoracic, 6 lumbar, 3 sacral and 24 (±1 to 3) vertebrae. Although the vertebral column is capable of assuming a nearly straight configuration under special circumstances (such as when climbing on vertical surfaces), the usual posture involves pronounced dorsoventral curvatures. The cervical series normally descends from the skull at angles of 35 to 55° to the horizontal (Figs. 2 and 5). The reversed thoracic inclination of 30 to 40° creates a sharp flexure in the cervicothoracic region (Figs. 2 and 5). The lumbar series usually inclines posteroventrad at angles of 15 to 25° from horizontal, an orientation shared with the sacrum and proximal caudals.

The tail functions in a variety of postures but is not prehensile. At relatively slow gaits, the tail is held off the ground with the proximal half at approximately the same horizontal level as the sacrum. The distal half is highly mobile and is carried either in an upwardly curved arc, or more or less straight with ventral inclination (Fig. 8), or with distal tip turned downward, the remainder of the tail then being straight. Tail-flicking is frequent, particularly during excitement or at the onset of locomotion; in this movement the distal half is brought forward over the base of the tail in a single rapid twitch. When perched on a branch, *T. glis* frequently holds the tail in a downwardly curved arc. During rapid locomotion, the tail apparently trails flaccidly and undulates dorsoventrally in response to body movement (Fig. 4).

Pronounced flexion and extension of the trunk accompanies locomotion. The region of maximum mobility lies at the T11–T12, T12–T13, and T13–L1 intervertebral articulations. The remainder of the thoracic spine has a small amount of mobility in comparison, and the lumbar series does not appear to engage in any significant flexion. The striking feature of *Tupaia glis* in a slow, exploratory walk is the apparent elongation and contraction of the trunk; the head and shoulders appear to move ahead and stop, followed by the hindquarters in similar pattern. During "contraction," the back in the middorsal region is noticeably arched (Fig. 8B and D). Flexion at T11–T12, T12–T13, and T13–L1 is principally responsible for this pattern, and is related to limb movement. As the left hindlimb begins to advance during early phase IV, flexion in the middorsal region is only slight (Fig. 8A). By late phase IV (or early phase I) on the left side, the right hindlimb is completing its propulsive thrust (phase III); simultaneous flexion at the middorsal spine permits the pelvis to advance without corresponding movement at the forelimbs (Fig. 8B). Spinal extension returns as one forelimb moves through phase IV and the contralateral limb enters phase II (Fig. 8C). During the bounding run used for maximum speed, hind- and forelimb pairs function more or less synchronously; spinal flexion, although more pronounced, still appears to be principally restricted to the intervertebral articulations between T11 and L1. The lumbar series remains rigid and does not contribute to even the most extreme flexion observed. Maximum flexion approximately coincides with the initial contact of the hindfeet with the substrate (Fig. 8C), and maximum extension with phase III of the hindfeet (Fig. 8A).

The anatomical adaptations whereby truncal flexion is concentrated largely at the T11–T12–T13–L1 intervertebral articulations are not obvious. The diaphragmatic and the anticlincal vertebra (T10), which commonly has been regarded as the site of maximum spinal mobility, in fact, is only peripherally associated.

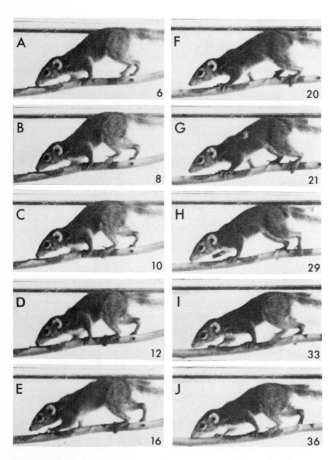

Figure 7 Tupaia glis *walking along a branch 14 mm in diameter and using independent movements of the hallux to secure the pedal grip. For further details, see text. This record is taken from a 16-mm movie film exposed at 64 frames/second; individual frame numbers are indicated.*

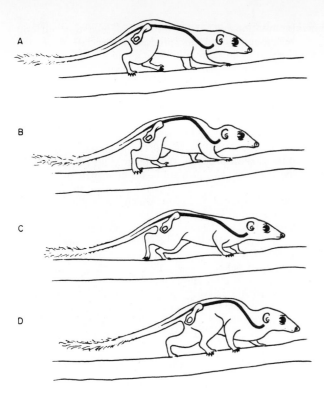

Figure 8 *(A–D) Sequential phases of the walk in* Tupaia glis *illustrating the coordination of flexion–extension of the vertebral column (heavy dark line) and limb movements. Based on radiographical and photographical records.*

There are, however, four features which relate to the dorsoventral mobility of this region.

1. In lateral view, the vertebral bodies of T12, T13, and L1 are trapezoidal, the dorsal (neural canal) aspect of the body being longer than the ventral length (Figs. 5 bottom and 9). Elsewhere in the vertebral column the bodies in sagittal section are either rectangular or resemble a parallelogram; apparently flexion is in part restricted by the compression of the intervertebral disc that accompanies approximation of the ventral edges of adjacent vertebral bodies. However, if the ventral edges of adjacent bodies are recessed and the intervertebral disc is correspondingly wedge-shaped, flexion is facilitated by an increase in compressible substance (disc) and a decrease in noncompressible substance (ventral edges of adjacent vertebral bodies). This type of mechanism, well known in the human spine, appears to be operating at the T11–T12, T12–T13, and T13–L1 intervertebral joints of *Tupaia glis.*

2. Another feature related to spinal flexion at T11 through L1 is the fact that the distances between the centers of pre- and postarticular surfaces of each vertebra are greater than the body length (measured between the centers of adjacent nuclei pulposi). Based on measurements made from single X-ray films of stationary individuals, this difference appears to be on the order of 10 to 20% at T12 and T13, but is usually less and sometimes nil at T11 and L1. Measurements from articulated as well as disarticulated vertebral columns of embalmed individuals yielded similar ratios. In one specimen this difference was approximately 25% at T13. The significance of this feature is that in the neutral position, with the centers of the articular surfaces on pre- and postarticular processes apposing, the vertebral column is in slight flexion.

3. At most intervertebral articulations in *T. glis* the anteropos-

terior length of a zygapophyseal prearticular surface is comparable to that of a postarticular surface of the preceding vertebra and the displacement between them (and the intervertebral movement) is relatively small. Dissection and ligamentous preparations, however, reveal that the prearticular cartilaginous surfaces are normally about 25% longer than the postarticular surfaces at the L2–L1, L1–T13, and T13–T12 joints. The T12–T11 articular surfaces vary between specimens in this regard. The functional result of this differential length is increased mobility; the smaller postarticular surface has greater potential for anteroposterior movement on the larger prearticular surface than it might were the dimensions of the two surfaces essentially the same.

4. The anterior margin of the prearticular surfaces at L1, T13, and T12 are more ventral than the posterior margin. The resultant convexity, not found elsewhere, is related to the greater degree of flexion at these joints. The curved surface represents an arc, the center of which lies within the nucleus pulposus. During flexion, the articular processes, in the course of "separating," trace an arcuate path which has been incorporated into the form of these prearticular surfaces.

HYPOTHESES OF ARBOREAL ANCESTRY

Ever since Huxley (1880) proposed that marsupials descended from an arboreal stock, the possibility of arboreal adaptation in mammalian ancestry has been repeatedly investigated. Did the early primates merely exploit the primitive mammalian mode of life or did they represent an adaptive innovation? In short, were the early placental mammals arboreal or terrestrial in adaptation? Huxley's original idea was based on the fact that members of most families of living marsupials possess an apparently prehensile pes by virtue of an abductable hallux; those species with a vestigial or absent hallux Huxley supposed possessed a "reduced prehensile pes" by virtue of their taxonomic relationships. Dollo (1899) developed the evidence in more detail, particularly in regard to the problem of the atrophied hallux. Dollo argued that in marsup-

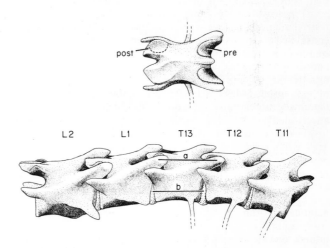

Figure 9 *Vertebral specializations of* Tupaia glis *that contribute to localization of spinal flexion and extension at the L1–T13, T13–T12, and T12–T11 joints. Below, vertebrae L2 through T11 are shown in lateral view. Note the wedge-shaped bodies at T13 and T12; the greater distance between articular surfaces (a) than that between adjacent nuclei pulposi (b). Top, the thirteenth thoracic vertebra is shown in dorsal view. Note the anterior edge of the prearticular surface (pre) is curved sharply ventrally, and the anteroposterior length of this surface is longer than that of the postarticular (post) surface.*

ials with tetradactylous feet (the hallux reduced or absent), features such as syndactyly, reduction of toes II and III, and enlargement of toe IV were common to forms with opposable halluces, and therefore indicated an arboreal heritage for all marsupials. Bensley (1901a, b) accepted the hypothesis as well.

Matthew (1904, 1909, 1937), using paleontological evidence, inferred that placental mammals also had an arboreal ancestry. Although Matthew's 1904 paper contained brief reference to such features as a gracile build and "probably prehensile" tail among early Tertiary mammals, the argument was based principally on the interpretation of an opposable or semiopposable digit I. In particular, he regarded the feet in creodonts (early Tertiary carnivores) as representing the basic placental morphology adapted to arboreal life. No other pedal structure or function was given as careful consideration as prehensility, and it is apparent from Matthew's last paper that his interpretation was founded on the identification of this ability. Subsequently, the theory of an arboreal ancestry for mammals gained wide acceptance and has been reinforced to a limited degree by some neontological evidence (see discussion in Haines, 1958, and Martin, 1968b). An early paper by Gregory (1910) probably also contributed to the genesis of the theory.

Only two authors have challenged Matthew's theory. Gidley's (1919) objection that mere pollical or hallical divergence was inadequate evidence of semiopposability was answered by Matthew (1937) with a detailed description of the manus in *Claenodon*. In *Claenodon* (and presumably other creodonts with the presumptively "primitive" mammalian manus), Matthew interpreted the pollical carpometacarpal and carpotrapezium joints as permitting a degree of pollical opposability. Haines (1958), in a careful reexamination of this material, contradicted Matthew's interpretation, and indeed the evidence for opposability appears insubstantial. However, Haines went on to suggest that the feet of the Egyptian mongoose (*Herpestes ichneumon*), a terrestrial carnivore, adequately represent the primitive placental plan and that, consequently, early mammals were probably terrestrial.

Previous theories on primitive mammalian arboreal or terrestrial adaptation have several shortcomings; first, the fossils considered are too late in time (Paleocene-Eocene) or too specialized in form (Creodonta) to represent common ancestral forms, and second, there has been inadequate appreciation of the adaptations among living mammals which bear on the interpretation of fossil forms. Matthew (1937) had available only early Tertiary mammals, rather than Cretaceous forms which he suspected were ancestral. Haines (1958), on the other hand, virtually ignored fossil evidence in his selection of a primitive mammalian analog among living species; without reference to fossil material, *Herpestes* is no more credible an analog of an early mammal than is, for example, a marsupial. Both appropriate fossils and relevant experimental evidence from living mammals are necessary to make interpretations of the anatomical adaptations of early mammals. Although the evidence is far from complete, tree shrews appear to offer at least one viable model for such interpretations on the basis of skeletal similarity to that of early mammals.

THE ORIGINS OF PRIMATE ARBOREALISM

Discussions of the evolutionary origins of major taxa usually focus on adaptive innovations (e.g. anatomical, physiological, or behavioral features) and new adaptive zones (e.g. in habitat or food source). For the Primates, the adaptive theme is arborealism. Although not all living primates are tree-living, they all are derived phylogenetically from arboreal ancestors. The question naturally follows as to what selective forces initially favored

arboreal adaptation and habitat; escape from predation and food source are among the commonly proposed answers. However, the circumstances in which trees became an accessible and viable adaptive zone for early primates remains unknown. Two alternatives seem plausible. The theory that early mammals were basically terrestrial implies that the origin of primates represented an adaptive innovation both in anatomical form and in habitat. Thus, the earliest primates are believed to have developed arboreal specializations which set them apart from other early mammalian groups. Since teeth constitute the major part of the fossil record of the earliest primates, the tendency has been to regard dietary specialization as a primary factor in primate origins (see Szalay, 1969). However, implicit is the recognition that other arboreal features were also acquired which less specialized mammals, both fossil and living, do not possess. This viewpoint, of which Haines (1958) is the most recent proponent, treats "arborealism" and "terrestrialism" as necessarily discrete phenomena. An alternative theory, proposed here, is based upon the implications of tree shrew behavior and habitat in which "terrestrialism" and "arborealism" are not discrete phenomena.

By emphasizing arboreal versus terrestrial behavior, published accounts of tree shrews appear to have misrepresented the true nature of their habitat and adaptations. With the possible exception of a few species, all tree shrews can and do move freely between ground and trees. Even the most "terrestrial" species are known to be good climbers. This ability relates not simply to climbing trees, but more realistically to the fact that the forest offers, to a small mammal, an extremely uneven and disordered substrate for locomotion. The forest floor, traversed by roots and littered with plant debris (including fallen trunks), grades upward through secondary growth (including vines, bushlike growth and small trees) to tree trunks and the lower story. In the forest habitat of tree shrews, the distinction between "arboreal" and "terrestrial" locomotion is artificial in the sense that the substrate everywhere requires basically the same locomotor repertoire for this sized mammal.

The locomotor repertoire of tree shrews is adapted to two major topographical features of their habitat. The first is that most surfaces, in terms of the actual plantar area of tree shrew feet, are not level. The ability to invert-evert the hindfeet, and pronate-supinate the forefeet, is critical in this situation. Combined with movements of plantar flexion and dorsiflexion, the feet are able to appose a nonlevel surface while the animal retains its upright posture. Postural stability is likewise important on uneven surfaces, and may be achieved in part by maintaining a relatively low center of gravity while carrying the body on as wide a base as practicable. In terms of limb posture, this is accomplished by employing the elbow and knee joints in flexed rather than extended positions, and by maintaining the limbs in abducted rather than parasagittal postures. A second feature of the forest habitat is the disordered spatial arrangement of locomotor surfaces. In the perspective of a small mammal, the forest floor, the trees, and all the interconnecting secondary growth are not a smooth continuum but rather a ragged network of substrate possibilities. Irregular spacing of footfalls is as appropriate for moving about among the debris of the forest floor as it is for climbing through forest vegetation. Both situations favor a highly flexible and versatile locomotor pattern, and may account for the predominance of asymmetrical gaits among tupaiids. In tree shrews, such versatility of foot placement is in part a function of the lateral mobility of the limbs (significant abduction at the shoulder and hip is common, particularly when climbing) and the flexibility of the vertebral column. The latter's function in lengthening the stride in some cursorial mammals has often been emphasized. A similar function is served for the tree shrew leap and bounding run. However, in other situations spinal flexion and extension also allows considerable variation in the distance between pectoral and pelvic girdles, and consequently between fore- and hindfeet. In terms of foot placement along an irregularly spaced substrate, the ability to make gross adjustments in the length of the quadrupedal stance pattern is important. Such adjustments during walking have already been discussed (Fig. 8). Variations in quadrupedal stance are common in other situations and always involve distinctive vertebral column postures. For example, when positioned transversely on a branch, tree shrews must closely approximate all four feet and to do so use maximal spinal flexion. On steep or vertical surfaces, where, in terms of leverage, it is advantageous to keep the body (center of gravity) close to the surface, the feet are spread widely and the vertebral column is maximally extended.

The locomotor pattern of *Tupaia glis* is significant both in a comparative and phylogenetic context. The basic characteristics of the pattern are not peculiar to tree shrews, but are shared by other relatively generalized, noncursorial mammals (Jenkins, 1971). Although some specialization has undoubtedly occurred among tupaiids, such fundamental features as limb posture and excursion are similar to those in the rat (*Rattus norvegicus*) or the Virginia opossum (*Didelphis virginiana*). Thus, a common locomotor pattern for noncursorial, relatively generalized mammals is recognizable, and a similar functional pattern—if osteology is a reliable guide—appears to have been present among early mammals.

A new hypothesis on the origin of mammalian posture and locomotion, as well as the origin of primate arborealism, is now possible. Mammalian limb posture traditionally has been interpreted in terms of the presumed biomechanical efficiency of upright, parasagitally oriented limbs; it is now known that this arrangement is not present in noncursorial mammals (Jenkins, 1971). Instead, the primitive mammalian locomotor mode may have evolved as an adaptation to moving on uneven, disordered substrates (which, of course, are not found only in forest habitats). Flexion and extension of the vertebral column, the mobility of the feet, and the flexed, abducted limbs may be interpreted as mechanisms to permit versatility in stance and locomotor pattern. Most Mesozoic mammals were approximately tree shrew sized, and the importance of a versatile locomotor repertoire to a small mammal has already been emphasized. Accordingly, the question of an arboreal or terrestrial ancestry for mammals no longer is relevant. The locomotor niche which ancestral mammals exploited undoubtedly included both "terrestrial" and "arboreal" surfaces. Ancestral primates very likely occupied forest habitats in the manner of some living tree shrews. The adaptive innovation of ancestral primates was therefore not the invasion of the arboreal habitat, but their successful restriction to it.

This study was supported by grants from the National Science Foundation (GB-13662 and GB-30724) and National Institutes of Health (5-S05-RR-07046-06). The author is grateful to M. W. Sorenson and J. G. Fleagle for reviewing the manuscript, to L. Meszoly for preparation of Figs. 2 and 7, and to A. H. Coleman for photographic assistance.

2

Rethinking Primate Origins
M. CARTMILL

If you asked a student of human evolution to explain why human beings, unlike other mammals, walk around on only two legs, you would be baffled and unhappy if he answered, "Because in man's ancestral lineage, individuals who could not run away from predators left fewer offspring." You would be justified in retorting that the same remarks apply equally to thousands of other species of mammals, yet none of these have developed upright bipedal locomotion. The purported explanation, you would properly conclude, may be a true proposition, but it is worthless as an explanation.

An explanation is a hypothesis of a complex sort. Ordinarily, to explain one fact in terms of another requires that there be an a posteriori rule which allows us to deduce the first from the second, and which warrants testable expectations other than the one in question (Hospers, 1966). We reject the foregoing "explanation" of human bipedality because we sense that its explanatory force depends on the lawlike generalization, "Natural selection favors bipedal locomotion in any mammal species that has predators," and that this generalization is false. Yet some evolutionary biologists and philosophers of science (Scriven, 1959; Lewontin, 1969; Feduccia, 1973) have argued that evolutionary explanations do not involve any such generalizations, and hence are not subject to refutation by counterexamples. In this view, we have no grounds for dismissing the "explanation" with which I began; the objection that the same remarks apply to species which have remained quadrupedal is beside the point.

I have suggested elsewhere (Cartmill, 1972) that this and similar objections are very much to the point; that, when valid, they demonstrate the inadequacy of the explanation in question; and that such objections must be raised systematically if we wish to arrive at adequate explanations of historical processes. These assumptions underlie the following reassessment of what has been called the arboreal theory of primate evolution.

THE ARBOREAL THEORY AND ITS BACKGROUND

The Linnean concept of the order Primates, which included the bats and colugos, was still current as late as 1870 (Gray, 1870). In 1873, Darwin's antagonist Mivart proposed ordinal boundaries which excluded these animals, but which (unlike the taxonomies then advocated by Milne-Edwards, Grandidier, and Gervais) included the prosimians as a suborder of Primates (Mivart, 1873). Mivart also proposed a list of traits that distinguished prosimians and anthropoids from other placental mammals. These traits included a complete bony ring around the eye, a well-developed occipital lobe of the cerebral cortex, and a grasping hind foot with an opposable, clawless first toe.

In the second decade of the 20th century, G. E. Smith and his pupil, F. W. Jones, put forth the first systematic attempts at explaining these and other characteristic primate traits in terms of natural selection. Smith, a comparative neuroanatomist, was principally concerned with explaining the distinctive features of primate brains. He proposed (Smith, 1912) that the remote ancestors of the primates were shrewlike terrestrial creatures that entered upon an arboreal way of life. In the complex networks of tree branches through which these early primates moved and foraged, the olfactory and tactile receptors in the snout did not provide adequate guidance; snuffling blindly along in hopes of scenting something edible, as most living insectivores do, was no longer a viable foraging pattern. Accordingly, vision gradually replaced olfaction as the dominant sense. In correlation with this the hand assumed the tactile and grasping functions primitively served by the mouth and lips; eye-hand coordination replaced nose-mouth coordination. Arboreal life also required more precise and rapid motor responses. Thus, Smith was able to account for the primates' reduced olfactory centers and elaborated visual, tactile, motor, and association cortex in terms of the selection pressures exerted by the arboreal environment.

Wood-Jones' (1916) reinterpretation of these ideas reflects his professional interest in the anatomy of the hand and foot. Jones proposed that the arboreal habit led to a functional differentiation of the limbs. While the foot remained a relatively passive organ of support and propulsion, the hand, used by the primate ancestors for reaching out and grasping new supports when climbing about in trees, became specialized for prehension—and therefore preadapted to take over the mouth's functions of manipulation and food-gathering. As the snout lost importance as a sensory and manipulative organ, it dwindled in size; and the eyes were perforce drawn together toward the middle of the flattening face. The progressive specialization of the hind limb for support and propulsion led to a more upright posture, with correlated changes in the axial skeleton, gut, and reproductive organs. For Jones, most of the things that distinguish human beings from typical quadrupedal mammals were originally adaptations to living in trees.

The arboreal theory was open to the obvious objection that most arboreal mammals—opossums, tree shrews, palm civets, squirrels, and so on—lack the short face, close-set eyes, reduced olfactory apparatus, and large brains that arboreal life supposedly favored. Jones tried to account for these counterexamples. Accepting Matthew's (1904) thesis that primitive mammals had been arboreal creatures with opposable thumbs and first toes, Jones proposed that the absence of primate-like traits in other arboreal lineages resulted from a period of adaptation in each

Reprinted with permission from *Science*, Volume 184, pages 436–443, Washington, D.C. Copyright © 1974 by the American Association for the Advancement of Science, Washington, D.C.

lineage to terrestrial locomotion. During this period, the thumb and first toe became reduced, the primitive reptilian flexibility of the forelimb was lost, and the primitive flat nails were replaced by claws. These changes blocked the specialization of the forelimbs for prehension. Accordingly, in nonprimate mammals that had reentered the trees, the primate evolutionary trends did not materialize.

Stated thus boldly, Jones' thesis is obviously inconsistent. His treatment of the evolution of the brain, which he borrows from Smith, presupposes that primitive mammals were small-eyed terrestrial beasts that nosed their way through the world, guided by specialized olfactory and tactile receptors in the snout; but when the evolution of the limbs is in question, he assumes that arboreality is primitive and that early mammals were neither terrestrial nor typically quadrupedal.

The late W. E. Le Gros Clark's reformulation of the arboreal theory, which more skilfully conceals this inconsistency, has been almost universally accepted by other students of primate evolution. Much of Le Gros Clark's primatological work centered around the now-discredited (Van Valen, 1965; C. B. G. Campbell, 1966; Martin, 1966; Luckett, 1969) proposition that the tree shrews (Tupaiidae) are persistently primitive lemuroids that have somehow failed to develop the perfected adaptations to arboreal life seen in the other extant primates. Le Gros Clark believed that primitive Insectivora were tree-climbing beasts with clawed, non-prehensile hands and feet, small eyes and brains, and elaborate olfactory apparatus. The unspecialized, squirrel-like climbing habit of tree shrews (and ancestral primates) is invoked by Le Gros Clark to explain their incipiently primate-like morphology; tree shrews have a complete bony ring around the orbit, a relatively extensive visual cortex, a highly differentiated retina, some simplification of the olfactory apparatus, and a few minor grasping adaptations of the joints and muscles of the hind foot. More perfect arboreal adaptations, of the sort seen in lemurs, involve the replacement of sharp claws by flattened nails overlying enlarged friction pads, the divergence and enlargement of the first toe and thumb to produce effective grasping organs, and the approximation of the two eyes toward the center of the face. This last change, in Le Gros Clark's view, had a positive selective advantage for acrobatic arboreal mammals; it produced a wide overlap of the two visual fields, allowing stereoscopic estimation of distance in jumping from branch to branch (Le Gros Clark, 1970)[1].

THE COMPARATIVE EVIDENCE

If progressive adaptation to living in trees transformed a tree-shrew-like ancestor into a higher primate, then primate-like traits must be better adaptations to arboreal locomotion and foraging than are their antecedents. This expectation is not borne out by studies of arboreal nonprimates. The diurnal tree squirrels (Sciurinae) provide the most striking counterexample. The eyes of squirrels face laterally, the two visual fields having only about a 60° arc of overlap (Kaas, Guillery, and Allman, 1972); the olfactory apparatus is not reduced by comparison with terrestrial rodents (Le Gros Clark, 1925); all the digits (except the diminutive thumb) bear claws, which are sharper and more recurved than those of terrestrial sciurids (Peterka, 1937), and the marginal digits of the hand and foot are not opposable or even very divergent (Fig. 1). Yet squirrels are highly successful arboreal mammals, and seem to have little difficulty in accomplishing the arboreal activities in which primates might be expected to excel. Despite their laterally directed eyes (and presumed lack of stereoscopy), squirrels of several genera may leap from 13 to 17 body lengths from tree to tree (Fig. 1D) (D'Souza, 1974; Hill, 1949;

Harrison, 1951; Hatt, 1929; Shorten, 1954), which compares favorably with the 20 body lengths reported for the saltatory lemuroid *Propithecus verrauxi* (Jolly, 1966)[2]. Although squirrel hands and feet are not adapted for grasping, squirrels easily walk atop or underneath narrow, sloping supports, and can forage for long periods in slender terminal branches hanging by their clawed hind feet (Fig. 1, A to C, F). Clearly, successful arboreal existence is possible without primate-like adaptations.

A partisan of Le Gros Clark's form of the arboreal theory might still postulate that tree squirrels are under selection pressure which favors their developing primate-like morphology, but have not undergone a long enough period of adaptation to arboreal life for them to have converged markedly with primates. Accepting this, we would still expect that arboreal squirrels would differ in primate-like ways from terrestrial sciurids, at least to a slight extent. We would have similar expectations about arboreal members of other nonprimate families.

The facts do not bear out these expectations. Virtually the only features of the hands and feet which systematically distinguish arboreal from terrestrial squirrels are the longer fourth digits and generally larger carpal pads of the former; the arboreal genera show no tendency toward enlargement of the thumb, reduction of claws, or development of a wide or deep cleft between the first and second digits (Pocock, 1922). Orbital convergence in all sciurids is slight, and is actually greater in the more terrestrial species (Fig. 2E), although the optic axes of ground squirrels' eyes are not more convergent than those of tree squirrels'.

Since small mammals have relatively large eyes, orbital-margin convergence in most mammals varies inversely with size, other things being equal (Cartmill, 1972). For a given skull length, this convergence is somewhat greater in higher primates than in lemurs (Cartmill, 1971)[3]. When convergence is plotted against skull length for several families of arboreal mammals and the lemuriform and haplorhine regressions are traced on the plot (Fig. 2), it is evident that arboreality (or saltatory arboreal locomotion, in wholly arboreal taxa) does not correlate with proximity to the primate regressions. The slow-moving lorises have, for their size, more convergent orbits than the saltatory galagos (Fig. 2A). Among feloid carnivores (Fig. 2B), the terrestrial *Felis bengalensis* approaches the primate regressions most closely. Both arboreal and terrestrial procyonids (Fig. 2D) fit a regression parallel to those of the primates, from which the semiarboreal coatimundi is widely displaced away from the primate lines.

Certain primate-like specializations of the visual pathways of the brain may perhaps represent adaptations to arboreal life per se. Diamond and his coworkers (Diamond and Hall, 1969; Harting *et al.*, 1973) have found that the common tree shrew and the Carolina gray squirrel resemble *Galago senegalensis* in having little or no overlap between the projection from the retina to the occipital visual cortex (relayed via the lateral geniculate) and a significant visual projection to the temporal cortex from the superior colliculus (via the pulvinar). This is not the case in the cat, in which these areas overlap widely and the temporal cortex is given over to projections from the medial geniculate. Since arboreality is about the only thing that tree shrews, squirrels, and galagos have in common, the suggestion that this represents a specifically arboreal adaptation (Diamond and Hall, 1969) may be correct. However, its adaptive significance is obscure. The expectation that "*any* mammalian line that relies heavily on visual cues" will develop a visual temporal lobe (Harting *et al.,* 1966) is clearly unwarranted; cats rely heavily on visual cues, and in fact show several primate-like features of the visual system that are absent or unknown in squirrels and tree shrews—for example, parallel optic axes, substantial ipsilateral radiations of each optic

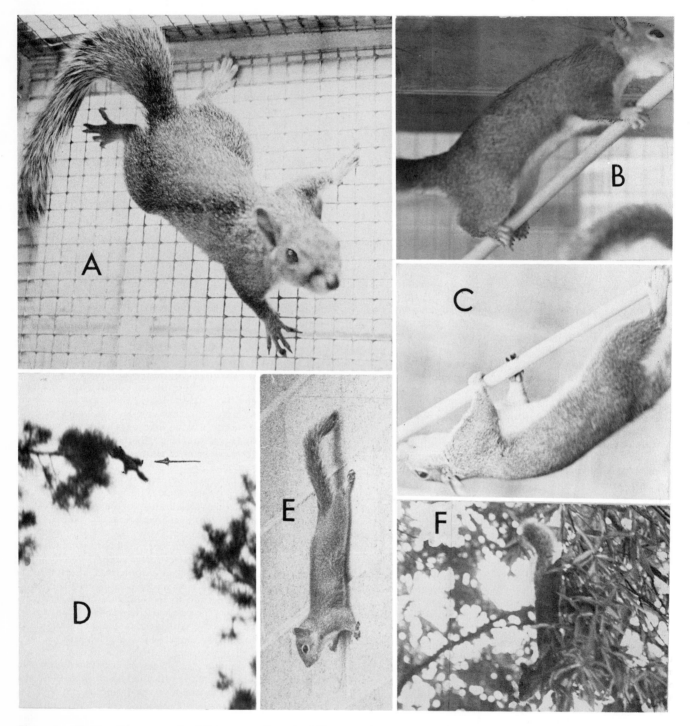

Figure 1   *The Carolina gray squirrel,* Sciurus carolinensis, *(A) hanging from wire grid, showing nonopposable first digits; (B) climbing thin sloping support; (C) descending underneath thin sloping support; (D) (squirrel shown by arrow) leaping across gap in the canopy, about 20 m above the ground; (E) clinging to vertical cinder block wall; and (F) foraging in terminal branches of a willow oak (*Quercus phellos*) hanging bipedally.*

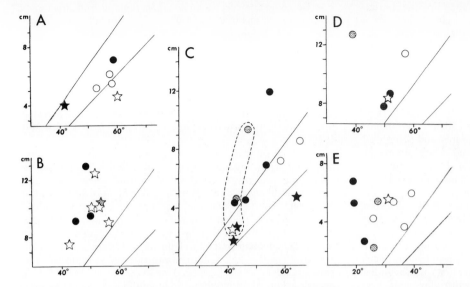

Figure 2   *Five bivariate plots of species mean values of skull lengths (prosthion to inion, centimeters) and orbital convergence (dihedral angle between orbital and midsagittal planes, degrees): (A) lorisiform prosimians, (B) feloid carnivores, (C) didelphids (dashed line) and diprotodont marsupials, (D) procyonid carnivores and (E) sciurids. White symbols represent terrestrial animals (such as* Monodelphis*) or slow-moving arboreal forms (such as* Phalanger*); stippled symbols represent semiarboreal animals (such as* Didelplis*); stars represent predominantly carnivorous animals (such as* Monodelphis*). In each plot, the diagonal lines represent the least-squares regression of convergence on skull length for Madagascar lemurs (upper line) and haplorhine primates (tarsiers and anthropoids: lower line). Data from Cartmill (1971)[3].*

nerve, and the presence of "binocular depth cells" in the striate cortex (Polyak, 1957; Johnson, 1901; Hubel and Wiesel, 1970). These features are all functionally related to steroscopic depth perception. Since most of the projection from the retina to the lateral geniculate body seems to correspond to the binocular portion of the visual field (Kaas, Guillery and Allman, 1972; Glickstein, 1969), the relative de-emphasis of the older tectopulvinar system in cats can even be described, from a different perspective (Polyak, 1957), as a special similarity to higher primates.

The comparative evidence, then, does not support the idea that the selection pressures of arboreal life favor the replacement of tree shrew-like morphology by primate-like morphology. In many respects, the first sort of morphology is actually of superior adaptive value. Clawed fingers and toes are superior adaptations for locomotion on nonhorizontal surfaces with large radii of curvature—including vertical walls (Fig. 1E) as well as tree trunks (Cartmill, 1974a). Like marmosets (Rothe, 1972), squirrels tend to avoid very thin branches in normal arboreal locomotion, but can walk on them easily enough, relying on the largely passive grip of the proximal volar pads when the support is horizontal and (unlike marmosets) gripping with opposed hands and opposed feet when the support is sloping (Fig. 1B). Primate-like approximation of the orbits increases visual field overlap, but decreases parallax, reducing the distance over which visual field disparities can provide distance cues. In a leaping arboreal animal, selection should act against the extreme orbital approximation seen in tarsiers and higher primates. This expectation is borne out by a comparison of lorises with galagos; the slow-moving *Loris* and *Nycticebus* have more convergent and closely approximated orbits than the saltatory galagos [4], whose wide interorbital space allows stereoscopic ranging over greater distances.

Evidently, the close-set eyes and grasping extremities typical of extant primates are adaptations to some activity other than simply running about in the trees; arboreal life per se cannot be expected to transform a primitive tree shrew-like primate into a lemur. Le Gros Clark's version of the arboreal theory is not adequate.

WERE PRIMITIVE MAMMALS ARBOREAL?

Jones's version of the arboreal theory holds, not that the primate characteristics will be selected for in any arboreal mammal lineage, but that they all result from the primates' unique preservation of the grasping hands and mobile forelimbs supposedly found in the arboreal ancestors of the Mammalia. This conception of what early mammals were like can be traced to several sources. Huxley (1880) and Dollo (1899, 1900) proposed that the last common ancestor of the living marsupials had a grasping hind foot, but they thought this represented an arboreal specialization and that early mammals were terrestrial. Matthew (1904), following Cope (1885b), reinterpreted this trait as a primitive retention, and suggested that Eocene and Paleocene placental mammals (and early ungulates in particular) showed features indicating derivation from an arboreal ancestor.

Most of the supposedly arboreal features identified or inferred for the ancestral mammals by Matthew (1904) and his inheritors (Klaatsch, 1923; Morton, 1935; Martin, 1968b) can be shown (Cartmill, 1970; Gidley, 1919; Haines, 1958; Romer, 1955) to be either chimerical or irrelevant to arboreality. Others represent specializations fixed at various points along the reptilian lineage leading to mammals (such as the loss of all but two phalanges in the thumb and first toe, the "anomalous" arrangement of the thumb's extrinsic muscles, and the appearance of a tuber calcanei). Some are mere amphibian retentions (for example, persistence of the clavicle) that were lost in later mammalian lineages that developed cursorial specializations. Most of those who have believed that primitive mammals were lemur-like arboreal animals have also thought that terrestrial habits select for cursorial locomotion and thus for simplification and stabilization of the limbs; that "the final stage of this process is exemplified in the horse" (Wood-Jones, 1916), and that primates could therefore not be descended from ancestors that had long been terrestrial. However, the fact that placental ancestors could not have been very much like horses does not imply that they were very much like lemurs. The same suite of primitive retentions seen in the primates is also seen in many terrestrial Insectivora. Most extant insectivores manifest no ungulate-like trends toward simplifying the limb skeleton—apart from a general but not universal tendency toward distal tibiofibular fusion, which can also occur in arboreal primates (*Tarsius*) and marsupials (*Marmosa*) (Barnett and Napier, 1953). Cursorial specializations are adaptations for rapid visually directed pursuit of prey or rapid and prolonged flight from predators, and are best developed in large mammals inhabiting open country. They would have had little or no selective advantage for the small, shrewlike mammals of the Mesozoic, and their absence does not imply arboreality.

In support of Matthew's hypothesis, Lewis (1964) points out that in reptiles the peroneal muscles arising from the fibula insert on the fifth metatarsal, but in mammals part of this musculature forms a peroneus longus muscle, whose tendon runs across the sole to insert on the first metatarsal. Lewis suggests that peroneus longus originally acted to adduct a divergent first toe in arboreal grasping. However, in extant mammals with rudimentary first toes, the peroneus longus typically persists, shifting its attachment one toe over to the base of the second metatarsal. This demonstrates that it has some important function unrelated to adduction of the first toe. An alternative explanation of its original adaptive value is that it acted to evert the foot against resistance. If the earliest mammals walked with their feet pointing somewhat sideways, as echidnas do (Jenkins, 1970), eversion would have added propulsive thrust at the end of the stance phase, and would have worked more efficiently if part of the everting musculature exerted its force through an attachment at the anterior (preaxial) edge of the foot. Intermediate stages in the shift of this attachment across the sole would yield progressively more efficient eversion, whereas, if its original function had been to adduct the first toe, selectively advantageous intermediate stages would not be possible.

In short, there is no reason to believe that the Triassic ancestors of the Mammalia had clawless, grasping extremities, as Jones's version of the arboreal theory requires. The point may be settled by forthcoming studies of the virtually complete skeleton of the Triassic mammal *Megazostrodon* (Crompton and Jenkins, 1968; Wood *et al.,* 1972). There is in any event ample evidence to show that late cynodont reptiles and their mammalian descendants progressively developed a more elaborate olfactory apparatus than is found in other reptilian lineages (Simpson, 1928; Hopson, 1969), and that the earliest mammals had relatively small and degenerate eyes, in which the sauropsidan mechanisms of accommodation and nictitation had been lost (Stibbe, 1928; Walls, 1965). These facts suggest that the earliest mammals were shrewlike terrestrial creatures, guided largely by olfactory and tactile stimuli. This does not mean that early mammals were incapable of climbing branches that presented themselves as supports or obstacles; as Jenkins (1974) points out, any small mammal needs this ability in a forest community.

THE VISUAL PREDATION HYPOTHESIS

If primate traits cannot be interpreted either as the products of a primitive arboreality retained only in primates, or as specializations necessarily selected for in any lineage of arboreal mammals, then neither form of the arboreal theory can explain why primates differ from squirrels or opossums, and an alternative set of explanations is needed. One recently proposed alternative (Cartmill, 1972, 1974a) has been induced from a survey of the distribution of primate-like traits in other taxa.

Grasping hind feet with a divergent first toe are characteristic of marsupials, chameleons, and certain arboreal mice and rats. Their adaptive significance varies. In at least some climbing mice, the grasping hallux is an adaptation to locomotion on the large siliceous stems of bamboos (Medway, 1964; Musser, 1972), on which claw grip is useless. In chameleons, grasping extremities represent a predatory adaptation, permitting prolonged and stealthy locomotion on slender terminal branches in pursuit of insects, which these specialized lizards stalk in the dense marginal undergrowth and lower canopy of tropical forests (Schmidt, 1919).

The notion that ancestral marsupials had a grasping hallux remains generally accepted. In the smaller South American opossums like *Marmosa robinsoni,* this trait correlates with a chameleon-like way of life involving visually directed predation on insects "in the intricate interlacing of vine and branch that characterizes the second growth which abounds around the edges of clearings" (Enders, 1935). Insects, which these small didelphids require for adequate nutrition (Hudson, 1932), are seized either in the hands or the mouth, bitten, and eaten held in one or both hands (Enders, 1935, Eisenberg and Leyhausen, 1972)[5]. The occasional use of the hands by didelphids in seizing prey becomes the most frequent pattern in small bush-frequenting Australian marsupials, including diprotodonts like *Cercartetus* as well as polyprotodonts like *Antechinus* (Eisenberg and Leyhausen, 1972; Hickman and Hickman, 1960). *Cercartetus* and related small insect-eating diprotodonts like *Burramys* differ from other arboreal marsupials and resemble primates in having much-reduced claws (Wood-Jones, 1924; Warneke, 1967). When allowance is made for allometry, insectivorous diprotodonts also have more convergent orbits than other marsupials (see Fig. 2C).

These comparisons suggest that the close-set eyes, grasping extremities, and reduced claws characteristic of most post-Paleocene primates may originally have been adaptations to a way of life like that of *Cercartetus* or *Burramys,* which forage for fruit and insects in the shrub layer of Australian forests and heaths. By this interpretation, visual convergence and correlated neurological specializations are predatory adaptations, comparable to the similar specializations seen in cats and owls, and allowing the predator in each case to gauge its victim's distance accurately without having to move its head. The grasping feet characteristic of primates allow insectivorous prosimians like the smaller cheirogaleines and lorisiforms to move cautiously up to insect prey and hold securely onto narrow supports when using both hands to catch the prey. Although claws are advantageous in most arboreal locomotor situations, they are actually a hindrance for a bush-dwelling animal that grasps slender twigs by opposition of preaxial and postaxial digits, and has little occasion to climb on larger supports (Cartmill, 1974a).

Olfactory regression has not been characteristic of most arboreal mammals. The slight simplification of the olfactory apparatus seen in strepsirhine prosimians, and the marked regression found in haplorhines (tarsiers and higher primates), are necessary results of the approximation of the medial walls of the two orbits; since the optic nerve leaves the base of the skull and the orbital openings lie in the dermal bones of the skull roof, the olfactory connections between braincase and snout must necessarily be constricted if the orbital cones draw closer together. This effect is evident in a comparison of small felids with canids: in the former, the interorbital space is generally narrower, and the olfactory bulbs are correspondingly smaller and have constricted connections with the olfactory fossa (Radinsky, 1969). In *Tarsius,* the close approximation of the huge eyeballs reduces the interorbital volume (filled, in typical mammals, by olfactory scrolls of the ethmoid) to a single plate of compact bone, the interorbital septum, over the top of which a few olfactory fibers arch to reach a much-reduced nasal fossa (Cartmill, 1972; Cave, 1967; Spatz, 1968, 1970)[6]. Small ceboids and cercopithecoids resemble *Tarsius* in these respects. Other lineages of visually directed predators have achieved comparable degrees of visual field overlap without pronounced olfactory constriction; in marsupials (cover photograph), optic convergence is produced by the coexistence of a low frontal region with a broad and high zygomatic arch (Cartmill, 1972), while in lorises the eyeballs come together around and outside the olfactory connections, which reach the nasal fossa between the optic nerves (Cartmill, 1972; Spatz, 1968, 1970)[6]. The unique arrangement seen in the smaller extant haplorhine primates probably reflects derivation from a big-eyed Eocene prosimian like *Pseudoloris* (which appears to have had a *Tarsius*-like interorbital septum); it does not represent perfected adaptation to arboreal life. Marsupial lineages which have evidently been arboreal since the Cretaceous have undergone no olfactory regression; arboreal life per se does not encourage loss of olfactory acuity.

Most of the distinctive primate characteristics can thus be explained as convergences with chameleons and small bush-dwelling marsupials (in the hands and feet) or with cats (in the visual apparatus). This implies that the last common ancestor of the extant primates, like many extant prosimians (for example, *Tarsius, Microcebus, Loris, Arctocebus,* and smaller galagines), subsisted to an important extent on insects and other prey, which were visually located and manually captured in the insect-rich canopy and undergrowth of tropical forests.

THE FOSSIL RECORD

Like any other evolutionary explanation, the visual-predation theory must be tested against the relevant paleontological data. Here it encounters difficulties. However we choose to define the order Primates, its early representatives differ from the earliest placentals in several features of the molar teeth, including reduction of the stylar shelf and associated cristae and decrease in the size and height of the trigonid. Since similar changes are seen in the earliest rodents and ungulates (Fig. 3), Szalay (1969, 1972a) has proposed that the differentiation of the Primates from the Insectivora involved an adaptive shift from an insectivorous diet to a predominantly herbivorous one. If true, this vitiates the visual-predation hypothesis.

Szalay's thesis has recently been challenged by Simons (1974c), who suggests that, in at least four of the six families of early Tertiary mammals usually assigned to the order Primates, the earliest representatives have molars functionally similar to those of the carnivorous prosimian *Tarsius*. Although it has been said that the carnivorous diet of *Tarsius* could not be inferred from the morphology of its dentition (Szalay, 1972b), my colleague R.F. Kay has recently developed a multivariate biometric statistic which is over 90 percent accurate in "predicting" the dietary habits of the extant primates, including *Tarsius*. Despite the reduction of the stylar shelf in extant prosimians, at least some of them have recognizable dental adaptations for masticating prey; other shearing mechanisms have replaced the primitive shear of trigonid against paracrista and metacrista (Kay, 1973b). The application of Kay's procedure to early primate dentitions will permit us to test certain aspects of the visual-predation theory.

The plesiadapoids of the Paleocene (Plesiadapidae, Paromomyidae, Carpolestidae) are assigned by paleontologists to the order Primates, although they show none of the diagnostic primate traits listed by Mivart (1873). Where known, plesiadapoid orbits are small and widely set, there is no postorbital bar, the braincase is small relative to the facial skeleton, and there is no apparent reduction of the olfactory apparatus; *Plesiadapis,* at

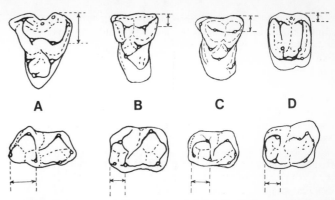

Figure 3   *Upper left (above) and lower right (below) molar teeth of (A) the Cretaceous opossum* Alphadon wilsoni, *(B) the mid-Paleocene plesiadapoid* Palenochtha minor, *(C) the Late Cretaceous ungulate* Protungulatum donnae, *and (D) the early rodent* Paramys copei. *In the latter three, the stylar shelf (vertical arrows, above) and trigonid (horizontal arrows, below) are reduced by comparison with the more primitive condition seen in* Alphadon.

least, also had clawed digits resembling those of a squirrel or dermopteran (Szalay, 1969; Russell, 1964; Wilson and Szalay, 1972). The plesiadapoids are assigned to the primates on the basis of minutely detailed resemblances between their molars and those of later undoubted primates; where known, the ear region of plesiadapoids also shows certain diagnostic primate features (Szalay, 1969; Szalay, 1972a).

There is little doubt that the plesiadapoids are close collateral relatives of the Eocene prosimians. There is also little doubt that the known plesiadapoids are not directly ancestral to the Eocene prosimian families, since at least one genus in two of the three Eocene families (Adapidae, Anaptomorphidae, and Tarsiidae) retained teeth that had been lost in known plesiadapoids [7]. Plesiadapoid lineages that can be traced through time did not converge with the early lemurs and tarsiers of the Eocene, but developed progressively more specialized dentitions displaying loss of canines and anterior premolars, hypertrophy of the fourth lower premolar, enlargement and complication of the anterior incisors, and other peculiarities. The fossil evidence suggests that the (unknown) lineages leading to the Eocene "primates of modern aspect" (Simons, 1972) must have branched off from the plesiadapoid lineages at least by the Torrejonian (mid-Paleocene).

The radiation of the phalangeroid diprotodont marsupials in Australia provides suggestive parallels with the plesiadapoid radiation. Plesiadapoid-like dental specializations, including reduction of the stylar shelf and hypertrophy of the lower central incisors, must have characterized the last common ancestor of the diprotodonts. Of the three extant diprotodont superfamilies (Kirsch, 1968), the phalangeroids have been the most successful. The ancestral phalangeroids were probably small arboreal mixed feeders; from these are derived not only the kangaroos, but also a complex radiation of arboreal marsupials. These include many forms with plesiadapoid counterparts (Fig. 4). The larger and more herbivorous phalangeroids like *Trichosurus* and *Pseudocheirus* have strong claws and a post-incisor diastema, and are roughly comparable to the plesiadapids. Smaller phalangeroids have retained varying amounts of insect prey in their diets, and generally more complete dental formulas; they can be compared to the early paromomyids. From such an ancestry there have arisen the gliding omnivore *Petaurus*, likened by Gingerich (1974c) to the specialized paromomyid *Phenacolemur*, and the

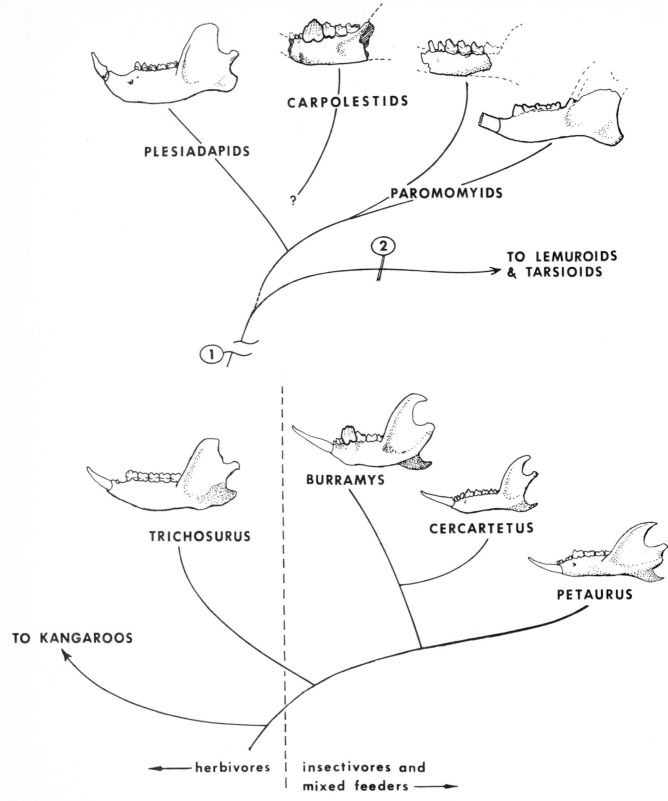

Figure 4   *(Above) Representatives of the plesiadapoid radiation (left to right:* Plesiadapis tricuspidens, Carpodaptes aulacodon, Palaechton alticuspis, Phenacolemur jepseni). *(Below) Possibly comparable extant representatives of the phalangeroid marsupial radiation: phylogenetic relationships after Kirsch (1968). The morphological shift at ①, which established the dental traits shown in Fig. 3B, is usually taken as the boundary of the order Primates. The inferred shift at ②, here considered to be a shift toward visually directed predation, could (if monophyletic) serve as the boundary of a more coherent primate order.*

mountain pygmy possum *Burramys,* whose enlarged, serrated third lower premolars, used in cutting open seeds and hard insect cuticles (Dimpel and Calaby, 1972), find a parallel in the carpolestid plesiadapoids. Unspecialized paromomyids like *Palaechthon* (Fig. 4) may prove to be plesiadapoid counterparts of *Cercartetus.*

As shown above, the adaptations of *Cercartetus* for visually directed predation among fine branches represent plausible structural antecedents for the traits that distinguish the extant primates. However, *Cercartetus* is considerably more primitive in these respects than superfically similar prosimians like *Microcebus murinus.* This is equally true of *Palaechthon,* which resembles *Plesiadapis* and differs from typical Eocene prosimians in having widely separated orbits, an unossified postorbital ligament, and a relatively small braincase (Russell, 1964; Wilson and Szalay, 1972).

Since early plesiadapoids had not acquired the traits (considered here to be adaptations to visually directed predation in forest undergrowth) that distinguish primate families from the Eocene on, and since later plesiadapoids did not converge with the true prosimians of the Eocene, it has been suggested (Cartmill, 1972) that the order Primates would be more coherent if the plesiadapoids were relegated to the Insectivora, and the postorbital bar and clawless, divergent hallux were taken as diagnostic primate traits, as Mivart considered them 100 years ago. It has been objected that "this diagnostic simplification certainly would not justify the resulting loss of phylogenetic information" (Gingerich, 1973b). Similar objections could be made to the exclusion of the therapsid reptiles from the Mammalia, or of the rhipidistian fishes from the Amphibia. Taxonomic boundaries must reflect more than mere phylogenetic affinity; they must also mark important adaptive shifts that underlie the evolutionary trends characteristic of a radiating higher taxon.

SUMMARY

Clawed digits, nonopposable thumbs and first toes, and wide-set eyes are primitive mammalian traits. For an arboreal mammal, the adaptive value of these traits is equal or superior to that of primate-like grasping extremities and closely apposed eyes. The loss of the primitive traits in the order Primates therefore cannot be explained by merely invoking the putative selection pressures imposed by arboreal locomotion per se. Visually directed predation on insects in the lower canopy and marginal growth of tropical forests is characteristic of many living prosimians, and also of small marsupials and chameleons. Primate-like specializations of the visual apparatus and extremities occur in all these groups. This suggests that grasping extremities were evolved because they facilitate cautious well-controlled movements in pursuit of prey on slender supports; and that optic convergence and stereoscopy in primates originally had the same adaptive significance they have in cats. The arboreal theory of primate differentiation, proposed in two incompatible forms by G. E. Smith and F. W. Jones, can be shown to be inadequate by counterexamples drawn from other lineages of arboreal mammals. Although some evolutionary biologists and philosophers regard such counterexamples as irrelevant, their relevance must be admitted if we want to work toward genuinely explanatory accounts of historical processes.

I thank K. Brown, P. D. Gingerich, W. C. Hall, W. L. Hylander, F. A. Jenkins, Jr., R. F. Kay, and V. T. Lukas for their help. Several of the drawings of early Tertiary mammals are redrawn after materials furnished by Dr. Kay. Portions of the research underlying this paper were supported by grants-in-aid from the Society of the Sigma Xi and the Wenner-Gren Foundation for Anthropological Research, Inc.

Notes

[1] This idea seems to have originated with T. Collins (1921).

[2] Leaps of 12 to 24 body lengths are reported for *Lemur macaco* by J. Buettner-Janusch (1973); *Galago senegalensis* is reported to leap more than 36 body lengths by J. M. Watson (1971).

[3] The data points yielding these regressions are pictured in Cartmill (1971).

[4] This is true for the three Asian lorisine species; but *Perodicticus potto* and (judging from one specimen) *Arctocebus calabarensis* have galago-like interorbital breadths. Character divergence from the sympatric galagines has resulted in a specialized dependence on olfaction in *Perodicticus* and *Arctocebus;* their retention of a broad interorbital space probably reflects this, rather than any need to maximize parallax. See P. Charles-Dominique (1971) and A. Bishop (1964).

[5] P. Murray, personal communication. Eisenberg and Leyhausen (1972) report that the hands are never used to seize prey; but the observations of Enders and Murray contradict this.

[6] Chameleons have a tarsier-like rostral configuration.

[7] Four upper and four lower premolars on each side are still present in the adapid dentition, and (in Simons' interpretation) in the dentition of the probable anaptomorphid *Teilhardina.* The lower dental formula of Eocene tarsiids may be 0:1:4:3, but this is debated by Simons (1972).

3

Where to Draw the Nonprimate-Primate Taxonomic Boundary

F. S. SZALAY

During the recent past Cartmill (1972, 1974c) has presented articulate and detailed series of arguments for a new delineation of the order Primates. In suggesting the realignment of the nonprimate-primate boundary he concluded by saying that, 'Taxonomic boundaries must reflect more than mere phylogenetic affinity; they must also mark important adaptive shifts that underlie the evolutionary trends characteristic of a radiating higher taxon' (Cartmill, 1974c, p. 442). Although Cartmill and I disagree on where to draw this line, which is artificial as all taxonomic boundaries are, we are in firm agreement on the philosophical foundations of evolutionary systematics.

It is perhaps no accident that students whose research heavily emphasizes the determination of branching among lineages and their subsequent classification are preoccupied with phylogeny *(sensu stricto)*, viewing anagenetic aspects of evolution as a matter of secondary importance compared to cladistic relationships. This school of systematics (Hennig, 1966) would increasingly abandon classificatory practices which attempt to reflect important anagenetic aspects of evolutionary history and still maintain a monophyletic system of classification (Simpson, 1961).

There is a need for the determination of well-understood cladistic and, when possible, phylogenetic relationships, and for monophyletic taxa, points on which both 'classical' evolutionary and 'cladistic' systematists agree (with some minor differences of opinion on what should constitute monophyly). Thereafter the most difficult and important decision systematists face centers on where boundaries (generic, ordinal, etc.) are to be drawn. There is a general desire to accomodate organismic classification within a Linnean system of hierarchies, yet it has long been believed that to construct a classification with a reasonable number of hierarchies, employing only a cladistic framework of reference (particularly as advocated by Hennig and followers), is just not possible (see particularly Løvtrop, 1973). What remains for systematists is to increasingly concern themselves also with phenetic relationships of taxa in an evolutionary frame, given an agreement on need for monophyly.

Once cladistic relationships between a host of taxa are understood, and the shared derived characters present in the morphotypes of various categories are known, systematists should attempt to decipher the adaptational significance of diagnostic features as, for example, has been attempted for primates, although of differing concepts of this higher category (Cartmill, 1974c; Szalay and Decker, 1974). Once this process is completed for the various morphotypes of several monophyletic taxa, the taxonomist, hopefully a biologist in the holistic sense, must judge the primary and relative importance of various adaptations. How can this be accomplished with some semblance of harmony when this is an area where most differences of opinion arise among students, as illustrated by Cartmill's (1972, 1974c) stance against including the predominantly Paleocene Paromomyiformes in the Primates?

Given sufficient morphological divergence from the ancestral group, many higher categories are set up on a model of 'primary adaptations'. Birds as a class are defined on the flight adaptations of the common ancestor; so are pterodactyls as a group. Among mammals, the whales, dermopterans, bats, and aardvarks are defined based on adaptations to certain substrates and to particular types of locomotor patterns.

The arguments for delineation of the order Primates do not differ from those suggested to be a part of a philosophical system that unites all cetaceans or all bats. The evidence (Szalay and Decker, 1974) points to the facts that the common ancestors of all primates (the Paromomyiformes included) had acquired adaptations related to habitual arboreal existence from a terrestrial ancestry, and this adaptational complex is the basis of further elaborations in this milieu. One may argue then that all subsequent adaptations have been added to this broad complex. In fact, these characters of the postcranial anatomy of early primates, in addition to such 'markers' of the cranium as a petrosal bulla, are the shared derived features (the rigorous recognition of which is the outstanding contribution of Hennigean systematics) delimiting the primates from their ancestors and contemporaries.

Cartmill (1972, 1974c), however, advocates a new diagnosis of the order Primates along grounds which are not of cladistic origin but rather involve the patristic relationships of living Primates. Cartmill (1972, p. 121) suggests that primates should be delineated at the precise point where, he believes, 'the ancestral primate adaptation involved nocturnal, visually-directed predation on insects in the terminal branches of the lower strata of tropical forests' and states that '...a monophyletic and adaptively meaningful order Primates may be delimited by taking the petrosal bulla, complete postorbital bar and divergent hallux or pollex as ordinally diagnostic.'

I have no quarrel with Cartmill's statement that some of the characteristic traits of living primates cannot be explained simply as adaptations to arboreal life. Features of any arboreal group are specific simply because of their unique transformation of ancestral inherited features for their own purposes.

The only character, however, which is common to the skull of 'real primates' (*sensu* Cartmill), the orbital ring and its loose relation to orbital convergence, is so pervasive in so many different mammals that it appears unlikely that its adaptive significance is anything but similar from group to group. No one would argue that the adaptive modifications of the skulls of, e.g., *Varecia, Megaladapis, Palaeopropithcus, Ateles, Daubentonia, Pongo,* and *Hylobates* are shockingly divergent, and that furthermore these modifications have some patristic similarity to the last common ancestor of these taxa that might very well have been a small, visual, nocturnal predator. All these and other primate taxa, however, bear the stamp of various modes of arboreal adaptations on the postcranium!

Excluding the arboreal Paromomyiformes from the Primates, however, when in fact we know that the arboreal adaptations of the order have been well established before the last common ancestor of the Strepsirhini and Haplorhini, would seem to be a violation of a practice which takes into account the broadest adaptive significance of shared derived characters in a monophyletic grouping of lineages. Most primate radiations are known to have been primarily arboreal and show extremely varied feeding adaptations and catholic dietetic preferences from one species to another. In addition, whole radiations are known to be primarily phytophagous. The delineation of the order Primates in an arboreal milieu, but based on the alleged adaptations for visual-manual predation of the last common strepsirhine-haplorhine ancestor, as Cartmill (1972, 1974c) suggests, therefore seems unsound. Were this practice to be followed widely we would be left with the Paleocene primates (many probably having been visual predators!) sharing a host of derived characters with other primates but considered either as a group in limbo or allocated to another, adaptively dissimilar group of Eutheria.

The causal explanation for the evolution of orbital convergence for the ancestor of strepsirhines and haplorhines runs into some difficulties when taxa other than those Cartmill has examined are scrutinized. It is a curious fact that the most convergent orbits among the Malagasy strepsirhines, found on the long-snouted, large *Palaeopropithecus,* are not the result of an allometric factor, as the even larger, closely related *Archaeoindris* does not show the former's extreme of orbital convergence. Admittedly, fossil taxa are difficult to assess in terms of their feeding adaptations, but that *Palaeopropithecus,* a large, forearm-dominated locomotor, was most likely a facultative herbivore seems extremely likely from its entire feeding mechanism. At least in this instance, the predation hypothesis correlates negatively with increased orbital convergence.

It follows from Cartmill's views that evolution of the complete postorbital bar and the divergent hallux or pollex may be causally related to visually directed predation. It is not my aim to dispute Cartmill's hypothesis (for which the evidence is scanty) on the adaptations of the common ancestry of the Strepsirhini and Haplorhini. In fact I would probably agree with him in his assessment of this particular morphotype and for several others for other taxa of primates.

Cartmill's assertion that the grasping feet and postorbital bar of living primates may be causally related is not supported by other Mammalia. In *Vandeleuria oleracea,* the long-tailed climbing mouse, for example (Maser and Maser, 1973), toe opposability (that of V) is well developed but the animal apparently prefers nuts and cereals as its main diet staple. A host of arboreal marsupials, and very likely the marsupicarnivoran morphotype, although probably possessing grasping feet, did not develop a postorbital bar. On the other hand, tupaiids, herpestine viverrids, suids, camels, hippos, cervids, bovids, giraffids, numerous other ungulates, and even a sirenian, *Manatus senegalensis,* have complete postorbital bars. No grasping feet, however! As far as I know, a petrosal covered auditory bulla is unique to *all* primates, *in combination* with a lost medial carotid artery and a complex of derived postcranial characters attesting to arboreal adaptations. This complex might have been derived from an arboreal insectivore or frugivore (the differences in feeding regimes are minor in these two categories), but a paromomyiform with a full dentition (see Clemens' [1974] account on the lower jaw of *Purgatorius*) did not possess a postorbital bar. Recent intensive research on feeding adaptations by Hladik *et al.* (1971) points out that frugivores, as well as folivores (the former more than the latter), take a large percentage of insects. Because insects contain protein, carbohydrates, and lipids, but leaves only protein and carbohydrates and fruits largely carbohydrates, insectivory is a necessary occupation for most arboreal mammals.

A host of paromomyiform genera were probably insectivorous (e.g., *Navajovius, Palenochtha, Palaechthon, Saxonella,* etc.); we have no knowledge, however, as to the nature of their orbits or the opposability of the pollex or hallux. What are we to do in the future if we discover that, e.g., two sister lineages of the Paromomyidae differ from one another in that one possesses an incipient postorbital bar (as in fact *Palaechthon* does indeed) and perhaps some other shared characters with a strepsirhine-haplorhine morphotype? Are we then to include this ancestral twig in a semi-Hennigean fashion and leave out the other, along with other paromomyids, picrodontids, plesiadapids, and carpolestids? This might solve the needs of primatologists and anthropologists whose primary concern is with man and primates (in increasing importance proportional to their man-like qualities) and who clearly regard all other considerations having to do with biological balance as secondary. Textbook writers for introductory courses in human evolution might find matters easier but I doubt that systematic research on the Mammalia has much to gain from the exclusion of the Paromomyiformes from the Primates.

Conclusions and Summary

The Paromomyiformes is the sister group of the Lemuriformes. Excluding it from the Primates, this in spite of the fact that its members possess most of the key, readily identifiable primate diagnostic characters, is unadvisable. It would not only cause great difficulties in classifying the archaic primates in relation to other eutherians, but such a taxonomic change would deprive primatologists from appreciating the biological and historical information that is conveyed by the retention of that archaic group within the order.

I am grateful for both discussion and helpful comments on the manuscript by two friends and colleagues, Dr. Eric Delson and Dr. Patrick Luckett.

4

Cranial Morphology and Adaptations of *Palaechthon nacimienti* and Other Paromomyidae (Plesiadapoidea, ? Primates), with a Description of a New Genus and Species

R. F. KAY AND M. CARTMILL

INTRODUCTION

The extant primates are derived from early representatives of the predominantly Paleocene superfamily Plesiadapoidea, regarded as a suborder of primates by most recent authorities (Simpson, 1935, 1945; Simons, 1972, 1974c; Szalay, 1969, 1972b, 1974a; Szalay *et al.,* 1975; Gingerich, 1974a, 1975b). The earliest plesiadapoid known from remains other than those of teeth and jaws is the Middle Paleocene paromomyid *Palaechthon nacimienti,* described by Wilson & Szalay (1972) from a skull and maxillary and mandibular fragments collected as part of the Angel's Peak faunule from the Nacimiento Formation, New Mexico, of Torrejonian age (Wilson, 1951). The cranial anatomy of *Palaechthon* is more primitive than that known for other plesiadapoids, and its dental morphology approaches that of the Early Paleocene paromomyid *Purgatorius unio,* which is primitive enough to be a possible common ancestor of plesiadapoids and primates of modern aspect. A detailed functional understanding of *Palaechthon*'s cranial and dental anatomy is therefore important to an understanding of the ecology of the ancestral primate stock. Some of our conclusions about *Palaechthon* have been presented in an earlier study (Kay & Cartmill, 1974). We wish here to support and refine these conclusions, describe a new genus and species of paromomyid that was included in the original hypodigm of *Palaechthon nacimienti,* and present an analysis of the ecology and phylogeny of paromomyids and other plesiadapoids.

The details of the geologic setting of *Palaechthon nacimienti* were published by Wilson & Szalay (1972). Specimens of this species came from a layer of reddish silt about 160 feet below the rim of Kutz Canyon, which is about 12 miles south of Bloomfield in the San Juan Basin, New Mexico. The specimen UKMNH (University of Kansas Museum of Natural History) 7903, tentatively referred to *Palaechthon* by Wilson & Szalay, comes from 70 feet lower in the Kutz Canyon Section and was found ¼ mile southeast of the other material. UKMNH 7903 forms the type for the new genus and species of paromomyid described below. Both Kutz Canyon species come from low in the *"Deltatherium* Zone," while *Torrejonia* (Wilson & Gazin, 1968) is from the succeeding *"Pantolambda* Zone" (Wilson & Szalay, 1972).

Direct correlations between structure and function can be made only for living organisms. Therefore much of what we have to say about the probable adaptations of *Palaechthon nacimienti* is based on field or laboratory information about living animals with similar morphology. Critics of this mode of interpretation have pointed out that the adaptation of any species must differ in some respects from the adaptations of its ancestors, and that the ancestral way of life influences the morphology of the descendants. The point is put concisely by Szalay (1975b):

The phenotype . . . reflects not just the adaptedness to present demands, but is clearly a compromise between the adaptedness of its ancestors and of its own. . . To understand adaptations of various species one must have some understanding of the adaptations of the inferred or actual common ancestor. Assessments of the ancestral condition, therefore, have important consequences in the evaluation of adaptations in the descendant species. Finally, diagnostic inferences on the mode of life of a given fossil taxon must necessarily be based on derived character states, as these are the ones differentiating a given form from its ancestor. Primitive character states may only represent adaptations of a precursor, although this . . . clearly does not mean that functions of the primitive morphology . . . were not part of the adaptedness of the species analyzed.

This seems sensible enough, but there are epistemological difficulties with it. The most obvious is the fact that it involves us in an infinite regress. Granting, for the sake of argument, that a species' adaptations cannot be understood without understanding the adaptations of ancestral species, how are we to gain understanding of those ancestors' adaptations? Apparently, by gaining understanding of *their* ancestors' adaptations; and so *ad infinitum.* Inferring function from the morphology of a putative common ancestor (or morphotype) is in no respect different from inferring function from the morphology of an actual fossil or living animal. We have not resolved the difficulty by ending our regress with a hypothetical construct; morphotypes have ancestors, too.

It might be urged that considerations of physics and mechanics allow us to make many reliable inferences from form to function without looking for living analogs. This is true for some inferences leading to negative conclusions; there can be no doubt, for instance, that brontosaurs were unable to fly. But positive conclusions based on biomechanical theory need to be tested for living animals before being applied to fossils. Suppose, for instance, that a study of the hand skeleton from Bed I at Olduvai Gorge, using photoelastic techniques like those applied by Oxnard (1972a) to the hands of extant hominoids, showed that stresses in this hand would have been maximized in a hanging posture, and minimized in a knuckle-walking posture. The obvious conclusion, that the animal represented by the fossil hand bones was a knuckle-walker, would have to be rejected if, for example, a similar analysis of a gibbon hand yielded the same conclusion. Such an analysis would demonstrate, not that gibbons are poorly adapted for brachiating, but that the data and methods used in the analysis can yield false conclusions. Although the laws of physics apply universally, positive inferences

Reprinted with permission from *Journal of Human Evolution,* Volume 6, pages 19–35, Academic Press Inc., Ltd., London. Copyright © 1977 by Academic Press Inc. (London) Ltd.

from mechanics to function are biological, not physical, hypotheses, and must accordingly be tested by application to living organisms.

We propose that, for any species *S*, with a complex of traits *T* whose function is not known from observation of living specimens of *S*, the inference that *T* has function *F* in *S* is warranted if:

1. There are some extant organisms (other than those in *S*) which have *T*. (No positive inferences are allowable for traits which have no analogs in extant organisms, unless the overall adaptation of *S* is reliably inferrable from other aspects of its morphology.)

2. In all extant organisms which have *T*, *T* has *F*. (An opposable hallux in *Gigantopithecus* would therefore not imply habitual arboreality, even probabilistically, though it might imply that *Gigantopithecus* had arboreal ancestors.)

3. There are no reasons for believing that *T*'s fixation in any lineage preceded *T*'s assumption of *F*. (The most important function of the projecting bridge of the nose in *Homo sapiens* today is holding up eyeglasses; but this is not the trait's original function, since it appears in the fossil record long before the invention of eyeglasses.)

4. All the features specified in the definition of *T* have some functional relationship to *F*. (All extant organisms with a petrosal bulla and continually growing incisors use the incisors to expose wood-burrowing insect larvae; but it does not follow that an extinct organism meeting this description would have done so, since the configuration of the bulla is irrelevant to the incisors' function, and such incisors serve many different functions in extant organisms.)

We have tried to hew to these principles in the following attempt to reconstruct the way of life of *Palaechthon nacimienti*. The principal difference between our approach and that recommended by Szalay resides in our conviction that a morphological trait which correlates perfectly with some function in a variety of extant animals (and satisfies rules 3 and 4 above) may be assumed to have been correlated with that function in extinct forms, even if the trait in question is one of heritage. For a trait of habitus, a perfect correlation with some function shows that acquisition of the trait is required for the appearance of its function; but for a heritage trait, such a correlation shows that *loss* of the function in question has always been accompanied by the trait's disappearance. Relatively few traits of either sort are so perfectly correlated with function, but the few that are can be taken as reliable guides to function in animals for which direct observation of function is no longer possible.

DENTITION OF *PALAECHTHON NACIMIENTI*

Anterior dentition

Fragmentary incisor alveoli are preserved in two partial mandibles of *Palaechthon nacimienti*, both of which are catalogued as UKMNH 9559. (The same catalog number is shared with an isolated premolar and a piece of mandible with several tooth roots, but these pertain to some other species.) The I_1 was considerably larger than I_2, and its root probably extended back below the canine. One of the mandibular fragments shows an I_2 socket about 0.6 mm in diameter. Judging from the alveoli, the relative size and cross-sectional shape of I_1 differed little from their counterparts in the closely related paromomyids *Palaechthon alticuspis* and *Plesiolestes problematicus*. The lower anterior dental battery of *Plesiolestes* is similar to that of extant erinaceines (Figure 1), but *Plesiolestes* has somewhat larger central incisors

than erinaceines. We infer that this comparison probably also applies to *Palaechthon nacimienti*.

The lower canine socket of *P. nacimienti* is slightly smaller than that of I_1. Its mesiodistal diameter at the alveolar border is about 1.8 mm. The canine socket is relatively slightly larger than that of *Plesiolestes* and much larger than that of *Palaechthon alticuspis* (cf. USNM 9481).

The crown of P_2 is not known, but a small socket between the canine and P_3 alveoli demonstrates the presence of a single-rooted tooth in that position. The P_2 of *Palaechthon alticuspis* is larger than P_3 and ranges from being two-rooted to having barely separable roots (Simpson, 1937). The single-rooted P_2 of *P. nacimienti* was considerably smaller than P_3, a condition approximated in *Plesiolestes*. The only known P_3 of *P. nacimienti* is heavily worn; it is two-rooted, with a large protoconid and a narrow talonid. P_4 is preserved, though heavily worn, in both of the UKMNH 9559 specimens. It gives some indications of having had a small paraconid, but the condition of the metaconid cannot be ascertained (*contra* Wilson & Szalay, 1972). The talonid of P_4 is anteroposteriorly elongated, with a distinct hypoconid supporting an oblique cristid [1], and an entoconid from which a crest runs anteriorly bordering a shallow talonid basin. On the average, the P_4 of *Plesiolestes* is more molariform than its homolog in either species of *Palaechthon*. Wilson & Szalay (1972) published measurements of the P_4 of UKMNH 9560, but this tooth is absent from the specimen and in their figure.

No upper incisors are preserved in any specimen of the genus *Palaechthon*. The type skull of *P. nacimienti*, UKMNH 9557,

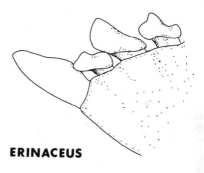

ERINACEUS

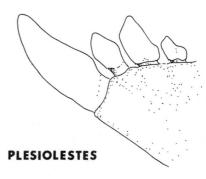

PLESIOLESTES

Figure 1 *Anterior lower dentition of* Plesiolestes problematicus *(Princeton Univ. 14149) and* Erinaceus europaeus *(not to same scale).*

preserves the right canine root in its socket, though the socket has been enlarged by post-mortem rostrad displacement of the root. The left canine has come out of its socket intact, and is lying on the surface of the palate. Its crown is oval in cross-section and triangular in lateral view. The root is between three and four times as long as the height of the crown from cemento-enamel junction to tip. There is an indistinct groove in the canine's root near the cemento-enamel junction. Behind the canine is a single-cusped, two-rooted P^2, whose crown is slightly recurved in a mesiodistal direction. Neither canine nor P^2 cusps support sharp shearing edges.

P^3 is a simple, probably three-rooted tooth with a large paracone. A trenchant shearing blade runs from this cusp back to the posterior margin of the tooth, where it bears a small cuspule. A cingulum runs around the lingual margin of P^3.

P^4 has a prominent protocone medially and a larger paracone laterally. A small but distinct metacone is joined by crests to the paracone anteriorly and a small metastyle posteriorly. A distinct parastyle is present anterior to the paracone. There is a small posterointernal basin-like expansion of the P^4 margin. The P^4 is much the same as that of *Palaechthon alticuspis*, but both species contrast markedly with *Plesiolestes problematicus*, which has a much more molariform P^4 bearing a large metacone and a prominent paraconule.

As Wilson & Szalay (1972) point out, the most marked differences between *Palaechthon nacimienti* and *P. alticuspis* are in the proportions of the antemolar teeth. In *P. nacimienti*, a large lower canine is followed by a small single-rooted P$_2$, whereas in *P. alticuspis* (e.g. AMNH 35474) the canine and double-rooted P$_2$ are subequal in size. Since Cretaceous placental mammals generally have large canines, we regard the condition in *P. alticuspis* as more derived, resulting from a specialized reduction of the canines and slight enlargement of P$_2$.

Molar morphology and occlusion

The molars of *Palaechthon nacimienti* are very similar to those of *P. alticuspis*, but the former has smaller third molars, with a small three-lobed M$_3$ (preserved in UKMNH 9561) and a small M^3 metacone. The M^3 metastyle of *P. nacimienti* is at the margin of the tooth, and its posterointernal basin is very small.

The M$_1^1$ and M$_2^2$ of *Palaechthon nacimienti* are illustrated in Figure 2. Their crown morphology is reconstructed from UKMNH 7752, 9557, 9560, and 134481.

Table 1 summarizes the events of occlusion in *Palaechthon nacimienti*. The nomenclature used in describing the chewing cycle and the teeth is that used by Van Valen (1966), Szalay (1969), Crompton (1971), Hiiemae, and Kay (Hiiemae & Kay, 1973; Kay & Hiiemae, 1973, 1974a). Shearing is *en echelon*. Phase I begins with contact between the leading edges of crests situated on the lateral margins of the molar crowns. Each of these leading edges is slightly concave, so that a lozenge-shaped space forms between each matching pair of upper and lower edges, producing double point-point shear (Crompton & Sita-Lumsden, 1969)—or, for those who prefer Every's (1974) terminology, scissorial occlusion between diakidrepanons and diakidrepanids—as the teeth are brought together and the spaces between their edges narrow. The lateral shearing crests on the lower molars continue to move upward and medially, engaging a second set of shearing blades supported by well-developed conules on the upper molars. Further upward and medial movement of the lower molars brings the crests bordering the posteromedial edges of the upper molar crowns into shearing engagement with those along the anteromedial edges of the lower molar crowns. Phase I ends with the protocone's lateral and posterolateral faces lying flush against the talonid basin and the tip of the paraconid.

In Phase II, the lower molars are moved anteromedially downward, producing a grinding action between the protocone and matching surfaces of the paraconid and talonid. The importance of this grinding phase in plesiadapoids is shown by the prominently developed third lobe on the M$_3$ of almost all species. Butler (1972, 1973) and Kay & Hiiemae (1974a,b) present comparative analyses of plesiadapoid molar occlusion.

Dental adaptations

Ingestion. Two separate stages are recognized in the oral portion of the digestive process. Food acquisition, manipulation, and separation of the bite are collectively called *ingestion*. The preparation (physical reduction) of bites of food prior to swallowing is called *mastication*. Ingestion is carried out with some combination of incisors, canines and premolars; mastication involves the premolars and molars.

Table 1 *Occlusion of second molars of* Palaechthon nacimienti. *Time is represented on the horizontal axis; sequence of occlusal events proceeds from left to right. For example, lower molar protocristid shears across 1a and 1b in Phase I and then is no longer active. Again, shearing facet 9 is only in contact with upper molars in centric occlusion and Phase II of occlusion. The postprotocingulum is also known as the "Nannopithex-fold."*

Lower molar surface or crest and wear facet	Shearing, Phase I (dorsal and anteromedial movement)	Crushing, centric occlusion (isometric contraction)	Grinding, Phase II (ventral and anteromedial movement)
Protocristid (1)	Postmetacrista (1a) → Postmetaconule crista (1b)		
Paracristid (2)	Preparacrista (2a) → Preparaconule crista (2b)		
Oblique cristid (3)	Postparacrista (3a) → Postparaconule crista (3b)		
Hypocristid (4)	Premetacrista (4a) → Premetaconule crista (4b)		
Postmetacristid (5)	Preprotocrista (5)		
Pre- and Postmetacristid (6)		Postprotocingulum (anterior part) (6)	
Preparacristid (7)		Postprotocingulum (posterior part) (7)	
Premetacristid and accompanying wear facet (8) absent			
Lateral surface of hypoconid (9)			Lateral surface of protocone
Lateral surface of protoconid (10)			Postero-lateral surface of protocone

*The anterior cingulum on each molar provides an "occlusal stop" for a protoconid and metaconid; the posterior cingulum provides the same function for the paraconid.

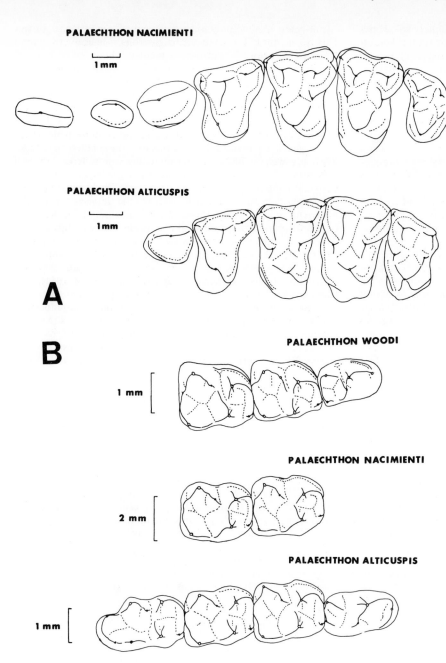

Figure 2   *Upper (A) and lower (B) dentition of* Palaechthon nacimienti *(Univ. Kansas Mus. Nat. Hist. 9557 & 7752) compared with known dentitions of* Palaechthon alticuspis *(American Museum of Natural History 35472, 35487; U.S. National Museum 9430, 9484) and* P. woodi *(Museum of Comparative Zoology 18740), the other two species of the genus. Scale shown for each dentition.*

Incisor occlusion among plesiadapoids is well known only in the highly derived genera of plesiadapids. Gingerich (1974a) inferred from a study of wear facets that incisor occlusion among plesiadapids involved a movement of the lateral edge of the lower incisor tips along the angle between the medial edge of the upper incisor crown and the anteromedial edge of its prominent medial basal cusp, creating a triangular shearing chamber which could efficiently shear interposed food. Incisor occlusion among paromomyids is more poorly understood, since upper incisors are unknown. The caliber of the first lower incisor in *Plesiolestes* and *Palaechthon* was much smaller than in plesiadapids, and it is unlikely that shearing incision utilizing the outer shearing edge of the lower incisors could have been very powerful, since the incisor roots are horizontally oriented.

Among living species, the incisor morphology of the *Palaechthon-Plesiolestes* group most clearly resembles that of erinaceines (figure 1) and to a lesser degree that of caenolestids. The first lower incisors of erinaceines resemble that of *Plesiolestes,* but are shorter and less pointed; those of caenolestids are longer and less pointed. Information about incisor use in these extant forms is scant. Captive caenolestids (Kirsch, unpublished manuscript) will tug on lumps of meat to pull off smaller bits, but this operation generally fails and pieces of food are usually chewed off between the cheek teeth. This is similar to the sort of "ingestion by mastication" practiced by *Tupaia* and *Didelphis* (Crompton & Hiiemae, 1970; Hiiemae & Kay, 1972). Kirsch's captive caenolestids also used the incisors in a "stab-bite," in which the prey was pinned to the ground by the fore paws and the incisors repeatedly thrust into it. We know much less about incisor use among erinaceines. Examination of the incisors of erinaceines at the United States National Museum shows that many have heavily worn upper and lower incisors but that the wear was not usually produced from tooth-tooth contact. Sharp incisor edges suitable for shearing are not maintained. Hedgehogs do not have a killing bite in which the prey is seized and shaken to death (Poduschka, 1969), and their incisors presumably serve mainly for stabbing and holding prey, as in caenolestids.

Thus, direct and indirect evidence on incisor use in extant mammals with incisors resembling those of early paromomyids suggests that these Paleocene species did not utilize their incisors for cutting up foods by shearing. A manipulative grasping or stabbing function is more probable. The use of the incisors for powerful shearing or crushing, inferrable for some later plesiadapoids like *Plesiadapis*, is a specialization not seen in *Plesiolestes* or *Palaechthon*.

Mastication: molar morphology, body size and diet. Inferring diet from dental morphology depends to a considerable extent on knowledge of the body mass. Our estimate of body mass for *Palaechthon nacimienti* rests on the assumption that extant animals of known weight and skull dimensions provide accurate models for predicting body weight in fossil species with similar skull shapes and dimensions. A tupaiid model was chosen, since *Palaechthon* shares with this group relatively small orbits, broad interorbital dimension, and a rather long dental arcade. Two methods of estimation were used. A multiple regression (ln body mass = 3.917 (ln dental arcade length) — 1.577 (ln skull length) —1.578), based on relationships between body mass and skull dimensions in 5 tupaiid species, yields an estimated body weight of 106 g for *P. nacimienti*, with a multiple *R* of 1.00000 and a standard error of 0.0008. When separate body weight estimates are generated using regression equations between body mass and each cranial dimension for tupaiids, the average of 8 estimates is 100 g with a 95% confidence interval of ±51.6 g in *Palaechthon*. The measurements yielding these estimates (Table 5) all involve reconstructed areas of the skull, but the possible error is not large.

Kay (1973b) and Kay & Hylander (1978) have shown that, for their body masses, living insectivorous primates and marsupials tend to have larger molars than their fruit- and sap-eating relatives. Insect-eaters also have higher and more acute cusps (enhancing their puncturing functions), longer molar shearing blades, and enlarged crushing surfaces on the molars. Table 2 presents the expected average tooth dimensions of an insectivorous primate or tree shrew weighing 100 g, the estimated body weight of *Palaechthon*. Also shown are the expected average molar dimensions for 100 g fruit- and sap-eating primates (see Kay, 1973b, for details). *Palaechthon* most closely approaches the insect-eater model. We are confident that *Palaechthon nacimienti* was not primarily a

fruit-eater. Our 95% confidence interval for body mass in this species is 48–152 g. A living frugivorous primate with similar tooth dimensions would have a predicted body mass of about 1100 g (Kay, 1975, Table5). This is more than 19 standard deviations away from the body mass predicted from skull dimensions of *Palaechthon,* and makes it highly improbable that *Palaechthon's* diet was mainly fruit. There is a possibility that *Palaechthon* was a leaf-eater, since the molars of folivorous species also tend to be large with a pronounced emphasis on shearing. However, body size practically rules this out; the smallest folivorous primate weighs 700 g (12 standard deviations from our best estimate of *Palaechthon's* body mass). Large size is advantageous for mammals with high-cellulose diets. Small mammals pass food through their digestive tracts faster than large ones. Since cellulose can only be broken down if it remains in the gut for some time, increased body mass allows more efficient digestion of leaves or other plant substances which are high in cellulose (Parra, 1978). A few exceptional small rodents are folivores. The microtine *Phenacomys longicaudus* weighs less than 100 g and eats conifer needles (Howell, 1926; Benson & Borell, 1931). This ability may be related to the extremely specialized microtine molar dentition, which probably allows food to be reduced to particles sufficiently small to allow digestive processes to act on them more efficiently, thus compensating for the food's rapid passage through the gut. *Palaechthon's* molars are insufficiently specialized to permit folivory in a mammal of its size.

DENTAL FUNCTION IN PLESIADAPOIDS

We have considerable confidence in our estimation of the diet of *Palaechthon nacimienti* because we know its tooth dimensions and probable body weight range, and because it has been demonstrated (Kay, 1973b) that living insect- and fruit-eating species can be distinguished from each other if we know tooth dimensions and body weight. If we could estimate the body weights of the other Paleocene species, we could determine their diets with equal confidence. But rigorous body weight estimates are not possible given the fragmentary evidence for most plesiadapoid species.

An alternative procedure utilizing ratios was undertaken. Table 3 presents ratios of the M_2 oblique cristid shearing length to M_2 length, for samples of five predominantly insectivorous and five predominantly frugivorous or gummivorous species. The sample sizes of the two groups are small, and statistical tests do

Table 2 Palaechthon nacimienti. *Second molar dimensions compared with expected for insectivorous and frugivorous models based on data from Kay (1973b). Arrows indicate which model is most closely approximated. Body mass = 100g.*

	Insectivorous model (mm)	P. nacimienti (mm)	Frugivorous model (mm)
Tooth length	2.26	← 2.41	1.73
Oblique cristid	0.92	← 0.88	0.76
Crown height (hypoconid)	1.34	← 1.54	0.91
Phase I	0.87	← 1.13	0.59
Phase II	0.83	← 0.98	0.63
Crushing surface	1.36	← 1.22	0.64

Table 3 *Ratios of oblique cristid to molar length in extant species. Sample sizes in parentheses.*

Species	M_2 oblique cristid/M_2 length
I. Insectivorous species:	
Arctocebus calabarensis (1)	0.44
Galago demidovii (5)	0.41
Loris tradigradus (3)	0.50
Tarsius spectrum (8)	0.38
Tupaia glis (1)	0.41
Mean for 5 species	0.43
S.D.	0.05
II. Gummivorous and frugivorous species:	
Saguinus geoffroyi (8)	0.35
Perodicticus potto (5)	0.27
Euoticus elegantulus (5)	0.40
Phaner furcifer (1)	0.37
Cheirogaleus major (2)	0.25
Mean for 5 species	0.33
S.D.	0.06

not reveal significant differences, but it is noted that four of five frugivorous species have smaller ratios than any of the insectivorous species, and that the mean ratios of the two groups are almost two standard deviations apart. Thus, in all but one of the sampled species, a ratio of M_2 shearing to M_2 length permits an estimate of the diet.

The use of the oblique cristid as a standard estimate of the second lower molar shearing capacity, used above for living primates, presents some difficulties when plesiadapoids are involved. The process of trigonid reduction and talonid expansion so characteristic of Eocene and later primate species was not as far advanced in plesiadapoids. Thus, shearing crests supported by the cusps of the trigonid appear to be proportionately larger in plesiadapoids than in most living species, while the crest on the edges of the talonid (including the oblique cristid) are comparatively reduced. Using the oblique cristid alone as an estimate of shear for fossil species may underestimate the shearing capacity of the second lower molar. Therefore, an alternative ratio using the sum of the lengths of four shearing blades and M_2 length was used, since shearing is known to be more important for food preparation in insect-eaters than in fruit-eaters (Kay, 1975). Table 4 presents estimates of total shear on the second lower molars of representative Paleocene species. Total M_2 shear was estimated by summing the lengths of all shearing blades and dividing by M_2 length. *Palaechthon alticuspis, Palenochtha minor* and *Plesiolestes problematicus* have considerable amounts of second lower molar shearing, while shearing is reduced in *Paromomys maturus* and *Phenacolemur praecox*. Our tentative conclusion is that *Palenochtha minor* was extremely insectivorous; that *Palaechthon* and *Plesiolestes* were insectivorous but took a small amount of fruit, nectar, or gum; and that *Paromomys* and especially *Phenacolemur* fed predominantly on fruit, gum, or nectar.

Observations based primarily on figures and descriptions in the literature are combined with the above standard analyses in Figure 3, which indicates the probable feeding regimes of the better-known Paleocene genera. The dentally most primitive early and middle Paleocene genera, *Purgatorius, Palenochtha* and *Palaechthon*, were insectivorous. They probably did not use

Table 4 *Summed lengths of second lower molar shearing blades of some Paleocene paromomyids. Numbered blades are identified in Table 1. Sample sizes are shown in parentheses.*

	Total Shear $(2 + 3 + 5 + 6)/ M_2$ Length
Palenochtha minor (1)	1.76
Palaechthon nacimienti (1)	1.64
Palaechthon alticuspis (2)	1.59
Plesiolestes problematicus (3)	1.63
Paromomys maturus (1)	1.25
Phenacolemur praecox (1)	1.36

their incisors and canines to bite off pieces of resistant food items for mastication; this function was relegated primarily to the cheek teeth. Incisors may have been used to grasp prey and subdue it by puncturing. The progressive paromomyid *Phenacolemur* has molars suggesting a diet composed of fruit, nectar or gum. *Phenacolemur's* long and slender incisors were suitably specialized for use as gouging, prying and grasping organs. *Petaurus australis*, the marsupial sugar glider, has similarly specialized incisors which it uses to make V-shaped gouges in the bark of preferred food trees, much like a man tapping a sugar maple; the resulting sap flow is licked up periodically from the apex of the V (Wakefield, 1970). The enlarged and simplified P_4 of *Phenacolemur* probably functioned in puncture-crushing and incision by mastication (Gingerich, 1974c). The less specialized paromomyids *Torrejonia* and *Paromomys* have reduced molar shearing, suggesting a mixed fruit-insect diet.

Pronothodectes, the earliest plesiadapid known, had stout anterior incisors which were almost certainly used for cutting up resistant food into pieces before mastication. The horizontal ramus of the mandible is deep, allowing it to resist the forces generated in powerful incisor biting. (This functional system was maintained in other plesiadapids, and is seen in an exaggerated form in *Chiromyoides*.) *Pronothodectes* was a small animal, and its molars retain well-developed shearing features. We infer that *Pronothodectes* and other (unknown) primitive plesiadapids fed

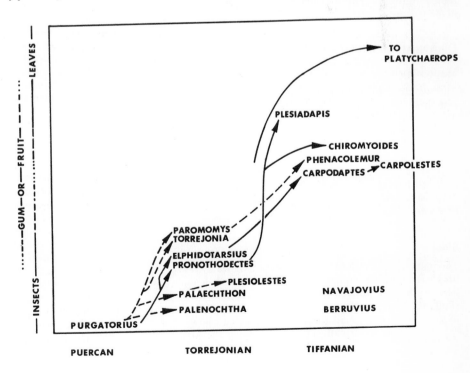

Figure 3 *Evolution of feeding regimes in plesiadapoids. Dashed arrows indicate paromomyid lineages.*

chiefly on insects, with some fruit. *Plesiadapis* and *Platychaerops* were larger than *Pronothodectes,* but maintained or increased the emphasis on molar shearing. This combination of features is observed today in leaf-eating species. In *Chiromyoides,* shearing features of the molars appear somewhat reduced, suggesting a tendency toward fruit-eating.

Apparently evolving from a *Pronothodectes*-like ancestor, Paleocene carpolestids develop progressively longer and slenderer incisors (Rose, 1975), indicating a decrease in the importance of these teeth in powerful incising. This function was assumed by the premolars. Carpolestid premolars have been modified into an ingestive shearing mechanism in which a lower molar "knife" occludes against an upper molar "butcher block" (Rose, 1975; Butler, 1973). Rose (1975) has pointed out that this shearing system does not resemble the shearing system found in some macropodid marsupials like *Bettongia;* in the latter, a lower premolar shearing blade is moved past a single upper premolar shearing blade. The molars of carpolestids are short anteroposteriorly, producing some reduction in molar shearing capacity and probably reflecting an increasingly frugivorous diet.

In summary, dental structure indicates that most middle Paleocene North American plesiadapoids were, like *Palenochtha* and *Palaechthon,* primarily insectivorous. A few, like *Torrejonia* and *Paromomys,* had a mixed diet including insects and fruit or gum. By late Paleocene times, a range of plant-eating adaptations had evolved. Some of the Plesiadapidae (e.g. *Plesiadapis*) had become leaf eaters, and some (e.g. *Chiromyoides*) were fruit eaters. Carpolestids (*Elphidotarsius, Carpodaptes* and *Carpolestes*) were probably fruit and nut eaters. The paromomyid *Phenacolemur* may have been feeding primarily on tree sap. The late Paleocene genera *Berruvius* and *Navajovius* were apparently insectivorous, but are poorly known and may not have been plesiadapoids in any case. This study does not support the conclusion of Szalay (1972b) that the origin of the order Primates was marked by an important dietary shift to plant eating. Early and middle Paleocene plesiadapoids appear to have been insectivorous. Specializations for eating plants do not appear among plesiadapoids until as much as 6 million years after the earliest known representatives of the group, and the lineages leading to primates of modern aspect branched off from plesiadapoids long before the appearance of these dietary specializations.

CRANIAL MORPHOLOGY

Mandible

The mandible of *Palaechthon nacimienti* exhibits an unfused symphysis which extends back to a point underneath P_3. The jaw's horizontal ramus is moderately and uniformly deep underneath the dental alveoli, as in *Palaechthon alticuspis* and other dentally primitive plesiadapoids, but is much shallower than that of any of the genera which have undergone enlargement of the incisors (e.g. *Phenacolemur, Plesiadapis*). The vertical ramus arises posterior to M_3, where a marked linea obliqua delineates a deep masseteric fossa. The coronoid, condylar and angular processes are missing on the known specimens.

Skull

The referred skull (UKMNH 9557; Wilson & Szalay, 1972, Figures 5-7) has been compressed dorsoventrally, and all bones have been crushed and broken to some extent. The caudal end of the skull, the zygomatic arches, and both premaxillae are lost, together with other smaller elements mentioned below.

The palate is virtually intact apart from the loss of the premaxillae, but the left and right sides have been separated posteriorly along the midsagittal suture, which is discernible along its entire length and remains intact anteriorly. Each palatine is pierced by a palatine foramen roughly midway between the interpalatine suture and the posteromedial corner of M^2. The posterior edge of each palatine displays a notch which is considerably larger than the palatine foramen and lies lateral to the anterior root of the pterygoid process. The foramen and notch correspond respectively to the greater and lesser palatine foramina of *Homo,* and presumably transmitted homologous branches of the maxillary nerve and vessels, as in most mammals. Although no incisive foramina are discernible, the anterior edge of the preserved part of the palate presents a roughly bilaterally symmetrical curve which resembles the maxillary-premaxillary suture of *Erinaceus,* and it seems likely that *Palaechthon,* like *Erinaceus,* had relatively small incisive foramina lying far anterior to the canine alveoli. The same would be true of *Plesiadapis tricuspidens* if it had any canines (Russell, 1964). We now reconstruct *Palaechthon's* skull with a longer premaxilla than in our earlier reconstruction (Kay & Cartmill, 1974), to allow for the incisive foramina (Figure 4).

The anterior margin of the preserved part of the rostrum is nearly bilaterally symmetrical. We conclude that the premaxillae were broken off and lost without significant damage to the maxillae, and that the irregular edge bounding the maxillae anteriorly closely follows the maxillary-premaxillary suture. In *Plesiadapis,* this suture extended back to the frontal (Russell, 1964; Gingerich, 1974a). The premaxillae of *Palaechthon* must have been considerably smaller and shorter, and the upper central incisors correspondingly less enlarged, than their homologs in *P. tricuspidens.* This is supported by the mandibular morphology (see above), and by the relatively small premaxillae found in *Phenacolemur pagei* and *Carpolestes dubius* (Gingerich, 1974a; Rose 1975).

No sutures are clearly distinguishable on the dorsal surface of the skull of *Palaechthon,* but the sutures bounding the nasals laterally may be represented by a bilaterally symmetrical line of fractures. On the left side, this fracture line bounds a gap in the roof of the rostrum; this gap is bounded in the midline by a straight edge representing the internasal suture. We conclude that the left nasal has probably been lost *in toto,* and that the resulting gap and its symmetrical reflection on the right may reveal the shape of the nasal bones. If so, *Palaechthon's* nasals were shaped like those of *Didelphis,* broadening markedly near their posterior ends and then converging on nasion, which lay posterior to the anterior orbital margin (Figure 4). Szalay's (1969) reconstruction of *Microsyops* has similar nasals, but *Plesiadapis tricuspidens* has narrower and shorter nasals (Gingerich, 1974a), due at least partly to the specialized posterior expansion of the premaxilla.

In the lateral view, the most significant feature of the rostrum of *Palaechthon* is the large infraorbital foramen, situated above the P^4 paracone as it is in many short-faced mammals of similar size (e.g. *Erinaceus, Hapalemur*). The lacrimal foramen appears to have been intraorbital, as in *Adapis* (Major, 1901). This represents another difference between *Palaechthon* and *Plesiadapis,* the latter genus having an expanded lacrimal bone with a large *pars facialis* pierced by the lacrimal foramen (Russell, 1964; Gingerich, 1974a). The position of the lacrimal foramen of *Palaechthon* is probably indicated on the left side by a rostrad bulge at the very front of the sharp orbital margin; this bulge defines the anterior edge of a shallow lacrimal fossa, posteriorly demarcated from the orbit proper by a crest between it and the posterior opening of the infraorbital canal (Plate 1). A similar configuration is seen in solenodontids, tupaiids, didelphids, and carnivorans, and is probably primitive for Theria, as suggested by Major (1901), Gregory (1920*b*), and Butler (1956).

The left orbit of *P. nacimienti* (Plate 1) is reasonably well preserved, although its superior margin has been crushed ventrally and the upper and lower post-orbital processes are missing. The existence of a short superior post-orbital process can be inferred from the fact that the superior temporal line is more prominent anteriorly than posteriorly. The preserved portion of the superior orbital margin is clearly defined, forming a smooth curved edge away from which the skull roof sloped medially and horizontally, and the medial orbital wall sloped medially downward. Although there was not a discrete supraorbital shelf formed by a simple lamina of bone projecting laterally, as in *Sciurus* or *Tupaia*, the superior orbital margin of *Palaechthon* was accentuated and laterally protrusive to a degree exceeding that seen in *Plesiadapis* or extant erinaceines. This is one respect in which *Palaechthon macimienti* seems less primitive than *Plesiadapis tricuspidens,* unless an allometric factor is involved.

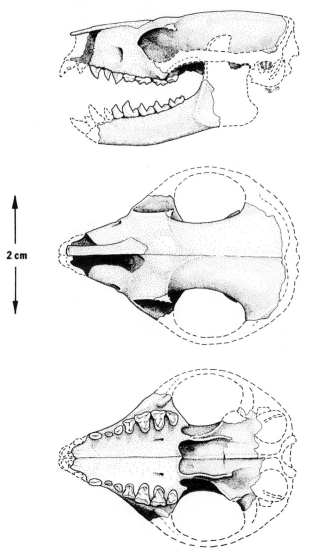

2 cm

Figure 4   *Reconstructed skull of* Palaechthon nacimienti, *based chiefly on UKMNH 9557 and 9559. The left nasal is restored in the lateral view (top).*

The possibility of a postorbital bar in *Palaechthon* cannot be conclusively ruled out, since the postorbital processes and zygomata have been lost. If present, the bar would have resembled that of *Tupaia*. However, we consider this unlikely for several reasons. The base of the right superior postorbital process was vertically shallow, and did not extend ventrally down onto the side of the skull even for a short distance, as it does in *Tupaia*. The observed morphology is more like its counterpart in *Sciurus*. Furthermore, the small size of the orbits of *Palaechthon* (see below) and their lateral orientation would have rendered a postorbital bar unnecessary, whether the function of the bar is to protect a laterally exposed periorbita (Prince, 1953), resist tension in the temporal fascia (Cartmill, 1972), or, as has been suggested for ungulates, to allow the orbit to project laterally beyond the temporalis for panoramic vision (Collins, 1921). Finally, the absence of a bar in *Plesiadapis* argues against its presence in an earlier and more generalized plesiadapoid like *Palaechthon*.

The anterior orbital margin of *Palaechthon* extends forward to a point above P[4] when the skull is oriented on the palate. This is a slightly more anterior position than that seen in most didelphids and insectivorans (Butler, 1956; McDowell, 1958; Cartmill, 1970). Since the orbit of *Palaechthon* is both small and displaced anteriorly, it follows that the anterior temporalis was larger than in extant prosimians. As can be seen from the reconstructed skull (Figure 4), the anterior edge of the mandible's ascending ramus lay well posterior to the orbit, and contraction of the temporalis as a whole must have produced a roughly vertical resultant. By contrast, the crest marking the origin of the superficial masseter at the anterior end of the zygomatic arch was slightly more anterior than in most extant prosimians (Kay & Cartmill, 1974), and this muscle must have had a correspondingly more horizontal orientation, as in *Nycticebus* or *Tupaia*.

The cranial base is badly fragmented, and no foramina or sutures can be identified. The most unambiguous feature of the cranial base is a midventral ridge or keel, which extends backward from the vomer along the entire length of the basicranium, becoming somewhat more pronounced posteriorly. We have seen nothing much like this in any other mammals, fossil or extant, and therefore cannot offer any testable hypotheses concerning its significance. A similar but lower and shorter crest is seen between the pterygoid laminae of *Plesiadapis tricuspidens* (Russell, 1964).

The caudal half of the midventral crest is displaced to the right, and it is evident from this and other landmarks that the posterior end of the skull has been twisted clockwise as viewed from the front. The posterior end of the undisplaced anterior half of the midventral crest is exposed, and the surface of the break is clean and smooth at this point. We infer that this marks the location of the synchondrosis between the basisphenoid and presphenoid; it is too far anterior to be the spheno-occipital synchondrosis.

Two ridges with broken ventral edges run backward along the cranial base on either side of the midventral crest. These ridges converge anteriorly and end in the pterygoid processes of the sphenoid, lateral to the choanae. We conclude that the ridges represent the bases of the lateral and medial pterygoid laminae, enclosing a large and probably shallow pterygoid fossa between them. The lateral laminae appear to have been crushed upward and laterally on each side. An alternative interpretation would regard the ridges as representing the outlines of the tympanic cavity on each side. We consider this highly improbable for several reasons. Such an interpretation implies that the choanal region, braincase and zygomatic arches were extraordinarily foreshortened; that the clean break in the midsagittal crest represents the spheno-occipital synchondrosis, lying in an exceptionally

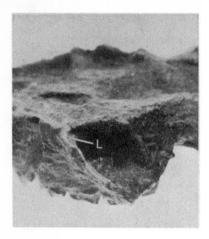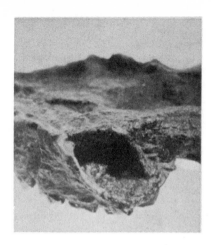

Plate 1 *Left orbit of* Palaechthon nacimienti, *UKMNH 9557, dorsolateral aspect
(stereo pair). The large infraorbital foramen is visible at the left. L, supposed
lacrimal fossa.*

posterior position between the centers of the two bullae; and that the basi-occipital extended far back behind the posterior margins of the bullae. All of this is not impossible, since similar morphology is seen in some specialized didelphids (Reig & Simpson, 1972), but it seems unlikely. Interpretation of the divergent ridges as bullar walls is further countermanded by the absence of any marked contours or other details on the floor of the fossa enclosed by the ridges, and by the fact that the medial ridge comes to an abrupt and clearly visible posterior termination on the left side, which the medial wall of the tympanic cavity would not do. We conclude that the divergent ridges represent the pterygoid laminae, and that the pterygoid fossa and musculature were correspondingly large.

The skull base terminates posteriorly in a ragged break that may mark the approximate location of the spheno-occipital synchondrosis. The basioccipital and petrosals are wholly or almost wholly lost, though some of the most posterior basicranial fragments probably represent pieces of these bones. Nothing can be said about the morphology of the bulla or occiput.

The roof of the braincase is crushed, distorted and virtually featureless, the only unambiguous landmarks being the superior temporal lines. These converge on a point about 7 mm posterior to the superior postorbital processes, and continue backward as a faint midsagittal line or ridge. A symmetrical pattern of fractures midway between nasion and the point where the two temporal lines meet may represent bregma, but this is dubious. The region between nasion and the supposed bregma forms a depressed concavity, which we interpret as the crushed roof of the chamber housing the olfactory bulbs.

Inion is not preserved. This is important, since some of our metrical comparisons of *Palaechthon* use prosthion-inion length as the independent variable in computing regressions. We regard the position of prosthion as less dubious than that of inion, but both are hypothetical, and we stress the possibility that we have underestimated skull length. Our reconstruction (Figure 4) places inion less than two millimeters behind the edge of the preserved part of the skull roof. We believe that the skull roof is virtually complete for two reasons. First, we believe that an ambiguous medially-directed ridge on some of the left parietal fragments may represent the posterior edge of the left temporal fossa. The second reason is intuitive: increasing the length of the skull roof in the reconstruction produced a disproportionate-looking result. Since an increase in 10% in prosthion-inion length would shift *Palaechthon* toward prosimians in some comparisons and away from them in others, and would produce no significant change in its position in any comparison, the question is not especially critical.

Functional inferences

The skull of *Palaechthon nacimienti* is too poorly preserved to permit a realistic reconstruction of the masticatory musculature, but, as pointed out in a preliminary report (Kay & Cartmill, 1974), the anterior temporalis and medial pterygoid appear to have been especially well-developed. Although *Palaechthon's* relatively small premaxilla makes it unlikely that the large incisors were used habitually in prolonged gnawing, *Palaechthon* was clearly well-adapted for powerful incisal biting.

An earlier study by the second author (Cartmill, 1970) employed a suite of cranial measurements designed to reflect primate-like evolutionary trends in the morphology of the orbits and rostrum of 63 species of therian mammals. Homologous measurements were taken or estimated for *Palaechthon* (Table 5), and extended to a number of fossil and extant primates and insectivorans not examined in the earlier study. A few new measurements were also added to the original set. Not all measurements were obtained for all specimens examined, due to the difficulties of returning to distant collections to take new or revised measurements. Some of the measurements taken on damaged fossil specimens involved guesswork; where possible, both authors took independent measurements on these specimens, and average estimates were used in constructing the figures. The species examined (and the sample sizes for the relevant measurements) are listed in the captions of Figures 5–7.

Most of the measurements employed (e.g. those reflecting rostral length) lack any single or simple functional correlate. We present here results for only one of these ambiguous measurements; minimum interorbital breadth. Figure 5 is a log-log plot of this dimension against prosthion-inion length. In those groups where the two variables are significantly correlated, the average slope of the least-squares regression is 1.25. Thus, larger animals can be expected (other things being equal) to have a relatively

Table 5 *Estimated skull dimensions of* Palaechthon nacimienti *(UKMNH 9557). Figures with asterisks are based on the skull reconstruction.*

Orbit diameter	6.5 mm
Interorbital breadth	14.5 mm
Prosthion-nasion	16.3 mm
Internasal suture length	14.5 mm
Dental arcade length	21.5 mm*
Preorbital rostrum	14.0 mm
Dental arc. width (maximum)	16.4 mm
Prosthion-inion	40.2 mm*

broader interorbital region. This is partly or entirely a result of negative allometric change in eyeball diameter.

As expected, the taxa sampled fall into a series of parallel regressions, with aberrant outliers in some groups. When allometry is taken into account (i.e. by measuring each taxon's distance from some line with a slope of 1.25), *Loris tardigradus* is seen to have the narrowest interorbital region of all mammals studied (Cartmill, 1972). Cercopithecines have an only slightly broader interorbital region; *Cebus* and *Nycticebus pygmaeus* lie on or near the upper edge of the cercopithecine cluster. Colobines and *Tarsius* lie still further away from the cercopithecine regression. The lemuriform regression comes next, overlapped at its upper end by the carnivoran cluster and at its lower end by the lorisiform cluster. Marsupials (didelphids and phalangeroids) have still broader interorbital regions, overlapping the upper parts of the lemuriform and carnivoran polygons. Tupaiids form a still higher parallel regression, with which *Echinosorex, Cynocephalus,* and *Daubentonia* are collinear. The greatest interorbital breadths are found in sciurids, *Rhynchocyon,* and *Phascolarctos.*

Most of the fossils measured fall into appropriate places on this plot. *Adapis parisiensis, Leptadapis magnus, Pronycticebus gaudryi, Notharctus tenebrosus,* and *Rooneyia viejaensis* (measured from a cast, and therefore not to be trusted overmuch) have interorbital dimensions resembling those of Madagascar lemurs. So does *Megaladapis. Tetonius* falls in line with the upper edge of the lemuriform distribution, roughly midway between *Eudromicia* and *Microcebus. Plesiadapis,* the Oligocene leptictid *Ictops,* and the erinaceoid "*Anchomomys latidens*" (Szalay, 1974b) fall into the insectivore regression. Three species, however, have somewhat unexpected positions. *Archaeolemur majori* and *A. edwardsi* have colobine-like interorbital dimensions (corrected for allometry); further study might show this to be true of other large extinct indriids as well. Finally, *Palaechthon nacimienti* falls at the extreme upper edge of the distribution, having a relatively wider interorbital region than any of the mammals studied except four of the tree squirrels.

The functional significance of great interorbital breadth is complex and not always clear. In some mammals (e.g. *Nasua* as compared to other procyonids), great interorbital breadth is clearly the result of an increase in size and complexity of the olfactory fossa, with consequent lateral displacement of the medial orbital walls. In other mammals (e.g. *Dactylonax* as compared to other phalangers—Cartmill, 1974b), increased interorbital breadth is associated with incisor hypertrophy and probably serves to dissipate compressive stress engendered in powerful incisal biting. Both factors may be involved in some cases—e.g. *Daubentonia.* Still other mammals have accentuated interorbital breadths for more obscure reasons. Cases in point are those mammals with laterally-facing orbits whose upper edge is drawn laterally to form a thin supraorbital shelf, of no evident relevance

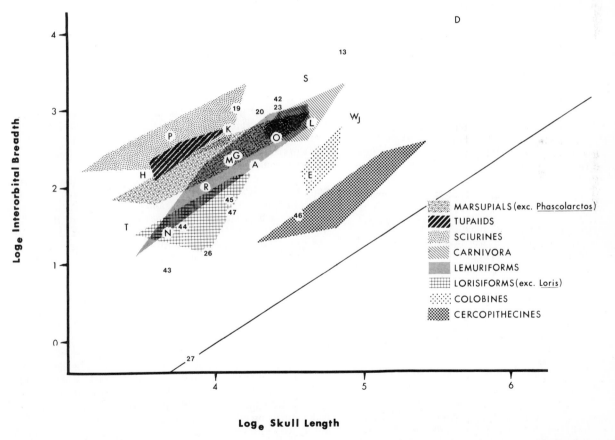

Figure 5 *Skull length and minimum interorbital breadth in* Palaechthon nacimienti *(P) and other mammals (measurements in mm). Species sampled, sample sizes, specimen numbers and code letters and numbers are listed in the caption of Figure 7, except* "Anchomomys" latidens, *(Mus. nat'l. d'Hist. nat. Qu 11012), indicated here by the letter H. Only species mentioned in the text are indicated by code symbols. The diagonal line through the* Loris tardigradus *datum (27) has a slope of 1.25, the average for groups with significant correlation between interorbital breadth and skull length. Aberrant outliers (*Phascolarctos, Loris*) were included in their taxa in computing correlations, but are set off here. (Data partly from Cartmill, 1970).*

to stress transmission or olfactory hypertrophy. This sort of arrangement is seen in typical tree squirrels (as contrasted, for instance, with *Tamias*), tupaiines (as contrasted with *Ptilocercus*), leporids (as contrasted with *Ochotona*), elephant shrews, and many ungulates. In some of these cases, the shelf may help to increase panoramic vision to the rear, by allowing the eyes to protrude laterally beyond the anterior temporalis. It is not clear, however, that *Sciurus* (for instance) has any real advantage over *Tamias* in this respect. Although supraorbital ossification correlates with arboreality in some groups (Hershkovitz, 1974a), it does not in others (e.g. lagomorphs, tupaiids). Extant mammals with well-developed supraorbital shelves are usually macrophthalmic; the combination of small orbits with extreme interorbital breadth, seen in *Palaechthon* and *Plesiadapis*, is not found in any of the extant mammals examined except *Phascolarctos*, where it is associated with an extraordinary invasion of the frontal region by the maxillo-turbinal (Wegner, 1964). This combination is approached, however, in *Ictops* (YPM 14041, 14046). Lacking extant analogs, we can only suggest that the morphology seen in *Palaechthon* certainly allowed a large olfactory fossa, and may also have served to dissipate compressive stresses produced in the skull roof during incisal biting.

Orbital diameter is plotted against skull length in Figure 6. Orbital diameter was not measured in the same way in all the species studied, since the superior or lateral orbital margins remain unossified in some animals, but we are persuaded that the results are morphologically comparable from one group to another. In the smaller animals examined, orbit diameter is positively correlated with skull length, but displays negative allometry—i.e. larger animals have absolutely larger, but relatively smaller, orbits. This is due to the fact that (other things being equal) visual acuity depends on the absolute number of receptor cells in the retina, so that small animals cannot afford to have proportionately reduced eyes. Three parallel but overlapping distributions can be discerned in the smaller placental species (up to prosthion-inion lengths of about 75 mm) shown in Figure 6. The central one of these three comprises small diurnal animals that rely to a considerable extent on vision: sciurines, tupaiines, *Rhynchocyon*, *Callithrix*, and *Callicebus*. A parallel distribution is seen among relatively large-eyed nocturnal animals which rely heavily on vision: lorises, galagos, cheirogaleids, tarsiers, *Lepilemur*, *Aotus* and *Galeopithecus*. The third parallel distribution comprises relatively small-eyed nocturnal forms, including *Erinaceus*, *Hylomys*, *Echinosorex* and many of the marsupials.

The marsupial distribution is worth discussing in some detail. All the sampled marsupial species are nocturnal, like all marsupials except *Myrmecobius* and a few macropodids. The marsupial distribution parallels the placental distributions, but shows less internal variability in relative orbit diameter. Small-eyed, rather terrestrial didelphids (*Didelphis, Metachirus, Monodelphis*) have orbits relatively as small as the erinaceids', reflecting similar terrestrial foraging habits. Didelphids and phalangeroids that rely more on vision in foraging (*Eudromicia, Marmosa, Caluromys*) have relatively larger orbits, but not so large as those of comparable nocturnal placentals; they fall at the edge of or into the diurnal placental distribution.

When skull length exceeds 75 mm, orbit size begins to predict activity patterns less reliably. At the upper end of the scale, there is fairly complete overlap of all distributions, and one finds nocturnal marsupials (*Phascolarctos*), dim-sighted nocturnal viverrids, and diurnal monkeys clustered together on the bivariate plot. Cercopithecoids of virtually identical skull lengths may differ considerably in orbit diameter, or the converse may be true. The most important factor producing this breakdown in correlation between relative orbit size and activity patterns is probably negative allometry in eyeball diameter. In small mammals, the eye is relatively large, and fills the available orbital space. In large anthropoids (Schultz, 1940; Washburn & Detweiler, 1943), the orbit is often considerably larger than the eyeball, and orbital dimensions must therefore be determined to a large extent by other factors, perhaps including the demands of stress transmission through the postorbital bar and temporalis fascia. Similar factors may be affecting relative orbital dimensions in the larger carnivorans and lemuriforms included in our sample. Whatever the final explanation proves to be, activity pattern cannot be inferred reliably from relative orbit size in large mammals. It follows that Walker's (1967a) inference of diurnality from relatively small orbit size in the large extinct Madagascar lemurs is not warranted. (Even if only the extant lemuriforms are considered, we are unable to discriminate perfectly between nocturnal and diurnal species, unless our data for *Daubentonia* are rejected because of our inadequate sample and *Hapalemur griseus* is considered nocturnal. The discrepancy between our results and Walker's reflects widely differing estimates of orbit diameter for certain species, especially *Avahi laniger* and *Phaner furcifer*.)

The extinct species examined fall into appropriate places on the graph, with a few exceptions. *Necrolemur, Tetonius, Pronycticebus* and *Mioeuoticus* appear to have been large-eyed nocturnal animals. *Rooneyia* has relatively smaller orbits, and may have been diurnal. *Palaechthon*, like *Ictops*, has extremely small orbits, relatively no bigger than those of *Hylomys* or *Monodelphis*. We conclude that vision was relatively unimportant to its way of life. The same inference can be made, though much more tentatively, for *Phenacolemur;* the orbits are not preserved on the American Museum skull, but the preserved parts allow an estimate of the maximum diameter of the orbital fossa, which could not have been much more than 8 mm.

The larger fossil species exceed the size range in which activity pattern can be inferred from orbit diameter and skull length. It is perhaps worth noting that *Plesiadapis tricuspidens* seems to fall on the same regression as *Palaechthon* and *Phenacolemur*. *Aegyptopithecus* does not differ much from extant cercopithecoids in relative orbit size, nor do the large extinct Madagascar lemurs examined. *Leptadapis magnus* and *Notharctus tenebrosus* have orbits somewhat smaller than Madagascar lemurs of comparable size. *Adapis parisiensis*, however, has extremely small orbits for a primate of its size[2], and falls quite close to *Echinosorex* on the bivariate plot. This does not demonstrate that *Adapis* was a nocturnal animal with poor vision; the datum for *Herpestes*, a diurnal carnivoran with a cone-rich retina and color vision (Dücker, 1957), lies in the same part of the graph, demonstrating that inferences about activity patterns are becoming unreliable at this size range. Nevertheless, the position of both *Adapis* and *Herpestes* on the graph is anomalous. There is no doubt that mongooses have extremely small eyes for carnivorans, but have keener vision at any rate than erinaceoids or *Monodelphis*. Mongooses' visual acuity and the role played by vision in their foraging activity have not been investigated.

The infraorbital foramen of mammals transmits the infraorbital branches of the maxillary nerve and vessels, which supply the upper lip, rhinarium, and vibrissae (Miller *et al.*, 1964). We might expect that the size of this foramen (measured here as height times breadth) would be greater in animals with better-developed vibrissae. The few data we have collected (Figure 7) show this to be the case. Again, a series of parallel regressions is seen. Terrestrial Insectivora and Carnivora have relatively large foramina; phalangerids and *Didelphis* have somewhat smaller foramina; those of diurnal tree shrews and strepsirhines are smaller still; and the smallest foramina are found in extant haplorhines, colugos and koalas, in which vibrissae are poorly-developed or absent. For the sampled marsupials, foramen size roughly reflects number and development of the mystacial vibrissae (Lyne, 1959).

Tetonius and *Rooneyia* fall with the strepsirhines in this distribution, suggesting that Paleogene tarsioids may not have

achieved the haplorhine condition seen in *Tarsius* and the anthropoids, or (more likely) may have retained more functional vibrissae. Similarly, *Mioeuoticus* has a relatively larger infraorbital foramen than extant strepsirhines (other than *Daubentonia*) have. *Palaechthon nacimienti* and *Ictops* unequivocally resemble

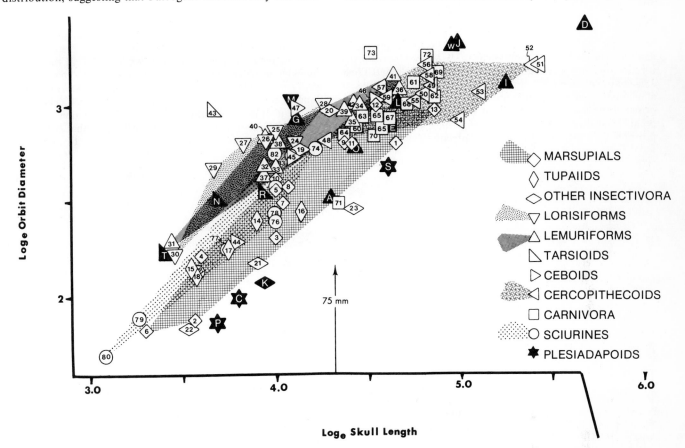

Figure 6   *Orbital diameter and prosthion-inion length (measurements in mm); sample means for various species of mammals. Species codes (and sample sizes); 1, Didelphis marsupialis (6); 2, Monodelphis brevicaudata (6); 3, Metachirus nudicaudatus (6); 4, Marmosa robinsoni (6); 5, Caluromys philander (6); 6, Eudromicia caudata (6); 7, Dactylonax palpator (6); 8, Dactylopsila trivirgata (6); 9, Trichosurus vulpecula (6); 10, Pseudocheirus (Hemibelideus) lemuroides (7); 11, Phalanger orientalis (6); 12, Phalanger maculatus (6); 13, Phascolarctus cinereus (6); 14, Tupaia glis (6); 15, Tupaia minor (6); 16, Urogale everetti (6); 17, Anathana ellioti (2); 18, Ptilocercus lowii (2); 19, Rhynchocyon stuhlmanni (6); 20, Cynocephalus sp. (4); 21, Erinaceus europaeus (1); 22, Hylomys suillus (3); 23, Echinosorex gymnurus (4); 24, Perodicticus potto (6); 25, Nycticebus coucang (7); 26, Nycticebus pygmaeus (5); 27, Loris tardigradus (6); 28, Galago crassicaudatus (6); 29, Galago senegalensis (4); 30, Galago demidovii (1); 31, Microcebus murinus (6); 32, Cheirogaleus major (4); 33, Phaner furcifer (2); 34, Lemur macaco (incl. fulvus) (6); 35, Lemur catta (6); 36, Lemur variegatus (2); 37, Lepilemur mustelinus (6); 38, Hapalemur griseus (8); 39, Propithecus verreauxi (6); 40, Avahi laniger (4); 41, Indri indri (6); 42, Daubentonia madagascariensis (1); 43, Tarsius syrichta (6); 44, Callithrix jacchus (6); 45, Callicebus moloch (24); 46, Cebus apella (6); 47, Aotus trivirgatus (6); 48, Cercopithecus talapoin females (2); 49, Erythrocebus patas males (5); 50, Macaca fascicularis males (7); 51, Mandrillus sphinx males (5); 52, Papio anubis males (5); 53, Theropithecus gelada males (3); 54, Theropithecus gelada females (1); 55, Colobus badius males (5); 56, Nasalis larvatus males (10); 57, Presbytis cristata males (5); 58, Rhinopithecus roxellanae males (1); 59, Simias concolor males (6); 60, Bassariscus astutus (6); 61, Procyon lotor (6); 62, Nasua narica (6); 63, Potos flavus (7); 64, Bassaricyon gabbi (6); 65, Paradoxurus hermaphroditus (7); 66, Genetta rubriginosa (6); 67, Viverricula indica (6); 68, Viverra tangalunga (6); 69, Arctictis binturong (7); 70, Nandinia binotata (4); 71, Herpestes edwardsi (6); 72, Cryptoprocta ferox (6); 73, Felis bengalensis (6); 74, Ratufa bicolor (6); 75, Sciurus carolinensis (behind Caluromys datum; n =6); 76, Rhinosciurus laticaudatus (1); 77, Dremomys everetti (7); 78, Callosciurus prevosti (6); 79, Sciurillus pusillus (6); 80, Myosciurus minutulus (4); 81, Tamias striatus (6); 82, Geosciurus inauris (6); 83, Cynomys ludovicianus (6). Fossils (black symbols); A, Adapis parisiensis (MNHN Qu 10873, MCZ 8886); C. Phenacolemur jepseni (AMNH 48005); D, Megaladapis edwardsi (MNHN 1936–660, AMNH 30024); E, Aegyptopithecus zeuxis Yale Peabody Mus. 23976; G, Pronycticebus gaudryi (MNHN Qu 11056); I, Palaeopropithecus ingens (MNHN 1913–16); J, Archaeolemur majori (MNHN 190721); K, Ictops sp. (YPM 14046); L, Leptadapis magnus (MNHN Qu 10825, 10870, 10872, 11002, PU 11481); M, Mioeuoticus sp. (Kenya Natl. Mus. RU 2052); N, Necrolemur antiquus (MNHN Qu 11013, 11058, 11060, MCZ 8879); O, Notharctus tenebrosus (YPM 142490); P, Palaechthon nacimienti (UKNMH 9557); R, Rooneyia viejaensis (AMNH plaster cast); S, Plesiadapis tricuspidens (MNHN R125); T, Tetonius homunculus (AMNH 4194); W, Archaeolemur edwardsi (MNHN Mad107). Data partly from Cartmill (1970).*

terrestrial insectivorans, and must have possessed a full complement of vibrissae resembling those seen in *Solenodon, Erinaceus* or *Echinosorex.* By contrast, the infraorbital foramen of *Plesiadapis tricuspidens* (Russell, 1964) is relatively small (not exceeding *Palaechthon's* in absolute height)[3], showing that some plesiadapoid lineages reduced the vibrissae in parallel with the ancestors of extant primates.

The cranial measurements used to compare *Palaechthon* with extant mammals were subjected to a principal-components analysis. The resulting distribution on the first three axes took the form of a series of parallel discoidal clusters lying along the first axis but tilted obliquely to it. Anthropoids composed the cluster at one end; tupaiids and other "menotyphlan" insectivorans comprised the cluster at the other end, with didelphids forming a parallel but divergent cluster. Phalangeroids, carnivorans and prosimians formed clusters between insectivorans and anthropoids, respectively and progressively approaching the latter; the carnivoran and prosimian distributions overlapped broadly. *Palaechthon* fell further from the anthropoid cluster than any other form except *Monodelphis,* but was closer to insectivorans than to didelphids. Since the original measurements were selected to quantify similarity to anthropoids (Cartmill, 1970), this distribution was neither surprising nor particularly informative. Other preconceptions were also confirmed by this analysis: *Aegyptopithecus* fell with anthropoids but very close to the edge of the prosimian cluster, *Tetonius* and *Rooneyia* fell near the strepsirhines but deviated toward the isolated position of *Tarsius, Leptadapis magnus* fell with the strepsirhines, and *Adapis parisiensis* (and to a lesser extent, *Notharctus tenebrosus*) fell near the primate end of the distribution but had deviant positions on axes II and III. The analysis demonstrated the obvious, that *Palaechthon* shows no resemblance in cranial morphology to primates of

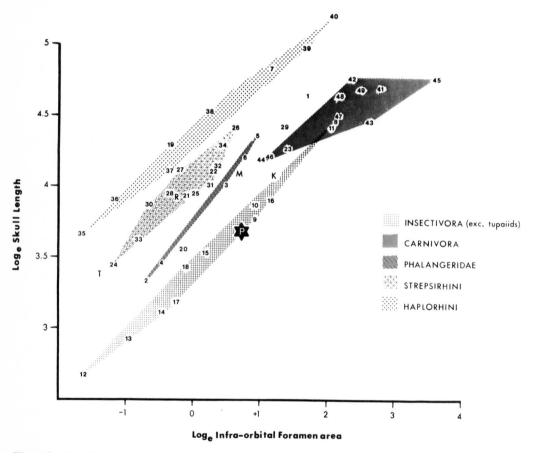

Figure 7 *Prosthion-inion length and area (height × breadth) of infraorbital foramen, for* Palaechthon nacimienti *(black star) and other mammals (measurements in mm); sample means. For three species, dial calipers were used in measuring the foramen; the other species were measured using a stereomicroscope with a reticle in the eyepiece. Species codes (and sample sizes): 1,* Didelphis marsupialis *(4); 2,* Eudromicia caudata *(2); 3,* Dactylopsila trivirgata *(2); 4,* Distoechurus pennatus *(2); 5,* Phalanger vestitus *(2); 6,* Pseudocheirus cupreus *(2); 7,* Phascolarctos cinereus *(2); 8,* Solenodon paradoxus *(2); 9,* Setifer setosus *(2); 10,* Hemiechinus auritus *(2); 11,* Echinosorex gymnurus *(2); 12,* Microsorex hoyi *(2); 13,* Sorex cinereus *(2); 14,* Blarina brevicaudata *(2); 15,* Talpa europaea *(2); 16,* Desmana moschata *(1); 17,* Chrysochloris villosa *(2); 18,* Elephantulus rufescens *(2); 19,* Cynocephalus *sp. (2); 20,* Ptilocercus lowii *(1); 22,* Tupaia glis *(2); 22,* Urogale everetti *(2); 23,* Pteropus medius *(2); 24,* Microcebus murinus *(2); 25,* Cheirogaleus major *(1); 26,* Lemur catta *(1); 27,* Hapalemur griseus *(2); 28,* Avahi laniger *(1); 29,* Daubentonia madagascariensis *(1); 30,* Loris tardigradus *(1); 31,* Nycticebus coucang *(2); 32,* Perodicticus potto *(2); 33,* Galago senegalensis *(2); 34,* Galago crassicaudatus *(2); 35,* Tarsius spectrum *(2); 36,* Saguinus nigricollis *(2); 37,* Aotus trivirgatus *(2); 38,* Cebus albifrons *(2); 39,* Pan paniscus *(cast); 40,* Homo sapiens *(1); 41,* Procyon lotor *(2); 42,* Nasua narica *(2); 43,* Potos flavus *(1); 44,* Mephitis mephitis *(2); 45,* Taxidea taxus *(2); 46,* Galidea elegans *(3); 47,* Herpestes brachyurus *(2); 48,* Ichneumia albicauda *(2); 49,* Viverra zibetha *(2). Fossil species are indicated by letters as in Figure 6 and represented by the same specimens, except that the* Ictops *sample includes a second specimen (YPM 14041).*

modern aspect; but it was of no help in attempting to interpret the functional significance of *Palaechthon's* morphology.

HABITS OF *PALAECHTHON*

The following points have been established about *Palaechthon nacimienti.*

(1) It was insectivorous.

(2) It had small eyes, which faced laterally; the plane of the orbital margin was no more than 35° out of a parasagittal orientation (Kay & Cartmill, 1974).

(3) It had a richly-innervated snout, and probably a full complement of mystacial and other facial vibrissae.

(4) It had a large olfactory fossa, and its eyes were very widely separated.

(5) It was a small animal, weighing between 48 and 152 g (95% confidence interval).

How would we expect an animal like this to make its living? Clearly, its hunting for insects must have been guided primarily by hearing, smell, and a sensitive whiskery snout. Although a few arboreal insect-eaters (e.g. *Perodicticus;* Charles-Dominique, 1971) locate their prey principally by smell, this is an inefficient way to search through a network of discontinuous branches for prey more active than caterpillars and snails. Typical arboreal insect-eaters may have well-developed facial vibrissae (e.g. *Marmosa, Ptilocercus*), but they also have more or less enlarged eyes and convergent orbits, reflecting the increased importance of vision in their foraging activities. We therefore consider it unlikely that *Palaechthon nacimienti* was primarily arboreal. We conclude that *Palaechthon* got most of its food by nosing about on the ground among rocks and plant debris, searching for concealed insects and other animal prey. Figure 8 presents a reconstruction of *Palaechthon* engaged in this activity.

The activity period of *Palaechthon* cannot be quite so confidently inferred. Extant forms that resemble it in orbit diameter and skull length are nocturnal, but the smallest mongooses (e.g. *Helogale*) probably exceed *Palaechthon nacimienti* only slightly in these dimensions. It is just possible that *Palaechthon* may have

been a mongoose-like diurnal terrestrial insect-eater. This possibility might be denied by pointing out that primitive placental mammals must have been nocturnal (Walls, 1942), and that *Palaechthon* is probably early and primitive enough to have preserved an ancestral preference for nocturnal activity; that even the diurnal strepsirhine prosimians preserve apparently symplesiomorph nocturnal features of the retina (Martin, 1972), which must also have been present in ancestral plesiadapoids; and that the diurnal adaptations of mongooses, like those of tupaiines and most other diurnal mammals, are probably relatively recent modifications of ancestral adaptations to nocturnal activity. However, these arguments are neither sound nor relevant. At best, they give us information about the adaptations of the ancestral plesiadapoid, but they do not preclude the appearance of diurnal specializations in later plesiadapoid lineages. We can say only that we have not seen any extant diurnal animal of *Palaechthon's* size that has such small orbits, and that there is accordingly some reason for thinking it to have been nocturnal and no reason for thinking it to have been diurnal.

Among living mammals, *Hylomys suillus* and *Monodelphis brevicaudatus* present two of the most likely ecological and behavioral analogs for *Palaechthon nacimienti.* Scanty field data indicate that both are primarily insectivorous, nocturnal and terrestrial (though *Monodelphis* is secondarily so); they also resemble *Palaechthon* in skull length. Both animals have been seen climbing in trees (Walker, 1964). The development of arboreal adaptations in later lineages descended from early plesiadapoids suggests that *Palaechthon* probably also had some limited tree-climbing ability, of the sort found in *Hylomys* and many other small and relatively unspecialized placentals. In Jenkins' (1974) words: "The adaptive innovation of ancestral primates was...not the invasion of the arboreal habitat, but their successful restriction to it." The available evidence implies that the ancestral plesiadapoids had not yet suffered this restriction of their activities.

A NEW GENUS AND SPECIES OF PAROMOMYID

Among the specimens tentatively assigned to *Palaechthon nacimienti* by Wilson & Szalay (1972) is one which we feel warrants separation at the generic level.

Superfamily PLESIADAPOIDEA
Family PAROMOMYIDAE
Talpohenach torrejonius [4], new genus and species
(Figure 9)

Type and only known specimen: UKMNH 7903 (R.W.W. 130), right maxilla with canine root and P²—M³.

Age and locality: University of Kansas locality 14, Kutz Canyon, Nacimiento Formation (70 ft below the level of the *Palaechthon nacimienti* specimens—Wilson & Szalay, 1972), Middle Paleocene of the San Juan Basin, New Mexico.

Figure 8 *Reconstructed appearance of* Palaechthon nacimienti *in life. General shape and appearance of the head are warranted, but pelage details, soft anatomy and the beetle are imaginary.*

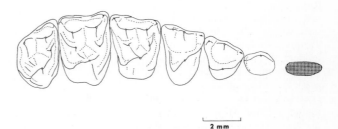

2 mm

Figure 9 *Upper dentition of* Talpohenach torrejonius *UKMNH 7903 (type). Canine root and socket stippled.*

Diagnosis

Size appreciably larger than the largest species of *Palaechthon (P. nacimienti)*. (Measurements and stereophotographs are given by Wilson & Szalay, 1972.) Dental formula ?-1-3-3. Canine root large, ovoid in cross-section with major axis anteroposteriorly oriented. P^3 unique among paromomyids in having a marked protocone, making the crown outline an equilateral triangle in occlusal view. Crest running posteriorly upward from the P^3 protocone, demarcating a posterointernal basin; smaller crest running anteromedially from the protocone along the edge of the crown. Pronounced P^3 metastyle, supporting the postparacrista, and much smaller parastyle supporting preparacrista. P^4 also with protocone and posterointernal basin. Small but pronounced P^4 metacone, twinned with larger paracone. P^4 parastyle larger than metastyle and set off from preparacrista. Intermediate conules absent on P^4. Molar stylar shelves well-developed; molar enamel somewhat crenulate. Ridge running posteriorly and dorsally from the molar protocone is blunt and raised, supporting a small but pronounced hypocone with its own wear facet (facet 8 of Kay & Hiiemae, 1974a) on M^2. A hypocone is also present on M^1 but is much less prominent. Internal outline of M^1 and M^2 two-lobed, reflecting hypocone development. Paraconules and metaconules on all molars, pronounced on M^{1-2}. M^3 with marked posterointernal basin, metacone slightly smaller than paracone.

Discussion

Talpohenach resembles *Palaechthon* in having a well-developed wear facet 5 on P^4, which suggests (but does not prove) the presence of a metaconid on P_4. Were it not for the P^3 protocone and the somewhat better-developed stylar shelf on the molars of *Talpohenach,* the genus *Palaechthon* could be stretched to include the new genus. Certainly the two are very closely related.

Talpohenach does not show tendencies in the direction of *Plesiolestes*; its P^4 structure is more primitive while its P^3 structure is more derived. A close relationship with *Torrejonia* is ruled out by premolar structure; *Torrejonia*'s small simple P_3 heel and simplified P_4 are incompatible with the more complex upper premolars displayed by *Talpohenach*. The reduced shearing edges of *Torrejonia*'s molars evince a trend opposed to the relatively increasing shearing indicated by the emphasized stylar shelf in *Talpohenach*.

EVOLUTION AND SYSTEMATICS OF PAROMOMYIDS

Enough is now known about the dentitions of early paromomyids and related Cretaceous and Paleocene Insectivora to reconstruct the ancestral dental morphology of paromomyids. When two early paromomyid genera of comparable age differ in one of the features mentioned below, we take that character state to be more primitive which more closely resembles typical Cretaceous placentals. The earliest paromomyids probably possessed a dental formula of 2-1-4-3 above and below (as in *Purgatorius*). The lower first incisor was probably procumbent, ovoid in cross-section, with a raised dorsal lateral edge (as in *Palenochtha* and *Plesiolestes*). I_2 was a small peg-like tooth. The lower canine was moderately large but smaller than I_1 (as in *Palenochtha*). P_1 and P_2 were simple and increased in size posteriorly (as in *Purgatorius*). P_3 had a slight talonid heel with a hypoconid and oblique cristid. P_4 had a paraconid but no metaconid. The heel of P_4 had a hypoconid and small entoconid (as in *Purgatorius*). Lower molars had reduced paraconids, and the M_{1-2} hypoconulids were small or absent. The trigonids were taller than the talonids. The talonid basins were somewhat expanded with two crushing areas for lateral and posterolateral surfaces of the protocone. Major shearing emphasis was on shearing crests 2, 3 and 4. There was an external cingulum on the lower molars. The M_3 heel had a simple third lobe.

Upper incisor structure is unknown. The canine was large, blunt, ovoid in cross-section, with smoothly rounded anterior and posterior crests (as in *Palaechthon*). P^1 is unknown; P^2 was anteroposteriorly ovoid with a slight internal cingulum (as in *Palaechthon*). P^3 was a larger version of P^2, with a better-developed internal cingulum. P^4 had a large protocone and a pronounced posterointernal basin bordered medially by a crest ("*Nannopithex*-fold") trending posteriorly. A small metacone was twinned with the paracone; the parastyle was well developed. The molars had reduced stylar shelves, well-developed conules, an anterior cingulum, and a pronounced posterointernal basin.

We visualize the following series of events in the evolution of Paleocene paromomyids (numbers corresponding to labelled points in Figure 10). (1) The last common ancestor of post-Puercan paromomyids lost its first premolars, and the process of stylar shelf reduction reached the stage seen in *Palenochtha*. (2) The lineage leading to *Palenochtha* lost the P_2. (3) An accessory cusp was added to the third lobe of the M_3. (4) The line leading to *Phenacolemur* and *Paromomys* developed an extension of the posterointernal basin on each upper molar; $P^{\underline{4}}_{\overline{4}}$ became enlarged and the $P^{\underline{4}}_{\overline{4}}$ metacone, paraconid, and metaconid were reduced and eliminated. I_2 was progressively reduced, and I_1 became a slender and elongate tooth, a process culminating in *Phenacolemur*. (5) The line leading to *Palaechthon* and *Plesiolestes* added a small but distinct metaconid on P_4, the condition seen in *Palaechthon*. (6) The differentiation of *Plesiolestes* (probably from an unknown species of *Palaechthon*) involved nearly complete molarization of $P^{\underline{4}}_{\overline{4}}$ through addition of a paraconule and expansion of the metaconid and metacone. The buccolingual breadth of M^3 was increased in *Plesiolestes*. (7) A protocone appeared on P^3 in the line leading to *Talpohenach*.

The systematic position of *Torrejonia* has been disputed. Like *Paromomys, Torrejonia* lacks a P_4 metaconid and has eliminated the P_4 paraconid. Other resemblances are enumerated by Gazin (1968) in his original description. Wilson & Szalay (1972) and Szalay (1973) state that *Torrejonia* shows no generic difference from *Plesiolestes*. *Torrejonia* certainly differs from the type of *Plesiolestes problematicus* in P_4 structure; *Plesiolestes* has a large distinct P_4 metaconid and paraconid while *Torrejonia* has neither. Assignment of *Torrejonia* to *Plesiolestes* has little to recommend it.

If our reconstruction of *Palaechthon nacimienti* is approximately correct, and the inferences we have made from that reconstruction are sound, then the question may be asked, as it in fact was at the Denver symposium where this paper was presented orally, "Why do you want to call *Palaechthon* a primate?" The question is embarrassing in several ways, not only because our own views diverge on this point, but also because there is no very happy solution to the problem of the plesiadapoids' systematic position. If we accepted Gingerich's (1974a, 1975b, 1976) notion, that plesiadapoids gave rise to tarsioids but not to other primates, then we would agree that inclusion of *Tarsius* in a monophyletic order Primates demands including *Palaechthon* as well. But we do not accept Gingerich's notion, for reasons largely agreeing with those advanced by Szalay (1977b). Therefore, we regard the ordinal status of *Palaechthon* and other plesiadapoids as essentially a matter of taste which depends entirely on one's preferences in the philosophy of systematics. We are of course in complete agreement that plesiadapoids are close collateral relatives of Eocene "primates of modern aspect" (Simons, 1972), and that they are linked to undoubted primates by many synapomorphies —e.g. the occlusal pattern of the cheek teeth (Gingerich, 1974b;

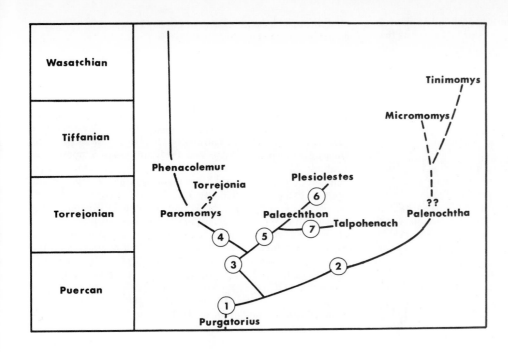

Figure 10 *A tentative phylogeny of Paromomyidae based on dental features. Circled numbers refer to comments in the text.*

Kay & Hiiemae, 1974a), the loss of the medial entocarotid (unless microsyopids are included in the superfamily), and the apparent absence of an independent entotympanic element in the bulla (although this has been denied, without supporting evidence, by Hershkovitz, 1974a). Our mutual agreement here is worth stressing, since the second author's suggestion that the plesiadapoids be returned to the Insectivora (Cartmill, 1972) has been taken by some as reflecting ignorance of these synapomorphies. This is not the case, and the lists of such synapomorphies published by defenders of the plesiadapoids' primate status are beside the point. Such lists are relevant replies to Martin's (1972) qualified suggestion that "...the Plesiadapidae are...quite unrelated to primates," but do not help to resolve the taxonomic problem that remains when the phylogenetic relationships have been agreed upon: at what point do we draw an ordinal boundary between ancestor and descendant?

"Phylogenetic systematists" following the system laid down by Hennig (1950) would, of course, reply that taxonomic boundaries cannot be drawn between ancestors and descendants, but must be drawn vertically, separating "sister groups" produced by a single cladistic event. Since "sister-groups must have the same rank in a phylogenetic system" (Hennig, 1965), this approach dictates that plesiadapoids must either (a) be included in Primates or (b) given an order of their own. The arbitrary or artistic element deplored by Wiley (1975) and other cladists is not removed by adopting a cladistic approach, however, since there is no reason to choose between these two alternatives—unless we invoke Hennig's (1950) dictum that rank orders of taxa must correspond to their chronological ages, in which case we should follow Crowson (1970) in lumping all primates into the family Hominidae so that this family will be comparable in age to invertebrate families.

Hennig's admirers in primatological circles have not, of course, accepted this aspect of his reasoning. Even Hennig's rejection of paraphyletic groupings, perhaps the most fundamental point of difference between evolutionary and phylogenetic systematics, has been abandoned by some cladistic taxonomists. One recent cladistic analysis of hominid phylogeny (Eldredge & Tattersall, 1975) includes sister-groups of differing rank, taxa distin-

guished from sister-groups by *absence* of derived features[5] (e.g. the genus *Ramapithecus*), and an unarguably paraphyletic grouping (the genus *Australopithecus*) comprising one sister-group and portions of another. So many of the basic principles of phylogenetic systematics have been discarded here that the effective result is a reconstitution of Simpsonian evolutionary systematics using Hennigian nomenclature, thus abandoning Hennig's undeniable virtues of elegance and conceptual rigor while retaining most of the vices inherent in a strictly vertical classification. The most important of these vices are inability to countenance ancestor-descendant relationships in the fossil record and incessant shifts in nomenclature due to the practical impossibility of resolving phylogenetic relationships with the degree of precision which the theory demands (Cartmill, 1975b). We feel that these shortcomings warrant the rejection of wholly cladistic procedures, and we favor the continued use of "paraphyletic" or "wastebasket" taxa defined by shared primitive retentions.

Use of paraphyletic taxa involves drawing horizontal taxonomic boundaries defined by phenetic criteria. In the mixed cladistic-phenetic systematics propounded in classical form by Simpson (1961) and Mayr (1969), the boundaries between wastebasket taxa and descendant taxa of the same rank are defined by the appearance in the descendant taxon of "some basic adaptation that evolved either coevally with the taxon itself, that is, at the base of what *later* becomes a higher taxon, or with more or less parallelism among its early lineages" (Simpson, 1961, p. 222). (The final qualification here involves Simpson's broad definition of monophyly, which we do not necessarily endorse.) If the plesiadapoids are to be included in the primates, an adaptive shift of some sort must be postulated which distinguishes the last common ancestor of plesiadapoids and *Homo* from earlier Insectivora. Szalay & Decker (1974), recognizing possible arboreal adaptations in tarsal bones attributed to *Plesiadapis,* suggest that a shift to life in the trees may represent the basal primate adaptation, warranting inclusion of the plesiadapoids in the order. Without dissenting from their analysis of the supposed *Plesiadapis* tarsals, we repeat that certain features of *Palaechthon nacimienti* and other primitive paromomyids suggest that their activity was predominantly terrestrial. Some of these features—the large infraorbital foramen, the apparently hedgehog-like anterior dentition—have been lost in *Plesiadapis*. This fact suggests that later plesiadapoid lineages and the ancestors of primates of modern aspect may have taken up arboreal life in parallel, adapting to rather different ways of making a living in the trees. We are unable to identify any significant adaptation in *Palaechthon*, apart from certain features of molar morphology, which would justify including it in Primates while excluding early erinaceoids. The problem becomes still more acute if we attempt to include *Purgatorius,* which is a reasonable structural common ancestor for plesiadapoids and later primates.

One weighty, if unscientific, reason for retaining the plesiadapoids in the order Primates is that they have been placed there by most paleontologists for four decades, and that nothing much would be gained by upsetting the systematic applecart once again. The order as presently conceived is apparently monophyletic, definable by synapomorphies, and paleontologically workable. These solid virtues must weigh heavily in the scale against the much less substantial inferences we can make concerning modes of life in fragmentary or imaginary ancestors. The problem could probably be solved to everyone's satisfaction by establishing a separate suborder to encompass all the primates of modern aspect; however, this step should not be taken until the phyletic relationships of tarsiers, lemurs, lorises, ceboids and catarrhines have been more satisfactorily resolved.

We are grateful for the generous assistance and co-operation extended to us by the staff at the various museums where much of this research was done. We thank Sidney Anderson, Malcolm C. McKenna, Hobart M. Van Deusen and Richard G. Van Gelder, at the American Museum of Natural History; Charles W. Mack, at the Museum of Comparative Zoology; Alan C. Walker, Harvard University; Richard Thorington, Jr., United States National Museum; Elwyn L. Simons and Friderun Ankel-Simons, Yale Peabody Museum; Philip Hershkovitz, Joseph C. Moore and William D. Turnbull, Field Museum of Natural History; Vincent Maglio, Princeton University; and Donald E. Russell, Daniel Goujet and C. Poplin, at the Muséum national d'Histoire naturelle in Paris. We are especially grateful to L. Martin for the loan of specimens under his care at the University of Kansas. We thank Kaye Brown for technical assistance and for comments on the manuscript, and Eric Delson and Milford Wolpoff for their criticisms. The research of one of us (R.F.K.) is supported by NSF Grant GS-43262; that of the other (M.C.), was made possible by NIH Grant 1-K04-HD00083-01 and by travel funds from the Wenner-Gren Foundation for Anthropological Research.

Notes

[1] The term *"cristid obliqua"* (Szalay, 1969), replacing *'crista obliqua"* of earlier authors (Gregory, 1920c; MacIntyre, 1966; Van Valen, 1966) is not acceptable, since *cristid* is not a Latin noun. Nevertheless, some way is needed to indicate that *"crista obliqua"* denotes a lower molar crest. We use the vernacular; alternative constructions would be "cristidum obliquum" (cf. German "das Hypoconid"), or "prehypocristid."

[2] Martin's (1973) estimates of 5.0 and 4.0 mm respectively for orbital height and breadth of the British Museum skull of *Adapis parisiensis* (BMNH M.1345) are incompatible with the accompanying radiographs, and are presumably misprints.

[3] Height, but unfortunately not breadth, was measured by one of us on the Paris skull before we discovered that both measurements were needed to produce an intelligible distribution.

[4] *Talpohenach,* from Welsh *talp o hen ach,* "Piece (lump, fragment) from an ancient lineage," anagram of *Palaechthon* (cf. *Palenochtha*). Trivial name from Torrejon beds; agreeing in (masculine) gender with W. *talp* (art. 30b-i, International Code).

[5] "The possession of at least one derivative (relatively apomorph) ground-plan character is a precondition for a group to be recognized at all as a monophyletic group. But it also follows from this that this same character in the nearest related group must be present in a more primitive (relatively plesiomorph) stage of expression. The exclusive presence of relatively plesiomorph characters is indicative of paraphyletic groupings; these are to be found only in pseudophyletic...and purely morphological systems...but not in phylogenetic systems" (Hennig, 1965). The existence of fossils like *Ramapithecus* (in which no characters can be detected that are apomorph relative to other hominids) makes it impossible to encompass them in a phylogenetic classification except by arbitrarily excluding them from the ancestry of later forms, thus saving the system at the cost of discarding one of the more important forms of evidence for the reality of organic evolution.

5

The First Radiation—Plesiadapiform Primates

K. D. ROSE AND J. G. FLEAGLE

The earliest primates, the Plesiadapiformes, were both diverse and common. Approximately 60 species in nearly 30 genera are currently recognized (see Table 1). In terms of taxonomic diversity this is nearly twice as many species as there are of extant prosimians. Plesiadapiform primates are the most common animals in many middle and late Paleocene faunas and are common in many early Eocene localities (Figs. 1, 2).

Plesiadapiform primates are known almost exclusively from fragmentary jaws and teeth. Definitely assignable skeletal remains are known of only two closely related genera, *Plesiadapis* and *Nannodectes,* and significant parts of skulls for only six of the genera (Fig. 3). Therefore, most definitions of Plesiadapiformes and of the genera and species that comprise the suborder are based on the dentition. The following dental features are characteristic of these early primates: molar teeth with relatively low, bulbous cusps (compared to contemporary or extant insectivores); relatively low trigonids and basined talonids on the lower molars; an unreduced third lower molar with an extended talonid; upper molars with prominent conules; poorly developed or absent stylar shelf and stylar cusps on the upper molars; a well-developed postprotocingulum (= nannopithex fold) or homologous wear facet on the upper molars; an enlarged procumbent lower central incisor.

These features indicate a functional shift towards more crushing and grinding in the cheek teeth as an adaptation towards increasing omnivory and herbivory. The prominent shearing surfaces on the teeth of many of these early primates (which often resemble the living *Tarsius* in this regard), as well as their small

size, suggest that many species were probably still largely insectivorous; others, however, seem to be highly specialized herbivores or even graminivores. The primitive dental formula for Plesiadapiformes, excluding *Purgatorius,* is $\frac{2.1.3.3.}{2.1.3.3.}$, but advanced members of all lineages show reduction and loss of antemolar teeth.

The identification of species and genera among the Plesiadapiformes and the arrangement of these genera into families varies somewhat from one author to another, depending upon their respective taxonomic philosophy and areas of expertise. We will follow the more inclusive systematic scheme of Bown and Rose (1976) in recognizing four or five families in North America.

MICROSYOPIDAE

The Microsyopidae are the most primitive known primates, so much so that many authorities consider them insectivores rather than primitive primates. They were quite successful, with a temporal range extending from the middle Paleocene through the late Eocene, and include both the smallest of all known primates (animals the size of a shrew) and the largest of the Plesiadapiformes (an animal the size of the opossum, *Didelphis*). Microsyopids share with other plesiadapiform groups the dental features described above. They comprise a relatively uniform group of Plesiadapiformes that have retained primitive features of cheek tooth morphology. However, so far as anterior teeth are known, all genera have a lower central incisor whose crown is narrow and lanceolate (spearhead-shaped), a feature which separates them

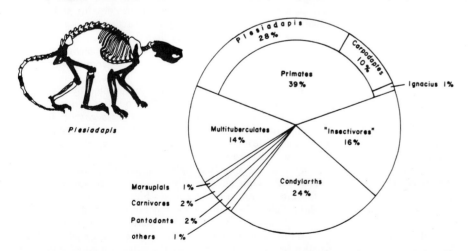

Plesiadapis

Figure 1 *Faunal composition for Cedar Point Quarry, a Tiffanian-aged quarry in northern Wyoming. Darkened parts of skeleton of* Plesiadapis *are elements currently known from fossils.*

Table 1 *Plesiadapiform Primates*

Suborder Plesiadapiformes Simons, 1972

Family MICROSYOPIDAE Osborn & Wortman, 1892 (M. Pal.–Late Eoc.)

Palaechthon Gidley, 1923 (M. Pal.) $\dfrac{2.1.3.3}{2.1.3.3}$

Pleisiolestes Jepsen, 1930 (M. Pal.)
(probably = *Palaechthon*)

Talpohenach Kay and Cartmill, 1977 (M. Pal.)
(probably = *Palaechthon)*

Torrejonia Gazin, 1968 (M. Pal.)

Palenochtha Simpson, 1935 (M. Pal.) $\dfrac{2.1.?3.3}{2.1.\ 3.3}$

Microsyops Leidy, 1872 (E.–Late Eocene) $\dfrac{2.1.3.3}{1.0.3.3}$
(includes *Cynodontomys* Cope, 1882)

Craseops Stock, 1934 (Late Eocene)

Berruvius Russell, 1964 (Late Pal.) (Eur.)

Navajovius Matthew and Granger, 1921 (Late Pal.) $\dfrac{?.1.3.3}{1.1.3.3\ \text{or}\ 2.1.2.3}$

Micromomys Szalay, 1973 (Late Pal.)

Tinimomys Szalay, 1974b (E. Eoc.) $\dfrac{?}{1.0.3.3}$

Niptomomys McKenna, 1960 (E. Eoc.) $\dfrac{?}{1.1.3.3}$

Uintasorex Matthew, 1909 (M.–Late Eoc.) $\dfrac{?}{1.0.3.3}$

Alsaticopithecus Hurzeler, 1947 (M. Eoc.) (Eur.)

Family PLESIADAPIDAE Trouessart, 1897 (M. Pal. – E. Eoc.)

Pronothodectes Gidley, 1923 (M. Pal.) $\dfrac{2.1.3.3}{2.1.3.3}$

from all other families. Microsyopids also seem to differ from other plesiadapiform genera and all other groups of primates in lacking an auditory bulla formed by the petrosal bone.

The earliest microsyopids are five closely allied genera: *Talpohenach, Palaechthon, Pleisiolestes, Palenochtha,* and *Torrejonia.* A crushed skull of *Palaechthon* (Wilson and Szalay, 1972; Kay and Cartmill, 1974, 1977) is the oldest and most primitive known primate skull. Unfortunately, the basicranium is poorly preserved, so the taxonomically more significant features of this region remain unknown.

Kay and Cartmill (1977) have recently provided a very thorough description of the preserved parts of the specimen in an attempt to reconstruct the habits of this animal. From the similarity of the anterior teeth with those of living erinaceid insectivores (hedgehogs) and caenolestid marsupials, they argued that *Palaechthon* (and other microsyopids) probably used these teeth for grasping and stabbing insect prey, rather than for ingesting plant food. Their analysis of the molar morphology also indicated that *Palaechthon* was largely insectivorous.

The skull (Fig. 3) has small, laterally directed orbits with a wide infraorbital region and a very large infraorbital canal. The first feature suggests that the visual abilities of *Palaechthon* were limited relative to those of more advanced primates, which have stereoscopic vision; the second suggests a large olfactory fossa; and the third indicates a richly innervated snout bearing sensitive

facial vibrissae. Kay and Cartmill concluded that *Palaechthon* probably hunted for concealed insects and other animal prey by "nosing around the ground" guided more by hearing, smell, and a sensitive whiskered snout than by visual activity. It is more likely that *Palaechthon* was nocturnal than diurnal.

Plesiolestes (Fig. 4) and its close relative *Palaechthon* probably lie near the ancestry of the two Eocene genera, *Microsyops* (Fig. 5) and its descendant *Craseops. Microsyops* is the best known member of the family, being represented by many jaws and two skulls. The opossum-sized skull of *Microsyops* although considerably larger, resembles that of *Palaechthon* in most known features. Like the earlier genus, *Microsyops* had procumbent, lanceolate incisors in the lower jaw. The single pair of upper incisors were large and set in very large premaxillary bones. As in *Palaechthon,* the orbits were relatively small, separated by a wide interorbital region, and laterally oriented. There was a very large infraorbital foramen and no postorbital bar (McKenna, 1966; Szalay, 1969). The auditory bulla is not preserved in any of the known skulls; however, it is clear that *Microsyops* lacked the petrosal bulla that characterized all other groups of living and fossil primates, and it probably had either an entotympanic or cartilaginous bulla, within which lay the free ectotympanic ring.

A less conservative group of microsyopids, placed in a separate subfamily, are the Uintasoricinae, comprised of three, tiny, shrew- to mouse-sized genera—*Navajovius* from the late Paleo-

Table 1 *Plesiadapiform Primates (continued)*

Nannodectes Gingerich, 1975d (Late Pal.)	$\dfrac{2.1.3.3}{1.0\ \text{or}\ 1.3.3}$
Plesiadapis Gervais, 1877 (Late Pal.–E. Eoc.) (Eur. and N. Am.)	$\dfrac{2.0\ \text{or}\ 1.3.3}{1.0.2\ \text{or}\ 3.3}$
Chiromyoides Stehlin, 1916 (Late Pal.–E. Eoc.) (Eur. and N. Am.)	$\dfrac{1?.?.3.3}{1.0.2.3}$
Platychoerops Charlesworth, 1855 (E. Eoc.) (Eur.)	$\dfrac{?}{1.0.2.3}$
Family CARPOLESTIDAE Simpson, 1935 (M. Pal.–E. Eoc.)	
Elphidotarsius Gidley, 1923 (M. Pal.)	$\dfrac{?}{2.1.3.3}$
Carpodaptes Matthew and Granger, 1921 (?M.-Late Pal.)	$\dfrac{?}{2.1.2.3}$
Carpolestes Simpson, 1928 (Late Pal.–E. Eoc.)	$\dfrac{2.1.3.3}{2.1.2.3}$
Family SAXONELLIDAE (Russell, 1964) (?Late Pal.)	
Saxonella Russell, 1964 (Eur.)	$\dfrac{?}{1.0.2.3}$
Family PAROMOMYIDAE Simpson, 1940 (M. Pal.-Late Eoc.)	
Paromomys Gidley, 1923 (M. Pal.)	$\dfrac{2.1.3.3}{2.1.3.3}$
Ignacius Matthew and Granger, 1921 (M. Pal.-Late Eoc.)	$\dfrac{2.1.2.3}{1.0.1\ \text{or}\ 2.3}$
Phenacolemur Matthew, 1915 (Late Pal.-M. Eoc.) (Eur. and N. Am.)	$\dfrac{?2.1.3.3}{1.0.1.3}$
Family PICRODONTIDAE Simpson, 1937 (M.-Late Pal.)	
Picrodus Douglass, 1908 (M. Pal.)	$\dfrac{?}{2.1.2.3}$
Zanycteris Matthew, 1917 (Late Pal.) (probably = *Picrodus*)	$\dfrac{?.1.3.3}{?}$
Plesiadapiformes, *Incertae Sedis* *Purgatorius* Van Valen and Sloan, 1965 (Late Cret.-E. Pal.)	$\dfrac{?}{?.1.4.3}$

Dental formulae provided where known. However, they may be based on only one species in the genus. All genera are known only from North America unless otherwise specified.

cene of the western U.S.A., and *Niptomomys* and *Uintasorex* from the Eocene. This subfamily might be derived from the middle Paleocene genus *Palenochtha*. Compared with other microsyopids, uintasoricines have very broad talonid basins and small trigonids on the lower molars; the last molar is often somewhat reduced. The last lower premolar is submolariform and the last upper premolar (known only in *Niptomomys*) is slightly enlarged and premolariform. Like other microsyopids, all three genera have the procumbent lanceolate lower incisor.

Two additional genera are provisionally placed in the Microsyopidae: *Micromomys*, from the late Paleocene of Wyoming and Saskatchewan and *Tinimomys*, from the earliest Eocene of Wyoming. These two genera are among the smallest of all known primates. While their molars resemble those of more conservative microsyopids, they are distinguished by a very large pointed last lower premolar. On the basis of their minute size and tarsier-like molar teeth, it seems likely that these genera were predominantly insectivorous.

PLESIADAPIDAE

Members of this highly successful family were among the commonest mammals of the North American Paleocene and were also prevalent in the late Paleocene and early Eocene of Europe. In North America, plesiadapids, particularly *Plesiadapis*, were sig-

nificant constituents of nearly all faunas ranging from Alberta and Saskatchewan in Canada to southern Colorado (Fig. 1). They are better known than any other plesiadapiforms. Five genera have been described, four of which occur in North America: *Pronothodectes* (middle Paleocene), *Nannodectes* (late Paleocene), *Plesiadapis*, and *Chiromyoides* (both late Paleocene and earliest Eocene). Plesiadapids were relatively large for archaic primates, ranging from about the size of a rat to the size of an opossum (*Didelphis*).

Compared to microsyopids, plesiadapids (Fig. 6) have cheek teeth with more bulbous cusps; in the lower molars the trigonids are relatively low and the talonid basins broad. The last two upper premolars are usually short but transversely broad with three prominent cusps. Plesiadapids have very large procumbent incisors. The lower central incisors are broader than those of microsyopids and the upper incisors have tricuspid, mitten-shaped crowns.

Recent work by Gingerich (1975d, 1976) has added substantially to our knowledge of plesiadapids. He recognizes 18 valid species of North American Plesiadapidae and has proposed a phylogeny based on their morphology and stratigraphic positions (Gingerich, 1976). The generalized *Pronothodectes* from Montana and Wyoming gave rise to two genera, *Nannodectes* and *Plesiadapis*. Only *Pronothodectes* retains the primitive plesiadapoid lower dental formula 2.1.3.3. Gingerich demonstrated that specimens formerly attributed to *Plesiadapis* actually represent two distinct lines of evolution, more appropriately regarded as separate genera. In *Nannodectes,* restricted to the late Paleocene, the dental formula is the same as in *Pronothodectes* except for loss of the small lateral incisor and, in the most advanced species, loss of the small canine as well. P₂ is invariably present. In the second line, *Plesiadapis,* there is further dental reduction. The lateral incisor and canine are always absent and, in later species, P₂ is present in only some specimens of some populations, or never present in others. In the late Paleocene, *Plesiadapis* branched into two lineages, one decreasing slightly in size, the other increasing considerably and culminating in the opossum-sized *Plesiadapis cookei,* one of the largest plesiadapiform primates. Derived from an early species of *Plesiadapis* was the fourth North American genus, *Chiromyoides,* until recently (Gingerich, 1973b) known only from Europe. It is distinguished from other plesiadapids by its deep, shortened mandible, very robust incisors, and broad cheek teeth probably associated with a graminivorous diet.

In contrast to some other plesiadapiform primates, plesiadapids all had generalized cheek teeth; the only modification of note is the distinctive form of the posterior upper premolars (P³⁻⁴). These teeth are short anteroposteriorly and in some species bear three prominent cusps, interpreted as an external paracone, internal protocone, and especially large and medial paraconule.

Figure 2   *Paleontological localities in North America yielding fossil nonhuman primates. Key: darkened circles = occurrences of plesia-dapiform primates; darkened triangles = omomyids; open squares = adapids.*

The molars of plesiadapids bear striking resemblance to those of *Pelycodus,* the earliest known adapid, and to those of early Eocene tarsioid primates.

The rather low-crowned, bunodont cheek teeth and the moderate size of plesiadapids suggest that they may have been more strictly herbivorous than most other plesiadapiforms. Gingerich (1974b) has postulated that *Plesiadapis* may have been a terrestrial grazer, feeding on small bits of vegetation, possibly stems.

Cranial fragments of *Plesiadapis* and *Nannodectes* have been recovered from late Paleocene beds in North America, but most of our knowledge of the skull morphology of plesiadapids comes from late Paleocene specimens of *Plesiadapis* from the Paris Basin (Russell, 1964; Gingerich, 1976) (Fig. 3). A notable feature is the long snout, consisting predominantly of the very large premaxillae and remarkable for the long diastemata between the large incisors and the cheek teeth in both upper and lower jaws. The interorbital breadth is narrow and there is no postorbital bar, a structure typical of later primates. The auditory region has been described in detail by Russell (1964) and a new, well-preserved specimen has been figured and discussed by Gingerich (1976). The bulla is apparently formed by the petrosal; no entotympanic element is evident, although one may have been present during initial states of ossification. Most important, the tympanic ring (ectotympanic) is fused to the lateral wall of the bulla and the ectotympanic extends laterally in the form of a tube as in tarsioids and hominids.

Although considerable skeletal material has been found for *Plesiadapis,* this material has never been thoroughly analyzed from a functional point of view. Consequently there is considerable disagreement regarding the probable locomotor habits of this early primate. Szalay and Decker (1974) and Szalay *et al.* (1975) have argued that *Plesiadapis* shares with all other primates a suite of characters in the talus and humero-radial articulation that are indicative of a basically arboreal adaptation. Simons (1967e) described *Plesiadapis* as an "arboreal scrambler, perhaps functioning like a large squirrel", but also suggested that some species may have been partly terrestrial in order to account for the widespread occurrence of the genus in both North America and Europe. He later cautioned (Simons, 1972) that the robustness of the skeleton casts doubt on the degree to which *Plesiadapis* might have been arboreal and suggested that its generalized anatomy indicates "semiterrestrial" habits. Gingerich (1974a) on the basis of both the abundance and widespread occurrence of *Plesiadapis* and skeletal similarities to the extant sciurid *Marmota,* has adopted the view of Teilhard de Chardin (1922) that *Plesiadapis* was probably a gregarious, terrestrial form. He has also noted (Gingerich, 1976) that the limb proportions of *Plesiadapis* are more comparable to those of terrestrial than arboreal sciurids.

Russell (1964) and Simons (1967e) noted similarities between the claws of *Plesiadapis* and those of dermopterans, but Russell felt that *Plesiadapis* was predominantly terrestrial. Finally, Walker (1974) has suggested that *Plesiadapis* may even have been a gliding form, but noted that there was no compelling evidence to indicate arboreal habits in the known material.

CARPOLESTIDAE

This strictly North American family contains three essentially time-successive genera constituting a well documented ancestor-descendant lineage. The earliest form, *Elphidotarsius,* from middle Paleocene beds, was a contemporary of *Pronothodectes.* The resemblances between these two genera, including detailed similarities of molar and upper premolar structure, are uniquely specialized features which indicate that the Carpolestidae and Plesiadapidae shared a common ancestor more recently than other archaic primates.

Carpolestids were small mouse-sized animals. The most distinctive aspect of the dentition is the remarkable specialization of the last lower premolar (P_4), which is hypertrophied, bladelike, and bears several cusps in line on its crest. The P_4 of *Elphidotarsius* is the least specialized and its morphology can be derived from a semimolariform tooth by opening the trigonid. By selecting for individuals with additional apical cusps, the proliferation of cusps characteristic of P_4 in the other two carpolestid genera, *Carpodaptes* and *Carpolestes,* can be explained (Fig. 7). Associated with the specialization of P_4 are progressive reduction of the teeth between P_4 and the enlarged incisor, and opening of the trigonid of M_1. The paraconid and metaconid of M_1 are widely separated and in *Carpodaptes* and *Carpolestes* the paraconid is aligned directly anterior to the protoconid (rather than anterolin-

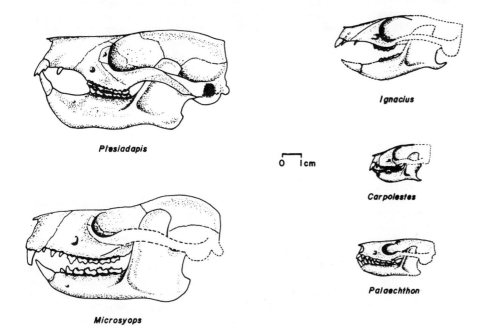

Plesiadapis

Microsyops

0   1cm

Ignacius

Carpolestes

Palaechthon

Figure 3   *Skulls of five genera of plesiadapiform primates.*

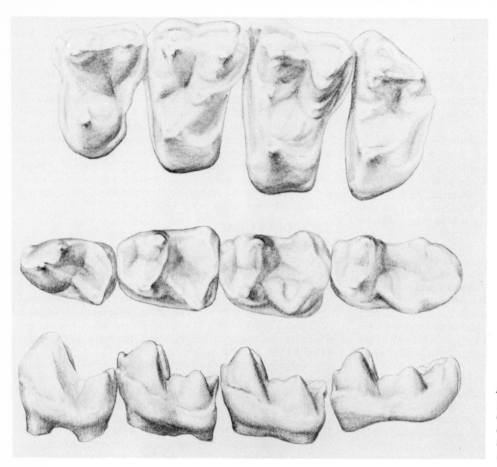

Figure 4   *Left P⁴₄–M³₃ of* Plesiolestes problematicus. *About ×12. In this and subsequent figures, tooth rows depicted are left side; top to bottom: upper teeth, occlusal view of lower teeth, lateral view of lower teeth.*

gual to it as in other primates), and both are in line with the cusps of P_4. Both *Carpodaptes* and *Carpolestes* have uniquely specialized upper premolars (P^3 and P^4) characterized by hypertrophy and development of accessory cusps.

A snout of *Carpolestes* preserving much of the dentition provides the only available evidence on cranial anatomy of carpolestids (Fig. 3). It reveals that the premaxillae are smaller and the snout shorter than in *Plesiadapis* or *Ignacius* (Rose, 1975).

Elphidotarsius is known from middle Paleocene sites in Wyoming, Montana, and Saskatchewan. The more specialized *Carpodaptes* succeeds *Elphidotarsius* in time and almost surely descended from it. *Carpodaptes* is the commonest and most widespread genus, with five species known from the late Paleocene of Wyoming, Montana, Colorado, North Dakota, Alberta and Saskatchewan. The further specialized *Carpolestes* (Fig. 8), derived from *Carpodaptes,* has been found in very late Paleocene beds of Wyoming and Montana and basal Eocene deposits in Wyoming. Rose (1975) presented a revision of carpolestid systematics and a detailed discussion of morphology and variability in these early primates. Like plesiadapids, carpolestid taxa had relatively short durations and, therefore, are often useful guides to the age of deposits in which they occur (Rose, 1977). In general, the presence of *Elphidotarsius* is indicative of Torrejonian age, *Carpodaptes* suggests late Paleocene (Tiffanian) age, and *Carpolestes* points to latest Paleocene or earliest Eocene age (late Tiffanian—Clarkforkian).

Dental morphology provides the primary evidence on the biology of Carpolestidae. In his detailed analysis of dental function of carpolestids, Rose (1975) found that dental anatomy and wear patterns on the teeth indicated capabilities for both herbi-

vorous and insectivorous feeding; however, a close parallel with *Tarsius* in molar structure may be indicative of insectivorous preferences.

The peculiarly modified lower fourth premolars were apparently more important as tearing blades, used in preparatory stages of mastication, than as shearing blades; this may be analogous to the function of the enlarged fourth premolar in *Phenacolemur* (Fig. 9). The mandibular dental configuration, including an enlarged procumbent incisor followed by reduced teeth and a hypertrophied trenchant tooth is a dental complex termed plagiaulacoidy, and it is not unique to carpolestids. It occurs also in some multituberculates, marsupials, and in the European plesiadapoid, *Saxonella.* The functional mechanism differs among these forms and may be associated with strict herbivority or, in some species, omnivority (including partly insectivorous tendencies). In other words plagiaulacoid dentition by itself is not clearly indicative of particular dietary propensities.

PAROMOMYIDAE

The Paromomyidae were a small but successful group of mouse- and rat-sized primates, probably the most herbivorous (frugivorous?) members of the Plesiadapiformes. On the evidence of dental and cranial anatomy, they are more closely allied to the Plesiadapidae and the Carpolestidae than to the more primitive Microsyopidae; hence these three families are sometimes grouped as the superfamily Plesiadapoidea. Paromomyids are characterized by relatively flat, low-crowned, rectangular lower molars with short squared trigonids and broad, shallow-basined talonids, and squared upper molars with expanded posterointernal basins.

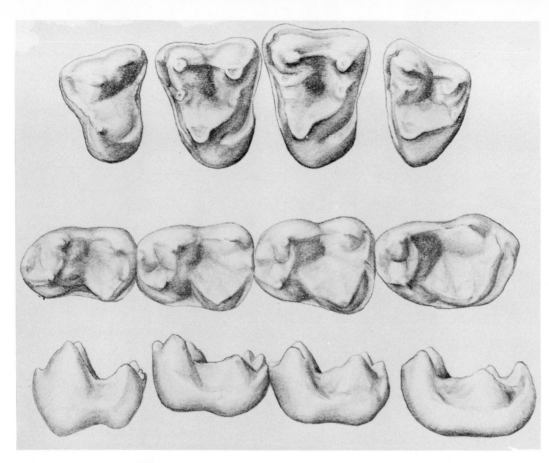

Figure 5   *Left P_4^4–M_3^3 of* Microsyops latidens. About x10.

Large crushing and grinding surfaces are developed on the molars, and the last molar above and below tends to be conspicuously elongated posteriorly. These features, particularly the breadth and low relief of the occlusal surfaces of the molars, are suggestive of a predominantly herbivorous diet. The pointed P_4, especially in more advanced species, may be tailored for powerful puncturing of food during the preparatory phases of mastication (Gingerich, 1974c).

Two lineages descended from middle Paleocene *Paromomys: Ignacius,* known from late Torrejonian into Uintan time, and *Phenacolemur* (Fig. 9), from late Tiffanian to Wasatchian deposits. Particularly characteristic of the two later genera is the development of a long diastema between the enlarged lower incisor and P_3 or P_4. The teeth in between have been lost, and P_3 is retained only in the earliest species of *Ignacius.* Others show diastemata separating the canine (and P^2 if present) from cheek teeth. Consistent dental differences between the species of these two genera indicate that they evolved independently since Torrejonian time, but overall similarity of the teeth reflects close affinity (Bown and Rose, 1976; Rose and Gingerich, 1976).

A single partial skull of each of the two later genera is known (Fig. 3), both revealing important aspects of paromomyid cranial anatomy. Szalay (1972a) described a crushed skull of early Eocene *Phenacolemur jepseni* in which part of the basicranium, including the auditory region, is preserved. From this specimen it can be seen that, as in *Plesiadapis,* the petrosal forms the auditory bulla (although the incorporation of an entotympanic element cannot be ruled out) and there is a large, tubelike ectotympanic lateral to the bulla. A skull of early Eocene *Ignacius graybullianus* (Rose and Gingerich, 1976), although lacking the basicranium, is

less crushed than the *Phenacolemur* specimen and preserves the front of the snout, missing in the other. Like *Plesiadapis, Ignacius* had very large premaxillae. The two incisors and canine in *Ignacius* are high-crowned, single-rooted teeth and P^2 is absent, whereas in *Phenacolemur* the two teeth preserved anterior to P^3 (P^2 and C?) are small, low-crowned, double-rooted teeth. As in *Palaechthon,* the infraorbital foramen is large, probably indicating a well-innervated snout with facial vibrissae.

PICRODONTIDAE

This aberrant, rare family is represented by two very closely allied monotypic genera, *Picrodus* from the middle and late Paleocene of Montana and Wyoming and *Zanycteris* from the late Paleocene of Colorado. *Picrodus* is known from lower jaws and isolated upper teeth; *Zanycteris* is known from a single crushed palate. The close resemblance between these two forms strongly suggests that they are congeneric.

Picrodonts are characterized by highly specialized low-crowned molars (Fig. 10). The first upper and lower molars are hypertrophied. In the lower molars the trigonids are small and the talonids large with crenulated enamel. In contrast to other plesiadapiforms, M_3 does not have a third lobe (expanded talonid). The anatomy of picrodontid molars suggests a diet consisting of fruit and nectar (Szalay, 1972b).

The affinities of picrodontids are unclear. Douglass (1908), describing *Picrodus,* referred it questionably to Ameghino's South American family Epanorthidae (=Caenolestidae). (He put *Megopterna,* later recognized as a synonym of *Picrodus,* in ?Insectivora). Matthew (1917) believed *Zanycteris* was a representative of the neotropical phyllostomatid bats. Later, Simpson

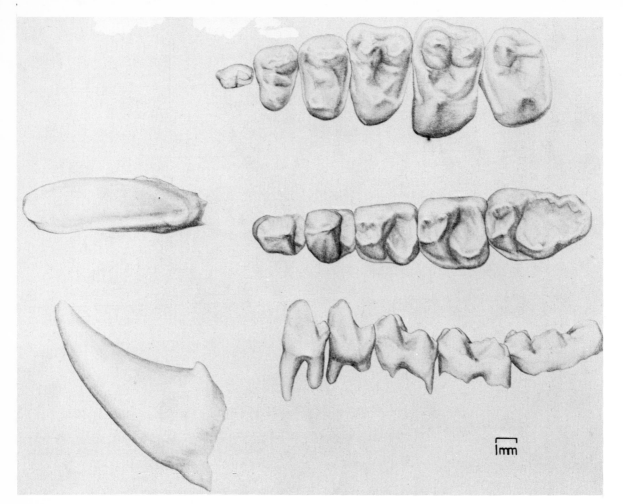

Figure 6   *P²–M³ and left I, P₃–M₃ of* Plesiadapis rex.

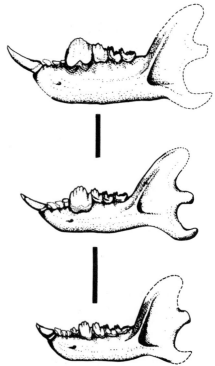

Figure 7   *Mandibular dentitions (bottom to top)* Elphidotarsius, Carpodaptes, Carpolestes, *showing evolutionary changes.*

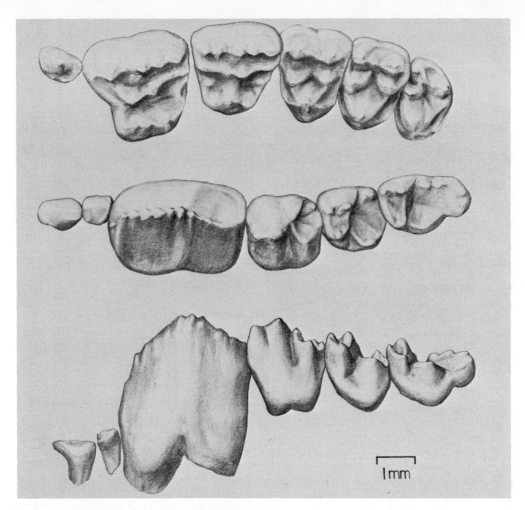

Figure 8 P^2-M^3 *and C,* P_3-M_3 *of* Carpolestes dubius.

(1937) and McGrew and Patterson (1962) argued against chiropteran affinities, allocating picrodontids to the Insectivora, with a query. Some recent authors (primarily Szalay, 1968b, 1972b) have included picrodontids in the Primates, but the evidence is equivocal. Using transformed coordinates, Szalay (1968b) depicted a possible derivation of picrodontid molars from a form like middle Paleocene *Palenochtha.* Although such a derivation is possible, no intermediate forms are known, and this remains an unproved pedigree. All other plesiadapiforms are characterized by comparatively conservative molars and a tendency to enlarge and sometimes specialize the posterior premolars. The situation is reversed in picrodontids. They show no particular resemblance to known plesiadapiform primates, except for the very general plan of their molars, and their allocation to this group must be regarded as dubious.

PALEOENVIRONMENT

Fossil primates of North America are known predominantly from the Rocky Mountain region (Fig. 2), where they are usually found in badland deposits in relatively arid basins. The environment in which they lived, however, was very different. The late Cretaceous and early Tertiary was a time of major orogeny in this area, when sediments eroding from uplifting areas accumulated rapidly in intermontane basins. Little has been preserved of life in the uplands during this time, for most fossiliferous beds seem to have formed at low elevation. In the early Tertiary these lowlands were dotted with lakes and swamps crossed by networks of meandering streams and rivers, some of which led to remnants of the epicontinental sea which had separated the western and eastern parts of the continent in the late Cretaceous. Prevalent in the streams were fishes such as bowfins (*Amia*) and garpikes (*Lepisosteus*), as well as turtles and crocodiles of many types. The latter, especially, suggest greater rainfall and higher mean annual temperatures as well as higher equability (less seasonal fluctuation in temperature) in the early Tertiary than occurs in the region today. Floodplains were forested, probably densely near the rivers. Hickory, cypress, ginkgo, and other plants indicative of a humid, subtropical or warm temperate regime were common. Some of the flora have living relatives known only in southeast Asia (Wolfe and Hopkins, 1967). The early Tertiary environment of the Rocky Mountain region is therefore sometimes compared to the present day environment of southern Asia or southeastern United States. One can readily appreciate that it provided hospitable habitats for the earliest primates.

Paleobotanical evidence indicates that a climatic deterioration occurred at the end of the Paleocene, with a resumption of warm, or even warmer, humid conditions in the Eocene (Wolfe

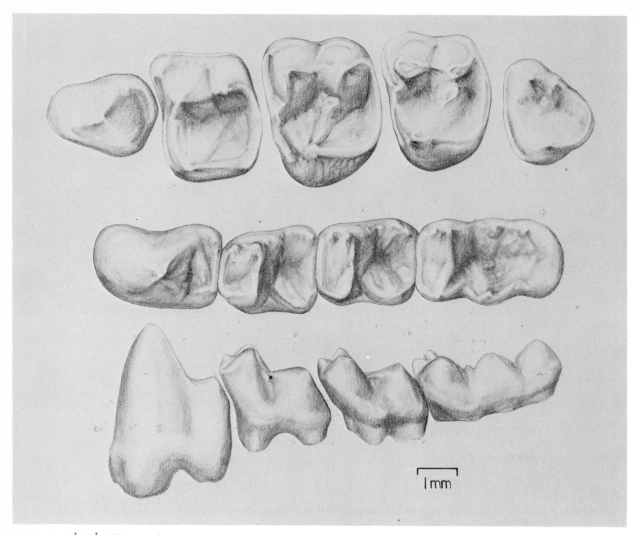

1 mm

Figure 9   *P³₄–M³₃ of* Phenacolemur praecox.

and Hopkins, 1967). Some groups of archaic Paleocene mammals were replaced by taxa of more modern appearance in the early Eocene, and this renewal of favorable climates was probably an important factor in the replacement. It seems likely that many mammals migrated northward into the Rocky Mountain region as the environment improved. The rather sudden appearance of tarsioid (omomyid) and lemuroid (adapid) primates in the early Eocene is probably associated with this climatic amelioration.

RELATIONSHIPS OF THE PLESIADAPIFORMES TO OTHER PRIMATES

Having summarized the systematics of these early primates and the most plausible reconstructions of their natural history, we must now confront the more difficult question of their relationship to other groups of primates, both living and fossil.

In assessing the phyletic relationships of plesiadapiform primates to more advanced primates, most authors have refrained from assigning any direct ancestor-descendant relationships between plesiadapiform and later primates. While numerous authors (Lemoine, 1878; Stehlin, 1916; Teilhard, 1921; Gidley, 1923; and Simpson, 1940) have noted similarities in dental morphology between plesiadapiform species and later prosimians,

both lemurs and tarsiers, there have been few serious suggestions that the plesiadapiforms were particularly closely related to any one group of primates. All workers generally agree that all known plesiadapiform genera (except *Purgatorius*) were too specialized morphologically (or showed too much dental reduction) to merit consideration as direct ancestors for later forms..

However, Gingerich (1976) has recently argued for a phyletic relationship between the plesiadapiforms and the tarsier-like omomyids of the Eocene. Like most other workers, Gingerich (1977b) agrees that only a plesiadapiform as primitive as *Purgatorius* retains enough of a generalized morphology to make a suitable ancestor for later forms.

Because plesiadapiforms are so distinct from later primates and because there are no known forms clearly linking this radiation with later groups, a number of workers have questioned why these animals should be considered as early aberrant primates rather than as insectivores or even as a separate order of mammals (Martin, 1968b; Hershkovitz, 1974a, 1977; Cartmill, 1972, 1974c, 1975a). The pros and cons of the systematic position of the plesiadapiforms have been discussed clearly by Cartmill (1975a), Szalay (1975a), and Kay and Cartmill (1977). As the latter authors pointed out, identification of the primate-nonprimate boundary

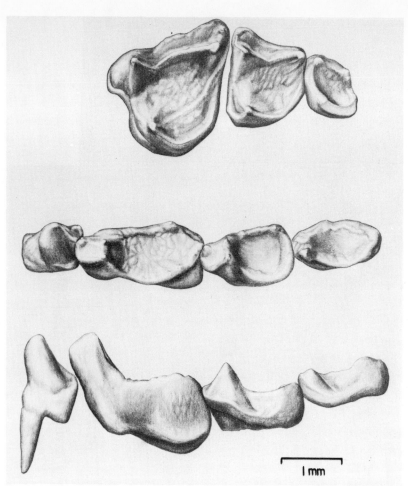

Figure 10   *M¹–M³ of* Zanycteris. *(Top).* P_4–M_3 *of* Picrodus silberlingi.

with respect to plesiadapiforms is, to some extent, a matter of taste or taxonomic philosophy. All plesiadapiforms share with later primates a basically similar and clearly derived pattern of cheek tooth morphology. All except the microsyopids (as far as known) share with later primates a petrosal bulla and loss of the medial entocarotid artery in the ear region. In addition, Szalay and Decker (1974) and Szalay *et al.* (1975) have argued that *Plesiadapis* shows features of the ankle, and perhaps other limb bones, that indicate unique affinities with primates as opposed to other mammal groups.

While there are many proposed solutions (almost as many as there are students of early primates), each has its flaws. Cartmill's (1972, 1974c) suggestion of placing all the Plesiadapiformes in the Insectivora has the elegance of making the order Primates clearly definable in terms of a large suite of unique characters that can be related to a major adaptive shift. At the same time, however, it overburdens the Insectivora, which is already something of a wastebasket taxon, by including one group (Plesiadapiformes) which clearly shares derived characters with primates.

The solution we have adopted here represents the other extreme—inclusion of all plesiadapiforms, including microsy-

opids, in the Primates. While this solution has the advantage of grouping all families with primate-like cheek teeth within the same order (and of being operable for the classification of fossil species known only by molar teeth in most cases), it does not formally recognize the cranial (and postcranial) similarities that plesiadapoids share with later primates, but that microsyopids lack. The solution adopted by Szalay is essentially an intermediate position which excludes those microsyopids lacking the cranial features (the petrosal bulla and loss of the medial entocarotid) which link plesiadapoids with later primates. This arrangement is based on his assessment that many of the genera (included here among the microsyopids) are more closely related to paromomyids than to later microsyopids (Bown and Gingerich, 1973; Bown and Rose, 1976; Szalay, 1973). At present the basicranial evidence which would clarify the relationships of these Paleocene genera to paromomyids and to microsyopids is unavailable. Until that becomes available, the more inclusive dental definition of the primate-nonprimate boundary seems the most acceptable.

We thank Luci Betti for preparing Fig. 1 and Karen Payne for Figs. 4–6 and 8–10. This work has been supported by NSF grants BNS 77-25921 and BNS 79-24149 to JGF.

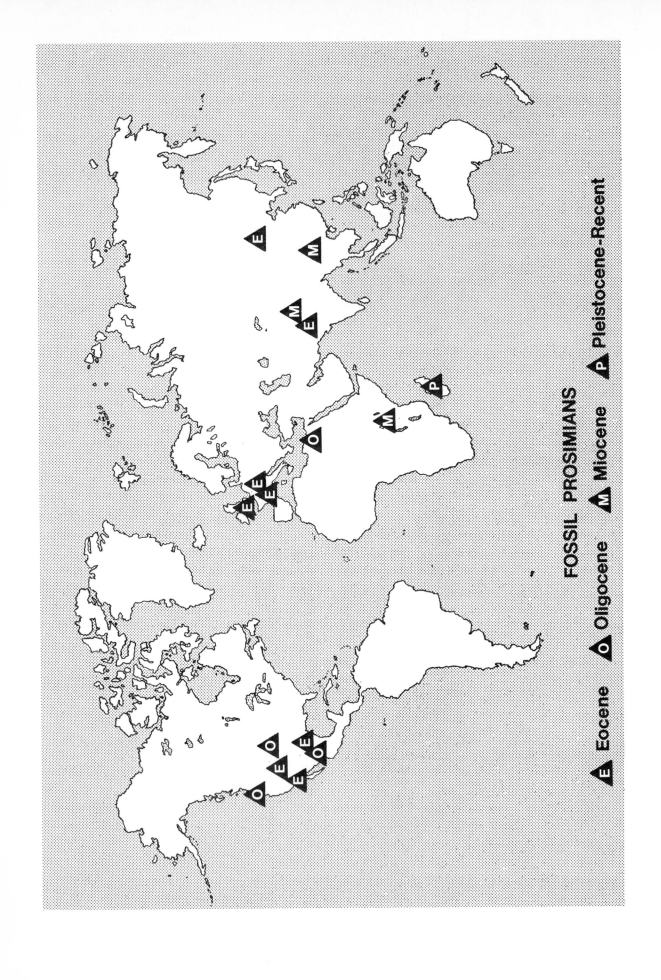

FOSSIL PROSIMIANS

△E Eocene △O Oligocene △M Miocene △P Pleistocene-Recent

Part II

Evolution of Prosimians

The chapters in this part are concerned with the evolution of prosimian primates (lemurs, lorises, tarsiers and their relatives) from their first appearance in the Eocene Epoch to the present.

Chapters 6 and 7, by P.D. Gingerich and K.D. Rose and J.G. Fleagle, respectively describe the adaptive radiations of the two main groups of fossil prosimians from the Eocene Epoch, the adapids and the omomyids (see also Chapter 12 by A.L. Rosenberger and F.S. Szalay in Part III). In contrast with the archaic primates of the preceding Paleocene Epoch, whose primate affinities were questioned by several authors, these relatively abundant primates of the Eocene have generally been recognized as relatives of the living prosimians. Elwyn Simons (1972) has appropriately dubbed them "the first primates of modern aspect."

The adapids, usually regarded as ancestral to living lemurs and lorises, were an extremely diverse group with a large adaptive radiation in Europe and a smaller one in North America. They averaged between about 1 and 8 kg and had relatively small brains when compared with living prosimians of a similar size. Some species seem to have been predominantly frugivorous; others were folivorous. The skulls of several species suggest that they were diurnal and at least one species shows evidence of sexual dimorphism. Gingerich emphasizes that although the Eocene adapids share a number of (probably primitive) phylogenetic similarities with living lemurs and lorises, adaptively they were very different from any living primate. In some ways they resembled living lemurs, in other respects they resembled living anthropoids.

The other group of Eocene prosimians, the omomyids, are considered by the majority of current workers to be relatives of the living *Tarsius* (e.g. Simons, 1961a, 1972; Szalay, 1975b, 1976; but see Cartmill, 1980; Schmid, 1981, 1983; for dissenting views). Like adapids, the omomyids underwent adaptive radiations in both North America and Europe, the former being far more diverse than the latter. In general, omomyids were smaller than adapids and many were probably nocturnal. Many researchers suggest that they were predominantly insectivorous and frugivorous in their dietary habits.

Chapter 8, by John Napier and A.C. Walker (1967), describes the locomotor adaptation of living prosimians and discusses the evolution of locomotor behavior among primates from the Eocene to the present. They argue that the "Vertical Clinging and Leaping" behavior of many living prosimians is the oldest locomotor adaptation found among living primates and probably characterized most of the prosimian primates known from the Eocene (see also Cartmill, 1972; Martin, 1972).

"Evolution of Lemurs and Lorises," by Pierre Charles-Dominique and R.D. Martin (Chapter 9), discusses the evolution of prosimians from the perspective of the behavior and morphology of living lemurs and lorises. They show that the living galagos from Africa and the cheirogaleids from Madagascar are very similar in many aspects of their behavior and cranial morphology (see also Szalay and Katz, 1973; Cartmill, 1975b) and suggest that these similarities between dwarf galagos and mouse lemurs may be ancestral primate characteristics. In contrast with most of the articles in this volume that emphasize analysis of the bony skeleton, this paper (and also Martin, 1972) is one of the best attempts to reconstruct the evolution of behavior using phylogenetic approaches.

Chapters 10 and 11, by Ian Tattersall (1975b) and William Jungers (1980), respectively, describe the adaptive diversity of the numerous lemurs that lived on Madagascar until the relatively recent arrival of humans during the last few thousand years. The extinct species were, in virtually all cases, larger than extant species, and, judging from their orbit size, most were diurnal. In their dental and locomotor adaptations, many of the extinct species resembled living monkeys and apes more than do any of the living Malagasy species. Others were strikingly different from any other known primates and document realms of primate evolution unknown among living species.

As the chapters in this part demonstrate, the biology of prosimian primates, both living and fossil, is one of the most exciting and active areas in all of primatology (see Cartmill, 1982), and there have been numerous recent symposia and books that treat in greater depth many of the issues touched on in this part (e.g. Martin, Doyle, and Walker, 1974; Tattersall and Sussman, 1975; Doyle and Martin, 1978; Chivers and Joysey, 1978; Tattersall, 1982; Niemitz, 1984). Many of the evolutionary relationships put forth in the chapters above have been questioned by other authors on the basis of new fossils, new comparative data, or different methods of analysis. In particular, several authors have questioned the evidence linking the Eocene omomyids with the living *Tarsius* (Cartmill and Kay, 1978; MacPhee and Cartmill, 1981; Schmid, 1981, 1983), whereas others have debated the systematics of living strepsirhines (e.g. Tattersall and Schwartz, 1974; Szalay and Katz, 1973; Cartmill, 1975b). Some of the topics treated in this part, such as the locomotor categories discussed by Napier and Walker (Chapter 8), have been widely debated issues in other areas of primatology as well (see Stern and Oxnard, 1974; Ripley, 1967; Mittermeier and Fleagle, 1977). Likewise, Tattersall's (Chapter 10) considerations of the mechanics of chewing and Jungers' (Chapter 11) concern over the structural correlates of size are topics of major importance for all areas of primate and human evolution.

6

Dental and Cranial Adaptations in Eocene Adapidae

P.D. GINGERICH

The primate family Adapidae is first known from the lower Eocene of Europe and North America. At first appearance, the European radiation is more diverse (including *Donrussellia, Protoadapis,* and *Pelycodus*) than that in North America (where only *Pelycodus* is represented). This suggests that Europe was possibly closer to the center of origin of the family than North America was. Other circumstantial evidence suggests that the origin and initial radiation of Adapidae was probably in Africa and/or South Asia during the late Paleocene or earliest Eocene. Some 17 genera and 48 species of adapids are known from the Eocene of Europe, Asia, Africa(?), and North America, compared with a single Oligocene genus *(Oligopithecus)* with one species from Africa, and two relict genera *(Indraloris* and *Sivaladapis)* with three species in the Miocene of South Asia (Gingerich & Sahni, 1979).

The phylogenetic relationships of Adapidae in Europe and North America have been reviewed recently (Gingerich, 1977c, 1979a; Gingerich & Simons, 1977). In this paper I would like to discuss in general terms some of the dental and cranial adaptations of Eocene Adapidae.

BODY SIZE

Body size is perhaps the most fundamental component of an animal's general adaptation. Morphological, ecological, behavioral, and life history parameters are influenced, if not determined, by body size. Fortunately body size is highly correlated with tooth size, and it can thus be estimated even from fragmentary fossil remains. The lower first molar is one of the teeth most commonly represented in fossil primate specimens. In a sample of 43 non-human living primates, ranging systematically from *Tarsius* to *Gorilla,* the log of crown area of M_1 has a coefficient of correlation $r = .95$, and coefficient of determination $r^2 = .90$. In other words, 90% of the variance observed in tooth size can be explained simply as a result of variation in body size in different species of primates. The following regression equation can be used to predict body size given tooth size:

$$\text{body weight} = \frac{[L \times W]^{1.62}}{48.4}$$

where body weight is measured in kg, and M_1 length (L) and M_1 width (W) are measured in mm (based mostly on data in Harvey et al., 1978; and Swindler, 1976). The evolution of body size in European Adapidae is illustrated in Fig. 1. Adapids were relatively small (200–1,000 g) when they first appeared in the European fossil record, but this distribution of body size in the family as a whole shifted rapidly to larger size. Most later adapids in Europe ranged from 600 g to 7–8 kg. All of the known radiation

of Adapidae in North American was in the 600 g to 8 kg range.

Kay (1975; see also Kay & Hylander, 1978) demonstrated that insectivorous and folivorous primates can be separated morphologically from frugivorous primates by the more crested structure of their molar teeth. Furthermore, insectivorous primates can be distinguished from folivorous primates on body size alone. The largest primate insectivore weighs about 300 g, whereas the smallest folivore weighs about 700 g. There is thus a threshold between the two adaptive zones at around 500 g body weight. Even frugivorous primates tend to fall on one or the other side of this threshold, depending on their secondary specialization for insects or leaves (Kay, 1975). The 500 g threshold between insectivorous-frugivorous and folivorous-frugivorous adaptive zones is shown in Fig. 1. This figure shows clearly that the European adapid radiation took place within the folivorous-frugivorous adaptive zone. Some adapids, such as *Cercamonius,* have relatively flat teeth indicating that they were feeding predominantly on fruit, while others, such as *Adapis,* have more sharply crested teeth indicating that they fed predominantly on leaves. Only one adapid lineage is known to have crossed the adaptive threshold and become insectivorous: the lineage from *Periconodon* to *Anchomomys.* The known North American radiation took place exclusively in the folivorous-frugivorous adaptive zone. By contrast with adapids, Eocene tarsiiform primates (Omomyidae) radiated primarily in the insectivorous-frugivorous adaptive zone. Fleagle (1978b) has previously discussed this important dichotomy in body size adaptation within Eocene primates.

CRANIUM

Brain Size and Arterial Circulation

Several genera and species of Adapidae are known from well preserved skulls. Best among these are the European *Adapis parisiensis* and North American *Smilodectes gracilis.* Comparing skulls of *Adapis* or *Smilodectes* with those of living lemurs or anthropoids, one of the most striking differences is in brain size. The fossil species have much smaller brains than those of living primates. The difference can be quantified using Jerison's (1973) encephalization quotient (EQ) as a measure of relative brain size corrected for differences in body weight. The endocranial volume of *Adapis parisiensis* can be estimated from two skulls, and these average about 8.8 cm³ (= 8.8 g). Using this brain size, and a body weight estimate of 2.3 kg, the relative brain size of *Adapis parisiensis* is estimated at EQ = .42 (Gingerich & Martin, 1981). A similar calculation for *Smilodectes,* using a brain weight of 9.3 g (Jerison, 1973; Radinsky, 1977) and a body weight of 2 kg (estimated from M_1 size), yields an encephalization quotient EQ = .49, which is slightly larger than the estimate for *Adapis.* These EQ

Reprinted with permission from *Zeitschrift für Morphologie und Anthropologie,* Volume 71, pages 135–142, Stuttgart. Copyright © 1980 by E. Schweizerbart'sche Verlagsbuchhandlung, Stuttgart.

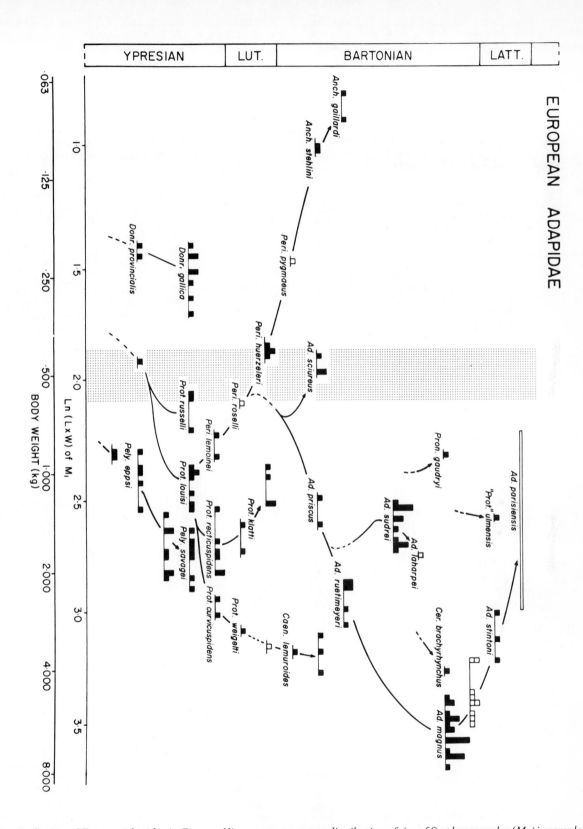

Figure 1 *Radiation of Eocene Adapidae in Europe. Histograms represent distribution of size of first lower molar (M₁) in samples from successive stratigraphic intervals. Abscissa is tooth size, and by inference body size (see text). Ordinate is time, based on biostratigraphic analysis of rodents and perissodactyls (see Gingerich, 1977c, and Godinot, 1978, for documentation). Stippling shows Kay's 500 g threshold separating insectivorous from folivorous primates. Note radiation of Adapidae is largely on folivorous side of threshold.*

values are below those of any primate living today, although they are relatively high for the Eocene. Among living primates these estimates are closest to the relative brain size of *Lepilemur*, which has an EQ = .60 (Jerison, 1973).

The blood supply to the brain in Adapidae was in part by way of the internal carotid artery passing through the middle ear. This artery includes two branches, the promontory and the stapedial. In *Adapis parisiensis* the bony canal for the stapedial branch is about four times the cross-sectional area of the canal for the promontory branch. In the one specimen of *Notharctus* where these diameters have been compared, the promontory branch was slightly more than twice the cross-sectional area of the stapedial branch (Gingerich, 1973a). Thus there appears to be significant variation in the relative size of different arterial branches supplying the brains of different Adapidae. The adaptive significance of these differences is discussed by Conroy (1980).

Dentition and Diet

The dentition of *Adapis parisiensis* is illustrated in Fig. 2. The upper and lower incisors are vertically implanted and have spatulate crowns very much like those of the living squirrel monkey *Saimiri*. The relative sizes of the incisors can be compared to those of living anthropoid primates using Hylander's (1975) regression of maxillary incisor width on body weight. Both *Adapis parisiensis* and *Adapis magnus* fall with the frugivorous primates on Hylander's figure, being about as far above his regression line as is *Saimiri*. Thus *Adapis* was not as highly specialized a folivore as, for example, modern colobines, even though cheek teeth of *Adapis* do suggest a folivorous diet.

The canine teeth of *Adapis magnus* are large, projecting,

interlocking teeth, with the back of the upper canine honed by the front of lower P_1 and P_2. Most adapids had similarly projecting canine teeth, but in *Adapis parisiensis* the canines are much reduced in size and functionally incorporated in the incisor series. This is like the condition in "short-tusked" callitrichids, which use their incisors to gouge the bark and superficial cambium of trees to harvest the exuded gum (Kinzey et al., 1975; Coimbra-Filho & Mittermeier, 1976). A similar dental adaption in *Adapis*, at this time when climate was becoming dryer and more seasonal in the latest Eocene (Wolfe, 1978) would permit exploitation of gums and resins, an important dry-season food resource. It is plausible, although certainly not demonstrated, that harvesting tree gum in this manner was an important stage in the evolution of the dental scraper or tooth comb in lemuriform primates.

The cheek teeth of adapids vary from having low rounded cusps *(Cercamonius, Amphipithecus, Pondaungia)* to being highly crested *(Adapis, Notharctus,* and *Sivaladapis)*. The former were probably largely frugivorous while the latter were more folivorous. Most adapids fall somewhere in between these two extremes, and probably fed on an intermediate combination of fruit and leaves. The small species of *Anchomomys* had sharp cusps and crests consistent with their insectivorous specialization.

Each of the largest adapids developed a fused mandibular symphysis, in most cases independently. *Caenopithecus, Adapis, Cercamonius* in Europe, *Northarctus* and *Mahgarita* in North America, and *Sivaladapis* in Asia all had at least partial symphyseal fusion. Beecher (1977) explains the evolution of symphyseal fusion in primates as an adaptation to leaf-eating, and suggests that folivory requires greater occlusal force during mastication than other diets do. Intuitively it is not clear why leaf-eating would require more force than fruit or seed-eating. Perhaps

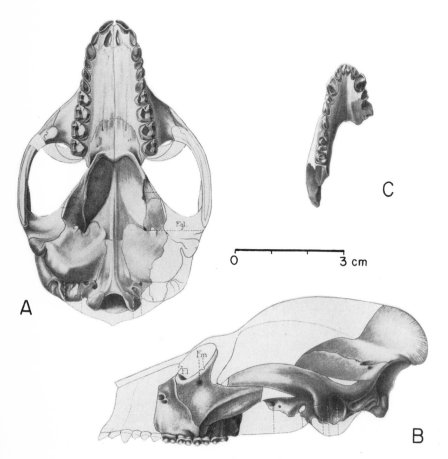

0 ⊢————————⊣ 3 cm

Figure 2 *Cranial anatomy of* Adapis parisiensis. *Judging from robust canines, large sagittal crest, and heavy flaring zygomatic arches, these specimens represent male individuals. A, cranium in palatal view. B, cranium in lateral view. C, anterior part of lower jaw in occlusal view. Short lower canines in this species functioned as a unit with the incisors, as in "short-tusked" callitrichids, possibly indicating an initial stage in the evolution of the dental scraper (tooth comb) characteristic of Lorisoidea and Lemuroidea.* **Figures from Stehlin (1912).**

larger primates generate greater occlusal forces than smaller primates for allometric reasons. In this case symphyseal fusion in primates could simply be a reflection of increased size and not necessarily an indication of increased folivory.

Sexual Dimorphism

Adapis magnus and *Adapis parisiensis* both have dimorphic crania, with the putative males having larger skulls, relatively enlarged sagittal and nuchal crests, and robust flaring zygomatic arches (Gingerich, 1979b, 1981a). *Adapis magnus* has dimorphic canine teeth, and the same may have been true for *Adapis parisiensis,* although the relatively reduced canines in the latter mask this relationship. Other genera have not yet been studied quantitatively, but *Notharctus* has been suggested to be dimorphic (Gregory, 1920c). Fig. 2 shows the cranial characteristics of a putative male specimen of *Adapis parisiensis*.

Orbital Diameter

The relative size of the eyes and bony orbits surrounding them sometimes gives an indication of whether an animal was nocturnal or diurnal. This is true for living Malagasy Lemuroidea (Walker, 1967a). Diurnal lemuroids fall below the regression of log orbital diameter on log cranial length, indicating that they have relatively smaller eyes than their nocturnal counterparts. *Adapis parisiensis*, *Adapis magnus*, and *Smilodectes gracilis* all fall well below this regression line, indicating that they were probably diurnal. By contrast, the omomyid *Necrolemur* falls above the line, suggesting that it was nocturnal.

SUMMARY

There is no animal living today that makes an ideal model for what an Eocene adapid was like in terms of its behavior and adaptations. Nor can one living model adequately represent a group as diverse as the Adapidae. Adapids possessed a mosaic of anatomical features, some primitive and some relatively advanced for their time. They share a few specializatons (chiefly postcranial) with lemuroids, and at the same time share a number of advanced specializations with anthropoids. Most adapids weighed between 0.6 and 8.0 kg, they had relatively small brains by modern standards, their dentition indicates a range in dietary adaptations from frugivory to folivory for most species, and at least one genus *(Adapis)* was sexually dimorphic. Considering all of these features, an average adapid may have resembled the hypothetical combination of a gentle lemur and a squirrel monkey.

7

The Second Radiation—Prosimians
K.D. ROSE AND J.G. FLEAGLE

Although a few lineages of the Plesiadapiformes continued into the early Eocene, overall this epoch marked a new phase in North American mammalian evolution. Faunas changed; their composition was different, and many of the new elements were more like modern animals than were those of the Paleocene. Many of the newcomers appear to have been immigrants, possibly arriving in the Rocky Mountain region from the south (Gingerich, 1976; Gingerich and Rose, 1977). The earliest prosimians were prominent members of the new Eocene faunas. Like modern prosimians, the primates which appeared in the early Eocene possessed a post-orbital bar completing the bony ring around their orbits, a divergent hallux, and nails rather than claws on their digits. Simons (1972) has therefore called them the "first primates of modern aspect". One group, the Adapidae, seems to represent the earliest relatives of the lemuriform primates (today including lemurs and lorises) while the other group, the Omomyidae, are the oldest relatives of the tarsier.

ADAPIDAE

Compared to the Plesiadapiformes, most adapids were rather large animals, comparable in size to many extant lemurs. If seen alive today, they would probably not differ greatly in appearance from extant Malagasy forms. They are divided into two subfamilies—the notharctines, a North American group whose earliest

genus also occurs in Europe, and the adapines, a predominantly European group with one North American representative (see Table I).

The Notharctinae are among the best known of all fossil primates. Many nearly complete specimens are known and have been described in great detail (e.g., Gregory, 1920c). They are best known from the central Rocky Mountain region, particularly the Eocene sedimentary basins of Wyoming. The earliest and most primitive member of this subfamily is the genus *Pelycodus*, with six recognized species from the early Eocene of the western U.S.A. (Gingerich and Simons, 1977) and one from Europe. The molar morphology is not unlike that of some plesiadapids, but the anterior dentition is quite different. *Pelycodus* (Fig. 1) has lower molars with broad basined talonids and three distinct cusps on the trigonid. In the earliest species the upper molars are basically tritubercular, but in later species a hypocone develops from the postprotocingulum (=nannopithex fold). All species have four premolars, prominent canines, and two small, vertical, spatulate incisors. The mandibular symphysis is unfused in *Pelycodus*. Little is known of the skull and skeleton of this genus.

Copelemur, with four species, is another early Eocene genus that is contemporary with and almost certainly derived from *Pelycodus*. Species of *Copelemur* were formerly included in the genus *Pelycodus*, but Gingerich and Simons (1977) showed that

Table I *North American Omomyidae and Adapidae*

Suborder: Prosimii, Infraorder Tarsiiformes Gregory, 1915

Family *OMOMYIDAE* Touessart, 1879 (E. Eoc.-Late Olig.)

Subfamily ANAPTOMORPHINAE Cope, 1883 (E. Eoc.-Late Eoc.)

Teilhardina Simpson, 1940b (E. Eoc.) (Eur. and N. Amer.)	$\frac{?}{2.1.3 \text{ or } 4.3}$
Anemorhysis Gazin, 1958 (E. Eoc.)	
Tetonoides Gazin, 1962 (E. Eoc.)	$\frac{2.1.3.3}{2.1.3.3}$
Tetonius Matthew, 1915 (E. Eoc.)	$\frac{2.1.3.3}{2.1.3.3}$
Chlororhysis Gazin, 1958 (E. Eoc.) (probably = *Tetonius*)	
Pseudotetonius Bown, 1974 (E. Eoc.) (includes *Mckennamorphus* Szalay, 1976)	$\frac{?}{1.1.3.3}$
Absarokius Matthew, 1915 (E. -M.Eoc.)	$\frac{2.1.3.3}{2.1.2 \text{ or } 3.3}$

Excerpt reprinted with permission from "The fossil history of nonhuman primates in the Americas," which appeared in *Ecology and Behavior of Neotropical Primates*, Volume 1, edited by A.F. Coimbra-Filho and R.A. Mittermeier, pages 111–167, Academia Brasileira de Ciencias, Rio de Janeiro. Copyright © 1981 by A.F. Coimbra-Filho and R.A. Mittermeier.

Table I   *North American Omomyidae and Adapidae (continued)*

Uintanius Matthew, 1915 (M. Eoc.)	$\dfrac{?}{2.1.3.3}$
Anaptomorphus Cope, 1872 (M. Eoc.)	$\dfrac{2.1.2?.3}{2.1.2.3}$
Trogolemur Matthew, 1909 (M.-Late Eoc.)	$\dfrac{?}{2.1.2.3}$
Aycrossia Bown, 1979 (M. Eoc.)	$\dfrac{2.1.2.3}{2.1.3.3}$
Strigorhysis Bown, 1979 (M. Eoc.)	$\dfrac{2.1.3.3}{2.1.2.3}$
Gazinius Bown, 1979 (M. Eoc.)	

Subfamily OMOMYINAE Trouessart, 1879 (E. Eoc.-L. Olig.)

Omomys Leidy, 1869 (E.-M. Eoc.)	$\dfrac{2.1.3.3}{2.1.3.3}$
Chumashius Stock, 1933 (Late Eoc.)	$\dfrac{?}{2.1.3.3}$
Ourayia Gazin, 1958 (Late Eoc.)	$\dfrac{2.1.3.3}{2.1.3.3}$
Shoshonius Granger, 1910 (E. Eoc.)	
Washakius Leidy, 1873 (M.-Late Eoc.)	$\dfrac{2.1.3.3}{2.1.3.3}$
Utahia Gazin, 1958 (E.-M. Eoc.)	
Hemiacodon Marsh, 1872 (M. Eoc.)	$\dfrac{?}{2.1.3.3}$
Dyseolemur Stock, 1934 (Late Eoc.)	$\dfrac{?}{2.1.3.3}$
Stockia Gazin, 1958 (Late Eoc.)	
Macrotarsius Clark, 1941 (Late Eoc.-E. Olig.)	$\dfrac{?}{2.1.3.3}$
Rooneyia Wilson, 1966 (E. Olig.)	$\dfrac{2.1.2.3}{?}$
Ekgmowechashala MacDonald, 1963 (Late Olig.)	$\dfrac{?}{2.1.3.3}$

OMOMYIDAE, Incertae Sedis

Loveina Simpson, 1940 (E. Eoc.)	$\dfrac{?}{2.1.3.3}$
Arapahovius Savage and Waters, 1978 (E. Eoc.)	$\dfrac{2.1.3.3}{2.1.3.3}$

Infraorder: Lemuriformes

Family *ADAPIDAE* Trouessart, 1879 (E. Eoc.-E. Olig.)

Subfamily NOTHARCTINAE Trouessart, 1879

Pelycodus Cope, 1875 (E. Eoc.) (Eur. and N. Amer.)	$\dfrac{2.1.4.3}{2.1.4.3}$
Notharctus Leidy, 1870 (M. Eoc.)	$\dfrac{2.1.4.3}{2.1.4.3}$
Smilodectes Wortman, 1930a (M. Eoc.)	$\dfrac{2.1.4.3}{2.1.4.3}$
Copelemur Gingerich and Simons, 1977 (E. Eoc.)	$\dfrac{2.1.4.3}{2.1.4.3}$

Subfamily ADAPINAE Trouessart, 1879

Mahgarita Wilson and Szalay, 1977 (Late Eoc.) (=*Margarita* Wilson and Szalay, 1976)	$\dfrac{2.1.3.3}{2.1.3.3}$

they tend to be more conservative than contemporary species of *Pelycodus* in showing less extensive development of the hypocone and mesostyles of the upper molars. The lower molars in all species have an open talonid with a distinct entoconid notch. In other aspects this genus resembles *Pelycodus*.

The middle Eocene genus *Notharctus* is derived from *Pelycodus* and differs from the latter genus in its larger size, possession of larger hypocones and mesostyles on the upper molars, greater reduction of the paraconid on the lower molars, and symphysial fusion. As in *Pelycodus* there are four premolars, large canines and small vertical incisors. Most of the features which characterize *Notharctus* are merely continuations of trends seen in time-successive species of *Pelycodus*, and the boundary between the two genera is essentially arbitrary. For taxonomic purposes, however, the genus *Notharctus* is currently defined by the presence of a fused mandibular symphysis. The teeth of *Notharctus* indicate that species of this genus were almost certainly herbivorous, and very likely folivorous.

The skull of *Notharctus* is well known (Fig. 2), and generally resembles that of the living *Lemur* (Gregory, 1920c). In contrast to the plesiadapiforms, there is a post-orbital bar, and the orbits are directed forward rather than laterally. There is a moderately long snout as in *Lemur*, but the lacrimal bone in *Notharctus* is located within the orbit rather than in front of it, and the lacrimal foramen thus lies within the orbital margin. There are prominent sagittal and nuchal crests on the cranium. The auditory region is well preserved in several specimens and reveals that *Notharctus* had a typically lemuroid ear region, with a tympanic ring lying free within the auditory bulla, and a large stapedial branch and small promontory branch of the internal carotid artery.

Several complete skeletons are known for *Notharctus*. In his classic study of this genus, Gregory (1920c) concluded that the Eocene taxon was generally similar to the living genera *Lemur* (including *Varecia*), *Lepilemur*, and *Propithecus*, but that most limb elements were shorter and more robust than those of extant forms. *Notharctus* had nails rather than claws on all digits, a large and divergent hallux, a long tail, and very long hindlimbs relative to the length of the forelimbs. Martin (1972) has argued that the calcaneus of *Notharctus* resembles that of the more quadrupedal lemur *Varecia variegatus*, and Oxnard (1973) has also suggested that in some skeletal proportions *Notharctus* is distinct from any living genus. Most authors, however have interpreted *Notharctus* as an arboreal animal primarily adapted for leaping and clinging on vertical supports (Fig. 3) (Gregory, 1920c; Simons, 1967e, 1972; Napier and Walker, 1967; Walker, 1974; Decker and Szalay, 1974).

Smilodectes was a middle Eocene adapid, closely related to *Notharctus*. Species of *Smilodectes* are generally distinguished from those of the latter genus by smaller, narrower molar teeth. The snout of *Smilodectes* is shorter than that of *Notharctus*, but in other aspects of the skull and skeleton the two genera seem very similar. The brain of *Smilodectes* is known from several complete

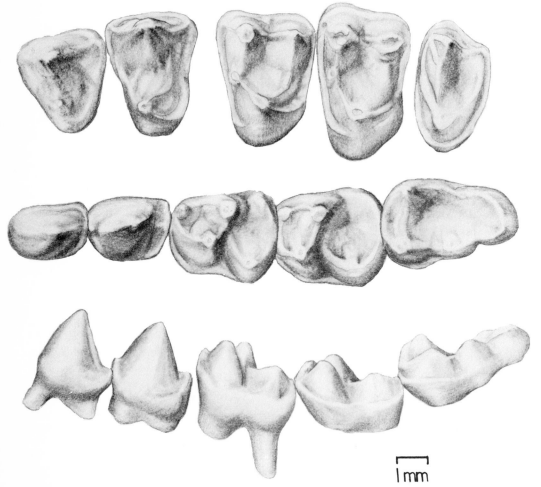

Figure 1   *Left P³–M³ (above) and left P₃–M₃ (middle: occlusal view; below: lateral view) of* Pelycodus mckennai.

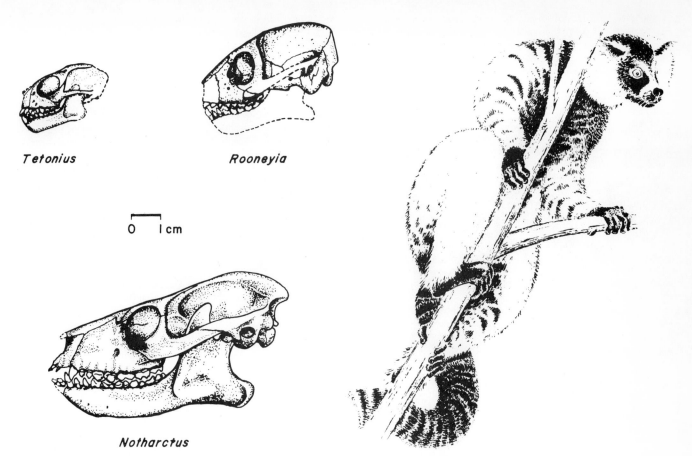

Figure 2 *Skulls of omomyid* (Tetonius and Rooneyia) *and adapid* (Notharctus) *primates from North America.*

Figure 3 *Artistic reconstruction of* Notharctus.

endocasts. Compared to other Eocene mammals, this primate had a relatively expanded visual cortex and reduced olfactory bulbs (Radinsky, 1975). The size of the brain relative to body size seems to have been less than that of extant prosimians (Radinsky, 1977).

The other subfamily of adapids, the adapines, are common and diverse in Eocene and Oligocene faunas of Europe. Recently, a single genus and species, *Mahgarita stevensi* (Wilson and Szalay, 1976, 1977), has been described from the late Eocene of Texas. In contrast to the notharctines, the adapine *Mahgarita* had reduced premolars (loss of P_1, and only a tiny P_2). Like the European adapines, but unlike notharctines, it possessed a hypocone arising from the lingual cingulum and pronounced crests on both the upper and lower molars. Wilson and Szalay (1976) argued that *Mahgarita* was probably insectivorous or carnivorous, but such dental adaptations seem to us more indicative of herbivory or folivory.

OMOMYIDAE

The tarsier-like omomyids also make their first North American appearance in the early Eocene. These smaller primates appear to be less common than adapids in most sedimentary environments, but show a much greater taxonomic diversity. North American omomyids are classified in two subfamilies, the Anaptomorphinae and the Omomyinae (Table I). Since most omomyids are known only from jaws and teeth, these are the primary elements that are used to diagnose the family as well as the genera and

species. Both subfamilies and especially the early, smaller genera have basically primitive teeth not unlike some of the microsyopids or the living *Tarsius*. The primitive dental formula for the family appears to be $\frac{2.1.3.3}{2.1.3.3}$ although there is possible evidence for the presence of four premolars in some specimens of European *Teilhardina* (Bown, 1976), a genus which may be ancestral to the entire family. The upper molars are broad transversely and in many species have a prominent postprotocingulum (nannopithex fold) joining the protocone to the posterolingual border of the tooth, but the hypocone in most species is small or absent. Lower molars have broad, deeply basined talonids and relatively small, low trigonids that are compressed anteroposteriorly. Nearly all omomyids have a pair of incisors on each side of the jaw. Where known, the medial incisors usually are enlarged and more or less procumbent, and the second incisor small. The canines tend to be relatively small, but not absent as in some Paleocene forms.

There is some debate over interrelationships among omomyid genera. While the make-up of the two subfamilies is relatively stable, there are some controversial genera. Both anaptomorphines and omomyines first appear in the early Eocene. The anaptomorphines (Fig. 4) are less diversified than the omomyines and are usually characterized by a large pointed P_4, reduced M^3, and, in many forms, reduction in the number of premolars from 3 to 2. The early Eocene genera, *Teilhardina* and *Anemorhysis,* are the most primitive anaptomorphines and their teeth are among the most primitive known among primates. They probably lie near the base of the entire omomyid radiation.

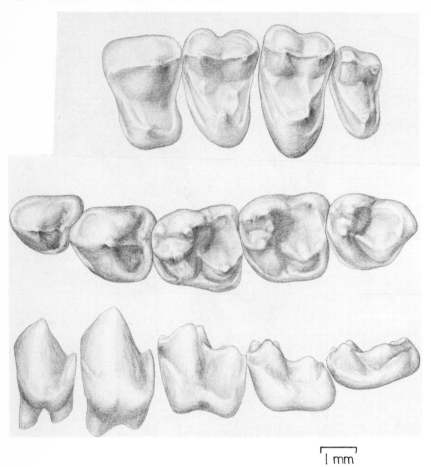

1 mm

Figure 4 *Left P⁴-M³ (above) and left P₃-M₃ (middle and below) of* Tetonius homunculus.

Tetonius, also from the early Eocene, is probably derived from a form like *Teilhardina* and in turn gave rise to *Absarokius.* No anaptomorphines are known beyond the Eocene.

The omomyines were more diversified and of longer duration, one form surviving into the latest Oligocene (early Arikareean). Whereas anaptomorphines were all very small animals, some omomyines reached moderate sizes perhaps comparable to the ceboids *Aotus* or *Pithecia* (Szalay, 1976). In contrast with anaptomorphines, this subfamily is characterized by less specialized upper and lower premolars, retention of three premolars, and retention of unreduced last molars. Many of the later genera of omomyines developed accessory molar cusps and crenulated enamel. Because of the diversity of this subfamily, the interrelationships among allocated genera are more difficult to determine than in the anaptomorphines.

While the morphology of omomyid teeth has been well documented (see especially Szalay, 1976), analyses of dental function in this group have been complicated by the great diversity and lack of comparable dentitions among living primates (but see Kay, 1977a). Thus in his recent monograph, Szalay (1976:281) could be no more specific than to suggest that *Ourayia,* one of the most completely known omomyids, was "probably mainly frugivorous-omnivorous, with of course a probably significant portion of its diet coming from insect prey". In general it seems that the larger omomyines with crenulated enamel and accessory cusps *(Ourayia, Macrotarsius, Ekgmowechashala, Rooneyia)* were more frugivorous, while the smaller omomyines and anaptomorphines with sharper cusps were more insectivorous. However, Szalay (1976) suggests that some of the smaller species were predominantly herbivorous and that at least one, *Washakius insignis,* was folivorous.

The skull is known in only two taxa of North American omomyids, *Tetonius,* a primitive anaptomorphine, and *Rooneyia,* an advanced omomyine from the Oligocene (Fig. 2). The skull of *Tetonius* is strikingly tarsier-like, with relatively large eyes (though not nearly so large as in *Tarsius*) surrounded by a bony ring, a short snout, and a globular braincase. The orbits are, however, less frontally directed than in the extant *Tarsius.* In relative orbit size, the small anaptomorphine most closely resembles living cheirogalines and was almost certainly nocturnal in habits. In both *Tetonius* and *Rooneyia,* the lacrimal foramen is large and is situated outside of the orbit, and the infraorbital foramen is smaller than in the Paleocene primates *(Palaechthon),* but larger than in extant *Tarsius.*

Natural endocasts are known for both *Tetonius* and *Rooneyia.* Compared with living primates, *Tetonius* had relatively large olfactory bulbs and small frontal lobes, but the occipital and temporal lobes were expanded more than in contemporary insectivores, suggesting a visually, rather than olfactorily, oriented animal. In relative brain size *Tetonius* lies below the range found among extant prosimians (Radinsky, 1977). Like *Tetonius,* *Rooneyia* has expanded visual cortex and reduced olfactory cortex relative to early insectivores, but smaller frontal lobes than those of extant prosimians. Its brain size lies within the lower half of the range found in extant prosimians (Radinsky, 1977).

The skeleton of omomyids is poorly known. Most of our knowledge is based on a number of hindlimb bones associated with teeth and jaws of the middle Eocene omomyine *Hemia-*

codon. These bones, which include parts of the pelvis and femur and a number of pedal elements, have usually been considered more similar to the counterparts in the lorisiform *Galago* than to those of *Tarsius* (Simpson, 1940b; Walker, 1974; Szalay, 1976). The shape of the femoral head, the distal femoral condyles, and the elongate tarsal elements all indicate leaping adaptations. However, the tarsal elongation was more general and not restricted to extreme elongation of the calcaneus and navicular as seen in the extant *Tarsius*. In addition, Szalay has attributed several isolated elongate calcanei to the early Eocene anaptomorphine *Tetonius*.

RELATIONSHIPS OF EOCENE PRIMATES

In the beginning of this century, most workers were confident in identifying adapids as early lemuriform primates and omomyids as early tarsiers (e.g. Wortman, 1903a; Gregory, 1920c). More recent workers have voiced reservations (e.g., Simpson, 1955; Gazin, 1958; Simons, 1963a) and cautioned that many of the features linking adapids with lemurs (e.g. carotid circulation, middle ear morphology, and general skull conformation) are likely to be shared primitive features, and that many of the features linking omomyids with *Tarsius* (e.g. tubular ectotympanic, tooth morphology, large orbits) are possibly parallel acquisitions. Nevertheless, this division of Eocene families is still accepted by the majority of workers today (Simons, 1972; Cartmill, 1975b; Szalay, 1975b, 1976; Gingerich 1976; Hoffstetter, 1974).

Prior to the description of cranial material of plesiadapiform primates, the phyletic relationships of the Eocene families were further confused by attempts to link them with various Paleocene groups with which they shared diet similarities. Indeed, it was this problem of sorting both Paleocene and Eocene genera into a lemuriform-tarsiiform dichotomy that seems to be behind much of the frustration expressed by Simpson (1955), leading him to conclude that fossil primates could not be so divided. Subsequent new finds and redescriptions of cranial and postcranial material from plesiadapiform, adapid, and omomyid genera have, however, demonstrated the distinctiveness of the earlier radiation with respect to both Eocene and modern forms of primates and supported the dichotomy of the Eocene families.

Despite the distinctiveness of the Plesiadapiformes, there is some disagreement whether the two groups of Eocene prosimians are monophyletic with respect to the Paleocene forms or whether one group, the omomyids, is more closely related to Plesiadapiformes than to the lemur-like adapids. Most workers (e.g. Cartmill, 1972, 1974c, 1975b; Simons, 1972; Szalay, 1975b, 1976; Bown, 1976; Kay and Cartmill, 1977; Cartmill and Kay, 1978) have interpreted the features which unite both adapids and omomyids with extant primates—e.g. nails rather than claws, divergent hallux, large eyes with post-orbital bar, a relatively large brain—as evidence that they are more closely related to each other than either is to more primitive plesiadapiform primates. By contrast, Gingerich (e.g. 1976, 1977b) has argued that the omomyids are more properly considered as the direct lineal descendants of the Plesiadapiformes and that similarities between omomyids and adapids are either primitive retentions from the earliest primates or parallel acquisitions. In particular, he points to the small size, presence of a large procumbent central incisor, reduced premolar formula, and a tubular ectotympanic as specialized characters linking omomyids with plesiadapiforms.

Current evidence would seem to support a closer relationship between the Eocene families than between omomyids and plesiadapiforms. Nevertheless, Gingerich's point is well made that the differences between adapids and omomyids are such that only *Purgatorius* (or possibly *Teilhardina*) is generalized enough to serve as a common ancestor for the two Eocene families.

We thank Karen Payne for drawing Figs. 1 and 4 and Mark Orsen for Fig. 3. This work has been supported by NSF grants BNS 77-25921 and BNS 79-24149 to JGF.

8

Vertical Clinging and Leaping—
A Newly Recognized Category
of Locomotor Behavior of Primates

J. R. NAPIER AND A. C. WALKER

This paper provides a preliminary account of a natural locomotor group among the primates. All the living members of the group are prosimians. The animals concerned are arboreal, have a vertical clinging posture at rest and are well adapted to a leaping mode of progression during which the hindlimbs, used together, provide the propulsive force. The special interest of this locomotor group of Vertical Clinging and Leaping is that it appears in a preliminary study to constitute the only known locomotor adaptation of Eocene primates; possibly it is to be regarded as the earliest locomotor specialisation of primates and therefore pre-adaptive to some or possibly all of the later patterns of primate locomotion.

The living forms comprising the Vertical Clingers and Leapers are as follows:

Tarsiidae	Indriidae	Lemuridae	Lorisidae
Tarsius	Indri	Lepilemur	Galago
	Propithecus	Hapalemur	Euoticus
	Avahi	simus	

A brief summary of original observations of the locomotion of these animals together with our own field observations and studies of film sequences of wild and captive animals is given below.

Tarsius

Observations of activities in the wild made by Le Gros Clark (1924a), Harrison (1963) and on captive animals by Cook (1939), Polyak (1957), Hill, Porter and Southwick (1952) and Ulmer (1963); the leaping activity in tarsiers in captivity at the London Zoo has been recorded on film[1], and Walker (1948) has published still photographs of various phases of locomotion.

In nature, when at rest, tarsiers cling in a bunched position to a vertical stem, with the tail pressed against the support. Hill (1955) in his summary of the locomotor activities of *Tarsius* states, "The hallux is always strongly opposed to the other toes, but the position of the pollex is variable and not necessarily one of opposition." Most accounts state that the animals do not habitually come to the ground, but, when placed on a flat surface, will walk clumsily in a quadrupedal manner. The resting posture on the ground is very often erect, the tail acting as the third leg of a tripod (Ulmer, 1963). During more active locomotion on the ground they may proceed bipedally in a series of long leaps or quick hops. Leaps of up to six feet have been recorded. During normal arboreal progression the leap is from one vertical stem to another and may involve a twist of the body through 180°. Analysis of single ciné-frames shows that the hind-feet reach the support ahead of the forefeet in any leap of moderate or large

dimensions (Fig. 1). The tail is inactive during the leap and trails inertly, acting as a counterweight.

Indri

The most recent field observations of this rare lemuroid are those of Petter (1962) and Attenborough (1961), both these observers having recorded the locomotion on ciné-film. From their summaries and from study of their film records[2] it can be seen that *Indri* rests in a vertical position and progresses by a series of leaps from vertical trunk to vertical trunk, contact being made first by the hindfeet (Fig. 2). *Indri* is rarely on the ground but when so the animal makes a series of bipedal hops until it regains the safety of the trees.

Propithecus

Many observers, including Petter (1962), Attenborough (1961), Rand (1935) and Webb (1953) have recorded the leaping habits of this animal. Petter and Attenborough have both recorded leaping activity in the wild on ciné-film. One of us (A.C.W.) also has observed and taken ciné-film of *Propithecus verreauxi* in the Berenty reserve of the de Heaulme family near Fort Dauphin. Petter states "Les Propithèques paraissent cependant ne quitter de plus de 100 m loin de leurs trajets journaliers pour chercher des graines de tamarin, alors que quelques sauts sur le sol leur auraient permis de raccourcir considérablement leurs itinéraires." Rand (1935) observed them in the wild and in captivity: "Their mode of progression in the trees is like that of *Indri* rather than

Tarsius

Figure 1

Reprinted with permission from *Folia Primatologica,* Volume 6, pages 204–219, S. Karger Ag, Basel, Switzerland. Copyright © 1967 by S. Karger, Ag, Basel, Switzerland.

Lemur; they cling to the sides of vertical branches and spring from upright to upright. Largely arboreal animals, they sometimes descend to the ground...". "On the ground they travelled upright, on the hind feet, the legs fully extended. Progression was by a series of hops, of about four feet each...."

From a study of the film material, it is clear that the tail plays only a minor rôle in the action of the leap, being trailed loosely behind for most of the trajectory (Fig. 3). Its function may be to aid in the body's rotation back to the upright landing position at the end of the leap. Petter has recorded leaps of over 10 m.

Avahi

The locomotion of this indrisine has been recorded by Rand (1935) and Petter (1962). Both reports make it clear that its locomotion is very similar to that of *Propithecus. Avahi* may tend while resting to curl its tail like a watchspring, a posture which is reported but rarely in *Propithecus*.

Lepilemur

Petter (1962) has now confirmed the suspicions of Rand (1935) that *Lepilemur*, although a lemurine, moves as do the indrisines. It has a method of locomotion almost identical to that of *Propithecus;* the principal difference being that *Lepilemur* is perhaps more clumsy on the ground ("Un *Lépilémur*, posé sur le sol, se déplace gauchement sur les pattes postérieures par une série de bonds." Petter [1962, p. 74]). A captive specimen in the Tsimbazaza Zoological Gardens in Tananarive hopped quite quickly when placed on the ground.

Galago and Euoticus

Hill (1953b, p. 208) summarizes the observations made by many authorities from 1863 onwards.

"They are very nervous, active animals and move from place to place in the trees by leaping frog-fashion, using their elongated hind-limbs to obtain the initial momentum. On the ground they tend to assume an erect or semi-erect posture, and progress by leaping on the hind-limbs like a jerboa or kangaroo. This is especially marked in the smaller forms of the *G. senegalensis* type, whose leaping powers are quite amazing, distances of upwards of 10 ft in an oblique direction being quite usual (*vide* especially Bartlett, 1863). *Euoticus* can cover 40 ft chasms in an upward direction (Sanderson). The larger, more heavily built forms of the *G. crassicaudatus* type are more quadrupedal, though even these become bipedal on the ground."

Our own direct observations on a captive *G. senegalensis* and on *G. crassicaudatus* on film, as well as high-speed ciné-film of *G. senegalensis* made available through the kindness of Dr. E. C. B. Hall-Craggs, support this summary; the observations of Sanderson (1940) however seem rather unlikely. While moving at slow speeds on the ground all Galaginae tend to quadrupedalism but revert quickly to a bipedal leaping gait for more speed when pressed (Fig. 4). A marked preference for vertical supports is exhibited by *G. senegalensis* in captivity.

Observation on a captive *G. demidovii* indicates a greater quadrupedal component in the locomotion than in the other species. This species does revert, however, to the hopping mode of locomotion when moving quickly on the ground.

Hapalemur simus

No observation of the arboreal locomotion of this species has been made so far as we are aware. According to Webb (1953) and Shaw (1879), this animal is narrowly confined to areas of reed-beds around Lake Alaotra in Eastern Maagascar and the lakes of Betsileo. They are sluggish, progressing very clumsily on the ground (Hill, 1953b, p. 374), a feature not shared with *H. griseus,* for Petter (1962) is content to include this species in the same locomotor division as *Lemur*. Lamberton (1965) gives a description of the locomotion of *H. simus* in which he states they "sautant de toufe en toufe avec beaucoup d'agilité et d'adresse". Webb (1953) describing how a tame *H. simus* greeted its mistress states "this intelligent little animal hopped (sic) down the stairs to greet her". A pregnant female specimen of *H. simus* is now under observation by one of us (A.C.W.), but at the time of writing only two or three short hops have been observed. The animal does, nevertheless, spend almost all of its time sitting in an upright position. The fishermen of the village of Andreba on the East shore of Alaotra report leaps of up to 5 m in this species.

The fact that this species is confined to a habitat in which the vertical stem is almost the only support, would suggest together with some morphological differences between the two species, in particular the greater hindlimb length in *H. simus,* a locomotor type similar to that of *Lepilemur*. Until adequate field and captive observations have been made, however, this species is only tentatively included in this newly recognized locomotor group.

MORPHOLOGICAL CHARACTERS OF THE VERTICAL CLINGING AND LEAPING GROUP

The following account of this locomotor group includes a variety of animals, some of which are taxonomically separated at a subordinal level. They therefore show many morphological dif-

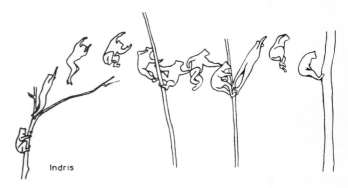

Indris

Figure 2

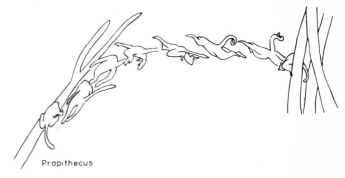

Propithecus

Figure 3

Galago

Figure 4

ferences. However, insofar as these animals have responded phenotypically in a similar way to the same selective pressures imposed by the environment, they share a number of common morphological characteristics which can be recognized as the correlates of their particular locomotor habit.

All vertical clingers and leapers are small to medium-sized primates. All have long tails with the exception of the largest, *Indri,* where the tail is rudimentary.

The hindlimbs are greatly elongated relative to the forelimbs (see Intermembral Indices, Table I). The hindlimb skeleton provides evidence that the limb is capable of an extreme range of flexion and extension. The femur is straight and the femoral head is cylindrical rather than spherical; the femoral condyles are projected posteriorly and the narrow patellar groove projects anteriorly to the long axis of the shaft thus displacing the patella well anterior to the shaft of the femur. The tibia is straight; in the tarsier the fibula is fused with the tibia in its distal half but is separate in the other members of the group. The forelimb is short; the humerus is shorter than the radius and, thus, the Brachial Index is high (Table I).

The vertebral column shows evidence of adaptation to the vertical posture. For instance the weights of the vertebral bodies (Petter, 1962, pp. 126–127) clearly shows that among Madagascan vertical clingers at least, the difference between the weights of the thoracic and lumbar regions, apparent in the quadrupeds, has largely disappeared.

The pelvis in this locomotor group shows a long iliac portion and a short ischial one—quite the opposite of that specified by Howell (1944) for jumping mammals of the jerboa-type. An allometric function is revealed in a study of the ilia; the larger animals have widely splayed iliac blades but in the smaller vertical clingers the iliac blades are narrow.

The grasping pes is remarkable for the size and strength of the hallux and the reduction of digits II and III. This reduction, which is due in part to the retention of grooming claws or claw, gives the foot both the appearance and function of a pincer. A further allometric effect is seen in the tarsus. The calcaneus and navicular are elongated in small vertical clingers such as *Galago* and *Tarsius* but are short in the bigger forms such as *Indri.*

The accent on the peroneal border of the pes is mirrored in the manus, where the ulnar border is dominant. This appears to be an adaptation associated with the vertical resting posture whereby the ulnar and peroneal borders of the extremities take the brunt of weight bearing function. The hands are long, due principally to the elongation of the phalanges (see Hand L. Index and Phalangeal L. Index, Table I). Large vertical clingers such as *Indri* and *Propithecus* which require a greater degree of prehensile power in the hand to support their relatively greater bulk have evolved the pincer-like arrangement between the first digit and the rest of the hand superficially resembling an "opposable" thumb but lacking in the rotatory element inherent in the concept of true opposability (Napier, 1961).

Associated with the vertical posture, the foramen magnum is placed in a more or less central position on the base of the skull. An orthognathous facial region is also seen, particularly in the smaller forms such as the tarsier. This cannot be considered to be wholly a function of the vertical posture; at least in part it can be related to great enlargement and frontality of the orbits associated with a nocturnal activity rhythm.

LOCOMOTOR ADAPTATIONS IN EOCENE PROSIMIANS

The morphology of such forms as *Propithecus* and *Tarsius* is generally regarded as a specialisation derived from a primitive quadrupedal primate morphology. However, when the known postcranial material of early fossil Eocene prosimians is considered in the light of vertical clinging and leaping adaptations, a different picture of the locomotor adaptations of early primates emerges. Of all the fossil postcranial bones reputably assigned to Eocene prosimians, none show the morphological feature of quadrupeds; all the skeletal characters point to these animals having been Vertical Clingers and Leapers. The post-cranial material of Eocene primates will be fully discussed in a later communication by one of us (A.C.W.), but in the meantime a list of Eocene and lower Oligocene fossil primates is given together with some of the principal features shown by these forms comprising the Vertical Clinging and Leaping type of morphology:

Low Intermembral Index	*Notharctus, Smilodectes.*
High Phalangeal L. Index	*Notharctus, Smilodectes.*
Femur with cylindrical head	*Hemiacodon, Necrolemur,*
and high, narrow patellar groove	*Notharctus, Smilodectes.*
Fused fibula and tibia	*Necrolemur,* perhaps *Nannopithex.*
Elongated calcaneus and os naviculare	*Hemiacodon, Microsyops. Nannopithex, Necrolemur, Omomys, Teilhardina.*
Very stout hallux and large peroneal tubercle	*Hemiacodon, Notharctus, Pseudoloris, Smilodectes.*
Tarsier-like pelvis	*Hemiacodon, Nannopithex, Notharctus, Smilodectes.*
Centrally placed foramen magnum	*Tetonius, Necrolemur.*
Widely divergent ramni mandibuares indicating a shortened facial region	*Aeolopithecus, Parapithecus, Amphipithecus, Microchoerus, Necrolemur, Tetonius.*

It can be seen that representatives of four separate families of Eocene primates figure in this list; they include (after Simons, 1963a): Adapidae, Microsyopidae, Omomyidae, Tarsiidae.

Doubt has been cast by several authorities on the primate character of the post-cranial material assigned by Filhol (1873) to the genus *Adapis.* Le Gros Clark (1959, p. 197) states, "only a few, rather uninformative, fragments of limb bones of the European fossil lemur *Adapis* are known, and there is even some doubt whether they have in all cases been correctly attributed to the genus". In view of the possibility that these Eocene specimens may not be primates, they have not been taken into account. Similarly, the post-cranial material assigned by Simpson (1931) to *Anagale* and transferred to the Insectivora by Saban (1958), has been omitted.

LOCOMOTOR CLASSIFICATIONS

Petter (1962) recognizes the existence of the vertical clinging among Madagascan lemurs as his 'specialised type', but does not expand his classification to include the tarsier or the African Galaginae. His account of the locomotion of *Lepilemur* (p. 74) serves as a description of that of the group:

"(a) Locomotion normale—Les *Lépilémures* sont bien adaptés au saut. C'est presque leur seule manière de progresser dans les arbres. Ils se déplacent ainsi d'un support vertical ou presque á un autre. Ils cheminent souvent au milieu d'arbustes ou d'arbres de moyenne grandeur. Dans ce cas ils sautent de tronc en tronc. La progression rapide se fait par une série de bonds successifs; les animaux ne touchent les troncs que pour y prendre l'élan d'un nouveau bond. Pendant le saut la queue semble inerte et pend derrière eux.

(b) Lorsqu'ils se déplacent sur les branches horizontales, généralement pour se nourrir, les *Lépilémures* peuvent adopter une démarche quadrupède normale, mais seulement pour un instant assez bref et pour un court déplacement. Leur allure semble gauche et hésitante.

Table I   *Shows upper and lower limb indices (based on longbone length) of vertical clingers. Indices of* Lemur *and of* Cercopithecus *are included for comparison. Indices from Mivart (1873), Mollison (1910), Napier (unpubl.) and Schultz (1953; 1956).*

	Rel.l.L.[1] Length	Rel.u.L.[2] Length	Brachial[3] Index	Crural[4] Index	Inter-membral[5] Index	Hand L.[6] Index	Phalan-geal L.[7] Index	Thumb L.[8] Index
Avahi	141	110	117	81	56	34	58	42
Propithecus	146	118	109	86	64	33	56	47
Indri	147	129	121	84	64	33	55	45
Lepilemur	—	—	114	89	64	32	56	42
Hapalemur	—	—	110	86	65	30	60	47
Galago	126	112	109	94	62	36	63	50
Tarsius	178	148	128	99	55	40	64	45
Lemur	111	106	106	93	70	29	55	46
Arboreal quadrupeds (*Cercopithecus*)	110	116	91	97	78	29	54	47

[1]Relative lower Limb-Length = Total length of lower limb as a percentage of trunk height (Schultz's [1956] method).
[2]Relative upper Limb-Length =
        Total length of upper limb as a percentage of trunk height (Schultz's [1956] method).
[3]Brachial Index   = Total length of radius in percentage of humeral length.
[4]Crural Index   = Total length of tibia in percentage of femoral length.
[5]Intermembral Index   = Total length of humerus and radius in percentage of total length of femur and tibia.
[6]Hand Length Index   = Total length of hand (carpus, metacarpus and phalanges in the line of digit III) in percentage of total arm length (humerus, radius and hand).
[7]Phalangeal Length Index = Total length of phalanges of digit III in percentage of total hand length.
[8]Thumb Length Index   = Total length of thumb ray (phalanges and meta-carpal) in percentage of total hand length.

(c) Lorsqu'ils ont à grimper ou à descendre sur une liane ou une branche verticale, par exemple pour changer de niveau, les *Lépilémures* le font par bonds successifs dus à la détente simultanée de membres postérieurs ou bien ils grimpent plus lentement, déplaçant un membre après l'autre à la manière d'un homme montant à l'échelle.

(d) Lorsqu'ils sont en train de manger dans les fines branches ils peuvent adopter des positions très variées, mais toujours pour un bref instant.

(e) Un *Lépilémur* captif, posé sur le sol, se déplace gauchement sur les pattes postérieures par une série de bonds; le plus souvent il saute mollement, se laisse retomber sur les pattes antérieures et rapproche ensuite les pattes postérieures. Il peut aussi, plus rarement, se déplacer lentement en position quadrupède normale."

Ashton and Oxnard (1957–1963; 1964) on the basis of fore-limb morphology divide the Prosimii into Quadrupeds and Hangers; they include *Avahi*, *Indri* and *Propithecus* with the slow moving Lorisidae in the latter group; and *Hapalemur*, *Lepilemur*, *Galago*, *Euoticus* and *Tarsius* (although the last three mentioned have a secondary designation of 'hoppers') in the former. Wood-Jones (1929) also lumped the Indriidae with the Lorisidae in one locomotor group.

Napier's (1963) classification was principally concerned with the brachiators and semibrachiators. Erikson's classification (1963) groups New World monkeys into Springers, Climbers and Brachiators. The following classification is that devised by the present authors, who regard it as a "working plan" which will

undoubtedly be subject to considerable future amendment as the behaviour and morphology of primates becomes better understood and the grades between apparently "discrete" locomotor groups, are revealed. The category, quadrupedalism, for instance, is inadequate for the variety of four-footed gaits met with among the primates; undoubtedly this category bridges the gap between "vertical clinging and leaping" and "brachiation". Some quadrupeds have a large component of vertical clinging and leaping in their locomotor pattern (e.g. *Lemur*) and others (e.g. *Colobus*) show an element of arm-swinging. The classification is based on those aspects of the observed locomotor behavior of primates that the authors regard as significant both phylogenetically and neontologically (Table II).

LOCOMOTOR EVOLUTION OF THE PRIMATES

A tentative synthesis of primate locomotor evolution is here put forward.

The post-cranial remains of *Plesiadapis* (Simpson, 1935), show that in the Palaeocene there were present small rodent-like primates having short limbs with non-prehensile clawed extremities. Their locomotion is inferred to have been quadrupedal and rather like that of a tree-shrew or possibly a squirrel. By the middle of Eocene, a large number of primate forms had appeared for which the vertical clinging and leaping type of locomotion has been inferred on osteological grounds. This is in agreement with the theories of Gidley (1919) who argued forcibly for the primary development of a grasping pes and the much later development of a grasping manus. The morphology and proportions of the limbs

Table II   *Locomotor categories*

(1) Vertical clinging and leaping		Leaping in trees and hopping on the ground
(2) Quadrupedalism	(i) Slow climbing type	Cautious climbing—no leaping or branch running
	(ii) Branch running and walking type	Climbing, springing, branch running and jumping
	(iii) Ground running and walking type	Climbing, ground running
	(iv) New World semi-brachiation type	Arm-swinging with use of prehensile tail; little leaping
	(v) Old World semi-brachiation type	Arm-swinging and leaping
(3) Brachiation	(i) True brachiation	Gibbon type of brachiation
	(ii) Modified brachiation	Chimpanzee and orang-utan type of brachiation
(4) Bipedalism	—	Human type of walking

and hands of vertical clingers is such that Eocene representatives of this locomotor group could have been capable of providing the basic stage from which other primate locomotor groups have evolved. Thus, the pincer-like hands and feet and the emphasis on the ulnar and peroneal borders of the extremities in the group as a whole, can be taken to be one of the basic prerequisites of preadaptation for the slow-climbing Lorisinae.

The locomotor evolution from vertical clingers and leapers to arboreal quadrupeds with prehensile extremities, long legs and relatively short arms, is no great step. Indeed, among the true lemurs there is structural and behavioural evidence to support the existence of "intermediate" types living today. *Lemur* possesses an Intermembral Index only just above the value of that found in vertical clingers. Under captive conditions and in the wild *L.catta* has been observed to adopt the vertical clinging posture, the leaping proclivities and even, for a step or two, the ground-hopping behavioural characteristics of the vertical clingers. A selection pressure leading to greater forelimb dominance associated perhaps with a change in diet and minor habitat, would lead to the evolution of arboreal "quadrupeds" such as the Cebidae and the Colobinae which retain the long hindlimbs of vertical clingers. Even among brachiators (in the sense of Napier, 1963) the long hindlimbs of gibbons for instance have been retained but are overshadowed by the tremendous length of the forelimbs. It would seem, in the light of these provisional considerations of locomotor phylogeny, that long legs relative to vertebral column length are an almost universal primate characteristic; and that evolutionary change leading to locomotor diversity has largely been a matter of phylogenetic adjustment of arm length.

Modern members of the vertical clinging and leaping group represent stable offshoots of the primitive Eocene type remaining in sanctuary areas. This hypothesis would help to clarify a rather surprising ecological anomaly. Petter (1962) has pointed out, the leaping type of locomotion is well fitted to predator avoidance but there are no large arboreal predators in Madagascar. If this hypothesis is correct then the Indrisines and *Lepilemur* exhibit a behaviour which must be regarded as a left-over from late Eocene times when in American, Asian and European forests, predator avoidance must have been highly necessary. This adaptation does allow, however, the use of certain restricted habitats where only vertical supports are found, such as the Alluaudia shrub of S.W. Madagascar, occupied by *Lepilemur* and *Propithecus* and the reed-beds of L. Alaotra which provide a refuge for *Hapalemur simus*. The limitations imposed by the vertical clinging and leaping habit upon feeding behaviour are also quite marked; the large forms, especially, are at a disadvantage when feeding in a small branch milieu. The implications of truncal uprightness, the key factor in the Vertical Clinging hypothesis, for the origin of human bipedalism are considerable, as Straus (1949; 1962) has long insisted.

SUMMARY

This paper provides a preliminary account of a natural locomotor group among the Primates. All the living members of the group are prosimians, they include: *Tarsius, Indri, Propithecus, Avahi, Lepilemur, Hapalemur simus, Galago, Euoticus*. The animals concerned are arboreal and have a vertical clinging posture at rest, and are well adapted to a leaping mode of progression during which the hindlimbs, used together, provide the propulsive force. The special interest of this locomotor group of Vertical Clinging and Leaping is that it appears in a preliminary study to constitute the only known locomotor adaptation of Eocene primates such as *Notharctus, Necrolemur, Smilodectes*, etc. Possibly it is to be regarded as the earliest locomotor specialisation of primates and therefore preadaptive to quadrupedalism, brachiation and bipedalism, the principal locomotor categories of living primates.

Notes

[1] The Tarsier. Distributed by British Film Institute. 16 mm. Black and White. Silent.

[2] Dr. J. J. Petter's film "Les lémuriens de Madagascar" is available from Service du Film de Recherche Scientifique, 96, boulevard Raspail, Paris. Mr. D. Attenborough's film is the copyright of the British Broadcasting Corporation.

9

Evolution of Lorises and Lemurs
P. CHARLES-DOMINIQUE AND R.D. MARTIN

Classifications of the order Primates attract great attention because of their special relevance to human evolution. Many different schemes have been proposed; but most work has been concentrated on relationships within major divisions and on the ranking of divisions within the Primates. The divisions themselves have remained largely unquestioned over the past 50 years, following widespread acceptance that there are six "natural groups" of living Primates: (i) Malagasy lemurs (lemur group); (ii) bush-babies, pottos and lorises (loris group); (iii) tarsiers; (iv) New World monkeys; (v) Old World monkeys; and (vi) apes and man.

PREVIOUS CLASSIFICATIONS

The chief points of contention have been: inclusion of the tree shrews (Tupaiidae) with the Primates (Simpson, 1945; LeGros Clark, 1962; Martin, 1968b); ranking of the Tarsiiformes (tarsier group) with either the prosimians (lemur group + loris group) (Simpson, 1945) or the simians (monkeys, apes and man) (Hill, 1953; Wood-Jones, 1929); the inclusion of various fossil groups (Simpson, 1945; Simons, 1963a; McKenna, 1963; Van Valen, 1965); and the fine details of hominid (human) evolution. All authors seem to agree on a basic division between prosimians and simians and on the existence of six groups. Indeed, the tendency is to relate various fossils to one or other of these groups, without considering the evolution of the Primates as a whole. In particular, fossils and living forms of dubious Primate status (namely the tree shrews) are typically placed with the lemurs, widely regarded as the most primitive living Primates.

Modern classifications, besides providing a useful reference system, should be consistent with supposed evolutionary relationships. The assumption that the lemur group and the loris group represent two distinct categories [variously ranked as families (Van Valen, 1969), superfamilies, infraorders (Simpson, 1945), or suborders (Hill, 1953; Pocock, 1918)] is clearly linked to the supposition that each of the two groups contains a range of forms derived from a distinct ancestral stock. This fact is not always clear. Confusion arises, first, because of the lack of clear guidelines in considering primate evolution (Martin, 1968b) and, second, because the published classifications are seldom accompanied by a hypothetical phylogenetic tree. Many authors hesitate to append an evolutionary tree, because much speculation is involved; but, on the other hand, the value of a classification is limited without some outline of the way a particular group has supposedly evolved.

The "ancestral Primate stock" (Martin, 1968b) is generally thought (or implied) to be located in the Upper Cretaceous and/or Palaeocene, while the existing and subfossil Malagasy lemurs are thought to have been derived from a single stock isolated on Madagascar in early Eocene times (McKenna, 1967), primarily as a result of continental drift. Thus a definite Malagasy lemur stock is generally assumed to have existed in the Eocene, perhaps persisting for some time. No such clear-cut date has been suggested for the occurrence of a loris group stock, but Walker would apparently locate this in the Miocene (Walker, 1969). In any case, according to much current reasoning, the distinct stock giving rise to all the Malagasy lemurs (living and subfossil) must logically have been separate from the line leading to the loris group since the early Eocene.

POSITION OF THE TREE SHREWS

This picture of prosimian evolution is made more complex by inclusion of the tree shrews in the Primates (Simpson, 1945; Le Gros Clark, 1962), because Simpson and Le Gros Clark both rank these mammals together with the Malagasy lemurs in the infraorder Lemuriformes, distinguishing these sharply from the loris group (infraorder Lorisiformes). If one follows this classification without the aid of a phylogenetic tree, several interpretations are possible; different patterns of relationship are consistent with the classification. One possible corollary is that tree shrews are more closely related to lemurs than the latter are to lorises, though neither author actually advocates this conclusion. An outline phylogenetic tree closer to the published opinions of these authors is given in Fig. 1, which shows that some features of the classification may be inconsistent with the probable evolution of prosimians. For example, Simpson places *Plesiadapis* as a Palaeocene relative of Malagasy lemurs in the superfamily Lemuroidea, apparently implying that many typical lemur characters (opposable digits, post-orbital bar, tooth-comb, and the like) had not been developed in the ancestral lemur/loris stock. Yet there can surely be no doubt that lemurs and lorises are far more closely related to one another than either group is to *Plesiadapis* or the tree shrews. In fact, much recent evidence shows that the tree shrews probably resemble Primates only in the retention of ancestral placental mammal characters and in the convergent development of adaptations found in many arboreal mammals. Because Simpson included *Plesiadapis* in the Primates at least partly on the grounds of comparison with (some) living tree shrews and (some) living Primates (Simpson, 1945), this allocation should also be reviewed.

ARE THE DISTINCTIONS BETWEEN LEMURS AND LORISES VALID?

One must now ask whether the traditional distinction between lemurs and lorises is really valid; that is, whether Fig. 1, or something close to it, represents their phylogeny accurately. If there were, in fact, a separate lemur stock and a separate loris stock, one should be able to list characters distinguishing the two.

Reprinted with permission from *Nature (London),* Volume 227, pages 257–260. Copyright © 1970 by Macmillan Journals Ltd. London.

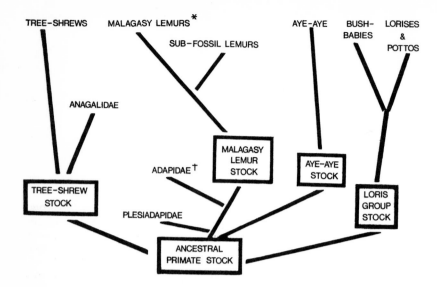

Figure 1 *One of several possible evolutionary trees for the lemurs and lorises derivable from the classification used by Simpson and Le Gros Clark. (Ancestral stocks indicate implied major subdivisions.) Note that the tree shrews, lemurs and aye-aye are placed in one infraorder (Lemuriformes), separate from the loris group infraorder (Lorisiformes).*

*"Malagasy lemurs" refers to living Madagascar prosimians, excluding the aye-aye.

† The family Adapidae is frequently split into two groups: the family Adapidae (European forms) and the family Notharctidae (American forms).

It must be assumed that these distinctions are traceable to specific, novel characters developed in one or the other stock. Published lists of distinctions are almost exclusively based on living forms (Le Gros Clark, 1962; Hill, 1953; Pocock, 1918; Hill, 1936). A summary of the commonest recognized distinguishing characters is given by Weber (1928) (see Table 1). Hill (1936) gives a far more extensive list, but the same general comments apply.

Such a list should apply to all known lemurs and lorises. In fact, the list is chiefly based on comparison of living Lorisinae (loris or potto) with living Lemurinae (usually *Lemur*). The Galaginae (bushbabies) and Cheirogaleinae (mouse and dwarf lemurs) have rarely been compared with one another, and subfossil lemurs have generally been ignored. Whenever the distinction between lemurs and lorises is examined with respect to the Galaginae and Cheirogaleinae, the initial clear separation evaporates.

Table 1 *List of Characters Distinguishing the Lemur Group and the Loris Group (from Weber, 1928, vol. 2)*

Lemuridae	Lorisidae
1. Ring-shaped ectotympanic bone within auditory bulla	1. Ectotympanic bone involved in formation of bulla wall
2. Jugal in contact with the lachrymal bone	2. Jugal separated from the lachrymal bone to varying degrees by the maxilla
3. Os planum of ethmoid bone lacking in the orbit wall	3. Os planum of ethmoid bone incorporated in the orbit wall
4. Ethmoturbinal I small; does not cover the maxilloturbinal	4. Ethmoturbinal I large; covers the maxilloturbinal
5. Internal carotid artery passes through the posterior carotid foramen into the bulla, continues through bulla and enters brain cavity through the basisphenoid bone	5. Internal carotid artery passes external to the bulla; enters the brain cavity anterior to the bulla, through the anterior lacerate foramen

To take characters from Weber's list (Table 1): in living Galaginae, a ring-like ectotympanic within the auditory bulla has been reported (Wood-Jones, 1929), while our examination of skulls of the common mouse lemur *(Microcebus murinus)* shows that the ectotympanic is sometimes fused to the bulla wall. Van Kampen, usually quoted as the authority on the structure of the auditory bulla in mammals, gives no description of Cheirogaleinae or Galaginae (Van Kampen, 1905). It has also long been known that

the Cheirogaleinae show the typical "lorisid" condition in the arrangement of the internal carotid artery and its foramina (Le Gros Clark, 1962; Hill, 1936). Our examination of material in the British Museum (Natural History) and in our own collections shows that in Demidoff's bushbaby (*Galago demidovii*), at least, the relationship between the jugal and the lachrymal is indistinguishable from that in *Microcebus murinus;* in both, the jugal generally contacts the lachrymal. *M. murinus* and *Cheirogaleus major* and *C. medius* have an os planum in the orbit, as Le Gros Clark pointed out in a footnote (Le Gros Clark, 1962). Finally, there is no significant difference between *M. murinus* and *G. demidovii* in the arrangement of the ethmoturbinals and the maxilloturbinal in the nose. In both, ethmoturbinal I is large and more or less covers the maxilloturbinal. Weber's list, and conclusions based on such lists, must therefore be revised.

SIMILARITIES BETWEEN CHEIROGALEINAE AND GALAGINAE

The vital fact is that the Galaginae were not effectively compared with the Cheirogaleinae in establishing such lists. For example, Pocock (1918) did not study *Microcebus* in preparing his monograph, and his conclusions regarding the Cheirogaleinae were apparently based on a small number of male *Cheirogaleus*. Our field studies of *Microcebus murinus* (R. D. M.; Madagascar, 1968) and *Galago demidovii* (P. C.-D.; Gaboon, 1965–69) have shown, however, that there are many ecological and behavioural similarities between these two species:

(1) Both are nocturnal forms, sharing extensive adaptations.

(2) Preferred habitat: fine branch and creeper niche. Locomotion: mixture of horizontal running and vertical clinging and leaping (trees) or hopping (ground).

(3) Diet omnivorous: chiefly fruit and insects. Insects are typically trapped with one or both hands. There is a conspicuous predatory pattern of propulsion by the back legs, which retain their grasp, followed by rapid retraction.

(4) Both species exhibit "urine-washing"—impregnation of both hands and feet with urine, involving a complex behaviour pattern.

(5) Females associate in groups: males are usually solitary. Both sexes usually disperse for feeding. Nests used for sleeping are typically globular arrangements of leaves (interwoven), with a lateral entrance; hollow trees may be used by the mouse lemur. Home ranges around the nests are very stable.

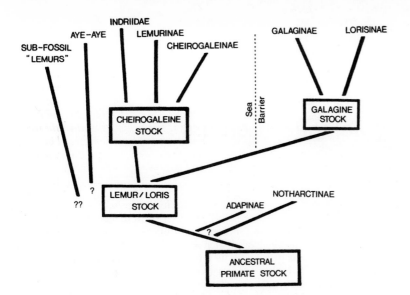

Figure 2 *A provisional outline of possible evolutionary relationships between lemurs and lorises. Note the exclusion of the tree shrews and the uncertain derivation of the subfossil lemurs* (Megaladapis, Archaeolemur, *and the like*). *The post-Eocene "sea-barrier" (= Mozambique Canal) may not have been impermeable.*

(6) Social and parental interactions include social grooming using the tooth-comb (identical structure in both species) and grasping fur in the hands. Grooming is most common in the head and neck area.

(7) The basic categories of adult vocalizations are identical.

(8) Male home ranges apparently overlap with those of females, thus facilitating reproduction. (When the female is in oestrus, the vulva passes through a cycle of expansion, reddening and opening; during periods of sexual inactivity it is closed.)

(9) The offspring—almost identical in appearance and state of development at birth in the two species—grow slowly and are weaned fairly late. The mother carries the babies in her mouth, not on the fur. Vocalizations of the offspring are generally comparable.

Two major factual conclusions emerge from these comparisons; first, the classical morphological distinctions between lemurs and lorises do not apply to *Microcebus murinus* and *Galago demidovii,* and perhaps do not apply to the Cheirogaleinae and the Galaginae in general; and, second, the two species are extensively similar in ecology, behaviour and general morphology. There are at least four possible interpretations open to discussion: (i) the two species have retained a great number of ancestral features present in all primates before the separation of Madagascar from the African mainland; (ii) convergent evolution for adaptation to a particular ecological niche has produced an enormous number of similarities in the two species; (iii) the Cheirogaleinae and Galaginae are far more closely related than is indicated by Fig. 1 (that is to say, invasion took place across the Mozambique Canal after the separation of Madagascar); (iv) some combination of (i), (ii) and/or (iii).

CHEIROGALEINAE AND GALAGINAE MAY SHOW ANCESTRAL PRIMATE FEATURES

Because many of the similarities between the Cheirogaleinae and the Galaginae seem likely to be ancestral Primate characteristics (or at least ancestral lemur/loris characteristics), it would seem that these two groups have remained very close to the original prosimian condition (although both would, of course, have undergone subsequent specialization). Convergent (or parallel) specialization many have emphasized the similarity based on retention of ancestral features, but the chief source of similarity is

probably the retention of ancestral Primate characteristics. This being the case, it is difficult to understand how one can trace the ancestry of the lemurs and lorises to anything remotely like *Plesiadapis* or ancestral tree shrews in the Palaeocene. (It should be pointed out that Simpson (1935) did not compare *Plesiadapis* with either the Galaginae or the Cheirogaleinae.)

We can provide a provisional evolutionary outline (Fig. 2), based on two assumptions: first, that no invasion of Madagascar by African prosimians (or vice versa) took place after the Eocene; and, second, that the similarity between Cheirogaleinae and Galaginae is so pronounced that explanations based exclusively on a convergence hypothesis are unsound. These suppositions may be questioned—indeed, the first, the non-invasion hypothesis, should be very closely examined—but we have attempted to give a clear statement involving wider consideration of lorises and lemurs. One possibility which we are not discussing here is that of diversification of extant and subfossil Malagasy lemurs before Madagascar split off from Africa. There could well have been several lemur types present in the African region during the late Palaeocene, and these might have given rise separately to the rather more extreme Malagasy lemurs (for example, *Daubentonia, Megaladapis* and *Archaeolemur*). Only new fossil evidence and extremely detailed comparisons can provide the answer.

The Eocene fossil Notharctidae, widely regarded as direct relatives of the Malagasy lemurs, probably came from a separate stock roughly contemporaneous with the hypothetical lemur/loris stock (Fig. 2).

The extremely provisional diagram of the supposed evolution of lemurs and lorises given in Fig. 2 is, in fact, fairly close to the classification given by Van Valen (1969). It is also in accordance with Romer's (1968) approach, whereby the lemurs and lorises are regarded as forming one "natural group", rather than two intrinsically separate groups. Significantly, both authors also agree in the exclusion of the tree shrews from the order Primates (Le Gros Clark, 1962; Romer, 1968). It must be emphasized, however, that Fig. 2 is provisional, and that post-Eocene migration across the Mozambique Canal cannot be ruled out.

The study of *Microcebus murinus* (R. D. M.) was supported by a NATO fellowship (SRC, London) and by a grant in aid for field study in Madagascar from the Royal Society. Finance and facilities for the study of *Galago demidovii* (P. C.-D.) were provided by the Mission Biologique au Gabon (CNRS; Professor P. P. Grassé).

10

Notes on the Cranial Anatomy
of the Subfossil Malagasy Lemurs

I. TATTERSALL

At a meeting of the Royal Society held on June 15, 1893, C.I. Forsyth Major described the skull of an extinct Malagasy primate, the first to come to scientific attention (Major, 1894). Since that time, the subfossil remains of some six genera and 12 species of extinct lemuroids have been recovered in Madagascar, many of them represented by quite abundant material.

The sites from which these forms are known are widely distributed over the western, central, and southern portions of the island (Fig. 1), and encompass the corresponding ecological zones. The central plateau sites, located in areas which are today almost entirely denuded of forest, appear to correspond to a forest environment containing many of the species, both floral and faunal, which typify the eastern humid forests. All the sites are of relatively recent origin: available ^{14}C dates range from 2850 ± 200 to 1035 ± 50 years B.P. (Mahé, 1965; Tattersall, 1973c; Mahé and Sourdat, 1972). The subfossil lemurs, whose extinction was almost certainly due to the activities of man (Walker, 1967a), must therefore be regarded as belonging to the modern lemur fauna (Standing, 1908), and as shedding no direct light on lemuriform phylogeny.

The relationships of the subfossil forms to the living lemurs are relatively clear, and a tentative phylogeny of the Malagasy primates is given in Fig. 2. The scheme is presented here without justification; this may be found elsewhere (Tattersall and Schwartz, 1974). As Gingerich (1975a) points out, their considerable dental (and, to a lesser extent, other) similarities with *Adapis* suggest strongly that the lepilemurines (*Lepilemur* and *Hapalemur*) are of adapid derivation. Adapids formed a significant proportion of both the Eurasian and North American prosimian faunas during the Eocene, and it is not unlikely that they also penetrated Africa (and Madagascar) during that epoch, if indeed they did not originate there.

At this point, it is necessary to define the proposition of an African origin (e.g., Charles-Dominique and Martin, 1970; Martin, 1972; Tattersall, 1973a,b), as opposed to an Asian/Indian one (Gingerich, 1975a) for the Malagasy prosimians. The argument against an Indian derivation rests partly upon the improbability of the presence of adapids (or, for that matter, of any primates) in India during the time period in question. The details of the breakup of Gondwana are still obscure. But whatever these may be it is virtually certain that India was well separated from Africa by the Jurassic (Dietz and Holden, 1970), while Madagascar, although perhaps isolated from the mainland by shallow seas at that time (Kent, 1972), remained in relatively close proximity. By the close of the Mesozoic, India was further isolated from Africa but was still remote from Asia. Madagascar's movement in the interim was evidently minimal (Kent, 1972).

The order Primates evidently had its origin in the latter part of the Cretaceous, but the similarities of certain Malagasy prosimians to characteristically Eocene forms suggests the descent of the former from representatives of the second major primate radiation, which originated and flourished during the time when India was perhaps at its farthest remove from any major land mass. It is thus far less likely that the adapids, a widespread Holarctic group, penetrated India during the latest Paleocene or earliest Eocene, than that they were in Africa, and subsequently Madagascar, at this time. The presence of an apparent adapid survivor in India during the Late Miocene (Tattersall, 1969b) provides no evidence of the presence there of adapids during the subcontinent's period of isolation, since by then contact with the Asian mainland had been established and faunal interchange via the Afghan corridor would have been possible. Thus an invasion of Madagascar directly from the Asian mainland and an introduction from Asia by way of India seem equally unlikely in view of the vast distances involved.

Perhaps a more important consideration is that the Paleocene Eurasian fauna contains no primates which may plausibly be regarded as sister groups of any of the Recent Malagasy lemurs besides Lemuridae. If multiple invasions of Madagascar by prosimian primates are accepted, as increasing evidence suggests they should be, we should expect to find forms related to the invading groups in the parent fauna. However, as we have already seen, these are signally lacking in the Eurasian early Tertiary record in all cases except one. It seems more reasonable to suggest that the unknown African primate fauna of the appropriate age may have contained such forms, including (less specialized) adapids, than to posit that Eurasia was the ultimate source of the modern Malagasy prosimian fauna.

RECENTLY EXTINCT LEMUROIDS

The three extinct lemur subfamilies include at least eight species grouped into five genera. In addition, there are two known extinct indriine, one (or possibly two) extinct lemurine, and one extinct *Daubentonia* species. Most extinct genera are known from sites throughout the sampled area of the island; however, there are, in most cases, specific differences between congeneric forms from the central plateau and from the coastal regions. There may also be such differences within some of the latter; thus, for instance, the single cranium of *Archaeolemur* from Amparihingidro, the sole northwestern site, appears intermediate in size and morphology between the southern and southwestern species, *A. majori*, and the plateau species, *A. edwardsi*. It is difficult to make any concrete judgment on this, however, because the differences between the two species are of degree rather than of kind, and only a single specimen is available from the northwest.

Reprinted with permission from *Lemur Biology* edited by I. Tattersall and R.W. Sussman, pages 111–124, Plenum Press, New York. Copyright © 1975 by Plenum Press, New York.

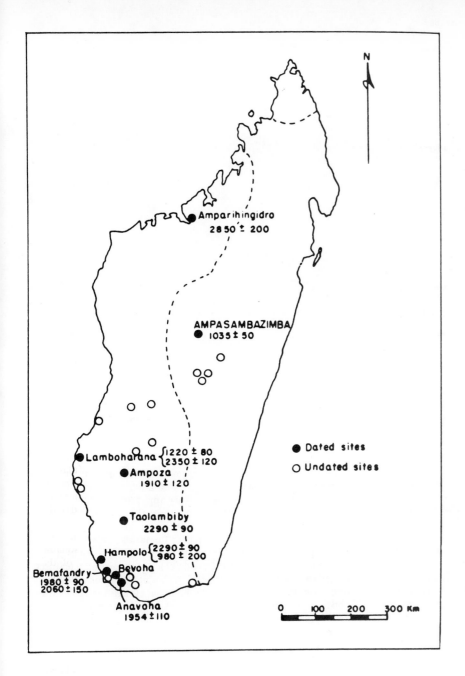

Figure 1 *Sites from which subfossil lemurs have been recovered, with dating where available. The dotted line indicates the western limit of the damp eastern flora. Dates from Mahé (1965), Mahé and Sourdat (1972), and Tattersall (1973c).*

Megaladapinae

Megaladapinae, a lemurid subfamily, contains the single genus *Megaladapis* (Fig. 3), represented by three species (*M. madagascariensis,* mean cranial length 240.5 mm, *n* = 3; *M. grandidieri,* mean cranial length 288.5 mm, *n* = 3; *M. edwardsi,* mean cranial length 296.0 mm, *n* = 10). As Major (1894) noted in describing the type of *Megaladapis*, "a *superficial* examination of the skull will certainly not suggest its classification among the Lemuroidea." In its highly elongated cranium, its peculiarly downturned nasal bones, its specialized ear region, and in many other characteristics, the skull of *Megaladapis* appears highly atypical for any lemur group. Yet an understanding of the functional anatomy of the skull together with attention to the details of dental and cranial morphology reveals the close relationship between this animal and the lepilemurines.

At the level of detail with which this chapter is concerned, the crania of the three species are sufficiently similar to be considered together, and all lend themselves admirably to analysis in terms of a model of the operation of the masticatory apparatus elaborated at some length elsewhere (Tattersall, 1973a, 1974; Roberts and Tattersall, 1974). Briefly, this model views the elevation of the mammalian jaw as being accomplished by a couple action between the two major groups of jaw-closing muscles, the masseter-internal pterygoid and the temporalis. In this system, the more anterior the point at which the bite force is exerted, the greater the effort required of the posterior (horizontal) portion of the temporalis muscle. Conversely, if biting takes place further posteriorly, the importance of the anterior (vertical) component of temporalis is emphasized.

According to this model, many of the unusual features of the

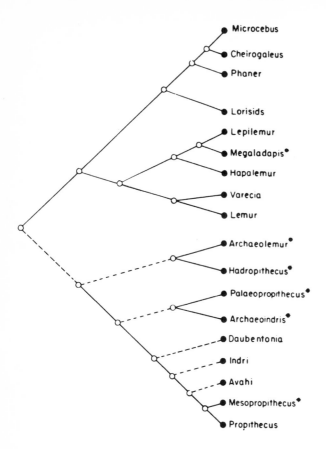

- Microcebus
- Cheirogaleus
- Phaner
- Lorisids
- Lepilemur
- Megaladapis*
- Hapalemur
- Varecia
- Lemur
- Archaeolemur*
- Hadropithecus*
- Palaeopropithecus*
- Archaeoindris*
- Daubentonia
- Indri
- Avahi
- Mesopropithecus*
- Propithecus

Figure 2 *Provisional theory of relationships of the Malagasy lemurs. Branching-points have no temporal or morphological significance. Extinct forms are denoted by asterisks.*

cranium of *Megaladapis* may be traced to the animal's possession of a greatly elongated face, reflecting an anterior concentration of dental activity. Most obvious among these features are extreme hypertrophy and posterior displacement of the posterior moiety of the temporalis, together with the reduction of the anterior portion of the muscle. Since relative to size, *Megaladapis* is extremely small brained, the posterior hypertrophy of the temporalis shows itself in the development of pronounced (posterior) sagittal and nuchal cresting. Further, the small size of the brain coupled with the need to shift the temporalis backward necessitated considerable spacing between the tiny neurocranium proper and the facial region. The development of truly enormous frontal sinuses correlates directly with this need. The failure of the brain itself to intrude anteriorly into the immediate postorbital area of the postfacial skeleton is reflected in the lack, noted by Radinsky (1970), of orbital impressions in the endocast, and in the inordinate length of the olfactory tracts.

The attenuation, for mechanical reasons, of the postfacial skeleton of *Megaladapis* also impinges on the structure of the cranial base, which exhibits a similar drawing-out. Beyond this, the atypical morphology of the cranial base can be attributed to two further major factors: first, the loss of the inflated auditory bulla, with the consequent rearrangement of the foramina of the cranial base (Saban, 1975); and second, the development of the paroccipital process into a large, elongate, protruding structure. This expresses the enlargement of the digastric muscle, both bellies of which were hypertrophied in the medium- and large-sized subfossil lemurs and which reached the zenith of their development in *Megaladapis*. Another somewhat atypical character of the cranial base in this animal is the highly robust postglenoid process, the broad anterior surface of which articu-

lates with a correspondingly extensive surface extending down from the posterior aspect of the mandibular condyle.

Modification of the masticatory apparatus in *Megaladapis*, largely in response to the great elongation of the face, thus accounts for many of its more striking differences in cranial construction from other members of Lemuridae. Why, then, this highly significant development of the face? As early as 1894, Major noted that *Megaladapis* possessed certain cranial features in common with the phalangerid marsupial *Phascolarctos cinereus*. Among such features are the facial elongation referred to above (although this is far more marked in *Megaladapis* than in *Phascolarctos*), the retroflexion of the facial skeleton relative to the plane of the cranial base, and the completely backward-facing foramen magnum, which is associated with an orientation of the occipital condyles perpendicular to the basicranial axis. Moreover, Walker (1967b) has recently shown that the locomotion of *Megaladapis* probably resembled that of *Phascolarctos*, the koala, which, although rather slow and clumsy, represents a form of vertical clinging and leaping.

Those characteristics of cranial construction common to both *Megaladapis* and the koala most likely represent adaptations to similar feeding patterns, since their effect is to transform the entire head into a long extension of the neck. *Phascolarctos* feeds, and *Megaladapis* presumably fed, by cropping leaves from branches pulled manually to within reach of the anterior teeth. The adaptations under discussion serve to maximize the radius within which feeding can take place from a single clinging position; this is advantageous for the relatively unagile koala, and would presumably have been yet more so for the vastly bulkier *Megaladapis*. The suggestion of a cropping habit for *Megaladapis* is reinforced by the loss in the adult of the upper incisor teeth

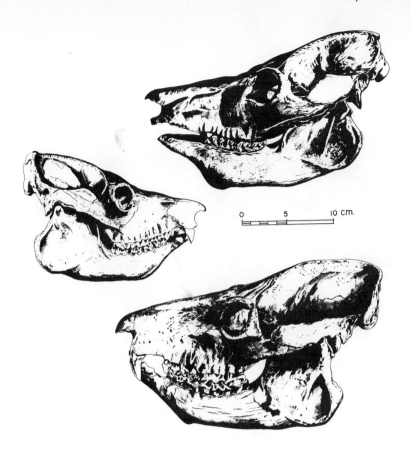

Figure 3 *Skulls of* Megaladapis *in lateral view. Above right:* M. grandidieri; *left:* M. madagascariensis; *below right:* M. edwardsi.

and their presumed replacement by a horny pad, an adaptation characteristic also of certain ruminants. Further, the possession of a mobile snout by *Megaladapis* may be indicated by the long nasal bones, downwardly flexed at their anterior extremeties and overlapping the nasal aperture, together with the highly vascularized nature of the bone covering the nasal region, which hints at a thick tissue covering.

Archaeolemurinae

Archaeolemurinae is an indriid subfamily containing three species *Archaeolemur majori* (mean cranial length 130.6 mm, *n* = 17), *A. edwardsi* (mean cranial length 147.0 mm, *n* = 17), and *Hadropithecus stenognathus* (two crania known, lengths 128.2 and 141.8 mm) (Fig. 4). Distinguished from the indriines both by morphological specialization of the dentition and by the possession of an additional premolar in each jaw, they nonetheless quite closely resemble them in cranial structure (Tattersall, 1973a,b). Indeed, one might establish a morphological sequence of skull form: *P. verreauxi* → *A. majori* → *A. edwardsi* → *H. stenognathus*, which largely expresses increasing specialization among the archaeolemurines (Tattersall, 1973a).

It is primarily from early studies of the archaeolemurines (e.g., Major, 1896; Lorenz von Liburnau, 1899; Standing, 1908) that the notion of "advanced" lemuroids has stemmed. More recent work, however, has shown that such views are untenable; the archaeolemurines are in no meaningful sense "advanced" over the indriines in any gradal characteristic (Lamberton, 1937; Piveteau, 1948; Tattersall, 1973a).

The most striking departures of the archaeolemurines from the ancestral indriid condition, most closely approximated today by *Propithecus*, are an increase in size and specialization of the masticatory apparatus. Dentally, the two species of *Archaeolemur* are extraordinarily similar; there are no consistent diagnostic differences which separate them, and there is even overlap in the size ranges of individual teeth. Both, then, are characterized by premolar rows modified into continuous shearing blades and by the possession of bilophodont molars. Such molar teeth are easily derivable from the quadricuspid indriid condition but are clearly distinct from it. If analogy with the bilophodont cercopithecids is permissible, the relatively low molar cusps in *Archaeolemur* suggest a predominantly frugivorous rather than folivorous diet. The central upper incisor is expanded, while the lateral, closely approximated to it, is small. The lower incisors are stout and somewhat forwardly inclined (possibly indicating derivation from a procumbent condition), the lateral exceeding the central teeth in size.

In *Hadropithecus,* the molar teeth are broad and high-crowned, with complex enamel folds replacing the simple cusp-and-loph pattern of *Archaeolemur*. Unlike the molars of the latter, these teeth wear rapidly. The posterior premolars are molarized, although the anterior ones, at least in their unworn state, retain a somewhat bladelike conformation. The incisors, canine and caniniform, are very greatly reduced; this is the primary, if not the only, reason for the remarkable orthognathy evinced by *Hadropithecus,* which in this respect contrasts markedly with the somewhat prognathous *Archaeolemur*.

The dental dissimilarities between *Hadropithecus* and *Archaeolemur* correspond quite closely to those described by Jolly (1970b) between *Theropithecus* and *Papio* (Jolly, 1970c; Tattersall, 1973a, 1974), and may quite plausibly be attributed to related selective pressures. In each case, *Papio* and *Archaeolemur,* respectively, may be viewed as approximating a primitive,

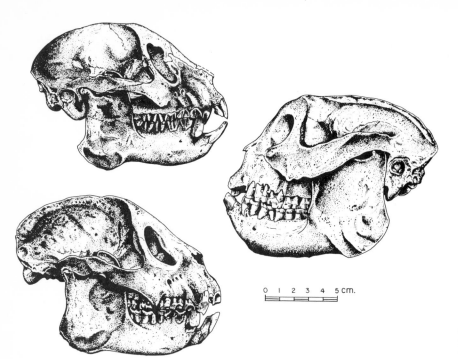

0 1 2 3 4 5 cm.

Figure 4 *Archaeolemurine skulls in lateral view. Above left:* Archaeolemur majori; *right:* Hadropithecus stenognathus; *below left:* Archaeolemur edwardsi.

dietarily more generalized, condition. A diet reminiscent of that of *Theropithecus,* i.e., of grass blades or tough, gritty morsels gathered with the fingers at ground level and placed directly between the cheek teeth, would satisfactorily explain the anterior dental reduction of *Hadropithecus,* together with the enormous expansion and complication of its posterior teeth. In turn, these opposite trends in the anterior and posterior dentitions are correlated with major modifications in cranial architecture.

The development of a powerful posterior grinding apparatus is intimately linked to that of the expanded masticatory musculature reflected in the sagittal and nuchal cresting characteristic of *Hadropithecus* and, to a slightly lesser extent, of *A. edwardsi.* Further, according to the model of masticatory mechanics mentioned earlier, the posterior concentration of masticatory activity in *Hadropithecus* is the cause of the facial deepening and the forward shifting of the muscles of mastication in this animal relative to the other archaeolemurines.

Also related to mastication is perhaps the most striking qualitative distinction between the archaeolemurines and the indriines: the fusion of the mandibular symphysis in the former, with which is correlated a considerable increase in mandibular robusticity. These allegedly "monkey-like" adaptations correspond merely to the adoption of a more powerful mode of chewing by the archaeolemurines: one in which, in particular, lateral mandibular movements could have been aided, if not primarily effected, by the contralateral muscles.

Palaeopropithecinae

Frequently considered inseparable from Indriinae, the two large-sized species *Archaeoindris fontoynonti* (length of single known cranium 269.0 mm) and *Palaeopropithecus ingens* (here taken to include *P. maximus;* mean cranial length 194.5 mm, *n* = 8) are in fact more divergent cranially from the indriines than are the archaeolemurines (Fig. 5). Dentally, however, their size differences notwithstanding, the indriines and palaeopropithecines are remarkably close, although the lower anterior teeth of the latter are nowhere near fully procumbent.

The primary characteristic distinguishing the palaeopropithecines from the indriids lies in the construction of the middle ear (Saban, 1975). Instead of possessing the "lemuroid" condition, with the tympanic ring free inside an inflated bulla, the palaeopropithecines lack the latter and the ectotympanic communicates with the exterior via a bony tube. I have noted elsewhere (Tattersall, 1973b) that the occurrence of a similar conformation in the remotely related *Megaladapis* suggests that the condition is size-related in lemuroids possessing large crania and relatively small brains.

The cranial anatomy of the palaeopropithecines has been described in some detail by Standing (1908) and Lamberton (1934), and in this brief review it is probably sufficient to note that the cranial differences between these animals and the indriines have been overstated. *Palaeopropithecus,* with its relatively long and low cranial contour, is not unreminiscent of *Indri* (although in certain features of its dentition, notably in the reduction of M^3 and in the disposition of the tooth rows it more closely resembles *Propithecus*). The relatively smaller brain and orbits of the subfossil form, possibly allometric manifestations, account for many of the major apparent distinctions between the two genera in overall cranial appearance. Thus, for instance, the more depressed profile of *Palaeopropithecus* in the region of the junction of the orbital and nasal capsules is probably to be traced to the relative reduction of its orbits. In *Indri,* conversely, the interorbital area, because it is flanked by the relatively large orbits, is filled out by the development of considerable frontal sinuses.

Propithecus and *Archaeoindris* both possess crania which are relatively abbreviated anteroposteriorly. Structurally, *Propithecus* is extraordinarily similar to *Indri;* virtually all the dissimilarities between the crania of the two genera are directly due to their differences in facial length. The flatter face of *Propithecus* is functionally linked to the shortening of its neurocranium and to the forward shifting of the origin of temporalis. Thus, although relative to skull length, the cranium of *Propithecus* is rather high when compared with that of *Indri,* this is not a real difference in functional terms. In *Archaeoindris,* however, deepening of the

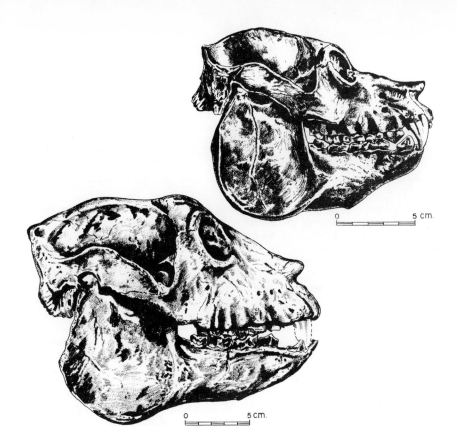

Figure 5 *Palaeopropithecine skulls in lateral view. Above:* Palaeopropithecus ingens *(= P. maximus); below:* Archaeoindris fontoynonti.

face and neurocranium appears to be biologically significant, although the relative longitudinal proportions of the splanchnocranium and neurocranium are determined by the requirements of the masticatory apparatus in a pattern very similar to that evinced by *Propithecus.* It seems plausible that the genuine facial deepening seen in *Archaeoindris* reflects the need to resolve large occlusal forces. The fusion of the mandibular symphysis in both palaeopropithecines, and particularly its foreshortening in *Archaeoindris,* would appear to support this suggestion.

Other Subfossil Forms

Subfossil indriines are limited to the genus *Mesopropithecus* (Fig. 6), with a central plateau species, *M. pithecoides,* and a southern and southwestern one, *M. globiceps* (Tattersall, 1971). Both species bear close resemblances to *Propithecus* in their crania, although *M. pithecoides,* in particular, is very much more robust, possessing, for example, both sagittal and nuchal crests. Whether or not these forms will ultimately prove congeneric with *Propithecus* depends partially on the acquisition of more knowledge of their postcrania; at present, despite their cranial similarties, *Propithecus* and *Mesopropithecus* are best regarded as generically distinct.

The two subfossil lemurines, *Varecia insignis* and *V. jullyi,* are close in cranial and dental morphology to their extant congener, although they are somewhat larger in size. *V. jullyi,* known from the plateau site of Ampasambazimba, possesses, on average, a slightly longer cranium than that of the southern *V. insignis,* but in general the differences between the two forms are no greater than those which, for instance, separate the larger eastern subspecies of *Avahi, A. laniger laniger,* from smaller western subspecies, *A. l. occidentalis.* I have therefore suggested (Tatter-

sall, 1973b) that the two subfossil forms be synonymized under *Varecia insignis.*

The remaining extinct lemur, *Daubentonia robusta,* differs in known parts (postcranial bones and a few teeth) from its living congener in size alone.

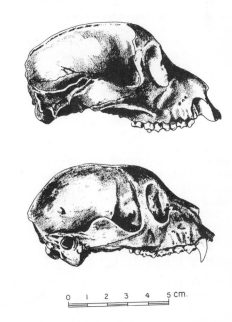

Figure 6 *Crania of the extinct indriines. Above:* Mesopropithecus pithecoides; *below:* M. globiceps.

11

Adaptive Diversity in
Subfossil Malagasy Prosimians

W.L. JUNGERS

The island of Madagascar exhibits an exceptional degree of topographical, climatic, and ecological diversity. This great internal complexity has been characterized by Martin (1972) as similar to "a closed group of (somewhat elastic) islands". The evolutionary implications of such a situation are readily apparent; the colonizing prosimian fauna no doubt encountered numerous opportunities for geographical separation of different populations, allopatric speciation, and subsequent competition leading to niche specializations. The considerable diversity in behavioral and morphological adaptations among living Malagasy prosimians strongly supports such an "adaptive radiation" scenario (Petter et al., 1977). Moreover, if recently extinct (subfossil) forms are taken into account, the entire remainder of the primate order strains to find suitable morphological and eco-ethological analogies to the Malagasy prosimians (Tattersall, 1973b; Walker, 1974).

It is possible to demonstrate quite close taxonomic relationships among various subfossil and extant Malagasy lemurs (Table 1). All recognized families except the Cheirogaleidae possess both living and extinct representatives. *Pachylemur* is the lone subfossil member of the Lemuridae; similarly, *Megaladapis* (three species) is the only extinct representative of the Lepilemuridae. *Daubentonia robustus* is little more than a larger, more robust form of the living aye-aye. Only within the Indriidae are there more known subfossil than living genera. Several of the subfossil deposits have yielded extinct specimens in association with living species (e.g., *Lemur catta* and *Propithecus verreauxi* at Andrahomana in association with *Archaeolemur* and *Megaladapis*), and radiocarbon dating of the various localities spans a narrow period of approximately 1,000–3,000 years B. P. (Tattersall, 1973c).

One of the most consistent differences between the living and the subfossil members of each family is that of size; the extinct forms tend to be larger than their extant counterparts, and some of the subfossil species are truly giant in comparison to their closest living relatives. As body size increases, specific shape transformations are frequently mechanically necessitated to maintain the constancy of some physical condition or ability. In other cases, changes in body size with accompanying alterations in geometry may reflect new adaptive potentials or novel functions (Gould, 1966; Jungers, 1978a; McMahon, 1973). It is the effects or adaptive consequences of changes in size (allometry in the most general sense) among the subfossil Malagasy prosimians that is considered in this report.

MATERIALS AND METHODS

The extant Malagasy prosimian sample of this study consists of over 100 adult skeletons (sources provided in Jungers, 1979). The subfossil specimens were measured by the author in the collections of the American Museum of Natural History, the British Museum of Natural History, the Museum National d'Histoire Naturelle, the Naturhistorisches Museum of Vienna, and the Académie Malgache.

Skeletal trunk length (Biegert & Maurer, 1972) was employed as the body size variable in the analysis of the Lemuridae and Lepilemuridae scaling trends. This osteological variable is isometric with body size in extant prosimians and is highly correlated with body mass ($r = 0.981$). In order to examine the allometric patterns of the Indriidae, it was necessary to substitute another size variable for skeletal trunk length because the vertebral column cannot be reliably reconstructed for *Palaeopropithecus*. The sum of pubis, ilium and ischium lengths was selected as the alternative size variable. This parameter is also highly correlated ($r = 0.995$) and isometric ($a = 1.06$) with body mass for extant prosimians.

The least-squares regression procedure was used in this analysis to estimate the exponent of the allometric relationship $y = \beta x^a$. This statistical technique assumes that long bone length is fundamentally dependent on body size (Gould, 1975; Jungers, 1978a) and tends to yield lower estimates of the exponent than other techniques which also consider error variance in the X variable (e.g., major axis). However, since we wish to examine the impact of body size on limb lengths, we feel the least-squares approach is reasonable (also cf. Goldstein et al., 1978). The standard anthropometric postcranial ratios (intermembral index, brachial index, and crural index) were also computed using maximum lengths of the elements. For the subfossil species, mean lengths of each appendicular element were used in the calculation of these indices.

As a base line for comparison to the subfossil species, only congeneric extant species or species of very closely related genera were initially grouped together in each of the interspecific analyses. Subsequently, the subfossil genera themselves were grouped interspecifically, either with other subfossil species or with related extant species of prosimians and/or other suggested non-primate mammalian analogues. In some cases (e.g. *Pachylemur*), the subfossils are not included in the regression analysis per se; rather their plots are compared to the values predicted by an extrapolation of extant regressions into expanded size ranges.

RESULTS AND DISCUSSION

Lemuridae

Fig. 1 illustrates the allometric patterning of body size and limb length in living lemurids and *Pachylemur*. The extinct species was not included in the computation of the scaling coefficient. Forelimb length scales negatively within the extant species ($a = 0.899$);

Reprinted with permission from *Zeitschrift für Morphologie und Anthropologie,* Volume 71, pages 177–186, Stuttgart. Copyright © 1980 by E. Schweizerbart 'sche Verlagsbuchhandlung, Stuttgart.

Table 1   *Malagasy prosimians.*

Family	Lemuridae	Lepilemuridae	Daubentoniidae	Indriidae	Cheirogaleidae
Extant Genera	*Lemur* *Varecia* *Hapalemur*	*Lepilemur*	*Daubentonia madagascariensis*	*Indri* *Propithecus* *Avahi*	*Microcebus* *Allocebus* *Cheirogaleus* *Phaner*
Subfossil Genera	*Pachylemur* (*Varecia*)	*Megaladapis*	*Daubentonia robustus*	*Mesopropithecus* *Palaeopropithecus* *Archeoindris* *Archaeolemur* *Hadropithecus*	No known novel subfossil genera

this implies that the larger species such as *Varecia* tend to possess relatively shorter anterior extremities. The forelimb length of *Pachylemur* appears to be somewhat longer than that predicted by size increase alone. Hindlimb length is also negatively allometric in extant lemurids ($a = 0.704$), only more so than the forelimb. The combination of such scaling trends should result in larger animals possessing relatively short anterior and posterior extremities, with relative hindlimb reduction outstripping relative forelimb reduction. This also implies higher intermembral indices among the larger forms. *Pachylemur* possesses hindlimbs that are somewhat shorter than would be expected on the basis of size alone. Consequently, the intermembral index of *Pachylemur* is the highest among all lemurids (Table 2). *Pachylemur* is quite similar to *Mandrillus* and *Papio* in intermembral, brachial, and crural indices (Napier & Napier, 1967). These proportions alone might suggest increased terrestriality for *Pachylemur;* however, *Pachylemur* exhibits short extremities relative to body size whereas the limbs of *Papio* are long relative to body size (Biegert & Maurer, 1972). *Pachylemur,* therefore, more closely conforms to the extant lemurid scaling trends, and its short, subequal limbs may have served to place the center of gravity close to horizontal arboreal supports during quadrupedal travel (Jungers, 1979). As Walker (1974) has pointed out, there is little in the postcranial osteology of *Pachylemur* to suggest a commitment to terrestriality.

Lepilemuridae

Inspection of Fig. 1 reveals that when only the species of *Megaladapis* are considered, forelimb length scales positively ($a = 1.341$) while hindlimb scales negatively ($a = 0.891$). A similar result was obtained in an earlier analysis of allometry in *Megaladapis* which utilized a different size variable (Jungers, 1978a); as body size increases, relative and absolute forelimb length increases whereas relative hindlimb length decreases. This results in quite high intermembral indices in the expanded size range of *Megaladapis* (Table 2). The mechanical significance of such an allometric pattern has been considered in detail elsewhere (Jungers, 1978a), and it will suffice here to note that such a morphological strategy conforms to the biomechanical expectations associated with climbing large vertical tree trunks (Carleton, 1936; Jungers, 1977). Pooling the *Megaladapis* values with those of its small relative, *Lepilemur,* and its probable behavioral analogue, the koala (*Phascolarctos*) (Jungers, 1977; Walker, 1974), generates another intriguing possibility for the proportional changes associated with increasing size. The forelimb scaling coefficient of this new grouping is slightly negative ($a = 0.860$), but that of the hindlimb is highly negative ($a = 0.455$). The intermembral index would still be predictably high since the rates of relative limb decrease are so disparate, and large forms would also be characterized by an exceptional degree of hindlimb shortening. Both of these conditions characterize *Megaladapis* (Jungers, 1978a) and the theoretical fit of such limb proportions to the expectations of the vertical support model remains quite good.

Indriidae

The adaptive diversity exhibited by the indriids qualifies them for their own "adaptive radiation" (Fig. 2). The extant indriids are usually classified as vertical clingers and leapers (Napier & Walker, 1967), but are also known to possess a very diversified positional behavioral repertoire (Petter et al., 1977; Richard, 1978). Among the best known of the subfossil Indriidae are *Archaeolemur* and *Hadropithecus* due to their prominent roles in models of early hominid craniodental evolution (Jolly, 1970c; Jungers, 1978b; Tattersall, 1973a). *Palaeopropithecus* possesses one of the most colorful histories of discovery and description, with its postcranial elements initially attributed to a hypothesized Malagasy sloth *"Bradytherium"* (Lamberton, 1944–1945). Analogies have also been drawn between *Palaeopropithecus* and the orang-utan (Standing, 1909; Walker, 1974), but the most convincing functional comparison remains with the sloths, the three-toed sloth (*Bradypus*) in particular (Carleton, 1936).

Fig. 3 illustrates that the forelimb of extant indriids is related in a positive allometric fashion to body size ($a = 1.251$) whereas the hindlimb is virtually isometric ($a = 1.076$). These results are supported by an independent analysis using skeletal trunk length as the body size variable (Jungers, 1979). Body size increase is accompanied by relatively and absolutely longer forelimbs, and, consequently, higher intermembral indices (Table 2). The forelimb plot of *Palaeopropithecus* falls remarkably close to the forelimb regression of the extant species. In other words, extrapo-

Table 2   *Relative limb proportions in extant and subfossil prosimians and selected nonprimate mammals.*

Taxon	Intermembral Index	Brachial Index	Crural Index
Varecia variegatus	72	90	93
Lemur catta	70	110	97
Lemur macaco	72	106	96
Hapalemur griseus	64	111	99
Lepilemur mustelinus	65	113	91
Lepilemur leucopus	59	108	96
Indri indri	65	122	88
Propithecus verreauxi	60	108	88
Avahi laniger	58	121	87
Daubentonia madagascariensis	71	101	98
*Daubentonia robustus**	85	101	99
*Magaladapis edwardsi**	120	87	73
*Megaladapis grandidieri**	115	88	85
*Megaladapis madagascariensis**	114	94	90
Pachylemur sp.*	94	110	90
*Palaeopropithecus** (central)	144	99	110
*Palaeopropithecus** (south)	138	87	102
*Mesopropithecus globiceps**	-	-	87
*Archaeolemur** sp.[1]	89	107	91
Phascolarctos cinereus	98	114	81
Bradypus tridactylus	172	92	87
Bradypus infuscatus	171	96	90
Choloepus didactylus	114	114	96

*subfossil prosimians
[1]Limb proportions of *Archeolemur* from Walker (1974).

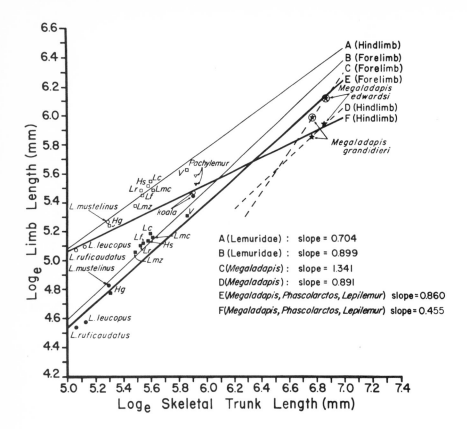

A (Lemuridae) : slope = 0.704
B (Lemuridae) : slope = 0.899
C (Megaladapis) : slope = 1.341
D (Megaladapis) : slope = 0.891
E (Megaladapis, Phascolarctos, Lepilemur) slope = 0.860
F (Megaladapis, Phascolarctos, Lepilemur) slope = 0.455

Figure 1 *Bivariate regressions of body size versus limb length in subfossil and extant prosimians. Abbreviations:* V, Varecia; Lc, Lemur catta; Lmc, Lemur macaco; Lf, Lemur fulvus; Lr, Lemur rubriventer; Lmz, Lemur mongoz; Hs, Hapalemur simus; Hg, Hapalemur griseus. *Solid symbols are forelimb plots, open symbols are hindlimb plots.*

lation of the extant trend into the size range of *Palaeopropithecus* would yield exceptionally long forelimbs. The plots of sloth forelimbs also fall quite close to this line. However, the hindlimb plot of *Palaeopropithecus* is far removed from the extant hindlimb plot (i.e., the hindlimb is substantially shorter than would be predicted by size alone). Again, the sloths parallel this condition. Presumably, the hindlimb of *Palaeopropithecus* was not under the length constraints imposed by a leaping adaptation (Hildebrand, 1974), and leaping abilities would have been greatly reduced. At the same time, *Palaeopropithecus* does not appear to be a scaled-up version of true sloths. The scaling coefficient for the forelimb of sloths with *Palaeopropithecus* is very close to 1.0, whereas that of the hindlimb is substantially positive ($a = 1.282$). This would imply decreasing intermembral indices with increases in body size. This does not appear to be the case for *Palaeopropithecus*. Central plateau representatives of this genus are considerably larger than their southern counterparts (unpublished observations). If the sloth-*Palaeopropithecus* trends are accurate, the smaller sized members should have higher intermembral indices than their larger members. This projection is not supported (Table 2); the southern group has a value of 138 compared to 144 for the central group. In this case, it would appear that sloth-like limb proportions could have been produced primarily by a large deviation from the scaling trends of the extant indriid hindlimb in combination with a quite close correspondence to that of the extant forelimb. Sloth-like feeding postures are known to occur in extant indriids (Richard, 1978). In a real sense, then, the raw behavioral variation upon which natural selection could focus and derive a sloth-like creature is still present in living relatives of *Palaeopropithecus*.

If one accepts the sloth model for *Palaeopropithecus*, the available possibilities for the positional adaptation of its close

relative, *Archaeoindris*, become most interesting. I accept Walker's (1974) synonymy of *Archaeoindris* and *"Lemuridotherium"*, the latter being a femur from Ampasambazimba of exceptional proportions (Standing, 1909). Both cranial and postcranial skeletal elements suggest that *Archaeoindris* was substantially larger than *Palaeopropithecus*. Yet several of the features which distinguish the latter from other indriids are present in *Archaeoindris* as well (e.g., the collodiaphyseal angle is 158 degrees in *Palaeopropithecus*, 152 degrees in *Archaeoindris*, but only 100–105 degrees in *Indri indri*). Perhaps *Archaeoindris* was Madagascar's "ground sloth" equivalent. Lamberton (1946) initially proposed the idea that *"Lemuridotherium"* might share behavioral and anatomical similarities with smaller representatives of the Megatheriidae. The paucity of skeletal remains attributable to *Archaeoindris* prevents pursuing the comparison much further.

In reference to Fig. 2 again, I have suggested that *Mesopropithecus* was an arboreal pronograde form. I base this largely on the high humero-femoral index of both species (*M. globiceps*, 90.6; *M. pithecoides*, 90.7) in comparison to the extant indriids (*Indri*, 54.8) and all lemurids (*Varecia*, 70.8; *Pachylemur* 85.3). Other osteological features such as transverse expansion of the trochlea of the distal humerus also support a more quadrupedal reconstruction for *Mesopropithecus*.

In suggesting that the Archaeolemurinae are arboreal/terrestrial, I have probably been too conservative. Walker (1974) has documented many anatomical features of the postcranial skeleton (especially the elbow) which strongly suggest an increased frequency of ground quadrupedalism. As with *Mesopropithecus*, the humero-femoral index of the Archaeolemurinae is quite high (e.g., *A. edwardsi*, 88.4), and pronogrady seems almost certain. While I do not have the necessary data to compare the scaling trends of the Archaeolemurinae to those of the extant indriids,

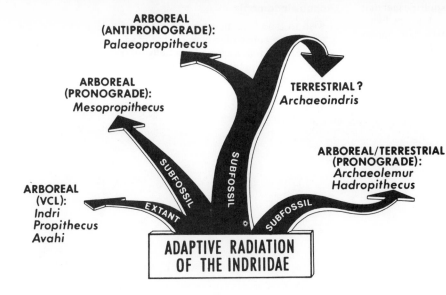

Figure 2 *Adaptive radiation of living and subfossil indriids. Probable substrate preference and habitual positional behavior are indicated.*

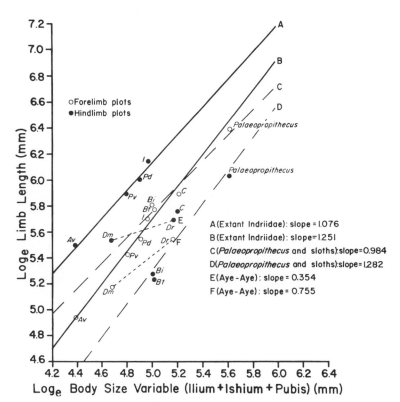

A(Extant Indriidae): slope = 1.076
B(Extant Indriidae): slope = 1.251
C(*Palaeopropithecus* and sloths):slope = 0.984
D(*Palaeopropithecus* and sloths):slope = 1.282
E(Aye-Aye): slope = 0.354
F(Aye-Aye): slope = 0.755

Figure 3 *Bivariate regressions of body size and limb length in subfossil and extant prosimians (and the sloths). Abbreviations:* I, Indri indri; Pd, Propithecus diadema; Pv, Propithecus verreauxi; Av, Avahi laniger; Dm. Daubentonia madagascariensis; Dr, Daubentonia robustus; C, Choloepus didactylus; B, Bradypus infuscatus; Bt, Bradypus tridactylus.

both fore- and hindlimbs appear to be quite short relative to the trunk (mounted skeleton of *Archaeolemur majori* at the Académie Malagache), and a relatively low center of gravity seems likely. This is quite unlike most terrestrial cercopithecoids (Biegert & Maurer, 1972), and I therefore hesitate to depict the Archaeolemurinae as strictly terrestrial primates.

Daubentoniidae

As mentioned before, *Daubentonia robustus* is very much like the living aye-aye except for its substantially larger body size and

increased robusticity. Like the living aye-aye, *D. robustus* was no doubt the local ecological equivalent of the woodpecker, foraging for wood-boring grubs in the forest (Cartmill, 1974b). Despite overall similarities in anatomy and ecological specialty, the manner in which *D. robustus* scaled up from *D. madagascariensis* is noteworthy. Both fore- and hindlimbs scale negatively in the aye-aye, but the hindlimb has an allometric exponent less than half that of the forelimb (Fig. 3). Both limbs are relatively shorter in *D. robustus,* but the intermembral index is higher. Paradoxically, these scaling trends could also be construed to conform to the predictions of the vertical support model, despite the fact that this model is relevant only to clawless mammals. Like the living aye-aye, *D. robustus* was probably equipped with claws rather than nails. Perhaps the adaptive significance of such scaling trends in the aye-aye once again resides in shifting the center of gravity closer to the substrate during both climbing and branch quadrupedalism.

Acknowledgments

I wish to thank the numerous curators of paleontological and mammalogy collections who made the specimens available for study. Special thanks go to Joan Kelly for typing the manuscript, to Luci Betti for preparing the illustrations, and to G. McDermott for photographical assistance.

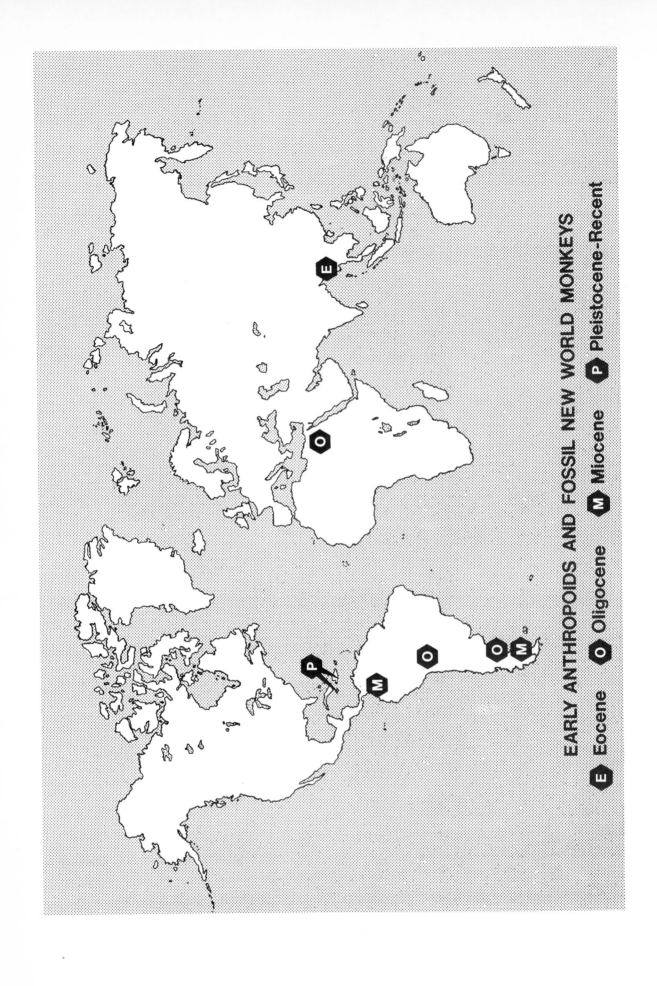

EARLY ANTHROPOIDS AND FOSSIL NEW WORLD MONKEYS

E Eocene   **O** Oligocene   **M** Miocene   **P** Pleistocene-Recent

Part III

Anthropoid Origins and New World Monkeys

The origin and early evolution of the Higher Primates or Anthropoidea constitute a series of complex interrelated issues. The chapters in Part III focus on this group's origin and on the evolution and diversification of the platyrrhine or New World monkeys. The evolution of Old World anthropoids, the central theme of Part IV, is also briefly considered in these chapters.

One of the primary issues in anthropoid origins is the question of phyletic relationships. From which group of Eocene prosimians did the Anthropoidea evolve? Was it from the lemur-like adapids or the tarsier-like omomyids, or was it from a tarsier-related group that evolved independently from either the Adapidae or the Omomyidae? A second issue concerns the relationship of the living New World monkeys to the Old World monkeys and apes. Are living anthropoids monophyletic—that is, did they share a common ancestor that was a "Higher Primate"—or did they evolve their remarkable similarity independently from one or more prosimian groups?

It is at this point that we can begin to focus our discussion on the origin of the New World monkeys. First, from what geographical source (North America, Africa, or Asia via North America) was the Platyrrhini derived? Second, from what phyletic group (adapid, omomyid, early anthropoid, or other) did they evolve? Third, what does the subsequent diversification of the Platyrrhini on the South American continent tell us about this group's early evolutionary history? For instance, which subgroup of New World monkeys is the more primitive and which is the more derived? All of the preceding issues are discussed at some length in the chapters of this part.

In the first two chapters, A. L. Rosenberger and F. S. Szalay (1980) and Phillip Gingerich (1980a) present detailed anatomical discussions of two major opposing viewpoints concerning the phylogenetic origins of the Anthropoidea in general and the Platyrrhini specifically. In Chapter 12, Rosenberger and Szalay argue that the stem "protoanthropoid" species was derived from an omomyid and that the Anthropoidea represents a monophyletic group. On the topic of New World monkey origins, Rosenberger and Szalay conclude that North America was the most likely continent of origin, but they do not rule out the possibility of a trans-South Atlantic dispersal of early Cenozoic anthropoids into South America.

In Chapter 13, Gingerich comes to an opposite conclusion concerning anthropoid phylogenetic affinities. Using a strato-phenetic approach and relying heavily on paleogeographic data, he reasons that anthropoids (=Simiiformes) are derived from Eocene Adapidae. He considers southern Asia to be the most likely area of origin of the earliest anthropoids and identifies *Pondaungia* and *Amphipithecus* (see Ba Maw et al., 1979, and Savage and Ciochon, 1985) as potential adapid-simiiform intermediates. Regarding platyrrhine origins, Gingerich hypothesizes that "protosimians" may have dispersed into North America from Asia across the Bering Land Bridge eventually spreading into South America via a Carribean volcanic island arc route. In

support of his model, Gingerich cites the close resemblance of the Eocene adapid *Periconodon* to the living squirrel monkey and to the earliest fossil platyrrhine, *Branisella*.

The geographical and phyletic origins of the Platyrrhini are the primary issues discussed by R. Hoffstetter (1972) in Chapter 14. In this now classic paper he argues that the anatomical similarities uniting platyrrhines and catarrhines are far too significant to have evolved in parallel from separate prosimian groups. Since no record of fossil or living anthropoids is known from North America, Hoffstetter (1972) looks to the African continent for the direct ancestors of the Platyrrhini. Hoffstetter supports his argument by noting that the distribution of hystricomorph rodents in South America and Africa (see also Lavocat, 1980) dramatically parallels the distribution of living platyrrhines and catarrhines. Furthermore, he draws together a wide variety of data ranging from parasitological evidence to early Cenozoic continental configurations and paleocurrent flow to bolster his assertions. The effect of Hoffstetter's (1972) pioneering conclusions are readily recognizable in that they are cited in each of the first five chapters of this part.

The origin of the New World monkeys from a variety of paleobiogeographic perspectives is explored in Chapter 15 by Russell Ciochon and Brunetto Chiarelli (1980). They begin with a cladistic analysis of the Higher Primates that supports anthropoid monophyly and further indicates that both platyrrhines and catarrhines are each monophyletic in their own right (see also Delson and Rosenberger, 1980). With the cladistic relationships (branching order) established, they review each potential geographical origin model and deployment scenario that has been proposed (i.e., North American-origin, Asian-origin, African-origin, and Vicariance-origin) and develop a maximum-parsimony model of platyrrhine origins. Their model, employing cladistic biogeographic techniques, supports an African origin of the New World monkeys from a preplatyrrhine-precatarrhine early anthropoid stock via dispersal across an equatorial south Atlantic inter-island route. This origin model proposed by Ciochon and Chiarelli (1980) fully corroborates the work of Hoffstetter (1972) mentioned above.

The next chapter by Ken Rose and John Fleagle (1981) presents a complete review of the fossil evidence of the Platyrrhini from the earliest fossils found in Oligocene beds through the late Pleistocene and Recent representatives. Rose and Fleagle also review arguments for both the phylogenetic and geographical origins of the New World monkeys. They compare and contrast the views of Gingerich (1973a, 1975c) on an adapid origin with those of Szalay (1976) on an omomyid origin concluding that the evidence slightly favors a tarsiiform connection. Concerning the geographical origin of platyrrhines, Rose and Fleagle observe that the evidence is not particularly compelling for any scenario.

In Chapters 17 and 18, Rosenberger (1980) and Ford (1980) each attempt to interpret the evolutionary history of the New World monkeys primarily from a consideration of the living genera only. Rosenberger employs a phylogenetic analysis concluding that living platyrrhines can be divided into two distinct groups: (1) cebids, which include marmosets, tamarins, *Cebus,* and *Saimiri,* and (2) atelids, which include the atelines and pithecines. He suggests that these two groups have pursued very different adaptive strategies, the former being insectivorous and the latter folivorous. Rosenberger points out there is no reason to consider the platyrrhine level of organization more "primitive" than the catarrhine grade as has often historically been the case. Instead, both of these groups of living anthropoids should be considered derived in their own right. Ford also surveys the Platyrrhini from a phylogenetic perspective specifically confronting the question of whether the callitrichids (marmosets and tamarins) represent the most primitive group of anthropoids or whether their primitive features are secondarily derived. She demonstrates that five supposedly primitive traits that characterize callitrichids are better interpreted as derived specializations associated with an evolutionary reduction in body size, which she calls "phyletic dwarfing."

Over the last decade there has been resolution of some issues concerning anthropoid origins and the phylogenetic derivation and geographical source of origin of the Platyrrhini, although there is still disagreement over certain points. For instance, a well-established case can now be made for anthropoid monophyly (see papers in Ciochon and Chiarelli, Eds., 1980). However, the issue over which Eocene prosimian group gave rise to the Anthropoidea remains unresolved. Recently, several authors have questioned the phyletic relationships between omomyids and living tarsiers (see Schmid, 1981, 1983; Cartmill, 1980) and have argued that anthropoids could share a phylogenetic relationship with living *Tarsius* without being derived from Eocene omomyids. Regarding the geographical origin of the Platyrrhini, there has been little new evidence presented since the appearance of the symposium volume by Ciochon and Chiarelli, Eds. (1980). The effects of continental drift on the evolution of the Earth's biota are evident from that volume and many other sources, and the strength of this evidence should certainly cause future defections from the North American origin model in favor of Africa. New fossil evidence that Oligocene platyrrhines resemble Fayum catarrhines in a number of dental features (Fleagle and Bown, 1983) add further support to an African origin. Ultimately, new fossil discoveries will be the major arbitrator of these controversies.

12

On the Tarsiiform Origins of Anthropoidea

A. L. ROSENBERGER AND F. S. SZALAY

INTRODUCTION

Many systematic papers of the past decade have employed a cladistic approach. Reasons for this lie largely in the superfically rigorous appearance of this method and, therefore, the impression of a powerful tool. An unvarying commitment to the operational underpinnings of cladism has led to the acceptance of a systematic methodology which perforce only recognizes, and can only search for, sister-group relationships (e.g., Engelman and Wiley, 1977). The notion of ancestor-descendant relationships, or to put it differently, that phena *transform* from antecedent phena, has been simply set aside. This fundamental aspect of real evolutionary history has come to be ignored *at the expense of a method* because hypotheses of descent are claimed to be unfalsifiable, whereas sister-group relationships are thought to be easily refutable.

More recently, the testability of ancestor-descendant relationships and the problems posed by a simplistic methodology which ignores the theoretical and empirical foundations of evolutionary biology have been treated by Bock (1977b), Szalay (1977a) and Naylor (manuscript). All of these authors make the point that a rigorous application of cladistics (considerably removed from Hennig's original formulation of phylogenetic systematics), as advocated by Cracraft (1974b, 1978), Engelman and Wiley (1977), Tattersall and Eldredge (1977), Nelson (1973), and many others, in fact deprives a systematic inquiry of the most precise (hence most easily corroborated) and therefore the most easily falsifiable of phylogenetic hypotheses—ancestry and descent. Discovery of single characters or fossils in a new stratigraphic position supplies ready and immediate tests of an ancestor-descendant hypothesis (Szalay, 1977a; Naylor, manuscript), either corroborating or forbidding it. As Naylor (manuscript) aptly puts it, "In this event, the hypothesis has been falsified and must be replaced by a *less precise* and *less readily tested* hypothesis of sister-group relationships." The discovery of a single autapomorphy in a postulated ancestral species immediately falsifies that hypothesis, and automatically relegates an explicit, boldly stated (ancestor-descendant) hypothesis to a more general one of lower order (a sister-group hypothesis). On the other hand, as Naylor further argues, the relative levels of primitiveness of characters, an empirically recognized consequence and attribute of evolution, offer irrelevant tests of sister-group relationships and are thus ignored by cladistics. but in postulating ancestor-descendant hypotheses, the recognition of additional primitive states can provide a strong corroboration of a transformation sequence, given that other features are not contradictory.

It should also be noted that character analysis, the cornerstone of phylogenetic studies, and particularly cladistics, relies heavily on inferences founded on vertical and horizontal comparisons (see Bock, 1977a), even though the stratigraphic position of the relevant morphologies must often be ignored because of the incompleteness of the record. These comparisons are transformational (in a sense ancestor-descendant) by nature and are required and justifiable simply because evolution is "descent with modification," whether this occurs during the history of an intact lineage or gene pool or at the time of splitting. Thus a polarity inference based on vertical comparisons is eminently falsifiable and equivalent, if not higher, in order than one based on horizontal comparisons.

Clearly, the synthesis resulting from a rigorously cladistic approach will be much less satisfactory biologically, both in terms of characters and organisms, than one which employs transformational hypotheses. With these foregoing remarks we proceed to employ and test ancestor-descendant hypotheses *and* sister-group hypotheses, preferring to allow the developing evidence determine which of these is more appropriate. Inasmuch as we treat even living animals as taxa and as characters (see especially Simpson, 1963), our operational concept of ancestry is an abstract one, realizing fully that our results can only be a first approximation of history. The nature of the problem dictates it.

THE PROBLEM

The radiation of Eocene euprimates (including strepsirhines and haplorhines) produced two rather successful groups of species that inhabited parts of North America and Eurasia, the Omomyidae and Adapidae [taxonomic and vernacular terms follow Szalay and Delson (1979)]. Each have been implicated as possible ancestors of the anthropoids. More specifically, some students have suggested direct phyletic ties of either omomyids or adapids with the South American platyrrhine primates. While these hypotheses may be tested and refuted, we believe that fruitful pursuit of that question, searching for the true origins of the platyrrhines, is beyond today's power of resolution. The New World monkeys are still poorly known paleontologically and serious studies of their morphological diversity and genealogical interrelationships are only now being undertaken. Furthermore, the evidence seems to point to a common ancestry of platyrrhines and catarrhines, requiring more broadly based comparisons. The emphasis of this contribution, therefore, is the bearing of omomyids and the extant tarsiiforms upon anthropoid origins, which is a phylogenetic question rather than a purely taxonomic one. As such, we will not restrict our examination to omomyids alone.

The morphology and relationships of omomyids have been considered by several workers during the past three decades. Essentially all are in agreement that the lineage leading to the

Reprinted with permission from *Evolutionary Biology of the New World Monkeys and Continental Drift*, edited by R. L. Ciochon and A. B. Chiarelli, pp. 139–157, Plenum Press, New York. Copyright © 1980 by Plenum Press, New York.

living *Tarsius* stems directly from an omomyid or some closely allied group, hence the widespread recognition of the taxon Tarsiiformes. But the higher relationships of tarsiiforms remain in dispute. A balance of cranial, dental, soft tissue and molecular evidence (see reviews in Luckett and Szalay, 1975, 1978) support the hypothesis that tarsiiforms and anthropoids are sister-taxa, together comprising the Haplorhini. This implies that higher primates are descendant of some species that would perhaps be classified as tarsiiform, or perhaps omomyid. Cartmill and Kay (1978) offer a bold refinement of this hypothesis, specifying the *Tarsius* lineage as the actual sister-group of Anthropoidea. A divergent view, one that we regard as highly unlikely, is that tarsiiforms are actually more closely related to the predominantly Paleocene plesiadapiform primates (e.g., Gingerich, 1975b, 1976, 1978b; Schwartz, 1978; Schwartz et al., 1978; Krishtalka and Schwartz, 1979). This view usually entails the corollary proposition that adapids gave rise to anthropoids. A variety of arguments specifically countermanding much of the rationale for this twin hypothesis has been provided by Szalay (1977b), Szalay and Delson (1979) and Cartmill and Kay (1978); see also Archibald (1977). Our chief purpose here is to review what we consider to be the most compelling morphological and phylogenetic arguments favoring a derivation of anthropoids from tarsiiforms such as omomyids and denying their origin from *bona fide* tarsiids and adapids.

OMOMYIDS AND ANTHROPOIDS

The overall construction of the omomyid cranium compares favorably with anthropoids, as Le Gros Clark (1963) and others have ably demonstrated, particularly in details of the facial skeleton and some of the derived aspects of the ear region. Compared to the proportions encountered among strepsirhines, the facial skull is reduced in length, though not especially abbreviated as in some modern anthropoids (e.g., colobines, callitrichines), and is hafted low on the neurocranium. The nasal fossa is apparently diminished in overall size but is dorsoventrally deep. The internal architecture of the nasal capsule is still unknown but the closely approximated orbits (Kay and Cartmill, 1977), which fuse posteriorly to form an apical interorbital septum in all the known omomyid skulls (Cartmill, 1975b; Cartmill and Kay, 1978), suggest that its posterior, olfactory components were correspondingly reduced. This is characteristic of the microsmatic *Tarsius* and the anthropoids (see Haines, 1950), both of which have a reduced system of olfactory scrolls and also lack a moist rhinarium (Pocock, 1918; Cave, 1967). Perhaps correlated with this is the expansion of the occipital lobes, which come to moderately overlap the cerebellum in *Tetonius* and *Necrolemur*, and the relative reduction of the olfactory bulbs noted in *Necrolemur* and *Rooneyia* (Radinsky, 1970). The fossilized brain of *Rooneyia* also lacks a coronolateral sulcus, common to strepsirhines but replaced in anthropoids by the central sulcus, possibly signaling a close relationship of omomyids with anthropoids (Radinsky, 1970).

Generally this organization stands in sharp contrast with that observable or inferrable for living strepsirhines and for adapids and plesiadapiforms: a long, shallow face, widely spaced orbits, deep olfactory recess situated between the orbital fossae, well-developed nasal turbinates, naked rhinarium, large olfactory bulbs, and small occipital lobes. Cartmill and Kay (1978) and Kay and Cartmill (1977) propose that the relatively large size of the infraorbital foramen of *Tetonius* and *Rooneyia* suggests that omomyids ". . . retained well-developed vibrissae and perhaps even a naked rhinarium" (1978, p. 212). While mosaicism and primitive retentions would not be surprising, particularly since

some callitrichines (Hershkovitz, 1977) present a fairly rich complement of vibrissal tactile hairs (but far fewer than visually oriented carnivorans), much of the evidence appears to reflect an advanced dominance of the visual system. Furthermore, because the foramen is or tends to be multiple in haplorhines (e.g., most anthropoids, *Tarsius, Tetonius, Washakius*, and *Rooneyia*) rather than single as in most of the species examined by Cartmill and Kay, we suspect that a variety of functions, biological roles, and conditioning factors are being sampled by their measure. The apparent reduction in *Tarsius*, for example, may reflect an encroachment of the hypertrophic eyeballs and orbits upon the posterior maxilla and the infraorbital canal. We therefore regard such implied similarities to strepsirhines to be of dubious systematic value.

The morphology of the omomyid ear region is well preserved in *Rooneyia* and *Necrolemur*, yielding much information. But there remain vexatious questions concerning the homologies and morphocline polarity of the tubular external auditory meatus (see Szalay, 1972a, 1976; Archibald, 1977), possessed by all known tarsiiforms, plesiadapiforms, some strepsirhines, and modern catarrhines. Also, the functional significance of this and other characteristics, and the several pattern variations, is unclear. Still, tarsiiforms present a number of features whose polarities can be inferred with reasonable assurance, and they also present a transformation series which illustrates the manner in which the anthropoid auditory region may have evolved. Detailed recent discussions of this region can be found in Cartmill (1975b), Szalay (1976), Archibald (1977), and Szalay and Delson (1979).

Omomyids typically present an enlarged promontory artery, which enters the bulla posteromedially, and a relatively small stapedial artery branching from it. The location of the carotid foramen and the arterial proportions contrast with the presumed primitive eutherian and primate pattern (Gregory, 1920c; Szalay, 1975d; Bugge, 1974; Archibald, 1977) and are thus derived conditions. Anthropoids present an essentially similar pattern, with the stapedial atrophying in adults. Although Cartmill and Kay (1978) caution that such features are liable to undergo convergence when lineages are independently evolving enlarged forebrains, it seems to us that the detailed similarities of haplorhine intrabullar circulation strongly increases the likelihood that these derived conditions are indeed homologous. This is especially so if the reduction of the stapedial correlates with the development of an anastomosis between the meningeal and the maxillary arteries to supply the dura in place of the stapedial, as is known in *Tarsius* and anthropoids (Hill, 1953a; Bugge, 1974). If so, Cartmill's (1978) view that this meningeal provision is possibly a haplorhine synapomorphy cannot be easily reconciled with his notion that *excessively* reduced stapedial arteries occur convergently in *Rooneyia* and *Tarsius*.

The general pattern of bullar inflation may also differentiate tarsiiforms and anthropoids from adapids, most lemuriforms and plesiadapiforms. In the latter groups, the tympanic cavity and hypotympanic sinus combine to form an expansive diverticulum ventral to the promontorium and middle ear. The bullar capsule is a spherical, inflated pocket. The major volume of the hypotympanic sinus is situated medial to the promontorium and extends posterolaterally around it. In haplorhines, on the other hand, the inflated bulla is diagonalized and the hypotympanic sinus extends anteromedially before the promontorium. The posterior recess of the hypotympanic cavity is much narrowed (related to the migration of the carotid foramen, see above) but the anterior portion is much enlarged. Thus the petrosal bone and the hypotympanic sinus of anthropoids becomes pneumatized elliptically in the very area enlarged by tarsiiforms, and almost to its anterior apex, quite unlike other primates. This is undoubtedly a derived morphology.

While *Rooneyia* is probably relatively primitive in retaining a capacious tympanic cavity and a laterally expansive, deep subtympanic recess (Cartmill, 1975b; Szalay and Wilson, 1976), *Necrolemur* presents a narrowed tympanum (as does *Tarsius*) and a dorsoventrally reduced subtympanic space, much as in anthropoids.

The anterior dentitions of omomyids and adapids are still poorly known. Nevertheless, they have figured prominently in recent debates on higher primate phylogeny. Gingerich (e.g., 1975b, 1978b) and Schwartz (1978; see also Schwartz et al., 1978; Krishtalka and Schwartz, 1979) have stressed the morphological similarity of tarsiiform and plesiadapiform lower incisors. They interpret this as evidence that tarsiiforms and plesiadapiforms comprise a monophyletic group that shared only a relatively remote common ancestry with adapiforms and anthropoids, which are presumed to represent another monophyletic group. Based upon the dental development and eruption patterns of a number of living primates and a small sample of nonadult fossils, Schwartz has explicitly argued that the enlarged anterior teeth (upper and lower) of plesiadapiforms and tarsiiforms are homologous *bona fide canine* teeth (see also Le Gros Clark, 1963), hence a synapomorphy linking these taxa phyletically. Cartmill and Kay (1978) strongly refuted his argument, exposing its shaky biological foundation and its failure to accord with the facts of *Tarsius'* dental development. We fully endorse their criticism.

Arguing in a more traditional vein, Gingerich (e.g., 1976, 1977b) has presented a dichotomous picture of incisor evolution in the primates, exemplified by enlarged apically pointed lower incisors appearing in plesiadapiforms and tarsiiforms on the one hand, and vertical spatulate incisors appearing in adapiforms and anthropoids on the other. He strongly implied that each pattern is independently derived relative to some unspecified ancestral primate condition (e.g., 1976; Fig. 42), but elsewhere (1977b) indicated that the similarity which he finds between the primitive omomyid *Teilhardina* and plesiadapids represent ancestral retentions.

The morphological evidence of the anterior dentition advanced by Gingerich at first seems compelling, but his analysis can be questioned at several levels. There is indeed a striking similarity between the lower incisors of certain plesiadapiforms (*Plesiadapis*) and some omomyids (*Trogolemur*). But as Szalay (1976) demonstrated, the morphology and proportions of omomyid antemolar teeth are variable within the group, and incisor hypertrophy, one of the most common independent trends among mammals, might well be expected to occur convergently (see also Szalay, 1977b; Cartmill and Kay, 1978). From another perspective, we consider the differences between plesiadapiform, omomyid, and anthropoid anterior teeth to have been generally overemphasized in the literature. All three groups share in common rather high-crowned, buccolingually thick and mesiodistally narrow upper and lower medial incisors (see Fig. 1a,b). This basic pattern (although not the exaggerated form of *Plesiadapis* or *Phenacolemur*) may well be primitive for the order, regardless of the number and conformation of peripheral crown conules seen in the uppers of derived plesiadapids (some of which however may also be ancestral for primates). This interpretation is supported by the distribution of large, procumbent, cylindriform lowers and correlatively robust, tall uppers in closely related archontans such as the tupaiids, mixodectids and microsyopsids (see Szalay, 1969, 1976; Gingerich, 1976; Szalay and Drawhorn, 1980). We envision various lineages modifying this primitive pattern according to the selection demands of the differing diets and varying harvesting roles encountered in an arboreal milieu. The secondary spatulateness of anthropoid I_1s (and I_2s) is correlated with a major reorganization of their antemolar dentition, no less dramatic than

the development of a toothcomb in living strepsirhines. Thus, irrespective of the real size and proportions of the ancestral primates' incisor teeth, the genetic substrate determining at least 1_1^1 shape may have been inherited essentially unchanged by tarsii-

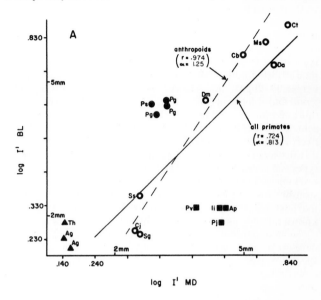

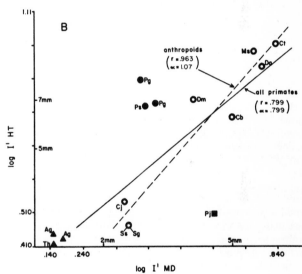

Figure 1   *Bivariate plots of (A) mediodistal length versus buccolingual breadth and (B) mediodistal length versus crown height (apical margin to cementoenamel junction) of the upper medial incisor. Regression based on sample means. Nonanthropoid fossils are represented by individual specimens. Conventions and sources: (triangles, omomyids) Ag, Arapahovius gazini, Th, Tetonius homunculus (Rosenberger, 1979b); (rings, anthropoids) Cb, Colobus badius, Ct, Cercocebus torquatus, Ms, Macaca sylvanus (Delson, 1973); Da, Dryopithecus africanus (Andrews, 1978); Cj, Callithrix jacchus, Sg, Saguinus geoffroyi, Ss, Saimiri sciureus (Rosenberger, 1979b); (squares, strepsirhines) Ap, Adapis parisiensis (Rosenberger, 1979b); Ii, Indri indri, Pv, Propithecus verreauxi (Gingerich and Ryan, 1979); Pj, Pelycodus jarrovii (Gregory, 1920c); (circles, plesiadapids); Pg, Plesiadapis gidleyi, Ps, Plesiadapis sp. (? gidleyi) (Rosenberger, 1979b). Strepsirhines contrast haplorhines and plesiadapids in their gracile, broad, low-crowned I¹s.*

forms and anthropoids from a euprimate common ancestor, which derived its morphology from some plesiadapiform.

A possible implication of this polarity hypothesis is that the much heralded similitude of adapid and anthropoid antemolar teeth, the spatulate incisors, is synapomorphic as Gingerich (e.g., 1976) indicated. We regard this as highly unlikely for there are important morphological and occlusal features distinguishing the two. For example, the upper incisors of the anthropoid morphotype, like all modern anthropoids, were robust and heteromorphic, I^1 being quite high-crowned and I^2 rather conical (Rosenberger, 1979b). Adapids such as *Pelycodus, Notharctus, Europolemur* and *Adapis,* in contrast, have low-crowned, buccolingually thin and transversely broad I^{1-2} (Figs. 1 and 2). The breadth of I^1 is acutely enhanced by a strong mesial process, while I^2 has a relatively simple, triangular shape. This gracile morphology is quite distinct from the postulated primate and anthropoid morphotypes and is clearly apomorphic. Furthermore, the pattern closely resembles all the essential details seen among the tooth-combed strepsirhines that have not drastically reduced the uppers (e.g., indriids). This very possibly represents a shared derived feature of known adapids and living strepsirhines.

The mandibular incisors of some adapids are also weakly developed, although their crowns are known from only a few specimens (see especially *Europolemur klatti*). Gregory (1920c) described at least three kinds of lower incisors: what he postulated to be the most primitive form found in *Pelycodus,* ". . . of small size, not chisel-shaped, not strongly procumbent . . ." (p. 229) but "semi-erect"; the more spatulate variety seen in *Notharctus;* and the progressively derived pattern of *Adapis parisiensis.* He suggested that the apices of *Notharctus* $I_{1,2}$ would have presented "bluntly pointed" or "rounded truncate" tips rather than the chisellike edge which typifies those of anthropoids. In *Adapis* the apically broadened lower incisors are ranked closely with the canine to form a cropping mechanism. Gregory (1920c), Gingerich (1975a), Szalay and Seligsohn (1977) and others have indicated that this is clearly a highly derived condition. It also involves a unique, derived occlusion of C_1 palatal to C^1. This articulation differs from *Notharctus* (Gregory, 1920c) and virtually all anthropoids (Rosenberger, 1979b), where C_1 occludes mesial to C^1 and into the precanine diastema. Furthermore, C_1 is exceptionally robust and low crowned in the known *Adapis,* quite unlike notharctine or protoadapin adapines such as *Mahgarita* and the presumed anthropoid ancestral condition. Thus, vertical spatulate incisors are not at all characteristic of adapids and bear only superficial resemblance to anthropoids. That genus which exhibits the most detailed similarities, *Adapis,* appears to also present the most derived incisor mechanism, evolving in a direction altogether different from that taken by anthropoids.

As a final point we wish to mention those omomyids which do conform with the projected ancestral anthropoid pattern. For example, the upper and lower median incisors of *Ourayia* are very similar to modern anthropoids. I^1 is robust, moderately high crowned and transversely broad at the apical margin (Fig. 3). I_1 is similarly shaped and vertical in orientation (Fig. 4). Upper central incisors assigned to *Tetonius* by Szalay (1976), and especially others allocated to the newly described *Arapahovius* (Savage and Waters, 1978), are also highly similar to anthropoid I^1s (Fig. 2). Simple metrical comparisons bear this out (Figs. 1a,b) and dramatically contrast a common omomyid-anthropoid pattern with that found in adapiforms and strepsirhines. These similarities themselves do not offer indications of descent but underscore the important point that the variability exhibited among omomyids includes a detailed transformation series ranging from what might be near the ancestral primate morphology (e.g., *Teilhardina*) to ones approaching the more derived condition found in anthropoids.

The cheek teeth of omomyids are exceptionally diverse when compared with the Adapidae (Szalay, 1976). Consequently, many genera present derived morphologies and tooth proportions that would exclude them from having direct bearing on anthropoid ancestry. In many features, however, including conule and cingulum development, overall tooth shape and occlusion, anaptomorphines such as *Teilhardina* and *Anemorhysis* are clearly primitive and are probably close to the ancestral euprimate molar morphology. More advanced forms such as *Omomys* and *Chumashius* may well represent conditions intermediate between an ancestral omomyid pattern and an anthropoid or platyrrhine condition (see Szalay, 1976). Kay (1977a) has recently discussed how a *Hemiacodon*-like molar occlusion may be reasonably transformed into a primitive catarrhine pattern. Clearly, there was a sufficient pool of appropriate variation from which higher primate molar and premolar teeth may be derived (see Kay, 1980). But the great primitive as well as convergent similarity of many adapid and omomyid cheek teeth, reflecting their common euprimate ancestry, diminishes the significance of a transformational approach in this case.

One area of morphological analysis which still is inadequately studied is the postcranium. Part of the reason is due to its poor representation in the fossil record (Szalay, 1976). However, a recent comparative study by Szalay and Dagosta (1980) on the distal humerus of Paleogene and living primates uncovered some interesting examples of special similarity between omomyids and primitive anthropoids (Fig. 5). The omomyid sample is relatively good: there are at least three different kinds of North American and at least two different European omomyids represented. It was found that the medial half of the trochlea curves distally in a highly diagnostic manner in omomyids, Fayum anthropoids, and a number of platyrrhine genera which are supposed to have retained this feature from a primitive anthropoid stage. The functional significance of this subtle but recognizable

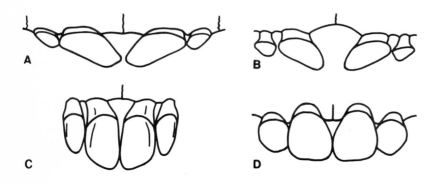

Figure 2   *Anterior views of the upper incisors of (A)* Adapis parisiensis, *(B)* Propithecus verreauxi, *(C) a compound of* Tetonius homunculus *($I^{1.2}$) and* Arapahovius gazini *(I^1), and (D)* Saimiri sciureus. *Not to same scale. Note the comparatively high-crowned incisors of haplorhines and the low-crowned shape of the strepsirhines; the strong medial process or angulation of the I^1 crown in the strepsirhines and their diminutive I^2s.*

Figure 3 *Lingual view of upper medial incisors of* Ourayia uintensis, *showing their relatively broad apical margin and anthropoid- or haplorhine-like crown shape. Compare with Fig. 2.*

character is not clear. We suspect, however, that this similarity is a shared, homologously derived one, as it does not occur among known adapids and lemuriforms.

TARSIUS AND ANTHROPOIDS

In their recent, cladistically slanted contribution on the affinities of *Tarsius,* Cartmill and Kay (1978) denied Simons' (1972) hypothesis of a close relationship with microchoerine omomyids (as did Szalay, 1975d) and proposed that "Derived features shared by *Tarsius* and anthropoids, but not by any omomyids, are more numerous (p. 211)." Subsequent to the statement of their hypothesis of *Tarsius*-anthropoid monophyly, they characterized the morphology of omomyids in such ways as to cast doubt upon, but not dismiss, their "haplorhine" status. However, we find little disagreement in the character analyses presented above and that produced by Cartmill and Kay. Both point to the inescapable conclusion that omomyids are members of the monophyletic Haplorhini. Perhaps we are less struck by the potentially primitive tarsiiform characters exhibited by omomyids. The numerous cranial features shared between omomyids and anthropoids does not allow the genealogical association of omomyids with the Strepsirhini even if the latter are to be defined by shared patristic (and not primitive) features.

The hypothesis advocated by Cartmill and Kay (1978) implies that anthropoids shared an immediate common ancestor with tarsiids *sensu stricto* rather than with some more primitive omomyid or haplorhine. The features itemized in support of this hypothesis are: the presence of an intrabullar transverse septum partitioning the hypotympanic sinus from the tympanic cavity; loss of the subtympanic recess beneath the ectotympanic; verticalized, anterolateral placement of the internal carotid canal; and

presence of an alisphenoid contribution to the postorbital septum. As discussed below, we consider these features to be either convergent similarities or indications of broader relationships within the haplorhines.

The restricted communication between hypotympanic and tympanic cavities in platyrrhines, such as *Cebus,* is due to the approximation of the promontorium and promontory canal with the lateral wall and ventral floor of the bulla rather than to the interposition of a transverse septum. This is at least partly conditioned by the lack of a subtympanic recess and a limited hypertrophy of the hypotympanic chamber, a function of the negative allometry between the bullar cavities and body size (see Cartmill, 1975b). The carotid canal of platyrrhines (representing the ancestral anthropoid state) retains its primitive haplorhine position. It courses laterad across the promontorium from a medial entry into the bulla at the carotid foramen, which lies opposite the ectotympanic meatus. The primitive anthropoid condition, therefore, is distinctly less derived than the pattern shown by *Tarsius,* where the internal carotid ascends vertically into the bulla. In smaller platyrrhines, such as callitrichines, the hypotympanic sinus and the tympanum may be relatively larger in volume, but the partitioning of these cavities is still effected by a lamina of the *internally* pneumatized petrosal bone, connecting the bullar floor proper to the dorsal aspect of the carotid canal. This is perhaps the more important point in terms of potential homologies with *Tarsius,* for the trabeculate petrosal is a striking and fundamental synapomorphy of anthropoids (e.g., Gregory, 1920c; Szalay, 1975c).

A combination of several factors distinguishes *Tarsius* from the patterns just described and from other haplorhines. The internal carotid artery enters the bulla centrally and the canal is directed vertically through to the endocranium. The canal is situated anterolateral to the promontorium, not upon it, and anterior to the external auditory meatus. The hypotympanic sinus is markedly inflated, ventrad and anteriad. The enveloping petrosal is but a thin, single sheet of bone. This arrangement could produce a vast continuous chamber connecting tympanic and hypotympanic cavities. It does not because the vertical promontory canal is connected to the medial wall of the bulla by a transverse septum extending dorsoventrally between the promontorium and the ventrally displaced bullar floor. Recourse to the more primitive haplorhine patterns exhibited by omomyids and platyrrhines suggests that the development of the septum may be related to verticalization of the carotid canal. This is borne out by von Kampen's (1905) ontogenetic studies which show that the canal is initially more horizontal, with the carotid foramen opening medially into the bullar wall. It reorients as the vertical diameter of the bullar chamber increases, dragging behind it tissues that later ossify to become the transverse septum. Thus the

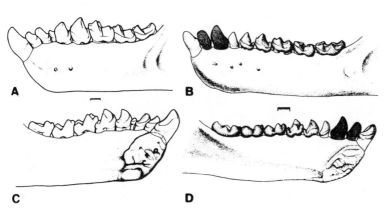

Figure 4 *Lateral (A, B) and medial (C, D) views of* Tetonius homunculus *(left) and* Ourayia uintensis *(right) lower jaw rami and tooth rows. (From Szalay, 1976, courtesy of American Museum of Natural History). Scales represent 1 mm. Although the* I_1 *of* Tetonius *is robust it is not disproportionately enlarged in height or thickness relative to the posterior premolars and molars. The proportions of anterior and cheek teeth in* Ourayia *may be even more similar to that of ancestral anthropoids.*

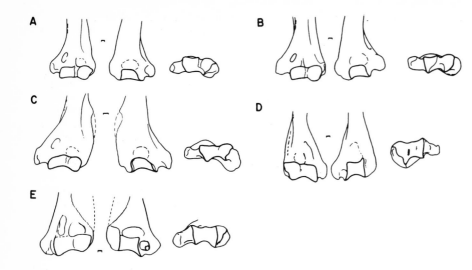

Figure 5 *Anterior (left), posterior (middle), and distal (right) views of the distal humeri of (A)* Hapalemur griseus, *(B)* Lepilemur mustelinus, *(C) a microchoerine (from the Phosphorites), (D)* Saimiri sciureus, *and (E) a Fayum anthropoid (YPM 24939). Scales represent 1 mm. All but D are from the left side.*

pneumatic foramen between tympanic and hypotympanic cavities lies lateral to the canal and promontorium somewhat as in anthropoids. However, neither the common presence nor the similar position of this foramen, actually representing the persistently primitive communication between bullar chambers, is sufficient evidence of synapomorphy since different morphologies and conditioning factors appear to be involved.

Similarly, these differences between platyrrhine and *Tarsius* bullae indicate that the central location of the carotid foramen and vertical orientation of the promontory canal of *Tarsius* are autapomorphic. The similar position of this foramen in catarrhines and larger platyrrhines like *Alouatta, Lagothrix* and *Cacajao,* relates to their large body size and/or transversely expanded petrosals. Despite this, no platyrrhine shows a verticalized promontory canal or a transverse septum eminating from its medial (i.e., dorsal) aspect.

The absence of a subtympanic recess, which is associated with a narrowing of the tympanic cavity, also need not be a synapomorphy of anthropoids and *Tarsius.* While *Rooneyia* does present a considerable ventrolateral inflation of the tympanic cavity beneath the ectotympanic, which is probably primitive for haplorhines and primates (Cartmill, 1975b), the recess is also reduced in *Necrolemur.* Convergence is therefore a possible explanation of this *Tarsius*-anthropoid similarity, as Cartmill (1975b) had previously advocated. However, we cannot exclude the possibility that this derived feature represents a phylogenetic marker within haplorhines, perhaps linking anthropoids with a tarsiiform lineage that does not include *Rooneyia* or linking *Necrolemur* and *Tarsius* to one another (see Szalay, 1976). This issue can only be resolved with additional data on this and other characters.

We also regard it unlikely that *Tarsius'* advanced degree of postorbital closure is synapomorphous with the postorbital septum of anthropoids. The participation of an alisphenoid element in the formation of the *Tarsius* partition (Le Gros Clark, 1963; Hershkovitz, 1977) may well be correlated with hypertrophic eyeballs, as is the expansion of the frontal bone (Starck, 1975; Cartmill and Kay, 1978) to which the sphenoid is sutured. Similar postorbital flanges occur in large-eyed strepsirhines such as *Loris,* and an incipient postorbital septum derived from the frontal is probably primitive for haplorhines (see *Rooneyia, Necrolemur*). Furthermore, the sutural details of the ancestral anthropoid postorbital septum are still in doubt because platyrrhines and catar-

rhines typically exhibit alternative mosaics (Ashley-Montagu, 1933).

As discussed above, we regard the proportions of the stapedial and promontory arteries of haplorhines, particularly the enlargement of the promontory artery in the protohaplorhine, as homologous. Because *Necrolemur* has a relatively larger stapedial than *Tarsius,* Cartmill and Kay (1978) regarded its morphology to represent the primitive haplorhine condition, implying that the *extreme* reduction in *Tarsius* and anthropoids is synapomorphic. We fail to see how this is a more parsimonious interpretation of this taxonomic distribution, especially when the great similarity of other aspects of the *Necrolemur* and *Tarsius* ear regions are taken into account. We prefer to regard the *extreme* reduction of the stapedial of *Tarsius* and of anthropoids to be convergent.

Finally, we advance one other criticism of the tarsier-anthropoid hypothesis. *Tarsius* is the only living representative of its immediate lineage. Other known tarsiiforms cannot be allied with *Tarsius* with much certainty [see Szalay (1975d) for a critique of the microchoerine-*Tarsius* link; see also Cartmill and Kay (1978)]. Therefore, the characters of *Tarsius* are the diagnostic features of the tarsiid morphotype. Nearly every important aspect of its craniodental and postcranial anatomy [see also Luckett (1975) on placentation] are thought to be derived relative to the conditions found elsewhere among tarsiiforms or haplorhines. To list but a few (see Szalay and Delson, 1979): hypertrophic eyeballs and orbits; markedly flexed basicranial axis; verticalized carotid canal; hypertrophied bulla; markedly reduced petromastoid and loss or absence of a pneumatized condition of the mastoid; loss of I_2; upper and lower incisor proportions and occlusion; low intermembral index; accessory toilet-claw. The extinct tarsiiforms known by cranial material are far less derived, as are many taxa based on dentitions. All these are therefore better suited as *structural ancestors of anthropoids.* It follows that the direct genealogical ties of such an "omomyid," cladistic semantics aside, is more relevant to anthropoid ancestry, whereas the *Tarsius* lineage has little direct bearing on the question of anthropoid origins.

The imposition of a rigorous cladistic philosophy, a search for the sister-group, is often less important than a search for a structural morphotype and a suite of character transformations within the context of an acceptable phylogeny. The exclusive use of the former approach, we think, has led to somewhat irrelevant, if not erroneous, results in this case. On neontological grounds, *Tarsius* has long been regarded by many as the immediate collateral relative of anthropoids, but this has limited bearing on *descent*. Even were our character analysis proven faulty, and that of Cartmill and Kay (1978) upheld, we would still refrain from recognizing *Tarsius* as the nearest actual relative of anthropoids for it would severely limit our phylogenetic and anagenetic hypothesis.

CONCLUSION

In light of the phyletic analyses presented above, we think it likely that the "protoanthropoid" species was derived from a haplorhine primate that might well be regarded as an omomyid. A more precise inference is possible only in terms of dismissing several genera and lineages of omomyids from possible ancestral status. This in turn increases the likelihood that forms such as the omomyids represent the nearest morphological approximation to such an ancestral lineage. But, thus far, they are known only by fragmentary dentitions. Crania of *Necrolemur* and *Tarsius* (to a much lesser degree) suggest possible, though highly tenuous, indicators of anthropoid affinities, but when other aspects of their anatomy are taken into account (Szalay, 1976) their lineages are also barred from direct ancestral status. Given our limited sample of the early tarsiiform radiation and our inadequate knowledge of the morphology of omomyids, postulating any known taxon as an anthropoid ancestor would be premature.

Even with a reasonably confident assessment of the monophyletic status of the anthropoids (Szalay, 1975d) and their position within the Haplorhini, a zoogeographic explanation of their current distribution, or for that matter that of the hystricognathous rodents, does not become much simpler. The tarsiiforms were apparently widely distributed across Laurasia during the Paleogene, and some survivors had at least a relic distribution within Eurasia during the Neogene. Since the Bolivian *Branisella* does appear to have direct platyrrhine affinities (Hoffstetter, 1969; Rosenberger, 1977, 1979b), contrary to Hershkovitz's (1977) remarks, the differentiation of anthropoids must have occurred prior to the earliest Oligocene. The moderately diverse Fayum faunas of Africa's late Oligocene attest to this (Simons, 1972; Szalay and Delson, 1979). Polyphyly aside, the major points arguing for an independent derivation of catarrhine and platyrrhine stocks from the north are: (1) the essential absence of recognizable platyrrhines and catarrhines in Laurasian Paleogene faunas, implying an *in situ* Gondwanan evolution of each group; (2) the presence of an Atlantic barrier making distant overwater dispersal unlikely (see especially Simons, 1976b) and overland dispersal impossible; (3) the lack of suitable, primitive forms on either continent that might be ancestral to collateral relatives on the other. At the moment, all of these reasons are essentially negative evidence, having neither falsifying nor corroborative powers.

Advocates of a vicariance explanatory model of anthropoid zoogeography (e.g., Brundin, 1966; Hershkovitz, 1977), attributing their disjunction to sea-floor spreading between Africa and South America directly or indirectly, have correctly recognized the implications of anthropoid monophyly but incorrectly assumed that platyrrhines and catarrhines were strictly endemic to South America and Africa, respectively. While the evidence for Cenozoic faunal endemism is good for South America, (e.g., Patterson and Pascual, 1972), the constitution of Paleogene African faunas indicates otherwise (Cooke, 1972). Whether or not the Burmese *Pondaungia* proves to be an Eocene anthropoid (Ba Maw et al., 1979), there is no reason to expect that catarrhines will not be found outside Africa in Laurasia during the Eocene and early Oligocene. For similar reasons, one cannot assuredly argue that archaic catarrhines gave rise to platyrrhines after island-hopping across a then narrower South Atlantic ocean. Certainly, the biological evidence counters Hoffstetter's (1977a) proposal that parapithecids were directly ancestral to platyrrhines. They are far too derived in their morphology (see Szalay and Delson, 1979; Kay, 1980) as are all *bona fide* catarrhines (Rosenberger, 1980). This leads us to the conclusion that it was an unknown early anthropoid which was ancestral to both infraorders, and that that species need not have inhabited either of these southern continents. This hypothesis does not, of course, escape difficulties. It too relies on negative evidence inasmuch as it needs a "home-land" and actual organisms which would serve as a putative ancestral phenon.

In sum, the evidence is still too incomplete to seriously contemplate the probabilities of a trans-South Atlantic dispersal by anthropoid primates during the early Cenozoic, even via archipelagos. The probability that this in fact occurred is a function of the availability of a suitable route, which can only be determined geophysically (see Tarling, 1980). Without relevant fossils, our estimation of the likelihood that primates followed this or a pair of parallel north-south dispersal routes across the Tethys, one into Africa and one into South America, should be based on the degree to which other Cenozoic animals of similar geographical situation are known to be in sister-group or ancestral-descendant relationship with one another. Either of the alternatives outlined for anthropoids is compatible with the evidence, and neither can be falsified by it. Our hesitancy to favor one or the other explanation reflects our own division on this issue. In either case, neither of us would think that the weight of the evidence would tilt the balance much toward either side.

13

Eocene Adapidae, Paleobiogeography, and the Origin of South American Platyrrhini

P.D. GINGERICH

INTRODUCTION

The origin of South American platyrrhine monkeys or Ceboidea is among the most interesting problems in primatology. This problem is basically an historical one, and geological evidence has special importance for any solution. Fossil primates, mammalian faunas, and paleogeography have a direct bearing on the origin of South American monkeys. Fortunately, much has been learned in the past twenty years about the fossil record of primate evolution. Several recent discoveries are particularly important for understanding the origin of higher primates. Furthermore, new evidence about climatic history and faunal migration during the early Cenozoic provides an improved background for interpreting the primate fossil record. Much remains to be learned, but the evidence available at present is sufficient to suggest a reasonably detailed hypothesis of ceboid origins.

SOUTH AMERICAN FAUNAS

Paleocene and Eocene mammalian faunas of South America (Riochican to Mustersan) include a diverse group of Marsupialia, edentates of order or suborder Xenarthra, and a variety of ungulates representing the orders Condylarthra, Notoungulata, Litopterna, Trigonostylopoidea, Xenungulata, and Astrapotheria (Patterson and Pascual, 1972). The major Cenozoic faunal events in South America are summarized in Fig. 1.

Marsupials, edentates, condylarths, and a notoungulate are all known from the late Paleocene and early Eocene of North America (Jepsen and Woodburne, 1969; Rose, 1978). Thus some faunal exchange between North America and South America must have occurred during the Paleocene, filtered by a discontinuous land connection and/or the intermediate zone of tropical climate. This evidence contradicts statements by some recent authors that Paleocene and Eocene faunal migration between North America and South America was improbable or impossible, based on the Eocene position of South America relative to North America published by Frakes and Kemp (1972). Frakes and Kemp's Eocene reconstruction has been widely cited in discussing the origin of South American primate and rodent faunas, but it was constructed for another purpose and not tested against known faunal distributions before being published. A more reliable reconstruction of continental positions during the Eocene, taken from the recent book by Smith and Briden (1977), is shown in Fig. 4. Here the connection between North America and South America more closely resembles the filtered route suggested by the known distribution of Paleocene and Eocene land mammal faunas.

A major event in the history of South American mammalian faunas was the appearance of both platyrrhine primates and caviomorph rodents in the early Oligocene (Deseadan), dated at about 35–36 million years (m.y.) before present (Marshall et al., 1977). The principal evidence of primates in this fauna is the type specimen of *Branisella boliviana* described by Hoffstetter (1969). Additional remains of primates from the Deseadan of Bolivia are fragmentary and all appear to represent *Branisella* as well. In contrast, the early caviomorph rodents known from the Deseadan are a diverse group including representatives of all five major suborders *Erethizontoidea, Chinchilloidea, Octodontoidea, Cavioidea,* and *Hydrochoeroidea* (Hartenberger, 1975). This diversity suggests that caviomorph rodents began radiating elsewhere before several different lines reached South America or, more probably, that they reached South America in the late Eocene. If primates arrived with rodents as part of the same faunal immigration, then primates too may have entered South America in the late Eocene. The late Eocene in South America, the "Divisaderan," is very poorly known and the Divisadero Largo fauna itself represents a peculiar facies difficult to date or relate to the mainstream of mammalian evolution (Simpson et al., 1962). Thus there is no real evidence that rodents and primates were absent, and there is some slight evidence favoring their entry into South America during the late Eocene. Early and middle Eocene mammalian faunas (Casamayoran and Mustersan) are well known, include abundant microfauna, but lack primates or rodents, and it is therefore very unlikely that primates and rodents entered South America before the late Eocene.

Another filtered interchange occurred in the late Miocene with the appearance of the procyonid *Cyonasua* in South America in the Huayquerian. Subsequently, in the Montehermosan or Chapadmalalan, a land bridge between North America and South America was established through Central America and the great American mammalian interchange began (Webb, 1976). The documented occurrences of faunal interchange between North America and South America in the early Tertiary and again in the late Tertiary and Quaternary, suggests that some limited faunal interchange in the middle Tertiary was at least a possibility.

BRANISELLA, APIDIUM, AEGYPTOPITHECUS, AND THE ORIGIN OF SIMIIFORM PRIMATES

Assuming that the earliest Platyrrhini and Caviomorpha entered South America in the late Eocene or earliest Oligocene, we can consider their relationship to primates and rodents in the late Eocene and early Oligocene elsewhere in the world. Deseadan primates and rodents are often compared with the Fayum Oligocene rodents and primates of northern Africa (Hoffstetter, 1972; Lavocat, 1974b; and others). Fayum primates and rodents are too young geologically to have given rise to Deseadan elements of these orders in the South American fauna, but they show such

Reprinted with permission from *Evolutionary Biology of the New World Monkeys and Continental Drift* edited by R.L. Ciochon and A.B. Chiarelli, pp. 123–138, Plenum Press, New York. Copyright © 1980 by Plenum Press, New York.

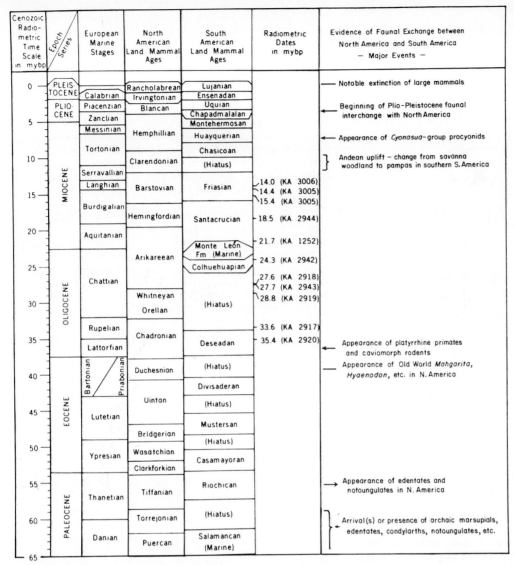

Cenozoic Radio-metric Time Scale in mybp	Epoch	Series	European Marine Stages	North American Land Mammal Ages	South American Land Mammal Ages	Radiometric Dates in mybp	Evidence of Faunal Exchange between North America and South America — Major Events —
0	PLEIS-TOCENE			Rancholabrean	Lujanian		— Notable extinction of large mammals
			Calabrian	Irvingtonian	Ensenadan		
	PLIO-CENE		Piacenzian	Blancan	Uquian		← Beginning of Plio-Pleistocene faunal interchange with North America
5			Zanclian		Chapadmalalan		
			Messinian	Hemphillian	Montehermosan		
			Tortonian		Huayquerian		← Appearance of *Cyonasua*-group procyonids
10				Clarendonian	Chasicoan		} Andean uplift – change from savanna woodland to pampas in southern S. America
	MIOCENE		Serravallian		(Hiatus)		
			Langhian	Barstovian	Friasian	14.0 (KA 3006) 14.4 (KA 3005) 15.4 (KA 3005)	
15			Burdigalian				
				Hemingfordian	Santacrucian	18.5 (KA 2944)	
20			Aquitanian				
				Arikareean	Monte León Fm (Marine)	21.7 (KA 1252)	
25					Colhuehuapian	24.3 (KA 2942)	
	OLIGOCENE		Chattian			27.6 (KA 2918) 27.7 (KA 2943) 28.8 (KA 2919)	
30				Whitneyan	(Hiatus)		
				Orellan			
			Rupelian	Chadronian		33.6 (KA 2917)	
35			Lattorfian		Deseadan	35.4 (KA 2920)	← Appearance of platyrrhine primates and caviomorph rodents
	EOCENE	Bartonian / Priabonian		Duchesnian	(Hiatus)		— Appearance of Old World *Mahgarita*, *Hyaenodon*, etc. in N. America
40					Divisaderan		
				Uinton	(Hiatus)		
45			Lutetian		Mustersan		
				Bridgerian			
50				Wasatchian	(Hiatus)		
			Ypresian		Casamayoran		
				Clarkforkian			
55			Thanetian	Tiffanian	Riochican		→ Appearance of edentates and notoungulates in N. America
60	PALEOCENE			Torrejonian	(Hiatus)		} Arrival(s) or presence of archaic marsupials, edentates, condylarths, notoungulates, etc.
			Danian	Puercan	Salamancan (Marine)		
65							

Figure 1   *Faunal succession and radiometric time scale for Cenozoic mammalian evolution in South America compared to sequences in North America and Europe. Major faunal events with a bearing on faunal migrations are indicated in the right-hand column. Data principally from Marshall et al. (1977) and Patterson and Pascual (1972), with additions from Wilson and Szalay (1976), Rose (1978), and others.*

similarity in structural grade that some reasonably close relationship is indicated. I am not sufficiently familiar with Eocene and Oligocene rodents to discuss the origin of Caviomorpha, but I have studied the original specimens of virtually all fossil primates relevant to the origin of Simiiformes (higher primates or "Anthropoidea"). I shall attempt to outline the nature of the paleontological evidence bearing on the origin of higher primates as simply as possible.

Branisella boliviana is known principally from the holotype maxillary fragment (Hoffstetter, 1969). In size and dental morphology this species corresponds closely to the living squirrel monkey (Fig. 2). The molars of *Branisella* in the holotype are somewhat worn, but they show the same trigon cusp and crest relationships, with a small hypocone on the internal cingulum, as seen in the living squirrel monkey. Virtually all of the fossil primates known from South America are similar to living genera and species of Cebidae, and it appears that living cebids do not

differ greatly in general structure from their South American ancestors in the Oligocene.

At least five genera of primates are known from the Fayum Oligocene of Egypt. These fall naturally into three groups: (1) the adapoid *Oligopithecus,* (2) the parapithecoids *Apidium* and *Simonsius,* and (3) the hominoids *Propliopithecus* and *Aegyptopithecus* (Simons, 1965a, 1972; Gingerich, 1978a). *Oligopithecus* is known only from a single mandible that resembles the Eocene adapid *Hoanghonius* from China (Gingerich, 1977c). The two genera that are best known anatomically and contribute most to our understanding of the morphology of Fayum anthropoids are *Apidium* and *Aegyptopithecus.* Cranially and postcranially *Apidium* and *Aegyptopithecus* resemble South American Cebidae to a remarkable degree (Simons, 1959, 1969c, 1972; Gingerich, 1973a; Conroy, 1976a; Fleagle, 1978b; Fleagle et al., 1975; Fleagle and Simons, 1978b). Thus, Oligocene *Branisella, Apidium,* and *Aegyptopithecus* taken together present a reasonably unified pic-

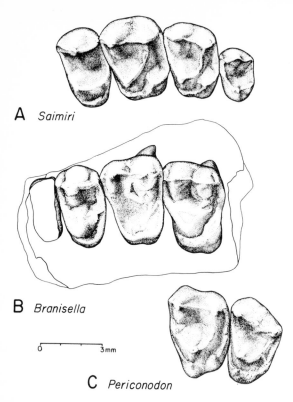

Figure 2 *Comparison of upper cheek teeth of primates related to the origin of South American primates, all drawn at same scale: (A) little worn left P^4M^{1-3} of the extant squirrel monkey Saimiri sciureus; (B) moderately worn left P^4M^{1-2} of the holotype of Oligocene Branisella boliviana; (C) little worn left M^{2-3} of the middle Eocene adapid Periconodon huerzeleri. Note close resemblance in overall size, and detailed similarity of trigon and hypocone cusps and crests in Branisella and Saimiri. Periconodon differs from these two principally in having a distinct pericone on the lingual cingulum, but otherwise it apparently represented an Eocene primate very similar in body size and dental adaptation to Branisella or even Saimiri. Branisella specimen is in the Muséum National d'Histoire Naturelle, Paris, and the Periconodon is in the Naturhistorisches Museum, Basel (Bchs. 640).*

ture of the anatomy of a truly primitive simiiform primate. Among living primates, primitive Oligocene Simiiformes most closely resemble cebids and not callitrichids, tarsiids, or lemurids.

Two large families of primates of modern aspect are known from the Eocene: the tarsiiform Omomyidae and the lemuriform Adapidae. Anatomical characteristics seen in Oligocene higher primates are listed in Table 1 for comparison with the characteristics of Eocene omomyid and adapid primates possibly ancestral to the simiiform radiation.

Paleontology and comparative anatomy furnish two complementary approaches to understanding the adaptations and evolutionary history of primates. In a group like the primates for which the fossil record is reasonably well known, it is possible to outline the phylogenetic history of the group based on hard parts preserved in the fossil record (Gingerich and Schoeninger, 1977; Gingerich, 1978b). Interpreting the distribution of anatomical traits of living members in light of this phylogeny yields information about the probable evolutionary pathways of other hard parts and of soft anatomical characteristics not preserved in fossils. Many possible phylogenetic trees showing the relation-

ships of primates can be suggested based on the comparative anatomy of living animals, but only one of these can reflect the actual historical pathway followed. Reversals, parallelism, and convergence are three well-documented evolutionary processes that cannot be detected by comparative study alone. For this reason, direct historical information about the actual stages of primate evolution is essential for reconstructing the evolutionary phylogeny of primates. In terms of the general question addressed in this chapter, the origin of higher primates, this reduces to two more specific questions: (1) What were the most primitive higher primates like?; and (2) what were possible precursors at an earlier stage like, and to which of these are primitive anthropoids most similar? In other words the general problem of the origin of higher primates focuses on the question of whether Oligocene Simiiformes more closely resemble Eocene Omomyidae or Eocene Adapidae.

Table 1 lists 16 anatomical characteristics preserved as hard parts in Eocene Omomyidae and Adapidae, and in Oligocene simiiform primates. Four of these are indeterminate, being shared equally by all, by Eocene lower primates but not Oligocene anthropoids, or by Oligocene anthropoids but not Eocene lower primates. Of the remaining twelve characteristics, eleven are similarities shared by Eocene Adapidae and Oligocene Simiiformes but not Eocene Omomyidae. Only one of the twelve diagnostic characteristics, relative brain size estimated by the encephalization quotient, favors Eocene Omomyidae as the ancestors of higher primates.

Kay (1975) has shown that insectivorous and folivorous primates differ in body size, with the former usually being smaller than 500 g and the latter being greater than 500 g in body mass. This size threshold at about 500 g may appropriately be called "Kay's threshold." Omomyids radiated on the insectivorous side of Kay's threshold, whereas adapids radiated at larger body size on the folivorous side of the threshold (Fleagle, 1978b; note that 500 g corresponds to an M_2 length of about 3.2mm, or ln M_2 length = 1.2, Gingerich, 1977a). The late Eocene and Oligocene radiation of simiiform primates was also on the folivorous side of Kay's threshold (Fleagle, 1978b).

The dental formula of omomyids and adapids is variable and by itself does not suggest special affinity of either group to early simiiform primates. On the other hand, virtually all other dental characteristics distinguish Adapidae and Simiiformes from Omomyidae. The mandibular symphysis of omomyids is never fused. Fusion occurred independently at least five times in adapids. There also appears to be a trend toward fusion in progressively smaller adapids through the course of the Eocene. Thus by the late Eocene even *Mahgarita stevensi* with a body weight estimated at about 1 kg had a solidly fused mandibular symphysis (Wilson and Szalay, 1976) like that of early Simiiformes. As discussed elsewhere (Gingerich, 1977b), the anterior dentition of adapids and anthropoids differs from that of omomyids in having vertically implanted, spatulate incisors with the lower central incisors smaller than the lateral ones. Omomyids, on the other hand, typically have enlarged central incisors and reduced lateral incisors and canines, with the central incisors forming an almost bird-like beak (Fig. 3). Adapids have projecting, interlocking canines honed by an anterior premolar as in primitive Simiiformes (Gingerich, 1975c). In addition, the canine teeth of some adapids appear to be sexually dimorphic (Stehlin, 1912; Gregory, 1920c; Gingerich, 1979b) like those of primitive simiiform primates. Canine dimorphism has never been documented in Omomyidae, and in most omomyid genera, the canines are greatly reduced in size relative to the central incisors (Fig. 4).

The earliest Omomyidae and Adapidae have molars that are very similar in morphology, the only diagnostic differences in the

Table 1 *Characteristics of Primitive Oligocene Simiiform Primates Compared to Those of Eocene Omomyidae and Adapidae*[a]

Morphological Characteristics	Eocene Omomyidae	Oligocene Simiiformes	Eocene Adapidae
Body size			
Radiation[d]	Below Kay's threshold	Above Kay's threshold	Above Kay's threshold
Dentition			
Dental formula[b]	2.1.4.3 to 2.1.2.3 or 1.1.3.3	2.1.3.3 or 2.1.2.3	2.1.4.3 to 2.1.3.3
Mandibular symphysis[d]	Unfused	Fused	Unfused to fused
Incisor form[d]	Pointed	Spatulate	Spatulate
Incisor size[d]	$I_1 \geq I_2$	$I_1 < I_2$	$I_1 < I_2$
Canine occlusion[d]	Limited	Interlocking	Interlocking
Canine dimorphism[d]	Absent	Present	Present
Canine/premolar hone[d]	Absent	Present	Present
Molar form[d]	Tritubercular	Quadrate	Tritubercular to Quadrate
Position of hypocone[b]	On basal cingulum	On basal cingulum	On postproto- or basal cingulum
Encephalization quotient[c]	EQ = 0.42 to 0.97	EQ = 0.85	EQ = 0.39 or 0.41
Postorbital closure[b]	None	Partial-complete	None
Ectotympanic[d]	Tubular	Free(?)-fused anulus	Free anulus
Stapedial artery[b]	Small	Lost	Large or small
Postcranium			
Calcaneum/navicular[d]	Elongated	Short	Short
Tibia/fibula[d]	Fused	Unfused	Unfused

[a]Where there is variation and a clear direction of evolution documented by the fossil record, the trend is written as primitive-to-derived. Variation is indicated wherever it is known. The complete dental, cranial, or postcranial anatomy is not known for any genus, and future discoveries may show that presently known genera do not adequately represent Omomyidae, Adapidae, or primitive Simiiformes. See text for discussion.
[b]Characteristics in which Adapidae and Omomyidae share equal similarity or dissimilarity with primitive Simiiformes.
[c]Characteristics in which Omomyidae are more similar than Adapidae to primitive Simiiformes.
[d]Characteristics in which Adapidae are more similar than Omomyidae to primitive Simiiformes.

dentition being in the morphology of the premolars and anterior dentition. Most omomyid genera retain a paraconid on the lower molars and a basically tritubercular molar structure (like that of *Tarsius*). Adapids, on the other hand, lost the paraconid on the lower molars relatively early and their molar structure is more quadrate than tritubercular. Oligocene simiiform primates have quadrate molars like those of adapids rather than omomyids. This is why genera like late Eocene *Amphipithecus* and *Pondaungia*, and early Oligocene *Oligopithecus* are difficult to classify. They have the molar structure of both Adapidae and Simiiformes

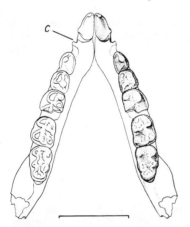

Figure 3 *Reconstruction of left and right mandibles of* Microchoerus erinaceus *showing the lower dentition in occlusal view. Note the large pointed central incisors (I_1), forming an almost birdlike beak. There are no second incisors (I_2) in* Michrochoerus, *and the lower canine (labelled C in the figure) is greatly reduced in size relative to I_1 or P_2. Specimen is in the British Museum of Natural History, London (M30345 and 30347). Scale bar is 1 cm.*

(Szalay, 1970, 1972c; Simons, 1971a; Gingerich, 1977c). The hypocone in most representatives of all three groups, Omomyidae, Adapidae, and Simiiformes, is a so-called "true" hypocone on the basal cingulum.

In cranial structure, the relative size of the brain can be measured using Jerison's (1973) encephalization quotient (EQ). Radinsky (1977) calculated EQ values of .42, .79, and .97 for the omomyids *Tetonius*, *Necrolemur*, and *Rooneyia*, respectively. He gives EQ values of .41 and .39 for *Smilodectes* and *Adapis*, respectively. *Aegyptopithecus* had an EQ of about .85 (Gingerich, 1977a), so in relative brain size omomyids are closer to *Aegyptopithecus* than adapids are. Postorbital closure separates primitive Simiiformes from both Omomyidae and Adapidae, and thus does not indicate any affinity with one family or the other.

The structure of the ectotympanic in Omomyidae is tubular as it is in *Tarsius*. Adapidae have a free ectotympanic within the auditory bulla like that of living Malagasy lemurs. The ectotympanic of both *Aegyptopithecus* and *Apidium* was ringlike, and it undoubtedly filled much of the lateral wall of the auditory bulla like it does in living Ceboidea. It is possible that this primitive anthropoid condition could be derived from the tubular ectotympanic of an omomyid, but I am not aware of any other examples of loss of the tubular extension of the ectotympanic in primate evolution. In addition, the squamosal of *Apidium* has a small cup-shaped depression that received the distal end of the ectotympanic anulus. The anulus itself is not preserved, but the presence of a distinct depression where its free end articulated with the squamosal suggests that the anulus was not solidly fused to the auditory bulla in *Apidium* (Gingerich, 1973a). Hershkovitz (1974b) has shown that the distal end of the ectotympanic sometimes does not fuse to the squamosal in *Tarsius* and in ceboids, but it is always solidly fused to the auditory bulla. The distal portion of the ectotympanic in *Tarsius* and ceboids is broad and flat, and it does not fit into a cup-like depression like that seen in

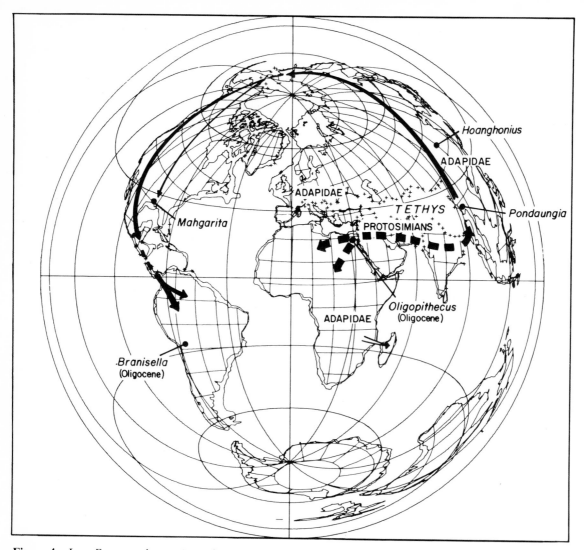

Figure 4   *Late Eocene paleocontinental map showing the position of South America relative to other continents. The geographic distribution of late Eocene adapid primates included Europe (Adapis, etc.), Asia (Hoanghonius), North America (Mahgarita), and almost certainly Africa and south Asia. By the middle or late Eocene the lemur fauna of Madagascar was probably isolated after derivation from African adapids. Late Eocene Pondaungia and Amphipithecus (both from the same general area of Burma) and early Oligocene Oligopithecus are transitional adapid-simiiform primates linking higher primates to an adapid origin. Early Oligocene Branisella is the earliest record of Ceboidea in South America. Note that the Burmese localities yielding Pondaungia and Amphipithecus were north of Tethys and part of Laurasia in the late Eocene. The evidence available at present favors origin of Simiiformes from an advanced adapid "protosimian" stock in south Asia or Africa or both. Part of the protosimian stock radiated in Africa, giving rise to the earliest Hominoidea by the middle and late Oligocene (Propliopithecus, Aegyptopithecus). Plausibly another part of the protosimian stock accompanied Mahgarita, Hyaenodon, and other Asian mammals across the Bering route into southern North America in the late Eocene. The protosimian stock then crossed from North America into South America by island-hopping via the route of present Central America or the West Indies. This hypothesis is shown by solid lines superimposed on the map. It is also possible that the adapid-proto-simian stock ancestral to Ceboidea crossed the South Atlantic directly from Africa to South America. Base map is a Lambert equal-area projection from Smith and Briden (1977).*

Apidium. Obviously, more complete specimens of *Apidium* are required to determine the detailed relationship of the ectotympanic to the auditory bulla and squamosal, but evidence at hand indicates that neither *Apidium* nor *Aegyptopithecus* had an omomyid-like tubular ectotympanic fused to the auditory bulla. The facet for the distal articulation of the ectotympanic with the squamosal in *Apidium* suggests that the primitive simiiform configuration may have included a partially free ectotympanic anulus more similar to that of adapids.

Simiiform primates differ from all Eocene lower primates in lacking a stapedial branch of the internal carotid artery. Omomyids and at least some adapids have a relatively reduced stapedial and enlarged promontory branch of the internal carotid artery (Gingerich, 1973a). Carotid circulation does not indicate any special similarity of either omomyids or adapids to early higher primates.

The postcranial skeleton of omomyids, adapids, and primitive Simiiformes is not yet sufficiently well described to permit a detailed comparison, but two aspects of hind limb anatomy deserve mention. The calcaneum is known in a number of different genera of omomyids, including *Hemiacodon, Teilhardina, ?Tetonius, Necrolemur, Nannopithex,* and *Arapahovius,* and in every case it is relatively elongated compared to generalized primates (Szalay, 1976; Savage and Waters, 1978). The tibia and fibula have been described in two omomyids, *Necrolemur* and *Nannopithex* (Schlosser, 1907; Weigelt, 1933; see also Simons, 1961a; Le Gros Clark, 1962), and in *Necrolemur* at least the fibula appears to be reduced in size and fused to the tibia. The conformation of the fibula in *Nannopithex* is less certain (Simons, 1961a). Calcaneal elongation and fibular fusion are resemblances of omomyids to living *Tarsius,* but they distinguish this group postcranially from both Adapidae and from Simiiformes.

Primitive Oligocene simiiform primates resemble Eocene Adapidae much more than they do Eocene Omomyidae. The most parsimonious interpretation of this evidence is that higher primates evolved from Adapidae and not from Omomyidae. It is generally accepted that living lemurs are derived from Eocene Adapidae and the living *Tarsius* from Eocene Omomyidae. Consequently, anthropoid primates and lemurs are probably more closely related to each other than either is to *Tarsius.* The implications for comparative anatomy are several. Anatomical characteristics such as the reduced rhinarium and nasal fossa (Cave, 1973), and the hemochorial placenta (Luckett, 1975) shared by *Tarsius* and Simiiformes but not Lemuriformes may be parallel evolutionary acquisitions (or possibly retained primitive states). The reliability of phylogenetic distances inferred from immunology and protein sequences (Goodman, 1975) appears somewhat questionable when these distances span a total temporal separation on the order of 80–100 million years (40–50 m.y. in each lineage compared). I doubt that placing lemurs and lorises slightly closer to anthropoids than *Tarsius* is would significantly decrease the parsimony of the immunological or protein sequence result.

There is disagreement regarding the major phyletic relationships of Tarsiiformes, Lemuriformes, and Simiiformes, with different results depending on whether one attempts to trace phyletic groups through the fossil record or to infer history from the comparative anatomy of living forms. This means on the one hand that our evidence regarding primate phylogeny is still far from complete, and on the other hand that we need to take a more critical look at different methods being used to reconstruct primate history. Parallelisms and reversals are common evolutionary phenomena. For this reason I tend to trust a phylogeny based on closely spaced historical records preserved as fossils rather than one based on selected comparisons of animals living, so to speak, 40 or 50 m.y. after the fact.

PALEOBIOGEOGRAPHY

The approximate distribution of continental land masses during the late Eocene, when simiiform primates evolved from their adapid ancestors, is shown in Fig. 4. Superimposed on early Cenozoic paleogeography was a series of major worldwide climatic changes documented paleobotanically on the continents (Wolfe and Hopkins, 1967; Wolfe, 1978) and isotopically in the oceans (Savin et al., 1975). The late Paleocene was generally a time of climatic cooling, followed by a definite warming trend at the end of the epoch that continued into the Eocene. After several fluctuations in the Eocene, there was a sharp drop in temperature worldwide at the end of the Eocene corresponding to Stehlin's (1909) *"grande coupure"* in European mammalian faunas. Climate strongly affects the distribution of mammalian faunas, and there is evidence that high latitude land bridges like the Bering route between Asia and North America were effectively opened or closed during the early Cenozoic by changes in climate as well as sea level.

Modern primates, more than most other orders of mammals,

are sensitive to climate. Thus it is probably no accident that the introduction of Omomyidae and Adapidae into Europe and North America coincided with early Eocene climatic warming, and the reduction in diversity of both families on northern continents also coincided with climatic cooling. The *grande coupure* marks the final exit of both Eocene families from Europe. Simpson (1947) made an extensive analysis of mammalian faunal similarity between North America and Eurasia. He showed that the greatest faunal interchange between North America and Eurasia took place during the late Eocene just before the *grande coupure.*

The major faunal interchange between North America and Eurasia in the late Eocene assumes special importance in explaining the distribution of Adapidae at this time. The principal radiation of adapids documented to date was in the Eocene of Europe, but finds in the poorly known early Cenozoic faunas of Asia *(Hoanghonius)* and Africa *(Oligopithecus)* suggest that major radiations of Adapidae may have taken place there as well. The notharctine adapid radiation in North America apparently became extinct early in the late Eocene, but one adapine genus, *Mahgarita,* is known from the late Eocene of Texas (Wilson and Szalay, 1976). It presumably reached North America as part of the late Eocene invasion from Asia that included the Old World creodont *Hyaenodon,* anthracotheres, etc. (Gingerich, 1979a). As a result, Adapidae apparently enjoyed a virtually worldwide distribution in the late Eocene.

Hoffstetter (1972, 1974) has advanced the hypothesis that tarsiiform primates radiated north of Tethys in Laurasia while their "sister-group" the simiiform primates radiated south of Tethys, initially in Africa and then in South America. It is true that omomyids are unknown outside of Laurasia, but it is difficult to see how Simiiformes could be derived from Tarsiiformes given this geographical exclusivity. A more reasonable hypothesis, I think, is that higher primates were derived from a group that shared a similar geographical distribution. For this reason, and all of the anatomical reasons discussed above, Adapidae as a group are a better candidate for simiiform ancestry than tarsiiform Omomyidae.

The most likely area of origin of higher primates, based on present evidence, appears to be Africa and/or South Asia. This is the region labelled "Protosimians" in Fig. 4, which lies between the known distribution of *Amphipithecus* and *Pondaungia* in Burma and *Oligopithecus* in Africa. All three of these genera have the distinction of being ambiguous adapid-simiiform intermediates at the time when simiiform primates were first differentiating.

The remaining problem is how the ancestors of Ceboidea reached South America if they originated in Africa or South Asia. There are two possibilities: (1) they crossed the Bering land bridge from Asia during a warm interval in the late Eocene and, with *Mahgarita,* colonized the southern part of North America, then crossed one of two possible volcanic island arcs bordering the Carribean Plate (see Gingerich and Schoeninger, 1977) and entered South America (Fig. 4); or (2) they crossed the South Atlantic directly, either by rafting or by island-hopping across the Walvis-Rio Grande rise (see Tarling, this volume). Of these two hypotheses, I favor the former because of the difficulty primates would have crossing large tracts of open ocean on rafts. It was no doubt necessary to cross some ocean by island-hopping in either case, but this would be minimized in crossing from Central America to South America. Unfortunately, there is little evidence available as yet to test either hypothesized route.

A third possibility deserves mention, although I do not yet think the evidence is sufficient to warrant serious consideration. The new mandible of *Pondaungia* (Fig. 5) recently described by Ba Maw, et al. (1979) from the late Eocene of Burma bears a surprising resemblance to *Notharctus*. Pilgrim (1927) also compared *Pondaungia* extensively with *Pelycodus* and *Notharctus*. The dentition of the new specimen differs in being adapted for a more frugivorous diet, whereas *Notharctus* has more crested folivorous cheek teeth, but the basic plan of trigonid and talonid cusps is very similar. If *Pondaungia* is a notharctine, it is possible that higher primates originated in southern North America and subsequently migrated in the late Eocene from North America into South America and also from North America across the Bering route into south Asia and ultimately Africa. This is not a serious hypothesis at present, but it is a possibility.

Much has been learned in the past twenty years about the evolution and geographical development of primates, and continued recovery of new fossil specimens at the current rate will undoubtedly contribute in the next twenty years to a better understanding of the origin of South American primates. The most critical tests of phylogenetic hypotheses are new fossils.

Acknowledgments

I would like to thank many museum curators for generous access to the fossil specimens discussed here, especially Drs. Elwyn L. Simons, Duke University Primate Center, Durham; D. E. Savage, University of California, Berkeley; M. C. McKenna, American Museum of Natural History, New York; R. Hoffstetter and D. E. Russell, Muséum National d'Histoire Naturelle, Paris; R. A. Reyment, Paleontologiska Institutionen, Uppsala; B. Engesser and J. Hürzeler, Naturhistorisches Museum, Basel; R. Wild and E. Heizmann, Staatliches Museum für Naturkunde, Stuttgart (Ludwigsburg); Darwish Alfar and Bahar El-Kashab, Cairo Geological Museum, Cairo; and M. V. A. Sastry and A. K. Dutta, Geological Survey of India, Calcutta. I also thank Drs. R. L. Ciochon and D. E. Savage for permitting me to reproduce a figure of one of their most important new specimens from Burma in Fig. 5. Ms. Karen Payne drew the specimens in Figs. 2 and 3. Ms. Karna Steelquist assisted with photography and Ms. Anita Benson typed the manuscript.

This research has been supported principally by a NATO Postdoctoral Fellowship at the Université de Montpellier (1975), grants from the Smithsonian Foreign Currency Program for museum research in Egypt and India, and NSF grant DEB 77-13465 for research on Paleocene and Eocene faunas in North America.

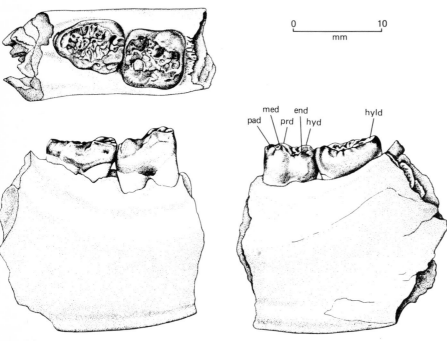

Figure 5   *Recently discovered right mandible of* Pondaungia *sp. with* M$_{2-3}$ *from the late Eocene of Burma. Specimen resembles the adapid* Notharctus robustior *in size, placement of trigonid cusps and crests, development of talonid cusps and crests, and to some extent in enamel crenulation. The molars are relatively broader and flatter in* Pondaungia *indicating an anthropoid-like frugivorous adaptation similar to that of* Aegyptopithecus, *whereas* Notharctus *has more crested cheek teeth indicating a predominently folivorous dietary adaptation. In spite of this difference in adaptation, among Eocene primates the molars of* Pondaungia *are most similar in structure to those of notharctine Adapidae, as Pilgrim (1927) stated over fifty years ago based on less well-preserved specimens. Figure reproduced from Ba Maw et al. (1979).*

14

Relationships, Origins, and History of the Ceboid Monkeys and Caviomorph Rodents: A Modern Reinterpretation

R. HOFFSTETTER

This modest essay is dedicated to Dr. George Gaylord Simpson, to acknowledge my admiration for the man and his work. The reading of his publications has been for me very enriching. But, more than his discoveries and the results of his studies, both considerable, I appreciate his rigorous and penetrating mentality, his courageous attitude with respect to ideas received, his scientific honesty which obliges him to reconsider syntheses in the light of new facts. This is why I have chosen a theme where, following the line of conduct of which he is an example, I have voiced some opinions which diverge from those he has defended and which I myself have long adopted.

INTRODUCTION

For a long time, biogeographical studies have drawn attention to groups that are essentially austral in distribution. Among tetrapod vertebrates, Leptodactylidae, Pipidae, Chelidae, Ratites, and Marsupialia can be mentioned. Other examples no less significant could be chosen among fishes, invertebrates, and plants.

To account for these distributions, former authors have been led to imagine an ancient austral continent, fragmented now by large subsidences. Such subsidences seeming unaccountable, in consideration of their width, some people have advocated the temporary emerging of transoceanic "bridges", whose number has increased to absurdity, in order to account for the geographical distribution of each animal or vegetable group. It is to the credit of W. D. Matthew and G. G. Simpson to have pointed out the improbability of those hypothetical constructions.

For a while Wegener's theory of continental drift raised the hope of explaining faunal exchanges between the large austral lands without appealing to such "land bridges." But the fragmentation of the Gondwana continent and the subsequent drift of its parts, while being denied by some authors, have been considered by others as too ancient a phenomenon to take a part in the biogeographical history of modern groups such as angiosperms, frogs, lizards, birds, or mammals.

As a natural reaction, and considering the ideas that prevailed then in geological circles, Matthew and Simpson and most authors after them did agree that oceans and continental masses practically have not changed since the middle Cretaceous, and especially in the austral hemisphere. Owing to that fact, direct faunal exchanges between austral continents were considered as highly improbable, not to say impossible, even by rafting, in view of the hugeness of interposed marine barriers. Invertebrate zoologists and botanists showed some reluctance in admitting such ideas. But as regards mammals, it was generally agreed that the center of origin and dispersion of most groups could only be the Holarctic region (North America and Eurasia). The different austral continents would have been populated by migrants of northern origin. Local differentiations subsequently gave to each

of these austral faunas its peculiar characters. Resemblance between some African, South-American, and Australian faunal elements could be explained by phenomena of parallelism or convergence. The same opinion still appears in the conclusion to the Symposium on The Evolution of Mammals on Southern Continents, which took place in Washington (1963) during the XVIth International Congress of Zoology. According to Simpson (1966a): "the southern continents Africa, Australia and South America have in common that they are dead ends as far as expansion of land mammals is concerned . . ." This clearly implies that mammals would never have reached Antarctica.

This statement was the logical outcome of the paleogeographical premises that discarded in fact any other hypothesis. Moreover, it has the agreement of nearly all the paleomammalogists. It was, in any case, a fruitful working hypothesis, as it instigated research in various branches in view of corroborating or invalidating it.

As a consequence of these researches in chorology, paleontology, comparative anatomy, parasitology, etc., a certain number of divergent opinions have been expressed during the last few years. But it was modern geophysical studies, above all, which brought new support to the theory of continental drift, with the intervention of sea-floor spreading. At present it has been practically demonstrated that the width of the oceans and, in consequence, relative positions of continental masses, have undergone considerable variations during geologic time. Some land areas may have changed in latitude and, consequently, passed from one climatic zone to another (this is particularly important for Antarctica). Such a phenomenon does not concern only old geological periods but also Cretaceous, Cenozoid and Recent times; thus, it must be taken into account in the history of modern animal and plant groups.

In fact, the whole history of the mammals, their migrations and consequent geographical segregations, must be reconsidered (see Hoffstetter, 1970a, b, 1971).

In the present essay, we consider only the mammalian exchanges that may have occurred have occurred between the austral continents whose paleontological documentation is richest, that is to say, South America and Africa. The problem as posed here concerns essentially the platyrrhine monkeys and caviomorph rodents, which suddenly appear in South America in the early Oligocene. First of all I shall briefly present the faunistic successions observed in South America, the arrival of groups here considered representing only an episode of this history. I shall sum up then the various pieces of information that shed some light on the question of phyletic and geographical origin of the Caviomorpha and the Platyrrhina.

As we shall see further on, this is the most striking example of the conflicts that arose between zoological classification (as given

Reprinted with permission from *Evolutionary Biology, Vol. 6*, edited by Th. Dobzhansky, M.K. Hecht, and W.C. Sterre, pages 323–347, Appleton-Century-Crofts, New York. Copyright © 1972 by Plenum Publishing Corporation, New York.

by the study of living and fossil forms) and restraints introduced by paleogeography (or at least the accepted ideas at a given time on this subject).

MAJOR FEATURES OF SOUTH AMERICAN MAMMALIAN HISTORY

The paleogeographic history of South America is not yet perfectly known. During a large part of the Tertiary, at least from the middle Paleocene to the middle Pliocene, it was isolated from the rest of the world by oceanic barriers. Patterson and Pascual (1968) think that this isolation goes back much further, but this is not confirmed by stratigraphical observations on isthmian Central America (Costa Rica and Panama), where the late Eocene rests nearly everywhere in discordance upon the late Cretaceous volcanic and marine sediments. But it is true that a branch of the sea may have existed at this time in northwestern Colombia, where stratigraphical data are still rather obscure.

As far as South American mammals are concerned, Simpson distinguishes three major episodes corresponding to three faunal strata [1], which he considers as resulting from immigrations of northern origin, some by terrestrial ways, others by "sweepstake routes." The first immigration, which he believes to have occurred at the beginning of the Tertiary or at the extreme end of the Cretaceous, could have brought marsupials, edentates, and ungulates: this is the ancient stock. The second one, which took place roughly at the Eocene-Oligocene boundary, corresponds to the arrival of "island hoppers"—caviomorph rodents and platyrrhine monkeys. In Neogene layers, formerly thought to be late Miocene but recently recognized as Pliocene in age, procyonid Carnivora appeared, also coming by rafting; they are sometimes connected with the second stratum, but it is more satisfactory to consider them as forerunners of the third one. Finally, the third stratum starts at the end of the Pliocene and increases through the Quaternary; it corresponds to faunal exchanges, massive and bilateral, that occurred between North and South America, thanks to the emersion of the Panamanian isthmus.

ANCIENT MAMMALIAN STOCK AND ARRIVAL OF THE EARLIEST NEOTHERIDA IN SOUTH AMERICA

The ancient mammalian stock of South America is known from fossiliferous localities of different geological ages: Laguna Umayo (Peru) for the latest Cretaceous; Rio Chico (Patagonia) and Itaborai (Brazil) for the late Paleocene; Casamayor and Musters (Patagonia) and maybe Chiococa (Peru), Tama and Gualanday (Colombia) for Eocene. Contrary to what is currently admitted, I think that these faunas are the results of long local histories. As far as mammals are concerned, there is little probability that South America was an empty continent until the latest Cretaceous. Indeed, some presently unknown (or unpublished) Theria, and non-Theria as well, may have inhabited it before this time. Strictly indigenous faunas may have existed originally. Moreover, one or several migrations have probably brought allochthonous elements which have undergone local differentiation and may have eliminated some indigenous forms. Therefore it seems better, when mentioning the known portion of this ancient stock, to avoid such usual terms as "initial stock," "first faunal stratum," "ancient immigrants," and "cenochron," some of these terms denying the possible existence of previous faunas, others postulating a single biogeographical history for all the constituent elements.

Be that as it may, during the time from which mammal-bearing deposits are known, this ancient stock consists only of marsupials

and "paleotherid"—or better, "henotherid" (nomen novum) [2]—placentals, the latter including xenarthran edentates, and a complex of ungulates and paenungulates. All of them belong to groups where the basicranial axis includes only three bony elements.

With the Deseadan (Early Oligocene), two placental orders, previously unknown for that continent, suddenly appear in South America: Rodentia and Primates, respectively represented by the Caviomorpha and the Platyrrhina (= Ceboidea). These are Neotherida whose basicranial axis includes four bony elements. It should be pointed out that the Neotherida seem to represent a monophyletic assemblage apparently originating in Laurasia from ancestral placentals ("insectivores" with three bony elements in the basicranial axis). But they undergo rapid diversification in the holarctic region. Successively there appear true Insectivora, Carnivora s.l. = Ferae (first represented by the Deltatheridia), and Primates in the late Cretaceous; Palaeanodonta (ascribed to Pholidota by Emry, 1970), Rodentia, and Lagomorpha in the Paleocene; Chiroptera and Tubulidentata in the Eocene.

The Neotherida are completely absent in the old South American stock. On this continent, their first representatives are, indeed, caviomorph rodents and ceboid primates, known from the Deseadan but whose arrival may have been a little earlier (see below). Anyway, at that time South America was geographically isolated by oceanic barriers. Therefore it was by passive transportation, probably thanks to natural rafts and maybe by means of island relays, that these immigrants ("island-hoppers" of Simpson) found it possible to reach South America. Their place of origin has been and is still controversial. In any case, it cannot be elsewhere than North America or Africa (the latter more or less connected with Eurasia).

The problems raised by the affinities and origin of both groups are astonishingly similar and have probably the same explanation. Similar, too, are the successive approaches that have been made to these questions, so that it is possible to consider them together.

FORMER CLASSIFICATIONS

Caviomorpha all possess the so-called hystricomorph structure (a very enlarged infraorbital canal giving passage to the deep layer of the masseter; Figure 1, *Myocastor*). Thus it has been natural to include them in the suborder Hystricomorpha Brandt 1855, together with similar Old World forms. This is still the position of Simpson (1945), who keeps the classical threefold division of rodents, and divides Hystricomorpha into one or two strictly African superfamilies (Hystricoidea and perhaps Bathyergoidea); three South American (Erethizontoidea, Cavioidea, and Chinchilloidea); and one composite (Octodontoidea, in which families from both continents are assembled).

In the same way, since the earliest zoological studies, South American primates were considered as monkeys but authors have distinguished them under the names of Platyrrhina Hemprich 1820 or Ceboidea Simpson 1931. Simpson's classification still remains conservative since his suborder Anthropoidea (in opposition to Prosimii) groups together Ceboidea, Cercopithecoidea, and Hominoidea.

However, it seems convenient to emphasize that in his explicative notes (Review of Mammalian Classification), Simpson (1945) states that he was obliged to adopt for these groups a morphological classification which does not necessarily reflect phylogenetic relations. Nevertheless, the same author (Simpson, 1961, Fig. 28, p. 213) proposes a phylogenetic scheme for Primates, in which the Anthropoidea (including Ceboidea) clearly appear as a monophyletic group.

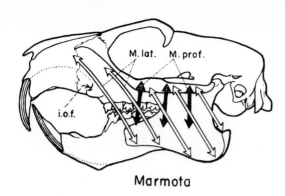

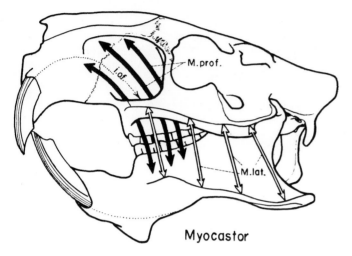

Figure 1 *Comparison between the skulls of the sciuromorph* Marmota marmota *and the caviomorph* Myocastor coypus, *x 4/5. The latter displays the "hystricomorph" disposition; the Masseter profundus (M. prof., black arrows) has spread forward through an enlarged infraorbital foramen (i.o.f.) onto the snout.*

TRANSOCEANIC MIGRATIONS OR PARALLELISM?

The classical systematic arrangements led very naturally to a search for an African origin for the Caviomorpha and the Platyrrhina, since Africa was the place where the nearest forms, at least morphologically, were found [3]. It implies transoceanic migrations from Africa to South America, raising serious problems. The first hypotheses (the emerging of continents or of transoceanic "bridges" followed by subsidences) has been at last abandoned: not only were they lacking geological support, but they did not take into account the characteristics of the migrations considered, which appear to be unilateral and limited to two mammalian groups. On the other hand, we have said already that geophysicists and geologists did not favorably receive Wegener's theory. Thus, Matthew being led to formulate his own fixist conception of paleogeography, there remained no more than two hypotheses to try to explain the resemblances between the primates and the rodents on both sides of the Atlantic. Either both groups crossed the Atlantic by means of natural rafts (this possibility was still considered in the case of Hystricomorpha by Matthew himself in 1915, and by Romer in 1945), or monkeys and Hystricomorpha arose respectively from Holarctic stocks and the likeness between New and Old World forms is attributable to parallelism and not to direct phylogenetic relations.

The immensity of the Atlantic barrier, in its present dimensions, makes very improbable a successful transportation by rafts; thus most authors have rallied to the second hypothesis (which, let us repeat, also postulates the rafting of immigrants from North America, or at least from nuclear Central America, to South America).

In order to support this hypothesis, workers have approached this problem in two ways: (1) by searching in North America for possible ancestors of the Caviomorpha and the Platyrrhina; (2) by trying to discover some anatomical criteria allowing a demonstration of the phylogenetic independence of both neotropical groups with respect to the Old World ones.

THE SEARCH FOR NORTH AMERICAN ANCESTORS

If the Caviomorpha did not arise in Africa, we are obliged to seek their ancestry in North America. But the only recent North American Hystricomorpha are a few late immigrants originating from South America, and the rich Tertiary faunas of North America have never yielded hystricomorph rodents. It was consequently among nonhystricomorph forms that workers have been led to search for ancestors of the Caviomorpha. The results have been disappointing. Of course there is no objection to deriving the Caviomorpha from the Paramyidae, a Holarctic family generally considered as the ancestral group of all the rodents. Wood (1949, 1950, 1962) has even pointed out in this family several genera *(Reithroparamys, Rapamys)* exhibiting some "hystricomorph tendencies," but taking into account their geological age, these genera are not enough advanced in the hystricomorph way to take place in the direct line leading to Caviomorpha.

The problem raised by ceboid monkeys is astonishingly similar. There are no monkeys, alive or extinct, in North America. The only extinct primates discovered there are Prosimii. Consequently, it is among these latter that workers were led to look for direct ancestors for the Ceboidea. It has been suggested that the stem stock could correspond to the Omomyoidea, the Holarctic group from which Old World monkeys are equally supposed to have arisen.

In both cases, no intermediate has ever been found between the supposed North American ancestral stocks and the first known representatives of the Caviomorpha and the Ceboidea. It was then assumed that these unknown intermediates lived in Central America during the Eocene epoch. Yet this is but a weak hypothesis. One cannot understand why these hypothetical Central American groups, which are supposed to have crossed southwards a large marine barrier in order to colonize all of South America, would not have spread northwards. And this is particularly unlikely for the rodents, which are well known for their aptitudes in adapting to different environments.

In fact, there is no ground for asserting a direct filiation, for either the Ceboidea or for the Caviomorpha, from North American stocks. Yet we need conclusive arguments to support an hypothesis which, once adopted, would lead to profound changes in classification. The concept of monkeys (and even of Simiae = Anthropoidea) should be abandoned in zoology if it were proved that it includes two stems independently originated from Prosimii. In the case of rodents, the situation is still worse since it results in putting at the end of the list a series of superfamilies and families incertae sedis! (See Wood's classification as adopted by Romer, 1966). That state of things seems to betray the inadequacy either of the criteria or of the phylogenetic conceptions used.

ANATOMICAL ARGUMENTS FOR AND AGAINST AN AFRICAN ORIGIN OF THE CAVIOMORPHA AND THE PLATYRRHINA

Comparative anatomical studies dealing with Caviomorpha and Old World Hystricomorpha were undertaken by various researchers in order to choose between affinity and parallelism as an explanation of the observed similarities.

As the result of his studies on the Deseadan genus *Platypittamys,* Wood (1949, 1950) thought he had discovered a relevant odontological criterion. According to him, the caviomorph cheek teeth would arise from a tetralophodont pattern, whereas the Old World Hystricomorpha are fundamentally pentalophodont. On the other hand, Stehlin and Schaub (1951) and later Schaub (1953, 1958) still classified the Caviomorpha (= Nototrogomorpha) among the Pentalophodonta, while Wood (1955, 1959, 1965) and Wood and Patterson (1959) keep their own position. New material from the Deseadan of Bolivia (Hoffstetter, 1968) gives strong evidence in support of a fundamental pentalophodonty in the Caviomorpha (Hoffstetter and Lavocat, 1970). Besides, it shows that *Platypittamys* does not deserve the meaning ascribed to it by Wood: several features considered as primitive by him do in fact result from secondary simplifications, and are found in other South American and African forms as well. Of course these grounds are not sufficient to infer a relationship between the Caviomorpha and the African Hystricomorpha (since pentalophodonty is found also among some nonhystricomorph rodents). Yet, the feature put forward by Wood is not relevant for excluding the possibility of such a relationship.

Landry (1957) noticed an impressive number of features common to the Caviomorpha and the Old World Hystricomorpha, but his conclusions are not always convincingly supported. Nevertheless, his contribution was real and deserved a reception quite different from the one that was given to it.

Brundin (1966), when adopting Hennig's ideas on phylogenetic systematics, has violently attacked the classical conceptions mainly from a methodological point of view. All of his assertions are not acceptable; but Hennig's ideas contribute a certain number of positive elements that Patterson and Pascual (1968) should not have discarded without discussion.

In fact our knowledge has considerably increased during the last years. African fossils have been studied by Wood (1968) for the Oligocene, and by Lavocat (1973) for the Miocene. We are now well acquainted with the Deseadan rodents in Patagonia (Wood and Patterson, 1959) as well as in Bolivia (Hoffstetter and Lavocat, 1970, and works in progress).

Presently there is no doubt that the hystricomorph structure of the orbitotemporal region arose among various independent branches and that the former suborder Hystricomorpha must be dismembered. In 1899, Tullberg already suggested that, from a phylogenetical point of view, the mandibular structure is far more relevant. Lavocat took up and strengthened this idea by supporting it with solid arguments. The Caviomorpha are both hystricomorph (by their infraorbital foramen; Figure 1, *Myocastor*) and hystricognath (a mandibular specialization [see Figures 2, 3] well studied by Waterhouse (1839, 1848), Tullberg (1899), Landry (1957), and Lavocat (1971), to which can be added more or less correlative features such as the perforation of the pterygoid fossae). To the Old World forms showing both features, i.e., the Phiomyidae and other related extinct families, the Thryonomyidae (including Petromurinae), the Hystricidae, and the Bathyergidae [4], Lavocat (1971) gives the name of Phiomorpha. The other Old World "hystricomorph" forms (Theridomorpha, Ctenodactylidae, Pedetidae, Anomaluridae), all of them sciurognath, obviously belong to distinct lineages.

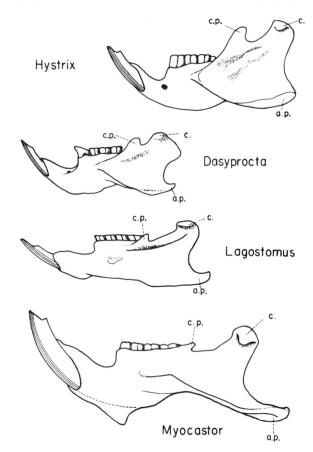

Figure 2   *Lateral view of hystricognath lower jaws of Old World Phiomorpha* (Hystrix longicauda) *and New World Caviomorpha* (Dasyprocta aguti, Lagostomus viscaccia, Myocastor coypus), x 4/5. *These few examples show various degrees of lowering of the condyle (c.), reduction of the coronoid process (c.p.), as well as backward and lateral expansion of the angular process (a.p.) See also Fig. 3.*

So it is between Caviomorpha and Phiomorpha, and more especially between Caviomorpha and Thryonomyoidea (Hystricidae and Bathyergidae being branches too specialized), that a comparative study ought to be done. Lavocat undertook it with, at first, the hope of supporting Wood's thesis (which he shared at the time) of a parallel evolution of two independent groups. His results, therefore, are all the more significant. In spite of a thorough study, he was not able to disclose any character, osteological or dental, that allowed a clear distinction between the two groups. Still better, he observed among South American and African forms such anatomical similarities (including the otic region) as can hardly be explained without appealing to a true relationship. He concludes, therefore, that the Caviomorpha and the Phiomorpha do constitute a natural group (suborder), for which he utilizes the term Hystricognatha as created by Tullberg in 1899.

The study of the present problem might take into account researches concerning the soft anatomy (myology, circulatory apparatus, endocrine glands, fetal membranes, and so forth), the histology, cytogenetics, serology [5], and physiology. Unfortunately the works till now published, on the whole, do not concern a sufficient variety of taxa in order to lead to clear conclusions. It must be noticed, however, that Guthrie's studies (1963, 1969) on

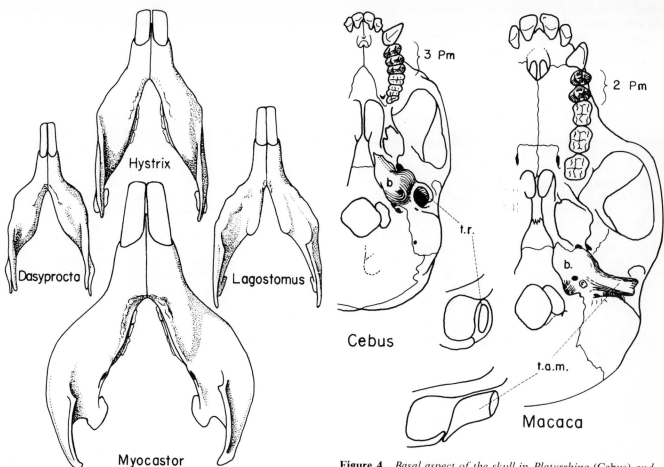

Figure 3 *Ventral view of hystricognath lower jaws, x 4/5: same specimens as in Fig. 2. The hystricognathy is characterized by the lateral position of the angular process in relation to the plane defined by the outer side of the incisor.*

Figure 4 *Basal aspect of the skull in* Platyrrhina (Cebus) *and* Catarrhina (Macaca), *x1. In the modern Catarrhina, the following features are characteristic: only two premolars (Pm) instead of three; auditory bulla (b) not inflated; ectotympanic bone (or tympanic ring: t.r.) developed in the form of a tubular auditory meatus (t.a.m.). The latter differentiation is illustrated too in the diagrammatic section (below), modified after Le Gros Clark, 1959.*

the arterial system disclose relevant similarities between the Caviomorpha and the Phiomorpha. On the other hand, Mossman and Luckett (1968) state: "Nidation and fetal membrane morphogenesis in *Bathyergus* (a fossorial African phiomorph) are, except in minor details, like those of the New World hystricomorphs *Erathizon, Cavia, Dasyprocta, Chinchilla,* and *Myocastor,* and not like those of any known sciuromorph or myomorph." They conclude: "The Bathyergidae are closely related to New World hystricomorphs. Is there a geologic explanation?"

In summary, it seems that no anatomical argument opposes a possible relationship between primitive Phiomorpha and the Caviomorpha.

The problem raised by the Platyrrhina is similar but more simple. A comparative study shows some anatomical differences between the living Platyrrhina and the Recent Catarrhina, differences which are also found in the Neogene representatives of both groups. The former are distinguished principally by the presence of three premolars, by their simple ectotympanic not elongated in the form of an ossified external auditory meatus, and by the retention of an auditory bulla (Figure 4). All of these features give to the Platyrrhina a relatively primitive character, which for a long time has prevented them from being considered as descended from the Catarrhina.

Nevertheless, Simons' work (1967c, d) has disclosed the existence of extinct African monkeys provided with three premolars (*Apidium, Parapithecus*) and shown that the earliest Catarrhina lacked a completely ossified auditory meatus. He has also noted among the primates from the Fayum (Egypt) other ceboid features concerning the humerus, the tail, etc. Simons (1967a) concludes that although these African primates are already true monkeys in their general features, they nevertheless correspond to a rather primitive stage more or less comparable to that of the living South American monkeys. He did not hesitate to qualify the Parapithecidae as a "primitive New-World monkey-like" family. It would certainly be rash to infer a direct filiation from Parapithecidae to Ceboidea. But these observations of Simons already quoted lead rather to suppose the existence of Eocene African monkeys, which gave rise to the Neotropical Platyrrhina and the African Catarrhina.

PARASITOLOGICAL EVIDENCE

Cameron (1960, p. 83) gives the following conclusion, based on a study of the parasites of the Platyrrhina and the Catarrhina: "The evidence suggests that both groups of monkeys evolved in common in Africa until a comparatively advanced stage—possibly until the second half of the Tertiary [sic]". Such an affirmation cannot be defended, but its basic premises have found some echo (Reig, 1968, p. 225). However, according to A. G. Chabaud and J. C. Quentin (pers. comm.), these bases themselves are very debatable and correspond perhaps to accidental infestations in captive animals. In fact, it is not possible to conclude anything from the distribution of the filarid *"Acanthocheilonema"* (= *Dipetalonema*) and of the cestode *Bertiella;* both of them attack many mammalian groups in the Old and New Worlds. The oxyures that infest the Platyrrhina do not belong to the genus *Enterobius* (which is localized in the Old World), but to a very different genus *Trypanoxyuris.* Cameron reports the same species of the nematode *Subulura* in both groups of monkeys; this indication is erroneous or else it rests upon a case of accidental infestation. In fact, the Subuluridae parasitizing monkeys belong to the genus *Primasubulura* (distinct from *Subulura*), subdivided into two subgenera, the more primitive of which *(Platysubulura)* attacks only the Platyrrhina, but the other (*Primasubulura s.s.)* is present only in the Catarrhina. Such a distribution does not allow one to go so far as Cameron does, but it speaks in favor of a common origin for the Platyrrhina and Catarrhina, because the genera that infest the Prosimii are different: *Subulura,* in Lemuroidea, and *Tarsubulura,* in Tarsioidea.

In the parasitological research concerning the Caviomorpha, an outstanding work, still in progress, on heligmosomid nematodes is being done by Durette-Desset (1971). A phylogenetic interpretation leads her to distinguish in this family several natural groups ("lineages"), of which two infest the Caviomorpha. One of these groups (Heligmonellinae—Pudicinae of Durette-Desset), issued from African forms like *Heligmonella* (parasitizing the phiomorph genera *Thryonomys* and *Atherurus*), is found in five families of Caviomorpha (Erethizontidae, Echimyidae, Myocastoridae, Capromyidae, and Dasyproctidae) and also in Neotropical Sciuridae. The other group (Viannaiinae of Durette-Desset), strictly South American, infests Neotropical marsupials and (secondarily) four families of Caviomorpha, distinct from the first ones (Caviidae, Hydrochoeridae, Cuniculidae, and Chinchillidae), as well as some cebid monkeys. Such a distribution is interesting. The presence, in five families of Caviomorpha, of Heligmosomidae closely related to African parasites of Phiomorpha, suggests an African origin for these Caviomorpha. In this hypothesis, the latter would have arrived in South America bringing their own parasites. The second group of the Caviomorpha, on the contrary, would have been contaminated, after its arrival, by parasites already present, at that time, in South American marsupials. If we consider that these two groups of rodents are completely distinct from one another, and if we note that the second group corresponds to the forms that are the most hypsodont [6], we may suppose that they would derive from two distinct colonizations, independent but not necessarily different in age. This is an important point in need of confirmation (see following text.

DATES AND NUMBER OF PALEOGENE COLONIZATIONS IN SOUTH AMERICA

The arrival of the rodents and the primates in South America goes back at least to the early Oligocene (Deseadan), since they are found beginning at this date in Patagonia (rodents), in Bolivia (rodents and primates), and possibly in Colombia (Peneyita rodent). Nevertheless, the initial diversity of the Caviomorpha, in Bolivia as well as in Patagonia, suggests an earlier colonization,

perhaps in the late or even middle Eocene. Wood and Patterson (1959) and also Patterson and Pascual (1968) are of the opinion that this arrival occurred during the interval between the Mustersan and the Deseadan. But oddly enough, rodents fail to appear in the Divisadero Largo fauna, which is situated precisely in this interval. Were the Caviomorpha confined for a time to the intertropical zone owing to their climatic requirements? Were the first immigrants cut off by a geographical barrier? Maps proposed by Schaffer (1924, Fig. 544, p. 443; reproduced in Kurtén, 1966) and by Cameron (1960, Fig. 2) suppose the existence of arms of the sea that in the Paleogene would have fragmented the South American continent. But there is no evidence of continuous marine sediments that would justify such an hypothesis. Landry (1957) suggests the existence of Antillean relays that could have served as centers of diversification prior to the Oligocene. Only the discovery of Eocene intertropical fossil localities will allow an elucidation of this problem.

It is possible that the Platyrrhina arose from a single stock, since the teeth of the Hapalidae as well as of the Cebidae may be derived from those of *Branisella* (Hoffstetter, 1969).

In the same way, Wood and Patterson (1959, p. 396) suppose that the Caviomorpha, as a whole, have diversified from a single invasion. But we have seen that parasitological evidence conveys more the impression of two different colonizations of South America. If this is proved to be the case, the Caviomorpha could no longer be regarded as a monophyletic group. But we must not make hasty conclusions. Maybe both the groups of Caviomorpha distinguished by Durette-Desset by their parasites, are in fact the result of an *ecological* segregation from a single South American stem-stock. Certain paleontological observations seem to support this last hypothesis. In Bolivia I know two Deseadan fossiliferous localities. All the rodents from one of these—Salla-Luribay—possess brachyodont or feebly hypsodont teeth (see Hoffstetter and Lavocat, 1970) and belong to the first of Durette-Desset's groups. The other locality—Lacayani (still unpublished)—is of the same age (presence of *Trachytherus, Glyptatelus,* etc.) but with a different facies (absence of *Pyrotherium, Rhychippus,* etc.). The rodents from the latter locality are very hypsodont and all belong to one or another of the families Chinchillidae and Eocardiidae (unknown from Salla); that is to say, to the second group distinguished by Durette-Desset. These two associations or rodents suggest distinct ecological control: it seems that the food sources supplied by the vegetation of Lacayani were more abrasive than that of Salla.

POSSIBILITY OF TRANSATLANTIC MIGRATIONS DURING THE PALEOGENE

It is therefore in Africa where we are obliged to seek the direct ancestors of the Caviomorpha and the Platyrrhina. We come back then to the former conceptions which have been dismissed for a long time because of paleogeographical reasons. Modern geophysical studies have restored with solid evidence the theory of continental drift, which is now admitted as a fact. During the Paleogene the Atlantic was far from having reached its present width, yet it already constituted a considerable barrier (see Le Pichon, 1968; Funnell and Smith, 1968; Dietz and Holden, 1970). It must be stressed, however, that the probability of a successful rafting depends perhaps more on marine currents than on the distance covered. During the Paleogene, the Atlantic and Pacific oceans communicated by a large branch of the sea, which was probably swept from east to west by the equatorial current. The latter rendered very improbable the transportation by rafting between the two Americas. On the contrary, the same equatorial current could carry African rafts westward only, toward the coasts of Brazil (thus accounting for the one-way direction of observed migrations).

Romer (1945, p. 504) at first admitted that "it is hard to escape the conclusion that the hystricoids, by some means or other, made a successful western crossing of the South Atlantic." Later on, he came to consider such a migration as impossible because, according to him (1968, p. 225), "the South Atlantic is a long swim for a necessarily pregnant female." Presented in this form, the argument is peremptory. Indeed, nobody will ever admit that it should be possible for a gravid female to cross the Atlantic (even if it were of a reduced width) by swimming. It is clear that, in order to give an ironical touch to an objection he wanted impressive, Romer deliberately limited himself to the most improbable hypothesis. In fact we can conceive other modes of transportation and colonization. Tropical rivers sometimes do carry veritable floating islands made up of entangled trees. Such rafts can harbor small populations of rodents and/or primates and, under propitious circumstances, may successfully transport them from one continent to another. Moreover, some island relays, made up by volcanoes such as are known since the Cretaceous in the South Atlantic ocean (see Deitz and Holden, 1970; maps, pp. 37–38), may have facilitated this crossing.

Nevertheless, even in this form, the success of such rafting remains very problematical. In most cases, the raft will disintegrate and the harboured animals die of thirst, starvation, or by drowning. However in the case considered, a feeble probability would be sufficient to account for one single successful crossing (by means of a raft bearing together rodents and monkeys) or even two or three successes (as distinct implantations of founding colonies, one of primates, one or two of rodents). The chances of success, still appreciable at the beginning of the Tertiary, diminished as the Atlantic Ocean grew wider. This is in accordance with the failure of any African faunal element to settle in South America since the beginning of the Oligocene, that is to say, during the last 40 millions of years.

CONCLUSIONS

In brief, the African origin of the Caviomorpha and the Platyrrhina is by far the most probable hypothesis. It is the only one to account satisfactorily for the totality of the evidence presented by paleontology, comparative anatomy, parasitology, and chorology. Immigrants probably arrived by rafting towards the end of the Eocene. Such an origin, if we accept it, requires a reconsideration of the phylogenetic and biogeographic history of the rodents and the primates.

The rodents are known since the late Paleocene in North America (Paramyidae). We know neither their ancestors—probably neotherid Insectivora—nor their place of origin.

I would be inclined to believe that the Sciurognatha and the Hystricognatha are the result of an early dichotomy caused by geographical segregation. As a matter of fact, the former are essentially Laurasian, the latter, essentially African. The two groups are clearly distinguished one from another by their mandibular structure. Hystricognathy represents an excellent apomorph character (according to Hennig's terminology), and makes it possible to contrast the group defined by Tullberg (1899) and Lavocat (1971) with the other rodents as a whole, which are all sciurognath and include the oldest representatives of the order.

In fact, the Hystricognatha are known only since the Oligocene, but, on the one hand, they are the only rodents described from Africa of that epoch, and, on the other hand, no other representative of them is known at that time in the rest of the Old World. Their ancientness is not known exactly, but they could have differentiated in Africa at the beginning of the Tertiary, our ignorance of their first representatives being due to the lack of African fossil-bearing deposits of adequate age. Toward the end of the Eocene, primitive Phiomorpha (Phiomyidae ?) succeeded in crossing the Atlantic ocean and settling in South America to give rise to the Caviomorpha [7]. These, first confined to the South American continent, reached the West Indies probably in the Neogene, and some of them passed to Central America, then to North America, after the emersion of the Isthmus of Panama.

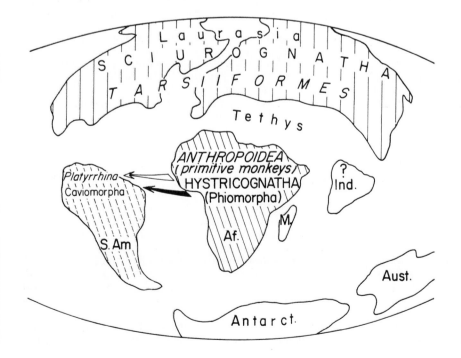

Figure 5 *Place of origin and early expansion of the hystricognath rodents and the anthropoid primates, following the present interpretation. In the early Tertiary both groups differentiated in Africa (Af.: oblique lines), whereas their respective sister-groups—Sciurognatha and Tarsiiformes—differentiated in Laurasia (vertical lines). Towards the late Eocene, both African groups crossed the South Atlantic (arrows) giving rise to Caviomorpha and Platyrrhina in South America (S. Am.: dashed oblique lines). The sketch indicates approximately the position of the land masses in the Eocene.*

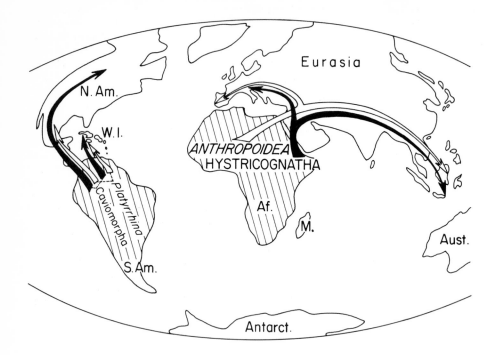

Figure 6   *Later migrations of the Hystricognatha and the Anthropoidea (excluding man): starting from Africa, penetration into Eurasia during and after the Miocene (after the Alpine orogenic phase of the late Oligocene); from South America, (a) rafting to the West Indies during the Neogene, (b) invading of Central America (and North America as regards the Caviomorpha) during the Pleistocene, after emergence of the Panamanian isthmus. Geographic disposition of the land masses as at present.*

In a parallel fashion, the Phiomorpha that remained in Africa gave rise to the central group Thryonomyoidea and also to the specialized families Hystricidae and Bathyergidae, the history of which is poorly known. Two families, (Thryonomyidae and Hystricidae) passed to Eurasia near the Miocene, but did not reach Australia or even Madagascar. (See Figures 5, 6).

It is more difficult to draw a picture of the history of the Sciurognatha, the initial radiation of which was extraordinarily rapid and led to precociously individualized branches, whose interrelationships are obscure. Convergences and parallelisms appear (groups with a "hystricomorph" orbitotemporal region, with pentalophodont molars, etc.), which explains the mistakes committed by certain authors. It is worth observing that the Sciurognatha, first confined to Laurasia, penetrated several times into Africa, at least since the Miocene, and the earliest colonizations gave rise to strongly endemic groups such as the Anomaluridae and the Pedetidae. Only later, some modern groups reached Madagascar (Cricetidae), Australia (Muridae), and South America (numerous Cricetidae since the Pliocene; a few Sciuridae since the latest Pleistocene; and one recently arrived heteromyid).

Thanks to the numerous works which have been devoted to the primates (see Valois, 1955; Piveteau, 1957; Le Gros Clark, 1959; Genet-Varcin, 1963, 1969; Wilson, 1966; McKenna, 1967; Simons, 1967c, d, and references cited by these authors), and taking in account the monophyletism of monkeys here advocated, the history of this order becomes rather clear, at least in its general features.

The primates are known since the latest Cretaceous of North America *(Purgatorius)*. They arose from Insectivora probably soon after the acquisition of the neotherid disposition by the latter. They gave rise precociously to an offshoot, the Plesiadapoidea, an aberrant group with chisel-like incisors, known only in the Paleocene and the early Eocene of North America and Western Europe.

The main stock of the primates seems to have undergone an early dichotomy, giving rise to the two groups for which Hill

(1953) reinstates the old names of Strepsirhini and Haplorhini.

The Strepsirhini, better known as Lemuriformes or Lemuroidea, represent the plesiomorph group of this dichotomy (Hennig's terminology). They spread over all the continents, except Australia and South America. But their history is only fragmentarily known. In fact, they have been found: (1) in the Eocene of Europe, North America, and Eastern Asia (Adapidae, incl. Notharctidae); (2) in the Pleistocene and Recent of Madagascar and neighboring islands (Lemuridae, incl. Indriidae; and Daubentoniidae, related to the former, but showing a very specialized dentition recalling that of the Plesiadapoidea); (3) in the Miocene of Africa and in the Recent of Africa and Southern Asia (Lorisidae, the very peculiar otic structure of which indicates a very early separation from the common trunk). All this suggests a long history, insufficiently illustrated, in which there acted climatic requirements (abandonment of the Holarctic zone after the Eocene), geographic segregations (origin of the Lorisidae), substitutions, and also refuges of which Madagascar is a typical example. But, in that last case, we do not even know whether the Lemuridae reached Madagascar from Africa or from India.

The Haplorhini, apomorph sister group of the former (Hennig's terminology) underwent in their turn an early dichotomy which seems to correspond to a geographic segregation. The Tarsiiformes (plesiomorph branch), which include the Omomyoidea and the Tarsioidea, differentiated in Laurasia; they lasted up to the lower Miocene in North America, up to the Eocene in Europe, and they survive nowadays with the genus *Tarsius,* which took refuge in islands of Indonesia and Philippines. The Anthropoidea = Pithecoidea = Simiae = Simii (apomorph branch) were certainly born in Africa but one unfortunately does not know their first representatives. Again, they subdivided by geographic segregation. Primitive monkeys, having crossed (I suggest) the Atlantic ocean by rafting towards the end of the Eocene, gave rise to the Platyrrhina = Ceboidea = New World monkeys (plesiomorph subdivision), which tardily reached Central America and also Jamaica *(Xenothrix).* The Catarrhina (apomorph subdivision) that remained in Africa include a primitive group (the Parapithecidae, which Genet-Varcin, 1969, refers to as "Protocatarhiniens"), still nearly related to the Platyrrhina; and two progressive groups, the Cercopithecoidea (Old World monkeys) and the Hominoidea (apes, *Oreopithecus,* and hominids). These two groups reached Eurasia: the apes in the Miocene, probably at the same time as the mastodonts did, the monkeys in the Pliocene (or before) and the Hominidae in the Quaternary. Finally, man conquered the entire world.

As we can see, the genealogic tree of the Primates is rather clearly drawn, but there is no general agreement about taxonomy. One will notice, in particular, that the phylogenetic relationships brought out above lead to the suppression of the concept Prosimii, a composite (or better, "paraphyletic" in Hennig's terminology) assemblage, which are not correct in contrasting with that of the Anthropoidea, if we want the classification to express the phylogeny.

The present interpretation, which seems coherent, brings out an astonishing analogy between the history of the Hystricognatha and that of the Anthropoidea. These two groups are of African origin. Both arise from a dichotomy resulting from a geographic segregation, their respective sister-groups being at the beginning localized in Laurasia. Both succeeded in crossing the Atlantic, maybe simultaneously, and in settling in South America. They penetrated only later on into northern continents: about the Miocene coming from Africa, the Pleistocene coming from South

America. They did not reach Madagascar any more than Australia (we except the special case of man). So many coincidences concerning two groups, independent phylogenetically, could not be fortuitous. It is clear that both groups, all along their history, have been stopped by the same biogeographic obstacles and have availed themselves of the same opportunities of expansion.

ADDENDUM

After the writing of the present essay, Wood and Patterson, published a paper (dated 1970, but published in 1971) answering Lavocat's note (1969). This paper would deserve a detailed analysis. I only shall express my agreement with the following conclusion of the authors: "...we must continue to stress the importance, in dealing with fossil rodents, of making no assumptions of the relationships of fossil forms (especially from areas widely separated, geographically) without a transitional series of fossils." However, this statement is valuable too for the other zoological groups, and must apply to the various possible hypotheses. Now, let us repeat that the respective ancestors of the Caviomorpha and the Platyrrhina had to cross a wide barrier, wherever they came from: Africa or North America. On the other hand, the lack of transitional series of fossils" is particularly flagrant in the hypothesis of a filiation from the North American Paramyidae and Omomyidae, respectively, to the South American Caviomorpha and Platyrrhina. The anatomical gap is less in the assumption of an African origin for the latter groups.

Notes

[1] Reig (1962) proposes to use instead the term *cenochron,* to distinguish the immigrant groups that became integrated, at a given time in a given area, into preexisting communities.

[2] The term "Palaeotherida" as created by Broom in 1927 in opposition to "Neotherida" applies to mammals that have but three bony elements in their basicranial axis. This term seems unsuitable for two reasons. Firstly, it may be mistaken, especially when in adjectival form, for its near homonym "Palaeotheriidae" (a family based on the genus *Palaeotherium*). Secondly, in its original meaning, this taxon includes monotremes, marsupials, several placental orders (Xenarthra, Artiodactyla, Perissodactyla, Proboscidea, Sirenia), and probably some extinct mammalian groups. This is certainly not a natural grouping. Actually the neotherid disposition (acquisition of a fourth basicranial axial element) seems to result from an early dichotomy of placentals (see Hoffstetter, 1970a); if this hypothesis proves to be right, the two resulting stems (superorders) may be designated Neotherida Broom (1927) and Henotherida nom. nov. (from Greek 'ενος 'old,' θηρ 'beast'), the latter for placentals that have not yet acquired the neotherid structure (excluding therefore nonplacental mammals). Furthermore, the affinities and taxonomic position of some aberrant groups (Chrysochloridae, Hyracoidea, Cetacea) still have to be debated.

[3] Some authors, in particular Ameghino, have considered an opposite filiation, from South American to African forms, but such an hypothesis encounters so many difficulties that we won't discuss it here.

[4] In fact, the Bathyergidae present a secondarily reduced infraorbital canal but, as was showed by Tullberg (1899), Landry (1957), and Lavocat (1971), on the whole their characters are clearly consistent with their belonging to Hystricognatha.

[5] As far as serology is concerned, the limited researches made by Moody and Doniger (1955) are not conclusive. According to them the North American and African procupines [*Erethizon* and *Hystrix*] "have but slight serological similarity," and "that similarity is of about the same order of magnitude as that shown by either procupine to the guinea pig [*Cavia*] and the agouti [*Dasyprocta*]". These first results emphasize the ancientness of these various phyletic lines, which are zoologically separated at least since early Oligocene. They do not exclude a common ancestry for the whole. More definite conclusions would need a deeper research, with crossed tests, using antisera prepared from a great variety of rodents (Caviomorpha, Phiomorpha, and non-hystricognath groups).

[6] Unhappily, we lack information about some families of the Caviomorpha, notably the Octodontidae, Abrocomidae, and Ctenomyidae. Moreover, referral of the Dasyproctidae to the first group and the Cuniculidae (pacas) to the second one needs a confirmation or a particular explanation if these two families are closely related, as is generally accepted. It is true that some authors, especially Ellerman (1940) and Schaub (1958), do not share this opinion. According to them, the Dasyproctidae are not closely related to the Cuniculidae; they may have some support from parasitology.

[7] If I interpret Lavocat's observations correctly, the terms Phiomorpha and Caviomorpha enable a distinction to be made between two biogeographical assemblages, but these cannot be separated by anatomical diagnoses. So, the distinction would not be valid from a taxonomic point of view.

15

Paleobiogeographic Perspectives
on the Origin of the Platyrrhini

R.L. CIOCHON AND A.B. CHIARELLI

INTRODUCTION

This chapter will explore the various hypotheses and scenarios that have been advanced concerning the paleobiogeographic source of origin of the Platyrrhini. The specific questions to be addressed include: (1) From what geographical region or regions were the immediate ancestors of the New World monkeys derived? (2) When and by what means did these first platyrrhines reach the island continent of South America? (3) What influence did continental drift have on the geographic source of origin and mode of dispersal of the Platyrrhini? Nearly all of the preceding chapters in this volume (see Ciochon and Chiarelli, Eds., 1980) have attempted to specifically answer one or more of these questions. It is our purpose herein to provide a background for the discussion of all these paleobiogeographic proposals and then to present a consensus-oriented maximum-parsimony model of platyrrhine origins and dispersal.

CONTINENTAL DRIFT

The changing positions of the continents throughout the Mesozoic and Cenozoic Eras have had a major impact on the past and present biogeographic distribution of life on this planet. The true significance of this impact has only recently come to light, resulting in a whole series of reinterpretations of the evolutionary history and biogeography of many of the earth's living and extinct organisms. The concept of continental drift and plate tectonics has now been successfully applied to the evolution and distribution of angiosperm plants (Axelrod, 1978), laroniinine spiders (Platnick, 1976), marine invertebrates (Valentine, 1973; Schram, 1977), ostariophysan fishes (Novacek and Marshall, 1976), early tetrapods (Milner and Panchen, 1973), fossil reptiles (Colbert, 1973; Galton, 1977), fossil birds (Cracraft, 1973; Rich, 1978), vertebrate distribution (Cracraft, 1974), Mesozoic mammals (Lillegraven et al., 1979), marsupials (Tedford, 1974), neotropical floras and faunas (Keast, 1972, 1977; Hershkovitz, 1972), hystricomorph rodents (Lavocat, 1974a, 1977), and fossil primates (Walker, 1972; Hoffstetter, 1972, 1977b; Szalay, 1975b), to name only a few. In almost every instance, the application of this "new paleogeography," to borrow the phrase of Tedford (1974), has resulted in a much more understandable and justifiable synthesis of previously existing data.

It is the synthetic and unifying nature of the "new paleogeography," which has truly revolutionized, and perhaps forever changed, our conceptualizations of the evolution and distribution of the earth's organisms. It is therefore not at all surprising that most of the contributors to this volume have used the concept of mobile continents to better decipher the evolutionary origins of the New World monkeys.

HISTORICAL BIOGEOGRAPHY

The application of continental drift, plate tectonics, and sea-floor spreading to paleobiogeographic problems has also brought about the introduction of several new principles of historical biogeography. Terms such as "Noah's arks" and "beached Viking funeral ships" (McKenna, 1972a, 1973) will be added to the list of such Simpsonian principles as sweepstakes routes, waif dispersal, filter bridges, and land corridors (Simpson, 1940a, 1953a, 1978). Indeed, some of these principles of historical biogeography bear directly on how the first primates reached the South American continent.

The nearly universal acceptance of the continental-drift paradigm in earth history and the recent desire by many researchers to develop more testable or falsifiable [in the sense of Popper (1959, 1963)] models of past biogeographic events has brought about a sort of minirevolution in the field of historical biogeography. No longer are many workers content with the analysis of their data following the traditional or narrative approach to biogeography as conceptualized by Darlington (1957, 1959, 1965), supported by Mayr (1965) and Briggs (1966) among many others, and vigorously defended by Darlington (1970). In place of the traditional approach to biogeography, two more rigorous methods have been developed. One can be referred to as "phylogenetic or cladistic biogeography" and the other as "vicariance biogeography."

Cladistic biogeography as we wish to term it here is a new, more formalized theory of biogeography (see Brundin, 1972; Ross, 1974; Cracraft, 1974a, 1975; Ashlock, 1974; Morse and White, 1979) which has as its primary tenet the construction of the most parsimonious hypothesis regarding the location of an ancestral group based on a definable, testable concept of phylogenetic relationship. Thus, reconstruction of the center of origin and pathways for dispersal of a group is deduced (a deductive inference) from prior phylogenetic analysis. Historical biogeography, in this sense, "... endeavors primarily to reconstruct the phylogeny of organisms, especially of higher taxa, and to place this phylogeny in geographic perspective" (Cracraft, 1974a, p. 215). Not all biogeographers feel that phylogeny should play such a central role in biogeographic reconstruction (for example, see Briggs, 1966; Darlington, 1970). In the past, many have reconstructed the center of origin of a particular group based primarily on patterns of species or generic diversity and on present-day diversity gradients, paying little if any attention to the group's phylogenetic history. Since the biogeographic event concerning the origin of the Platyrrhini, which we are attempting to reconstruct, occurred many millions of years in the past (certainly prior to the early Oligocene), we feel that it is absolutely essential to use all potential sources of biogeographic data, especially those derived from phylogenetic analysis. In this regard, we feel that the cladistic biogeographic approach has considerable merit.

Vicariance biogeography, according to Nelson and Platnick (1978), is as different from traditional biogeography in approach as cladistic analysis is from phenetic analysis. In other words, vicariance biogeography represents a nearly complete reformulation of the basic principles of historical biogeography. The vicariance paradigm, as currently formulated, owes its existence to the energetic writings of Leon Croizat (for example, see Croizat, 1958, 1962; Croizat *et al.,* 1974). For more than 30 years, he has expressed the view that the distribution of animal and plant life on the earth is not necessarily the result of innumerable dispersal events from various centers of origins but rather the result of geologic or geographic changes in the environment (vicariant events) on a worldwide basis which in themselves bring about episodes of allopatric speciation. Thus, tectonic change, not dispersal, is the basis of vicariance biogeography. When one considers that Croizat [a complete bibliography of his works appears in Nelson (1973)] formulated the vicariance paradigm during a period when most biologists considered the concept of mobile continents a near impossibility, it is a credit to his perseverance as a scientist that vicariance biogeography has become an acceptable (and in some cases all-inclusive) alternative in historical biogeographic analysis today.

During the past five years, a small group of workers (Nelson, 1974, 1975, 1978; Rosen, 1975, 1978; Platnick, 1976; Platnick and Nelson, 1978) have discussed, refined, and applied the vicariance paradigm in a number of useful ways. For instance, Rosen (1978, p. 187) has pointed out the importance of recognizing "... that geology and biogeography are both parts of natural history and, if they represent the independent and dependent variables respectively in a cause and effect relationship, that they can be reciprocally illuminating." Nelson (1974, p. 557) has suggested that "estimating the time of splitting of lineages through a study of vicariance in relation to dated barriers is an alternative to the traditional approach through paleontology, which tends to underestimate the absolute age, or age of origin, of lineages." Finally, Platnick and Nelson (1978, p. 1) contrast dispersal with vicariance by stating that "dispersal models explain disjunctions by dispersal across preexisting barriers, vicariance models by the appearance of barriers fragmenting the ranges of ancestral species." The concept of vicariance insofar as it can be applied to the origin of the Platyrrhini will be discussed further in an upcoming section.

PHYLOGENETIC BACKGROUND

Before any potential biogeographic models of platyrrhine origins are presented, the evidence derived from the various morphological and phylogenetic studies of this volume must be considered. Delson and Rosenberger (1980), in the preceding chapter, have summarized and interpreted much of this information in a well-thought-out and reasonable fashion. We will briefly summarize their interpretations here.

Nearly all of the morphological and biochemical evidence presented in this volume supports the concept of a monophyletic origin of the Anthropoidea. There is also a considerable body of evidence favoring both catarrhine and platyrrhine monophyly. This relationship is depicted in the cladogram presented in Fig. 1. Note how platyrrhines and catarrhines are each uniquely derived in their own right from a precatarrhine-preplatyrrhine anthropoid stock. The existence of this basal anthropoid group can be reconstructed from the large number of synapomorphies uniting Platyrrhini and Catarrhini in the Anthropoidea. Figure 1 illustrates the existence of at least 12 derived features that can be used to characterize the Anthropoidea as a whole, whereas platyr-

rhines and catarrhines are each distinguished by 5 and 6 derived features, respectively. All of these derived characters are presented in Table I of Delson and Rosenberger (1980). They have been assembled from a careful analysis of the contributions [1] in this volume (see Ciochon and Chiarelli, Eds., 1980) together with the addition of several outside sources. They are also supported by the biochemical studies of Baba et al. (1980) and Sarich and Cronin (1980). This list is by no means complete; as Delson and Rosenberger (1980) state, it is only a first effort. Yet, we feel that their analysis, even in its preliminary form, points out some very interesting trends.

The anthropoid synapomorphies depicted in Fig. 1 are twice as numerous as and are more significant in terms of level of organization (structural-functional transformations) than the synapomorphies characterizing either the Platyrrhini or the Catarrhini. We feel that these basic relationships will remain the same even as new characters are discovered and added to the list. Furthermore, we interpret this arrangement to mean that the transition from lower primates to higher primates was a fundamentally more significant evolutionary event than the origin of either of the two major divisions of the Anthropoidea. Following this rationale and its related graphic depiction in Fig. 1, we are compelled to view the evolutionary origin and early ancestry of the platyrrhines and catarrhines as very closely linked. This relationship has interesting implications for the biogeographic origin of the New World monkeys which will be discussed later.

From arguments presented above and throughout this volume, we may now infer that the paleontological origins of the New World monkeys will most likely be traced to a preplatyrrhine-precatarrhine early anthropoid stock. One of the more prominent issues in primate evolutionary biology today concerns the origin of this early anthropoid stock from the lower primates. A number of contributors to this volume support the view that the Anthropoidea are most closely related to the tarsiiform primates and can be grouped together in the monophyletic taxon Haplorhini [see Delson and Rosenberger (1980) for arguments and summary]. This would imply an origin of the anthropoids from an early tarsiiform group such as the Omomyidae (Rosenberger and Szalay, 1980). A contrary view is expressed by Gingerich (1980a), who suggests derivation of the anthropoids from early lemuriform primates such as the Adapidae (see also Gingerich, 1973a, 1975c, 1977c). Kay (1980), in his review of the dental evidence for platyrrhine origins, suggests that the arguments favoring an omomyid derivation vs. an adapid derivation are not compelling in either case.

We conclude that the balance of evidence (especially the soft anatomy) presented by the contributors to this volume and reviewed by Delson and Rosenberger (1980) does favor the omomyid or tarsiiform hypothesis of anthropoid origins. Figure 1 depicts this relationship as we view it. We do not choose to argue as strongly for this position as Delson and Rosenberger (1980) have done. Rather, we suggest that the apparent discrepancies between the two opposing viewpoints (adapid origins vs. omomyid origins) may be due as much to the differing methods of analysis employed, such as the stratophenetic approach (Gingerich, 1979c) vs. the cladistic approach (Szalay and Delson, 1979), as to the actually somewhat fragmentary fossilized data base.

Resolving the issue of platyrrhine, catarrhine, and anthropoid origins would naturally be greatly aided by the recovery of a more complete fossil primate record. Unfortunately, this fossil record, as it now stands, does little more than sketch the paleohistory of these groups in the broadest possible terms. The first record of a platyrrhine primate in South America occurs in the early Oligocene deposits of Salla-Luribay, Bolivia (Hoffstetter, 1969, 1974). Although rather extensive sediments of Paleocene and

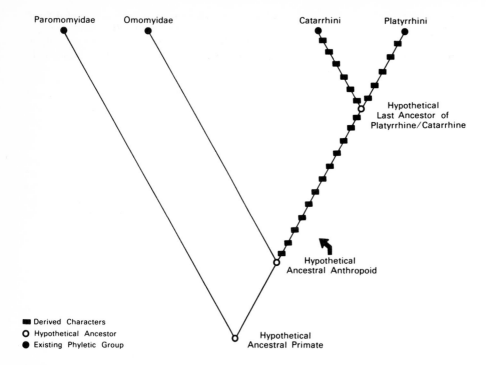

Figure 1 *Abbreviated cladogram of the primates showing the relationship of platyrrhine and catarrhine primates and the existence of a preplatyrrhine-precatarrhine ancestral anthropoid stock. Derived characters depicted in this cladogram have been compiled by Delson and Rosenberger (1980) from the various analyses presented in this volume (see Ciochon and Chiarelli, Eds., 1980) and their own respective research efforts (see their Table 1). The establishment of the derived anthropoid, platyrrhine, and catarrhine characters and their graphic depiction in this cladogram are put forth as a preliminary first attempt at understanding the relationships of the various higher-primate groups. No derived characters have been graphically presented for the sister group of the Anthropoidea, the Omomyidae, or for the Paromomyidae, though their relationship in this cladogram is confirmed by many chapters in this volume and by additional sources (for example, see Szalay and Delson, 1979). It should be pointed out that Gingerich (1980a and elsewhere) has cogently argued for considering the Eocene Adapidae the sister group of the Anthropoidea. This viewpoint is not represented in this cladogram for reasons discussed in the text.*

Eocene age are known from South America, no earlier record of primates has yet been documented, which has been interpreted by some to mean that none was in fact present. In Africa, the first record of a catarrhine primate likewise occurs in the early Oligocene deposits of the Fayum region of Egypt (Simons, 1962a, 1971). Unlike South America, the African Paleocene and Eocene fossil record is very poorly known. At present, the only documented recovery of a somewhat questionable primate (certainly a nonanthropoid) comes from the Eocene deposits of Algeria (Sudre, 1975). In North America and Eurasia, primates abound in the Paleocene and Eocene, yet all are thought to represent lower (nonanthropoid) primates with the probable exception of two late Eocene taxa from Southeastern Asia. These two primates from Burma, *Amphipithecus* and *Pondaungia,* have for a long time been considered potential early anthropoids (Colbert, 1937; Pilgrim, 1927). Recent discoveries of new, more complete specimens have reinforced that opinion (Ba Maw *et al.,* 1979; Savage and Ciochon, 1985). Neither *Amphipithecus* nor *Pondaungia* can be considered a potential direct ancestor of any known platyrrhine or catarrhine (in fact, *Pondaungia* may itself *be* an aberrant catarrhine). Yet, both of these taxa can be regarded as the only known occurrence of Eocene anthropoids. This places the earliest record of the higher primates in Asia. It may also be an indication that even earlier basal precatarrhine-preplatyrrhine anthropoids were evolving on the continent of Asia in the latter part of the Eocene.

DEPLOYMENT SCENARIOS OF THE PLATYRRHINI

There are but two basic geographic sources [2] from which to derive the immediate ancestors of the Platyrrhini: (1) the North American continent (see Fig. 2) and (2) the African continent (see Fig. 4). There are also a number of peripheral or intermediate geographical regions which have been suggested as possible centers of evolution for the ancestral New World monkeys, but in each case, the pathway of eventual dispersal would certainly have passed through either North America (including what little was present of Central or Middle America) or Africa. Based on the presence of the first platyrrhine in South America in the earliest Oligocene, we surmise that this dispersal probably occurred sometime during the Eocene and therefore involved the crossing of some sort of an oceanic barrier (since South America remained an island continent until the Pliocene). Apart from these basic points of agreement, divided opinions exist over exactly how the first anthropoids might have reached the South American continent. Contributors to this volume have supported source areas of origin in North or Middle America (Orlosky, 1980; Perkins and Meyer, 1980; Rosenberger and Szalay, 1980; Gantt, 1980; Wood, 1980; McKenna, 1980), Asia via the Bering Bridge to North America (Gingerich, 1980a; Delson and Rosenberger, 1980), and directly from Africa (Tarling, 1980; Lavocat, 1980; Hoffstetter, 1980; Maier, 1980; Bugge, 1980; Luckett, 1980; Martin and Gould, 1980; Chiarelli, 1980; Sarich and Cronin, 1980).

When considering the various platyrrhine sources of origin and dispersal scenarios outlined in the preceding chapters, it is important to distinguish exactly how each author has defined what constitutes an ancestral platyrrhine. The chapter by Delson and Rosenberger (1980) has done much to clarify this in a summary fashion. With consensus on the demonstration of anthropoid monophyly, we feel that platyrrhine origins and catarrhine origins should no longer be considered as two isolated independent events. Any potential source area of one group should probably be tied to the source area of the other. With these points in mind, we will now review the various paleobiogeographic models of platyrrhine origins.

North-American-Origin Model

The traditional geographic source of origin for the ancestors of the New World monkeys has been the North American continent.

The dispersal scenario usually associated with this model involves island-hopping by waif dispersal [3] across the Caribbean Sea to South America (Fig. 2). Simpson (1945) was the first to formally advocate this model, though in concept it existed a good deal earlier (see Matthew, 1915; Gregory, 1920c).

Most advocates of the North-American-origin model suggest that the Eocene Omomyidae, a tarsiiform primate family, Holarctic in distribution, comprise the basal stock of both the Platyrrhini and the Catarrhini (McKenna, 1967; Szalay, 1975b; Orlosky and Swindler, 1975; Simons, 1976b; Szalay and Delson, 1979). In the early to middle Eocene, continuity between the North American and European populations of omomyids, which would eventually become ancestral platyrrhines and catarrhines, was maintained by gene flow across the North Atlantic Bridge. Some time before the middle Eocene, this bridge was submerged, probably due to a combination of crustal rifting and marine

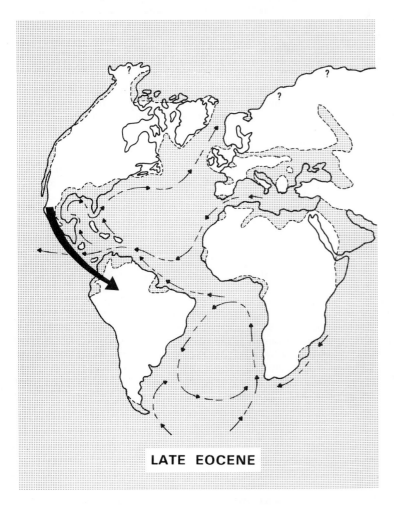

LATE EOCENE

Figure 2   *Model illustrating derivation of the Platyrrhini from the North American continent in the late Eocene. Stippled areas represent oceans or land surfaces transgressed by epicontinental seas. The heavy arrow indicates the direction of potential dispersal of the ancestral platyrrhines, not necessarily the precise pathway for this dispersal, which could have occurred via any of the reconstructed island arc chains in the Caribbean. Note the distance between the southernmost extent of land on the North American continent and the northernmost extent of land on the South American continent. Also note the direction of oceanic circulation patterns (small arrows) in and around the Caribbean. Configuration of continents in this reconstruction is after Sclater et al., (1977); position of epicontinental seas based on Bond (1978) and references therein; surface oceanic circulation patterns derived from Berggren and Hollister (1974); island arcs in the Caribbean reconstructed from Tomblin (1975) and Malfait and Dinkelman (1972). To focus primary emphasis on the North-American-derivation model, note that none of the potential island chains in the South Atlantic Ocean which subsequently appear in Fig. 4 has been included in this reconstruction.*

transgressions (see Pitman and Talwani, 1972; Fitch et al., 1974; McKenna, 1975). The dissolution of this transcontinental land bridge would have at first severely impaired and later brought a cessation of gene flow between the North American and European Omomyidae. The possibility that gene flow was also maintained via the Bering Bridge both during and after the closure of the North Atlantic Bridge exists, but is complicated by the imposing barrier of the Turgai Strait (also termed the Ouralian Sea), which bisected the Eurasian land mass throughout much of the Eocene.

As Simons (1976b) has argued, if one accepts on morphological grounds that the Omomyidae (perhaps in the form of a *Teilhardina* species) represent the ancestral populations from which the platyrrhines and catarrhines were independently derived, then on geological grounds, the submergence of the North Atlantic Bridge marks the point of the platyrrhine-catarrhine dichotomy. McKenna (1972b, 1975) has pinpointed this interruption in the North Atlantic Bridge as occurring late in the early Eocene or about 50 million years B.P. All subsequent evolution of the ancestral Platyrrhini in North America and their eventual dispersal into South America would therefore have occurred as an independent parallel event to the evolution of the Catarrhini in Europe and their separate dispersal into Africa.

The North-American-origin model of the Platyrrhini is grounded in a long tradition of North American paleontology dating back at least to Matthew (1915). Proponents of this traditional view attempt to derive all the major South American faunal elements from North American ancestors and disperse them southward in a series of successive "invasions." In concept, this fact alone constitutes no reasonable basis for rejection; yet, the fact that this model was conceived prior to the acceptance of continental drift does present some interesting conceptual difficulties.

It is now considered virtually certain that during the Eocene, the continent of South America occupied a position considerably more distant from North and Middle America than it does today. In fact, the global position of South America in the Eocene has been demonstrated to lie equidistant between the continents of North America and Africa (Sclater et al., 1977; Tarling, 1980; Lavocat, 1977) (see also Figs. 2 and 4). Furthermore, Ladd (1976) presents evidence that during the Eocene, South America may have temporarily reversed its direction of drift and actually moved away from North America for a period of time. Therefore, any colonization of South America from North or Middle America during the Eocene would have been over a considerable oceanic distance. The existence of volcanic island arcs in the Eocene Caribbean is documented (see Tomblin, 1975; Malfait and Dinkelman, 1972), which makes a variety of island stepping-stone routes possible. Yet, the exact location of these routes and the distance between islands in the reconstructed chains remain quite conjectural (for example, compare Lloyd, 1963; Malfait and Dinkelman, 1972; Weyl, 1974; Tomblin, 1975). The presence of oceanic circulation patterns in the Eocene which flowed both east to west *across* these potential Caribbean dispersal routes and south to north *against* the routes also seems well substantiated (see Figs. 2 and 4—based on Berggren and Hollister, 1974) (also see Frakes and Kemp, 1973; Holcombe and Moore, 1977; discussion by Lavocat, 1980). These patterns of paleocurrent flow in the Caribbean would argue strongly against rafting as a means of dispersal between preexisting volcanic island arcs. Since no direct land connection between Middle America and South America is known to have existed until the Pliocene, we are left with a rather improbable situation. Nevertheless, we feel that McKenna (1980) has very expertly argued that in spite of these paleogeographic

factors, some interchange between North America and South America did occur prior to the development of the Pliocene Panamanian Bridge. Whether or not the ancestral platyrrhines were among this faunal interchange remains to be demonstrated.

The North-American-origin model as we see it attempts to derive the Platyrrhini from middle Eocene North American omomyids (perhaps *Teilhardina*) which do not share, or are not known to share, the derived dental and skeletal features characteristic of the ancestral anthropoid stock as defined by contributors to this volume (see Fig. 1). The absence of a potential platyrrhine ancestor in North America coupled with the considerable oceanic distance separating Middle America and South America and the unfavorable prevailing oceanic circulation patterns, in our opinion, very nearly falsify this model of platyrrhine origins.

Asian-Origin Model

A variant of the North-American-origin model depicts the ancestral stock of the Platyrrhini developing in Asia and dispersing to North America via the Bering Bridge in the late Eocene and ultimately reaching South America by waif dispersal across the island arcs of the Caribbean (see Fig. 3). The ancestral catarrhines also disperse from a similar source area in Asia via a land corridor in the south and cross the reduced western extent of the Tethys Sea into Africa. The conceptual framework for this model was originally suggested by Gingerich (1977c) and has recently been further refined and substantiated by him (see Gingerich, 1980a). As a deployment scenario, it has also been proposed by Delson and Rosenberger (1980) and Szalay and Delson (1979, p. 519), though aspects of their version differ somewhat from that presented here. It appears to us that the unifying conceptual basis for all these authors' proposals stems from the viewpoint that the ancestries of the platyrrhines and catarrhines are very closely linked. Since it no longer seems tenable to derive these two groups from widely separate and isolated geographical areas, hypotheses such as the Asian-origin model provide a logical and testable alternative.

As previously indicated, Gingerich (1980a) favors derivation of the ancestral anthropoid stock from the Eocene Adapidae. In his view of the Asian-origin model, the presence of transitional adapid-anthropoid primates in the late Eocene of Burma (*Pondaungia* and *Amphipithecus*) and in the early Oligocene of Africa (*Oligopithecus*) provides evidence that the origin of the higher primates from an advanced "protosimian" (preplatyrrhine-precatarrhine) stock occurred in either southern Asia or Africa or perhaps simultaneously in both areas. Deriving the ancestral platyrrhines from a source area in southern Asia and dispersing them to North America is then based on (1) the more primitive and generalized appearance of the transitional adapid-anthropoids, *Amphipithecus* and *Pondaungia;* (2) the apparent open nature of the high-latitude Bering Bridge during a warm interval in the late Eocene; (3) the existence of a marked mammalian faunal similarity between Eurasia and North America in the late Eocene; and (4) the unexpected occurrence of *Mahgarita,* an adapine primate of Eurasian aspect, in the late Eocene of Texas.

Gingerich (1980a) considers the Asia-based model of platyrrhine origins (see his Fig. 4) the most parsimonious hypothesis based on current data. However, as a second only slightly less likely alternative, he suggests that "...it is also possible that the adapid-protosimian stock ancestral to Ceboidea crossed the South Atlantic directly from Africa to South America" (p. 131). Finally, as a third considerably less likely alternative, Gingerich (1980a, p. 135) cautiously states that "...it is possible that higher primates originated in southern North America and subsequently migrated in the late Eocene from North America into South

America and also from North America across the Bering route into south Asia and ultimately Africa." We feel that presentation of alternative models in this manner sets the stage for future hypothesis-testing which will eventually result in the development of a single falsifiable model.

Delson and Rosenberger (1980) take a different tack from Gingerich in their apparent support of the Asian-origin model of the Platyrrhini. They favor derivation of the ancestral Anthropoidea from the Omomyidae, a view which we also support. Instead of specifically stating that the ancestral anthropoids differentiated in southern Asia as we present in Fig. 3, they offer the scenario that these ancestral higher primates actually occupied a single biotic community which spanned the circum-Pacific region across the Bering Bridge all the way from Asia to North America. In the late Eocene, the formative catarrhine branch of this com-

munity then entered Africa from Asia, while the protoplatyrrhine branch dispersed to South America from the north via a Caribbean island-hopping route (Delson and Rosenberger, 1980). In support of their model, Delson and Rosenberger emphasize (1) the significant morphological resemblance between Asian and western North American anaptomorphine omomyids in the Eocene, (2) the potentiality of the Bering Bridge as a mammalian migration route throughout the Eocene, (3) the possibility of a phyletic link between *Pondaungia* and the Fayum catarrhines, and (4) the strength of faunal links between African Fayum mammalian taxa, including the nonanthropoid *Oligopithecus*, with Mediterranean and south Asian mammalian taxa, including *Hoanghonius*.

We suggest that both of these versions of the Asian-origin model as proposed by Delson and Rosenberger (1980) and Ginge-

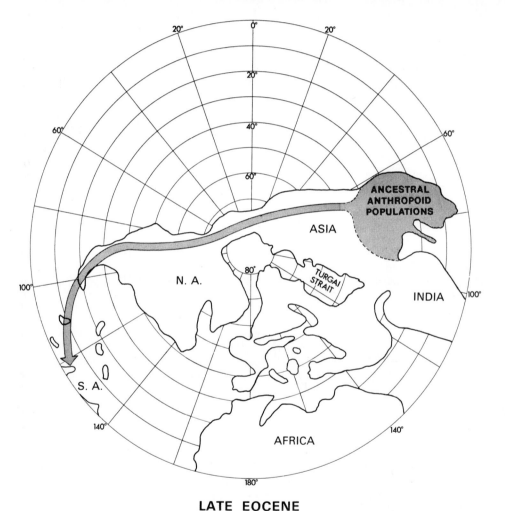

LATE EOCENE

Figure 3   *Model illustrating the derivation of the Platyrrhini from the Asian continent in the late Eocene. Areas below sea level in this polar projection are indicated by the presence of longitude and latitude demarcations. The stippled area represents the approximate range of ancestral anthropoid populations in Asia, and the arrow indicates the potential pathway of dispersal across the Bering Bridge through western North America to South America via an island-hopping route through the Caribbean. Note that the over-water distance from the southernmost extent of North America to the South American continent is still quite large despite the presence of inferred island arcs in the Caribbean. Configuration of continents in this reconstruction is derived from a North polar stereographic projection provided by A. G. Smith and Briden (1977); position of the Indian subcontinent is after Johnson et al. (1976) and Bingham and Klootwijk (1980); location and extent of the Bering Bridge based in part on McKenna (1972b); position of epicontinental seas derived from Bond (1978), Sahni and Kumar (1974), and Vinogradov (1967); island areas in the Caribbean reconstructed from Tomblin (1975) and Malfait and Dinkelman (1972).*

rich (1980a) have considerable merit, since each supports the existence of a preplatyrrhine-precatarrhine stem anthropoid stock (as depicted in Fig. 1) and each version either directly or indirectly supports a post-middle Eocene platyrrhine-catarrhine divergence date. The phylogenetic arguments reviewed in an earlier section, in our opinion, make these two assumptions a requisite basis for any viable paleobiogeographic model of platyrrhine origins. However, both Gingerich and Delson and Rosenberger still must contend with the arguments presented under the North-American-origin model indicating the potential difficulties to be encountered in dispersing the ancestral platyrrhines across the Caribbean via an island sweepstakes route.

African-Origin Model

The African-origin model of the Platyrrhini attempts to derive the ancestral New World monkeys directly from the African conti-

nent via a sweepstakes route across a much less expansive Eocene South Atlantic Ocean (see Fig. 4). Given the long-acknowledged morphological similarity of platyrrhines and catarrhines, it would seem that such a hypothesis, at least in principle, would have been proposed many years ago. However, it was not until the nearly universal acceptance of the continental-drift paradigm during the last decade that this model of platyrrhine origins began to receive serious consideration. Its list of adherents now includes Sarich (1970), Hoffstetter (1972, 1974, 1977b), Cracraft (1974a), Cronin and Sarich (1975), Thorington (1976), Lavocat (1974b, 1977), Hershkovitz (1972, 1977), and Washburn and Moore (1980), among others.

Many supporters of the African-origin model favor derivation of the ancestral anthropoid stock from Eurasian omomyid tarsiiforms in the middle to late Eocene. Soon after this stem anthropoid stock had populated the African continent either as a

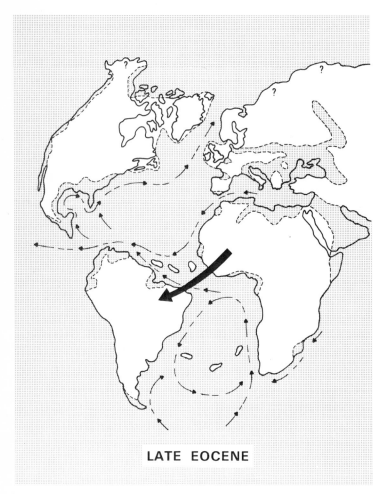

LATE EOCENE

Figure 4 *Model illustrating derivation of the Platyrrhini from the African continent in the late Eocene. Stippled areas represent oceans or land surfaces transgressed by epicontinental seas. As in Fig. 2, the heavy arrow indicates the direction of potential dispersal of the ancestral platyrrhines, not necessarily the exact pathway for this dispersal, which was evidently via island-hopping and most probably across the more northern portion of the South Atlantic. Compare the distance between the closest adjacent points of Africa and South America with the oceanic distance separating South America and North America. Also compare the direction of oceanic circulation patterns (small arrows) in the northern South Atlantic with the patterns of circulation in the Caribbean. Configuration of the continents in this reconstruction is after Sclater et al. (1977); position of epicontinental seas based on Bond (1978) and references therein; surface oceanic circulation patterns derived from Berggren and Hollister (1974); position of oceanic islands in the South Atlantic based on Tarling (1980), Sibuet and Mascle (1978), and Van Andel et al. (1977). As was the case with Fig. 2, to focus primary emphasis on the African-derivation model, note that none of the potential island arcs in the Caribbean which appeared in the former figure has been included in this reconstruction.*

precatarrhine-preplatyrrhine early anthropoid or possibly as a protoanthropoid, it dispersed to South America through rafting and island-hopping across the South Atlantic in the late Eocene or earliest Oligocene. Since no Paleocene or Eocene fossil record of any consequence exists over the entire expanse of the African continent, there is no way to substantiate this scenario. The early to middle Oligocene deposits of the Fayum region of Egypt have produced a rich primate fauna, yet these all appear to be bona fide catarrhines. Nevertheless, Hoffstetter (1980 and elsewhere) has attempted to derive the ancestral platyrrhines from one group of the Fayum primates, the Parapithecidae. As Hoffstetter, among others, has argued, the parapithecids may indeed represent the most primitive (least derived) members of the Catarrhini, but this alone does not make them appropriate ancestors for the earliest Platyrrhini (see arguments by Delson and Rosenberger, 1980; Kay, 1980; Szalay and Delson, 1979). If the existence of a precatarrhine-preplatyrrhine anthropoid as depicted in Fig. 1 is proven correct, then an African-based branch of this stock would be a more appropriate structural ancestor for the Platyrrhini. Without the aid of a Paleocene or Eocene primate fossil record in Africa, there is, of course, no way to test this hypothesis. It could just as easily be argued (and indeed has been) that anthropoids originated in Africa from unknown Eocene tarsiiforms or lemuriforms and dispersed to South America via the South Atlantic route to become ancestors of the New World monkeys and to south Asia via a crossing of the Tethys to become ancestors of the presumed anthropoids of Burma's Pondaung Hills.

Before being completely carried away with proposals for hypothetical platyrrhine-deployment scenarios and for equally hypothetical platyrrhine ancestors, we feel that it is necessary to mention one potential source of information that is *not* highly disputed and which *does* appear to lend credence to the African-origin model. Current evidence now indicates that the Platyrrhini and Catarrhini represent two equal subdivisions of the Anthropoidea that arose from a common early anthropoid ancestral stock (see Fig. 1). Therefore, it is proper, and indeed essential, to consider the Platyrrhini as the sister group of the Catarrhini and vice versa. The current predominant distribution of these two groups in the continents of South America and Africa, respectively, targets these regions as their potential areas of differentiation. This disjunct distribution and sister-group relationship takes on added significance when it is recalled that another group of mammals, the South American caviomorph rodents and African phiomorph rodents, exhibit a nearly identical distribution and phyletic relationship (Lavocat, 1980). When these similarities are further explored utilizing the method of Hennig (1966b) for determining the biogeographic origin of a group called the "search for the sister group" and the concept of Croizat *et al.* (1974) of "generalized patterns of biotic distribution" or "generalized tracks," a pattern emerges which supports a wholly southern-continents origin of platyrrhines, catarrhines, caviomorphs, and phiomorphs irrespective of hypothetical deployment scenarios or of hypothetical ancestors.

Historically, the main obstacle that has been used to argue against derivation of the platyrrhine primates and caviomorph rodents from a source in Africa is the reported great size of the marine barrier separating that continent from South America in the early Tertiary. Actually, the exact size of the equatorial South Atlantic during the later Eocene-earliest Oligocene (the most probable time a crossing would have taken place) is still a matter of some debate. There does appear to be a consensus that the final separation of the contiguous subaerial land masses of South America and Africa occurred in the early late Cretaceous (Larson and Ladd, 1973; Sclater *et al.*, 1977; Sibuet and Mascle, 1978; Tarling, 1980). Initially, this separation appears to have been in

the form of a series of continuous rift valleys which were subjected to periodic marine incursions. Similar marine incursions are known to have flooded the subsiding continental margins of South America and Africa as early as Albian times (ca. 105 million years B.P.), though these episodes did not represent the final rifted separation (Förster, 1978). The final contact of land between the two continents, probably in the form of an uplifted land bridge between Brazil and West Africa, occurred in the Turonian (ca. 92 million years B.P.) (Kennedy and Cooper, 1975; Reyment *et al.,* 1976). After that point, progressive development of the intracontinental rift coupled with a worldwide rise in sea level firmly established a pattern of continuous circulation between the Central and South Atlantic.

Because the rate of sea-floor spreading in the South Atlantic has not been constant since the final separation of South America and Africa (Herz, 1977; Sibuet and Mascle, 1978), precise estimates of the size of the marine barrier separating the closest approach of the two continents in the late Eocene-earliest Oligocene vary somewhat. Based on calculations from scaled paleocontinental maps, Funnell and Smith (1968) estimate the oceanic barrier at 1000 km, A. G. Smith and Briden (1977) at 1450 km, Sclater *et al.* (1977) at 1600 km, Tarling (1980) at 1650 km, and Ladd *et al.* (1973) at 1750 km. For comparison, the current size of the marine barrier separating South America and Africa is approximately 3200 km. The sheer size of this marine barrier even in the late Eocene would have acted as a nearly impenetrable filter if it had not been for the presence of a series of oceanic islands which very possibly acted as stepping stones.

The existence of volcanic islands in the early Tertiary South Atlantic is based on the cyclic occurrence of tectonically active processes associated with sea-floor spreading in the region. McKenna (1980, p. 53) has suggested that island chains in the equatorial and southern South Atlantic were very possibly brought about by (1) excessive production of Mid-Atlantic Ridge lavas, (2) temporarily elevated sections of fracture zones, and (3) off-ridge volcanic activity. The geographical distribution and petrographic-geochemical nature of existing islands in the South Atlantic Ocean (see Baker, 1973) bear out McKenna's proposals. The exact location of these potentially subaerial island chains would have been along the Ceará and Sierra Leone Rises in the equatorial South Atlantic and along the Rio Grande-Walvis Ridge further to the south (see Fig. 4). The structure and geologic history of the Ceará and Sierra Leone Rises is detailed by Kumar and Embley (1977) and Sibuet and Mascle (1978). Evidence is provided which indicates that both these features of the newly formed equatorial South Atlantic were volcanic in origin and potentially above sea level sporadically since their formation about 80 million years B.P. Tarling (1980, p. 34) suggests that both the Ceará Rise and the Sierra Leone Rise were "...strong positive features in the early Tertiary." He further states that "it seems probable, therefore that oceanic islands, some hundreds of square kilometers in size existed offshore from Brazil and West Africa, and that the mid-oceanic rise was also subaerial for much of this period" (p. 34). Distances between islands in this oceanic chain may have been as little as 200 km (Tarling, 1980).

In the southern portion of the South Atlantic, oceanic island chains potentially existed along the Rio Grande Rise and Walvis Ridge spanning the distance between present-day southern Brazil and South-West Africa. Studies of the depositional history and geochemistry of this region (Van Andel *et al.,* 1977; P. J. Smith, 1977) firmly establish the barrier nature and volcanic origin of these features and suggest the occurrence of subaerial volcanic islands or seamounts. The formation of the Walvis-Rio Grande Rises was in the early Cretaceous (Sclater *et al.,* 1977), which suggests that they predate the equatorial Ceará-Sierra Leone

Rises by at least 40 million years. This factor, among others, led Tarling (1980, p. 33) to state that "the Walvis-Rio Grande Rise...is likely to have been a much more rugged, intermittent migration route with the gradual decrease in its elevation leading to greater difficulty in 'island-hopping,' but the occasional volcanic eruption creating temporary new land and providing improved connections for short periods." McKenna (1980) offers the analogy that a portion of these rises might be considered a southern version of Iceland. The subsidence of the Rio Grande Rise-Walvis Ridge barrier in the earliest Oligocene established deepwater circulation in the southern South Atlantic for the first time (Van Andel *et al.*, 1977) and brought about the end of any potential island-hopping route in this region.

Even with the probable existence of volcanic island chains in both the Eocene South Atlantic and Caribbean, the colonizing ancestors of the Platyrrhini would have still been forced to make some sort of oceanic crossing. Arguments for and against the concept of an oceanic crossing have been particularly heated with regard to the African-origin model (compare Hoffstetter, 1974; Simons, 1976b). The obvious mechanism for the crossing of a substantial marine barrier would be by means of rafting. Darlington (1957, p. 15) quite simply characterizes rafting as the "... dispersal of land animals across water on floating objects." Without question, rafting as a mode of dispersal has occurred in the past. The existence of land mammals on islands separated by large marine barriers from all other land masses can often be explained by no other means. Several examples from the literature also document the existence of long-distance rafting as a mode of dispersal. Ridley (1930) cites the occurrence of a raft consisting of driftwood and living clumps of lalang grass and bamboos which was apparently ridden by a crocodile from the coast of Java to the Cocos Islands, some 1100 km in distance. Powers (1911, pp. 304–305) provides an extraordinary example of a floating island seen in 1892 off the coast of North America hundreds of miles out to sea in the Atlantic:

> When first noticed in July in latitude 39.5° N., longitude 65° W. (about 400 km west of Cape Cod) the island was about 9,000 square feet in extent, with trees thirty feet in height upon it, which made it visible for seven miles. It had apparently become detached from the coast of this country and been carried out to sea by the Gulf Stream. It was again seen in September in latitude 45° 29 N., longitude 42° 39 W., (about 600 km west of Newfoundland) after it had passed through a severe storm. By this time it must have traveled over 1,000 nautical miles (= 1850 km approx.), and it may have eventually arrived at the coast of Europe.

Such examples of floating islands are rare, but given the probability of chance dispersal over long periods of geologic time (see Simpson, 1952), such unlikely events as the rafting of small primates across marine gaps separating oceanic islands become possible at least in theory. It should be noted here that physiological and behavioral objections have been presented to primate rafting over long distances, [4] yet these would appear to apply equally to the Caribbean dispersal routes from North America and to the equatorial South Atlantic dispersal route from Africa.

Rafts like those described by Ridley (1930) and Powers (1911) do not necessarily require the action of great flooded rivers undercutting their banks during storm conditions to start them on their way. Rather, they can be formed and floated into the ocean anywhere that marshes or shallow lakes drain seaward (Darlington, 1957). Paleogeographic data indicate that such conditions almost certainly prevailed along the coast of West Africa in the late Eocene-earliest Oligocene. Potential rafts therefore may very

possibly have existed. Once the seaward drift of a raft had begun, the paleocurrents which flowed from along the coast of West Africa in a westerly direction across the South Atlantic (Berggren and Hollister, 1974) (see also Fig. 4) could have carried the raft or floating island to one of the equatorial volcanic islands or possibly all the way to the coast of South America. If this raft or floating island had vertically standing trees on it like the one described by Powers (1911), these could have acted as a sail, since prevailing surface winds in the Eocene and Oligocene blew in a westerly direction off the African continent (Frakes and Kemp, 1973). Even today, at a distance more than twice the size of the Eocene South Atlantic, floating drift from the Niger and Congo Rivers carried by the South Equatorial Current across the South Atlantic has been reported thrown up on the coast of Brazil (Darlington, 1957). Scheltema (1971) has shown from studies of pelagic marine larvae that such a passive crossing of the present-day South Atlantic can take as little as 60 days. Perhaps some time in the late Eocene-earliest Oligocene, such a waif-dispersal scenario via a rafting and island-hopping route across the South Atlantic was responsible for the introduction of the ancestral New World monkeys into South America.

In summary, we feel that the African-origin model of the Platyrrhini as presented here has much to offer, since it not only satisfies preexisting phylogenetic requirements of deriving both the Platyrrhini and the Catarrhini from a common geographical source but also provides a deployment scenario for dispersal of the Platyrrhini to South America over a potentially viable and ultimately verifiable oceanic sweepstakes route.

Vicariance-Origin Model

An alternate form of the African-origin model has been proposed by Hershkovitz (1972, 1977) as a possible alternative scenario for the origin of the Platyrrhini and Catarrhini. This model, depicted in Fig. 5, is described here by Hershkovitz (1972, p. 323):

> Prosimians arose in the rifted South American-African continents, possibly during the Cretaceous, and spread across both continents during early drift stages. Platyrrhines and catarrhines then evolved independently in isolated South America and Africa, respectively, perhaps during the early Tertiary from closely related, possibly congeneric prosimian stocks. This hypothesis accounts for basic similarities without the need for initial convergence.

Hershkovitz (1977, p. 67) later described an extension of the aforedescribed model:

> This hypothesis assumes that the unknown haplorhine forerunner of platyrrhines and catarrhines had already evolved to simian grade in the rifted but not yet widely drifted South America-Africa. The timing of the geological events may be questioned, but nothing in the fossil record denies the sequence of events outlined here.

The essence of the Hershkovitz model is his belief that platyrrhines and catarrhines appear so fundamentally similar that they must have had a common center of origin and were most probably derived from a pre-simian (tarsioid, haplorhine) stock (see Hershkovitz, 1977). Given the present-day disjunct distribution of platyrrhines and catarrhines in the continents of South America and Africa, respectively, Hershkovitz reasoned that their obvious area of origin would be in the combined Afro-South American continents. Unfortunately, as Hershkovitz (1977) himself admits, this Afro-South American-based derivation model of the Platyrrhini-Catarrhini does not appear to fit the current

LATE CRETACEOUS

Figure 5   *Model adapted from the Hershkovitz (1977) illustrating the origin of the Platyrrhini and Catarrhini in the rifted but not yet widely drifted southern continents from a currently unknown lower primate stock in the late Cretaceous. This model stresses the independent evolution of platyrrhines and catarrhines in South America and Africa, respectively, from a "pre-simian" (tarsioid, haplorhine) primate that had a pan-Afro-South American distribution toward the end of the Cretaceous. Since no actual dispersal of the New World monkeys into South America is implied by this model and since tectonic change in the form of an active Mid-Atlantic Rift can be viewed as the independent variable determining the origin of the Platyrrhini, this model appears to fit the vicariance paradigm. Indeed, proponents of vicariance biogeography, such as Nelson (1974) and Croizat (1971, 1979), have used arguments very similar to these proposed by Hershkovitz (1972, 1977) to explain the disjunct distribution of certain South American and African faunal elements. For further discussion, see the text. Configuration of the continents in this reconstruction is based on Sclater* et al. *(1977) and Tarling (1980); note that no oceans or epicontinental seas are indicated.*

paleocontinental configurations as proposed by Ladd *et al.* (1973) or Sclater *et al.* (1977), nor is there any direct fossil evidence to support it.

An inherent implication of the Hershkovitz model is that the tectonically active process of intracontinental rifting brought about the isolation of ancestral platyrrhines in South America and ancestral catarrhines in Africa. We therefore feel that his proposals can be interpreted as support for a vicariance model of platyrrhine origins. Though Hershkovitz has made no statement to this effect, we have nonetheless taken the liberty of including his model under the present subheading. In support of this, we would like to quote one final statement by Hershkovitz which, in

our view, bridges the gap between his innovative proposals of early 1970's and those currently made by Rosen (1978), a proponent of the vicariance paradigm. Referring to the hypotheses of platyrrhine-catarrhine origins presented above, Hershkovitz (1972, p. 324) states that they are "...based on an idealized construction of evolutionary and geological sequences, (which) may be inconsonant with the supposed chronology of either primate evolution [= cladistic sequences] or of continental drift [= vicariant events], but not both." Rosen (1978) would certainly argue that they should be consonant with both! (Phrases appearing in brackets in the quote were added by us for emphasis.)

Utilizing arguments presented by Croizat (1971, 1979) and Nelson (1974), a vicariance model of platyrrhine origins can be constructed. Following the depiction by Nelson (1974, p. 556, Fig. 1) of a vicariant event fostering the development of the current disjunct distribution of a hypothetical South American-African taxon, the following scenario can be visualized: The ancestral platyrrhine-catarrhine species, either an early anthropoid or a lower primate (Species A), exhibited a pan-South American-African distribution at some point in the past before the breakup of these continents. The descendant species of Platyrrhini (Species A1) and Catarrhini (Species A2) appeared when a marine barrier developed between South America and Africa (the result of seafloor spreading and continental drift), causing the splitting (vicariance) and allopatric speciation of the ancestral species. This vicariance scenario would fit the distributional and limited paleontological evidence of platyrrhines and catarrhines. Unfortunately, as Nelson (1974), Croizat *et al.* (1974), and Rosen (1978) have argued, the timing of the development of the barrier is the causal geographic factor in any vicariance model of speciation. Geological, geophysical, and invertebrate paleontological data have already been presented (for example, see Sclater *et al.*, 1977; Tarling, 1980; Sibuet and Mascle, 1978; Kennedy and Cooper, 1975) indicating that the final separation (development of the marine barrier) between South America and Africa occurred in the Turonian, at least 90 million years B.P. Since the oldest known primate, *Purgatorius,* thought to be very near the basal radiation of that order, does not appear in the fossil record until the latest Cretaceous of North America, at most 70 million years B.P. (Van Valen and Sloan, 1965; Clemens, 1974), it is difficult to argue for the vicariance origin of platyrrhines and catarrhines in the southern continents *much* earlier in time. It is indeed possible that lower primates older than *Purgatorius* may one day be found in the late Cretaceous of South America or Africa or both, which would then lend support to the vicariance model. However, as was presented in Fig. 1, results of a cladistic analysis performed by Delson and Rosenberger (1980) indicate that the Platyrrhini and Catarrhini shared some period of common ancestry as early anthropoids before their divergence. Use of this phylogenetic evidence as support for the vicariance model would mean that anthropoids (= higher primates) had developed in the prerifted South American-African continent some 20-25 million years prior to the earliest known occurrence of a lower primate anywhere else in the world. At present, this does not seem a likely possibility.

We conclude that evidence derived from the primate paleontological record, the cladistic relationships of platyrrhines and catarrhines, and the timing of intracontinental rifting of South America from Africa tends to falsify the vicariance model of platyrrhine origins. As we present in the following section, we do support the view that the origin of the platyrrhines and catarrhines was very likely a southern-continents event. However, waif dispersal from Africa in the late Eocene, not vicariance at a much earlier date, is the most probable mechanism for explaining the arrival of the first platyrrhines in South America. As Ferris (1980,

p. 72) suggests. "...vicariance biogeographers do not deny the possibility of dispersal over barriers, but they consider it a random event." In essence, then, we are discussing the reality of oceanic dispersal vs. the reality of vicariance of the pan-Afro-South American biota. In a similar discussion concerning the congruence between biological and geological relationships, Rosen (1978, p. 186) concludes that "a decision concerning the nature of this correspondence, whether by dispersal or vicariance or some combination of the two, must be a parsimony decision concerned with minimizing the number of separate assumptions entailed by the different types of explanations." Therefore, we suggest that a vicariance origin of the Platyrrhini based on the megavicariant event of an intracontinental rifting is not the most parsimonious decision at present. We would nevertheless support the view that microvicariant events, such as the transgression-regression of the Tethys Sea and Turgai Strait and the opening-closing of the Bering and North Atlantic Bridges, rather than dispersal *per se*, had a major impact on other phases of Paleogene primate evolution.

MAXIMUM-PARSIMONY MODEL OF PLATYRRHINE ORIGINS

The principle of parsimony dictates that a particular model is preferable to all others provided that the known data do not call for its rejection. Furthermore, following the hypotheticodeductive method as conceived by Popper (1959, 1963) and recently analyzed by Kitts (1977), it is now possible to formalize the process of verification of a parsimony model. To this end, we have selected a recently proposed technique for historical biogeographic analysis which meets these and other criteria, and we will apply it here to the development of a maximum-parsimony model of platyrrhine origins and dispersal.

Morse and White (1979, p. 357) have presented the following methodological outline for historical biogeographic analysis in comparative biology:

(1) Infer sequential hypothetical ancestors of the taxa (ultimately species) under study according to the "ex-group comparison" method of Ross (1974) or the "sister group" method of Hennig (1965, 1966) and Brundin (1966).

(2) Note the geographical distribution of each non-hypothetical taxon studied.

(3) Infer the distribution of each hypothetical ancestor, beginning with the most recent ancestors and working backward in time, according to the principles of phylogenetic compatibility....

(4) Having completed this hypothesis forming phase through the oldest ancestor of the group, trace the implied vicariances (instances of allopatric speciation) and dispersals in a narrative summary....

(5) Test the hypotheses as appropriate. If any are falsified infer alternate ones by repeating the necessary steps.

In essence, this technique of Morse and White has already been applied in various ways throughout the preceding reviews of the paleobiogeographic models of platyrrhine origins. What we plan to present here is a shortened but more formalized version of the model which our studies indicate is the most parsimonious one.

Cladistic Biogeography

Following the conclusions reached by Delson and Rosenberger (1980) and supported by many contributors to this volume, the Anthropoidea can be viewed most parsimoniously as a monophyletic group with the Platyrrhini and Catarrhini as two equal subdivisions each derived with respect to one another and to their hypothetical early anthropoid ancestor (see Fig. 1). Platyrrhines

and catarrhines can therefore be considered sister groups of one another. Hennig (1966b) has adopted a cladistic-based method for deciphering the biogeographic origin of a group which he has termed the "search for the sister group" (see also Ashlock, 1974). He suggests that the first critical question to be posed concerning the biogeographic origin of any group is: what is the group's closest sister group and where does that sister group occur? If this question were asked concerning the Platyrrhini, the answer would be the Catarrhini and Africa. Naturally, the answers would be the Platyrrhini and South America if a similar question were posed concerning the Catarrhini. As we see it, the central point to be grasped from this discussion concerns the application of a strictly cladistic approach to a problem in historical biogeography. None of Hennig's other biogeographic methods such as the Progression Rule, the Phylogenetic Intermediate Rule, or the Multiple Sister-Group Rule (for an excellent summary, see Ashlock, 1974) can be applied to the wholly disjunct distribution of platyrrhines and catarrhines. Furthermore, application of the new proposals of Platnick and Nelson (1978) for historical biogeographic analysis does little to resolve the issue of the biogeographic origin of the Platyrrhini.[5] In our opinion, the fact that the platyrrhines inhabit South America distinguishes them in principle from the Catarrhini of Africa in a fashion similar to that of a series of derived morphological features. Thus, the biogeographic distribution of taxa in itself can be an important source of information.

In our construction of a maximum-parsimony model of platyrrhine origins, we will employ the cladistic biogeographic approach (see Morse and White, 1979). As previously discussed, this method is based on the prior development and utilization of a testable model of phylogenetic relationship. As Brundin (1972, p. 74) has stated, "... a careful establishment of strict monophyly and sister-group relationship is a necessary prerequisite for a realistic interpretation of a distribution pattern." The cladogram we presented in Fig. 1 fits the phylogenetic requirements proposed by Brundin (1972) and Morse and White (1979). To place this cladogram in a geographic perspective, we have applied it to a paleogeographic model of the world in the late Eocene (see Fig. 6). Ross (1974) has discussed this method for inferring distribution and direction of dispersal from phylogenetic analysis, though in concept it is based on the work of Kinsey (1936). It is also remarkably similar to procedures outlined by Hennig (1965, 1966b) and Ashlock (1974). It is, of course, possible to argue that there are other ways to apply this cladogram to the late Eocene paleogeographic map, yet we conclude that this is the most parsimonious application based on its congruence with Hennig's (1966) "search for the sister group" and the occurrence of the earliest documented record of anthropoids in southeastern Asia.

Following the procedures outlined by Morse and White (1979) coupled with the geographic interpretations derived from Fig. 6, we have been able to construct a maximum-parsimony model of platyrrhine origins and dispersal (Fig. 7). This model combines phylogenetic relationships, areal relationships, and probable dispersal patterns. The apparent congruence (see Fig. 7) between the biological cladogram of taxa and the geological cladogram of areas corroborates this maximum-parsimony model. Rosen (1978) has recently argued that there should be nothing fundamentally different in cladograms expressing biological relationships vs. geological relationships of a particular group. We agree completely with his assessment.

Dispersal Scenario

It is easily discernible from information presented in Figs. 6 and 7 that this maximum-parsimony model supports an African origin of the Platyrrhini from a preplatyrrhine-precatarrhine early anthropoid stock via waif dispersal across an equatorial South Atlantic inter-island sweepstakes route. Arguments in support of

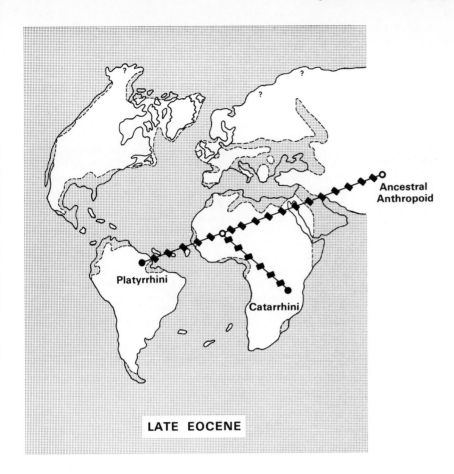

LATE EOCENE

Figure 6 *Cladogram of the higher primates originally presented in Fig. 1 overlaid on a paleogeographic model of the world in the late Eocene. Applying the cladogram to the model in this particular manner represents the most reasonable and falsifiable combination of phylogenetic and paleogeographic information attainable at present. This wedding of phylogenetic and geographic data results in a paleobiogeographic scenario in concert with an origin of the Anthropoidea in Asia, the existence of a precatarrhine-pre-platyrrhine early anthropoid stock in Africa, and the derivation of the Platyrrhini directly from the African continent via sweepstakes dispersal route through island-hopping across the South Atlantic Ocean. Thus, the African-derivation model of the Platyrrhini presented in Fig. 4 is supported by this scenario. For background information concerning the paleogeographic reconstruction depicted here, see the references in the Figs. 2 and 4 captions.*

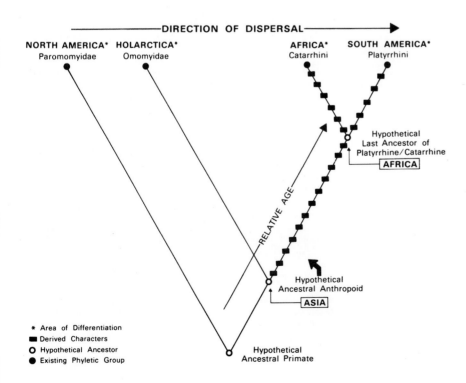

Figure 7 *Maximum-parsimony model in graphic form of the phylogenetic and biogeographic relationships of the primates as outlined in this chapter. The basic phylogenetic relationships expressed in this cladogram were originally presented and discussed in Fig. 1, adapted in part to a late Eocene paleogeographic model of the world in Fig. 6, and here combine the results of systematics, phylogeny, and geography in a synthetic and testable model. The horizontal arrow at top of figure points out inferred overall direction of dispersal for the existing phyletic groups, and the oblique arrow indicates decreasing relative age. Note that the ancestral platyrrhines appear most parsimoniously derived from a precatarrhine anthropoid from the African continent. A study by Brundin (1972) provides the conceptual basis for this maximum-parsimony model.*

this proposal have already been presented in the section on the African-origin model. Since any major opposition to this maximum-parsimony model will almost certainly center around the likelihood of the proposed dispersal scenario, we feel it is necessary to present once again the comments of the geophysicist D. H. Tarling (1979):

> The implication of these models for migration routes in the Eocene-earliest Oligocene is that uplifted oceanic islands probably existed between Africa and South America, with marine gaps of probably 200 km or less between them. There is much less likelihood of island arc connections between North and South America at this time, as the Antilles, formed at 80 m.y., only lead toward the marine conditions of the Gulf of Mexico, and the southward growth of Central America had only just commenced. Migration from Africa would also be aided by westward flowing ocean currents, but these would inhibit migrations from North to South America. The palaeogeography, based on the palaeomagnetic record, therefore strongly supports trans-South Atlantic connections between islands in both the equatorial Atlantic and possibly across the Walvis-Rio Grande Rises in Eocene-earliest Oligocene times and suggests that North to South American crossings are much less probable, although not impossible.

Earlier, J. Tuzo Wilson, developer of the transform fault concept of plate tectonics, had also argued for the existence of island chains in the formative equatorial Atlantic. Wilson (cited in Chace and Manning, 1972, p. 6) states:

> Ascension Island is only the latest in a series of islands whose remains form scattered seamounts and ridges from Ascension Island to the Cameroons in one direction (The Guinea Rise) and in the other direction to the north-east corner of Brazil.

Due to the nature of these arguments, we think it is incumbent upon these researchers who wish to argue against this maximum-parsimony model of platyrrhine origins to attempt initially to falsify the accumulating body of geological data on which it is based.

Falsifiability of the Model

No matter how many data can be assembled in support of a particular hypothesis, it will remain unproven with the criterion of falsifiability being the ultimate test of its adequacy. In its present form, this maximum-parsimony model of platyrrhine origins is not fully falsifiable. We suggest, as others have in the past, that no paleobiogeographic model can be fully falsifiable. However, if a single well-preserved higher primate jaw were recovered from Eocene deposits in West Africa, this discovery alone might lend direct biological support to our model. For example, if the jaw proved to be a preplatyrrhine/precatarrhine anthropoid or possibly an incipient platyrrhine, then this maximum-parsimony model would become testable and falsifiable on the basis of both paleontological *and* geological data. With this example in mind, it is interesting to note that Conroy (1980a, p. 450), in a study of auditory structures and primate evolution, suggests that the last common ancestor of platyrrhines and catarrhines has yet to be found because "...it still lies buried in the sand of the African Paleogene!"

To conclude, most aspects of this maximum-parsimony model do represent the consensus of contributors to this volume (see Ciochon and Chiarelli, Eds., 1980). It is directly supported by Lavocat (1980), Hoffstetter (1980), Maier (1980), Bugge (1980), Luckett (1980), Martin and Gould (1980), Chiarelli (1980), Sarich and Cronin (1980), and Tarling (1980) and discussed as an alternative proposal in Gingerich (1980a), Rosenberger and Szalay (1980), and McKenna (1980). We fully believe that its list of supporters will expand just as the continents themselves continue to drift.

Acknowledgments

We would like to thank Ms. Joy Myers, who greatly assisted one of us (R. L. C.) throughout all stages of the preparation of this chapter. We would also like to acknowledge Ms. Doris Carter, who typed and retyped the manuscript, and Ms. Evelyn Oates, who drew the figures. Finally, we wish to thank Dr. Eric Delson for his many useful comments throughout the course of this study. This research was supported, in part, by a grant to R. L. C. from the L. S. B. Leakey Foundation.

Footnotes

[1] One minor criticism which some may wish to level at the analysis of Delson and Rosenberger (1980) concerns the methods they use for selecting individual characters from all those presented by the contributors. Little effort has been made by the authors to present exactly how the primitive vs. derived characters were sorted out. Our insistence on brevity is perhaps the major contributing factor to this situation. In any case, the reader is still left with the choice of accepting their summary conclusions based on the degree of confidence that the reader chooses to have in their authority. Fortunately, it was our confidence in their authority which prompted us in our editorial capacity to solicit this contribution.

[2] To our knowledge, no one has yet proposed a derivation of the Platyrrhini from the continent of Antarctica. As Tarling (1980) has shown, some sort of land connection existed between Antarctica and South America throughout the Cretaceous and possibly into the early Tertiary. The absence of any living or extinct record of primates (except *Homo sapiens*) on either Antarctica or Australia coupled with the generalized patterns of biotic distribution which characterize these southern continents (see Brundin, 1966; Keast et al., 1972; Craw, 1979) would, in our opinion, constitute falsification of this proposal.

[3] Waif dispersal as defined by Simpson (1978, p. 321) "...refers to the occasional occurrence of one or a few members of a species outside its usual or previous range." Waif dispersal is commonly visualized as occurring along a sweepstakes route such as a series of widely separated oceanic islands (Simpson, 1953a). Sweepstakes dispersal is very nearly the equivalent of waif dispersal; Simpson (1978, p. 321) suggests that it should refer to "...geographic spread of a group of organisms across a barrier, such as an ocean or strait, for terrestrial organisms where the probability of such spread is very small but not zero, analogous to the probability of holding the winning ticket in a sweepstakes." McKenna (1973, p. 297) considers that "sweepstakes dispersal is inherently improbable although not impossible at any one time, but with sufficient time sweepstakes dispersal becomes probable, although random and unpredictable." Since South America remained an island continent until the Pliocene, the ancestral Platyrrhini had to reach this continent via an oceanic barrier. For terrestrial mammals in the size range of the ancestral New World monkeys (probably squirrel-sized), the crossing of this barrier was almost certainly made by waif dispersal across an oceanic sweepstakes route of tectonically active island arcs (island-hopping). Near plate margins, these volcanic island chains appear to have formed intermittently and occasionally coalesced into temporary land bridges, only to later drop back below sea level. Initial transport from the mainland to an island and any subsequent over-water inter-island migrations probably

occurred via a series of successive rafting episodes. It is possible that other dispersal mechanisms such as the innovative "Noah's ark" concept of McKenna (1972a, 1973) or a rather fanciful mechanism we propose of occasional passive transport via large predatory birds could also have been responsible. Whatever the actual dispersal mechanism, we envision the process of spread along the sweepstakes route to have taken many generations, with the probable result of a single colonizing group reaching the South American continent some time in the late Eocene.

[4] Simons (1976b, pp. 51-52) presents the following excellent critique: "The physiological and behavioral objections to primate rafting over long distances are great. The most crucial objection would be dehydration due to lack of water, accelerated by heat, sun, and lack of shelter. Exposed to salt spray, vegetation would wither and cease to provide a food or water source, thus upsetting the water/salt balance. Small monkeys do not normally utilize tree holes and thus would be exposed during the day to heat and wind stresses. Platyrrhines are known to be sensitive to heat stress and are easily susceptible to sun stroke. Lack of food and water, salt imbalance, and the aforementioned environmental stresses would render most small primates comatose or unconscious in 4-6 days. Primates would likely jump off a raft when it was forming, or, if not, the probable isolated rafting group would be only a few individuals. Mother-son and sister-brother incest avoidance occurs in marmosets and cebids and might further impede colonization. Also, because they are social animals, small group size and the characteristic slow birth rate might lead to abnormal behavior. Finally, the sudden adjustment to a new environment long-distance rafting necessarily requires (new predators, new foods) argues against a successful colonization."

[5] Platnick and Nelson (1978) argue that a two-taxon, two-area pattern like the one exhibited by platyrrhines and catarrhines is not immediately resolvable by their method of analysis for historical biogeography. Instead, they suggest that the three-taxon statement should be regarded as the most basic unit of analysis in biogeography. Since the Omomyidae are the sister group of the Platyrrhini-Catarrhini (see Fig. 1), it is therefore the third taxon available for analysis. Unfortunately, it has a documented Holarctic (pan-European-Asian-North American) distribution which does little to resolve the paleobiogeographic problem at hand. The Omomyidae could be the ancestral stock from which the potential early anthropoids *Amphipithecus* (see Simons, 1971) and *Pondaungia* (see Szalay and Delson, 1979) were derived. The absence of omomyids from the Paleocene-Eocene fossil record of Africa is a moot point, since virtually no fossiliferous deposits are known; in contrast, their absence from the Paleogene of South America appears to be well documented. Thus, the occurrence of Omomyidae in all the northern continents and possibly in Africa cannot be used to falsify the paleobiogeographic model outlined above. Insofar as the omomyids can be judged ancestral to the Burmese early anthropoids, this evidence can add support to the maximum-parsimony model depicted in Fig. 7.

[6] The aspect of this model which may be most easily falsified concerns the origin of the Anthropoidea in Asia. As previously stated in the text, many authors do regard *Amphipithecus* and *Pondaungia* as probable anthropoids, and these taxa have been recovered *only* from the South-East Asian late Eocene deposits of Burma. Unfortunately, the fossil record of Africa during this same period of time is almost entirely unknown. It is possible that the Eocene anthropoids of Burma enjoyed a pan-Asian-African distribution. However, paleogeographic evidence in the form of the epicontinental transgressions of the Tethys Sea and Turgai Strait effectively isolated Africa from faunal interchange with either Europe or Asia throughout most of the Eocene. Therefore, if anthropoid origins were in Africa, the Tethys Sea and Turgai Strait would have restricted any range expansion into Asia until the late Eocene. Since the lack of any Eocene African fossil evidence can neither support nor deny this possibility, it is not falsifiable at present. With regard to the Pondaung primates of Burma, their presence in the late Eocene of Asia does provide falsifiable evidence in support of an Asian origin of the Anthropoidea as depicted in Figs. 6 and 7. We feel that this is the most parsimonious hypothesis until evidence is produced in the future to deny it.

16

The Third Radiation—Higher Primates
K. D. ROSE AND J. G. FLEAGLE

The higher primates (Suborder Anthropoidea), represented in the New World by the Ceboidea, or New World Monkeys, are the third major group of Primates in the western hemisphere, and the only group to survive into the present. Like the two North American groups discussed previously (see Chapter 7), the New World monkeys appear abruptly in the fossil record with no indication of evolution *in situ* from an earlier radiation. Indeed, the origin of this group is one of the most fascinating and controversial problems in primate evolution today.

Higher primates as a group are distinguished from prosimians and plesiadapiform primates by a variety of cranial and dental characters. Although many of the features which distinguish higher primates from prosimians are also found in particular lower taxa of other (non-primate?) groups, the suite as a whole is nevertheless characteristic of higher primates as a taxon and probably represents shared derived characters possessed by the last common ancestor of all anthropoids.

In most groups of living higher primates, incisors, canines, premolars, and molar teeth are all present and each tooth group possesses a fairly distinctive morphology. All higher primates possess two small, vertically implanted incisors in each quadrant. Although these may be somewhat procumbent as in *Callithrix* (lowers) or *Pithecia* (uppers), they are never markedly so as in lemurs or plesiadapids. In higher primates the canine is always larger than the incisors. Although the canine may be laterally splayed (*Pithecia*), somewhat procumbent (*Callithrix*), or even relatively short (*Homo*), it is never absent, drastically reduced, or markedly procumbent (as in Lemuriformes), and it usually retains a large root even when the crown is reduced (*Homo sapiens*). All higher primates possess at least two premolars, and the primitive condition is clearly three premolars as in extant New World monkeys. The morphology of higher primate premolars varies considerably, from a caniniform puncturing tooth, to a molariform tooth with a differentiated trigon (id) and talon (id) or a bicuspid condition. The first lower premolar behind the canine is frequently a sectorial tooth that functions as a honing blade for the upper canine. In no extant higher primates is there any marked diastema between the anterior dentition and the cheek teeth. Extant higher primates possess three molars, except for callitrichids which have two. Compared with the upper molars of prosimians, those of higher primates are relatively longer and have a larger hypocone making for relatively square upper molars. The lower molars are characterized by reduced trigonids, loss of paraconid and a relative broadening of the crowns.

In most prosimians and placental mammals in general, the two halves of the mandible are joined in the center by a fibro-cartilaginous symphysis which permits considerable mobility between the two halves of the jaw during mastication. In all higher primates, the two halves of the mandible fuse prenatally in a bony symphysis. This development is presumably associated with the development of vertically implanted incisors.

In prosimians (but not Plesiadapiformes), the frontal joins with the jugal to form a lateral bar and thus complete a bony ring around the orbit. In higher primates, the frontal, jugal and maxillary bones combine to form a thin bony septum of variable completeness between orbital and the temporal fossa, a condition known as *post-orbital closure*.

In most prosimians, the metopic suture between the two halves of the frontal bone remains open postnatally and is visible as a suture in the adult. In most higher primates, the metopic suture fuses prenatally giving rise to a smooth frontal without an obvious suture.

The blood supply to the cranial cavity of most mammals is through two branches of the internal carotid, which divide within the middle ear—a stapedial branch which runs through the stapes bone and provides the major blood supply to the orbit and anterior part of the cerebrum, and a promontory branch which runs through a canal in the middle ear to join with the anterior communicating branch of the basilar artery to form the circle of Willis. In higher primates and *Tarsius* the stapedial branch of the internal carotid is greatly reduced and the promontory branch enlarged to provide the major supply of blood to the cerebrum. In addition to these specific traits which distinguish higher primates from prosimians, there are a number of more qualitative differences. Higher primates tend to have relatively more complex brains than do prosimians. Most of them, with the notable exception of baboons, have reduced snouts. Higher primates tend to have more frontally directed orbits. These three traits are related to the greater importance of stereoscopic vision and hand-eye coordination among higher primates compared with prosimians.

Among living higher primates, New World monkeys are distinguished from the Old World higher primates (monkeys, apes, and man) by two main characteristics: presence of three premolars and possession of an auditory bulla in which the tympanic ring forms the external boundary but is not extended to form a tube (the external auditory meatus).

FOSSIL CEBOIDS

In contrast with the extensive fossil record of Paleocene and Eocene primates from North America, the paleontological record of ceboid evolution is meager. Instead of evidence for major radiations of extinct animals allied to the extant New World

Excerpt reprinted with permission from "The fossil history of nonhuman primates in the Americas," which appeared in *Ecology and Behavior of Neotropical Primates*, Volume 1, edited by A. F. Coimbra-Filho and R. A. Mittermeier, pages 111–167, Academia Brasileira de Ciencias, Rio de Janeiro. Copyright © 1981 by A. F. Coimbra-Filho and R. A. Mittermeier.

Figure 1 *Map of Central and South America showing paleontological localities yielding fossil primates.*

monkeys, we have only a scattered array of isolated specimens ranging in age from Lower Oligocene to Recent, from sediments as far south as southern Argentina and as far north as Jamaica (Fig. 1). In the following section we will treat these specimens in chronological order from oldest to youngest.

Branisella boliviana

Branisella boliviana (Hoffstetter, 1969) from Lower Oligocene (Deseaden) sediments in Salla, Bolivia, is the oldest fossil monkey from the New World. The type specimen is a left maxillary fragment preserving P^4 through M^2, the roots of P^3 and P^2, and the edge of alveolus indicating the presence of M^3 (Fig 2). The most distinctive features of the molars of *Branisella* are the relatively low, bulbous cusps, reduced conules, and presence of a hypocone—all of which suggest higher primate affinities. The trapezoidal shape of the molars with the hypocone situated lingual to the protocone, the styles of the molars and especially the premolars, and the presence of a single root for P^2 have been suggested as primitive features reminiscent of some tarsiiform

primates, specifically omomyids. Among extant higher primates the teeth of *Branisella* resemble most closely those of *Saimiri* in tooth formula and structure. While Hoffstetter regards *Branisella* as definitely ceboid, he has not placed it in either family of extant New World monkeys.

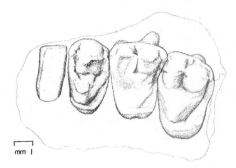

Figure 2 *P^3—M^2 of* Branisella boliviana.

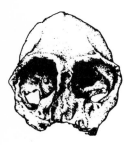

Figure 3   *Lateral (top), frontal (left), and basal (right) view of* Tremacebus harringtoni.

In disagreement, Hershkovitz has argued that the small root of P² and the quadritubercular molars of *Branisella* indicate a "highly evolved primate"...that was "neither platyrrhine nor ancestral to platyrrhines" (Hershkovitz, 1974a:25). Neither of these conclusions seems particularly justified on the basis of present evidence. As Orlosky (1973) points out, only the apical tip of the root of P² is preserved in the specimen of *Branisella* and it is perhaps rash to estimate the size and shape of the crown of the tooth from such minimal information. Certainly coalesced roots are common in *Saimiri sciureus.*

Furthermore, while the presence of cebid (quadritubercular) molars in an early Oligocene form seems unexpected, it need not necessarily indicate that *Branisella* is unrelated to later forms. Rather it more likely indicates that callitrichid (tritubercular) molars are a result of secondary reduction.

Tremacebus harringtoni

Originally described as a species of *Homunculus, H. harringtoni* (Rusconi, 1933, 1935), the single specimen attributable to this species from the Upper Oligocene (Colhuehuapian) of Sacanana, Chubut, Argentina, has recently been redescribed by Hershkovitz as the type specimen of a new genus, *Tremacebus* (Hershkovitz, 1974a) (Fig. 3). As the best preserved skull of an early ceboid, it is particularly important for our understanding of the evolutionary history of higher primates in the New World.

The skull of *Tremacebus harringtoni* is comparable in size to the smaller cebids (*Callicebus, Aotus* or *Saimiri*). The specimen has enlarged, obliquely directed orbits and a particularly wide interorbital region. The position of the canine sockets and the morphology of the maxillary bone suggest that the snout was remarkably broad for a New World anthropoid, and somewhat prognathic. Like most New World anthropoids, *Tremacebus* had a long, relatively narrow braincase. The temporal lines apparently joined to form a low sagittal ridge posteriorly. Although the occipital bone is almost completely missing from the specimen, both Rusconi (1935) and Hershkovitz (1974a) have argued that this region was callitrichid-like rather than cebid-like in showing a vertical occiput and posteriorly-directed foramen magnum. The

upper dental arcade appears to have been relatively broad, short, and U-shaped, rather than V-shaped as in many extant cebids.

Hershkovitz has argued that the skull of *Tremacebus harringtoni* possessed minimal post-orbital closure, and he weighted this character very heavily in his assessment of the specimen. We have been unable to confirm his observation regarding the absence of a post-orbital plate in the specimen. *Tremacebus* clearly shows a large orbito-temporal fissure in its present condition; however, there seems good reason to suspect (as Rusconi noted) that this condition is the result of post-mortem breakage either prior to fossilization or even during preparation. Although Hershkovitz cites the "rough symmetry of posterior, lateral and anterior openings of both orbits" as evidence for their natural proportions, the presence of an obviously broken margin on the orbit with the smaller orbito-temporal opening argues against the naturalness of the present condition. In addition, under high power magnification, we were unable to confirm that the left lateral orbital ring is definitely undamaged. On present evidence, the extent of post-orbital closure in *Tremacebus harringtoni* must remain an open question.

The skull preserves partial crowns of three molars on the left side and roots of all premolars and the canine on both right and left. Like the molars of *Branisella*, those of *Tremacebus* are remarkably cebid-like in showing quadrangular crowns with a well-developed hypocone and low bulbous cusps. There is a small lingual cingulum on all molars and a small pericone lingual to the protocone on M¹, similar to that described in *Branisella*. The last upper molar is reduced in size. The molars of *Tremacebus harringtoni* are most similar to those of *Callicebus moloch* among extant cebids. The last two premolars in *Tremacebus* were double rooted and presumably had crowns that were transversely broad. The P² of *Tremacebus* was clearly single-rooted like that of *Branisella boliviana*, although the root that is preserved in the Argentinian species appears slightly larger than that in the Bolivian species.

The broad molars and convergent temporal lines of *Tremacebus harringtoni* suggest an herbivorous diet for the fossil monkey. *Callicebus moloch,* the extant species that shares the most dental resemblance with the fossil, has been reported to be predominantly frugivorous. On the basis of the orbit size, which is intermediate between that of living diurnal forms and the only nocturnal cebid, *Aotus,* Hershkovitz (1974a) has proposed that *Tremacebus* was crepuscular in habits.

Comparison of *Tremacebus* with *Branisella* seems to have been overlooked by most authors. All of the features which have been described as advanced in *Branisella*,—such as rectangular molars, low bulbous cusps, large lingually placed hypocone and even a slight internal cingulum with a pericone on M¹—are present in *Tremacebus* as well. Indeed, the two features which Hershkovitz cited as evidence that *Branisella* is unrelated to later platyrrhines (the quadrangular molars and single-rooted P²) are characteristic of both Oligocene fossils.

In terms of the major divisions of living New World monkeys, *Tremacebus* shows cebid rather than callitrichid affinities in almost all features except the reconstructed shape of the occiput. The suggested absence of post-orbital closure which would distinguish it from either family remains conjectural on present evidence. Rusconi allied the specimen with later specimens of *Homunculus* from Argentina; Hershkovitz has outlined the numerous cranial differences between this skull and the partial face of *Homunculus patagonicus* and placed *Tremacebus* in its own subfamily within the Cebidae.

Dolichocebus gaimanensis

The badly crushed and edentulous skull of *Dolichocebus gaimanensis* (Kraglievich, 1951) from the upper Oligocene (Colhuehuapian) of Gaiman, Argentina, is difficult to compare with other fossil anthropoids of South America because of the discrepancy of the preserved parts. The skull apparently lacked the large orbits or wide interorbital regions seen in *Tremacebus*. In Hershkovitz's (1970b) reconstruction, the shape of the braincase is comparable to that of *Tremacebus*. It is similar in size to that genus but possessed a narrower and probably longer snout. It definitely had nearly complete post-orbital closure.

The dental formula was presumably 2.1.3.3. All premolars were double-rooted. The dental arcade appears to have shown more anterior convergence than seen in *Tremacebus*. The molars were unusually wide as in early Miocene *Homunculus*.

Most workers have tended to align *Dolichocebus* with *Homunculus*. However, Rosenberger (1979a) has argued for a close relationship between *Dolichocebus* and the extant *Saimiri* primarily on the basis of the presence of an interorbital fenestra in both taxa.

Homunculus

This genus was named by Florentino Ameghino, in 1891, who described a right mandible with symphysis and the crowns of I_2-M_1 as the species *Homunculus patagonicus*. In later years, a nearly complete mandible preserving crowns of most teeth (originally described as *H. ameghinoi*), an edentulous cranial fragment preserving much of the left side of the face, several other mandibular fragments, and several limb elements were assigned to the genus by other workers (Bluntschli, 1931). All of the specimens were found in early Miocene sediments (Santacruzian) near the mouth of the Rio Gallego, Santa Cruz, Argentina.

The dental formula of *Homunculus* is $\frac{2.1.3.3}{2.1.3.3}$. The mandible is V-shaped, with the lateral pair of incisors placed behind the central pair rather than beside them as in most higher primates. With a postcanine tooth row 23 mm in length, *Homunculus* was a monkey about the size of *Cebus albifrons*. The canine is relatively small in both the type specimen and the most complete mandible. The three premolars are relatively small and narrow compared to

the size of the molars. The first two molars are subequal in size and the last molar is slightly smaller. The lower molars seem to be quadritubercular with an additional crest outlining the anterior border of the trigonid. The expanded talonid basins are much larger than the enclosed trigonid, and the greater width of the talonid gives them a trapezoidal shape. Bluntschli (1931) and Scott (1928) felt that the dentition of *Homunculus* was indicative of a relationship with *Aotus*, whereas Gregory (1920c) and Stirton (1951) pointed out the numerous similarities between the molar structure in the Patagonian genus and that of the howling monkey, *Alouatta*.

The edentulous cranial fragment from Rio Gallego is assigned to this genus primarily on the basis of locality since there are no teeth available for comparison. The specimen shows moderate sized orbits, suggesting a diurnal monkey. The interorbital region is broad and the lacrimal foramen lies within the orbit. The canine root is prominent. Published illustrations of the specimen indicate postorbital closure.

The distal end of a humerus, a complete radius, and a complete femur have also been assigned to this genus. The radius and femur suggest an animal with the body size and limb proportions of *Cebus*. However, studies of the individual bones provide a less clear picture of the locomotor abilities of the Miocene monkey. Bluntschli (1931) and Scott (1928) felt that the limb bones indicated affinities with *Aotus* and *Callicebus*, two basically quadrupedal monkeys which are also effective leapers (Stern, 1971). In a morphometric analysis of the femur, Ciochon and Corruccini (1975) concluded that the affinities of *Homunculus* were with the callitrichids.

There is certainly little indication of a close phyletic relationship between the Patagonian material attributed to the genus *Homunculus* and any one genus or subfamily of extant ceboids. Despite the conclusions of Ciochon and Corruccini, the dental and cranial material indicates cebid rather than callitrichid relationships. On the other hand, Hershkovitz (1970b) rightly questions the justification of forcing Miocene fossils into a family classification based on extant species, and argues that *Homunculus* and *Dolichocebus* should be placed in a new family, the Homunculidae. A more conservative position was adopted by Orlosky and Swindler (1975), who placed these two genera in their own subfamily within the Cebidae.

Stirtonia tatacoensis

Originally described as a new species of *Homunculus* from the Miocene (La Venta) of Colombia, this species was later placed in a new genus by Hershkovitz (1970b). It is the largest fossil ceboid. The type and best specimen is a nearly complete mandible preserving both canines and most of the cheek teeth (Figs. 4, 5). Several isolated teeth are also referred to this species. *Stirtonia* is approximately 25% larger in most dental dimensions than the Argentinian *Homunculus*, and the mandibular tooth rows are less divergent. The lower dental formula was 2.1.3.3. The alveoli indicate that the incisors were diminutive, but aligned side by side in contrast with the situation seen in *Homunculus*. The canines were more robust and the premolars rather broader than those of the Argentinian genus. The first and second lower molars of *Stirtonia* are trapezoidal in shape with a relatively small trigonid and a greatly expanded talonid. The second molar is slightly larger than the first and, in the type mandible, the last molar is unerupted, despite the erupted and worn condition seen in other teeth. Hershkovitz has noted that such late eruption of M_3, if normal for the genus, is unique among New World monkeys. Alternatively it may indicate a trend toward loss of this molar in *Stirtonia*.

Stirton (1951) believed that this species showed definite resemblances to the howling monkey, *Alouatta,* but he was originally uncertain as to whether they should be interpreted as purely adaptive similarities related to leaf-eating (a view espoused by Gregory for the Argentinian *Homunculus*), or whether they were also indicative of a phyletic relationship.

He later placed the Miocene monkey in the same subfamily with the howling monkeys (Stirton, 1953), a view also adopted by Orlosky and Swindler (1975). While agreeing that many features of *Stirtonia* could be antecedents of those found in *Alouatta,* Hershkovitz (1970b) suggested that the reduced and (?) disappearing third molar and the divergent shape of the mandible are unique traits militating against placement of the Miocene genus in the same subfamily with any extant genus. He allocates it to its own subfamily, the Stirtoninae.

Cebupithecia sarmientoi

This is another taxon from the Upper Miocene La Venta fauna of Colombia. It was originally described by Stirton and Savage (1951) and discussed and figured in more detail by Stirton (1951). The type and only known specimen preserves maxillary and mandibular portions of the face with the canines and P_3–M_1, P^3–M^2 intact, and alveoli for most teeth (Figs. 7, 7, 8). Many of the articular ends from postcranial elements are also known. In his detailed description of the specimen, Stirton (1951) argued that the genus is definitely allied to the extant pithecines and shows the most similarities to *Pithecia*. He placed particular diagnostic emphasis on the large and procumbent upper incisors as indicated by the alveoli and the root of a central incisor, and on the size and splayed orientation of the upper and lower canines.

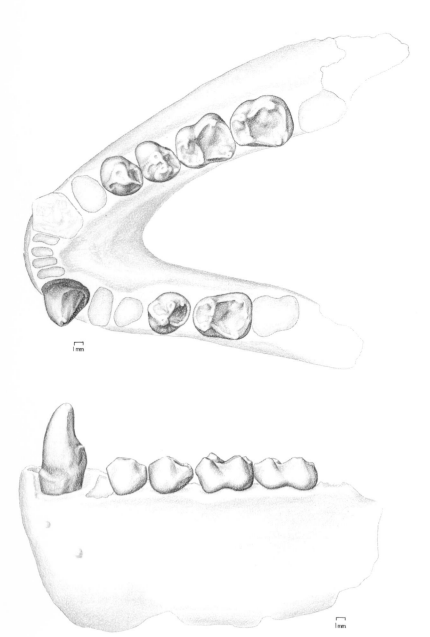

Figure 4   *Occlusal view of the lower dentition of* Stirtonia tataconensis.

Figure 5   *Lateral view of the mandible of* Stirtonia tataconensis.

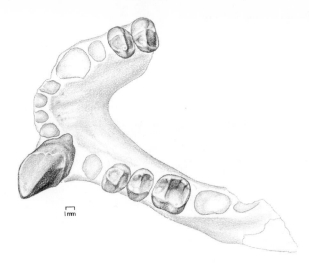

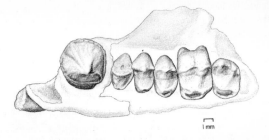

Figure 8   *Occlusal view of the upper dentition of* Cebupithecia sarmientoi.

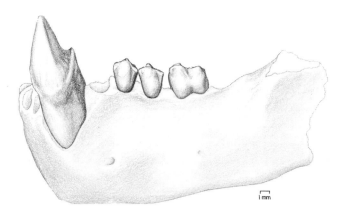

Figure 6   *Occlusal view of the lower dentition of* Cebupithecia sarmientoi.

Figure 7   *Lateral view of the mandible of* Cebupithecia sarmientoi.

Hershkovitz (1970b) has argued that these pithecine features of the anterior dentition of the Miocene fossil are all the result of distortion due to breakage. This interpretation, and the fact that the fossil lacks the wrinkled enamel characteristic of *Pithecia* and *Chiropotes,* shows reduction of the last molars not seen in extant pithecines, and has canines with rounded rather than triangular cross sections, led Hershkovitz to dissociate *Cebupithecia* from the pithecines and place it in its own subfamily within the Cebidae.

Orlosky (1973) and Orlosky and Swindler (1975) have questioned Hershkovitz's reassessment of this fossil. While agreeing that there is some distortion of the anterior region of the maxilla, Orlosky (1973) suggested that incisor procumbency and canine splaying were nevertheless present in the intact palate, and that distortion has merely exaggerated these conditions. The lower canines, according to Orlosky, are both splayed and pithecine in shape. He also pointed out that wrinkled enamel is not a universal characteristic of pithecines, frequently being absent in *Cacajao*. Finally, he noted that the flat molars with low relief are distinctly pithecine in morphology. Assuming that Stirton's original interpretation of *Cebupithecia,* and Orlosky's (1973) reassessment are correct, what can we say about the dietary habits of this fossil? Although dietary information on pithecines is still scarce, Mit-

termeier and Van Roosmalen (1981) have suggested that the procumbent incisors and robust splayed canines of these animals have evolved for opening hard nuts and fruits and removing the seeds in them. Orlosky has suggested that the low relief on pithecine cheek teeth may be adaptations for grinding hard nuts and fruit seeds, a suggestion that accords well with the preliminary dietary observations of Mittermeier and Van Roosmalen.

In his original description of *Cebupithecia,* Stirton (1951) noted that the skeletal elements were, overall, most comparable to those of *Pithecia* among the ceboids he examined, although some individual limb elements were unlike any extant species. In their morphometric analysis of the head of the femur, Ciochon and Corruccini (1975) argued that this part of the skeleton was more callitrichid-like than cebid-like, a feature which Stirton had noted about the femur as a whole. There seems to be no real disagreement here. Although the skeletal elements of *Cebupithecia* certainly warrant further comparisons and functional analysis, most joint surfaces seem to indicate a leaping rather than suspensory or quadrupedal locomotor habit and, in that regard, it was probably not unlike *Pithecia.*

Neosaimiri fieldsi

Of all the fossil ceboids, this third genus and species from the Upper Miocene La Venta fauna of Colombia (Stirton, 1951) is the only extinct form which is generally agreed to be closely related to an extant genus. The type and only known specimen is a nearly complete mandible preserving, on one side or the other, all of the dentition except I_1 and M_3 (Figs. 9, 10). The lower lateral incisor of the fossil differs from that of *Saimiri* in having a more distinct mesial edge and more procumbent orientation. However, the morphology of the canine and all postcanine teeth is quite similar to that in the extant genus, the main difference being the relatively narrower molars of the fossil. In *Neosaimiri,* the mandibular rami are more divergent than in extant squirrel monkeys. All authors agree that the fossil genus should be placed in the same subfamily with *Saimiri.*

Xenothrix mcgregori

Xenothrix mcgregori Williams and Koopman (1952) is known from a partial mandible with two teeth that was found in a cave deposit in Jamaica. The age of the deposit is either Pleistocene or Recent and the jaw may have been associated with a human kitchen midden. Unfortunately all evidence regarding the age of the fossil relative to the human remains is equivocal (Williams and Koopman, 1952).

Xenothrix was moderately large monkey with a mandible larger than that of *Cebus* and relatively large molars intermediate in size between those of *Ateles* and those of *Alouatta.* Although the Jamaican monkey had the lower dental formula of a callitrichid, 2.1.3.2, all who have studied it agree that *Xenothrix* was

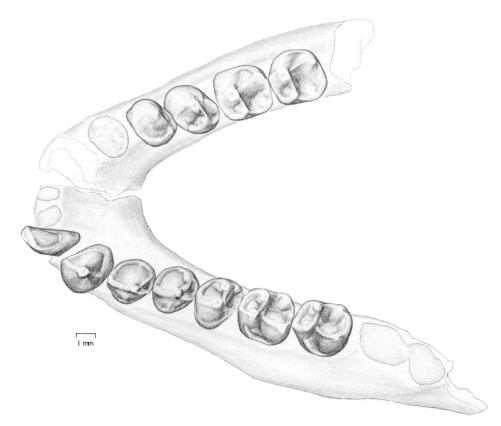

Figure 9   *Occlusal view of the lower dentition of* Neosaimiri fieldsi.

neither a callitrichid nor particularly closely related to any other genus of ceboid. The two left molars, the only teeth preserved, are relatively quite large for the size of the mandible. Compared with the teeth of other New World monkeys they are very bunodont and lack a distinct demarcation between the trigonid and the talonid. The alveoli for the three premolars suggest that those teeth increased in size from front to back and that all were relatively broad. Judging from the size of the alveolus, the canine

in *Xenothrix* was a very small tooth. The relative size and orientation of the two incisors cannot be reliably estimated from the morphology of the alveoli (Williams and Koopman, 1952; Rosenberger, 1977). The mandible had a fused symphysis and a greatly expanded angle posteriorly, similar to that seen in *Callicebus*.

In their original description, Williams and Koopman (1952) found that *Xenothrix* resembles marmosets (callitrichids) in den-

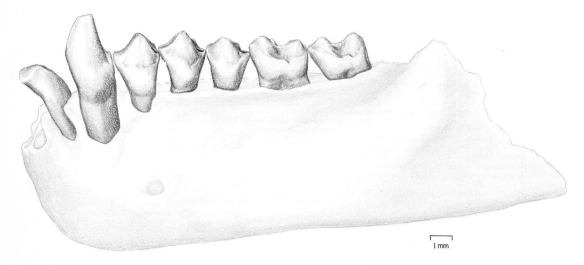

Figure 10   *Lateral view of the mandible of* Neosaimiri fieldsi.

tal formula, symphysial shape, and incisor orientation. In their estimation, the molars are most like those of *Cebus*, and the shape of the mandible like *Callicebus*. Although they placed *Xenothrix* in the Cebidae, their disenchantment with the distinctiveness of the family Callitrichidae indicates that they have felt that all ceboids should be included in a single family. Simpson (1956) and Hill (1962) accepted the cebid assignment of *Xenothrix* at face value, but later Simpson (1969) formally combined all ceboids into a single family, thus avoiding the question of the affinities of *Xenothrix*. Hershkovitz (1970b, 1977) considered *Xenothrix* unique enough to merit its own family, a view adopted by Simons (1972) and Hoffstetter (1974). As Rosenberger (1977) properly points out, however, this sidesteps the issue of the affinities of the Jamaican monkey. In a rigorous and thorough analysis, Rosenberger concluded that the Jamaican monkey shared a number of derived characters with extant forms within the Cebidae and that addition of *Xenothrix* to that family was well within the adaptive and phylogenetic breadth displayed by extant members. He did not speculate on the probable habits of the Jamaican genus, but suggested that its closest phyletic affinities lay with either *Aotus* or *Callicebus*.

Saimiri bernensis

Rimoli (1977) recently described *Saimiri bernensis* on the basis of a single maxillary fragment preserving $P^4 - M^2$ from Cuera de Berna, Dominican Republic. The specimen has a radiometric date of $3,850 \pm 135$ years B.P. Although he placed the new taxon in the genus *Saimiri*, Rimoli noted that *S. bernensis* is approximately twice the size of any other species of *Saimiri*. The most important implication of this new taxon is that along with *Xenothrix*, it suggests the presence of a distinctive ceboid fauna living in the Caribbean until very recently.

SUMMARY OF FOSSIL CEBOIDS

Our knowledge of the evolutionary history of New World monkeys, based on less than a dozen fossils from six localities spanning a vast area and 30 million years, is exceedingly scanty compared with that derived from the thousands of fossils, thick sedimentary sequences, and numerous quarries in the Paleocene and Eocene intermontane basins of North America. Despite the paucity of specimens, the ceboid fossil record does provide a number of interesting clues about the history of this group.

The question of whether callitrichids or small cebids (or some form with a combination of features) represent the ancestral ceboid morphology has been extensively debated on the basis of comparative anatomy of the extant forms, with no clear resolution to date (see Gregory 1920c, 1922b; Hershkovitz, 1977; Rosenberger, 1977). Gregory (1920c, 1922b) believed that, dentially, a quadritubercular molar such as that of *Callicebus* represented the primitive condition and, similarly, that small canines and bicuspid premolars were also the ancestral condition. Remane (1960) argued that, overall, *Saimiri*, in its molars and its anterior teeth, represented the ancestral dental condition. Both authors agreed that the callitrichid dentition (as well as other features) showed regressive changes from those present in the ancestral platyrrhine. Thomas (1913) and Hershkovitz (1977), however, have maintained that the small callitrichids (and especially *Cebuella*) display the ancestral condition in virtually all morphological features.

Although the presence of *Saimiri*-like molars in the earliest ceboids (*Branisella* and *Tremacebus*) does not prove that this dental morphology is ancestral, it certainly confirms its antiquity and its widespread occurrence (Orlosky and Swindler, 1975). Similarly, the cranial morphology of *Tremacebus* seems to resemble a small cebid like *Aotus* or *Callicebus* rather than a callitrichid.

ORIGIN OF NEOTROPICAL MONKEYS

In contrast to the large numbers of Paleocene and Eocene primates known from North America, there are no indications of primates in South America prior to the Oligocene appearance of *Branisella*, despite a rich record of late Paleocene and Eocene mammals. At present there is no clear answer to the question of geographic origin of ceboids, and considerable controversy exists regarding both the phylogenetic and geographical origins of New World higher primates.

While virtually all students of primate evolution would agree that monkeys and apes have evolved from ancestors that would be classified as prosimians, it is not yet clear whether those ancestors were tarsier-like or lemur-like prosimians, since higher primates share some anatomical features with both groups. The anatomical similarities between higher primates and living tarsiers include numerous features of the placental membranes, the internal (Cave, 1973) and external structure of the nasal region, the pattern of arterial blood supply to the brain, and the bony morphology of the ear region. The latter features can also be identified in various tarsier-like fossil primates such as the omomyids (Szalay, 1976). By contrast some authors, particularly Gingerich (1973b, 1976), have emphasized features such as the reduced, vertically-implanted incisors, relatively large canines, and fused mandibular symphysis, which higher primates share with the lemur-like adapids from the Eocene and Oligocene of North America, Europe, and Asia. Thus, currently available evidence seems to show that living higher primates share more features with the extant *Tarsius* than with extant lemuriform prosimians; however, among fossil primates the lemuriform adapids seem more like higher primates than do the tarsioid or tarsiiform omomyids. It seems likely that only recovery of more clearly intermediate fossil forms linking higher primates with one or the other of the prosimian groups is going to resolve the question.

Equally problematical and as controversial as the question of the phyletic origin of the New World higher primates is the question of their geographical origin and route to South America. As noted above, higher primates first appear in the early Oligocene of South America with no evidence of a history on that continent. Because South America was an island continent throughout the early Tertiary, the first appearance of primates was almost certainly the result of a chance rafting from some other continental area. Two source areas have been proposed—Africa and North America (Fig. 11).

Traditionally, paleontologists have generally presumed that New World monkeys are derived from an ancestral group that reached South America through North America. Since there is no record of any higher primates ever inhabiting North America, such a scenario generally assumes that ceboids are derived from a group of prosimians from North America and that the similarities between New and Old World higher primates are parallelisms rather than characters indicative of a common higher primate ancestor (e.g. Patterson, 1954; Romer, 1966). Such a derivation would mean that Anthropoidea is a grade, not a clade.

More recently, Hoffstetter (1972, 1974) has argued for an African origin of all higher primates with the ancestral ceboids rafting across the South Atlantic to South America in the late Eocene. In support of his hypothesis, he points to the presence of undoubted higher primates, many of which indeed resemble New World monkeys in a number of features, in the Oligocene of Egypt (see Simons, 1972; Fleagle *et al.*, 1975; Conroy, 1976a). Likewise, the South American caviomorph rodents, another

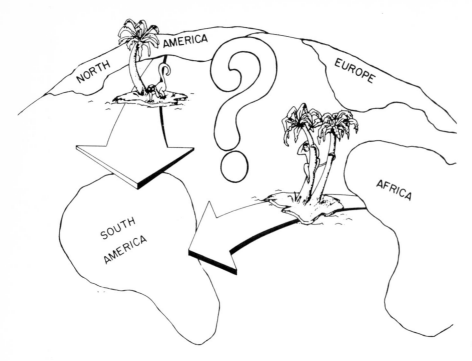

Figure 11   *Where did New World monkeys come from?*

group of Oligocene immigrants, also have close relatives in the African Oligocene and Miocene (Lavocat, 1969). These authors also argued that in paleogeographical reconstructions Africa was as close to South America as was North America and that the prevailing oceanic currents were more favorable for a South Atlantic crossing from East to West.

Hoffstetter and Lavocat's proposal for an African origin of both ceboid monkeys and caviomorph rodents has been subjected to a variety of criticisms. Wood and Patterson (1970) have reinforced their earlier arguments for a North American origin for the caviomorph rodents, and Wood (1972, 1974) has subsequently provided possible, but by no means convincing, evidence for ancestral caviomorphs in the early Tertiary of North and Central America. In an extensive review of historical evidence for rafting, Simons (1976b) has argued that the probable distances between Africa and South America in the early Tertiary were almost certainly much larger than those originally proposed by Lavocat and Hoffstetter, and he doubts that rafting from the Western coast of Africa is a feasible explanation for the origin of New World monkeys. Finally, Gingerich and Schoeninger (1977) have provided evidence (from Malfait and Dinkleman, 1972; also Barr, 1974) that the distances between South America and North America in the early Tertiary were not so great as previously suggested

and that, in addition, the two western continents were connected by at least two volcanic chains providing a pathway for animals traveling between them. As an example of faunal interchange during the time in question, they note the occurrence of endemic South American notoungulates in the Eocene of both North America and Asia (Gingerich and Rose, 1977).

In their current state, both the North American and African hypotheses for the origin of New World monkeys have generally complementary strengths and weaknesses, and the ultimate answer to this problem is not in sight (Fig. 11). While the idea of an African origin from a higher primate stock indigenous to that continent is particularly appealing in eliminating the need for parallel evolution of the various features common to all extant and fossil higher primates, the proposed mechanism—rafting over large expanses of open ocean—requires a considerable leap of faith. While the connections between North and South America seem more reasonable, there still remains the obstacle that the well-known early Tertiary fauna of North America does not contain any higher primates or any satisfactory higher primate ancestors. However, the recent finding of a Eurasian adapid, *Mahgarita,* in the Eocene of Texas certainly opens the possibility that other Old World groups, perhaps even higher primates, will eventually be found in North America.

Acknowledgments: We thank Luci Betti for preparing Figs. 1, 3, and 11, and Karen Payne for all the figures of dentitions. Dr. Jose Bonaparte, then of the University of Tucuman, Argentina, kindly provided access to the skull of *Tremacebus.* This work has been supported by NSF grants BNS 77–25921 and BNS 79–24149 to JGF.

17

Gradistic Views and Adaptive Radiation of Platyrrhine Primates

A. L. ROSENBERGER

Because they are poorly represented in the fossil record, and because systematists have largely relied on the fossil record for motivating phylogenetic hypotheses and evolutionary interpretations, the neontologically-based concept of grade or level of organization has strongly influenced evolutionary and taxonomic studies of the platyrrhine primates. This is clearly evident in the widely accepted, highly split classifications of Napier & Napier (1967), Hershkovitz (1977) and others. In another sense, gradistic argument has apparently substituted for phyletic analysis in appraising the relationships and histories of Old and New World monkeys and the category Anthropoidea generally. This has led to the often implied or cited, but never documented, conclusion that anthropoids are not strictly monophyletic (Cachel, 1979) but represent a convergently achieved grade of organization instead. A more influential example of this approach is the Huxleyian alignment of primate morphology into a series of "trends" and primate lineages into a sequence of "grades" (e.g., Le Gros Clark, 1963; Napier & Napier, 1967; Hershkovitz, 1977), yielding the picture of an ascending adaptive scale of primate evolution that posits platyrrhines as the primitive link between the catarrhines and "prosimians."

Several such applications of the grade concept, as they pertain to issues of platyrrhine evolution, will be examined in this paper. Additionally, I will present an abstract of an alternative framework for examining the adaptive radiation of the ceboids. Given the many recent advances in evolutionary biology, both in theory and in method, the philosophical basis of the grade concept clearly warrants a serious reconsideration. While this cannot be attempted here, it is hoped that critical analysis of its use will reveal some of its weaker points and suggest other, perhaps more heuristic, ways of looking at the same data. Unless indicated otherwise, the classification adopted is from Szalay & Delson (1979) and Rosenberger (1979a,b) and groups the living Ceboidea as follows—Cebidae: Cebinae *(Cebus, Saimiri),* Callitrichinae *(Callithrix, Leontopithecus, Saguinus, Callimico);* Atelidae: Atelinae *(Ateles, Brachyteles, Lagothrix, Alouatta),* Pitheciinae *(Pithecia, Cacajao, Chiropotes, Callicebus, Aotus).* A full elaboration of many of the points presented, and more complete documentation, is given elsewhere (Rosenberger, 1979b).

DISCUSSION

The Platyrrhine Grade

As a taxon, the platyrrhines were usually negatively defined relative to modern cercopithecoids (and hominoids) by 19th century authors, who noted such features as: their complement of three premolars; ring-like ectotympanic bone; inflated auditory bulla; broad internarium and laterally facing nostrils; lack of ischial callosities and cheek pouches; pendulous or prehensile tail. This practice continues today (Le Gros Clark, 1963) and has become further emphasized in the interests of paleogeographic theories of platyrrhine origins. The form of this comparison goes hand in hand with the assumptions that New and Old World monkeys, as monkeys, represent succesively higher levels of organization and that apes evolved from monkeys. This phyletic conception of the catarrhines is contradicted by cladistic analyses which demonstrate that ceropithecoids are highly derived in dental, postcranial and soft anatomy, and are a comparatively recent lineage which differentiated from a hominoid-like stock (Szalay & Delson, 1979). The recovery of good Oligocene catarrhine material (Simons, 1972) contests such a gradistic ranking by illustrating that early undoubted catarrhines cannot be distinguished by the traditional suite of craniodental characters and are in fact more similar to some platyrrhines than to eucatarrhines in certain postcranial features (Conroy, 1976a). An even more basic objection is that the analysis is based on a minimal number of subjectively chosen morphological criteria and taxa. Given the expansive range of morphologies evident among platyrrhines (and catarrhines!), I doubt that one can easily define a platyrrhine adaptive grade that would not approach a meaningless generalization. At best we might attempt a reconstruction of some ancestral platyrrhine characteristics and, by functional analysis and analogy, postulate what some of their biological roles and adaptive significances might have been. But this has yet to be done.

Some (e.g., Conroy, 1978) regard the similarities of Fayum catarrhines and ceboids as an affirmation of a platyrrhine level of organization and as evidence that that grade was primitive for Anthropoidea. But such abstractions are always possible when comparing isolated attributes of closely related taxa. There is no reason to suspect that early catarrhines or protoanthropoids should especially resemble living platyrrhines in form or adaptation. New, anatomically unique, parapithecid pelves offer a cautionary caveat in this regard (Fleagle & Simons, 1979). The fossil record is revealing what comparative anatomy has always suggested: that we are dealing with transformation series of characters which are ultimately expressed as mosaics of primitive, derived and autapomorphous states in an assortment of taxa. In at least some respects protoanthropoids are likely to recall catarrhines more than platyrrhines. This appears to be the case in the occlusal anatomy of upper molars, for example. Fayum catarrhines retain paraconules and metaconules, widespread among Paleogene primates and certainly an ancestral condition of Anthropoidea, but these are essentially absent in all ceboids.

Given that platyrrhines will inevitably retain the primitive conditions of characters which become modified in catarrhines,

Reprinted with permission from *Zeitschrift für Morphologie und Anthropologie,* Volume 71, pages 157–163, Stuttgart. Copyright © 1980 by E. Schweizerbart'sche Verlagsbuchhandlung, Stuttgart.

and vice versa, it remains to be shown that any of those features portray qualitative differences in goodness of adaptation. That they do is another fundamental premise of the gradistic view, at least in its early formulations ("...from the crown and summit of the animal creation down to creatures, from which there is but a step, as it seems, to the lowest, smallest, and least intelligent of the placental Mammalia"; Huxley, 1863). Most of the morphological differences between these infraorders probably represent chance paradaptive differences (Bock, 1969) rather than actual adaptive improvements. This contrasts with other cases in primate history where anagenetic advances are referable to certain features that distinguish higher taxa, such as the enhanced frugivorous capabilities of anthropoids versus tarsiiforms or the improved arboreal locomotor aptitude of euprimates versus plesiadapoids.

Grades of Platyrrhines

The classical family divisions of Ceboidea, separating the clawed "callitrichids" from the nailed "cebids," reflects a classical segregation of supposed grades rather than a documentable cladistic distinction (Rosenberger, 1979b). Hershkovitz (1977) has elaborated this system by introducing a third intermediate "marmoset-like cebid" grade, composed of *Saimiri, Aotus* and *Callicebus,* and provided some four independent gradistic rankings for comparing each of the six marmoset (and tamarin) genera that he maintains (p. 406; Fig. VII.3). Hershkovitz also recognized grade distinctions between the saki-uakari group and the "larger and prehensile-tailed cebids," *Cebus, Alouatta, Lagothrix, Ateles* and *Brachyteles.*

Of these groupings, I fully endorse the associations of callitrichines, saki-uakaris and atelines. These all appear to share specific, unique clusters of morphological and behavioral attributes which tie them genealogically and indicate a common, though internally diverse, ecological adaptation. Hershkovitz' (1977) other assignments are far too heterogeneous. While *Cebus* is the largest nonateline ceboid, it still weighs only half as much as an average ateline (Bauchot & Stephen, 1969). Though it sports a powerful, manipulable tail, in craniodental morphology, body proportions, locomotor mode, behavior and general habitus, *Cebus* is quite unlike atelines (Erikson, 1963; Hladik, 1975). The "marmosetlike cebid" assemblage unites three phenetically disparate genera and seems even more arbitrary. It includes species with diurnal and nocturnal activity rhythms, polygamous and monogamous social systems and dietary adaptations of fundamentally different sorts. In essence, Hershkovitz' justification for defining and ranking these grades is based on similarities in body size, features correlated with it, and the morphological "stage of evolution" attained by parts of the brain, skull, dentition and postcranium of the various genera as they evolve along what he considers to be predictable, practically unswerving pathways. In principle, this seems highly unlikely. In practice, my own character analysis suggests that evolution is far more plastic. Finally, in view of the model of ceboid differentiation and diversification outlined below, I question the value of recognizing a "cebid" grade altogether.

As Hershkovitz' (1977) recent historical review shows, opinions have long been divided on the interpretation of marmoset morphology and the marmoset "grade": do they represent a conservative or highly modified platyrrhine stock? Evidence supporting the hypothesis that they are essentially primitive is detailed by Hershkovitz, who maintains this view. But a variety of studies contradict this rather emphatically in suggesting that some or all of the callitrichines are characterized by features that are uniquely derived for primates as well as ceboids. Among them are: the combination of twining and simplex uterus; extended-family monogamy involving female-female reproductive inhibition; secondarily derived claws, not homologous with primitive eutherian claws; a clawed thumb; reduction or loss of M_3^3 and reduction of M_2^2; and in *Callithrix*—V-shaped modified incisal occlusion; hypertrophied C_1 and P_2 in both sexes; extensive mandible; staggered incisor-canine emplacement; modified canine occlusion; increased $I_{1,2}$ crown height; hypertrophy of buccal enamel and reduction or loss of lingual enamel on $I_{1,2}$. The emerging evolutionary interpretation of the callitrichines is that they are a rather specialized lineage which secondarily occupies a canopy-subcanopy spatial niche, thereby reducing competition with sympatric, larger bodied species of the canopy strata. They are probably secondarily small in size though selection has perhaps favored a subsequent increase in body size in some sublineages, e.g., *Leontopithecus*. Members of genus *Callithrix*, particularly the smallest, most derived species, *C. pygmaea* and *C. jacchus*, have become marvelously adapted to a highly gumivorous diet, which is quite an unusual strategem for a full-fledged anthropoid but may be an important feature of the callitrichine radiation.

Adaptive Radiation and the Fossil Record

An ecological approach to evolutionary interpretations of the platyrrhines avoids the many assumptions inherent in the purely morphological, gradistic approach and provides testable hypothesis at various levels. Observational (Hladik, 1975) and mechanical (Kay, 1975) studies of feeding and masticatory adaptations have shown that living ceboids are generally classified as frugivore-insectivores and frugivore-folivores with a minimal amount of taxonomic overlap between categories. These and other studies (e.g., Rosenberger & Kinzey, 1976) suggest that dietetic specializations of certain species are also discernible and with the inclusion of additional relevant data, e.g., other parts of the feeding mechanism, foraging modes, locomotor and manipulative behaviors, body size and social organization, adaptive inferences may be substantially refined. Following this rationale, I suggest that living ceboids, as a class of arboreal frugivores, occupy two semi-exclusive adaptive zones, a Frugivorous-Insectivorous Zone (FIZ) and a Frugivorous-Folivorous Zone (FFZ). This ecological division corresponds with a basal phyletic dichotomy and the family-group classification I employ. Initial zonal segregation centered on selection for efficient exploitation of alternative primary protein resources and is reflected in fundamentally contrasting organizations of the masticatory apparatus (Rosenberger, 1979b). To generalize, cebids primitively inhabit the FIZ and display a light-weight feeding mechanism that deemphasizes molar processing; atelids primitively inhabit the FFZ and are characterized by a heavy-duty system designed for powerful molar occlusion.

Radiation within each zone, i.e., finer niche partitioning, is reflected in the derived morphologies of zone members. In FIZ, for example, an increase in body size and suspensory behaviors and a modification of occlusal morphology may have facilitated evolution of the impressive dietetic opportunism of *Cebus,* which genus may be justifiably regarded as an advanced omnivore. Callitrichines, as previously indicated, evolved a vertically ranging foraging mode and, in more derived lineages, a gum-harvesting dentition. *Saimiri,* by anthropoid standards, may prove to be a rather specialized insectivore, small in size, equipped with acutely designed puncture-crushing postcanines large premolars and highly convergent, frontated orbits.

Among FFZ constituents, *Aotus* altered its dyadic rhythm. Within the size range of frugivore-insectivores, yet removed from their competitive sphere of influence, the nocturnal *Aotus* partakes in a relatively large proportion of leaves (P. Wright, pers. comm.), perhaps conditioned more by its heritage than anything else. *Callicebus, Pithecia, Chiropotes* and *Cacajao* commonly share a very modified ensemble of dental features, particularly evident in the harvesting incisors that appear to be specifically fruit-adapted, possibly to some highly exclusive resource. In postcranium and prehensile tail, atelines evince adaptations to a unique, though not fully understood, foraging mode. For *Ateles,* Cant (1977) suggests that rapid suspensory locomotion may be time-and-energy saving in traveling between widely dispersed patches of preferred fruit sources. *Ateles* and *Alouatta* also exhibit dental adaptations that are quite specific to fruits and leaves, respectively.

The fossil record of Cenozoic platyrrhines suggests that these dual radiations were already underway during the Oligocene-Miocene. The morphology of the early Oligocene *Branisella* is still poorly known but it does not appear to show the derived gnathic or molar features of FFZ atelids. There is very strong evidence of a close cladistic relationship between the late Oligocene *Dolichocebus* and *Saimiri,* whose crania bear several hallmarks of the light-weight FIZ feeding mechanism (Rosenberger, 1979a). This implies that both *Cebus*, the closest living relative of *Saimiri,* and callitrichines, the sister-group of cebines, had already differentiated as lineages. The late Oligocene and early Miocene *Tremacebus* and *Homunculus,* collateral relatives of *Aotus* and the other pitheciines, respectively, document the roughly contemporaneous presence of atelids. *Tremacebus* appears to exhibit *Aotus*-like orbital expansion while *Homunculus* preserves indications of high-crowned, narrow-calibered incisors and stout canines, harvesting specializations of *Callicebus* and saki-uakaris. *Neosaimiri, Cebupithecia* and *Stirtonia,* known from middle Miocene material, each present dental features and adaptations that resemble their modern closest relatives, *Saimiri,* saki-uakaris, and *Alouatta,* in very fine details.

Thus, however meager the record may be, there are good indications that the four major sublineages of living ceboids, and several of the adaptive modalities they represent, were established quite early. It also shows that several generic lineages (i.e., *Dolichocebus-Neosaimiri-Saimiri; Tremacebus-Aotus; Stirtonia-Alouatta)* are particularly long-lived. The presence of such ancient generic lineages is suggestive of a common pattern of anagenetic advances, offsetting lineage extinction by increasingly fine habitat differentiation within relatively stable adaptive zones and heritage parameters. This may explain the surprising similarity of middle Miocene and modern species, the relatively large phenetic gaps between the modern genera, and, in part, their disproportionately large number of monospecific or narrowly varying genera.

This scenario is unlike the pattern inferred for catarrhines, which are even better sampled paleontologically. Their generic lineages do not extend anterior to the middle Miocene (E. Delson, pers. comm.) and a large number of Neogene genera have differentiated, possibly developing into a series of successional adaptive replacements. One speculative explanation for these long range contrasts is that they reflect the greater endemicity of the South American island continent, its reduced surface area and environmental homogeneity placing a premium on directional selection and character divergence and minimizing dispersal possibilities.

Acknowledgments

I thank the organizers and hosts of the VII Congress of the International Primatological Society, Drs. P. D. Gingerich, R. L. Ciochon and B. Chiarelli for their generosity and for making my participation possible. I am also grateful to the various museum officials who have allowed me access to their collections.

18

Callitrichids as Phyletic Dwarfs, and the Place of the Callitrichidae in Platyrrhini

S. M. FORD

INTRODUCTION

The Callitrichidae, which includes the marmosets and tamarins of South America *(Cebuella, Callithrix, Leontopithecus,* and *Saguinus,* but excluding, for now, *Callimico),* are quite distinct from the rest of the South American primates. The major distinguishing characteristics of these monkeys are reproductive twinning, absence of third molars, tritubercular upper molars (lacking a hypocone), the presence of claws instead of nails on all digits except the hallux, and small body size. Most of these characters have been interpreted as representing the primitive condition for all New World monkeys, most recently by Hershkovitz (1970a, 1972, 1977). Evidence will be presented to support the argument that they are all derived features within Platyrrhini and would not have been present in the last common ancestor of all of the New World monkeys. It will be further argued that these features are the result of phyletic dwarfing, i.e., a reduction in body size within a lineage through time.

CALLITRICHIDAE AS A DERIVED CLADE

Reproductive Twinning

Schultz (1948) and more recently Hershkovitz (1977) have argued that the first character, multiple births, is a primitive retention. This argument was based primarily on the fairly widespread distribution of multiple births in Mammalia and on the tendency of smaller mammals to have larger litters. However, as Schultz (1948, 1969c) noted, all other platyrrhines and most other primates generally have single offspring. The only other primates for which multiple births are the norm rather than a relatively rare exception are *Cheirogaleus, Microcebus,* and *Varecia* (or *Lemur*) *variegatus* (Boskoff, 1977; Harrington, 1978; Hill, 1973; Petter-Rousseaux, 1964; Schultz, 1948, 1969c). Hamlett and Wislocki (1934) found that for all mammals, dizygotic twinning (as opposed to litters of varying size or single births) is characteristic of only the callitrichids, the euphractine armadillos (*Euphractus* and *Chaetophractus*), the cape mole (*Chrysochloris*), the platypus (*Ornithorhynchus*), and possibly the African hyrax (*Dendrohyrax*). These data suggest that twinning in callitrichids may be derived.

This suggestion is supported by the fact that marmosets and tamarins have a simplex uterus. A simplex uterus appears to be ". . . designed for the accomodation of a single embryo" (Hamlett & Wislocki, 1934: 94), and is otherwise found only in primates that have single births (all other anthropids). This led Hamlett and Wislocki (1934), Leutenegger (1973), and Hampton (1974) to conclude that callitrichids evolved from ancestors that had single births, and that diovular twinning is a specialization. The only other mammals to have evolved a simplex uterus are the phylos-

tomatid bats (Carter, 1970) and the edentates (Goffart, 1971; Minoprio, 1945; Wislocki, 1928). In both, it appears to have evolved as an adaptation to single births, as in primates. There are two groups of edentates that normally have multiple births (Hamlett & Wislocki, 1934; Wislocki, 1928), but as for the callitrichids, a strong case can be made that single births are primitive and multiple births derived. *Dasypus* accomplishes multiple births by polyembryony, with up to 12 young from a single ovum, which is clearly derived. *Euphractus* and *Chaetophractus* normally have diovular twins (as do the callitrichids), which would appear to be an independent derivation of multiple births convergent with *Dasypus.* These analogous cases strengthen the argument that the simplex uterus evolved as an adaptation to single births and that its occurrence in Callitrichidae strongly argues that reproductive twinning in this family is a derived trait. Hershkovitz' (1977) comment that a bicornuate uterus occurs both in animals with single and with multiple births is then irrelevant to the argument.

In addition, both Schultz (1948) and Boskoff (1977) suggested a link between the number of nipples and number of offspring. Schultz found it ". . . surprising that so far only 1 case of polymastia [supernumerary nipples] has been found in marmosets which usually have twins" (1948: 18) because he considered their twinning to be a primitive retention. The very low frequency of polymastia in callitrichids would be expected, however, if they evolved from ancestors with single births and a reduced number of nipples.

Comparative data on litter size in other primates and other mammals combined with the evidence of the reproductive system of callitrichids and their reduced number of nipples strongly suggest that reproductive twinning in callitrichids is a derived trait within Primates and within Platyrrhini in particular.

Absence of Third Molars

A second major character that links the callitrichids is the absence of third molars (there are occasional exceptions, as in a specimen of *Leontopithecus* at the National Museum [USNM 269705] which has very tiny third molars on both sides). While no one, to my knowledge, has ever suggested that this trait is primitive for New World monkeys, it is well known that loss of a character can occur many times independently. It is here suggested that loss of the third molar has occurred at least twice in South American primates. All New World monkeys have reduced upper third molars (although less markedly so in *Lagothrix, Pithecia,* and *Chiropotes*) and reduced lower third molars, with the exception of some specimens of *Lagothrix* and *Alouatta* (although again somewhat less reduced in *Pithecia, Chiropotes,* and *Callicebus*). Both upper and lower third molars are present but greatly

---

reduced in *Callimico*. In the fossil ceboid *Xenothrix* from Jamaica, third molars are absent. Rosenberger (1977) concluded that *Xenothrix* has independently lost its third molars, based on other highly divergent features of *Xenothrix*, such as molar and mandibular morphology, overall size, and relative tooth size within the tooth row, characters which do not serve to align it with the callitrichids. Since reduced third molars are widespread among New World monkeys, it seems clear that either relatively small third molars are primitive for Ceboidea, or the trend to their reduction is almost universal within the group.

In either case, loss of the third molar is a derived trait which probably developed independently more than once in Ceboidea (*Xenothrix* and the callitrichids) and which may be developing independently in other ceboids now, as shown by the extreme reduction of the third molars of *Callimico, Cebus,* and *Saimiri.* These three genera may, however, be exhibiting a shared derived condition of extremely reduced third molars, a trend which may have culminated in the loss of the third molars in callitrichids (Rosenberger, pers. comm.). Loss of the third molar should not be considered conclusive evidence for the classification of fossil ceboids, nor even weighted as heavily as other characters such as molar morphology. In the case of the fossil monkey *Dolichocebus* and the debate over the presence or absence of a third molar (e.g., Kraglievich, 1951; Hershkovitz, 1970b, 1974a; Hill, 1960), even if and when a final decision is reached, this trait alone should not be considered as conclusive for the placement of *Dolichocebus* in one taxon within Platyrrhini over another.

Tritubercular Upper Molars

A third characteristic of callitrichids is the absence of a hypocone on the upper molars. One occurs very rarely on M^1 (Kinzey, 1973). Hershkovitz (1970a, 1972, 1977) and Kinzey (1973, 1974) both claimed that this is the primitive condition for New World monkeys, and that absence of the hypocone in callitrichids therefore represents a primitive retention. However, the widespread distribution of hypocones on upper molars both within Platyrrhini and within all of the strepsirhine-anthropoid primates (the Simiolemuriformes of Gingerich [1976]) suggests that the ancestral anthropoid had a hypocone which was retained in the ancestral ceboid. This conclusion is supported by the fossil evidence. The earliest known anthropoids from the Fayum of Egypt all have hypocones (Simons, 1972), as does *Branisella,* the earliest known fossil primate from South America, which led Hoffstetter (1969, 1974) to the conclusion that presence of a hypocone is primitive for New World monkeys. Gregory (1922b), Hill (1957), and Rosenberger (1977) also reconstructed the primitive New World monkey as having a hypocone, implying that callitrichids have secondarily lost this cusp. All other known fossils of New World monkeys have hypocones. Therefore, callitrichids appear to have undergone a simplification of their molar morphology, resulting in their present tricuspid upper molars, from a ceboid ancestor that retained a hypocone.

Claws

The presence of claws instead of nails on all digits except the hallux is the fourth character shared by the callitrichids. Cartmill (1974a) clearly demonstrated that nails and apical pads appear to be a functional complex and are not better adapted than claws to most arboreal locomotor activities. He suggested that nails and pads are merely an alternative mode of adaptation to life in the trees, not a superior mode derived from arboreal clawed animals. Rosenberger (1977) pointed out that callitrichids have well developed apical pads lying ventral to the distal phalanx, a position that would seem to exclude these pads from any major role during locomotion in clawed callitrichids such as the one they play in other primates with nails. It is difficult to explain the presence of these pads in callitrichids, unless the callitrichids are descended from an ancestor that had nails.

The presence of a terminal matrix and a deep stratum are generally considered characteristic of "true claws" (Le Gros Clark, 1936). The occurrence of these features in the claws of callitrichids has led to the suggestion that callitrichid claws are a primitive trait (Cartmill, 1974a; Le Gros Clark, 1936). Thorndike (1968) discovered a vestigial form of the terminal matrix and deep stratum in the hallucial nail of callitrichids and in the nails of *Cebus albifrons.* She noted that the connection between these features and claws is not as direct as previously assumed, and that claws and the terminal matrix-deep stratum complex may be lost independently in primates. Therefore, it is clear that callitrichid claws could be derived from nails such as those of *Cebus albifrons.* This supports the findings of Rosenberger (1977), who concluded that the claws of callitrichids are part of a derived adaptive complex within Platyrrhini, since the terminal matrix and deep stratum, as well as the digital flexors, are reduced in callitrichids, whereas none of these features are reduced in other mammals with claws, and since only *Daubentonia* and the callitrichids among all living primates have a claw rather than a nail on the pollex.

The redevelopment of claws in callitrichids, and probably in *Daubentonia,* from an ancestral primate with nails would appear to be at least as parsimonious as four or five independent losses of claws and acquisitions of nails in other primates, as suggested by Cartmill (1974a). In fact, as Cartmill pointed out, other primates appear to be independently approaching a clawed condition, as shown by the development of "*specialized* pointed nails" (Cartmill, 1974a:58, emphasis mine) in *Galago elegantulus, Phaner,* and *Microcebus coquereli.*

Thus, it appears that the presence of claws in callitrichids may better be interpreted as a derived condition within Platyrrhini.

Small Body Size

The fifth and last major characteristic shared by these four genera is their small body size. The largest callitrichid is smaller than the smallest cebid, although their body weights overlap (Napier & Napier, 1967). Hershkovitz (1970a, 1972, 1974a, 1977) has continually advocated the hypothesis that this small size is primitive, and that there is a trend of increasing size and derivedness within Platyrrhini—"...from the smallest and most primitive to the largest and most advanced" (Hershkovitz, 1972:377). This is based on his unsupported assertions that, "All mammals, it appears, tend to become larger in time. . . . Where selective pressures for small size should prevail, the trend toward larger size would be held in check, but not reversed. The small species remains small. It does not become smaller." (Hershkovitz, 1972:372). These assertions would seem to deny the evolution of pygmy elephants and an extensive literature on island faunas.

The smallest callitrichid, *Cebuella,* which Hershkovitz would have as the most primitive New World monkey, is smaller than any other living primate except *Microcebus. Leontopithecus,* the largest callitrichid, is smaller than all other non-callitrichid anthropoids and is larger than only seven other extant non-callitrichid primate genera (Napier & Napier, 1967). If size is treated as any other morphological character, then distribution of this trait within Platyrrhini, Anthropoidea, and Primates as a whole would suggest that the small body size of the callitrichids is derived within Platyrrhini. This suggestion is supported by the fact that all platyrrhine fossils, in fact all anthropoid and most or all strepsirhine fossils, are larger than the callitrichids (Simons, 1972). Hershkovitz (1977: 73) stated, ". . . the size grade between the largest callitrichid and the smallest of the marmosetlike cebids is slight but marks the crossing of a crucial threshold in the form of diet, method of arboreal support, locomotion, and reproduction." There may indeed be a "crucial threshold" in size change which then necessitates various correlated changes is morphology and behavior, but in the platyrrhines, it appears to have been crossed in the opposite direction. The callitrichids have decreased in size through time and are phyletic dwarfs.

CALLITRICHIDS AS PHYLETIC DWARFS

Thus, all five traits which are characteristic of the four callitrichid genera—reproductive twinning, absence of third molars, tritubercular upper molars, claws, and small body size—appear to be derived features within Platyrrhini. These five traits may all be linked to a single adaptive complex, as suggested by Moynihan (1976). Included in this complex would be reduction in body size, or phyletic dwarfing. I suggest that the first four traits have developed as a direct consequence of the fifth—reduction in body size through time.

Gould (1975) posited a "negative allometry" in dwarfed fauna, in that changes associated with decreases in size are not the exact reverse of changes caused by increases in size through time. Marshall and Corruccini (1978) found in marsupials that the allometric relationship in dwarfing lineages exhibits a trend opposite to that among synchronous species of increasing size. It appears that different changes and patterns may be associated with size decrease in lineages than are seen with size increase through time, and that the two processes should be studied separately. However, most of the studies of allometry deal with changes as animals increase in size, rather than decrease.

A small number of studies of dwarfism in mammals exist, primarily of examples from the Pleistocene. However, most of these are concerned with documenting examples of dwarfism and suggesting its possible causes, not its allometric (or "negatively allometric") effects (e.g., Davis, 1977; Edwards, 1967; Kurtén, 1965; Tchernov, 1968). A few studies of dwarfism do investigate its correlated effects on the morphology of the animals. Most of these document the common occurrence of two major changes in the dentition both of which are derived features found in the callitrichids.

First, Gould (1975) suggested that in dwarfing lineages, body size decreases much more rapidly than the size of the post-canine dentition. Evidence tending to support this hypothesis has been reported by both Maglio (1973) and Marshall and Corruccini (1978). Maglio (1973) found that the post-canine teeth of the dwarf fossil elephants of the Mediterranean (*Elephas falconeri*) are proportionately larger than in their mainland ancestors (*E. namadicus*). Marshall and Corruccini (1978) showed an increase in relative molar size within Pleistocene/Recent marsupial lineages which were decreasing in body size. Most of the change in relative tooth size was expressed transversely, or bucco-lingually. If the callitrichids have reduced in size through time, reduction in the size of the jaw would probably have occurred more rapidly than reduction in tooth size. This would result in a *relative* enlargement of the molar battery (although the molars would decrease in absolute size, along with a more rapid decrease in general body size), allowing the dwarfed animal to maintain a large functional chewing surface. Selection for two relatively large molars in a rapidly reducing jaw may have led to the "crowding out" and loss of the third molar, which was most likely already greatly reduced as in *Callimico* and *Saimiri*. Third molar loss in *Xenothrix* may have occurred for the same reason, that is crowding due to an

increase in the relative size of M_1^1 and M_2^2, but by a different evolutionary pathway. As Rosenberger (1977) pointed out, the size of the molar teeth relative to the rest of the dentition is exceedingly large in *Xenothrix*, more so than in any other known New World monkey. In this case, the absolute and relative enlargement of M_1^1 and M_2^2, without a concomitant increase in the size of the mandible, may have independently resulted in the "crowding out" and loss of the third molar.

Boekschoten and Sondaar (1966, 1972), Sondaar and Boekschoten (1967), and more recently Sondaar (1977) have extensively studied nanification or dwarfism of island faunas in the Mediterranean. *Phanourios minor*, the dwarfed hippo, lost the fourth premolar. Sondaar and Boekschoten suggested that this loss may have been the result of changes in occlusion, in turn possibly related to a major dietary and ecological shift from the semi-aquatic habitats of other hippos to one of dry land and different food resources, in association with dwarfing. However, the changes in occlusion may have been due to the loss of the premolar as a result of dwarfism, as suggested for the loss of the third molar in callitrichids, and not vice versa. It appears, then, that a relative enlargement of the post-canine dentition, often resulting in the loss of one or more post-canine teeth as in the Callitrichidae and *Phanourios minor*, is a result of phyletic dwarfing.

The second dental specialization is a simplification of the molar morphology (as exhibited by a loss of the hypocone in callitrichids). Pygmy proboscideans occurred on islands in the Mediterranean, the Celebes, and the Channel Islands of California, and the substantial literature on them was summarized by Maglio (1973). The dwarfing has occurred many times independently, as is evident from the evolution in isolation of these small proboscideans on islands which were (and are) well separated geographically (Europe, Southeast Asia, and North America). All of these pygmy elephants have a reduced number of plates on the molars. Maglio (1973), Sondaar and Boekschoten (1967), and Sondaar (1977) have all suggested that this modification is a direct result of dwarfing.

Moore (1959), in a revision of the Sciurinae, found similar trends in the four genera of pygmy squirrels. They are found in three widely-separated geographical areas, two in Southeast Asia, one in West Africa, and one in South America. These genera are convergent, as is evident from biogeographical and geological data (Moore, 1959; Patterson & Pascual, 1972; Simpson, 1950, 1953a; Webb, 1976; although Hershkovitz [1972] presented an unsupported dissenting view) as well as from their morphology. These four genera differ in 15 major, consistent characters, which in each case align the genera phylogenetically with other, non-dwarfed tree squirrels in that particular geographic area (Moore, 1959). Moore (1959) concluded that dwarfing has occurred independently at least three times. Among the convergent characters shared by these pygmy squirrels is the reduction from four transverse ridges to three on the upper molars, with the third often reduced to a minute cusp. Moore (1959:190) stated that "the twelve skull characters that appear to unite the pygmy genera are the results of dwarfing." Some of the features characteristic of the pygmy squirrels also occur in other genera of tropical tree squirrels that are diminutive, leading Moore to suggest that these other animals may be "incipient dwarf squirrels" (p. 189).

Larry Marshall (pers. comm.), in his study of dwarfing in Pleistocene/Recent kangaroos, has discovered in the *Macropus-M. giganteus* lineage there is often an accessory cuspule on the upper mid-link. This cuspule has a high frequency of occurrence in the large fossil species, while it is only incipiently developed or absent in the "dwarfed" recent kangaroo. Finally, Jon Baskin (1980) had found a simplification of the molar cusp pattern in association with a decrease in size through time in the fossil rodent *Mylogaulus*. All of these occurrences support the correlation between dwarfing and simplification of the molar cusp pattern.

Three of the derived features of callitrichids (loss of the third molar, loss of the hypocone, and reduced body size), then, appear to be direct results of dwarfing. This leaves the last two characters, reproductive twinning and claws. As to the first of these, Leutenegger (1973) argued strongly that the derived development of twinning in callitrichids is a direct response to dwarfing, in particular to the reduction of maternal body size and of the size of the birth canal. Relative to the claws, Cartmill (1974a:59) showed that "...for small mammals, claws are adaptively superior to flattened nails in climbing on large vertical trunks or branches." Callitrichids primarily inhabit dense secondary growth and the lower story of the forest (Hershkovitz, 1977; Moynihan, 1976). If the callitrichids have reduced in size through time, the development of sharp claws would appear to be an important adaptation to allow these small primates (small due to dwarfing) to cling to the large branches and tree trunks of this habitat, which their hands and feet may be too small to grasp between opposed digits (as suggested by both Moynihan, 1976, and Rosenberger, 1977).

Therefore, all five of the characters that unite the callitrichids may be interpreted as being the direct result of phyletic dwarfing (reduction in body size through time).

IMPLICATIONS FOR PHYLOGENY RECONSTRUCTION

There are two major alternate hypotheses of relationship that could be derived from this evidence. One is that these four genera form a monophyletic group, and that the five characters which they share are homologous and retained from a common ancestor that underwent dwarfing at some point in the past. The other is that these characters are not all homologous, and that there have been two or more independent dwarfing events within Platyrrhini. As was noted earlier, dwarf proboscideans and dwarf squirrels developed many times independently. Yet in each case, the dwarfed species share many convergent features. *Saguinus* and *Leontopithecus* may share some further derived traits, and *Cebuella* and *Callithrix* certainly share derived morphological and behavioral traits (such as diet, locomotion, and incisor/canine relationship [Kinzey, Rosenberger, & Ramirez, 1975; Moynihan, 1976; Rosenberger, 1977]). However, since all of the known characters which unite all four genera are apparently the result of dwarfing, and since dwarfing has occured independently in other groups of closely related species and resulted in the development of many seemingly identical traits (many of which are the same ones that occur in callitrichids), the evidence at present is not sufficient to falsify either one of the possible hypotheses of relationship. Until such evidence has been found, neither possibility should be totally ruled out. It is not sufficient in this case merely to claim that multiple dwarfings are "unparsimonious." Apparently, just such an unparsimonious hypothesis best reflects the evolutionary history of both the dwarf squirrels and the dwarf proboscideans.

Callimico goeldii has often been considered intermediate between the cebids and the callitrichids. It normally has single births, retains a very small third molar, and has a tiny hypocone, but it also has claws like those of the callitrichids and is intermediate in body size, approaching that of *Saguinus*. All of these traits, with the exception of claws and small body size, appear to be primitive retentions. The presence of claws and small body size (and perhaps the extreme reduction of the third molars and the hypocone) are most reasonably interpreted as parts of a dwarfing

complex not as far advanced as in the other four genera. All five genera could form a monophyletic group, with *Cebuella*, *Callithrix*, *Saguinus*, and *Leontopithecus* forming a further derived clade in association with greater decreases in size. Or *Callimico* could be converging on some or all of the others, representing an "incipient dwarf platyrrhine" like Moore's "incipient dwarf squirrels."

The one line of evidence that at present may provide an independent (i.e., non-dwarfing related) test of these hypotheses is the serum albumin studies of Cronin and Sarich (1975). They concluded that these five genera form a clade within Platyrrhini. However, they also concluded that *Callimico* is most closely related to the *Callithrix/Cebuella* sister-group, which would necessitate either the independent development of reproductive twinning, loss of the third molar, and loss of the hypocone in *Saguinus/Leontopithecus* and in *Callithrix/Cebuella* (that is, independent dwarfing events), or a reversal in all of these characters in *Callimico*. On the other hand, Moynihan (1976), based on behavioral data, suggested that the *Cebuella/Callithrix* group and the *Saguinus/Leontopithecus* group did develop independently, but that *Callimico* is either related to the latter or independently converging on all four. Thus, there still is not sufficient evidence to allow a reasonable choice between these alternatives.

DISCUSSION

Cebuella Callithrix, Leontopithecus, and *Saguinus* share five derived character states, all of which are related to and caused by their reduction in size through time. The question left unanswered is why these monkeys underwent dwarfing. Most known cases of dwarfing occurred during the Pleistocene, with the most striking examples being size changes (both dwarfism and gigantism) in island populations (e.g., Boekschoten & Sondaar, 1966, 1972; Foster, 1965; Hooijer, 1967; Kurtén, 1972a; Maglio, 1973; Sondaar & Boekschoten, 1967; Sondaar, 1977). Several possible causes of phyletic dwarfing have been proposed, and these have been summarized and discussed by Case (1978), Heaney (1978), and Marshall and Corruccini (1978). These suggested causes have included:

1. Dwarfism serves to maintain a large breeding population under conditions of severely limited space and limited resources (Hooijer, 1967; Kurtén, 1965, 1968; Sondaar & Boekschoten, 1967). As Marshall and Corruccini (1978) noted, this is a group selectionist argument. Heaney (1978) criticized this hypothesis, pointing out that it does not provide a mechanism for dwarfism (unless it be "inbreeding, loss of genetic variability, and subsequent loss of overall fitness" [p. 37]).

2. Dwarfism results from selection against individuals at the large end of the size range, perhaps coupled with selection for those at the small end, under conditions of limited resources (Marshall & Corruccini, 1978). Individuals of smaller body size can survive on absolutely less food (although they require more relative to body weight [Bourlière, 1975]), and they can successfully breed at lower body weights, which would be adaptive under these conditions (K-selection) (Heaney, 1978; Kellogg & Hays, 1975; Kurtén, 1972a; MacArthur & Wilson, 1967). Others have noted that a fluctuating, as well as a limited, resource base may cause dwarfism, as larger-bodied species tend to be more stenotopic and thus more prone to extinction in periods of extreme environmental perturbations (Boucot, 1976; Hallam, 1975).

3. Dwarfism is due to predation. Both increased predation (Edwards, 1967; Grant, 1965; Guilday, 1967; Walker, 1967a) and reduced predation (Sondaar & Boekschoten, 1967) have been suggested as causes of dwarfism. Interestingly, in a study

of North American squirrels, Heaney (ms.) found that body size did not correlate with mortality, indicating that at least in this case body size does not appear to change with or be affected by predation.

4. Dwarfism (and enlarged body size) are adaptive for greater physiological efficiency in relation to climate (Mayr, 1963a; see Heaney, 1978, for review). Heaney (ms.) found that body size of North American squirrels is correlated with various climatic parameters, which he suggested might be measures of productivity and seasonality of the food supply. Therefore, size change in association with climate may merely be a reflection of size change due to limitations and/or extreme fluctuations in the source base (hypothesis 2).

5. Dwarfism is a result of character displacement due to interspecific competition (Davis, 1977; Kurtén, 1965; McNab, 1971; Schoener, 1974). Heaney (1978:38) suggested that, "If an increase in number of competitors in a community allows new species only at the small end of the range of body size [as shown by Brown, 1975], and if no small species are available to fill the niche, selection might act to lower the size of some or all members to take advantage of the new niche space."

6. In his study of the relationship between island area and body size, Heaney (1978) concluded that all of the last four factors listed above (resource limitation, predation, physiological efficiency in particular climates, and interspecific competition) interact to cause body size changes, and that the importance of any particular variable will differ with each individual case (depending on the species, the climate, and the size and degree of isolation of the region). Thus, he suggested that resource limitation may be more important on small islands, predation on medium-sized islands, and interspecific competition on large islands and continents.

These hypotheses have not been well tested, and it is unclear at present which, if any, of them best explains dwarfism in mammals. The problem addressed here is what caused dwarfing in callitrichids. Since we lack a fossil record of Callitrichidae, there is no way of knowing what ecological and climatic conditions were prevalent when they underwent dwarfing. One cannot assume that they dwarfed during the Plio/Pleistocene merely because most of the other known cases of dwarfing occurred during that time period; this may be an artifact of the better fossil record for the Pleistocene. No minimum date can be set for the occurrence of dwarfing event(s) other than post-origination of the Platyrrhini. It is possible to suggest a fairly narrow location: forested areas of continental South America. This would seem to be supported by the fact that all extant ceboids live, and probably all extinct ones occurred, in forested regions (Moynihan, 1976; Simons, 1972). Also, the only occurrences of ceboids outside of continental South America appear to be the result of relatively recent (Plio/Pleistocene) dispersals of genera into Central America (Patterson & Pascual, 1972) and the Greater Antilles (Rimoli, 1977; Williams & Koopman, 1952).

The tropical forest of South America is a varied region that supports one of the richest and most diverse faunas the world (Eisenberg & Thorington, 1973; Hershkovitz, 1972; Müller, 1973). The forest has undergone many changes during the Pleistocene and Recent, and possibly much earlier. Climatic oscillations during the late Tertiary and Quaternary caused periodic arid phases, during which the tropical forest was broken up and dis-

crete patches may have acted as refuges or "islands" (Haffer, 1969, 1974; Müller, 1973). Even now, rivers form very effective barriers to dispersal of much of the lowland neotropical fauna, and changes in river courses cut off populations to form "islands" in the forest (Müller, 1973). Moynihan (1976) suggested that tropical environments may fluctuate more and be less predictable than temperate environments. While the larger ceboids are found primarily in gallery forest, the smaller ones (especially the callitrichids) inhabit dense secondary growth and scrub forest environments. It is possible that the periods of aridity and of reduced and widely separated forest regions, combined with the normal perturbations of the tropical environment, may have caused resource limitations which provided a push toward dwarfism in callitrichids.

Either subsequent to this initial push to dwarfism, or along with it in independently dwarfing lineages, the callitrichid genera developed divergent dietary specializations, with *Leontopithecus* being primarily insectivorous, *Saguinus* omnivorous (insects and up to 60% fruits, a diet much like that of *Saimiri*), and *Callithrix* and *Cebuella* being primarily gummivorous, or sap-feeders (Izawa, 1978; Kay, 1973b; Kinzey, 1974; Moynihan, 1976). It is possible that these divergent diets are partially due to competition. However, as Müller (1973:203) pointed out, "The competition hypothesis does not clearly separate features which can be explained by competition from those acquired much earlier, when forms that now compete with each other were distributed allopatrically."

SUMMARY

1. *Callithrix, Cebuella, Saguinus,* and *Leontopithecus* are all characterized by reproductive twinning, loss of the third molars, tritubercular upper molars (lacking a hypocone), claws on all digits except the hallux, and small body size.

2. All of these traits are best interpreted as derived character states within Platyrrhini. Multiple births are most likely a derived condition in armadillos as well.

3. All of these traits are the result of the monkeys' reduction in body size through time; i.e., these four genera are phyletic dwarfs.

4. Phyletic dwarfing causes a different set of changes than result from increase in size through time. Two changes that seem to be correlated with dwarfing in mammals are: (a) an increase in the relative size of the post-canine dentition, often resulting in the loss of post-canine teeth; and (b) a simplification of the molar cusp pattern.

5. It is equally possible that these four genera underwent dwarfing in two or more separate lineages, or that they are the result of a single dwarfing event within Platyrrhini and form a monophyletic group.

6. *Callimico goeldii* has apparently undergone dwarfing, although not as extensively as have the other four genera. *Callimico* may form a sister-group with some or all of the other four genera, or it may be converging on all of them.

7. No definite conclusion is reached as to the cause of dwarfing in New World monkeys. However, it is suggested that resource limitation due to the formation of "islands" habitats, as a result of the normal perturbations of the neotropical environment and of periods of increased aridity, may have played some role.

Acknowledgments: I wish to thank Dr. Leonard Kristalka, Dr. Lawrence Marshall, Dr. Jeffrey Schwartz, Lawrence Heany, Alfred Rosenberger, Gary Morgan, Jon Baskin, and Simon Davis for advice on earlier drafts of this article. I am especially grateful to Dr. Richard Thorington, Jr. and Stuart Peters for discussions, encouragement, and criticisms. This work was completed while under the support of a Smithsonian Predoctoral Fellowship. An earlier version of this paper was presented at the Annual Meeting of the American Association of Physical Anthropologists, Toronto, April 1978.

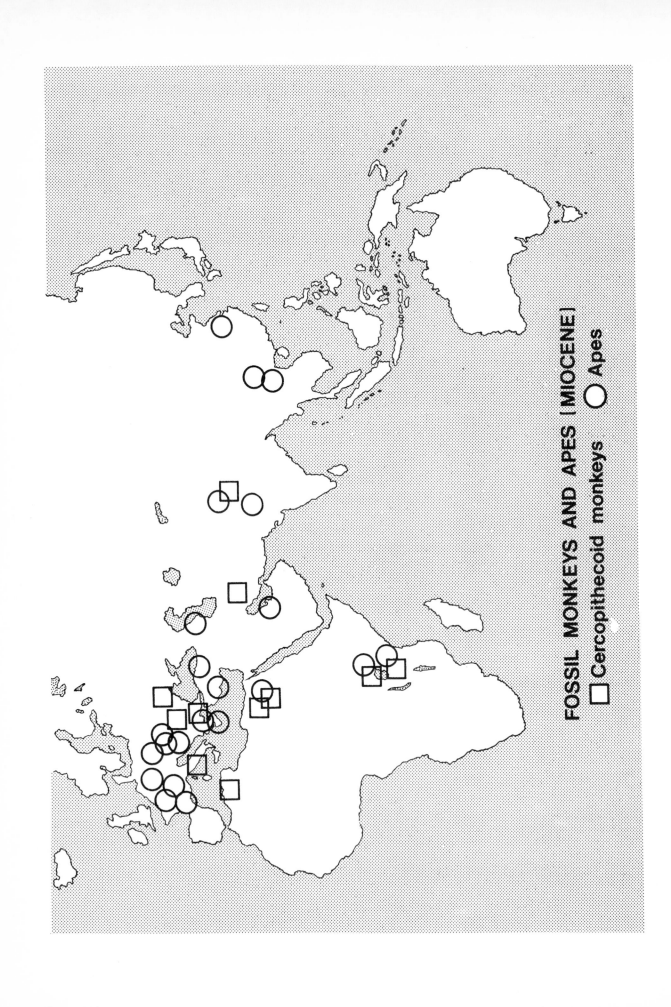

FOSSIL MONKEYS AND APES (MIOCENE)

□ Cercopithecoid monkeys ○ Apes

Part IV

The Evolution of
Old World Monkeys and Apes

The fossil record of higher primates in the Old World begins in the Oligocene of Egypt and extends through the Miocene, Pliocene, and Pleistocene of East Africa, Europe, and Asia. This aspect of primate evolution has been one of the major hotbeds of debate over the past twenty years and is the subject of both part IV and Part V.

In Chapter 19, Elwyn Simons (1965a) discusses several of the many fossil apes discovered by his expeditions in the Fayum Province of Egypt. Simons describes two new genera and species from the early Oligocene (over 31 million years old)—*Aegyptopithecus zeuxis*, which he felt was an early pongid, and *Aeolopithecus chirobates*, which appeared to be a primitive gibbon. When these early anthropoids were known only from their jaws and teeth, they seemed to provide not only ancestors for the more abundant fossil apes from the succeeding Miocene Epoch, but also evidence for the divergence of lineages leading to living apes. He further suggests that another Fayum ape, *Propliopithecus haeckeli*, may be the earliest hominid.

Chapter 20, by Richard F. Kay and E. L. Simons (1980), describes the behavior and ecology of the early anthropoids from Egypt based on their dental and cranial anatomy. They suggest that most were arboreal fruit-eaters, but one species, *Parapithecus grangeri*, may have been more folivorous and perhaps also more terrestrial than the others. In addition to the information it provides about the habits of the Egyptian early anthropoids, this paper demonstrates quite clearly the type of functional analysis of dental structures that have been used to reconstruct the dietary habits of many other fossil primates (see also Kay, 1975, 1977a,b,c, 1978, and Chapters 4 and 31 in this volume).

Chapter 21, by J. G. Fleagle and R. F. Kay (1983), provides a summary of recent views regarding the Fayum apes as well as a review of changing interpretations of these important fossils during the past century. With the recovery of more fossils, including cranial and dental remains, it has become increasingly clear that these Oligocene "apes" were more primitive than any living catarrhines. In much of their morphology they are more similar to platyrrhines than to either Old World monkeys or apes. It thus seems most likely that the Fayum apes occupy a more basal position in catarrhine phylogeny than was previously recognized. Rather than being uniquely ancestral to any one group of living apes, they are probably the common ancestors of all later catarrhines.

Chapter 22, excerpted from a much longer review paper (1970), presents E. L. Simons' early views on the phyletic position of another group of Fayum anthropoids, the parapithecids (see also Gingerich, 1978a, Kay, 1977a; Simons, 1972; Simons and Delson, 1978). On the basis of striking similarities in their molar morphology, he argued that these smaller anthropoids with three premolars were ancestral to cercopithecoid monkeys.

The next chapter, by Eric Delson (1975b), presents an alternative viewpoint on cercopithecoid origins. Delson challenges Simons' views on the role of parapithecids in the ancestry of Old World monkeys. In this view, the parapithecids are a specialized group of early catarrhines whose similarities to Old World monkeys are the result of evolutionary convergence or parallelism, rather than a close phyletic relationship. Instead, he argues for an origin of both apes and Old World monkeys from the larger Fayum "apes," the same view supported by Fleagle and Kay in Chapter 21.

Chapters 24 and 25 discuss the locomotor behavior and phyletic position of the numerous fossil apes from the early Miocene of Kenya—a topic that has generated considerable disagreement and debate during the past twenty years. In the primary descriptions of this important material, Le Gros Clark and Thomas (1951) and Napier and Davis (1959) argued that the early Miocene apes *Limnopithecus* and *Proconsul* were "dental apes" with skeletons intermediate between those of living apes and living monkeys. The conclusions of many subsequent workers were more polarized with respect to the ape-like or monkey-like nature of these primitive apes. Some authors in the late 1960s and early 1970s emphasized the ape-like features of the fossil primates in keeping with a view—based largely on dental characteristics and size—that the living pongids had probably diverged from one another during this period. Thus the early Miocene species of *Proconsul* were depicted as primitive gorillas and chimps, respectively (see Pilbeam, 1969a; Simons, 1967d; Zwell and Conroy, 1973). In direct contrast, Corruccini, Ciochon, and McHenry (1976 and elsewhere) presented a long series of multivariate analyses in support of their view that the Miocene apes were really just "dental apes" with skeletons like those of Old World monkeys. Chapter 24 summarizes the results from many of their studies.

Next, Michael Rose (1983) discusses the skeleton of one early Miocene species, *Proconsul africanus*. His analysis places particular emphasis on the value of joint articulations for reconstructing locomotor abilities in this fossil ape. He agrees with Corruccini et al. (Chapter 24) that the early Miocene apes had a skeletal morphology and locomotor behavior that was very different from that seen among living apes. However, he also finds that they lack many distinctive skeletal features that characterize living cercopithecoids. In his opinion, they had a more functionally and phyletically generalized skeleton lacking the specializations that characterize both groups of living catarrhines.

The final chapter in this part, by Peter Andrews (1981b), reviews the ecological adaptations of fossil apes and compares them with those of living apes and cercopithecoid monkeys. He documents that the diversity of hominoids decreased through the Miocene, while the diversity of cercopithecoids increased. He then discusses the ecological differences between living apes and Old World monkeys and examines the hypothesis that cercopithecoid monkeys are the ecological replacements for Miocene "dental apes."

The succession of chapters in Part IV clearly records an increasing awareness during the past twenty years that the evolution of monkeys and apes was far more complex than had been previously imagined. The fossil apes from the Oligocene and Miocene seem to document not just old relatives of living species but whole radiations of catarrhines that were distinctly different from the monkeys and apes of today. We are only beginning to appreciate the behavioral diversity, ecological preferences (e.g. Pickford, 1983), and biogeography of the numerous fossil monkeys and apes. Needless to say, these reinterpretations of ape evolution have also affected our understanding of hominid origins, the topics of Parts V and VI.

19

New Fossil Apes from Egypt and the Initial Differentiation of Hominoidea

E. L. SIMONS

The third Yale Paleontological expedition to Egypt (November 1963–March 1964) supported by a U.S. National Science Foundation grant in geology (P–433) concentrated fossil-collecting efforts on a locality (I) in the upper levels of the Qatrani Formation, Oligocene of Egypt, which had yielded a few surface finds of early Anthropoidea in the previous season (January 1963). Extensive excavation at this quarry last winter, primarily during December and January, produced a considerable series of continental vertebrates, particularly rich in primates. Eleven mandibles and more than thirty isolated teeth of primitive Anthropoidea were recovered in addition to two jaws found in the previous season.

These finds add considerably to our understanding of the origins of Old World Higher Primates including, potentially, the ancestors of man at that period. Among these new primate materials are several interesting juvenile specimens of *Apidium phiomense* Osborn (1908); another represents a new large species of *Parapithecus*. The primary purpose of this article, however, is to describe briefly two wholly new and somewhat unexpected genera and species of Primates which were recovered in the collections from Quarry I. These and some 90–100 other Fayum primate specimens from lower horizons are to be illustrated and discussed by me in a forthcoming monograph on early Cainozoic mammalian microfaunas of the Fayum. However, in view of widespread interest in higher primate origins it seems advisable to record taxonomically at the present time the two most significant new species.

GEOLOGICAL HORIZON

The locality which yielded the new primates described here is appreciably higher stratigraphically than the classic fossil-vertebrate localities of the Fayum region, and is approximately that of the "upper level" of Osborn (1908, 1909). From this upper fossil wood zone came specimens which Osborn described as the types of a rodent *Metaphiomys beadnelli*, a primate *Apidium phiomense,* and a creodont carnivore *Metasinopa fraasi*. Evolutionary advances in the fauna at Quarry I, compared to those of the lower levels, suggest that considerable time elapsed between accumulation of these faunas, but at present it is not possible to say more than that most species appear to be different from those of the lower zones. For example, examination of Fayum rodents by A. E. Wood, soon to be published, have shown that a common rodent of this upper level, *Metaphiomys beadnelli*, is not present at the next lowermost Quarry (G) in the section, but that a smaller, more primitive species of *Metaphiomys* that could be ancestral to *M. beadnelli* occurs in Quarry G. Similarly at Quarry G a small species of *Apidium*, *A. moustafai*, is abundant. *A. moustafai* is almost certainly ancestral to the much larger *A.*

phiomense of the Quarry I level. Apparently the type specimen of the archaic hominoid primate *Propliopithecus haeckeli* Schlosser (1911) came from about the level of Quarry G, or even lower in the section for a left M_1, indistinguishable from that of *P. haeckeli,* has been recovered from Quarry G. Other teeth of *Propliopithecus* sp. from Quarry G are appreciably larger and somewhat different in morphology from those of the type species, *P. haeckeli*. At the Quarry I level no teeth of *Propliopithecus haeckeli* have been found, but instead only those of a generally larger and more advanced primate species. Taken together these evolutionary differences between the faunas of the two upper quarries, G and I, their stratigraphical separation by more than 250 ft. of varying, partly cyclothemic sediments, including riverine sands, marine limestones and marls, indicate that the fauna of Quarry I represents the latest known Oligocene fauna of Africa and one which is distinctly different from the classic Fayum faunas examined by Andrews (1906) and Schlosser (1911).

SYSTEMATICS

Order, PRIMATES
Superfamily, HOMINOIDEA
Family, Pongidae
Genus *Aegyptopithecus* [1], gen. nov.
Type: Aegyptopithecus zeuxis [2], new species.

Generic characters: Lower dental formula 2?.1.2.3, size approximately that of a gibbon and 25 percent larger than the type of *Propliopithecus* in most comparable measurements. Differs from its contemporary *Propliopithecus* in showing relatively larger canine, premolar heteromorphy and relatively larger M_2 and M_3. Resembles *Proconsul* in marked molar size increase posteriorly, $M_1 < M_2 < M_3$, but differs from members of the latter genus, *Pliopithecus* and *Dryopithecus*, in possessing a more triangular M_3, narrowing posteriorly; unlike most *Proconsul*, entoconid and hypoconulid (rather than hypoconid and hypoconulid) joined by distinct crest. Resembles *Dryopithecus* in rounded outline of M_{1-2} but unlike later dryopithecines retains lower and more rounded molar cusps, as in *Propliopithecus*. Ascending ramus of mandible approximately 40 percent broader from front to back compared with depth of horizontal ramus at M_2 than in *Propliopithecus;* ascending ramus of mandible more vertical relative to the tooth row than in *Propliopithecus* or *Proconsul*, probably indicating comparatively short face.

While resembling *Propliopithecus, Aegyptopithecus* is clearly much closer to Mio-Pliocene dryopithecines particular to East African Miocene species of this group which it agrees with in showing unusually marked molar size increase posteriorly. At the symphysis, however, *Aegyptopithecus* is more typically pongid than some members of the East African species of *Proconsul*, and shows comparatively less anteroposterior thickening of the symphysis and a more backward-directed simian shelf. The characters of symphyseal robusticity and degree of development of simian shelf are notably variable in living hominoids (Schultz, 1963). In *Proconsul nyanzae* and *Proconsul major* this shelf is usually more ventrally directed than in *Aegyptopithecus*.

Aegyptopithecus is close in size to *Pliopithecus (Limnopithecus)* from East Africa but differs from it and from European *Pliopithecus* in showing more distinct labial cingula, relatively smaller premolar and molar trigonid, less distinct cross-crests between protoconid-metaconid and entoconid-hypoconid, relatively much larger M_3 with narrower M_3 talonid, and comparatively larger M_3 hypoconulid. Being much earlier and more generalized than *Dryopithecus* or *Proconsul*, *Aegyptopithecus* does show some likeness to *Pliopithecus (Limnopithecus)*, which Ferembach (1958) has maintained may have greater affinity with the great apes (pongines) than with the gibbons. However, resemblance between all the East African and Egyptian hominoids of the Oligocene and Miocene is great. In general, they exhibit the sort of morphological interrelationship to be expected of various members of a category comparatively close to the time of its initial radiation. These interrelationships which are reasonably apparent in early hominoids naturally affect all attempts to place given species in higher categories. Dental features now seen only in gibbons or only in the African apes *(Pan, Proconsul)* were then combined in one species as Ferembach's observations show.

Since it is characteristic of African *Pliopithecus (Limnopithecus)*, European *Pliopithecus* and modern gibbons to have lower third molars which are shorter or equal in size to M_2 (and other features cited here), this new Fayum primate would appear to relate more definitely to the ancestry of *Proconsul*. The comparatively enlarged M_3 of *Aegyptopithecus* and of *Proconsul* species is one of the few features which seem to differentiate species of the latter from the type species of *Dryopithecus, D. fontani*. Moreover, if some Fayum primates are near the ancestry of later Hominoidea, as certainly appears to be probable, another contemporary of *Aegyptopithecus* (described here) would seem to be a much better candidate for possible ancestry to the gibbons of Miocene-Recent times.

Aegyptopithecus zeuxis [2], new species
Type: C.G.M. 26901, left mandibular ramus with P_4–M_2; partial alveoli of C, P_3 and M_3 and symphisis.
Hypodigm: Type and Yale Peabody Museum (Y.P.M.) 21032, left horizontal and vertical ramus with broken P_4, M_1–M_3, American Museum of Natural History (A.M.N.H.) 13389, left mandibular ramus with broken C, P_3 roots and alveoli of P_4–M_3.

Horizon and *Locality:* Yale Expedition Quarry I, upper fossil wood zone, Qatrani (Fluviomarine) Formation, Oligocene, Fayum Province, Egypt, U.A.R.
Specific characters: Not distinguished from generic.
The three known mandibular rami complement each other in parts preserved although the reference of A.M.N.H. 13389 to this species remains slightly in doubt in view of its fragmentary condition and uncertain locality and horizon; it was collected in 1906.

Family, (?) Hylobatidae
Genus *Aeolopithecus* [3] gen nov.
Type: Aeolopithecus chirobates, new species.
Generic characters: Lower dental formula 2.1.2.3. Slightly smaller than *Propliopithecus* and much smaller than all other Oligocene-Recent hominoids. Differs from the contemporary genus *Propliopithecus* in showing marked premolar heteromorphy (sectorial P_3), comparatively much larger canines and more procumbent incisors and from both *Aegyptopithecus* and *Propliopithecus* in having tooth rows more divergent posteriorly and in showing relatively higher and deeper genial fossa, with horizontal ramus of mandible shallowing posteriorly. Differs from *Pliopithecus* in possessing relatively larger canines, and greater degree of premolar heteromorphy. Resembles some *Hylobates* and differs from *Aegyptopithecus*, European *Pliopithecus* and *P. (Limnopithecus)* in having much reduced M_3.

Although molar crown patterns are obscured by destruction of the enamel the observable characters of the mandible and teeth all suggest that this primate is related to the gibbons, living and fossil. Particularly pertinent to this conclusion are the large and long canines, reduced M_3 and posterior shallowing of the mandibular ramus. As in some *Hylobates* the genial fossa is remarkably high and deep.

In size *Aeolopithecus chirobates* is rather larger than *Oligopithecus savagei* and *Parapithecus fraasi*. It also comes from a higher horizon than either of the latter species. *A. chirobates* differs from *Oligopithecus savagei* not only in size and proportion of the mandibles (the mandible shallows less posteriorly in the latter), but also in that the mental foramen is located much farther forward, the canines are relatively larger and the premolars more heteromorphic. Molar crowns of *Aeolopithecus chirobates* also appear to lack the cross-crests of *Oligopithecus savagei*. Close relationship of *Aeolopithecus chirobates* to *Apidium phiomense, Apidium moustafai* or *Parapithecus fraasi* is ruled out by their differing dental formulae and wholly dissimilar tooth morphologies.

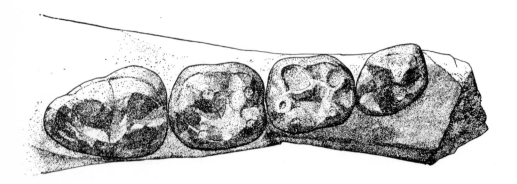

Figure 1   *Crown view of the left mandibular teeth of the type of* Aegyptopithecus zeuxis, *C.G.M. 26901, with* M_3 *restored from Y.P.M. 21032 (x 4.00).*

Measurement (mm) of Aegyptopithecus zeuxis *and* Propliopithecus haeckeli

Dentition	C.G.M. 26901 type	A. zeuxis Y.P.M. 21032	A.M.N.H. 13389	P. haeckeli S.M.N. 12638 type (left side)
Length of lower tooth series C–M_3 inclusive	—	—	40.5+	27.0
Length of series P_4–M_3	—	25.1	27.5+	18.5
Length of series P_4-M_2	17.7	16.8	18.0+	13.1
P_4 maximum transverse diameter (width)	4.2	—	—	4.0
P_4 maximum anteroposterior diameter (length)	4.4	—	—	3.5
M_1 maximum transverse diameter	5.2	5.4	—	4.5
M_1 maximum anteroposterior diameter	6.1	5.9	—	4.8
M_2 maximum transverse diameter	6.3	6.3	—	4.5
M_2 maximum anteroposterior diameter	7.1	6.8	—	4.8
M_3 maximum transverse diameter	—	6.8	—	4.1
M_3 maximum anteroposterior diameter	—	8.6	—	5.6
Mandible				
Depth beneath M_2	14.9	14.1	13.0+	12.4
Depth beneath M_3	16.0e	16.2	15.0+	14.0

Abbreviations: A.M.N.H., American Museum of Natural History, New York, C.G.M., Cairo Geology Museum, Cairo, S.M.N., Stuttgart Museum of Natural History, Stuttgart, Germany, Y.P.M., Yale Peaboy Museum, New Haven, Connecticut.

This is one of the most distinctive of Fayum primates known so far and it is regrettable that only one specimen exists. However, it is the second most complete specimen of a Oligocene primate mandible known, after the type of *Parapithecus fraasi* (S.M.N. 12639*a*), and it should provide data for useful comparative analysis when early gibbons become better known.

The most direct relationships of this species are with *Pliopithecus*, particularly sub-genus *Limnopithecus* which possesses a similarly large simian shelf and with which *Aeolopithecus* agrees in most other observable features except for its much smaller size. Conceivably *Aeolopithecus* is closer to the line that gave rise to modern *Hylobates* and *Symphalangus*, that is, *Pliopithecus*, in view of its third molar reduction, as is also observed in some species of *Hylobates*. Nevertheless, adoption of such a conclusion on the strength of this one feature alone seems inadvisable. It remains entirely possible that none of the known fossil gibbons is actually ancestral to present-day south-east Asian species.

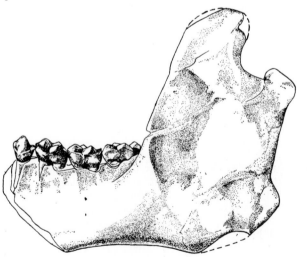

Figure 2 *Lateral view of the left lower jaw of* Aegyptopithecus zeuxis *Y.P.M. 21032 with P_4 restored from C.G.M. 26901 (x c. 1.50).*

In the early part of this century Fourtau (1918) described a species, *Prohylobates tandyi*, from the supposed Burdigalian (early Miocene) fauna of Wadi Moghara, west of Alexandria, Egypt. As the name implies, he concluded that this fossil was related to the gibbons. Consequently, it is necessary to consider the relation of this primate to *Aeolopithecus*. Examination shows clearly that *Prohylobates tandyi* is a cercopithecoid monkey, a fact missed by most previous commentators on its status (including Remane (1924), MacInnes (1943), although Le Gros Clark and Leakey (1951) did remark "the molars [of *P. tandyi*] appear to be more cercopithecoid in their cusp pattern". This is a typical example of the danger of publishing discussion on fossils which have not actually been seen and compared with other pertinent materials. *Prohylobates tandyi* is clearly congeneric and probably conspecific with a still undescribed ceropithecoid monkey which occurs in the Miocene Rusinga Island fauna of East Africa, and consequently bears no taxonomically significant relationship to *Aeolopithecus*.

Aeolopithecus chirobates [4] new species

Type: C.G.M. 26923, complete horizontal rami of mandible fused at symphysis with incisor alveoli and left and right C–M_3.

Hypodigm: Type.

Horizon and locality: Yale Expedition, Quarry *I*, upper fossil wood zone, Qatrani (Fluviomarine) Formation, Oligocene, Fayum Provence, Egypt, U.A.R.

Specific characters: Not distinguished from generic.

GENERAL DISCUSSION

In determining the position of given fossils in evolutionary trees it is often necessary to make use of inferences which cannot be tested in full but which often serve as a basis for further investigations. One of these is that many Egyptian Oligocene primates are likely to be in or near the ancestry of later monkeys, apes and men. Students of fossil mammals know that there is little evidence of the occurrence of higher primates (Anthropoidea) in the known Eocene Oligocene and early Miocene faunas of Eurasia; thus pointing to the continent of Africa (where Oligocene and Miocene Primates are abundant) as the most probable place of differentiation of monkeys, apes and men. Of course, were early and middle Cenozoic faunas better known in Asia, distribution of early Anthropoidea might be extended. Recent review of the early

Measurements (mm) of Aeolopithecus chirobates *in Comparison with other Fayum Primates of Similar Size* [5]

Dentition	Aeolopithecus chirobates type, C.G.M. 26923		Oligopithecus savagei type, C.G.M. 29627	Parapithecus fraasi type, S.M.N. 12639a
	Left	Right	Left	Left
Length of C–M_3 inclusive	28.3	28.0	—	23.5
Length of P_3–M_3 inclusive	22.0	22.0	—	21.0
Length of P_4–M_3	18.0	17.9	—	17.7
C transverse diameter	3.5	3.5	3.3	2.5
C anteroposterior diameter	5.9	5.6	3.7	3.2
P_3 transverse diameter	2.4	2.6	3.1	2.5
P_3 anteroposterior diameter	4.3	4.5	4.2	3.3
P_4 transverse	2.4	2.7	3.0	2.5
P_4 anteroposterior	3.5	3.6	3.3	3.3
M_1 transverse	3.7	3.6	3.4	3.2
M_1 anteroposterior	5.0	4.9	4.2	4.2
M_2 transverse	4.0	3.9	3.5	3.4
M_2 anteroposterior	5.2	5.0	4.2	4.3
M_3 transverse	3.1	3.0	—	3.3
M_3 anteroposterior	4.0	4.0	—	4.2
Mandible				
Maximum symphyseal thickness	6.8		—	—
Maximum symphyseal depth (along axis)	16.1+			9.0
Depth of mandible below P_3	11.7		10.4	7.5
Depth of mandible below M_1	9.8		9.4	7.6

Cainozoic primates of Europe (Simons, 1961a, 1962b) has shown that none of the known Eocene primate species of that continent is a good candidate for possible relationship to the ancestry of monkeys, apes or men. One species, *Alsaticopithecus leemani,* described by Hürzeler (1947) from middle Eocene deposits in Alsace, has been tentatively suggested as having molar resemblance to Anthropoidea by several authors, as indeed it does. However, McKenna (1960) has correctly identified this species as a member of the family Microsyopidae. It thus belongs to a doubtfully primate family known otherwise only from the Eocene of North America. Microsyopid molars parallelistically resemble those of certain hominoids, but it is unlikely that Old World higher primates are derived from Microsyopidae or anything like them.

The only remaining, possibly hominoid, primates of early age outside Africa are from the late Eocene of Burma: *Pondaungia cotteri* (Pilgrim, 1927) and *Amphipithecus mogaungensis* (Colbert, 1937). The original materials of both these species have been examined during the process of preparing this article, but the fragmentary condition of the types of these two forms precludes a definite statement that the origins and early geographic distribution of Old World higher primates were extended or were centred outside Africa.

The type of *Pondaungia cotteri,* which was quite inadequately figured and described by Pilgrim, consists of partial mandibular rami of both sides of the lower jaw carrying chemically corroded second and third molars. As nearly as can be judged, in size, outline and cusp placement, these molars resemble those of *Propliopithecus*. Associated with the two mandibular fragments of *Pondaungia* at the site of discovery was a left maxillary fragment with M^{1-2}, apparently belonging to the same individual.

Pilgrim's figures (Pilgrim, 1927) and discussion of *Pondaungia* left much to be desired and indeed, on the basis of the data he then provided, it was scarcely possible to make an ordinal placement for the form. Consequently, most publications on early Old World primates since 1927 either omit *Pondaungia cotteri* or question placement of this species as a primate.

Pilgrim noted that web-like chemical erosion of the enamel of lower right and left M_{2-3} obscured their crown patterns. He also supposed this to be the case for the upper molar surfaces. However, microscopic examination of the type at Calcutta indicates that, while the lower molars do appear to have suffered erosive damage, the grooves and ridges on the upper teeth represent natural surfaces except in one or two broken areas. Thus it is possible to observe upper molar structure in considerable detail in

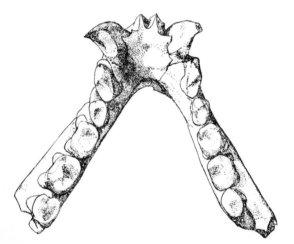

Figure 3   *Type mandible of* Aeolopithecus chirobates *C.G.M. 26923 (x c. 2.00).*

Pondaungia. These molars have the typical placement of the four primary cusps, pronounced lingual cingula and crenulate occlusal surface characteristic of most early hominoids, but multiple cusps exist in the region of paraconule and metaconule and there are no ridges connecting any of the main cusps.

In the combination of these molar features, *Pondaungia* somewhat resembles species of the European Eocene tarsioid genus *Microchoerus* and members of the North American omomyid primate genera *Hemiacodon* and *Ourayia.* It is tempting to interpret molar crown structure in *Pondaungia* as morphologically transitional between that characteristic of Omomyidae and that of Pongidae, but there are non-Primates which *Pondaungia* also resembles, particularly some species belonging to the hyopsodontid Condylarthra. This archaic mammalian family is not definitely known to have occurred in Asia, but its presence there in the late Eocene would not be particularly surprising. The evidence on the taxonomic relationship of *Pondaungia* is therefore equivocal but, rather than regarding it as *incertae sedis,* the balance would seem to weigh in favour of a questioned placement among *Hominoidea*—archaic members of which it resembles in known comparable parts.

From the Pondaung Formation of Burma is the type of *Amphipithecus mogaungensis* Colbert (1937) (Fig. 4). Colbert's reasons for placing this animal among the higher primates remain acceptable and have been strengthened by its dental resemblance to the recently discovered Fayum primates *Oligopithecus savagei* and *Aegyptopithecus zeuxis.* In its mandibular depth, foreshortening of the anterior dentition, and symphyseal cross-section this species is typically hominoid. Indeed, Colbert made a queried reference of the species to the taxonomic family of the apes—Simiidae, now Pongidae. Unfortunately, only two premolars and first molar of the left side are preserved in the type and only specimen. Roots of a canine and a P_3 are also evident and if his identification of these teeth is correct this means that three premolars were present. Although all later Old World Anthropoidea, with the exception of the Parapithecidae, possess only two premolars above and below, the occurrence of three here is presumably a primitive feature. Most other Eocene Primates had at least three premolars. Taken altogether, the structure of the teeth and jaw fragment of *A. mogaungensis* permits its close taxonomic association with the early apes of the Fayum, but it is regrettable that so little is known of this species. Since this is the case, understanding of its full significance must wait until more material is secured. It is clear, as Colbert quite adequately demonstrated, that this fragment is more like comparable parts of early catarrhines than like members of any other known order of Mammalia. The conclusion which must be drawn is that known materials of *Pondaungia* and *Amphipithecus* indicate, but do not prove, that early hominoids or hominoid-like primates existed in the Asian tropics before the end of the Eocene Epoch. The question whether these

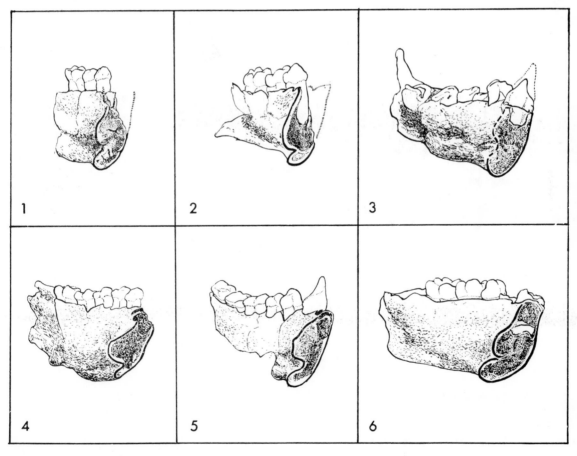

Figure 4   *Symphyseal cross-sections in several early primates, showing variations in symphyseal thickness and in expression of simian shelf (not to the same scale). (1)* Amphipithecus mogaungensis *A.M.N.H. 32520 type, Eocene of Burma; (2)* Aegyptopithecus zeuxis *C.G.M. 26901 type; (3)* A. zeuxis *A.M.N.H. 13389; (4)* Parapithecus *sp. nov. Yale Coll.; (5)* Sugrivapithecus salmontanus, *Y.P.M. 13811 type.*

species bear any relation to the latter radiation of apes, man and monkeys of the Old World cannot really be settled on the basis of evidence available at present.

Consequently, the Fayum primates of Egypt remain the only adequate basis for speculation as to the nature of the early stages of differentiation of the higher primates of the Eastern Hemisphere. The following tentative conclusions as to their phylogenetic meaning are consequently presented.

Propliopithecus haeckeli Schlosser (1911) has most often been treated as representing an ancestral gibbon or at least a good morphological antecedent for Mio-Pliocene *Pliopithecus* species. Although Schlosser (1911) in his original description stressed this point of view, his comparisons between *Propliopithecus* and *Pliopithecus* are not very specific and he equally considered a number of other phyletic relationships for *Propliopithecus*, including the possibility that it was ancestral to *Pithecanthropus* and modern man. Of course, at the time of his writing there were not many other fossil Higher Primates for comparison. Afterwards, with more comparative material available, Gregory (1916), Abel (1931) and Le Gros Clark (1952) also stressed resemblances between *Propliopithecus* and European or East African *Pliopithecus*. More recently, in the most comprehensive investigation of *Pliopithecus* published so far, Hürzeler (1954a) questioned that *Propliopithecus haeckeli* could be ancestral to the Mio-Pliocene gibbons. He noted that in this Egyptian species there is no P_3 sectoriality, that the anterior fovea of P_1 does not open on the lingual side as in *Pliopithecus* species and the metaconid crest of P_4 of *Pliopithecus* is lacking in the Fayum primate, which instead possesses a labial cingulum not seen in *Pliopithecus*. Moreover, in *Propliopithecus* a weak crest exists between M_3 hypoconid and hypoconulid while in *Pliopithecus* instead a crest joins entoconid and hypoconulid. In *Propliopithecus* the lower molar cusps are lower and more rounded and lack much of the crown-crenulation and development of accessory tubercules common in *Pliopithecus*. All these distinctions, when taken together with the now documented presence of a rather more *Pliopithecus*-like species, *Aeolopithecus chirobates*, contemporary with at least one species of *Propliopithecus* in the Fayum, suggest that *Propliopithecus haeckeli* should not be thought of as taxonomically close to the gibbons.

In its overall morphology, the type mandible and dentition of *Propliopithecus haeckeli* (S.M.N. 12638) is more reminiscent of what one would expect to see in an Oligocene forerunner of the family Hominidae rather than in an early dryopithecine or gibbon. The probability that this primate indicates that, by early Oligocene times, species ancestral to living man had already differentiated from those which led to *Dryopithecus* and subsequent great apes is thus increased. Confirmation of this interesting possibility will not come unless Oligocene-Miocene forms transitional between *P. haeckeli* and *Ramapithecus punjabicus* of the late Miocene of Africa and Eurasia are recovered by future field research in relevant areas of the Old World.

Notes

[1] Named in reference to the Egyptian provenance of the type.

[2] From Greek, yoke, join. In reference to the mandibular and dental morphology which shows intermediacy between *Propliopithecus* and *Dryopithecus*.

[3] Named with reference to the fact that the type specimen was exposed in Yale Quarry I by a wind-storm, for Aeolus god of the winds, and in analogy with *Oreopithecus, Limnopithecus*, etc.

[4] From Greek *cheir*, hand, and *bates*, treader or climber, with reference to the use of hands in arboreal progression presumed for early hominoids and in analogy with the modern gibbon, *Hylobates* (forest climber).

[5] All measurements are decreased by surface erosion or corrosion of enamel in the type and are therefore $1/2$ mm less than they would be in an undamaged individual. As nearly as can be judged, the removal of enamel has affected all tooth dimensions about equally, so that the same proportionate relations are retained as between any two given teeth.

20

The Ecology of Oligocene African Anthropoidea
R. F. KAY AND E. L. SIMONS

INTRODUCTION

... from deposits ... Egypt. Finds ... specimens of ... that substan- ... anial remains ... ions based on ... the available ... zations of the

... ni Formation, ... dstones with ... requent inter- ... r. The forma- ... senting rivers, ... ate is thought ... l, with dense ... e possibility of ...

... y overlies the ... f Late Barto- ... cene), and it is ... ically dated at ... esent (Simons, ... to the deposi-

... t of this centu- ..., and "Moeri- ... epresented by ... iomense, des- ... e Upper Fossil

... nd 1970s have ... l el Qantrani ... mes the oldest ... m a single left ... omes the type, ... s well as some ... morphology to ... cus species. A

... l Wood Zone, about 50 yd. above Quarry G, at Quarries I and M, including *Apidium phiomense, Parapithecus grangeri, Aegyptopithecus zeuxis,* and *Proplipithecus (= Aeolopithecus) chirobates* (Simons et al., 1981).

The Fayum primates appear to fall into two natural groups. One group, the Parapithecidae, includes *Apidium* and *Parapithe-*

cus. A second, dentally ape-like, group includes *Aegyptopithecus* and *Propliopithecus.* The affinities of *Oligopithecus savagei* are poorly understood, although it has the same dental formula as the fossil apes.

BODY SIZE OF FAYUM SPECIES

Estimation of Body Size

Dental, especially molar, remains are the most commonly recognized fragments of Fayum anthropoids. In fact, more than half of the recognized species are represented by dental fragments alone. For this reason it has been necessary to rely on dental dimensions to estimate the body weights of these species. A regression equation was formulated to express log body size (body weight) as a function of log second lower molar length for 106 species of living primates including lemurs, lorises, tarsiers, hominoids, cercopithecoids, and cebids. The New World monkey family Callitrichidae was not sampled because of extreme molar reduction in that group. The equation is $\log_{10}(B) = 2.86 \log_{10}(L) + 1.37$, where B is the body mass in grams, and L is the second lower molar length in millimeters. The coefficient of correlation for the equation is 0.949. Thus, body weight alone accounts for 90% (i.e., 0.949^2) of the variance in second molar length among extant primates. This regression equation is very close to one reported by Gingerich (1977a) for extant Hominoidea, where $\log_{10}(B) = 2.99(L) + 1.46$, with a coefficient of correlation of 0.942. Table I provides information on the second lower molar length of samples of 10 species of Fayum anthropoids, together with an estimate of their body mass, in grams, on the basis of our regression equation. The body size estimates provided here should be used with great caution. Even with the very high coefficients of correlation in the log-log regression equations, the 95% confidence intervals of estimated body size are extremely large.

Body size estimates for *Aegyptopithecus* can also be obtained from the dimensions of a nearly complete skull in the Cairo Geological Museum (Table II, Fig. 1). The mean estimated body weight for *Aegyptopithecus*, based on cranial dimensions, is about 5300 g, with a 95% confidence interval of 935 g. This estimate of *Aegyptopithecus* body size is virtually identical to that estimated from molar length.

Generally, parapithecids tend to be smaller than the apes at any given stratigraphic level. For example, *Parapithecus grangeri* and *Apidium phiomense* from the Upper Fossil Wood Zone are smaller than *Aegyptopithecus zeuxis* and *Propliopithecus chirobates.* From geologically older Quarry G, *Apidium moustafai* is smaller than contemporary *Propliopithecus* sp.

Similarly, there is a tendency for earlier (geologically older)

Table I   *Second Lower Molar Length and Estimated Body Mass*

	M_2 Length			Estimated Body Mass (g)	95% Confidence Range of Body Mass (g)
	X	SD	N		
Apidium phiomense	3.92	0.16	13	1177	1051–1318
Apidium moustafai	3.24	0.25	12	682	596–781
Parapithecus grangeri	4.55	0.18	7	1802	1632–1990
Parapithecus fraasi	3.94	—	1	1194	1067–1337
Oligopithecus savagei	3.38	—	1	763	668–872
Propliopithecus chirobates	5.43	0.33	4	2951	2698–3228
Propliopithecus markgraf[a]	4.48	0.07	3	1724	1561–1903
Propliopithecus haeckeli	5.00	—	1	2360	2147–2594
Aegyptopithecus zeuxis	6.90	0.47	12	5876	5346–6450

[a] Here, as in Tables V and VII, two isolated teeth from Quarry G have been assigned tentatively to *Propliopithecus markgrafi* Schlosser, although the type comes from an uncertain horizon.

Table II   Aegyptopithecus zeuxis: *Cranial Dimensions and Estimated Body Mass[a]*

Cranial Dimension	Value (mm)	Slope	Constant	R for Equation	Estimated Body Mass (g)
(A) Orbital diameter	18.0	5.55	−8.21	0.89	2,558
(B) Interorbital breadth	8.5	2.54	−3.32	0.79	6,420
(C) Prostion-nasion	38.5	2.14	+0.36	0.92	3,545
(D) Maximum dental arcade width	33.6	3.68	−4.29	0.96	10,038
(E) Anterior orbital edge to anterior point on maxillary-premaxillary suture	20.6	1.61	+3.95	0.91	6,705
(F) Prostion-inion	101.8	3.07	−5.85	0.94	4,149
(G) Maximum dental arcade length	39.4	2.58	−1.10	0.96	2,850
			Mean estimated mass:		5,281
			Standard error:		382
			95% Conf. interval:		+935

[a] Regression equations are derived from 16 Old and New World monkey species using the following equation: ln Body Mass (g) = Slope (ln Dimension) + Constant. The cranial dimensions are identified and defined by Cartmill (1970).

species to be smaller than their closest later (geologically younger) relatives. Thus, *Apidium moustafai* from Quarry G is smaller than *Apidium phiomense* from Quarries I and M, higher in the section. The same is true for ape material from Quarry G in comparison with Quarries I and M ape species.

Aspects of the Adaptive Significance of Body Size

Among living primate species, body sizes are broadly correlated with what a species eats and where it eats it. Thus, insectivorous species tend to be smaller than folivorous species. (Fruit-eating species overlap both groups.) Furthermore, terrestrially or semi-terrestrially adapted primates tend to be larger than their closest arboreal relatives.

Body Size and Diet   Body size may be viewed as part of a species' adaptation to exploit various types of food (Kay, 1973b). Histo-

grams of the body size distribution of arboreal species with insectivorous, folivorous, and frugivorous diets are shown in Fig. 2. (As used here, the terms "insectivore," "folivore," and "frugivore" are a shorthand notation for whether a species' diet is dominated by invertebrate foods, plant foods containing a high percentage of fiber, or plant foods low in fiber, respectively. Each term refers to the most common kind of food within each category.) Insectivorous primate species, with the doubtful exception of *Daubentonia* (Kay and Hylander, 1978), are smaller than folivorous species. Thus, it is possible to distinguish all extant primate insectivores

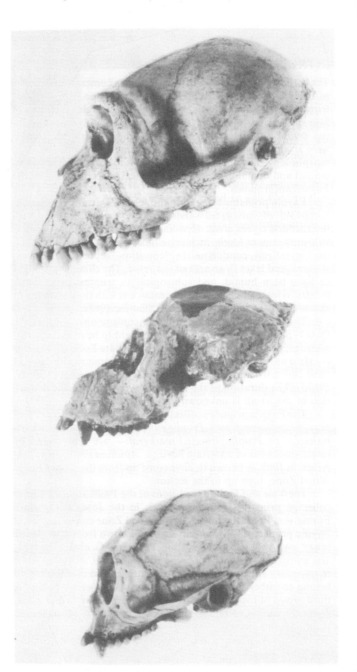

Figure 1   *Comparison of the skulls of* Macaca sylvanus *(USNM 398445) (top),* Cebus albifrons *(AMNH 19014) (bottom), and* Aegyptopithecus zeuxis *(cast of CGMU 40237) (center) to illustrate the approximate size of this skull.*

from extant folivores by body size alone. The adaptive reasons for this size difference are discussed by Kay and Hylander (1978).

The body size distribution of fruit-eating arboreal primates has three peaks (Fig. 2). This distribution may be explained by the limitation of fruit as a protein source. According to Hladik *et al.* (1971) and Hladik (1977), the fruits eaten by some groups of primates contain large amounts of readily available carbohydrates, but small amounts of protein. Primate fruit-eaters must eat other kinds of foods to get their protein. Two such protein sources are leaves and insects, each of which contains more than 20% protein by dry weight (Hladik *et al.*, 1971; Hladik, 1977; Boyd and Goodyear, 1971). The choice of leaves or insects as a source for protein in the diet of frugivorous primates is an important element in selection for body size. Frugivores which concentrate on insects as a source for their protein tend to be relatively small, whereas frugivores which obtain their protein from leaves tend to be relatively large. The intermediate peak in the histogram for frugivores in Fig. 2 is made up of fruit-eaters with a secondary specialization in insects and fruit-eaters with a secondary specialization in leaves.

The body size distribution among the Fayum species (Fig. 2), as estimated from their second lower molar lengths, tends to rule out insectivory as a plausible feeding strategy for these species. On the basis of body size, a leaf- or fruit-eating diet is indicated. The larger species, *Parapithecus grangeri, Oligopithecus, Propliopithecus,* and *Aegyptopithecus,* would most likely have used leaves as a source for their protein. The smaller species, *Apidium moustafai* and *Parapithecus fraasi,* could equally well have utilized insects as a source for protein.

Body Size and Terrestriality   Among extant primates, the Old World monkeys serve as the only available model for comparison of body size among closely related arboreally and semiterrestrially adapted species, since each of the major taxonomic groups of Old World monkeys contains species with both lifestyles. (As used here, the term "semiterrestrial" is intended to refer to a species which spends a considerable portion of its foraging time on the ground, even though it may be quite at home in an arboreal setting.)

Although arboreal and terrestrial Old World monkeys over-

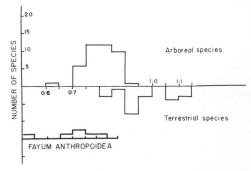

Figure 3.   *Histograms of $\log_{10}$ second lower molar length (as a measure of size) for arboreal species (above the line) and terrestrial species (below the line), with the size distribution of Fayum species shown separately.*

lap in size to a considerable degree (Table III, Fig. 3), terrestrial species tend to be larger than arboreal species. This phenomenon is especially marked when each subgroup of Old World monkeys is considered separately (Table III). Arboreal members of the tribes Cercopithecini and Papionini, and the subfamily Colobinae, are smaller than their more terrestrially adapted close relatives. Some overlap in body size occurs within the cercopithecines and the colobines, but none is found within African or Asian papionins. However, it should be noted that terrestrially adapted cercopithecins are smaller on average than terrestrially adapted papionins or colobines. Thus, while there is a tendency for arboreal species to be smaller than their close terrestrial relatives, body size, taken by itself, does not completely separate the groups.

Several possible reasons may be suggested for the size disparity between arboreal and terrestrial members of the major groups of Cercopithecidae. One possibility is that, for some reason, an upper body size limit is selectively maintained for arborealists. Such a limit might be related to a limited food supply or to the increased difficulty in locomotion in an arboreal setting for large animals. The removal of this limit may lead to evolutionary increases in body size. Thus, when any arboreal species becomes adapted to a more terrestrial habitus, body size increase will be favored. Accordingly, more terrestrially adapted species will tend to be larger than their arboreal close relatives, but not necessarily larger than arboreal members of more distantly related taxa. Another possibility could be that body size increase is favored in terrestrial species in response to increased predator pressure.

Fayum primates fall well within the body size range of arbo-

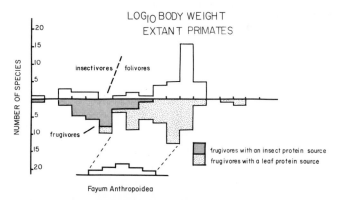

Figure 2.   *Histograms of $\log_{10}$ body weight for extant primate species. Each species is assigned to a dietary category—frugivory, insectivory, or folivory—according to the best behavioral evidence. Frugivorous species are assigned to one of two further categories according to whether they eat large additional amounts of leaves or insects. Each interval in the histogram represents a $0.2 \log_{10}$ interval. Fayum taxa are shown separately.*

Table III   *M_2 Length of Arboreal and Terrestrial Old World Monkeys[a]*

	Number of Species of Uncertain Habitus	Arboreal Species		Terrestrial Species	
		N	$\log_{10} M_2$ Length Range	N	$\log_{10} M_2$ Length Range
Cercopithecini (tribe)	3	18	0.60 to 0.83	4	0.81 to 0.88
African Papionini (tribe)	0	2	0.83 to 0.86	11	0.90 to 0.98
Asian Papionini (tribe)	2	4	0.83 to 0.88	7	0.90 to 0.98
Colobinae	3	19	0.74 to 0.91	1	0.91

[a] Within each group the more terrestrially adapted species tend to be larger than the more arboreally adapted species.

real anthropoids and outside the range of extant terrestrial anthropoids, suggesting an arboreal adaptation for the group (Fig. 3).

DENTAL MORPHOLOGY

The details of the functional adaptations of the molars of Fayum primates and the phylogenetic implications of these findings have been dealt with in detail elsewhere (Simons, 1967d; Kay, 1977a). We will concern ourselves here with the adaptive significance of the dental morphology of Fayum species, using the habits and morphology of living primates as a model.

Incisor Morphology

Recent studies of catarrhine incisor morphology have shown that the more frugivorous species have relatively larger incisors than do the more folivorous or graminivorous species (Hylander, 1975). This difference may be related to the extent of incisal preparation prior to mastication. Certain foods, including leaves, stems, berries, grasses, seeds, buds, and flowers, may not ordinarily require any extensive preparation prior to mastication. In contrast, large, tough-skinned fruits may require a considerable amount of incisor handling prior to mastication. A great deal of food preparation leads to a large amount of incisor wear. Thus, enlarged incisors may be an adaptive response to delaying the time when the incisors wear out.

The relative size of the incisors of any anthropoid may be expressed in relation to a regression equation of incisor size versus body weight for anthropoid species as a whole. Knowing the body weight, we can estimate the incisor size for an "average" anthropoid of that size. This estimate can be compared with the actual incisor dimension by means of a ratio. Hylander (personal communication) has kindly provided a regression equation to express maxillary incisor size as a function of body mass. Among Cercopithecidae, $\log_e$ (incisor size) $= 0.31167 \log_e$ (body mass in grams) $+0.09762$, where maxillary incisor size is the maximum linear distance between the distal cemento-enamel junction of the left and right lateral incisors (e.g., a chord dimension) in millimeters.

A maxillary incisor dimension of 17.8 mm may be estimated for *Aegytopithecus*, represented by a skull (Fig. 1) which preserved undamaged incisor sockets. The estimated body size of this individual is 5281 g, with a 95% confidence interval of 932 g. From these estimates, an average cercopithecid should have a maxillary incisor dimension of 15.95 mm, with a 95% confidence range of 15.01-16.78 mm. *Aegyptopithecus* resembles Old World monkeys which use considerable incisors preparation prior to mastication. Such a morphology is often associated with a frugivorous diet, and never found among catarrhine folivores. By inference from Hylander's model, *Aegyptopithecus* was probably frugivorous.

Molar Shearing

Recent studies have explored the relationship between dietary habits and molar structure, especially the relative length of the shearing crests on the second lower molar. Within various taxonomic groups, including strepsirhines, New World monkeys, Old World monkeys, and apes, more folivorous and insectivorous species have relatively longer molar shearing crests proportionate to molar length than do frugivorous and guminivorous species (Kay *et al.,* 1978; Kay, 1977c; Kinzey, 1978; Rosenberger and Kinzey, 1976; Seligsohn, 1977).

Because of their overall functional similarity to Fayum primate molars, the molars of extant Hominoidea were chosen as a model on which to base hypotheses concerning the diets of Fayum primates. Ten species of extant hominoids were studied: Eight are predominately frugivorous (species of *Hylobates, Pongo,* and

Pan) and two are more folivorous [species of *Gorilla* and *Hylobates (Symphalangus)*] (Kay, 1977b). A regression equation was calculated to express the summed length of eight shearing crests (see Kay, 1977b, for details) as a function of the second lower molar length for the eight frugivorous species: $S_E = 2.46$ (M_2 length)$^{0.95}$, where S_E is the expected summed shearing blade length. From this equation, the relative amount of shearing development (SQ, or shearing quotient) may be expressed by the equation SQ $= [S_O - S_E(100)]/S_E$, where S_O is the observed summed shearing blade length. As expected from the way this equation is formulated, frugivorous hominoids had SQ's clustering around zero (mean $= +0.23$, SD $= 3.40$). The two folivorous hominoids *Gorilla gorilla* and *Hylobates (Symphalangus) syndactylus* have mean SQ's of 7.06 and 9.04, respectively, exceeding the range of mean SQ's for frugivorous hominoids (Table IV). This indicates that folivorous hominoids have proportionately more molar shearing than do frugivorous species. To test the validity of this model for estimating dietary adaptations, SQ's were calculated for samples of *Cebus capucinus,* a small frugivorous New World monkey, and *Alouatta seniculus,* a more folivorous New World monkey. The SQ's are -2.00 and $+11.16$, respectively: These species would be correctly assigned to the frugivorous and folivorous categories on the basis of the hominoid model.

SQ's were calculated for the Fayum anthropoids (Table V). With the exception of *Parapithecus grangeri,* all Fayum species fall well within the cluster of extant frugivorous hominoids, suggesting a diet dominated by fruit. The relatively well-developed shearing crests on the M_2 of *Parapithecus grangeri* are suggestive of a more folivorous diet for this species (but not approaching that of folivorous extant hominoids). A shearing quotient of -14.01 is calculated for *Oligopithecus savagei,* extremely low for a hominoid species. We are not confident that our model is a realistic standard of comparison for *Oligopithecus;* the second lower molar of *Oligopithecus* lacks two shearing crests found in other catarrhines, the premetacristid (7) and postentocristid (8), rendering suspect the significance of the low SQ of *Oligopithecus savagei.*

Crown Height

Two aspects of terrestrial feeding may subject the dentitions of primates to a degree of wear not found among arboreal species. First, grasses provide a major potential food source available to terrestrial primates but unavailable to arboreal species. Grasses contain relatively large amounts of highly abrasive silica in their

Table IV *Total Shearing Capacity of the M_2 among Extant Hominoidea*

Species	N	Shearing Quotient *(SQ)*[a]	Range	SD
Hylobates lar	5	−0.69	−10.31 to 2.62	5.47
H. klossi	5	−2.28	− 8.90 to 2.36	4.44
H. moloch	5	−1.31	− 4.66 to 2.50	2.72
H. agilis	5	−0.14	−10.31 to 7.04	8.35
H. hoolock	2	6.76	3.74 to 9.78	4.74
H. syndactylus	3	9.04	7.82 to 11.37	2.02
Pan paniscus	11	−1.08	− 8.91 to 5.86	4.09
P. troglodytes	10	−3.37	− 9.94 to 1.79	3.99
Pongo pygmaeus	5	3.95	− 3.86 to 11.20	5.46
Gorilla gorilla	14	7.06	1.52 to 16.70	4.50

[a]Shearing quotient *(SQ)* is expressed as $SQ = 100$ $(S_T - S_E)/S_E$, where S_T is the summed length of shearing blades 1–8 for each species, and S_E is the shearing "expected" for a frugivorous hominoid based on the equation $S_E = 2.46$ (M_2 length)$^{0.9531}$. See text for further discussion.

Table V *Total Shearing Capacity of the M_2 among Oligocene Anthropoidea[a]*

Species	N	Shearing Quotient (SQ)	Range	SD
Lower Fossil Wood Zone species				
Oligopithecus savagei[b]	1	−14.01	—	—
Quarry G species				
Apidium moustafai	12	− 3.80	−9.03 to 1.39	3.51
cf. *Propliopithecus markgrafi[c]*	3	− 1.06	−3.67 to 0.64	2.29
Quarries I, M species				
Apidium phiomense	11	+ 1.50	−9.73 to 11.78	6.16
Parapithecus grangeri	4	5.73	1.65 to 8.02	2.82
Aegyptopithecus zeuxis	3	− 3.64	−6.14 to 0.86	3.91
Propliopithecus chirobates	2	− 3.86	−5.67 to 2.05	2.56
Species of uncertain horizon				
Parapithecus fraasi	1	1.79	—	—

[a]See text and Table IV for further discussion. *Propliopithecus haeckeli* is represented only by a single, heavily worn specimen.
[b]This species has only six shearing blades, rather than the eight always found among living and fossil catarrhines.
[c]The type of *Propliopithecus (= Moeripithecus) markgrafi* is from an uncertain horizon. Two isolated M_2's from Quarry G are tentatively assigned to this species. The *SQ* of the type is 0.64; those of the Quarry G specimens are −3.67 and −0.16.

stems and leaves, whereas the same parts of trees and shrubs contain relatively small amounts of silica (Eisenberg, 1978). Thus, the dentitions of grass-eating species are subjected to a relatively large amount of wear.

Second, one would expect to find more dust and other gritty substances on foods found near the ground than on foods taken from the forest canopy. This dust could have a highly abrasive effect on the dentitions of terrestrial herbivores.

Either or both of the above possibilities may explain why there is a general tendency for more terrestrial Cercopithecoidea to have considerably higher crowns than their arboreal close relatives (Table VI). (Similar cross-species comparisons are not possible for other primate groups, since they are virtually all arborealists.) In separate comparisons, when allowance is made for differences in tooth size, primarily terrestrial-adapted species of the tribes Papionini and Cercopithecini have relatively higher crowns than their more arboreal relatives. Relative crown height (C_R) is expressed as $C_R = 100 (C_T-C_E)/C_E$, where C_T is the sum of the hypoconid height and the metaconid height, and C_E is the summed crown height "expected" for a terrestrial cercopithecin from the empirically derived regression equation $C_E = 1.63 (M_2$ length)$^{0.92}$. For both tribes of cercopithecins, terrestrial species have relatively larger C_R's than arboreal species at the $p < 0.01$ level.

Relative crown height is not a very reliable indicator of the degree of arboreality or terrestriality in primates that lack close extant relatives, since it cannot be applied confidently in between-group comparisons. For example, although the more terretrial species of the tribe Cercopithecini have higher crowns than the arboreal species, they tend to have lower crowns than terrestrial papionins. Similarly, the molar crowns of arboreal papionins are higher in some cases than those of terrestrially adapted cercopithecins. Thus, the only conclusion which can be made is that, in a

Table VI *Relative Crown Heights of Arboreal and Terrestrial Cercopithecidae[a]*

Species Group	Terrestrial Species			Arboreal Species		
	N	X	SD	N	X	SD
Macaca spp.	5	2.38	4.51	2	−5.17	0.12
Cercopithecus spp., *Miopithecus* spp., *Erythrocebus* sp.	3	−4.26	1.03	16	−9.91	3.07
Cercocebus spp.	3	3.74	1.77	2	−5.60	2.15
Papio spp., *Mandrillus* spp., *Theropithecus* sp.	6	0.11	2.97	—	—	—
Colobinae (except *Presbytis entellus*)	—	—	—	19	−4.59	3.18
Presbytis entellus	—	0.71	—	—	—	—
Totals for all species	17	0.78	3.92	39	−6.85	3.92

[a]"Relative" crown height (C_R) is expressed as $C_R = 100 (C_T-C_E)/C_E$ where C_T is the sum of the hypoconid height and the metaconic height, and C_E is the summed crown height "expected" for a terrestrial cercopithecin from the regression $C_E = 1.63 (M_2$ length)$^{0.922}$. For each group, and for all species combined, the crown height of terrestrial species is higher than that for arboreal species at the $p < 0.01$ level.

group of extinct species, the more terrestrial members might be expected to have relatively higher crowns.

One species of Fayum primates, *Parapithecus grangeri*, stands out as having high crowns, by comparison with both other Fayum primates and the relatively higher-crowned papionins (Table VII). Thus, *Parapithecus* seems likely to have been more terrestrially adapted than its contemporaries.

CRANIAL MORPHOLOGY

From the skull of *Aegyptopithecus zeuxis* and a reconstruction of the face of *Apidium phiomense* presented by Simons (1970), we can get some idea of the relative size of the orbits of Fayum species. Using cranial length as a standard of comparison, it has been shown that *Aegyptopithecus zeuxis* had relatively small orbits by comparison with those of living cercopithecoids (Kay and Cartmill, 1977). Although it is not possible to make more than a rough guess, visual inspection suggests that *Apidium* has very small orbits as well. The presence of small orbits tend to be a relatively good indicator of activity pattern among small primates. Nocturnal species have relatively larger orbits than diurnal species. Among living mammal species with skull lengths greater than 75 mm, orbit size is a less reliable indicator of activity pattern (Kay and Cartmill, 1977). Unfortunately, *Aegyptopithecus* falls somewhat above the size range where relative orbit size yields useful information about activity pattern. *Apidium's* small orbits

Table VII *Relative Crown Height for Fayum Species[a]*

Species	N	Mean Relative Crown Height	Standard Error of Mean
Oligopithecus savagei	1	−19.93	—
Apidium moustafai	12	− 9.24	1.79
Apidium phiomense	11	− 8.93	1.11
Propliopithecus markgrafi	3	− 0.13	5.57
Aegyptopithecus zeuxis	3	−17.14	6.89
Propliopithecus chirobates	2	−21.21	2.95
Parapithecus grangeri	4	0.66	3.42
Parapithecus fraasi	1	−14.36	—

[a]See text and Table VI for clarification of the "relative crown height" measure.

suggest a diurnal mode of existence similar to that found among almost all extant anthropoids.

The relative size of the infraorbital foramen provides information about the vascularization and innervation of the snout. This foramen transmits the infraorbital branches of the maxillary nerve and vessels which supply the upper lip, rhinarium, and vibrissae. Among extant mammals, a relatively small foramen indicates that these snout structures are poorly developed (Kay and Cartmill, 1977). This is always the case among living anthropoids, for which these sensory modes are relatively unimportant. *Aegyptopithecus* and *Apidium* have very small infraorbital foramina, implying an anthropoid grade of organization.

BRAIN SIZE AND MORPHOLOGY

Radinsky (1973) estimates an overall cranial capacity of between 30 and 34 cm³ for the skull of *Aegyptopithecus*. More recently, on the basis of Jerison's (1973, p. 50) "double integration" method, Radinsky revised this estimate downward to 27 cm³ (Radinsky, 1977).

Given Radinsky's brain size estimates and our body estimates, it is possible to calculate Jerison's encephalization quotient (EQ), an expression of the relative brain size of a mammalian taxon, by comparison with that of an "average" living mammal. (An EQ of 1.0 would indicate a brain comparable in size to that of an average living mammal.) If *Aegyptopithecus* had a 32g brain (the midpoint of Radinsky's early estimates), body size estimates based on M_2 length would yield a mean EQ of 0.82, with a range of 0.43 to 1.53; an EQ of 0.87 would be predicted from body size estimates based on cranial dimensions. The EQ's for *Aegyptopithecus* would be somewhat lower given Radinsky's more recent brain size estimate of 27 cm³: M_2-estimated body weight yields a mean EQ of 0.69 (a range of 0.36 to 1.29); weight estimates from cranial dimensions would yield an EQ of 0.73. These estimates are in agreement with those of Gingerich (1977), who estimated body weight based on a regression equation of M_2 length against body weight for seven species of hominoids and used Radinsky's earlier estimates of brain size.

Based on any set of the foregoing brain and body size estimates, the brain of *Aegyptopithecus* is very small for an anthropoid. Jerison (1973) provides EQ's for 46 anthropoid species. All species, with the exception of two species of *Presbytis*, have EQ's exceeding 1.43. The most liberal estimate of relative brain size in *Aegyptopithecus* is smaller than that of all but one living anthropoid. On the other hand, *Aegyptopithecus* had a brain size well within the limits of those of living strepsirhines, which range from 0.60 to 1.73.

Radinsky (1973) studied the incomplete endocasts of the brain of *Aegyptopithecus*. He concluded that *Aegyptopithecus* had a brain with a relatively larger visual cortex and a relatively smaller olfactory bulb than is the case for most prosimians. However, *Aegyptopithecus* was more primitive than modern anthropoids in having a smaller frontal lobe. He concluded from this that *Aegyptopithecus* had developed an increased emphasis on vision, and a decreased emphasis on smell, in a fashion similar to that seen among living Anthropoidea.

INFERRED HABITS OF FAYUM SPECIES

The following points have been established about the Fayum primates.

(a) Fayum primates were primarily frugivorous. *Parapithecus grangeri* may have been somewhat more folivorous than the other Fayum primates. There is no evidence that any of these species were primarily insectivorous or primarily folivorous in their diet.

(b) A predominantly arboreal habitus is suggested for most species. *Parapithecus grangeri* is a possible exception; the high crowns of its molars are suggestive of semiterrestrial habits.

(c) The relatively small orbits inferred from the reconstruction by Simons (1970) suggest that *Apidium* was diurnal, similar to extant anthropoids. Because of the large body size of *Aegyptopithecus*, its comparatively small orbits warrant no interference concerning its activity pattern.

(d) The small infraorbital foramina of *Apidium* and *Aegyptopithecus* show that they had a poorly developed tactile sensory apparatus in the snout, a feature similar to extant Anthropoidea.

Probably the closest living ecological analogues of the Fayum species are found among living New World cebids. The latter are similar in their range of body size and have diurnal habits and a predominantly frugivorous diet. Our inference of a slightly more folivorous and slightly more terrestrial habitus for *Parapithecus grangeri* is particularly intriguing, since in this combination of features, *Parapithecus* may come closest among the Fayum species to filling the ecological niche predicted by Kay (1977c) for the ancestral Old World monkey.

ACKNOWLEDGMENTS

The specimens of living primates used in this report came from the American Museum of Natural History, New York; the British Museum of Natural History, London; the Institute of Comparative Anatomy and Division of Mammals and Birds, Paris; the Museum of the Congo, Tervuren, Belgium; the Museum of Comparative Zoology, Harvard University; and the Natural Museum of Natural History, Washington, D.C. The fossils examined are housed at the Yale Peabody Museum, the Duke University Primate Center, and the Cairo Museum. We thank the staffs of these institutions for their assistance. Current support is provided by NSF Grant BNS77-08938 to Richard F. Kay and NSF Grant BNS77-20104 and Smithsonian Institute Grants SFCP FC7086-9600 and FC-80974 to Elwyn L. Simons. This paper was originally prepared for presentation at the 1979 International Primatological Congress, Bangalore, India, but due to airline scheduling difficulties beyond our control, we were unable to present it.

21

New Interpretations of the Phyletic Position of Oligocene Hominoids

J. G. FLEAGLE AND R. F. KAY

INTRODUCTION

In attempts to trace the ancestry of living hominoids, the Miocene apes from Europe, Asia, and sub-Saharan Africa have long played a central role. First described over 125 years ago, the genera *Dryopithecus* and *Pliopithecus* were recognized from their initial descriptions as an ancestral pongid and hylobatid, respectively, a situation which Darwin (1871) acknowledged and which, until most recently, was universally accepted.

Since the first decade of this century, fossil evidence for the hominoid antecedents of these taxonomically diverse and geographically widespread Miocene apes has come almost exclusively from the (?Eocene—) Oligocene sediments of the Fayum province of Egypt. When Schlosser (1910) described the first fossil anthropoids from Egypt, he recognized this relationship by naming one genus *Propliopithecus*. Since Schlosser's original description, numerous additional species have been described and dozens of new fossil specimens have been recovered from these same deposits (e.g., Simons, 1967d, 1972; Kay *et al.*, 1981). Over the past 70 years, many divergent hypotheses concerning relationships with later apes and humans have been proposed for these early anthropoids. Various Fayum species have been identified at

one time or another as the earliest hominid, pongid, hylobatid, Old World monkey, the common ancestor of all hominoids, or the common ancestor of all catarrhines including Old World monkeys, apes, and humans. Many of the relationships were advanced when the Oligocene species were known only from a single type specimen. Now most are known from dozens of jaws, as well as cranial and skeletal material. With the greatly increased fossil record of these early anthropoids, paleoprimatologists can better evaluate the phyletic relationship of the Fayum apes with respect to later ape and human evolution.

SYSTEMATICS AND INTERRELATIONSHIPS OF THE FAYUM APES

The first recognized ape jaws from the Fayum Province were collected by Richard Markgraf, apparently in the spring of 1907 (Gingerich, 1978a). Schlosser (1910) named two taxa, *Propliopithecus haeckeli* and *Moeripithecus markgrafi*, each on the basis of a single specimen (see Figs. 1, 2). Simons (1967d) suggested that the two are congeneric and dropped *Moeripithecus*, a move which has been followed by all later workers. The distinctiveness of the two species is generally accepted. Unfortunately, no information

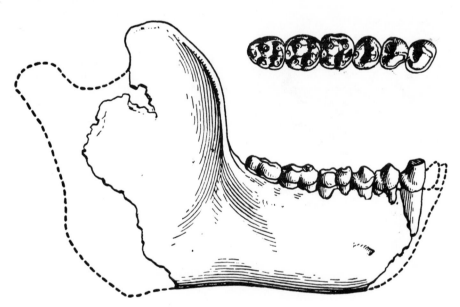

Figure 1 *Right lateral and left occlusal view of the type (SNM 12638) jaw and teeth of* Propliopithecus haeckeli *Schlosser (X c.2.0).* [*From Gregory (1916).*]

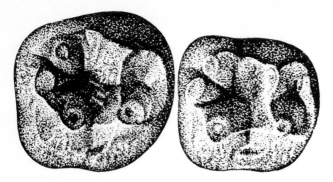

Figure 2   *Occlusal view of the type (SNM 12639G) of* "*Moeripithecus*" (= *Propliopithecus*) *markgrafi Schlosser, a right jaw fragment containing M_1 and M_2 (X c.8.0).* [*From Kalin (1962).*]

survives as to the stratigraphic position of either type specimen.

The type specimen of *Oligopithecus savagei* (Simons, 1962a) was recovered from the Lower Fossil Wood Zone in 1961. This animal is still known only from the type specimen preserving $C-M_2$ and a recently discovered unworn lower molar. It resembles extant catarrhines in having two rather than three premolars, the anterior of which bears a well-developed wear facet for contact with an interlocking upper canine. On this basis, and because it has a relatively deep jaw, Simons (1972) has taken the view that *Oligopithecus* should be tentatively linked to the Fayum apes. He notes that the Upper Fossil Wood Zone parapithecid *Parapithe-*

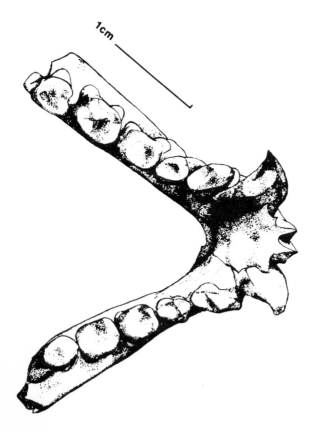

Figure 3   *Type mandible of* "*Aeolopithecus*" (= *Propliopithecus*) *chirobates, CGM 26923 (X c.2.0).* [*From Simons (1965a).*]

cus grangeri, which he feels stands closer to Old World monkeys, still retained three premolars, implying that the two-premolared condition of the Cercopithecidae was achieved separately at a later time. However, *Oligopithecus* lacks the peculiar derived molar occlusal pattern found in other Fayum apes, parapithecids, and all later catarrhines (Kay, 1977a). Thus, it is equally plausible that *Oligopithecus* is much more primitive and is an offshoot of the ancestral Old World anthropoid group that gave rise to parapithecids and apes, and that reduction from three to two premolars by loss of P_2^2 occurred in parallel in this genus (Szalay and Delson, 1979). An alternative view is that *Oligopithecus* is an adapid (Gingerich, 1977c).

Simons (1965) described two species of ape from the Upper Fossil Wood Zone, *Aegyptopithecus zeuxis* and *Aeolopithecus chirobates.* The former, based on four specimens, including an edentulus jaw, which had been collected for the American Museum of Natural History in 1906, is morphologically distinct from *Propliopithecus* in several taxonomically important ways (Kay *et al.,* 1981). In all specimens of *Aegyptopithecus zeuxis* M_2 is much larger than M_1, whereas in all *Propliopithecus* species M_1 is about the same size as M_2. The lower molar occlusal surfaces of *Aegyptopithecus* are compressed buccal-lingually, and the crown margins flare outward, giving the molars a bulbous appearance. In *Propliopithecus* species typically the lower molar crowns are not compressed and the crowns are more steep-sided. *Aegyptopithecus* also differs from *Propliopithecus* species in that the premolar cingulum tends to be less developed. *Aegyptopithecus* lacks a lingual cingulum on P_4, whereas a P_4 lingual cingulum is invariably preserved in *Propliopithecus.* The first upper molar of *Propliopithecus* is consistently broader than that of *Aegyptopithecus.* Finally, lower incisors in *Propliopithecus* are shorter and of smaller caliber in relation to M_1 size than are those of *Aegyptopithecus.*

On the basis of a single lower jaw, a second species of ape was described by Simons (1965a) as "*Aeolopithecus*" *chirobates* (Fig. 3). Because of chemical erosion [probably due to digestive enzymes in the gut of a crocodile (Fisher, 1981)], few features are preserved on the teeth of the type, but these seemed at the time to indicate an ape quite distinct from other known species. The type of "*Aeolopithecus*" is much smaller than contemporaneous *Aegyptopithecus zeuxis.* The large canines and premolar heteromorphy of the type initially set it apart from *Propliopithecus haeckeli.* Recent collections of material from Quarries I and M show, however, that the type of "*Aeolopithecus*" *chirobates* is a male; the females of this species have small canines (Fleagle *et al.,* 1980; Kay *et al.,* 1981; see Fig. 4). This species has been best referred to *Propliopithecus* (Szalay and Delson, 1979; Kay *et al.,* 1981). It resembles *P. haeckeli* and is distinctively different from *Aegyptopithecus* in having similar sized first and second molars with marginally placed lower molar cusps. In addition, lower molar crowns are steep-sided, and premolar cingula are relatively well developed.

Gingerich (1978a) tentatively suggested that, should more material of *P. chirobates* be found, it might prove to be conspecific with *P. haeckeli.* Currently available samples of the two taxa do not show sufficient overlap in morphology to confirm this. In premolar morphology, *P. chirobates* is more derived than *P. haeckeli,* which it closely resembles in overall size. The shape of the P_3 is oval and compressed in both presumed males and presumed females of *P. chirobates* (Fig. 5), leading to premolar heteromorphy similar to that of *Aegyptopithecus;* in *P. haeckeli,* the sole known specimen of which is probably a female, P_3 is practically round in occlusal outline. Our sample of *P. chirobates* mandibles shows that their mandibular corpora average considerably shallower than in *P. haeckeli.* Even so, jaw depth varies

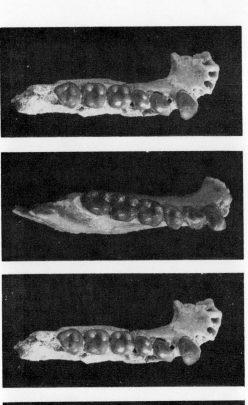

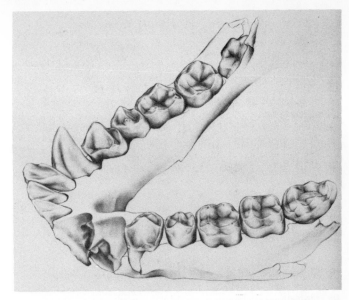

Figure 5   Propliopithecus chirobates, *oblique lateral view of DPC 1069 (X c.2.2).*

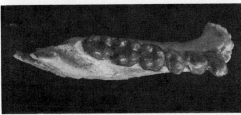

Figure 4   *Stereophotographs of mandibles of* Propliopithecus chirobates *from the Upper Fossil Wood Zone, Jebel el Qatrani Formation. Left, DPC 1029, a male. Right, DPC 1103, a female. (X c.1.6).* [*From Kay* et al. (1981).]

considerably in sexually dimorphic species; in body-size dimorphic species, jaw depth under M_{1-2} is usually greater in males than in females (Fleagle *et al.*, 1980). While the molars in the type specimen of *P. haeckeli* (a female as judged from canine size) are smaller than any in the sample of *P. chirobates*, its mandible is 30% deeper under M_{1-2} than are females of *P. chirobates*. Thus, in various slight differences, *P. haeckeli* is distinct from all known *P. chirobates* in having a less sectorial P_3 and a very deep mandible. We prefer to maintain two species at present.

We also cannot demonstrate that *"Moeripithecus" markgrafi* and *Propliopithecus haeckeli* are conspecific. In molar size and M_1 : M_2 proportions, *"M." markgrafi* resembles *P. haeckeli*, but the molar crowns of the former are very compressed (see Fig. 2), a feature of similarity shared with *Aegyptopithecus*. The larger samples of *P. chirobates* show that this degree of morphological distinctiveness would not be anticipated in a single species and we conclude that *"M." markgrafi* is a valid species. The question of whether *"Moeripithecus"* is a valid genus is more of a problem, given the lack of information about most of its anatomy. In a few known features, this species bridges the gap between *Aegypto-*

pithecus and *Propliopithecus*, with a few morphological similarities to the former and the size proportions of the latter. If more material of *"Moeripithecus"* were to become available, it might be necessary to conclude that it was congeneric with *Aegyptopithecus* on grounds of molar morphology. Such a step is not warranted at present, given the very incomplete nature of the type, and we take the more conservative course of including *"M." markgrafi* within *Propliopithecus*.

The presently available ape material from the Fayum is thus divisible into four species, *Propliopithecus haeckeli, P. chirobates, P. markgrafi,* and *Aegyptopithecus zeuxis. Aegyptopithecus zeuxis* and *P. chirobates* are known to be contemporaries of one another; the geological provenence of the other taxa is unknown. How are the four interrelated? Table 1 summarizes the distribution of 13 character states discussed more extensively elsewhere (Kay *et al.*, 1981). In terms of overall similarity, *Aegyptopithecus zeuxis* and *Propliopithecus haeckeli* are at opposite poles: the two differ in every characteristic of the dentition that shows variation among the four species. *Propliopithecus chirobates* falls in an intermediate position, but shares more features with *Propliopithecus haeckeli* than with *Aegyptopithecus zeuxis*. The dentition of *P. markgrafi* is so poorly known that it can be compared with the other three taxa in three characters only.

To assess correctly the phylogenetic relationships among these forms, it would be necessary to know the morphology of their last common ancestor. This would permit determination of which features of similarity result from the shared retention of primitive characteristics from the common ancestor and which result from the acquisition of new characters at various stages subsequent to the initial speciation event. Only the latter similarities might indicate relative relatedness within the group. If one assumes that other Oligocene anthropoids (parapithecids and *Oligopithecus*) demonstrate the primitive condition for the apes, all of the character states shown by *P. haeckeli* are probably primitive for the four species, with the possible exception of the degree of cingulum development (characters 6 and 10 in Table 1). If this interpretation is correct, by far the most parsimonious cladogram of the three best known forms shows an initial split of *P. haeckeli* from the common ancestor of *P. chirobates* and *A. zeuxis,* followed by the separation of the latter two. If *P. haeckeli*

Table 1   *Morphological Comparisons of Fayum Apes*

Propliopithecus haeckeli	*Propliopithecus chirobates*	*Propliopithecus markgrafi*	*Aegytopithecus zeuxis*
1. —	M^1 smaller than M^2	—	M^1 much smaller than M
2. —	M^x conules absent	—	M^2 conules variably present
3. P_3 rounded: no mesial-buccal flare	P_3 oval with well-developed mesial-buccal flare	—	P_3 oval with well-developed mesial-buccal flare
4. —	C -P_m sexual size differences	—	C-P_m sexual size differences
5. P_4 trigonid broad buccal-lingually	P_4 trigonid broad buccal-lingually	—	P_4 trigonid compressed buccal-lingually
6. P_4 lingual cingulum present	P_4 lingual cingulum is variable	—	P_4 lingual cingulum is absent
7. $M_1 = M_2$ in mesial-distal length	M_1 averages 5% shorter (M-D) than M_2	M_1 5% shorter (M-D) than M_2	M_1 averages 17% shorter (M-D) than M_2
8. Mandibular corpus very deep for female	Mandibular corpora broad, shallow under M_{1-2} when sexual dimorphism is controlled	—	Mandibular corpora broad, shallow under M_{1-2} when sexual dimorphism is controlled
9. Molar crowns broad, steep-sided	Molar crown broad, steep-sided	Molar crown compressed, sides flare	Molar crown compressed, sides flare
10. Molar buccal cingulum well developed	Molar cingulum moderate	Lower molar cingulum weak	Lower molar cingulum weak
11. M_3 distal fovea absent	M_3 distal fovea small to lacking	—	M_3 with large distal fovea
12. —	1_1 broad, low-crowned	—	1_1 narrow, high-crowned

proves to be primitive in all respects, and if it turns out to be geologically older than the other species, it would be an ideal ancestor for the others. Given the poor state of knowledge about *P. markgrafi*, it is not possible to decide which are its closest relations. If a well-developed molar cingulum were primitive for the last common ancestor of Fayum apes, a possible link with *A. zeuxis* is implied; otherwise, *P. markgrafi* is almost the same as *P. chirobates*.

EVIDENCE FOR RELATIONSHIPS WITH LATER ANTHROPOIDS

In his initial description of *Propliopithecus haeckeli*, Schlosser (1910, 1911) suggested that the species could represent an ancestral gibbon through the lineage *Propliopithecus → Pliopithecus → Hylobates*. He argued that *P. haeckeli* was gibbon-like largely through its similarities to *Pliopithecus*, especially in having molars with five rounded cusps located around the periphery of the teeth. On the other hand, he suggested that the small canine, deep jaw, and relatively broad anterior premolar were hominid-like features. At the same time, Schlosser was very impressed with many "primitive" similarities which he found between *Propliopithecus* and cebids—especially in size, in the shape of the jaw, and the morphology of the premolars. However, because of the catarrhine dental formula, he never made much of these resemblances.

Largely because of a lack of any new fossil material attributed to this genus for over 50 years, Schlosser's two main interpretations have persisted down to the present and are even found in some recent textbooks. Over the years various authorities have adopted one, the other, or both of these positions. For example, Keith (1923) argued that since *Propliopithecus* was gibbon-like, this was possible corroborating evidence that the earliest apes were brachiators. In contrast, some authors offered the human-like nature of the canines and premolars in *Propliopithecus* as well as its generally primitive molars and mandible as evidence for a very early separation of humans from the ape lineage (Osborn, 1927b; Simons, 1965a; Pilbeam, 1967; Kinzey, 1971; Kurten,

1972b; Wood-Jones, 1929). This phylogenetic scheme had the decided advantage of not requiring a reduction of canine size in hominid evolution. Others, such as Gregory (1922b), Kalin (1961), Simons (1967d, 1972), Delson and Andrews (1975), Andrews (1978), and Szalay and Delson (1979) argued that this taxon was probably best interpreted as a generalized ancestral hominoid which was broadly ancestral to most later apes, including the dryopithecines.

Schlosser's other ape species, "*Moeripithecus*" *markgrafi*, has had an equally diverse history of opinions as to its affinities. Schlosser originally suggested that his species might be an early ancestor of Old World monkeys. Abel (1931) felt that "*Moeripithecus*" *markgrafi* was related to *Apidium* and probably also to later cercopithecoids. Kalin (1961, 1962) considered the possibility of cercopithecoid affinities for this species, but decided otherwise. Most recent workers (Simons, 1967d, 1972; Delson, 1975b; Szalay and Delson, 1979) have tended to regard this taxon as just another species of *Propliopithecus* with no specific affinities to particular later taxa.

In 1965, Simons described *Aeolopithecus* as a possible early hylobatid because of the large canines, reduced third molar, supposed posterior shallowing of the mandibular corpus, and a high, deep genial fossa. He tentatively supported this relationship in a series of later publications (Simons, 1967d, 1972; Simons *et al.*, 1978), as did other workers (e.g., Howell, 1967), with the proviso that additional material of this species might necessitate a reevaluation of this relationship. As described above, additional specimens have indeed increased the knowledge of this species considerably and now cast considerable doubt as to its hylobatid affinities. Such hylobatid features as the posteriorly shallowing mandible and the marked reduction of M_3 in the type specimen are artifacts of the poorly preserved type specimen, and the large canines are found only in the males (Fleagle *et al.*, 1980; Kay *et al.*, 1981). Thus, unlike extant gibbons, *P. chirobates* had sexually dimorphic canines. Now that the molar morphology is known, this species is clearly very close to and possibly even conspecific with *Propliopithecus haeckeli*, and its phyletic affinities are best considered in conjunction with that taxon.

When Simons (1965a) first described *Aegyptopithecus zeuxis*, he argued that it was more closely allied with *Proconsul* species from the Miocene of East Africa and with the later pongids and hominids than with either the small apes from Africa, *Pliopithecus*, or hylobatids, primarily on the basis of its relatively large M_3. He later expanded the argument that *Aegyptopithecus zeuxis* was clearly more closely related to *Dryopithecus* sp. from the Miocene and to the Great Apes than to any other group of anthropoids (Simons, 1974a). Howell (1967) carried this possibility to the extreme by suggesting synonymy between *Aegyptopithecus* and *Dryopithecus*. In direct contrast, Andrews (1970) argued that *Aegyptopithecus zeuxis* was very similar to *Limnopithecus legetet* and *Dendropithecus macinnesi*. Indeed he even provisionally assigned specimens from Rusinga Island (early Miocene, Kenya) to a new species of *Aegyptopithecus* before reassigning them to *Dendropithecus macinnesi* (Andrews, 1978).

More recently Delson and Andrews (1975), Delson (1977a), and Szalay and Delson (1979) have argued that, on the basis of shared primitive features, all of the Fayum hominoids are most closely allied to *Pliopithecus* from Europe and *Dendropithecus macinnesi* from East Africa and that they should all be placed in a primitive family of hominoids, the Pliopithecidae, which is the sister taxa of all other hominoids. Likewise, Simons and Fleagle (1973) and Fleagle and Simons (1978a) noted the numerous similarities (in primitive features) between *Aegyptopithecus* and *Pliopithecus* (but not with *Dendropithecus*).

Although many authors have noted the many ways in which *Aegyptopithecus* (and *Propliopithecus*) are far more primitive than any Miocene to Recent catarrhines (Gregory, 1922b; Simons and Fleagle, 1973; Fleagle and Simons, 1978a, 1979; Fleagle, 1980; Delson and Andrews, 1975; Delson, 1975b, 1977a; Szalay and Delson, 1979), only a few (Remane, 1965; Groves, 1972b; Fleagle and Simons, 1978b; Fleagle, 1980; Kay *et al.*, 1981; Andrews, 1981b; Fleagle and Rosenberger, 1983) have actually argued that the Oligocene taxa are probably ancestral to both cercopithecoid monkeys and later hominoids.

In light of current knowledge about the very primitive dental, cranial, and skeletal morphology of the Oligocene hominoids *Propliopithecus* and *Aegyptopithecus*, there is no reason to believe that any single group of extant hominoids (either hylobatids, hominids, or pongids) can be traced back to an Oligocene divergence (Fleagle *et al.*, 1980). However, the issue of the relationship between these Oligocene hominoids and the numerous Miocene apes from Africa and Europe is very complex and controversial and can only be evaluated by a careful review of the available cranial, dental, and skeletal material of Oligocene and early Miocene species.

Cranial Morphology

The best material available for cranial comparisons with *Aegyptopithecus zeuxis* are the partial skulls (Figs. 6–8) of *Proconsul africanus* (East African early Miocene) (Le Gros Clark and Leakey, 1951; Napier and Davis, 1959; Davis and Napier, 1963) and *Pliopithecus vindobonensis* (Czechoslovakian middle Miocene) (Zapfe, 1960). The skull of *Aegyptopithecus* (CGM 40237; Fig. 6, 8) stands out for its primitiveness with respect to all later Anthropoidea. The large facial component of the skull, combined with the small brain case, the somewhat dorsally facing orbits, the sharp angulation of the nuchal region with the cranial vault, and the extreme postorbital constriction are reminiscent of the skulls of Eocene adapids, such as *Leptadapis*, and of some of omomyids, such as *Rooneyia*. On the other hand, *Aegyptopithecus* shows an anthropoid grade of organization in having a fused mandibular symphysis with both superior and inferior transverse

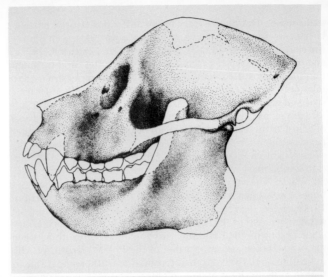

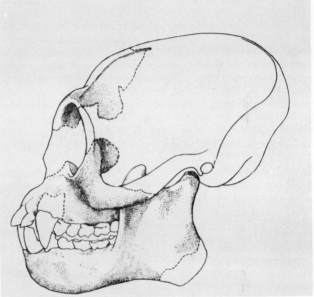

Figure 6 *Lateral views of skulls of* Aegyptopithecus zeuxis *(CGM 40237) (top) and* Pliopithecus vindobonensis (bottom), based on material described by Zapfe (1960). (X c.0.75).

tori (unlike Eocene—Oligocene adapids, which lack the superior transverse torus, even when the symphysis is fused) (see Figs. 9, 10). *Aegyptopithecus* also exhibits an advanced stage in postorbital closure, very close to the condition seen in extant gibbons, whereas no known Eocene primate has advanced beyond the "prosimian" grade of postorbital configuration.

A detailed comparison of the cranial morphology of *Aegyptopithecus zeuxis*, *Pliopithecus vindobonensis*, and *Proconsul africanus* (Figures 6–8, Table 2) shows that most of the similarities between *Aegyptopithecus* and *Pliopithecus* are primitive for the apes in general, whereas *Pliopithecus* shows a number of derived features with *Proconsul*. This suggests that the Miocene taxa may have shared a more recent common ancestor with one another in their derivation from a form like *Aegyptopithecus*.

Details of the facial skeleton of *Aegyptopithecus* (Fig. 8) are considerably at variance with those of *Pliopithecus* and *Proconsul*. The premaxilla of *Aegyptopithecus* is uniquely large for an anthropoid. This bone is very broad anteroposteriorly, particularly at its dorsal end, much in the same way as is seen among Eocene—Oligocene adapids (Simons, 1972; Gingerich, 1973a) and omomyids, such as *Rooneyia* (Table 2). (Here and throughout, the terms dorsal, ventral, anterior, and posterior are used in relation to the Frankfurt Horizontal). Anterodorsally it forms a sharp-edged margin for the nasal aperture. The infraorbital surface of the maxilla is smooth, lacking a pronounced canine fossa or a lip-like projection of the inferior orbital margin. The distance between the inferior orbital margin and the ventral root of the zygomatic arch increases slightly proceeding laterally. The inferior orbital margin is slightly posterior to the root of the zygomatic arch. The root of the zygomatic arch is above M^2. The nasal bones are broken away, leaving only the interorbital portion. However, it is clear from the configuration of the maxillae and the nasal aperture that the nasal bones were long and formed a very acute angle with the Frankfurt Horizontal. The breadth between the orbits, taken as a proportion of the P^3—M^3 length, is 0.30, indicating a relatively large interorbital breadth, comparable to that of gibbons, most extant colobines, and Eocene primates as well.

By comparison with *Aegyptopithecus zeuxis*, the premaxillae of both *Pliopithecus* and *Proconsul* are very narrow anteroposteriorly at their dorsal ends. The overall configuration of the premaxilla in the latter genera is similar to that found among extant catarrhines. The infraorbital surface of the maxilla of *Proconsul* resembles that of *Aegyptopithecus* in lacking a pronounced canine fossa or a well developed lip-like forward projection of the inferior orbital margin. Additionally, the inferior orbital margin is posterior to the plane of the root of the zygomatic arch. It differs from *Aegyptopithecus* in that the distance

between the inferior orbital margin and the root of the zygomatic arch shallows laterally. In *Proconsul* the root of the zygomatic arch is above M^{1-2}.

The conformation of the maxilla in the infraorbital region in *Pliopithecus* is distinct from both *Aegyptopithecus* and *Proconsul* in that the inferior orbital margin forms a lip-like projection anteriorly. As a result, the inferior orbital margin of *Pliopithecus* is situated well anterior to the root of the zygomatic arch. *Pliopithecus* is also different from either of the older forms in that the vertical distance between the inferior orbital margin and the root of the zygomatic arch becomes much greater laterally. These features are not seen in *Proconsul* but occur in recent hylobatids, colobines, and some cercopithecines.

In *Pliopithecus*, the root of the zygomatic arch is above M^{1-2}. The configuration and size of the surrounding bones give some indication of the size and orientation of the nasal bones of *Pliopithecus* even though they are missing. The greater part of the nasals are preserved in *Proconsul*. In both of the Miocene forms, the nasals must have been considerably smaller than in *Aegyptopithecus* and well within the range of modern hominoids. The long axis of the nasals of *Pliopithecus* must have been oriented quite obliquely with respect to the Frankfurt Horizontal. Those of *Proconsul* were apparently oriented somewhat more acutely than in *Pliopithecus* but were not as horizontal as in *Aegyptopithecus*. In the latter respect, the reconstructions of Le Gros Clark and Leakey (1951) and Davis and Napier (1963) are quite different.

The interorbital breadth, taken as a ratio of P^3—M^3 length, is 0.38 in *Proconsul*, approximately the same as in *Aegyptopithecus*. The same index is 0.60 in *Pliopithecus*, illustrating the extremely great distance between the orbits in that taxon.

Table 2   *Comparison of the Cranial Anatomy of* Aegyptopithecus zeuxis, Pliopithecus vindobonensis, Proconsul africanus, *and a Sample of Eocene Omomyids and Adapids, Including* Notharctus, Rooneyia, Adapis, *and* Tetonius

Eocene/Oligocene "Prosimians"	*Aegyptopithecus*	*Pliopithecus*	*Proconsul*
Premaxilla large	Premaxilla large	Premaxilla small	Premaxilla small
Infraorbital surface of maxilla smooth	Infraorbital surface of maxilla smooth	Infraorbital surface projecting, lip-like	Infraorbital region smooth
Variable	Infraorbital maxilla depth increases slightly laterally	Infraorbital maxilla depth increases greatly laterally	Infraorbital maxilla depth decreases laterally
Above either M_1 or M_2	Root of zygomatic arch above M_2	Root of zygomatic arch above M_2	Root of zygomatic arch above M_2
Nasals long, projecting	Nasals probably long, projecting	Nasals probably short, projecting	Nasals short, not projecting
Extremely large interorbital breadth	Large interorbital breadth	Extremely large interorbital breadth	Large interorbital breadth
Palate straight-sided or widens posteriorly	Palate widens posteriorly	—	Palate widens posteriorly
Orbit size variable	Small orbits	Orbits large	Orbits intermediate in size between *Pliopithecus* and *Aegyptopithecus*
Orbits face slightly laterally and dorsally	Orbits face slightly laterally and dorsally	Orbits face forward	—
Small brain compared to modern Anthropoidea	Small brain compared to modern Anthropoidea	Possibly modern sized	Brain in modern anthropoid size range
?Large visual cortex	Large visual cortex	—	Large visual cortex
Large olfactory bulbs	Small olfactory bulbs	—	Small olfactory bulbs
Marked postorbital constriction	Marked postorbital constriction	Little postorbital constriction	Little postorbital constriction
Variable	No external auditory tube	External auditory tube incomplete ventrally	Well-developed tubular ectotympanic

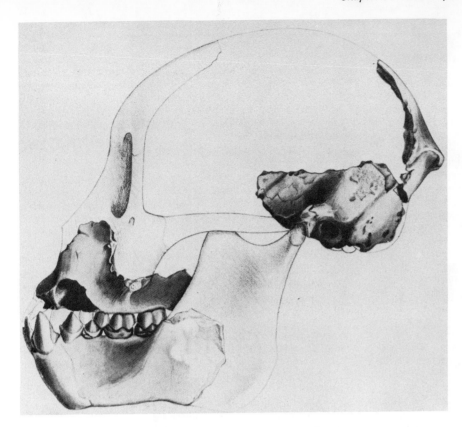

Figure 7 *Lateral view of reconstructed skull of* Proconsul africanus. *[From Davis and Napier (1963).] Note that illustration has been reversed to facilitate comparison with Fig. 6 (X c.0.75).*

The palate of *Aegyptopithecus* (Fig. 8) widens posteriorly such that the distance between the lingual margins of the M³'s is 1.33 times greater than the distance between the lingual margins of the P³'s. Data presented by Andrews (1978) show that the palates of Miocene African dryopithecines except for *Proconsul major* widen posteriorly but not to the degree seen in *Aegyptopithecus*. The palate of *Pliopithecus* is not well enough preserved for one to be certain of its shape. [Andrews (1978) provided measurements of the palate of *Pliopithecus,* apparently based on the reconstruction of Zapfe (1960). Unfortunately, the reconstructed palate in question is mostly made of plaster.]

The orbits of *Aegyptopithecus* (Figs. 8a, 8d) are quite small by comparison with extant cercopithecoids, using skull length or P³—M³ length as a standard for comparison (Kay and Simons, 1980). The ratio of orbit diameter to P³—M³ length is 0.63. The orbits of *Pliopithecus* are relatively much larger, comparable to those of many extant catarrhines. The ratio of orbit size to P³—M³ length is 0.89 for *Pliopithecus*. The orbits of *Proconsul* are proportionately smaller than those of *Pliopithecus* but larger than those of *Aegyptopithecus*. The orbit to P³—M³ ratio is 0.74. The relatively small orbit in *Proconsul africanus* compared with *Pliopithecus* is not out of line for the larger size of the former and the negatively allometric trend in orbit size among Old World anthropoids (Kay and Cartmill, 1977). However, the orbits of *Aegyptopithecus* are surprisingly small for an Old World anthropoid.

In *Aegyptopithecus*, the lateral orbital margin is more posterior than the medial orbital margin, and the superior orbital margin is more posterior than the inferior orbital margin. Thus, the orbits are directed somewhat laterally and dorsally. In *Proconsul*, the orbits are badly distorted so that their orientation is indeterminant. The orbits of *Pliopithecus* face forward in a manner similar to modern catarrhines.

The brain size of *Aegyptopithecus*, as estimated by Radinsky (1977) and allometrically corrected for body size, shows that this species had a smaller brain than extant anthropoids and within the range of extant strepsirhines (Kay and Simons, 1980). *Aegyptopithecus* was also more primitive than extant anthropoids in having comparatively smaller frontal lobes (Radinsky, 1973). On the other hand, the brain of *Aegyptopithecus* had a larger visual cortex and smaller olfactory lobes than is the case for most "prosimians," and within the modern anthropoid range.

The brain of *Proconsul africanus* (see Falk, 1983) was apparently within the modern anthropoid range (Gingerich, 1977a). All aspects of the morphology of the endocast resemble extant Old World anthropoids (Radinsky, 1977). No information is available about the configuration of the available parts of the inside of the brain case of *Pliopithecus*. A hint that *Pliopithecus* may have an essentially modern-sized Old World anthropoid brain is indicated by the lack of any great postorbital constriction. The small-brained *Aegyptopithecus* skull has a marked postorbital constriction and relatively flat-lying frontal bone. The large-brained *Proconsul* more closely resembles *Pliopithecus* in this feature.

The skull of *Aegyptopithecus* has well-developed ridges marking the origin of the temporalis muscles (Figs. 8c–8e). These converge posteriorly and medially to form a pronounced sagittal crest. The anteriormost extent of the sagittal crest is found in the same coronal plane as the maximum postorbital constriction. The crest continues posteriorly to inion, where it is continuous with flange-like nuchal crests. The pronounced development of this cresting is unique for an anthropoid of comparable body size, especially considering that this individual was probably a young adult (the canines are not quite fully erupted). This development is more likely to reflect the small size of the brain case than to be an indication of comparatively better developed jaw or neck musculature than seen in extant catarrhines. This conclusion is

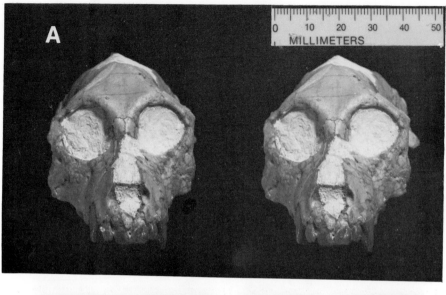

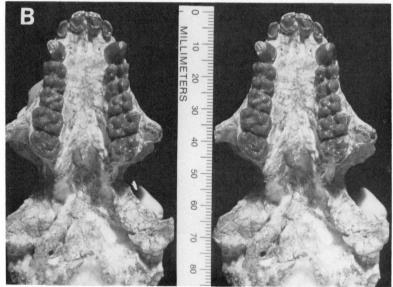

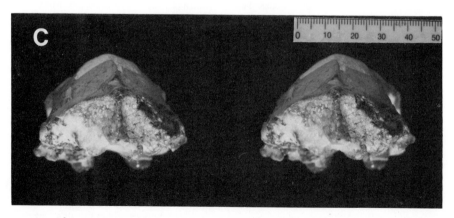

Figure 8   *Stereophotographs of* Aegyptopithecus zeuxis *skull (CGM 40237) (various magnifications, scales provided). A, frontal view; B, palatal view; C, posterior view.*

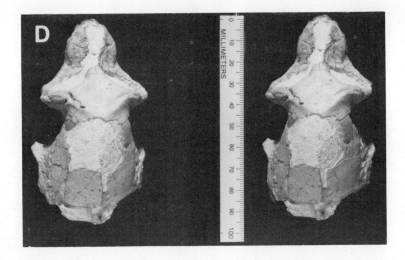

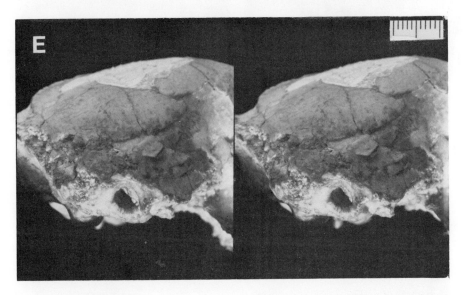

Figure 8 (continued)   *Stereophotographs of* Aegyptopithecus zeuxis *skull (CGM 40237) (various magnifications, scales provided). D, top view; E, right lateral view of ear region.*

drawn because the overall size of the nuchal plate, the temporalis fossa, and the surface area of the origin of the temporalis muscles are similar to those of comparably sized Old World monkeys [5000-5500 g (Gingerich, 1977a; Kay and Simons, 1980)]. The foramen magnum of *Aegyptopithecus* is located well under the cranial vault in a fashion similar to the position in extant cercopithecines.

The temporal lines of *Pliopithecus* converge to form a very low midsagittal ridge well back on the skull. These crests again diverge before reaching the back of the skull. Laterally they are confluent with the nuchal crest. The only specimen of *P. africanus* with the relevant areas preserved is a juvenile female, so the condition of temporal lines on the adult cranium is indeterminate.

The ectotympanic bone of *Aegyptopithecus* is fused solidly to the petrosal bone, but is unique among adult Old World

Anthropoidea in not being prolonged outward as a tube (Simons, 1972; see Fig. 8e). By contrast, *Proconsul* had a well-developed tubular ectotympanic (Davis and Napier, 1963). *Pliopithecus* is intermediate in this respect (Szalay and Delson, 1979); short ectotympanic flanges are prolonged laterally from anterior and posterior parts of the petrosal bulla, but the ventral portion is incompletely ossified (Zapfe, 1960, Figs. 28–30).

The mandibular tooth rows of *Aegyptopithecus* (Fig. 9a) diverge posteriorly, giving the jaws a V-shaped appearance in occlusal view. The ratio of breadth between P_4's to breadth between M_2's is 0.74 in one female *Aegyptopithecus*. A similar ratio is present in the type of *Propliopithecus chirobates*. The tooth rows of East African Miocene apes diverge posteriorly also, as do those of *Pliopithecus* (Andrews, 1978; Zapfe, 1960). In the Fayum apes, the ratio of mandibular depth under M_{1-2} to M_2

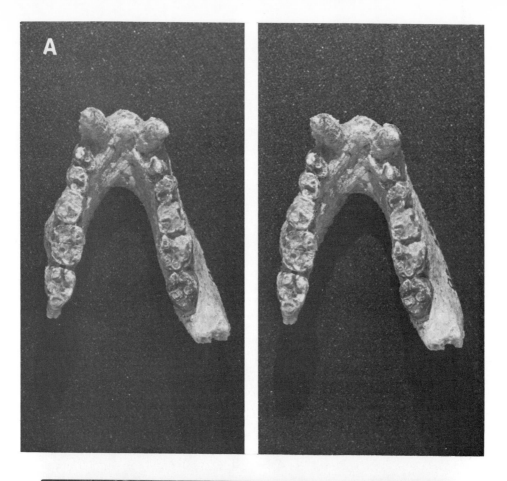

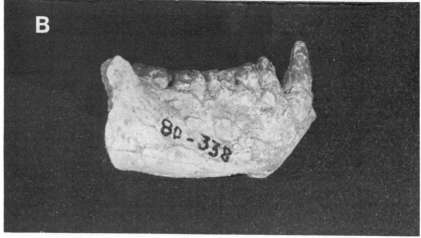

Figure 9   *A, stereophotograph of occlusal view of mandible of* Aegyptopithecus zeuxis, *female, newly recovered (1980) from Quarry M, Upper Fossil Wood Zone. B, Lateral view of the same (X c.1.5).*

length is extremely high for males by comparison with males of Miocene ape species [comparative data from Andrews (1978) and Zapfe (1960)]. Two male *Aegyptopithecus zeuxis* have ratio of depth to M_2 length of 3.03; two male *Propliopithecus chirobates* have a ratio of 3.57. Three male *Pliopithecus vindobonensis* average 2.17 and a *P. antiquus* male has 2.38 (Zapfe, 1960). Some male *Proconsul nyanzae* and *P. major* approach the relative jaw

depth of male Fayum apes, but the males of some other early Miocene species have shallower jaws.

The mandibular symphysis of both *Aegyptopithecus zeuxis* (Fig. 10c) and *Propliopithecus chirobates* is buttressed by well-developed superior and inferior transverse tori. This condition is found also in *Dendropithecus macinnesi* (Miocene), but East African *Proconsul* and *Limnopithecus* usually lack an inferior transverse torus (Andrews, 1978). *Pliopithecus* tends to have a well-developed inferior transverse torus; in *Pliopithecus vindobonensis* the superior torus is quite variable (Zapfe, 1960).

Male *Aegyptopithecus zeuxis* mandibles show an extremely heavy scar for the origin of the anterior belly of the digastric muscle. This scar extends on the ventral surface of the mandibular corpora from near the midline of the symphysis back to the middle of M_2. Such a broad origin for digastric muscle is typical of extant catarrhines (Hill, 1966; Raven, 1950).

Dental Morphology

All of the Fayum and Miocene ape taxa are known from dental remains, allowing a broader range of comparison than was possible with cranial material. The most complete reviews of the dental anatomy of Miocene apes on which the present comparison is based are the studies of Andrews (1978), Fleagle and Simons (1978a), and Zapfe (1960).

Andrews places considerable emphasis on central incisor shape and height as diagnostic criteria for recognition of East African Miocene taxa. In particular, *Dendropithecus macinnesi* is said to have strongly mesial-distally compressed, high-crowned incisors by comparison with the contemporaneous taxa. Using Andrews' (1978) criteria, the lower cental incisor of *Propliopithecus chirobates* (Fig. 11) is comparatively broad and low-crowned, resembling most East African Miocene taxa; those of *Aegyptopithecus zeuxis* are very narrow and high-crowned, resembling *Dendropithecus*. However, the sample sizes on which Andrews'

conclusions are based are very small and the variability in the measurements is very high. The largest sample of lower central incisors for a Miocene ape consists of six specimens of *Proconsul (Rangwapithecus) gordoni*; *Dendropithecus macinnesi* is represented by two specimens only, and unaccountably one specimen is eliminated from all of Andrews' sample statistics. The shape indices within each species vary widely. One sample of three I_1's of *Limnopithecus legetet* has buccal-lingual/mesial-distal ratios ranging from 104 to 148. This exceeds the range of species means for all six Miocene taxa analyzed by Andrews. In fact, the only statistically significant difference in lower incisor shape which emerges from Andrews' study is that *Proconsul (Rangwapithecus)* apparently had comparatively narrow incisors. [This would be consistent with a possibly more folivorous diet for this species (Kay, 1977b).] There are no associated upper central incisors for any of the Fayum apes.

The upper canines of the Fayum species (one each of *Aegyptopithecus zeuxis* and *Propliopithecus chirobates*) are nearly round in cross section at the base. The lower canines of these species are more bilaterally compressed. In *Pliopithecus* and East African Miocene hominoids, upper canine breadth is 70–80% of canine length (in the occlusal plane) and the lower canines are also bilaterally compressed. Delson and Andrews (1975) argued that "bilaterally compressed" upper and lower canines were primitive for Old World Anthropoidea, and that dryopithecines are characterized by rounded upper canines, but the evidence of the Fayum apes suggests that the upper canines of ancestral apes may have been more circular in cross section.

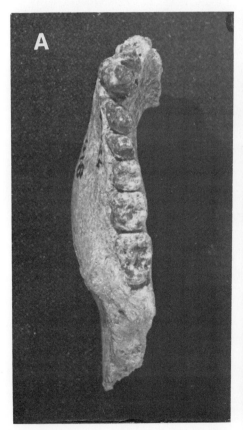 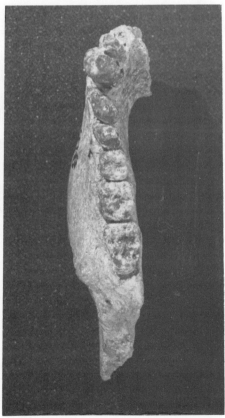

Figure 10   *A, stereophotograph of occlusal view of mandible of* Aegyptopithecus zeuxis, *male, (CGM 40137) from Quarry M, Upper Fossil Wood Zone (X c.1.6).*

The upper premolars and M$^{1-2}$ of *Propliopithecus* and *Aegyptopithecus* are quite broad buccal-lingually, M^1 of the former more so than that of the latter. The ratio of P^4 breadth to length exceeds 1.50 for these species; the ratio of M^2 breadth to length is greater than 1.20. In this respect, the Fayum species resemble species of *Micropithecus, Pliopithecus, Limnopithecus, Dendropithecus,* and *Proconsul. Proconsul (Rangwapithecus) gordoni* and *P. (R.) vancouveringi* differ from all contemporary and earlier apes in having much narrower upper premolars and nearly square molars [data from Andrews (1978), Fleagle and Simons (1978a), and Zapfe (1960)]. In both of these respects, *P. (Rangwapithecus)* very closely resembles later *Sivapithecus* and extant Great Apes.

The lower fourth premolar of the Fayum apes tends to be broad: the ratio of buccal-lingual breadth to mesial-distal length is greater than 1.00. In this respect there is a close similarity to *Proconsul.* On the other hand, P$_4$ tends to be longer (breadth/length < 100) in *Pliopithecus, Limnopithecus, Dendropithecus,* and *Proconsul (Rangwapithecus).*

As noted above, the M$_2$ is about 5% longer (mesial-distally) than M$_1$ in *Propliopithecus.* In *Aegyptopithecus,* M$_2$ is much longer than M$_1$. *Micropithecus, Limnopithecus, Dendropithecus,*

Pliopithecus, Proconsul, and *P. (Rangwapithecus)* all have much longer M$_2$'s than M$_1$'s, resembling *Aegyptopithecus.*

The upper premolar and molar lingual cingula of *Aegyptopithecus zeuxis* and *Propliopithecus chirobates* are very strongly developed. In African Miocene apes, the upper lingual cingulum is best developed on the upper molars and is progressively weaker toward the anterior upper premolar. The cingulum development reaches forward to include P^3 in *Proconsul (Rangwapithecus),* which resembles the Fayum apes in this respect. It is absent from P^3 but strongly developed on P^4—M^2 in *Dendropithecus, Pliopithecus* and *Limnopithecus.* It is poorly developed on P^4 and well developed on M$^{1-2}$ in *Proconsul.*

Postcranial Morphology

In the past 5 years, increasing numbers of skeletal elements attributable to the Oligocene hominoid have been recovered and described [Fleagle et al. (1975); Conroy (1976a,b); Fleagle and Simons (1978b, 1982); reviewed in Fleagle (1980)], including most of the humerus and ulna, a hallucial metatarsal, and a phalanx of *Aegyptopithecus zeuxis.* In addition to their value for reconstructing the locomotor habits of these early anthropoids, these

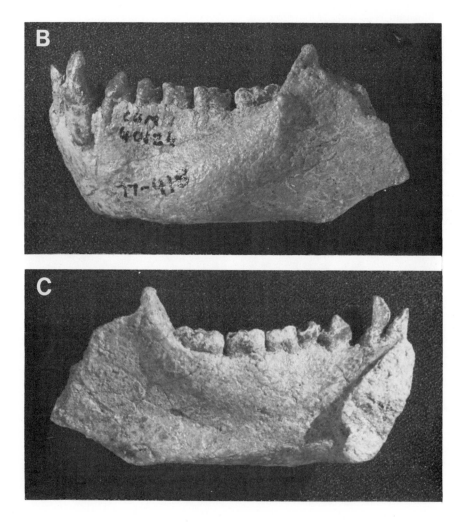

Figure 10 (continued)   *Side view of the mandible of* Aegyptopithecus zeuxis, *male, (CGM 40137) from Quarry M. Upper fossil Wood Zone, B, lateral view, and C, medial view of the same (X c.1.6).*

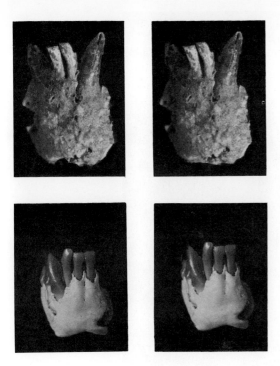

Figure 11 *Stereophotographs of incisors of* Aegyptopithecus zeuxis *(top) DPC 1112, and* Propliopithecus chirobates *(bottom) DPC 1069 (X c.1.5).*

skeletal elements provide important phyletic information of the relationship between Oligocene and later hominoids.

The humeri attributed to *A. zeuxis* (Fig. 12) are more primitive than the humeri of any extant higher primate, either platyrrhine or catarrhine (Fleagle and Simons, 1978b, 1982). The morphology of tuberosities, the bicipital groove, the deltoid plane, the brachialis flange, and the shaft as a whole are more "prosimian"-like than anthropoid-like. The distal aspect of the humerus in both *A. zeuxis* and *P. chirobates*, however, is more comparable to that of extant platyrrhines in the shape of articular surfaces and the possession of an entepicondylar foramen. This latter feature, primitive for mammals as a group, is found in *Pliopithecus* and about half the genera of extant platyrrhines (often variably), but is absent in all Miocene to Recent catarrhines, including all extant cercopithecines, and hominoids except as a rare variation in *Homo*.

The relatively wide trochlea, large medial epicondyle, and thin supracapitular region in the humeri of Fayum species are apparently primitive characteristics which link them with platyrrhines and later hominoids and contrast with the derived conditions seen in cercopithecoids. The Fayum taxa and *Pliopithecus* lack the spool-shaped trochlea seen in extant hominoids, and to a lesser degree in *Proconsul africanus*.

The ulna attributed to *A. zeuxis* (Fleagle et al., 1975; Preuschoft, 1975; Conroy, 1976a; Schön-Ybarra and Conroy, 1978) is similar to the same bone in the living platyrrhines such as *Alouatta* and unlike the ulnae of either extant cercopithecines or hominoids. Morphologically it is similar to (but more robust than) the ulna of *Pliopithecus* and neither shares any clearly derived features with any group of later catarrhines.

The first hallucial metatarsal attributed to *A. zeuxis* clearly demonstrates the presence of an opposable grasping hallux in this species (Preuschoft, 1975) and also shows a facet for a prehallux bone. A prehallux and the associated facet is not present in hominids, living pongids, or cercopithecoids, but is found in many platyrrhines, in *Hylobates*, in *Pliopithecus*, and in *P. africanus* (Conroy, 1976b). Its presence in *A. zeuxis* is certainly a primitive anthropoid feature. The phalanges are similar to those of other arboreal anthropoids with no remarkable features.

Summary of Morphological Features

In all known characteristics, with the possible exception of very small orbit size, the cranial anatomy of *Aegyptopithecus zeuxis* appears to present what might be expected of a primitive anthropoid. In fact, other than the diagnostic anthropoid features of postorbital closure and a fused mandibular symphysis, the cranial anatomy is very similar to that of adapids and advanced omomyids like *Rooneyia. Aegyptopithecus zeuxis* does not share any derived features of the cranium, suggesting a closer relationship to either *Pliopithecus, Proconsul africanus,* or any other single Miocene taxon. On the other hand, *Pliopithecus* and *Proconsul* share such advanced anthropoid cranial characteristics as dorsally reduced premaxillae, short, nonprojecting nasals, reduced postorbital constriction, relatively shallow mandibular rami, and at least partial development of a tubular ectotympanic bone. In this last feature, *Proconsul* is apparently more advanced than is *Pliopithecus*.

Dentally, both *Aegyptopithecus* and *Propliopithecus* are very similar to African Miocene hominoids and *Pliopithecus*. Lower incisor morphology is comparable. The lower canines of the Fayum species are very similar to those of Miocene apes, but the upper canines are more rounded (less bilaterally compressed) than in any of the Miocene species considered here. The upper premolars and molars of the Fayum apes are very broad buccal-lingually, which is true for the Miocene apes except *Proconsul (Rangwapithecus)*. M_2 is longer than M_1 in *Aegyptopithecus* and the Miocene apes. *Propliopithecus* in this regard differs strikingly in having M_2 about the same length as M_1. All taxa have a well-developed lingual cingulum on the upper molars, and in

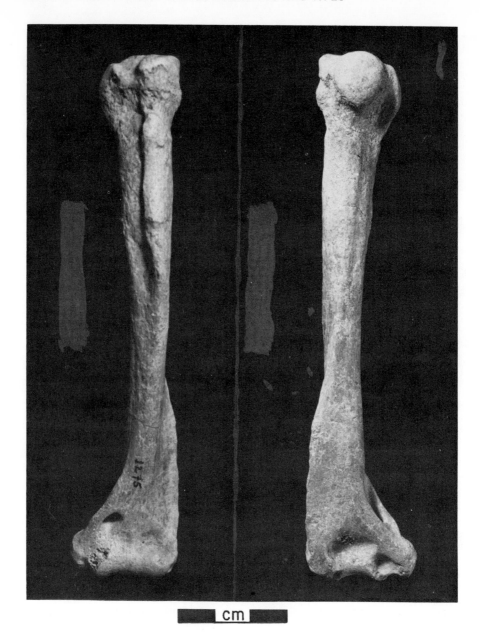

Figure 12 *Anterior (left) and posterior (right) views of the humerus of* Aegyptopithecus zeuxis *(DPC 1275). Approximate size.*

many species there is a comparable development of the cingulum on the upper premolar also.

In postcranial anatomy (see Fleagle, 1983) the Fayum hominoids are completely primitive with respect to both living hominoids and cercopithecoids (or even platyrrhines in many ways). There are no derived features linking the Egyptian genera with any one Miocene taxon or with either extant catarrhine group. Indeed, the Fayum genera show a primitive condition that could be ancestral to both groups. *Pliopithecus* is more "advanced" than *Aegyptopithecus* in the more gracile nature of the humerus and ulna, but retains the primitive entepicondylar foramen and generally primitive limb anatomy. All of the dryopithecines from the early Miocene of East Africa, including *Dendropithecus,* lack this foramen, as do all extant catarrhines. *Proconsul africanus* appears to be even more like a modern hominoid in that it has a

distal humeral articulation like a living ape and also apparently a very chimpanzee-like foot and shoulder (Walker and Pickford, 1983). However, both *Pliopithecus* and *Proconsul africanus* lack the characteristic hominoid wrist morphology with a very reduced ulnar styloid.

In summary, the preponderance of anatomical evidence indicates that *Pliopithecus* and the early Miocene hominoids from Kenya and Uganda, were more like modern hominoids and more derived than *Aegyptopithecus.* The Fayum apes are both sufficiently primitive and geologically old enough to have been the ancestors of all later hominoids.

So primitive is the anatomy of *Aegyptopithecus* and *Propliopithecus* that these Oligocene anthropoids might easily be ancestral to all later catarrhines (Fleagle and Simons, 1978b; Fleagle, 1980; Kay et al., 1981). Miocene to Recent catarrhines (both hominoids and cercopithecoids) share a number of probably derived features not present in *Aegyptopithecus,* including a tubular ectotympanic, shortened nasals, reduced premaxillae, and enlarged brains, as well as loss of the entepicondylar foramen and numerous modifications in the shape of the humerus. Having ape-like molars would not necessarily rule out the Fayum "apes" from Old World monkey ancestry. Certainly hominoid molars more closely resemble the primitive primate molar pattern than do those of Old World monkeys.

Other considerations have nevertheless led many authors to exclude the Fayum apes from the ancestry of Old World monkeys and to regard them as specially related to Miocene—Recent hominoids alone. There are possibly derived features of the molar dentitions which link *Aegyptopithecus* and *Propliopithecus* with Miocene hominoids and would tend to rule them out of the ancestry of Old World monkeys. The upper molar lingual cingulae are particularly well developed in Fayum apes and the M_{1-2} hypoconulids are very large and centrally or buccally placed. These features are the reverse of what would be expected in a cercopithecoid ancestor. A second, but independent, argument against deriving Old World monkeys from the Fayum "apes" is that the Parapithecidae, particularly *Parapithecus grangeri* from the Upper Fossil Wood Zone, show a considerable degree of derived dental similarity to the Cercopithecidae. Recently recovered dental specimens show that, although *Parapithecus grangeri* still had three premolars, the first of these was considerably reduced, and the lower molars are extremely high-crowned and incipiently bilophodont (Simons, 1970, 1972, 1974b), Unlike *Apidium,* upper molars of *Parapithecus* are quadrate, with four main cusps. This, and a combination of other derived dental features reviewed by Kay (1977a), support Simons' (1970) case that the parapithecids are ancestral to Old World monkeys. However, like the Fayum apes, parapithecids retain a number of very primitive cranial and postcranial features which suggest that they antecede the divergence of hominoids and cercopithecoids (Cartmill et al., 1981; Delson and Andrews, 1975; Szalay and Delson, 1979; Fleagle, 1980; Fleagle and Rosenberger, 1983).

PHYLOGENY AND SYSTEMATICS OF FAYUM ANTHROPOIDS

Any phylogeny linking Fayum primates with later catarrhines must involve considerable parallelism. The "prosimian"-like features in the humerus and skull of *Aegyptopithecus* must have been lost independently in platyrrhines and catarrhines. Furthermore, if *Aegyptopithecus* is a phyletic ape [that is, specially related to apes alone; as advocated by Simons (1967d, 1972), Simons and Pilbeam (1972), and Szalay and Delson (1979)], then the few osteological features proposed which unite extant catarrhines (e.g., tubular ectotympanic; loss of entepicondular foramen) are necessarily also parallelisms. Alternatively, if *Aegyptopithecus* and the other Fayum anthropoids phyletically precede the hominoid—cercopithecoid divergence, then the dental similarites between *Parapithecus* and cercopithecoids (Simons, 1974a; Kay, 1977a) are parallelisms. On the present evidence, we support the latter phylogeny because of the lack of convincing derived features linking *Aegyptopithecus* and *Propliopithecus* uniquely with later hominoids and their lack of many apparently derived features seen in all modern catarrhines.

If *Aegyptopithecus* and *Propliopithecus* are really primitive catarrhines belonging to a group ancestral to both cercopithecoids and living hominoids, rather than uniquely related to later apes, to what higher taxon of higher primates should they be assigned? They clearly should not be regarded as cercopithecoids; all fossil and living Old World monkeys are a very coherent group

showing numerous uniquely derived features in their dentition and skeleton (Szalay and Delson, 1979). *Aegyptopithecus* and *Propliopithecus* show none of these derived features and do not even show any inkling of special cercopithecoid affinities.

There are two remaining taxonomic alternatives. They can be left in the Hominoidea with the clear implication that Hominoidea is a wastebasket taxon, or they can be placed in a new taxon for noncercopithecoid, nonhominoid catarrhines. [Szalay and Delson (1979) have used a comparable taxon for the parapithecids.] We prefer to leave the Oligocene apes in the Hominoidea for several practical reasons.

First, and most important, is the very obvious fact that *Aegyptopithecus* and *Propliopithecus* are merely the oldest of numerous "dental apes" from the Oligocene and Miocene of Africa and Eurasia whose phyletic position *vis à vis* later apes is debatable. *Pliopithecus* shares only a few more derived features with living catarrhines than do the Oligocene taxa, and even the dryopithecines share very few clearly derived dental, cranial, or skeletal features with living pongids, hylobatids, or hominids. Expanding the Hominoidea to include *Aegyptopithecus* and *Propliopithecus* eliminates the additional difficulties of deciding what to do with the other dental apes who grade ever so tenuously into a more modern-looking hominoid appearance.

Second, a broad definition of the Hominoidea recognizes what anatomists have long realized (e.g., Wood-Jones, 1929; Le Gros Clark, 1934), that in many aspects of dental and skeletal anatomy, cercopithecoids are equally or more derived from the ancestral anthropoid condition than are apes. Furthermore, Old World monkeys appear in the fossil record "full-blown" in recognizably modern form, but it is impossible to identify a likely ancestor for cercopithecoids among the "dental apes" of either the Oligocene or Miocene (Delson, 1975b).

Within the Hominoidea, the most primitive taxa can reasonably be grouped in a single family, the Pliopithecidae (Remane, 1965; Szalay and Delson, 1979). In contrast with the latter authors, we would recognize two subfamilies: Propliopithecinae for the more primitive Fayum genera such as *Aegyptopithecus* and *Propliopithecus,* and Pliopithecinae for the somewhat more advanced *Pliopithecus. Dendropithecus* is more closely allied with the other dryopithecines of East Africa, and should be placed with them (Fleagle and Simons, 1978a).

SUMMARY

Four species of "dental apes" currently are recognized from the Fayum (Oligocene, Egypt). *Aegyptopithecus zeuxis* (Simons, 1965a) and *Propliopithecus chirobates* (Simons, 1965a) from the Upper Fossil Wood Zone are both known from numerous jaws and various skeletal elements. *Propliopithecus haeckeli* (Schlosser, 1910) and *Propliopithecus markgrafi* (Schlosser, 1910) are known only from their type specimens collected early this century from an unknown stratigraphic level and possible isolated teeth from Quarry G (Simons, 1972). Dentally, these taxa are similar to later hominoids, particularly *Pliopithecus* from the Miocene of Europe and *Proconsul, Limnopithecus,* and *Dendropithecus* from East Africa. In skeletal anatomy they share many primitive features with *Pliopithecus.* In contrast, in the cranial and skeletal anatomy they are more primitive than any later catarrhine and share no obviously derived features with any group of living catarrhines, cercopithecoid monkeys, hylobatids, pongids, or hominids. They are thus suitable phyletic ancestors for all later catarrhines. They should be retained within the superfamily Hominoidea as a separate subfamily Propliopithecinae of the primitive hominoid family Pliopithecidae. This arrangement implies that Hominoidea are the primitive catarrhines and that cercopithecoids are derived from a hominoid ancestor.

Acknowledgments

The research reported here was supported by NSF grant BNS77-25921 and BNS79-2419 to John G. Fleagle, BNS77-08939 to Richard F. Kay, and BNS77-20104, BNS80-16206, and Smithsonian Foreign Currency Grant FC0869600 and FC80974 to Elwyn L. Simons. The field work was carried out with the cooperation of the Egyptian Geological Survey and Mining Authority, and especially Ragi Eissa, Darwish el Far, Abed El Ghani Ibrahim, Bahay Issawi, Baher el Khashab, Galal Ali Moustafa, M. F. el Ramly, and Rushdi Said. We also thank W. L. Jungers for comments on the manuscript. We thank the curators of Mammals at the American Museum of Natural History, the Museum of Comparative Zoology, and the Smithsonian Institution for allowing us to study the primate specimens in their care. We are especially grateful to Prof. E. L. Simons for allowing us to participate in the Fayum expeditions under his direction.

22

The Deployment and History
Of Old World Monkeys:
Oligocene Cercopithecoidea
E. L. SIMONS

Editor's Note:

This short excerpt is taken from a comprehensive review of the evolution of the cercopithecoid (Old World) monkeys written by Elwyn Simons in 1970. In this article Simons, for the first time, outlines his view that the common ancestor of all later cercopithecoids can be traced back to the Parapithecids of the Oligocene of Egypt. This excerpt presents Simons arguments.

Apidium and *Moeripithecus* from the Egyptian Fayum were suggested by Gregory (1922a) and by others as possible ancestors of monkeys or of *Oreopithecus*. The many new finds of dryopithecine jaws and teeth from our Egyptian Oligocene quarries G and I, belonging to species of both *Propliopithecus* Schlosser (1910, 1911) and to *Aegyptopithecus* Simons (1965a), show that these animals should undoubtedly be considered dryopithecine pongids. They further show that the unique type mandible of *Moeripithecus markgrafi* belongs to a pongid species that is nearly intermediate between *Propliopithecus haeckeli* and *Aegyptopithecus* but apparently closer to the former. I have consequently referred the genus "*Moeripithecus*" to junior synonymy under *Propliopithecus* (Simons, 1968d). *P. markgrafi* is a distinct species but it clearly can no longer be considered a monkey forerunner. The Fayum dryopithecines are not very similar to their contemporaries among the parapithecines (Simons, 1969b) and consequently a common ancestry would have to have been much earlier in time than the Oligocene. In fact the parapithecines are so different from the Egyptian dryopithecines that several scholars (but see principally Hürzeler, 1968) have even challenged the fact that they are primates.

The many new finds of mandibles, maxillae, isolated teeth and a considerable number of isolated postcranial bones which can be referred to the Fayum Parapithecinae, establish the relatedness of these animals to the cercopithecoid monkeys. What is equally clear is that they are not some otherwise unknown group of prosimians—even in their most primitive features; judging from the numerous postcranials, parapithecines are at least as "simian" as many platyrrhine monkeys. Cranially also these animals have reached the grade of Anthropoids, for they exhibit pre-juvenile (perhaps pre-natal) fusion of the metopic and symphyseal sutures as well as the development of post-orbital plates. In 1967 I discovered part of an *Apidium phiomense* frontal associated with teeth of this species at Yale Quarry I and a right maxilla with the orbital margin and post-orbital bar was found at Quarry M. These show that the more complete frontal described

by Simons (1959) belongs to *A. phiomense*. Together with a mandible and isolated upper canines and incisors from Quarry I, these two specimens can be made the basis for the composite reconstruction of Figure 1.

To date, four species of parapithecines have been discovered in the Fayum, two of *Apidium* and two of *Parapithecus*. Species of these two genera resemble each other closely in morphology of the antemolar teeth. They have relatively small canines compared to primates other than the marmosets, which they approximate in size. Like platyrrhines, species of *Apidium* and *Parapithecus* possess 3 pairs of premolars above and below. Consequently loss of P_2^2 in Cercopithecoidea must have occurred independently, and later than it did among ancestral Hominoidea. In molar and mandibular morphology *Apidium* and *Parapithecus* species differ distinctly. Relative to tooth size, *Apidium* horizontal mandibular rami are typically deeper than in *Parapithecus*. In lower molars of both genera the paraconid is reduced to a crest and the M_{1-2} are long and block-like as seen from above (rather than nearly circular in outline as are the Fayum pongid M_1 and M_2's). This is because, in parapithecines, the trigonid and talonid are slightly separated from one another by median internal and external indentations.

Narrow or "waisted molars", are typical of most later Cercopithecoidea. Such indentations are not typical of the vast majority of hominoids but do occur in *Oreopithecus bambolii* and *Gigantopithecus blacki*. Species of *Apidium* differ markedly from those of *Parapithecus* in that there is a clear-cut size increase in the molar series posteriorly. The upper and lower increase in the molars of *Apidium* are distinctly polycuspidate for a primate while in *Parapithecus* species, cuspules beyond the basic five are almost never present, and M_3 is always distinctly smaller than M_2. In many cases (just as in the Talapoin) the M_3 of *Parapithecus* is considerably smaller. In *Parapithecus* the foveal height of the lower molars is great, and lower molars have four principal cusps (the paraconid and hypoconulid are very low and reduced in size) arranged in a quadrate pattern which is almost identical in form to that of teeth in the smaller species of *Cercopithecus*, or *Miopithecus talapoin* (see Figure 2).

At Quarry I in the Fayum the two most common mammalian species are *Apidium phiomense* and *Parapithecus sp. nov.* Since Quarry I definitely represents a river-lain sediment it is tempting to conjecture that the Oligocene parapithecines lived in the canopy along river banks and over swamps just as many small, arboreal, living cercopithecoids such as the Talapoin and Allen's swamp monkey do today.

Excerpt reprinted with permission from "The Deployment and History of Old World Monkeys (Cercopithecidae, Primates)," which appeared in *Old World Monkeys: Evolution, Systematics, and Behavior,* edited by J. R. Napier and P. H. Napier, pages 97–138, Academic Press, Inc., New York. Copyright © 1970 by Academic Press, Inc., New York.

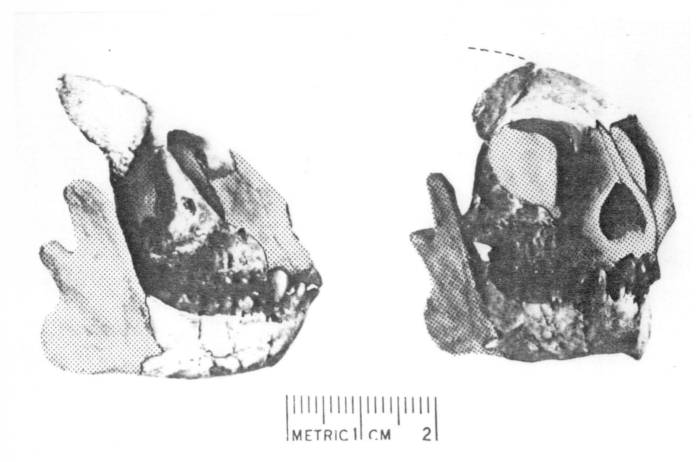

Figure 1 *Composite reconstruction of the face of* Apidium phiomense *based on separate finds; A.M.N.H. 14556 (frontal), C.G.M. 26920 (maxilla), Y.P.M. 21018 (mandible), and isolated upper canines and incisors associated in Quarry I. Stippled areas reconstructed from other specimens.*

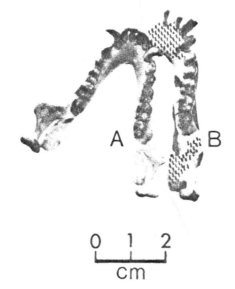

Figure 2 *Comparison between mandibles of the modern African swamp monkey,* Miopithecus talapoin *(A) and a new species of* Parapithecus *from the Fayum (B). The four anterior teeth of B are from the same site but do not belong to the same individual. In both these catarrhines there are four principal lower molar cusps and in the Oligocene species the hypoconulids are so reduced and flattened that they do not receive significant wear except in extreme dental age.*

23

Toward the Origin of the Old World Monkeys

E. DELSON

INTRODUCTION

The problems of applying Hennigian or cladistic methods in paleontology have been discussed by other speakers at this meeting and elsewhere. In brief, I consider that Hennig and his colleagues have made their greatest contribution in emphasizing the value of derived *versus* ancestral conditions in determining relationships among taxa. Other aspects of "cladism" are more open to question, but systematics and comparative anatomy have long accepted the importance of derived or "specialized" features, although there has until now been little attempt to employ them exclusively.

The Old World monkeys (family Cercopithecidae) are a group of higher primates whose living members share at least one derived "key character" of adaptation: a bilophodont dentition. This adaptation is already present in the fossil forms unequivocally assigned to the family (see Delson, 1975a). In fact, certain authors (Straus, 1953; Kälin, 1961, 1962) have argued that no animal can be termed a cercopithecid which does not possess this dental morphology. Although such an extreme view may not be necessary, similar criteria have been (and are still) used to define early Mesozoic mammals. What is of interest here, however, is to determine what can now be said about the possible ancestors of known Neogene Cercopithecidae.

Following Hennig (1966a), Schaeffer *et al.* (1972) and others, the study of morphologic diversity among the living and fossil members of a group may allow the determination of derived character states and by consequence those which are ancestral for the group. The combination of the latter conditions in a single model animal may be termed a postulated ancestral morphotype, representing those characters expected in the latest common ancestor of all known members of the group. Two additional procedures, to be considered here, are the extrapolation of this morphotype one step farther back in time and its subsequent comparison to known "primitive" fossils, which of course have the value of placing time and space constraints on our reconstructions of evolutionary history.

MORPHOLOGY OF THE ANCESTOR OF CERCOPITHECIDAE

In the development of an ancestral morphotype for cercopithecid monkeys, emphasis was placed on the variation observed in modern species, although all known fossils were considered as well, in light of the increased ranges of total morphological variation thus obtained. Of hard parts available for fossilization, the dentition and less distinctively the skull provide valuable characters, but the post-cranial skeleton is much more responsive to an analysis of habitus than of heritage.

Based on the dentition, four major taxonomic-morphologic groups can be distinguished among modern monkeys, and all but a few early and possibly still-evolving taxa fit clearly into one of these. As I have detailed elsewhere (1975a, 1973), it is thus possible to define morphotypes representing the shared (derived and ancestral) features of the Colobinae, the Cercopithecini, *Theropithecus* and the other Papionini (baboons, macaques, etc.). The last of these groups appears the least derived from a common cercopithecid ancestor in most dental characters, and the modern genus *Macaca* may be the least "specialized" of all.

The resultant ancestral cercopithecid dental morphotype would thus be rather similar to macaques in possessing high-crowned ("hypsodont") cheek teeth with transverse crests (as in all monkeys) linking relatively low, rounded cusps. The main notches between these cusps, especially lingually on lower molars, would not be very deep. The mesial fovea (trigonid basin—see Delson, 1975a for terminology and illustrations) of lower molars would be of similar size to the distal fovea, both smaller than the central fovea (talonid basin). Four cusps only would be present on dP_4–M_2, with a distal hypoconulid on M_3 and a mesial paraconid crest on dP_3. From dP_4 to M_2, the teeth would become longer and relatively wider, while the M_3 would be quite long but narrow. Four cusps and three foveas would characterize dP^3–M^3, while two subequally large cusps and a single loph (id) would be found on P^{3-4} and P_4. A honing mechanism involving C^1 and P_3 would involve a large mesiobuccal flange on the latter unicuspid tooth, especially in males; large upper canine teeth would also present a mesial groove from apex to root base. No cingulum would be present laterally on any cheek teeth; instead there would be a smooth "flaring" outward from the cusp apexes to the cervix on all lophodont teeth, most prominent lingually on uppers and buccally on lowers. The lower incisors would be completely sheathed in enamel and the four incisors of similar proportions, with I^2 possible conical. In general, these are the characters found in the teeth of all cercopithecids or in macaques and baboons when there is distinct variation; the incisor region, however, has been most changed in known papionins other than *Theropithecus,* while colobines have preserved a suite of characters closer to the ancestral pattern.

In terms of other anatomical systems, there is less broad variation in the family. My studies and those of Vogel (1966) suggest that the earliest cercopithecids had a facial morphology similar to that found today in colobines and in gibbons (Hylobatinae); the cercopithecine face is most derived among modern

Reprinted with permission from *Problèmes Actuels de Paléontologie-Évolution des Vertébrés, Colloq. Int. Cent. Nat. Rech. Sci.* 218: 839–850. Copyright © 1975 by Centre National de la Recherche Scientifique, Ministere de la Recherche et de L'Industrie, Paris, France.

catarrhines. This ancestral pattern includes such features as wide interorbital region, short and broad nasal bones and a lacrimal fossa extending beyond the lacrimal bone onto the maxilla. The face in general would have been relatively short, broad and perhaps high, most similar to a modern *Presbytis*. On less secure morphological grounds, it may be expected that neither the colobine stomach specialization nor the cercopithecine cheek pouches had yet been developed; the limbs would have been of subequal length; and discontinuous ischial callosities were probably present. It may even be suggested that this early monkey possessed the presumed ancestral catarrhine diploid chromosome number of $2n = 44$ (shared today by most colobines and gibbons). Such animals might have lived early in the Miocene, and following ecological hypotheses discussed by Delson (1975c), may have been African arboreal or semi-arboreal quadrupeds who ate fruit when possible but supplemented their diet with leaves when necessary.

The next and more interesting step backward in time is to attempt reconstruction of the ancestor of this first cercopithecid, that is a primate lacking the Old World monkey dental specializations altogether. Many previous workers have attacked this point directly, without the above intermediate stage on which to base hypotheses. Moreover, most have become fascinated with the question of how bilophodonty developed, especially in terms of the origin of the loph(id)s themselves and of the postero-internal cusp of cercopithecid upper molars. Voruz (1970) gave a good summary of these studies, many of which discuss the possibility that the cusp in question is not a "true" but a "pseudo-"hypocone, that is, not derived from internal cingulum. I consider that loph(id)s arose in monkeys as they did in other mammals, as an adaptation to processing of food requiring slicing as well as crushing—today bilophodonty is most strongly developed in leaf-eating colobines and grass-chewing *Theropithecus*. The rearrangement of cusps and their linkage by crests is a functional problem dependent on the relative positions of cusps (and crests) on opposing teeth. Thus, even if it can be shown (see Kälin, 1962, fig. 1) that the cercopithecid hypocone was never linked to the internal cingulum, I would still consider it a hypocone because of its functional relationship with trigonid cusps, so long as it cannot be shown to be a "migrated conule", as Hurzeler (1949) has suggested. Von Koenigswald (1969) has assigned to *Victoriapithecus*, an African Miocene monkey, an upper molar which shows remnants of the crista obliqua, linking protocone and metacone. This specimen and contemporaries are now under further study, but as they are most probably cercopithecid, it would appear that ancestral populations possessed this crest more commonly. The origin of loph(id)s is not solved, but I agree with von Koenigswald that their development might have been relatively rapid under selection pressure for slicing occlusion, especially if a switch from fruit to leaf eating were involved.

Upper molars of pre-cercopithecids might thus be expected to have four cusps and a crista obliqua, lowers five cusps without clear talonid crests—derivation of a protolophid by strengthening the metaconid-protoconid link is simple to envisage. The presence of a hypoconulid on all lower molars (at least dP_4-M_3) is accepted as reasonable by comparison with all living and fossil primates of anthropoid grade, as well as advanced prosimians.

In addition to bilophodonty, high crowns, high relief and lateral molar "flare" characterize known cercopithecids but might have been lacking in their ancestors. It is suggestive to recall that flare seems most pronounced lingually in upper molars and buccally in lowers, as this is the distribution of cingulum in other advanced primates. It seems likely that flare developed through the incorporation of cingulum directly into the tooth face, producing a smooth surface, slightly bulged out near the cervix.

Some cingulum would thus be expected in pre-cercopithecids. Because crown height and especially crown relief are greater in cercopithecid cheek teeth than in those of most other primate families, these features might be expected to be only partly developed in the ancestors of Cercopithecidae; most other characters would probably have been as described above for early cercopithecids. It is important to note that no early fossils were considered the development of this postulated morphotype. Thus it is reasonable now to compare known fossils with the morphotype in order to seek out common features, without fear of circularity.

PARAPITHECUS AS AN ANCESTOR OF CERCOPITHECIDAE

In a search for known fossil relatives of the ancestor of Old World monkeys, the only possible candidates at present are the early catarrhines of the Fayum Egyptian Oligocene. Szalay (1970, 1972c) has rather convincingly denied *Amphipithecus* of the Burmese Eocene any place in higher primate history. Simons (1972 and personal communications) has suggested that its contemporary *Pondaungia* shows closer morphological similarities to catarrhines, but the specimens have not been fully studied. Unfortunately, the same is true for the Fayum materials collected in the 1960's by Simons. However, he has published numerous short comments and has been most generous in allowing colleagues access to figured specimens, for which I gratefully thank him.

I must preface all further remarks by accepting the limited nature of the Fayum record, both in small numbers of specimens of most primate taxa and as representing only a fraction of the area of Africa, much less the Old World, during only part of the Oligocene. However, these several species existed and need to be interpreted, and they may aid in the search for the origins of the Cercopithecidae and other modern groups.

As described by Simons (e.g., 1972), the Fayum primates are known from all three fossiliferous horizons. The unique holotype of *Oligopithecus savagei* is from the oldest level, the Lower Fossil Wood Zone. The many specimens of *Apidium moustafai* and perhaps Schlosser's holotypes of *Parapithecus fraasi*, *Propliopithecus haeckeli* and "*Moeripthecus*" *markgrafi* are derived from an intermediate horizon. The youngest beds, known as the Upper Fossil Wood Zone, have yielded large series of *Apidium phiomense* and *Parapithecus grangeri* (Simons, 1974b; see also Fig. 1), as well as specimens of *Aegyptopithecus zeuxis* and the unique type of *Aeolopithecus chirobates* (and perhaps *Propliopithecus*, see below). The species of *Apidium* and *Parapithecus* differ by about 15% in size, the younger populations being larger.

Since the first finds early in this century (Osborn, 1980 and Schlosser, 1910, 1911), authors have claimed ape or monkey status for each Fayum primate, often without close study. Because of his major accomplishments in raising our knowledge of these animals from four specimens in 1959 to hundreds today, the opinions of E. L. Simons are the most important for further consideration. As expressed in several papers since 1967, Simons argues for special, indeed direct ancestral relationship between cercopithecids and the Fayum primate *Parapithecus*. It must be assumed that the total body of evidence in support of this view is eventually to be published as a coherent unit, perhaps in the manuscript cited as in press in Simons, 1972. As yet, however, these arguments have not been presented in full, but on the basis of previous papers by and communications from Dr. Simons as well as my observations of the fossils, a brief description of the characters of *Parapithecus* may be given (see also Kälin, 1961).

The lower molars of *Parapithecus* species each have five cusps with a relatively small but distinct hypoconulid on the midline. The molars are relatively high-crowned, but the lingual intercusp notches are shallower even than in cercopithecines. Paraconids are reduced to a crest perhaps confluent with the mesial margin in molars, but are distinct on dP_4 (and probably dP_3). There are no lophids crossing the teeth, but the median buccal notch does angle inward as in cercopithecids to constrict the tooth between trigonid and talonid; unpublished upper molars may functionally approach a lophodont condition through incorporation of conules. There is no indication of a cercopithecid-type distal fovea, set off between distal margin and rear cusps, in the lower teeth of *Parapithecus*, nor would one be expected because of the omnipresent hypoconulid. All three permanent molars are rectangular and rather wide, and the M_3 is slightly or even considerably shorter than M_2. Unworn teeth of *P. grangeri* are more crest-oriented than in the bulbous cusped *P. fraasi*, but rapidly flatten with wear. The lower premolars increase in size but decrease in complexity from P_4 to P_2: a talonid basin and low metaconid are present on P_4 and less clearly on P_3, but P_2 is a simple, unicuspid tooth which appears to have been incorporated in some form of canine honing, as in modern platyrrhines. These features may be discerned on Figures 1 A-C and 2 B, D.

As Simons has argued (1967a, p. 599 and later), the new finds clearly show that two incisors and accompanying alveolar bone were lost from the type mandible of *P. fraasi* (Fig. 2A). It is less certain that both remaining teeth are I_1; they seem slightly different in morphology, and as no specimen of *Parapithecus* (or *Apidium*) has yet been recovered with both I_1 and I_2 *in situ,* the question remains open. It may finally be noted that while the mandibular corpus of *Parapithecus* is rather shallow, the bone is much deeper in the closely related *Apidium,* which differs further in possessing a long M_3, premolars relatively smaller by comparison to molars and accessory cuspules on molars, notably a centroconid (mesoconid) on all lowers.

Simons has selected certain of the listed characters which are similar to those in cercopithecids in order to promote *Parapithecus* as an Old World monkey ancestor. It is of interest to review the way his views have changed with increasing information and study. At the last C.N.R.S. colloquium in 1966, he argued that the parapithecids (*Parapithecus* and *Apidium*) were anthropoidean and suggested them as possible ancestors for Old World monkeys, noting that this would imply independent loss of P2 in Cercopithecoidea and Hominoidea (1967a, p. 599). A later paper that year (1967b) compared unworn M_{1-2} of a new *Parapithecus* species to those of living *Cercopithecus (Miopithecus) talapoin;* this is still the closest similarity yet demonstrated. Simons wrote (1967b, p. 321): "preliminary work suggests that some of these creatures *(Apidium* and *Parapithecus* species) may be related to the Old World monkeys." Further study allowed the statements that : "*Parapithecus* is extremely close in a number of dental and mandibular details to the smallest (and possibly most primitive) living African monkey *C. talapoin.* . . . In order to derive the dental-mandibular morphology of the talapoin from that of *Parapithecus* it would only be necessary to reduce lower molar hypoconulids from a small cusp to a flat platform. . . and eliminate the P_2" (Simons, 1969b, p. 323). In terms of formal taxonomy, Simons proposed (1970, p. 100) that "*Parapithecus* and *Apidium* should no longer be considered as representing a distinct primate family, Parapithecidae, but instead should be classified as the Parapithecinae, a primitive subfamily of the Cercopithecidae."

In his most recent work (1972), Simons appears to waver somewhat in his assertions. In a formal classification (p. 288), the Parapithecinae is one of three subfamilies of Cercopithecidae, of equal rank with Colobinae and Cercopithecinae. In discussing reasons for the lack of Miocene cercopithecoid fossils (pp. 185–186), Simons argues for (and I concur in) a late diversification with only a few (Early and Middle) Miocene species. He doubts that apes were more successful than monkeys at this time, mostly because of the abundance of Fayum parapithecines by comparison to contemporary "apes". On a more plaintive note, he writes (p. 189): "it does seem unlikely that such abundant and adaptively successful primates as *Apidium* and *Parapithecus* were aberrant side branches, doomed to eventual extinction." Finally, Simons concludes his discussion by noting (1972, p. 191): "The similarity between the teeth of *Miopithecus* and *Parapithecus* is remarkable, but may be due to parallel evolution. In any case, *Apidium* and *Parapithecus,* with symphyseal and frontal fusion and with post-orbital closure, have reached the anthropoid grade and qualify to be considered monkeys."

It is at this point that a confusion is revealed which has affected all of Simons' argument. It is evident to anyone who has considered the evidence that the parapithecids (or -cines) are at a *grade* level equal to that of monkeys in the vernacular sense, perhaps more "progressive" in some characters than the marmosets (Callitrichidae) of South America (see Simons, 1970, p. 102). But this is not proof of taxonomic placement, much less ancestry, by comparison to other forms, living or extinct. Success in terms of numbers in one time interval does not guarantee evolutionary survival or advancement, as witness the extremely successful primate *Plesiadapis,* the equid *Hipparion* and innumerable other "dead-ends". If the parapithecids were not ancestral to the later cercopithecids, the success of the former is not (directly) relevant to the rarity of the latter. Only a more detailed analysis of the morphology involved can demonstrate or "prove" relationships, and even then, not necessarily ancestry.

It is to be recalled that Simons' most convincing display of morphological similarity between *Parapithecus* and a monkey was with *C. (M.) talapoin,* which he termed "possibly most primitive". Only brief comparison was made with the earliest fossil cercopithecids and none with the suite of characters expected in a pre-cercopithecid ancestor. Moreover, my analysis indicates that in both dental and cranial form, the talapoin is a member of the highly derived Cercopithecini, which have lost M_3 hypoconulids, reduced lateral flare and increased relative length in all cheek teeth. Further, as shown by Verheyen (1962) and others, the skull of the talapoin is probably derived (through neoteny ?) by comparison to typical *Cercopithecus.* Only the (unrelated) character of female sexual swelling may be an ancestral trait retained by *C. (M.) talapoin.*

A direct comparison of the morphology of *Parapithecus* and the pre-cercopithecid morphotype postulated above reveals that they share high-crowned molars and the less definite features of deep mandibular corpus (in *Apidium*) and slightly rotated P_4 (in some Miocene monkeys). The retention of rather larger hypoconulids and of a paraconid on dP_4 are probably ancestral catarrhine (Old World anthropoid) features, as is retention of three premolars. However, the development of canine honing on P_2 is derived, and it seems unlikely that such a specialization would be lost and then redeveloped on P_3 if P_2 were dropped, nor is it likely that the P_3 of *Parapithecus* might be lost, leaving only P_2 and P_4. The low metaconid on P_4, short M_3 and general lack of cingulum are additional important differences between this fossil and the hypothesized pre-cercopithecid condition.

It would seem best at present to consider that *Parapithecus* and *Apidium* represent a group of early anthropoids of "monkey" grade which may have been the ecological vicars of cercopithecids

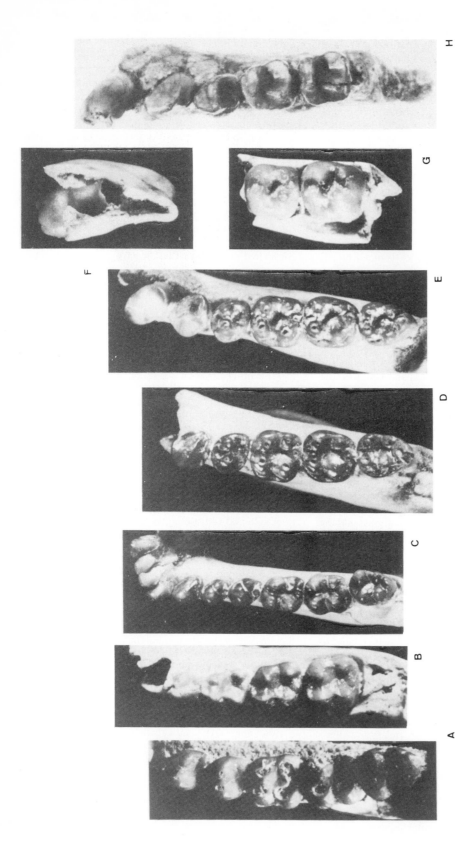

Figure 1 *Occlusal (and distal) views of selected Fayum primate specimens. A. Parapithecus sp., C.G.M. 26912, P₃–M₃. B. Parapithecus sp., Y.P.M. 23796, dP₃–M₂ (photographically reversed). C. Parapithecus fraasi, type, S.M.N.S. 12639a, R C₁ L M₃. D. E. Propliopithecus haeckeli, type, S.M.N.S. 12638, P₃–M₃ and C₁–M₃ (E photographically reversed). F, G. Propliopithecus (= "Moeripithecus") markgrafi, type, S.M.N.S. 12639b, M₁–₂; distal and occlusal view (G reversed photographically). H. Oligopithecus savagei, type, C.G.M. 29627, C₁–M₂. All at approximately three times natural size, except H about four times natural size. A is type P. grangeri Simons, 1974b [1].*

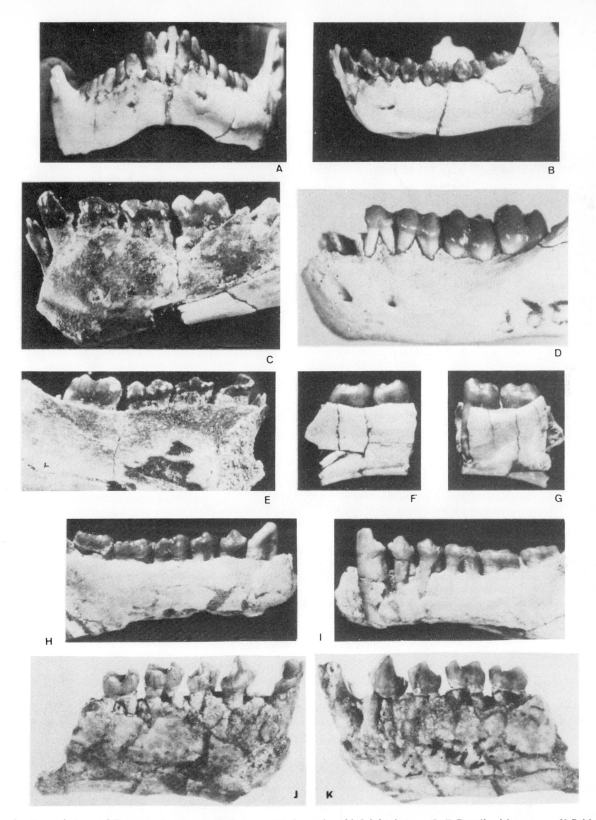

Figure 2   *Lateral views of Fayum primates. A, B. Same as 1C, frontal and left labial views C, E.* Propliopithecus *sp., Y. P. M. 23804, dC₁-M₁, photographically reversed right labial and lingual views. F, G. Same as 1F, G, photographically reversed right lingual and labial views. H, I. Same as 1E, photographically reversed right lingual and labial views. A B, F-I at approximately twice natural size; C-E, J-K at approximately three times natural size.*

and whose environment might have been specially similar to that of talapoins, causing further parallels. In some dental characters, notably increased crown height and perhaps a tendency toward lophodonty by a different route, this group converged on cercopithecid morphology, while in other ways they modified ancestral conditions in parallel. Several of Simons' above-cited statements are more easily interpreted in this light as well. Finally, the dangers of searching only for ancestors are clearly revealed, for if parapithecids are not the ancestors of cercopithecids, where are they to be placed in primate phylogeny?

AN ALTERNATIVE APPROACH TO MONKEY ORIGINS

Having rejected Simons' interpretation of cercopithecid ancestry as unsatisfactory, the most prudent course might be to claim that the incompleteness of the fossil record allows no further conclusions at this time. But hypothesis-formulation and prediction are among the essential aspects of the scientific method, and paleontology is built around fossil evidence. Therefore we may ask if any other known fossils are better candidates for having shared recent common ancestry with the progenitor of Cercopithecidae. All of the non-parapithecid Fayum primate species are at present considered "apes" by Simons (1972). One immediately obvious feature of this adaptive-taxonomic complex is the variation observed at what appears to be an early stage of catarrhine evolution. *Aegyptopithecus* is known from a skull, several partial mandibles and isolated teeth; of all the Fayum species, it seems most similar in morphology to early African species of *Dryopithecus* (and perhaps *"Limnopithecus"*, with which it has been compared by Andrews, 1970). *Aeolopithecus* is represented by a single mandible whose precise dental morphology cannot be evaluated due to its erosion, but it does afford further evidence of Fayum primate variability. The three remaining named species, including only five or six specimens with certainty, are of greater interest to this discussion.

Oligopithecus savagei, the oldest Fayum catarrhine (Figs. 1H, 2J, & K), is known from a single mandibular fragment with C–M$_2$, presumably of a male because of the well-developed C honing facet on P$_3$ (see also Simons, 1971a). *Propliopithecus haeckeli* has as holotype two portions of (probably) a single mandible, preserving both C–M$_3$ (except the left C, - Figs. 1 D & E, 2 H & I). Simons (1967a, 1971b) has reported but never described isolated teeth of similar morphology and size from the middle faunal horizon, suggesting a similar (or older) age for the type. Two incompletely described specimens from the upper horizon also present similar morphology: a juvenile mandible with dC–M$_2$ (Fig. 2 C & E; see 1967b) and a specimen with P$_3$–M$_3$, previously figured and identified as *Aegyptopithecus zeuxis* (1971a, b, 1972). Simons now (personal communication) identifies the latter as *Propliopithecus* sp. because of its short M$_3$; I suggest it to be a male, with a large P$_3$, while the type is presumably female. The final specimen of interest is that which Schlosser (1910, 1911) termed *Moeripithecus markgrafi,* a mandible fragment with M$_{1-2}$ (Figs. 1 F-G, 2 F-G). Simons (1967a, 1971b, 1972) has argued convincingly that this specimen is a juvenile best considered as representing a distinct species of *Propliopithecus.* It certainly appears referable to that genus, and its juvenile character is further emphasized by the open distal root on M$_2$ (see Fig. 1 F) [2].

If these several specimens are considered together as representing a pool of morphological variability known to be present in the Oligocene of northeastern Africa, it is possible to discern a mosaic of features similar to those postulated for a pre-cercopi-

thecid of similar age. These include: the presence of only two premolars, the P$_3$ with well-developed upper canine honing facet, the P$_4$ with subequally large metaconid and protoconid and two large foveas; a cingulum, especially buccally, around the lower molars, which have lost the paraconid and moderately enlarged the trigonid, separating the protoconid and metaconid but linking them by a crest (this feature only in *Propliopithecus* spp.); the tooth crowns are high, but the lingual notches relatively low; hypoconulids are present and of moderate size, but with evidence of appression to the entoconid—this is clearest in *Oligopithecus* and on M$_1$ more than M$_2$, while it is strongest on dP$_4$ of *Propliopithecus* sp. The molars are rounded, with flat wear of thin enamel in *Propliopithecus:* in *P. haeckeli,* the M$_3$ is narrow, but M$_{1-3}$ are subequally long; the M$_2$ is wider and longer than M$_1$ in *"Moeripithecus",* in which the buccal cingulum slopes outward almost in a "flare", and the trigonid basins are longer.

It has already been suggested that the lower buccal cingulae in P$_4$–M$_3$ could have developed into a cercopithecid flare, since the high crowns of these teeth are similar to those expected in premonkeys; shallow intercusp notches are also expected in ancestral forms, especially as they seem to have deepened independently in both subfamilies (compare colobines and *Theropithecus*). The large P$_4$ metaconids of Fayum "apes" are more cercopithecid-like than those of *Parapithecus,* and as Voruz (1970) has also observed, it is quite conceivable that the hypoconulid on dP$_4$–M$_2$ could have been lost through fusion with the entoconid due to interference with upper molar occlusion, in much the same way that many lineages lost the paraconid through interference with a developing hypocone. Hypoconulids were retained on M$_3$ to occlude with the M^3 distal fovea. Few upper molars of Fayum non-parapithecids are known (or published), and these are quite wide with lingual cingular shelves, but with cingular incorporation into the lingual surface, a moderately flared tooth would result.

In his original description of *Oligopithecus,* Simons (1962a) suggested possible similarities to cercopithecids, but later abandoned this view in favor of a parapithecid ancestry. Kälin (1961 and especially 1962) discussed the possibility that *"Moeripithecus"* might be a monkey ancestor, but rejected this relationship while concentrating on questions of cusp and loph(id) homology. He also agreed with Schlosser that the shallow jaw would imply a long face, but this shallowness is a function of its juvenile state, not a definitive character. Abel (1931) found *Moeripithecus* difficult to place phyletically, but thought it might be somehow linked to *Apidium,* which he considered a cercopithecid forerunner. Even earlier, Remane (1924) had noted some similarity between *Propliopithecus* and *Prohylobates tandyi,* an (early?) Miocene Egyptian fossil that Simons (1969a) has shown to be one of the earliest identified cercopithecids. Simons emphasized the distal wear facet on M$_2$ of the type specimen, suggesting it to be a true hypoconulid, on the midline and thus similar to *Parapithecus.* This may indeed be a remnant hypoconulid, as might be expected in light of the remnant crista obliqua in roughly contemporary *Victoriapithecus;* but similar facets are found on M$_2$ of some *Theropithecus* individuals, representing merely an accessory cuspule. Finally, Kurten (1972a) has recently argued that *Propliopithecus* could be an early hominid, a view held previously but since dropped by both Simons and Pilbeam. Much of Kurten's argument is based on the small P$_3$–C complex in the type specimen of *P. haeckeli*—the value of this point diminishes greatly if it is accepted that this individual is female, a previously unconsidered possibility.

It seems clear that known *Propliopithecus* spp. are morphologically close to *Aegyptopithecus,* and *Aeolopithecus* and *Oligo-*

pithecus probably also shared recent common ancestry with these genera. The purpose of the present discussion is not to prove that any known population is *the* ancestor of the Old World monkeys, but merely to clarify some of the geometry of phylogenetic relationships. The search for ancestors alone is generally counterproductive, more so in this case where the parapithecids may have been the ecological replacements of cercopithecids in the Fayum region. On the other hand, the variability seen in known Fayum "apes", especially *Oligopithecus* and *Propliopithecus* spp., suggests that perhaps in a different environmental zone, some early catarrhines combined in a single animal the mosaic of cercopithecid-like features found above in several species. This ancestor might be termed a "hominoid" on overall grounds of grade similarity, but a "cercopithecid" in terms of its descendants. In less precise terms, one can almost facetiously say that monkeys did not give rise to apes, but that (some) "apes" gave rise to (Old World) monkeys.

OTHER RELATIONSHIPS WITHIN ANTHROPOIDEA

This study has suggested that the Old World monkeys developed as a specialization from an Oligocene catarrhine stock already possessed of an "apelike" dentition. Additional derivatives of this stock became known forms of apes and men, among other groups. But then to return to an earlier question, what is the position of the Parapithecidae? If the above analysis is correct, they would seem to represent a group rather distinct from the living catarrhines and their ancestors. In cladistic terms, the Catarrhini *sensu stricto* (including *Oligopithecus,* etc.) and the Parapithecidae are sister groups descended from a common ancestor, unknown and of uncertain grade. The step of constructing a morphotype for this ancestor is of great interest, although beyond the scope of this paper (but see Szalay, 1970, 1972c). Such an ancestor, however, may not yet have developed a canine honing mechanism: that of parapithecids seems weak on the retained P_2, while the strong hone on *Oligopithecus* has cut through the as yet unthickened enamel, suggesting a recent, "unperfected" adaptation.

Classification at family-group level among the Anthropoidea is quite unsettled, but one possibility which merits consideration is to place Cercopithecidae, perhaps Oreopithecidae and one to three families of apes and men in a single superfamily [3]. Parapithecidae might be placed in such a superfamily by a classical systematist (e.g., Mayr or Simpson) on grade ("horizontal") grounds, but would certainly be excluded by a strict cladist. In the latter case, the infraorder Catarrhini could include two superfamilies, rearranging their contents while retaining the present concept of this monophyletic group.

Finally, the question of a possible Old World origin of South American Platyrrhini has been reviewed by Hoffstetter (1972). I remain unconvinced by either the the morphological or paleogeographical arguments presented to date, but recently Gingerich (1973a) has shown that *Apidium* had the same annular ectotympanic development as *Aegyptopithecus* (Simons, 1972), *Pliopithecus* and modern ceboids. This is presumably the ancestral condition for anthropoids, as opposed to the tubular meatus of later catarrhines. Parapithecids are again seen to be extremely conservative animals, and it is morphologically conceivable that a West African primate group may have existed which was related both to the known Fayum genera and to ceboids.

More African Paleogene fossils and detailed analysis of extant and extinct anthropoideans will provide the only meaningful solutions to these problems.

Note Added in Proof

In an article published while the present paper was in press, Simons (1974b) formally described and named *Parapithecus grangeri*—this name is therefore used in the text as it is no longer a *nomen nudum*. Unfortunately, the distinctions between the species of *Parapithecus* and between this genus and *Apidium* remain incompletely documented—the relatively larger premolars in *Parapithecus* are especially diagnostic. Essentially, Simons has reiterated his arguments as summarized above, concluding in part (pp. 9–10): "there can no longer be any doubt that Parapithecidae are monkeys, not prosimians; zoogeographic considerations ally them with cercopithecoids—not ceboids." This is still the grade argument I have discussed. Simons (1974b, p. 10) continues to classify *Parapithecus* and *Apidium* in the Cercopithecoidea, but has returned them to family rank. "Their ranking among the Cercopithecoidea need in no way imply that the parapithecids would or could have been directly ancestral to any surviving Old World Monkey group but does leave open the possibility that *Parapithecus* may well prove to have been such an ancestor." This purely phenetic approach to taxonomy is unsatisfactory, as no special relationship has been shown to exist between Cercopithecidae and Parapithecidae.

Footnotes

[1] Photographs by the author except 1H and 2J-K courtesy of E.L. Simons. Thanks are due to Dr. Simons and Dr. K. D. Adam for permission to study and photograph Yale and Stuttgart specimens, respectively. Abbreviations used for museums:

C.G.M.: Cairo Geological Museum;

S.M.N.S. Staatliches Museum für Naturkunde, Stuttgart;

Y.P.M. Yale Peabody Museum, New Haven.

[2] More detailed descriptions, illustrations and different interpretations of Schlosser's three types are given by Kälin (1961).

[3] The vicissitudes of nomenclatural priority lead to this superfamily being named Cercopithecoidea, a sidelight which may delay serious consideration of its utility as a concept.

24

The Postcranium of Miocene Hominoids: Were Dryopithecines Merely "Dental Apes"?

R. S. CORRUCCINI, R. L. CIOCHON, and H. M. MCHENRY

INTRODUCTION

Many of the earliest describers of Miocene hominoids felt that, on the basis of dental morphology, these apes were closely related or directly ancestral to individual living species (Pilgrim, 1915; Hopwood, 1933; Gregory & Hellman, 1926; Gregory, Hellman, & Lewis, 1938). Discoveries of postcranial material, however, led to the realization that these extinct taxa were monkey-like in anatomy and inferred locomotion (Le Gros Clark & Leakey, 1951; Napier & Davis, 1959; Napier, 1964, 1967b; Preuschoft, 1973; Tuttle, 1967). This information coincides with accumulating molecular and biochemical data regarding affinities of living primates (reviewed in Goodman & Lasker, 1975). Using many different techniques and various proteins, the biochemistry demonstrates that hominoids shared a significant period of common evolution following the cladogenesis of every living cercopithecoid and other lower primate. Thus in all probability man and great apes shared a common ancestor through much of the Miocene, and would not be expected to have evolved their postcranial distinctions early in that epoch. Furthermore, living hominoids (including man) share a host of important and detailed functional similarities in upper trunk and limb anatomy, as has been copiously documented by Schultz (1930, 1936, 1950, 1951, 1953a, 1963), Washburn (1963a, 1968b, 1972), Washburn and Moore (1974), Tuttle (1967), and Tuttle and Basmajian (1974). These authors feel such similarities can only be explained as a result of relatively recent common ancestry in an arboreal hominoid which was fully adapted to behaviors involving "arboreal climbing, hauling, hoisting, and suspensory behavior" (Tuttle & Basmajian, 1974, p. 87) and, occasionally, true brachiation. This behavior pattern departs radically from lower anthropoids and from the entire generalized mammalian pattern they represent. This evolutionary hypothesis is seemingly reinforced by its concordance with at least one interpretation placed on the large body of molecular information (Sarich, 1971, 1973). Since cercopithecoids have a more derived dentition (Von Koenigswald, 1968, 1969) but less derived postcranium than hominoids with regard to their latest common ancestor, monkey-like limb bones and hominoid dentition are the expected primitive condition in early Miocene apes. Thus the dryopithecines have been called "dental apes."

Recently, however, these interpretations have been controverted from both the molecular and anatomical viewpoints (Simons, 1967e, 1972; Pilbeam, 1969a; Zwell & Conroy, 1973; Lewis, 1971a; Lovejoy & Meindl, 1972; Uzzell & Pilbeam, 1971). The views of Simons and Pilbeam and their students (Simons, 1967e, 1972; Pilbeam, 1969a; Pilbeam & Simons, 1971; Simons & Fleagle, 1973; Conroy & Fleagle, 1972; Zwell & Conroy, 1973) may be summarized as follows:

(a) Living pongid species differentiated earlier than once thought.
(b) The early Miocene *Dryopithecus major* was probably ancestral to the gorilla. *D. africanus* was in "or close to the ancestry of" the chimpanzee (Pilbeam 1969a, p. 126). The Indian *Dryopithecus (Sivapithecus)* species are perhaps ancestral to the orangutan. *Limnopithecus, Pliopithecus* and possibly *Aeolopithecus* indicate an early separation of the gibbon lineage.
(c) Since "the radiation of living hominoids had already occurred in the Miocene" (Pilbeam, 1969a, p. 133) the African ape locomotor pattern "might go back a long way" (Simons, 1967e, p. 250), and whatever traits apes show in common which are lacking in Miocene fossils must have evolved in parallel.
(d) Since several specific postcranial fossils (of *D. africanus* or *D. fontani*, of *D. major*, and of *Limnopithecus-Pliopithecus*) are similar to particular living apes (*Pan troglodytes* or *P. paniscus, P. gorilla*, and *Hylobates,* respectively), the arm-swinging and suspensory capabilities of hominoids must have evolved by the early Miocene, not too long after the divergence of hominoids from cercopithecoids.
(e) Chimpanzee and gorilla lineages split more than 20 million years ago and from a common ancestor that was an incipient knuckle-walker.

In the present paper we wish to review the non-dental morphological configuration of dryopithecines with special reference to the hypothesis of linear relationships between certain fossil species and living analogues. We will place special emphasis on recent multivariate metrical analyses of the fossils and comparative extant primates. Since the gibbon is in many ways an integral ape (Washburn, 1963a), and its putative ancestors *(Limnopithecus* and *Pliopithecus)* are very similar to dryopithecines (Simons, 1967e; Andrews, 1973), we shall include reference to the fossil lesser apes as well.

THE WRIST

There is one important fossil dryopithecine wrist complex—the KNM-RU-2036 specimen from Rusinga Island. This fossil assigned to *D. africanus,* has been thought indicative of palmigrade quadrupedal monkey-like actions (Napier & Davis, 1959; Tuttle, 1967). Lewis (1971a, 1972a, b, 1974) recently challenged this opinion, citing evidence of an intra-articular meniscus between ulna and proximal carpals, midcarpal locking mechanism, and other carpal shape features similar to *Pan.* These points have in turn been controverted by Schön and Ziemer (1973), Morbeck (1972, 1975) and O'Connor (1974).

One way of attempting resolution of a controversy of this sort is to subject conflicting anatomical observations and traits to standardized metrical analysis. This has been done by Corruccini, Ciochon, and McHenry (1975). A clear multivariate discrimina-

Reprinted with permission from *Primates,* Volume 17, pages 205–223. Copyright © 1976 by the Japan Monkey Centre, Inuyama, Aichi, Japan.

tion between hominoids and various monkey taxa resulted. *D. africanus* was unequivocally affiliated with the morphological pattern of quadrupedal monkeys. Hominoids are distinguished from cercopithecoids and ceboids in having reduced ulnar and radial styloid processes, greatly expanded crescentic distal radial articular facet on the ulna with resulting broadened ulnar head, intra-articular meniscus separating the ulna from the reduced pisiform and pyramidal triquetral and changing the form of facets on these bones, a true hook on the hamate, extension of articular cartilage well onto the dorsum of metacarpal V, change in placement of Lister's tubercle, and other features (Lewis, 1969, 1972a; Washburn, 1968b; O'Connor, 1974; Morbeck, 1974; Corruccini et al., 1975). *D. africanus* shows none of these features so far as they can be measured (Fig. 1). There are no features evident in the actual original fossil specimen indicating presence of an intra-articular meniscus, though some have been surmised from examination of a cast.

For *D. africanus* to be lineally ancestral to the chimpanzee and phyletically independent of the line leading to the gorilla, we would have to assume: (a) both apes independently adapted to a suspensory arboreal behavioral repertoire, evolving many highly detailed wrist structures completely independently from a cercopithecoid phenetic source, or (b) a knuckle-walking adaptation evolved directly from the wrist structure of a palmigrade monkey, with the features of a "brachiator" somehow developing independently. Neither of these is a likely proposition (Cartmill, 1974; Napier, 1974; Tuttle, 1967, 1969b, 1970; Tuttle & Basmajian, 1974; Washburn, 1963a, 1968b, 1972; Lewis, 1969).

Several well-preserved postcranial remains exist of the Miocene genus *Pliopithecus,* including three partial wrist joints. Measurements of *Pliopithecus vindobonensis* II and III were also included in the analysis of Corruccini et al. (1975). Like *Dryopithecus,* the wrist of *Pliopithecus* shows no indication even of the beginning of the hominoid adaptation for increased supination and hanging ability (Fig. 2).

THE ELBOW

Like the wrist, the elbow joint in hominoids has adapted for increased freedom of movement in a variety of positions, charac-

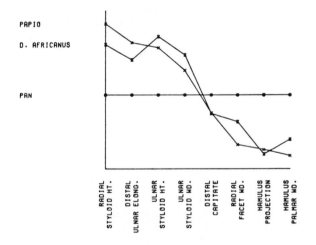

Figure 1 *Comparison of mean standardized shape values over key wrist measurements differentiating apes from monkeys.* Pan troglodytes, Papio, *and* D. africanus *are compared; values for* Pan *are held constant and the others adjusted to compare with this standard. The vertical scale represents approximately 4 standard deviation units.*

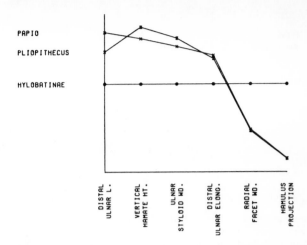

Figure 2 *Comparison of wrist measurements, as in Fig. 1 for* Hylobatinae, Papio *and* P. vindobonensis *(averaged for two specimens).*

teristically flexed. Hominoids can freely supinate the forearm in almost any posture. This is reflected in increased size of the flexor muscles as opposed to extensors, elongated trochlea for more stable articulation with the ulna, separation of trochlea and capitulum by a ridge, high, and rounded capitulum, increased biepicondylar and especially medial epicondylar width in the coronal plane, shallow olecranon fossa correlated with reduced olecranon process of ulna and importance of triceps, and differently oriented proximal radial facet on the ulna which renders radial rotation independent of humero-ulnar articulation and forearm position.

McHenry and Corruccini (1975) have subjected this morphological complex to metrical analysis based on the distal humerus. The KNM-RU-2036 specimen was again the major evidence of elbow morphology in *Dryopithecus;* additionally there was the new Fort Ternan (KNM-FT-1271) fragment of unknown classification, a probable medium-sized dryopithecine (A. Walker, personal comm.). One distal humeral specimen of *Limnopithecus* (KNM-RU-2097) was included. The analysis widely separated extant great apes from cercopithecine and colobine monkeys. Man was an integral member of the ape cluster. *Hylobates* and the Miocene hominoid fossils all fell in a somewhat intermediate position between apes and monkeys, but were phenotypically nearer monkeys (Figs. 3 & 4) and joined them in a cluster analysis. *Limnopithecus* and *D. africanus* were quite close to one another and to *Presbytis,* but not particularly near *Hylobates;* the Fort Ternan form was unique and differed from all other species. The metrical configuration of the capitulum height, trochlear transverse width and anteroposterior thickness, medial and lateral walls of the olecranon fossa, and the fossa's depth indicate that the elbow of Miocene hominoids may well have commenced the adaptive shift toward greater supination and mobility perfected in extant large hominoids, but that it was far from complete as Napier and Davis (1959) pointed out long ago. Likewise, other components of the elbow joint, particularly the form of the proximal ulna and olecranon process, and further fossil specimens (such as the European *"Austriacopithecus"* and *Pliopithecus*) confirm the generally primitive and monkey-like elbow motions of the known pre-Pleistocene hominoids (Morbeck, 1972).

THE SHOULDER

The shoulder joint is perhaps the most indicative of behavioral motions of the forelimb. Here, "brachiators" (i.e., hominoids)

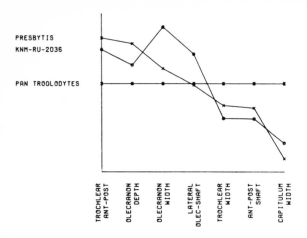

Figure 3   *Comparison of distal humerus shape variables between* Pan troglodytes, Presbytis rubicunda, *and* D. africanus.

have evolved decreased stability but increased mobility through flattening and rounding of the glenoid cavity, lateral restriction of articular cartilage on the humeral head, a relatively expanded and rounded humeral head, a narrow and deepened bicipital grove, elongated clavicle which holds the shoulder joint well to the side in combination with the broadened thorax, and enlarged coracoid process and acromion which project well laterally to the gleno-humeral articulation and anchor the coraco-acromial ligament. This ligament improves the leverage of the medially thickened deltoid in combination with the supraspinatus, and correlates with a number of other muscular changes such as insertion of pectoralis minor on the coracoid and secondary origin of pectoralis major from the clavicle as well as numerous changes in serratus and trapezius. As contrasted with the ancestral generalized mammalian pattern in monkeys, these traits allow the motions of hoisting, hanging, climbing, reaching, and swinging in branches.

Corruccini and Ciochon (1976) have subjected this morphological complex to multivariate morphometric analysis, based on 18 dimensions of 587 anthropoid specimens. The highly detailed similarity among hominoids, and the phenetic hiatus to all monkeys (whether arboreal or terrestrial, Old World or New World) increases the probability that this adaptation occurred only once

in a single common ancestral hominoid, not several times in parallel. Even the acrobatic cebid *Ateles* shows only an incipient convergence on apes in these diagnostic joint surface measurements (Ciochon & Corruccini, 1977a). The unique conformity among apes and man in the shoulder complements data from the elbow and wrist and from molecular biology in indicating a very small probability that Miocene apes had perfected this adaptation, or that several independent lineages were to evolve it in parallel.

However, there is no direct evidence of pectoral girdle morphology from dryopithecine fossils. The several humeri do not preserve the proximal head, and the supposed clavicles from East Africa are probably crocodile femora (Morbeck, 1972). There are several beautifully preserved shoulder girdles of East European *Pliopithecus vindobonensis*. Multivariate analysis of these in the matrix of extant Anthropoidea (Ciochon & Corruccini, 1977a) indicated that the putative hylobatine fossil shows no indication even of initial development of hominoid features (Fig. 5). This agrees with the conclusions of Zapfe (1958, 1960) that the *Pliopithecus* postcranium is monkey-like. If *Pliopithecus* is associated with gibbon ancestry, we must assume the many functional anatomical similarities between gibbons and great apes evolved *de novo* in parallel—an unlikely proposition. Similarly, much of the rest of the shoulder girdle and the thorax of *Pliopithecus* indicates the anatomy of a cercopithecoid (Zapfe, 1960; Morbeck, 1972). Ankel (1965) also presents evidence that *Pliopithecus* possessed a tail.

THE *D. AFRICANUS* FORELIMB

Thus far we have seen evidence of the primitive morphological and locomotor nature of Miocene hominoids, contrasting with the radical divergence shared by their extant relatives, and reducing the probability that the detailed hominoid specializations are mere parallelisms. In terms of multivariate metrical analyses, there has been one major exception to this viewpoint. Zwell and Conroy (1973) reprocessed the data provided by Napier and Davis (1959) on the KNM-RU-2036 forelimb bones, relying on principal component analysis. They concluded that *D. africanus* was very similar to the knuckle-walking *Pan troglodytes* and *Pan paniscus.* Apparently Zwell and Conroy used the covariance matrix rather than correlation matrix to compute principal com-

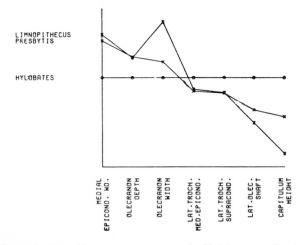

Figure 4   *Distal humerus comparison for* Hylobates lar, Presbytis, *and* Limnopithecus.

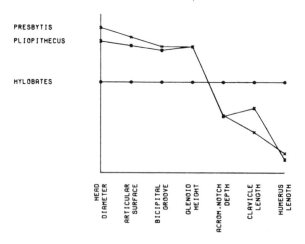

Figure 5   *Comparison of key shoulder joint shape variables discriminating apes from monkeys.* Pliopithecus vindobonensis *(composite of three fossils) is compared with* Presbytis rubicunda *while values for* H. lar *are held constant.*

ponents, an undesirable feature since variables would not automatically have equal weight in the analysis and one measurement expressed in different units might unduly influence results (Corruccini, 1975b). Their first component was dominated by the radial neck index, which out of all Napier and Davis's indices was the only one expressed as a simple ratio. We therefore conducted a principal component analysis of the correlation matrix between the 17 indices provided by Napier and Davis (1959). There was an important difference in the results, especially evident in the first component (Fig. 6). This component does not separate locomotor groups as the covariance component supposedly did, but merely separates the extreme terrestrial quadruped *Papio* from all others. *Pithecia, Cacajao,* and *Saimiri* fall in the range of variation of *Pan*. This component is determined by positive loadings for medial epicondylar, radial neck, third phalangeal, hand length, and humeral length indices and by negative loadings for brachial, third metacarpal, and radius length indices.

There are disadvantages to an analysis based on multiple indices such as those provided by Napier and Davis (Corruccini, 1975b). Corruccini (1975b) therefore attempted a more direct assessment of "total morphological pattern" by taking all 21 of the original linear measurements, converting them into allometrically adjusted shape variables, and performing a principal coordinates analysis. This analysis did succeed in separating *Pan* from cercopithecoids, thus providing a basis for judging the affinities and probable locomotor behavior of *D. africanus*. *Dryopithecus* was clearly aligned with the quadrupedal monkeys on the major principal coordinate, and was uniquely separated from all extant taxa on other axes. The fossil fell nearest *Cercopithecus* and *Presbytis* (Fig. 7). Therefore we can agree with Zwell and Conroy (1973, p. 374) that "there is no need to consider the modern postcranial and locomotor similarities of *Pan* and *Gorilla* as a puzzling case of parallelism," not because ancestors of both were knuckle-walkers in the Early Miocene, but rather because of recent common ancestry.

THE FOOT

Several tali and calcanei as well as other foot bones are known for East African dryopithecines. Le Gros Clark and Leakey (1951) described the tali, finding them more generalized and monkey-like than those of apes. Day and Wood (1968, 1969), in the course of a multivariate analysis designed to assess bipedality in australopithecines, included dryopithecine tali. These fell nearest samples of living chimpanzees and gorillas. Pilbeam (1969a, b) subsequently made much of this result, postulating special phyletic relationships between *Pan troglodytes* and the Rusinga talus (which Pilbeam felt represented *D. nyanzae*), and on the other hand between *Pan gorilla* and the Songhor talus (putatively *D. major*).

These conclusions now appear to be seriously in error. Day and Wood's research design was inappropriate for assessing dryopithecines, for it only included comparative samples of living African apes and men, but no monkeys. Thus interpolation of dryopithecine values into the discriminant functions left only two alternatives: for the fossils to be classified with bipedal hominoids or with knuckle-walking apes. Logically enough they fell with the latter. This analysis in no way tested the possibility of monkey-like affinities in dryopithecine tali. Another error lies in Day and Wood's method of interpolating fossil values into discriminant axes computed only on the basis of extant samples. Oxnard (1972b) shows that including the fossils in distance and discriminant computations, thus allowing them to have some direct influence on dispersion in the metric space, resulted in a marked separation of dryopithecines to a unique position in the morphometric space far removed from apes. Responding to this critique, Wood (1974) changed his research design, including *Papio* and colobine samples in canonical variate computations. This time the dryopithecines segregated with *Papio,* showing that Oxnard's (1972b) unique phenetic dimension in which dryopithecines differed from extant apes was in fact the cercopithecoid phenetic direction.

Supplemental evidence has been published by Lewis (1972c, p. 24) concerning the primate foot. He showed that "loss of the prehallux...did not occur until after the middle Miocene, as is evident from the bones of *Dryopithecus (Proconsul) africanus*. Only then was the basic pattern persisting today in the living great apes established." Lewis (1972c, p. 24) adds: "the obvious absence of a prehallux in the foot skeleton of *Pliopithecus vindobonensis* ...reinforces other anatomical evidence suggesting that this form could have occupied no place in the ancestry of modern gibbons."

PROXIMAL FEMUR

Le Gros Clark and Leakey (1951) and Napier (1964) have described probable dryopithecine proximal femora from Moboko and Songhor (KNM-SO-1011) as being essentially monkey-like.

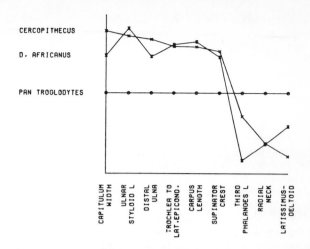

Figure 7. *Comparison of normalized shape variables separating apes from monkeys for the forelimb over* P. troglodytes, *Cercopithecus, and* D. africanus.

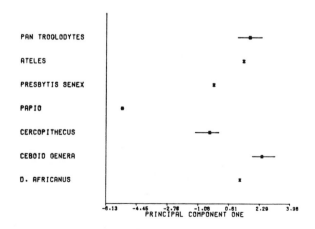

Figure 6. *Means and ranges for primate samples along the first principal component of the correlation matrix between 17 pairwise indices from Napier and Davis (1959).*

McHenry and Corruccini (1976) conducted a functional metrical analysis of primate femora including these fossils, the Eppelsheim femur, comparative samples of three genera of monkeys, six species of living Hominoidea including man, and fossil hominids. Monkeys were distinct from apes, the separation residing in a combination of the high greater and lesser trochanter positions relative to the depression on the superior surface of the neck, short neck lengths, and small head diameters of the monkeys. The three dryopithecine specimens were basically phenetically intermediate between monkeys and small apes, but much closer to monkeys than to *Pan gorilla* or *Pongo* and nearer to *Hylobates* than to *Pan troglodytes*. The Moboko and Songhor specimens, probably *D. africanus* or a close relative, do not appear to show any special relation to the chimpanzee (see Fig. 8).

THE SKULL

Although this paper emphasizes postcranial analysis, cranial morphology is also important to phylogenetic inference and relates indirectly to locomotor-behavioral conclusions based on the postcranium. For instance, Le Gros Clark and Leakey (1951) showed that several palato-facial features of the Rusinga *D. africanus* cranium were monkey-like, supplementing the other evidence as to the primitive phyletic position of dryopithecines.

Bilsborough's (1971) maxillary metrical analysis is of interest in this regard. He compared samples for *Pongo, Pan troglodytes, P. gorilla,* and *Homo sapiens* using the D^2 statistic, and computed evolutionary rates assuming *a priori* direct ancestry of *D. major* to *P. gorilla* and of *D. africanus* to *P. troglodytes*. A conflicting ancestor-descendant sequence could not have been yielded by Bilsborough's research design, for he included no monkeys in the comparison. Furthermore, he computed distances from raw measurements, allowing general size to be the major determinant of distance. Thus small dryopithecines fell relatively near small apes (*D. africanus* and *P. troglodytes*), and large dryopithecines appeared related to large apes (*D. major* and *P. gorilla*).

Corruccini (1975b) processed Bilsborough's mean measurements using shape distance, and found that in fact the large and small dryopithecine species were not only very similar to one another but also far removed from *Pan,* the species of which were likewise very close. Thus when general size is replaced by considerations of shape, it becomes necessary to postulate an enormous amount of parallelism to explain the similarity of extant chimpanzees and gorillas if indeed they evolved from distinct Miocene dryopithecines (Fig. 9). Since Bilsborough supplied no data for

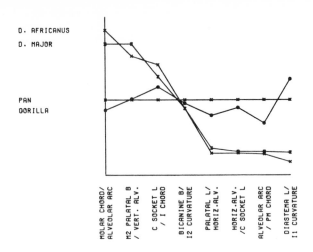

Figure 9   *Covariance between* P. troglodytes *and* P. gorilla *and between* D. major *and* D. africanus *over pairwise indices derived from Bilsborough's (1971) mean measurements.*

cercopithecoids, it is not possible to say whether the dryopithecine divergence from pongines (Corruccini, 1975b, Fig. 1) is in a monkey-like phenetic direction. Some indication is provided by comparing several pairwise indices of measurements given in Le Gros Clark and Leakey (1951) on *D. africanus* with specimens of *Papio* and *Pan troglodytes* (Fig. 10). There are several facial-palatal proportions in which *D. africanus* resembles the primitive rather than derived hominoid condition. This comparison is quite superficial and tentative and needs to be substantiated by a more carefully constructed metrical analysis; nevertheless, it serves at least to illustrate the desirability of including cercopithecoids in comparisons of dryopithecine metrical variability.

Similarly, the measurements provided by Pilbeam (1969a) for the Moroto *D. major* indicate some monkey-like as well as ape-like features (Fig. 11). Pilbeam (1969a) also gave measurements on pongid and dryopithecine mandibles and incorporated them in a principal component analysis. Here again, the absence of monkeys from the comparison and the use only of raw measurements virtually assured results in line with the *a priori*

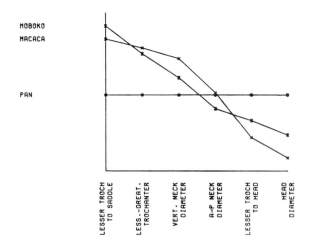

Figure 8   *Comparison of proximal femoral shape.*

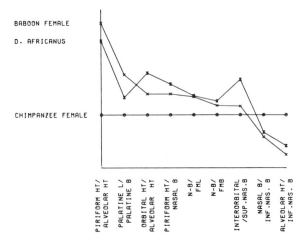

Figure 10   *Comparison of indices derived from Le Gros Clark and Leakey's (1951) description of the* D. africanus *cranium with* P. troglodytes *and* Papio hamadryas.

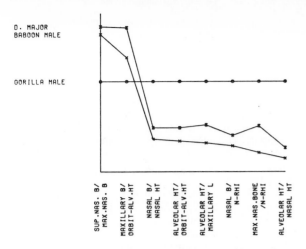

Figure 11. *Comparison between the Moroto* D. major, Pan gorilla, *and* Papio cynocephalus *based on data from Pilbeam (1969a).*

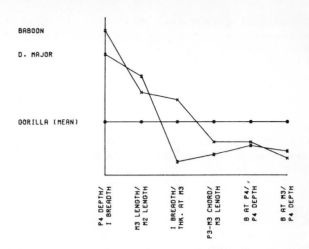

Figure 13 *Comparison as in Fig. 12 but between the Songhor (BMNH 190, 1)* D. major *mandible and* P. gorilla.

assumption of similarity between large dryopithecines and pongines (*D. major* and *P. gorilla*) and between small dryopithecines (*D. africanus* or *D. nyanzae*) and *P. troglodytes*. No other pattern could have resulted from a research design which only allows comparison of fossils with their putative descendants. Tentative introduction of a monkey specimen into some shape ratio comparisons (Figs. 12 & 13) does not support the above hypothetical lineages. Again, it is stressed that this is merely a preliminary finding based on a few specimens and subjectively chosen indices. More elaborate metrical analysis of dryopithecine crania is underway to verify these points. All that can be indicated with certainty is the necessity of including ceropithecoid samples, and of considering shape as well as size to assess accurately the affinities of dryopithecines in future analyses.

DISCUSSION

There are many cranial and postcranial morphological features in Miocene hominoids that recall an ancestral cercopithecoid-like condition rather than the derived condition characteristic of living hominoid species. While this certainly does not rule out the

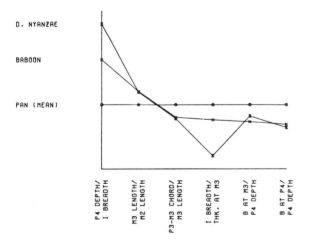

Figure 12 *Comparison of the* D. nyanzae *"CMH 1, 1" mandible with* Pan troglodytes *and* Papio cynocephalus *over measurements from Pilbeam (1969a).*

possibility that several fossil forms might be ancestral to extant species, since we cannot expect an ancestor to resemble its descendant in every way, it does seem to eliminate the consideration that specific dryopithecines were highly analogous to specific living hominoids in morphology, habitat, and behavior. All living hominoids have shared a strong common adaptive radiation, evident in their functional anatomy, in their behavior, and in their molecular biology. All this detail argues in favor of the recency of their own common ancestry relative to their last common ancestor shared with cercopithecoids. The hypothesis of early separation of gibbon, orang, gorilla, chimpanzee, and man "would demand an extraordinary parallel evolution in the viscera, trunks, teeth, skulls, and limbs. It may be fairly stated that these very early separations simply do not fit the evidence of comparative anatomy, behavior, or immunology" (Washburn 1972, p. 357).

Recent interpretations of the cercopithecoid fossil record provide more evidence regarding a late hominoid adaptation and radiation. It may be that none of the Oligocene Fayum primates are definite cercopithecoids, the earliest true monkeys appearing in the early to mid-Miocene (Delson, 1974b). Von Koeningswald (1968, 1969) presents convincing evidence that the Miocene *Victoriapithecus* is a stem cercopithecoid just recently derived from a hominoid stock with perhaps a gibbon-like dentition. Presumably a hominoid ancestor of cercopithecoids would still retain the primitive, generalized mammalian postcranial pattern, just as extant cercopithecoids do today. Thus moving the cercopithecoid-hominoid divergence up in time also indirectly necessitates upward revising of divergence dates within the hominoids.

If theories are true suggesting that living species of apes derived directly from specific Miocene hominoids, then one must invoke detailed and far-reaching parallel evolution in many lineages to explain similarities within dryopithecines and within extant hominoids and the vast differences between the two groups. For instance, Simons (1967e) and Pilbeam (1969a) argue that "brachiation" evolved five times in parallel from quadrupeds: in *Oreopithecus, Ateles, Hylobates, Pongo,* and *Pan*. While *Oreopithecus* does show arm elongation, so also do some large Pleistocene baboons (*"Simopithecus"*), and the Italian fossils are in too poor condition to really allow detailed analysis of *Oreopithecus* postcranial morphology and inferred locomotor capabilities. *Ateles* represents a very different approach, biomechanically and anatomically, to arm-swinging (Lewis, 1971b), and does not swivel while swinging like apes but instead looks straight ahead while using the prehensile tail as a third support—thus the thorax retains the monkey condition of being deeper than it is broad (M. Schön, pers. comm.). The fact that *Ateles* is convergent on, not homologous with, the hominoid pattern is easily evident from detailed morphometric examination (Corruccini et al., 1975; Ciochon & Corruccini, 1977a). This leaves the gibbon, orang, and African apes to be considered. The highly integrated similarities among these, and man, leave little doubt but that their common structures are due to common arboreal heritage (Schultz, 1963; Lewis, 1969, 1971b; Tuttle & Basmajian, 1974). The Hylobatinae were presumably a rather early offshoot of this lineage, but *Pongo, Pan,* and *Homo* probably shared a common ancestor in which forelimb joint modification had been fully accomplished allowing suspensory, climbing, reaching, swinging, and even throwing motions. Neither cercopithecoids nor ceboids are capable of these exact motions, and both retain the generalized primitive postcranial pattern. Although Pilbeam (1969a) and Simons (1972) infer that dryopithecines, if quadrupedal, were cebid-like quadrupeds as opposed to cercopithecoid-like, this makes no difference *vis à vis* their differences from hominoids. Thus consideration of whether dryopithecines were more similar to New World or Old World monkeys is immaterial.

The inference that postcranial similarities evolved in parallel in hominoids therefore must rest heavily on the dental evidence. Even here there is an alarming amount of parallelism that must be invoked. Subgenus *Proconsul* of genus *Dryopithecus* has a primitive dentition (Andrews, 1973; von Koeningswald, 1973), retaining similarities to Omomyid prosimians in the hooklike talonid, M_3 elongation and the strong cingulum. *Pan troglodytes, P. paniscus, P. gorilla,* and *Pongo* all show M_3 length reduction, hypoconulid reduction, reduced posterior dental arcade divergence associated with enlarged canines and incisors, and cingulae which, in the infrequent instances in which they are present, are thin enamel bands that wear off quickly. All these similarities must have formed independently according to Pilbeam (1969a). We agree with von Koeningswald (1973) that the cingulum in *Proconsul,* particularly on maxillary teeth, is fundamentally different from that in extant apes in that it is a wide shelf which persists in its expression through advanced attrition, extends distally from the protocone-hypocone furrow, and is not merely a superficial enamel feature. Most of the dental characteristics affiliating *D. major* with the gorilla (Pilbeam, 1969a) reside in size or expectably correlated allometric effects.

Andrews (1973, 1974) has proposed an ecological model of dryopithecine evolution which helps explain this. He sees the Miocene dryopithecine radiation as analogous to the Pliocene cercopithecoid radiation, which involved a spectrum of different-sized species filling a variety of niches (and perhaps expelling relict dryopithecines from those niches). Groves (1971) infers that *D. major* evolved in response to a Miocene environment not too different from that of the living mountain gorilla, and therefore finds it unnecessary to postulate direct ancestry of *D. major* to *P. gorilla* since their general similarities could be due to habitat convergence. The dryopithecine radiation preceding the cercopithecoid radiation also involved size deployment among closely related species. This necessarily entailed some superficial parallelism in size and associated allometric shape to unrelated modern ape species. Focusing attention on functionally important nondental anatomical characteristics, and on size-independent shape statistics, makes it seem unlikely that there were vast independent parallel developments from Dryopithecinae to Ponginae. The concept of parallel evolution is misused when invoked to dismiss every fact which does not agree with one's favorite phylogenetic speculation. Deriving chimpanzee and gorilla from separate dryopithecines implies many parallelisms, while their derivation from a later common ancestor implies parallelism in size only between particular dryopithecine and pongine species (von Koeningswald, 1973).

A further difficulty involves Pilbeam's (1969a) and Simons' (1972) practice of retaining the different *Dryopithecus (Proconsul)* species in subfamily Dryopithecinae, while recognizing subfamily Ponginae for *Pan* and *Pongo.* If indeed two or more species of dryopithecines evolved directly into known pongines, then subfamily Ponginae is biphyletic. A widely accepted principle of classification is that the immediate ancestry of a taxon should be included within a group of lower rank than itself (Simpson, 1935). We are surprised that eminent proponents of modern taxonomic concepts, particularly ones who have chastised physical anthropologists at length for their shortcomings in these matters, favor such an arrangement (Simons, 1963a, b, 1964a, b, 1967b, 1968b, 1972; Simons & Pilbeam, 1965; Simons, Pilbeam, & Ettel, 1969).

We prefer Andrews' (1974, p. 190) opinion that "the relationship of these fossil hominoid primates with the living ones remains uncertain....It is no longer feasible to suggest direct ancestral-descendant relationships between fossil and living species. Presumably one or more of the Miocene species was ancestral to the later pongids, but which this was...cannot be known from the available evidence." Geographically, we have no reason to believe that the fossil record has necessarily sampled the paleoenvironments containing all species ancestral to extant forms (Andrews, 1973).

Acknowledgments: We thank the late L.S.B. Leakey and the staff of the Center for Prehistory and Palaeontology, Kenya, R.E.F. Leakey and the staff of the Kenya National Museum, P. Andrews, A. Walker, H. Zapfe and the staff of the Naturhistorisches Museum, Vienna, J. Hürzeler of the Natur-Historische Museum, Basel, and D. Brothwell and the staff of the British Museum (Natural History), London, for permission to study the original Miocene hominoid material; L. Barton of the Powell-Cotton Museum, Birchington, W. W. Howells, B. Lawrence, C. Mack, R. Thorington, R.G. Van Gelder, and M. Poll for use of the comparative skeletal material in their charge; L. McHenry for helpful comments and editing; and the Wenner-Gren Foundation for Anthropological Research for partial funding of one of us (H.M.M.). E. Delson also commented. Travel support was provided by NIGMS training grant MH-1224 to R.L.C.

25

Miocene Hominoid Postcranial Morphology: Monkey-like, Ape-like, Neither, or Both?

M. D. ROSE

INTRODUCTION

It is not the purpose of this chapter to provide definitive answers to any of the questions asked in its title, even though various aspects of these questions have formed a large part of the lively debate that has been conducted in recent years concerning Miocene hominoid postcrania. The material available for investigation has been increased significantly recently by new specimens from the early Miocene of East Africa [KNM-RU 2036C and KNM-RU 5872 specimens (Walker and Pickford, 1983)], the later Miocene of Rudabánya, Hungary [Rud specimens (Morbeck, 1983)], and the Potwar Plateau of Pakistan [most GSP specimens (Pilbeam *et al.*, 1980)]. The main purpose of this chapter is to make some general comments on functional features of the morphology of some Miocene hominoid postcrania and on possible positional capabilities consistent with those features. Similarities to and differences from features of Miocene species evident in the postcrania of groups of living higher primates will be made purely in terms of function. Attention will be directed toward the larger bodied Miocene hominoids. Original specimens of all the East African and Asian material have been examined. The European material has been examined in cast form.

THE FORELIMB

Shoulder. The partial scapula, and proximal humeral shaft and metaphysis, KNM-RU 2036CH, of *Proconsul africanus* from Rusinga Island show a number of distinctive features. In posterior view the scapular spine is set fairly obliquely and the lateral half of the spine inclines somewhat cranially from its root. The acromion extends somewhat lateral to the cranially inclined glenoid fossa. As evident from the metaphysial surface, there was a greater amount of medial torsion of the humeral head than estimated by Napier and Davis (1959) for the KNM-RU 2036AH specimen. Details of the proximal shaft and the region of deltoid insertion indicate that there was some anterior inclination of the shaft, with relatively well developed deltotriceps and deltopectoral crests. The deltoid insertion may not have extended as far distally on the shaft as estimated by Napier and Davis. The proximal shaft features are present in the dryopithecine humeri from Maboko, BMNH M 16334 (Le Gros Clark and Leakey, 1951), and Klein Hadersdorf (Zapfe, 1960). There was evidently considerable mobility within the shoulder region of *P. africanus*. The scapular features imply that overhead positions of the forelimb could be achieved fairly easily. The relative importance of deltoid as an abductor at the gleno-humeral joint is equivocal, as is the positioning of the scapula on the thorax. This leaves the functional significance of the humeral head torsion explicable in more than one way. The torsion would be an obvious consequence if the scapula was relatively dorsally placed (Le Gros Clark, 1959). However, if the scapula was more laterally placed, the torsion might imply that some medial rotation of the forelimb at the glenohumeral joint was associated with commonly used forelimb positions.

No living primates show a pattern of features of the shoulder region that closely corresponds to that described above. The cercopithecoid, and especially the cercopithecine, shoulder region is specialized mainly in terms of features relating to parasagittal movements of the limb as a whole. Similarities with the Miocene hominoid pattern are only evident in the proximal humeral region, and not in features related to general mobility of the shoulder girdle. Some larger cebids, and living hominoids in which shoulder mobility is high, show similarities in scapular features and in the torsion of the humeral head, but the proximal humerus of living hominoids shows a different pattern of features.

Elbow and Forearm. The elbow region of *P. africanus* has received considerable attention (e.g., Napier and Davis, 1959; McHenry and Corruccini, 1975; Corruccini *et al.*, 1976; Morbeck, 1975, 1976). The humero-ulnar joint of *P. africanus* can be reconstructed from the specimens KNM-RU 2036AH and AK (distal humeri) and CF (proximal ulna). The spool-shaped humeral trochlea is relatively broad compared with the capitular region and has a fairly even transverse curvature throughout its extent. A moderately developed medial trochlear lip extends posteriorly and a sharply defined lateral lip continues as the lateral margin of a quite deep olecranon fossa. As seen in the proportions of the trochlear surface of the proximal ulna, the articular surface is relatively long compared to its width. The expanded olecranon beak of the ulna extends somewhat anteriorly and articulates proximolaterally with the lateral lip of the humeral trochlea and lateral wall of the olecranon fossa. The coronoid beak extends slightly further anteriorly than the olecranon beak and is moderately expanded where it engages medially with the medial lip of the humeral trochlea. In side view the articular surface of the coronoid beak slopes anterodistally.

Flexon–extension is obviously the predominant movement allowed at the joint. The combination of a moderately protuberant olecranon beak and fairly deep olecranon fossa indicates that considerable extension was allowed. Napier and Davis (1959) suggest a range of up to 180° on the basis of humeral morphology. The slope of the coronoid beak suggests that a common working position of the joint may have been one of semiflexion, when this surface would have been aligned horizontally. Preuschoft (1973) reaches similar conclusions on the basis of other morphological features of the KNM-RU 1786 *Proconsul nyanzae* ulna. The morphology

Reprinted with permission from *New Interpretations of Ape and Human Ancestry* edited by R. L. Ciochon and R. S. Corruccini, pages 405–417, Plenum Press, New York. Copyright © 1983 by Plenum Press, New York.

of the medial and lateral sides of the joint provide stabilization against medially or laterally directed forces throughout the flexion–extension range (Napier and Davis, 1959; Jenkins, 1973). Conjunct movements are minimal in this type of joint. A broad area for the origin of brachialis indicates that powerful flexion at the humero-ulna joint may have been possible. A complete ulnar olecranon process is present on the KNM-RU 1786 *P. nyanzae* ulna. The process is mediolaterally broad, quite short proximodistally, and inclined somewhat posteriorly. Triceps insertion was thus quite extensive, indicating that it may have been a quite powerful muscle. Posterior inclination of the olecranon process has been discussed by Jolly (1967), who relates it to efficient triceps action in elbow positions of nearly full extension. Harrison (1982) suggests that this and other features of the elbow region in *Proconsul* species indicate some terrestrial activity. As far as the olecranon process is concerned, this morphology is not incompatible with efficient triceps action at more flexed elbow positions, or as a synergist or fixator, for example, during activities in which the limb is held in an overhead position.

The KNM-RU 1786 *P. nyanzae* ulna is similar to that of *P. africanus* in most features of the trochlear surface. Although it is eroded, it is possible that the coronoid beak extended further anteriorly than in *P. africanus*, indicating that more extended elbow positions may have been common. However, the inclination of the surface is similar to that of *P. africanus*, indicating that more flexed elbow positions were also used. The Klein Hadersdorf ulna shows similar features. The possibly hominoid ulna KMN-FT 3381 from Fort Ternan has a coronoid beak that extends even further anteriorly and its articular surface slopes more anteriorly than anterodistally. These features are shared by the Rud 22 ulnar fragment. The emphasis in these specimens is therefore more on extension at the humero-ulnar joint. The later Miocene humeral specimens Rud 53, the medium-size GSP 6663 and 13606, and large-sized GSP 12271 have an even broader and more spool-shaped trochlea, and better defined medial and lateral trochlear lips than the earlier Miocene specimens. The olecranon fossa is relatively deep and its lateral wall is aligned so as to make extensive contact with the lateral part of the olecranon beak. These features, which are most marked in GSP 12271, indicate an emphasis on extension, and the stability of the humero-ulnar joint in this position.

The humeral capitulum of *P. africanus* is directed as much distally as anteriorly, has a strong mediolateral curvature that is more pronounced laterally, a fairly even anteroposterior curvature, and is separated from the lateral lip of the trochlea by a pronounced groove. The proximal radial head of KNM-RU 2036CE is complete and the dished area for articulation with the central part of the capitulum has a circular margin, surrounded by a flatter articular rim that is more extensive anterolaterally, giving the head an oval outline in proximal view. The head is slightly tilted proximomedially to distolaterally. There was thus good contact and free movement between the joint surfaces during forearm pronation–supination at any position of elbow flexion –extension. Contact would be maximal in full pronation, when the anterolateral area of the radial head would engage the more medial part of the capitulum, the tilt of the head having been effectively eliminated by medial movement of the distal radius. The articular area on the radial head for the proximal radio-ulnar joint extends completely round the head and is proximodistally narrowest dorsolaterally. The corresponding radial notch on the ulna faces anterolaterally. As with the humeroradial joint, the proximal radio-ulnar joint would be most stable in a position of full pronation. This position was thus probably a common working position of the forearm.

The form of the capitulum in other large Miocene hominoids largely conforms to the pattern described above. While the lateral trochlear region shows some peculiarities in the KNM-FT 2751 humerus (Morbeck, 1983), it is clearly separated from the globular capitulum. These features are even more pronounced in the later Miocene specimens GSP 6663 and 12271, and Rud 53. This may indicate that the radial head was more circular in outline and less tilted than in *P. africanus*. The shape of the radial head of Rud 66, certainly approaches circularity (see Fig. 2D in Morbeck, 1983). The *P. nyanzae* ulna KNM-RU 1786, the Fort Ternan ulna KNM-FT 3381, and the Klein Hadersdorf ulna all have radial notches similar in form to that of *P. africanus*. The radial notch of Rud 22 faces completely laterally.

As with the shoulder region, morphological patterns of the elbow and forearm regions are not closely matched by that of any group of living higher primates. The humero-ulnar joint of cercopithecoids, especially cercopithecines, show a number of specializations that indicate functional differences. The humeral articular surface is mediolaterally narrow and is buttressed mostly to withstand medially directed forces tending to adduct the ulna. Laterally directed forces are ultimately withstood by the shape of the humeroradial joint and the annular ligament of the proximal radio-ulnar joint (Washburn, 1951). Due to spiraling of the articular surface, flexion is accompanied by a conjunct medial translation of the ulna and attached proximal radius. Extension is limited mostly by the form of the ulnar olecranon beak. Larger cebids show some similarities in this region, especially to the morphology of the ulnar trochlea of earlier Miocene hominoids. It is the living hominoids that share most resemblances with the Miocene hominoids (Morbeck, 1975, 1976, 1983). Similarities are least to the earlier Miocene hominoids, where differences in the proportions of the articular surfaces, the degree of mediolateral curvature of the trochlea, and the degree of development of the lateral trochlear lip are evident (McHenry and Corruccini, 1975). Similarities are greatest to the partial humerus, GSP 12271, for features that can be compared (Pilbeam *et al.*, 1980; McHenry and Corruccini, 1983). These similarities all relate to an extensive flexion-extension range that includes marked extension with medio-lateral stability throughout the range, but especially in extension. The situation with the humeroradial and radio-ulnar joints is more complex. The ability of the radius to spin on the capitulum at different positions of the flexion-extension range of the humero-ulnar joint, the range of pronation—supination possible, and habitual working positions of the elbow and forearm are all involved. In cercopithecoids the morphology indicates that a habitual working position is one of the semiflexion at the elbow and full pronation of the forearm. In this position the oval tilted head of the radius is in a close-packed position at both the humeroradial and the proximal radio-ulnar joints. An anteriorly facing radial notch, bringing the radius into a position almost completely anterior to the ulna in pronation, is an additional feature in this complex.

Cercopithecoids differ from Miocene hominoids in the shape of the capitulum, which is flatter distally and medially in cercopithecoids, and in the positioning of the radial notch of the ulna. Larger cebids generally show more similarities than cercopithecoids to the Miocene hominoids. Similarities include a more globular capitulum and a more laterally facing radial notch, as well as details of radial head shape. Activities involving a fully pronated forearm are implied in the morphology of all three groups. Cercopithecoids are additionally specialized for stability at the humeroradial and proximal radio-ulnar joints in positions of semiflexion of the elbow. The relationships of the orientation of the radial notch of the ulna to pronation—supination range is not clear, but it is certain that the range is relatively great in cebids, where the notch faces anterolaterally, and extensive in hominoids, where it faces laterally (O'Connor and Rarey, 1979). In living hominoids, of course, the circular radial head, completely surrounded by articular cartilage, and the extensive distal radio-ulnar articular surfaces are the main factors relating to this range. The extent of the articular surface for the distal radio-ulnar joint is slightly greater in *P. africanus* than in most cercopithecoids and cebids (Corruccini *et al.*, 1975; Robertson, 1979). The indications are therefore that the pronation—supination range in *P. africanus* was at least as great as in some cebids. The orientation of the radial notch of Rud 22 and the form of the radial head in Rud 66 suggests that the range may have been even greater in those individuals.

One interpretation of some of the features of the shoulder, elbow, and forearm mentioned above is that during the later stages of the stance phase of quadrupedal locomotion in *P. africanus* the forelimb was placed so that the forearm was in full pronation and the semiflexed elbow pointed posterio-laterally due to medial rotation at the glenohumeral joint. Grand (1968a) has investigated this type of forelimb use in *Alouatta caraya*. In later Miocene hominoids the elbow may have been held in a more extended position during the stance phase. All Miocene hominoids could probably place the forearm over a reasonable pronation-supination range during a wide range of elbow positions. There are numerous activities other than quadrupedal progression in which these capabilities might be employed.

Wrist and Hand. The morphology of the wrist and to a lesser extent the hand of *P. africanus* has received considerable attention. The general consensus is that to the extent that quadrupedal progression was used, that quadrupedalism was palmigrade (e.g., McHenry and Corruccini, 1983). Two points concerning palmigrady are worth discussing. The first relates to the fact that the prime requirement in palmigrady is for adequate contact to be made between the palm and the substrate, and the second relates to the fact that in all types of quadrupedalism the joints of the wrist and hand take up their close-packed positions toward the end of the stance phase, as the propulsive thrust is made against the substrate. In cercopithecoids the palm of the hand does not reach a fully horizontal position when the forearm reaches the end of its limited pronation range. Wrist rotation is necessary to bring about a more horizontal palm position. This rotation mainly involves the trapezoid and capitate in the distal row, and the scaphoid and centrale in the proximal row. Jenkins (1981) has demonstrated that in *Macaca mulatta* at least, the embrasure formed by trapezoid and capitate is wide dorsally and narrower ventrally. Midcarpal pronation, during which the scaphoid and centrale rotate dorsally, is relatively free. Midcarpal supination is limited by the centrale engaging in the capitate-trapezoid embrasure. During quadrupedal progression passive midcarpal pronation produced by body weight acting during hand placement would effectively complete the rotation started by forearm pronation. Midcarpal supination occurring at the pushoff stage would lock the centrale and bring the "beak-like" process of the scaphoid into contact with the trapezium (Napier and Davis, 1959). This rotation lock would combine with extension locks within the carpus and between the carpus and metacarpus (O'Connor, 1975, 1976). In *Ateles* the midcarpal supination locking mechanism is lacking due to the fact that supination is extensive and is used during suspensory activities (Jenkins, 1981). During quadrupedal activities carpal locking is presumably achieved mainly by close packing in extension.

Schön and Ziemer (1973) have documented such a complex in *Ateles* and *Alouatta*. They point out similarities between it and a locking complex that can be reconstructed for the incomplete KNM-RU 2036 *P. africanus* carpus. Thus, in *P. africanus* at least, rotation required to bring the palm into contact with the substrate could probably have been generated within the forearm, and locking of the carpus could have been achieved mostly by close packing in extension. The presence or absence of an additional supination lock in *P. africanus* is problematic, as there are no trapezoid or centrale specimens. Evidence from the form of the KNM-RU 2036M *P. africanus* capitate is indirect and equivocal. The GSP 17119 capitate (Rose, 1984) shares some features of head shape with *Ateles* and so provides minimal evidence that a midcarpal supination lock may not have been present. The absence of the "beak-like" process on the KNM-RU 2036Q scaphoid also suggests the absence of a rotatory lock. If relatively free midcarpal rotation was present in *P. africanus*, it would be consistent with the rotatory capabilities suggested above for the shoulder and forearm. It should be emphasized that most groups of living hominoids, suspensory knuckle-walking, and quadrumanous, possess rotatory capabilities within the carpus, in at least some positions of flexion-extension of the carpus, that are considerably greater than those being discussed here, and that in these groups, as well as in cercopithecoids and various cebids, different patterns of rotatory and extension locking take place (Jenkins and Fleagle, 1975; Lewis, 1977; Jenkins, 1981).

Abduction—adduction within the wrist of *P. africanus* probably took place mostly at the midcarpal joint (Morbeck, 1975, 1977), although Robertson (1979) presents evidence from distal radius and lunate features indicating that there was an adduction (ulnar deviation) set to the hand of *P. africanus*. This is consistent with the possible positioning of the forelimb during quadrupedal activities that has been described above. Napier and Davis (1959) found a mixture of features within the metacarpus and phalanges of *P. africanus*, none of which are inconsistent with grasping in association with arboreal activities, a conclusion also reached by Preuschoft (1973). The New World monkeys are the only group of living higher primates showing similarities to *P. africanus* in the form of the trapezium—first metacarpal joint. The large proximal pollicial phalanx GSP 6664 shows features indicative of a powerful grasping capability. As with the large humeral fragment GSP 12271, it is the large living hominoids that share most features with this specimen.

In summary, the forelimb of the earlier Miocene hominoids is characterized by having moderately well-developed rotatory powers in all segments, a relatively mobile shoulder girdle, and extensive range of movement at the elbow, and a freely mobile hand. The forelimb is well designed for palmigrade quadrupedal activities performed in a fairly flexed-limbed way, and the general mobility suggests that a wide range of other general activities could have been used. There is no evidence of functional features related to highly specialized quadrupedal, suspensory, saltatorial, or other specialized positional activities.

THE HINDLIMB

Hip and Knee. The proximal femur shows a number of distinctive features that are common to most of the larger Miocene hominoids. These include a generally quite spherical femoral head with an articular surface that extends onto the neck anteriorly, and especially posteriorly, where there is a smooth transition between head and neck. These features are found in KNM-SO 399 from Songhor, probably from *P. africanus*, BMNH M 16331 (Le Gros Clark and Leakey, 1951), probably a *Proconsul* species, KNM-RU 1753 (Le Gros Clark and Leakey, 1951) of *P. nyanzae*, and GSP 11867 and 13929 from the medium-sized Potwar group. A relatively highly inclined neck is present in the KNM-SO 399, BNMH M 16331 and KNM-RU 1753 femora, and can be inferred for the medium-sized GSP 12654 specimen. The functional implications are that movement was relatively free about all axes at the hip joint, and that abduction and lateral rotation were favored. A tubercle on the posterior aspect of the femoral neck, present on KNM-SO 399, BNMH M 16331, KNM-RU 1753, and the Eppelsheim femur, has also been interpreted, by Walker in Napier (1964), as being related to a lateral rotatory capability.

Information on distal femoral morphology is sparse for Miocene hominoids. The KNM-RU 3709 and KNM-RU 5527 *P. nyanzae* distal femora are both incomplete and distorted. Some indication as to the proportions of the distal femur can be gained from the KNM-RU 2036 CQ *P. africanus* specimen, which lacks the epiphysis, and from the medium-sized GSP 13420 specimen, where some of the supracondylar region is present. The distal end of the Eppelsheim femur shows similarities with KNM-RU 5527. The articular region is fairly narrow mediolaterally, the condyles are subequal in size, and extend distally to the same extent. The patellar groove is relatively wide. Harrison (1982) suggests that for the KNM-RU 5527 specimen this combination of features is at least in part related to a placement of the knees directly beneath the hip joints in the quadrupedal posture and with the hip joints in the neutral position. The incomplete proximal tibial metaphysis of *P. africanus*, KNM-RU 2036CR, generally confirms the proportional features of the *P. nyanzae* distal femur. The angulation of the metaphysial surface in side view indicates that the surface would have been horizontal in a position of semiflexion. These features of hip and knee are consistent with a hindlimb position in which the hip is flexed, abducted, and laterally rotated, and the semiflexed knee points anterolaterally. This position is illustrated by Grand (1968b) for *Alouatta caraya* and is the hindlimb equivalent of the forelimb position during quadrupedal activities described above. A number of studies have concluded that the femur of Miocene hominoids is not clearly matched by that of any living primate (Le Gros Clark and Leakey, 1951; Napier, 1964; McHenry and Corruccini, 1976; Pilbeam *et al.*, 1980; Harrison, 1982). The pattern in cercopithecoids, which, as with the forelimb, relates to parasagittal movement, is similar for the distal femur, while cebids and large living hominoids share features relating to general hip mobility.

Ankle and Foot. The ankle region is relatively well represented for *Proconsul* species. *Proconsul africanus* specimens include a distal fibula, KNM-CA 1834 from Koru; distal tibiae KNM-RU 2036BA and CN and complete tali KNM-RU 2036BF and CO, KNM-RU 1744, and KNM-RU 1745. *Proconsul nyanzae* specimens include distal tibiae KNM-RU 1939 and KNM-RU 5872 and complete tali KNM-RU 1743 (Le Gros Clark and Leakey, 1951) and KNM-RU 1896 (Preuschoft, 1973). *Proconsul major* specimens include a distal tibia from Napak, Napak I 58, currently in the care of the British Museum (Natural History), and a complete talus KNM-SO 389 (Le Gros Clark and Leakey, 1951). Partial tali from later Miocene sites are Rud 27 and GSP 10785b. Features relevant to the talocrural joint include a relatively deep talar trochlea with more or less equally developed medial and lateral lips, with moderate wedging of the articular surface in dorsal view. The lateral trochlear lip is emphasized in Rud 27 (Morbeck, 1983). The medial and lateral surfaces of the talar body, articulating with the fibular and tibial maleoli, are relatively steep-sided. A distinctly developed, almost globular articular surface on the medio-anterior part of the tibial maleolus is matched by an area on the junction of the medial side of the talar body with the neck.

Details of the action of the talocrural joint in higher primates are complex (Barnett and Napier, 1952; Conroy, 1976a; Lewis, 1980a). They center around the degree to which conjunct abduction and inversion of the foot as a whole accompany dorsiflexion and mechanisms by which a stable position in dorsiflexion are achieved. These movements and positions are all involved during the later stages of the stance phase of quadrupedal climbing and walking. The features mentioned above are consistent with a reasonable degree of abduction and not a great deal of inversion (except in Rud 27) accompanying dorsiflexion, and good stability in dorsiflexion, where the trochlea and the maleolar-talar "subjoint" are both fully engaged.

The proportions within the tarsus in different groups of living primates vary considerably. This is reflected in the proportions of individual bones and in the amount of articular contact between neighboring bones. However, in all groups in which arboreal activities are at all important, the mechanisms for making effective contact with the substrate and for stabilization of the tarsus tend to be similar (Lewis, 1980b,c). The virtually complete tarsus of *P. africanus*, KNM-RU 2036, are used as the basis for the following discussion. Additional specimens include the *P. major* calcanei KNM-SO 390 (Le Gros Clark and Leakey, 1951) and KNM-SO 969 (Preuschoft, 1973), the small GSP 4664 calcaneus, and the medium-sized intermediate cuneiform GSP 6454 and lateral cuneiform GSP 17118 (Rose, 1984).

The calcaneus is relatively elongated, especially in its more distal part. There are confluent facets for the anterior talocalcaneal joint on the sustentaculum, and the axis of the subtalar joint is directed distomedially. The orientation of the facets of the posterior talocalcaneal joint are such that the calcaneus is advanced to make more complete contact with the cuboid during the considerable inversion movement taking place at the subtalar joint. There is extensive contact between the head of the talus and the navicular, especially laterally. The calcaneocuboid joint is characterized by a well-developed ventral beak on the cuboid, which articulates with a correspondingly deep surface on the distal calcaneus, and close packs when the forefoot supinates about the hindfoot to complete inversion taking place at the subtalar joint. There are relatively extensive areas of contact between the cuboid and the navicular, and between the calcaneus and the navicular. These contacts allow the navicular to move on the talus during inversion—eversion, and with the cuboid during supination-pronation, and also allow the calcaneus to fully contact the navicular when supination and inversion are combined. This pattern of articulation seems to be a consequence of the relative proximodistal proportions of the tarsus as much as being functionally required.

The three cuneiforms are all moderately elongated proximodistally. There is extensive contact distally between the medial and intermediate cuneiforms, minimal distal contact between the intermediate and lateral cuniform, and only dorsal contact between the lateral cuneiform and the cuboid. GSP 6454 and 17118 show similar features, except that in each case they are relatively shorter proximodistally than the earlier Miocene specimens. *Proconsul* metatarsals and such phalanges as are known are relatively gracile. The articulation between the metatarsus and the tarsus is relatively mediolaterally narrow. The first metatarsal-medial cuneiform joint indicates that hallux faced as much plantarward as laterally in its neutral position. The medium-sized partial hallux GSP 14046 is more robust than those of *Proconsul* species, and was capable of grasping powerfully.

Lisowski *et al.* (1976), Lewis (1980a), and Harrison (1982) have all noted that there are no consistent patterns of similarities between the morphology of the talus in living primates and the tali of Miocene hominoids, although the various authors have emphasized different partial sets of similarities in each case. This inconsistency is evident when comparisons are made to features of the tarsus and metatarsus of *Proconsul* species. The tarsus is relatively proximodistally longer in cercopithecoids and shorter in living hominoids. Cercopithecoids share features of the talocrural joint relating to dorsiflexion stability, but not features relating to the inversion set of the joint. Hominoids are the only group of living primates with a similar calcaneocuboid joint, while both cercopithecoids and hominoids differ from *Proconsul* species in details of the cubonavicular and calcaneonavicular joints, and in the pattern of articulations in the distal tarsus. More similarities are evident between cercopithecoids and *Proconsul* species in the metatarsus and in the digits, although the hallux faces more laterally than plantarward in cercopithecoids, and more plantarward than laterally in living hominoids. There are more resemblances between the hallux and cuneiforms of large living hominoids and the late Miocene Potwar specimens than with the *Proconsul* specimens. However, the more "monkey-like" Rud 27 talus maintains the overall inconsistency of the comparative picture.

Conclusions as to overall hindlimb function are similar to those made for the forelimb of larger Miocene hominoids. General mobility within the limb is high and features relating to very specialized hindlimb function are lacking. The limb is well adapted for arboreal quadrupedal activities during which, as with the forelimb, a fairly flexed limb position may have to be used. The morphology is also compatible with the utilization of a range of other general positional activities.

CONCLUSIONS AND DISCUSSION

The general conclusion of this study, that the functional morphology of Miocene hominoids relates to generalized arboreal quadrupedalism, does not differ from the assessment made by Napier and Davis (1959) for *P. africanus,* or many of the studies referred to above. Rather more material has been considered, and aspects of the function of some regions has been considered in a somewhat different way than in some previous studies. When considered *in toto* the Miocene material shows some interesting features. Size variation between different species at any particular time does not involve major differences in functional morphology. Thus the comparable features of the *Proconsul* species are remarkably consistent. Similarly, although the sample sizes are extremely small and only limited comparisons can be made, the medium- and large-sized Potwar material and the Rudabánya specimens do not show fundamental differences. Similarly, differences between earlier and later Miocene hominoids are mostly of degree rather than of kind. The basic plan upon which these variations were made was evidently a very successful one. This success was based on a generalized morphology underlying generalized capabilities.

The differences in degree between earlier and later Miocene species as shown in the elbow region relate to increased movement in the elbow and forearm and a greater independence of elbow and forearm function in the later Miocene forms. These features are shared with some living apes. Were later Miocene hominoids therefore more "ape-like" than earlier species, and were the earlier species more "monkey-like"? These labels have purposely been avoided in this chapter, except in its title, for a number of reasons. While cercopithecoids are similar to Miocene hominoids in showing a number of variations on a basic morphological plan, and while this plan relates to generalized capabilities, the morphology is nevertheless a relatively specialized one. The morphology is a conservative one, and as it relates to quadrupedal progression, it is expressed in predominantly parasagittal limb movement. Cercopithecoids are thus in a sense "specialized generalists" rather than "generalized generalists." Similarities with Miocene hominoids are greatest in those parts of the limb skeleton that are most generalized, for example, the wrist, and least in specialized regions, such as the humero-ulnar joint. Similarly, living hominoids share features, for example, in the shoulder, elbow, and hip, that relate to general abilities. Thus, in terms of special features Miocene hominoids are neither (cercopithecoid) "monkey-like" nor "ape-like." To the extent that some cebids are morphologically and behaviorally quite generalized, the early Miocene hominoids are more "New World monkey-like" than "Old World monkey-like." While information on the morphology and behavior of living primates is of great importance in understanding functional features of any fossil species, the importance of these features may be lost if labels such as "monkey-like" and "ape-like" are applied to them in any but very clearly defined and limited ways. In some ways it is better to ask how particular living primates resemble Miocene hominoids, as has been done above, rather than *vice versa.*

Many of the studies of Miocene hominoid postcrania have been concerned as much with phylogeny as with function, and in this context the demonstration of features that are, for example, more "monkey-like" than "ape-like" may be important. However, relative similarity does not necessarily correspond to absolute similarity. There are of course very important questions concerning features that are primitive or derived for various groups of catarrhine primates involved with these points, but they are not the main concern of this chapter. Miocene hominoids were a highly successful group, both in time and space, and were primarily "Miocene hominoid-like" rather than like any contemporary group.

Acknowledgments

I would like to thank D. Pilbeam for making the Potwar postcranial material available to me for investigation, R. E. F. Leakey for permission to study the Miocene hominoid material in the collections of the Nairobi National Museum, P. Andrews for permission to study the Miocene hominoid material in the collections of the British Museum (Natural History), and M. F. Morbeck for providing me with casts of the Rudabánya postcranials.

I have learned much from discussion relevant to this chapter with P. Andrews, J. G. Fleagle, T. Harrison, F. A. Jenkins, Jr., M. Marzke, M. E. Morbeck, D. Pilbeam, and A. C. Walker.

I would also like to thank R. Ciochon, A. Hill, B. Jacobs, L. Jacobs, M. Marzke, M. E. Morbeck, and C. Rose, who have at various times provided encouragement, assistance, and hospitality, or have shown great patience during the course of this study.

The research reported here was supported by a grant from the L. S. B. Leakey Foundation.

26

Species Diversity and Diet in Monkeys and Apes during the Miocene

P. ANDREWS

INTRODUCTION

At the present time the higher primate fauna of the Old World is dominated by monkeys. The apes are restricted in their distribution and diversity, so much so that they are often seen as evolutionary failures. This state of affairs has not always been so, and at their first appearance in the fossil record, during the early part of the Miocene period, apes were more abundant than monkeys and occupied greater ranges of habitats. This paper will document ape and monkey diversities throughout the Miocene and discuss the ecological background for the replacement of one by the other.

It has been recognized for some time that the fossil apes from the early Miocene of Africa were in many respects more like living monkeys than like apes. Le Gros Clark and Leakey (1951) observed that the brain and skull of *Proconsul africanus* were more like those of cercopithecoid monkeys than like living apes; and the postcrania too, especially the talus and calcaneus of the larger *Proconsul* species, were seen to be monkey-like. In subsequent descriptions of associated postcrania (Le Gros Clark and Thomas 1971, Napier and Davis 1959) reference was again made to the existence in *Proconsul* and *Limnopithecus* of a primitive cercopithecoid grade of evolution in terms of locomotor morphology and function, and comparison was also made with New World Monkeys. This has been confirmed more recently by a whole series of publications (Morbeck 1975, Wood 1973, McHenry and Corruccini 1975, 1976, Corruccini 1975a, Corruccini, Ciochon and McHenry 1975, 1976) which show that in terms of postcranial morphology and function, and thus in locomotor adaptation, the early Miocene apes resembled living monkeys more closely than living apes.

This concept was carried a stage further by Napier (1970), who stressed that the early Miocene apes were 'functionally and ecologically monkeys', and by Simons (1970) in the same volume, who proposed not only that the small apes were adaptively similar to monkeys but also that the disappearance of the small apes from Africa could be due 'to their failure in competition with a more successful radiation of arboreal monkeys such as the cercopithecines, which might just possibly be better fitted than apes to feed on fruits'. I took up some of these issues (Andrews 1973, 1978, Andrews and Van Couvering 1975), in particular the relative diversities of monkeys and apes, and showed that in modern monkey faunas four to six species could be present in a single patch of forest, while in fossil ape faunas presumed to have lived in comparable habitats, similar levels of diversity were achieved. This compares with extant ape faunas, which are restricted usually to no more than two species occurring together. The species diversity of Miocene apes was seen to be unusual in the context of living apes but more easily explicable if they are the ecological equivalents of living monkeys (Andrews 1973). This hypothesis was accepted by Kay (1977b), who further pointed out that the size range of the early Miocene ape species is similar to that of extant cercopithecoids (2 to 40 kg) and that they were similar in being arboreal quadrupedal frugivores, but he also introduced a note of caution that in terms of diet there was no strict equivalence between the monkeys and early Miocene apes. There is no evidence, for instance, that any of the latter fed on 'seasonally high proportions of leaves, bark, buds or grasses, as do some cercopithecines today' (Kay 1977b). Fleagle (1978b) has also shown that the size range of Miocene apes is identical to that of extant apes and monkeys, and since the replacement of apes by monkeys has thus taken place within the same size adaptive zone, it adds further support for their ecological replacement.

REVIEW OF THE FOSSIL EVIDENCE

The earliest known catarrhine primates have been recovered from late Oligocene deposits of the Egyptian Fayum. These had clearly reached catarrhine grade, but their relationship with later catarrhines is far from clear. Taxonomy is not a major issue in this paper, and for present purposes I will use the classification proposed by Groves (1972a) and Delson and Andrews (1975), with two primitive families of the Catarrhini, namely the Parapithecidae for *Parapithecus* and *Apidium,* and the Pliopithecidae for *Propliopithecus, Aegyptopithecus* and *Aeolopithecus.* There is no good evidence linking these fossil taxa with any of the extant families of the Catarrhini, and it is possible that their occurrence pre-dates the split between the Cercopithecoidea and the Hominoidea. *Parapithecus* species from the same deposits may (Simons 1974b) or may not (Delson 1975a) be considered early representatives of the cercopithecoid lineage.

Early Miocene

Nearly all of the early Miocene primate-bearing sites are located in East Africa. They have yielded rich samples of primitive anthropoids in association with hominoids and cercopithecoids. The East African sites are associated with volcanics so that the dates of many of the sites are relatively well known (Bishop, Miller and Fitch 1969, Baker, Williams, Miller and Fitch 1971), ranging between 18 and 20 million years ago. Three main groups of localities are known at present, each centred on a volcanic source. Rusinga, Mfwangano and Karungu are grouped around Rangwa; Songhor, Koru and Fort Ternan around Tinderet; and the Napak sites around the Napak volcanic centre (Le Gros Clark and Leakey 1951, Bishop 1958, 1963, 1968).

The fossil catarrhines from the early Miocene of Africa are divisible into three groups. Firstly there are the fossil apes grouped in the sub-family Dryopithecinae. Two genera at least are recognized, *Proconsul* and *Limnopithecus,* and at least six species (Andrews 1974, 1978), although there is some evidence for the presence of two additional species. Secondly there is a group of primates that were formerly considered to belong to the Hominoidea but which are now recognized to be a primitive group lacking the synapomorphies of the Hominoidea (Delson and Andrews 1975). *Dendropithecus macinnesi* is the best known species of this group, but a second as yet unnamed species from Koru and *Micropithecus clarki* from Napak, Uganda (Fleagle and Simons 1978a) may also belong. Finally there are three species of monkey, all undifferentiated cercopithecoids. A single molar is known from Napak (Pilbeam and Walker 1968), and this has the bilophodont specialization characteristic of the Cercopithecoidea. A second specimen from Napak was originally described by Pilbeam and Walker (1968), but this has since been shown to belong to *Micropithecus clarki* mentioned above (Fleagle and Simons 1978a). The second species is *Prohylobates tandyi* which is represented by three mandibles from Wadi Moghara in Egypt (Simons 1969a), and it appears different from the Napak specimen. Wadi Moghara is of uncertain age, however, and while there is a general consensus accepted here that it is early Miocene, it may well be later, and the age of the third species is even more doubtful. This is based on a single specimen from 'near Gebel Zelten' and this is attributed to a second species of *Prohylobates, P. simonsi* on the basis of its larger size than the type species (Delson 1979). Delson considers that both species are probably of early Miocene age.

Two species of dryopithecine are known from outside East Africa. These are two unnamed species from Ad Dabtiyah in Saudi Arabia (Andrews, Hamilton and Whybrow 1978). The affinities of the Arabian dryopithecines are not yet clear, but they appear most similar to *Proconsul* species without actually fitting into the species ranges of the East African forms. The associated fauna is sparse, and both the age of the deposits and their palaeoecology are uncertain. The age appears to be latest early Miocene (Hamilton, Whybrow and McClure 1978) based on fragmentary bovid and cricetodontid teeth.

Middle Miocene

There are two centres of radiation of Miocene primates in the middle Miocene, East Africa again, and central Eurasia. In East Africa primates are known from two localities, Fort Ternan and Maboko Island. Fort Ternan is fairly accurately dated at between 12.5 and 14 Ma, but there is some conflict of opinion as to which of these two bracketing dates the fossiliferous horizons are closer to (Bishop, Miller and Fitch 1969). The Maboko Island deposits consist of a series of clays and calcified siltstones passing conformably upwards into tuffs and lavas, which are correlated with an extensive series of lavas on the neighbouring mainland of Central Nyanza, all of which have yielded dates of between 11 and 13 Ma (Baker *et al.* 1971), so that on this reasoning the Moboko fossiliferous sediments are probably no more than 13–15 Ma, i.e. equivalent in age to the Fort Ternan deposits.

The primates found at these localities appear to have been less numerous, in terms of both species variety and numbers of individuals, than at many of the early Miocene sites. The commonest species at Fort Ternan is a species similar to *Limnopithecus legetet,* with additional fragmentary remains of *Proconsul africanus* and *Ramapithecus wickeri* (Andrews and Walker 1976). At Maboko there are at least two species of monkey (von Koenigswald 1969, Delson 1975a), one of which is more abundant than the other, and there are also three hominoid species, *Limnopithecus legetet, Ramapithecus* species, and *Proconsul (Rangwa-*

pithecus) vancouveringi (Andrews 1978). There is also said to be an *Oreopithecus*-like primate from Maboko (von Koenigswald 1969), but this does not appear to be a primate at all. The two monkeys at Maboko are considered by Delson (1975a) to show evidence of the differentiation of the colobines and the cercopithecines, and Maboko is the earliest site for which this is recorded. Finally, a third species of monkey was present at Ombo, Kenya, and this also shows affinities with the colobines.

Catarrhine primates appear in Eurasia for the first time in the middle Miocene. Primitive catarrhines of the family Pliopithecidae are represented by a number of species of *Pliopithecus* and *Crouzelia;* the hominoids are represented by two species of *Dryopithecus* in France and Spain and a third in Hungary, while a major new radiation of thick enamelled hominoids is present in eastern Europe, Greece, Turkey, India, Pakistan and China. A summary of the more important fossil localities is given in table 1, and it is important to note that in these and in all other known sites of the same age cercopithecoid monkeys are completely unknown in Eurasia.

The type species of *Dryopithecus, D. fontani* first appears in the middle Miocene of Europe. It extends from southern France and Germany into Spain, where a second species appears in the Vallesian *(D. laietanus).* The Spanish material has yet to be described in detail, and it is possible that a third species, *D. piveteaui* should also be recognized. It may also be that the new Hungarian specimens described by Kretzoi (1975) as *Rudapithecus hungaricus* and *Bodvapithecus altipalatus* may be conspecific with the Spanish dryopithecines: *Rudapithecus* is a small thin enamelled dryopithecine, but without good descriptions of the material the matter cannot be resolved.

The thick enamelled hominoids are the most widespread group in the middle Miocene. At least four species of the genus *Sivapithecus* and one of *Ramapithecus* are currently recognized. *S. sivalensis, S. indicus,* and *R. punjabicus* survive from Simons and Pilbeam's (1965) synthesis. An earlier and more primitive species from early middle Miocene deposits of Turkey is *S. darwini* (Abel 1902, Andrews and Tobien 1977). This has acquired the thick enamel characteristic of the group but still retains many primitive features lost by other species of *Sivapithecus.* Another new species from Greece is *Ouranopithecus macedoniensis* (de Bonis, Bouvrain, Geraads and Melentis 1974, de Bonis, Bouvrain and Melentis 1975, de Bonis and Melentis 1977b) which I consider conspecific with *Sivapithecus meteai* (Andrews 1978, Andrews and Tekkaya 1980).

At most of the Eurasian localities no more than two or three pliopithecid and hominoid primates are to be found at any one locality and by inference in any one habitat. Monkeys are not found in any of the large number of Eurasian middle Miocene localities. One of the characteristic species groupings that is found commonly consists of *Ramapithecus* and *Sivapithecus* species, and these are found together in many of the Siwalik localities and at Pasalar in Turkey. Another combination consists of *Dryopithecus* and *Pliopithecus* species, for instance at La Grive in France and Rudabanya in Hungary, but apart from these in primate species are found singly at most sites. There are few records of thin enamelled species of *Dryopithecus* or *Pliopithecus* being found associated with thick enamelled species of *Ramapithecus* or *Sivapithecus,* so that although it is not yet possible to identify habitat differences between them based on their associated faunas some such difference may well exist.

Late Miocene

This period sees the final disappearance from the fossil record of the pliopithecids and dryopithecines and the expansion of the monkeys both within Africa and in Europe and Asia. In Africa, hominoids are known from two localities, Lukeino and Ngorora in the Kenya Rift Valley, where they are represented by two species. The Ngorora deposits span an age of 11.8 to 9.2 Ma and the Lukeino deposits 6.7 to 7.4 Ma (Bishop and Chapman 1970, Pickford 1975). Their faunas indicate more open conditions than were seen to be present at Maboko and Fort Ternan: bovids and giraffids have become more common with time and the degree of hypsodonty of the teeth of these animals has increased. Equids appear for the first time in the African fossil record at Ngorora, and both localities appear to have had savanna habitats (Pickford 1979a,b).

The hominoid species present at Ngorora and Lukeino have not been named. Single isolated molars from the two sites do not provide much evidence on the nature of the species present (Bishop and Chapman 1970, Pickford 1975), and it can only be concluded that a chimpanzee-sized hominoid was living at that time. Also from Ngorora there are the remains of a smaller dryopithecine species that could have affinities with the early Miocene subgenus *Rangwapithecus*, but again the material is too scanty to say more. Monkeys are present at both these localities in slightly greater numbers than the hominoids, and appear to represent at least two species. Monkeys are also known from a number of other localities in Africa, namely a cercopithecine from Ongoliba (Hooijer 1963), a cercopithecine referred to *Macaca* from Marceau, Algeria (Arambourg 1959) and a colobine from the same deposits (Delson 1975a), a colobine, *Libypithecus,* from Wadi Natrun (Delson 1975a) and another macaque from the same deposits, and a new cercopithecine and *Libypithecus* again from Sahabi in Libya (Boaz, Gaziry and Ali el-El-Arnauti 1979).

In Europe and Asia the pattern is similar to that of Africa in the late Miocene. There are no records of pliopithecids at any locality. A single species of *Dryopithecus* is known, poorly represented by some isolated teeth from Germany (table 1). Only the thick enamelled hominoids are present in any numbers in the uppermost Nagri beds of the Siwaliks and in the succeeding Formation, the Dhok Pathan. A single specimen is also known from Pyrgos, Greece (von Koenigswald 1972a), and all of these species, with one exception, are indistinguishable from the middle Miocene species. The exception is *Gigantopithecus bilaspurensis*

(Simons and Chopra 1969), which together with *Sivapithecus indicus* and *Ramapithecus punjabicus* make up the thick enamelled species; the thin enamelled dryopithecines are represented by *D. fontani.*

Monkeys are more numerous in fossil deposits of the late Miocene of Europe and Asia than are hominoids. In particular, *Mesopithecus* is relatively abundant at Pikermi, although less so in other similar-aged deposits. Two species of *Mesopithecus* are probably present at Pikermi (Zapfe, personal communication), both large ground-living colobine monkeys apparently more closely related to Asian than to African colobines (Delson 1975a). A younger colobine from the latest Miocene of Italy is referred by Delson (1975a) to the same genus, as *M. monspessulanus,* and from similar aged deposits in Hungary is the earliest record of the mainly Plio-Pleistocene species *Dolichopithecus ruscinesis.* In Asia there is a small colobine species from Dhok Pathan deposits of the Siwaliks which Delson (1975a) refers to, *Presbytis sivalensis,* this being a combination of *Macaca sivalensis* and *Semnopithecus (= Presbytis) asnoti* which Simons (1970) originally recognized as representing the same small colobine species. Macaques are not well known from the late Miocene of Eurasia, apparently being absent from the Siwalik deposits, and the only record is from Montpellier in France.

SPECIES DIVERSITY OF APES AND MONKEYS IN THE MIOCENE

During the early Miocene the hominoids reached their greatest known levels of diversity. All this time there existed in Africa and Arabia eight to ten species of apes belonging to the Dryopithecinae (Andrews 1978). These ranged in size from that of the smallest Old World monkeys living today to the size of female gorillas. In the same deposits there were in addition two to three species of the Pliopithecidae. (The lower number of species numbers refers to that generally accepted in the literature, while the higher number includes reference to unpublished specimens.) Four species of dryopithecines and *Ramapithecus* but no species of pliopithecid survived into the middle Miocene of Africa (Andrews and Walker 1976, Andrews 1978). At the same time monkeys were uncommon in the early Miocene, with three species represented by a total of

Table 1   *Summary of Some of the More Important Middle and Late Miocene Anthropoid Localities in Europe and Asia (Hurzeler 1954a, Berggren and Van Couvering 1974, Delson 1973, Sickenberg 1975, Bernor et al. 1979).*

	Pliopithecus and *Crouzelia*	*Dryopithecus*	*Ramapithecus* and *Sivapithecus*	Cercopithecoidea
Late Miocene		Salmendigen	Pyrgos	Gascino
		Trochtelfingen	Dhok Pathan	Hatvan
(Turolian)		Ebingen		Pikermi
				Maragha
			Nagri (2)	Dhok Pathan
		Can Llobateres		
	Elgg	Can Ponsic	Sinap	
	Rudabanya	Rudabanya	Ravin de la Pluie	
(Vallesian)	Rumikon	Eppelsheim		
Middle Miocene	Goriach	St Gaudens	Chinji	
	La Grive	La Grive		
	Neudorf an der March		Candir	
	Sansan		Sandberg	
	Manthelan		Pasalar	
(Aragonian)	Pont Levoy			

five specimens, and they were not much more common in the middle Miocene, with three species again in Africa and none at all outside Africa (table 2).

Not only was the overall species diversity of the apes greater in the early Miocene than that of the monkeys during the middle Miocene, but their habitat diversity was greater also. At a number of localities in East Africa there are four or five species of ape known in association (with one or two pliopithecid species in addition) from single sedimentary levels (Andrews 1973, Andrews and Van Couvering 1975, Andrews, Lord and Evans 1979). These are interpreted as having occupied forest habitats, and there is no evidence of mixing of faunas to create artificially high diversity. This level of diversity is very different from that seen in living apes, where usually one and never more than three ape species occupy the same habitat today, but it parallels the diversity of living monkeys. In forest habitats today it is quite common for there to be four to eight monkey species living in the same area of forest, where they occupy different parts of the canopy and eat different food (Kingdon 1971, Struhsaker 1975). For example, in relic lowland forest areas in Kenya, such as Kakamega forest of the Tana River flood plain, there are four species found together, a colobine and three cercopithecines in both cases. Further west in Africa (that is closer to the forest refuge areas of the Congo Basin) two colobine species may occur together with up to six cercopithecines, raising the number of monkey species in single habitats to eight, as for instance in the Semliki lowland forest (Kingdon 1971). Similarly, forest habitats in West Africa can also have up to eight species, as at Seredou in Guinea (Roche 1971). In the Miocene faunas, therefore, the presence together of three dryopithecine species at Koru, four at several localities on Rusinga Island, and five at Songhor, with the further addition of one or two species of *Dendropithecus* in every case, suggests distribution patterns more like those of present day monkeys than of apes.

In the middle Miocene the overall species diversity of the apes had apparently increased. This may, however, be the result simply of their much greater geographical spread, for in addition to the African species they are known throughout Europe and Asia during the middle Miocene. Two or three species of *Dryopithecus* occurred in France, Spain (Simons and Pilbeam 1965) and possibly Hungary (some of the *Rudapithecus* specimens of Kretzoi (1975) may represent a small species of dryopithecine). These were all medium sized apes, and in addition four to five smaller species of pliopithecid also occurred in deposits throughout Europe (Hurzeler 1954a, Bergounioux and Crouzel 1959, Zapfe 1960, Ginsburg 1975, Kretzoi 1975). At the same time a third group of thick-enamelled fossil hominoids emerged, the Ramapithecinae (Pilbeam, Meyer, Badgley, Rose, Pickford, Behrensmeyer and Shah 1977a) with five or six species in Eurasia and one other species in Africa (Simons and Pilbeam 1965, de Bonis *et al.* 1974, 1975, Andrews and Tobien 1977, Pilbeam *et al.* 1977a). These were all medium to large sized apes characterized by molar teeth with relatively thick enamel.

Thus, totals of eleven to thirteen species of fossil ape and four to five species of pliopithecid are known during the middle Miocene of Africa and Eurasia, more than during the early Miocene of Africa when the numbers were eight to ten and two or three

respectively (see table 2). In accounting for this apparent increase in species diversity, two factors must be considered. First, the middle Miocene localities cover a greater geographical and temporal range than do the early Miocene ones. The latter are known from restricted parts of East Africa and Arabia, and with the possible exception of Moroto they seem to cover a time span of around three million years. No evolutionary stages are recognized in the primate (or other) species (Andrews 1978). By contrast, the middle Miocene localities are distributed across Africa, Europe and Asia, the African ones in East Africa again and the Eurasian forms ranging from France to China (Xu Qing-hua and Lu Qing-wu 1979), so that they cover far greater latitudinal and ecological ranges. The sheer distances involved are so much greater that increased speciation would be inevitable, and in addition to this, age-related evolutionary trends can be recognized in several groups. Hurzeler (1954a) has documented time-successive species for *Pliopithecus,* and I have provided evidence for evolutionary changes in the species of *Sivapithecus* and *Ramapithecus* (Andrews and Tobien 1977). All these factors lead to the apparently higher species diversity in the middle Miocene.

The second factor to be considered in this respect relates to hominoid species diversity in more limited geographical and depositional settings. Overall ape diversity has been seen to be eleven to thirteen species in the middle Miocene, but diversity within any one geographical region is lower, the highest figures being for East Africa, Europe, and India/Pakistan with four species each. Moreover, ape diversity and single localities is no more than two or three species in Pakistan, one to three species in Africa, three species in Hungary, and one or two species in France, Spain and Turkey. These values are all considerably lower than those given earlier from Kenya during the early Miocene, and, while this may be partly the effect of differences in habitat, it would appear that the range of fossil localities is sufficiently great to give a reasonable approximation of the true diversity levels and that the difference is a genuine one.

Taking these all together, therefore, the greatest species diversity of fossil apes is known from the early Miocene of Africa with slightly lower diversity in the middle Miocene of Africa and the rest of Europe and Asia. By the late Miocene the known numbers of species had diminished to three in Asia, and one in Europe. There were also two species in Africa, a possible dryopithecine at Ngorora and an unnamed hominoid from Ngorora and Lukeino, Kenya (Bishop and Chapman 1970, Pickford 1975). There is no shortage of fossil localities of this age in Africa and Eurasia, so this reduction in numbers appears to be a real event, not an artefact of site availability.

At the same time that the apes were becoming less common, the monkeys were expanding, although this event took place earlier in Africa than in Europe and Asia. The earliest fossil monkeys appeared in Africa during the early Miocene and the early part of the middle Miocene at the time when the numbers of fossil ape species were reduced from the early Miocene totals of 8–10 to four in the middle Miocene and to two in the late Miocene. Three species of monkeys are known in the early Miocene and three in the middle Miocene, all in Africa. The three monkey species in the middle Miocene increased to fourteen by the late Miocene (table 2), and these were distributed throughout Africa and Eurasia.

These changes in species diversity are summarized in Figure 1. This shows the trends in species numbers expressed as proportions of the total monkey and ape diversity throughout the Miocene. These are limited to Africa since monkeys do not appear in the fossil record of Eurasia until late in the Miocene. Proportions were used so that they could be compared with extant populations of monkeys and apes, and it can be seen from figure 1 that the

Table 2   *Distribution of Fossil Anthropoids in the Miocene.*

	Anthropoids		Hominoids		Cercopithecoids	
	Africa	Eurasia	Africa	Eurasia	Africa	Eurasia
Early Miocene	3	—	8–10	—	3	—
Middle Miocene	—	4–5	4	7–9	3	—
Late Miocene	—	—	2	4	8	6

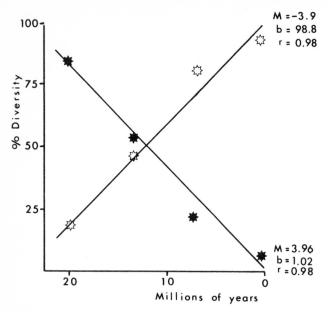

Figure 1 *Species diversity of hominoids (solid stars) and monkeys (open stars) in the Miocene of Africa. The figures are shown as percentages of the summed species numbers of hominoids plus monkeys for the period concerned so that they would be comparable with present day diversity figures (these are based on figures from Napier and Napier (1967) of 43 hominoid and monkey species living in Africa today). The fossil species numbers are given in table 2.*

relative proportions of extant species is close to that predicted by the trend in past diversity. (Extrapolating the regression line for the diversities of the fossil hominoids along predicts an x-axis intercept (time in millions of years) of 3.2 million years for $y=0$; that is the time when the hominoids would have become extinct if the trend shown during the Miocene had been continued to the present time.) The decrease in diversity of the hominoids since the early Miocene has been almost constant, with the trend expressed by the equation $y=4.18x+0.03$, but this does not give the full picture, as may be seen from the following considerations.

1. Reversal of trend

Although the hominoids as a whole have declined since the early Miocene, in two groups of hominoids the trend is reversed. These are the hylobatids and man, and it seems most likely that their success has been due in both cases to their adaptation to unusual ecological niches through the evolution of highly specialized forms of locomotion, bipedal walking in the case of man and bimanual brachiation in the case of the gibbons.

2. Sampling errors

It may also be questioned whether the Miocene diversity figures are sufficiently representative to be compared with present-day diversities. The number of living apes and monkeys in Africa today is 43 species excluding man (Napier and Napier 1967), so that the joint totals of monkeys and apes of 12 in the early Miocene, 7 in the middle Miocene, and 10 in the late Miocene (table 2) may provide only a poor indication of the probable species diversities existing at those times. How reliable these estimates are depends on how close they are to the actual diversity values in the areas concerned, and there are some grounds for supposing that they might be better than they appear. It is possi-

ble, for instance, that species diversity in the past has been less than it is at present, partly at first because both monkeys and apes were newly emergent in the early Miocene and would not have reached present diversity levels, and partly also because the present high diversity is largely influenced by a comparatively recent radiation, that of the *Cercopithecus* monkeys which make up more than half the present diversity.

Most important, however, is the fact that it is unrealistic to compare species diversity for Africa as a whole with Miocene values for relatively small parts of the continent. The early Miocene localities are restricted to a small area on either side of the Kenya—Uganda border and single localities in Egypt and eastern Saudi Arabia, and a survey of these areas today shows the presence of ten species of monkey (no hominoids) occupying the full range of habitats that is present now (man is excluded from the present analyses). The early Miocene diversity of 10–12 species is thus comparable with this. Considering East Africa alone, the extant higher primate fauna contains eight species, while the early Miocene fauna has six to eight species. Similarly, in the middle Miocene the fossil species are restricted to East Africa and the seven fossil species known compare with the extant fauna of eight species of monkeys. The late Miocene fauna comes from broader geographical limits in Africa, covering parts of East Africa, the western Rift Valley, North Africa and Egypt, and the extant fauna from this region is two hominoids and 14 monkeys, making 16 in all. This compares with 10 fossil species, so that in this case the past diversity is slightly lower than the present. It is concluded, however, that species diversity levels known from the Miocene fossil record are sufficiently similar to present levels to make past changes in diversity meaningful in both an evolutionary and an ecological sense.

3. Ecological similarity

A more important problem, which cannot be fully treated here is that the faunas from the different stages of the Miocene and from different geographical regions are not necessarily ecologically comparable. Indeed, there is some evidence that they are not. Many of the early Miocene faunas that have hominoids associated with them appear to be representative of tropical forest habitats, while the middle Miocene and late Miocene faunas appear to represent more open woodland to savanna tropical and subtropical habitats, and the patterns of ecological diversity for these different faunas would not be expected to be the same (Andrews, Lord and Evans 1979). As part of this overall difference, the primate species diversity is less in woodland habitats than in forest and is less in savanna habitats than in woodland (Napier and Napier 1967), and some allowance has to be made when comparing, for example, species diversity of primates in early Miocene forest habitats with those of middle Miocene woodland and savanna habitats. This is particularly true of the hominoids, and it is possible that the reduced species diversity in Africa from early to middle Miocene is partly the result of ecological differences in the localities available for study. The same factors, however, are operating for the monkeys as for the hominoids but with an increase in species diversity not a decrease, so that it does not appear that the changes in species diversity are caused only by ecological differences between localities. At this stage, with so few localities available for comparison, it is not possible to estimate the extent of the ecological differences and their impact on changes in primate species diversity.

4. Ecological change

On a wider issue, the possibility of ecological change between Miocene and the present must be considered. The early Miocene paleoecology of Kenya has been shown to have been dominated by forest ecosystems (Andrews and Van Couvering 1975, Andrews et al. (1979), and forest habitats in the Miocene appear to be similar to present day habitats in terms of vegetation (Chesters 1957), gastropod fauna (Verdcourt 1963), vertebrate faunas (Andrews and Van Couvering 1975) and also in terms of the community structure of the mammalian faunas (Andrews et al. 1979). The evidence from community structure is particularly important because it has been shown to be largely independent of the species composition of the faunas but relates directly to the niche-availability in the ecosystem, so that the degree of similarity existing between Miocene and the present may be taken to show ecological similarity between then and now.

5. Ecological replacement

Even when all these factors are taken into account it is not possible to show whether replacement of apes by monkeys took place as a result of direct competition or whether their relative changes in species diversity were due to some other, extraneous, cause. There are several other groups of animals in competition with both monkeys and apes, not least the insects but also larger fruit eating animals such as the fruit-eating bats, hornbills and squirrels. All these are known from the African Miocene, but their fossil record is too incomplete, in contrast of that of the primates, to form any idea of their importance as competitors. The anomalurid 'flying squirrels' have the best represented fossil record, and they appear to have had similar and even slightly higher levels of species diversity in the African early Miocene compared with the present, but bats and birds are poorly represented as fossils.

6. Changes in locomotor and positional behavior

Another factor that must be considered is the monkey-like morphology of the early Miocene hominoids. It was mentioned in the introduction that these early apes were more like monkeys than apes in their postcranial adaptations, so that it can be seen that the monkeys have not just replaced apes ecologically but have also apparently usurped part of their locomotor and positional behaviour as well. This implies that the early apes shared with the monkeys the same locomotor and positional behaviour, contrary to a recent suggestion by Ripley (1979) who put forward the theory that the differentiation of hominoids and cercopithecoids came about through changes in positional behaviour. Ripley (1979) attempts to correlate trends in positional behaviour, social organization, habitat and diet in the anthropoids, recognizing a basic dichotomy between a below branch hang-feeding line and an above branch feeding-and-moving line (Ripley 1979). These postural adaptations are considered characteristic of the hominoids and cercopithecoids respectively, and they are thought to be the result of differential adaptation to forest canopy stratification, which Ripley suggests became 'coarser grained' (i.e. greater differentiation of strata in the forest canopy) as a result of increasing climatic seasonality during the Miocene. Ripley envisaged the primitive anthropoid stock as consisting of small quadrupedal arboreal frugivores occupying both above branch and below branch positions, but, as a result of increasing climatic seasonality, greater stratification of the canopy and deciduousness of the tree species led to both the differentiation of two crown types, which she called monolayers and multilayers, and the bifurcation of the primitive anthropoids into below branch apes and above branch monkeys adapted to the two crown types. The impact of different feeding strategies on the structural types of canopy relates diet to posture and habitat, and Ripley considers that whereas the multilayers offer wider scope for folivorous species

the monolayers are much more limited, so that the folivore/frugivore dietary divergence is greater in the cercopithecoids (above branch feeders on multilayers) than it is in the hominoids. It would also appear from Ripley's theory that the adaptation of the above branch cercopithecoids to the multilayers would favour the cercopithecoids against the hominoids if it were really true that the increse in multilayers is the outcome of increased seasonality, so that her theory also provides an explanation for the increase in species diversity of the cercopithecoids during the Miocene, and the decrease of the hominoids.

Compelling as Ripley's (1979) synthesis appears to be, there are several problems arising from it. The early Miocene hominoids appear to show adaptations both to below branch hanging and to above branch quadrupedalism (Aiello, 1981), so that as yet there is no fossil evidence for the bifurcation in positional behaviour that Ripley predicts as distinguishing monkeys and apes. It may be questioned whether the evidence for 'coarsening of grain' of the forest canopy during the Miocene is valid, for increasing seasonality of climate generally leads to decrease in canopy complexity rather than increase (Eggeling 1947, Richards 1952, Langdale-Brown, Osmaston and Wilson 1964, Lind and Morrison 1974). This might then be expected to have had the opposite effect, according to Ripley's (1979) theory, in favouring the hominoids against the cercopithecoids. It must also be remembered, as Ripley herself points out in her evolutionary scenario, that the evidence for increasing climatic seasonality in Africa is mainly East African in origin but that this phenomenon in East Africa is mainly the result of tectonic activity producing rain shadow effects. To extrapolate the East African situation to cover the rest of Africa is not justified, for in the central African equatorial belt there is no evidence for the contraction of climatic (and therefore vegetation) zones above and below the equator as a result of climatic cooling, but there is no evidence that the nature of the vegetation within any one zone has changed ecologically since the Miocene.

7. Delay of the cercopithecoid radiation

One of the odd features of the origin and radiation of the cercopithecoid monkeys is that the two events are so far removed in time. The cercopithecoid monkeys today are such a successful group compared with the apes that it is hard to see why they did not become more immediately widespread following their initial divergence. Their origin has been shown to have occurred by the beginning of the Miocene period, for they are known from early Miocene deposits in Africa, but they still appear to have been rare in the middle Miocene and did not become more diverse until the late Miocene, at least ten million years later. The further back in time the origin of the monkeys is put the longer the gap between the two events. For instance, if *Parapithecus* from late Oligocene deposits of the Fayum is a fossil monkey as claimed by Simons (1974b), the divergence of the cercopithecoids from the ancestral anthropoid stock would have had to have been so much earlier, for *P. grangeri* is already highly specialized, resembling the specialized cercopithecines more than the generalized ones. It occurs in deposits at least 20 million years earlier than the late Miocene cercopithecoid radiation, so that in this event in time separating the origin and radiation of the monkeys would be still greater.

In the late Miocene the diversity of the monkeys is increased, both because they became more widespread in Africa and because they occur for the first time in deposits in Europe and Asia. It was suggested by Napier (1970) that the absence of monkeys from early and middle Miocene deposits in Eurasia was an artefact of the fossil record and that they were present but not preserved at least from the beginning of the Miocene. This does not seem likely, for, contrary to Napier's statement that the Miocene Eurasian sites occupy only a fraction of the total areas concerned, there are a great many sites of this age sampling a variety of different habitat types, and if monkeys were present it is likely that they would have been represented in the fossil record.

Napier's (1970) reasons for suggesting the early presence of monkeys in Asia were based on several misconceptions about their early adaptations. He considered that the colobines represented the ancestral cercopithecoid stock because leaf eating was the original specialization that distinguished monkeys from apes, and he considered further that the colobines have had an exclusively arboreal history, with the cercopithecines diverging into non-arboreal niches. Now for an arboreal forest animal to migrate from Africa to Eurasia there must have been present a forested migration route through North Africa, and as available evidence suggests that there has not been such a route since the end of the Oligocene their migration must have occurred by then. More recently, however, Delson (1975a) has shown that, far from being exclusively arboreal, early colobines were at least partly terrestrial, so that there is no need to predict their early migration on ecological evidence.

The second part of Napier's (1970) argument, that cercopithecines diverged from colobines by becoming more terrestrial, was based on the presence of cheek pouches in cercopithecines. These have been considered to be adaptations to ground foraging, whereby monkeys can store food obtained in a high risk situation on the ground and then consume it at leisure in the safety of a tree. Murray (1975) and my unpublished work on cheek pouches have shown that something like the opposite is the case, for in the cercopithecines as a whole the species that have the largest cheek pouches are the more arboreal ones, not the terrestrial ones. Even within a single genus such as *Cercopithecus,* the ground foraging vervet monkey *(C. aethiops)* has small cheek pouches while arboreal species such as *C. ascanius* have larger ones. Baboons have small cheek pouches, and, by measuring the extent to which the pouches become stretched through use, Murray (1975) has also shown baboons use their pouches infrequently. There is no reason, therefore, for maintaining a terrestrial origin for the cercopithecines on the basis of presence of cheek pouches.

The rest of Napier's (1970) argument concerns the latitudinal distribution of vegetation belts during the Miocene. He considers that the cercopithecines evolved from colobines in Eurasia because in a time of gradual climatic deterioration it was at the higher latitudes that the greatest scope was available for terrestrial animals. This explanation has been seen to be unnecessary, since the cercopithecines neither evolved in Asia nor were necessarily initially terrestrial. Much more plausible is Delson's (1975a) suggestion that the ancestral cercopithecoid morphotype combined cercopithecine features of the postcrania and teeth with colobine characters of the skull, and that the divergence of the two subfamilies occurred in the early middle Miocene of Africa, with the evidence of their separation to be seen in the two species present at Maboko Island in Kenya (von Koenigswald 1969, Delson 1975a). Pilbeam and Walker (1968) had suggested an earlier divergence of the colobines and cercopithecines on the basis of two specimens from Napak, but the specimen on which the presence of the colobine was established has since been shown to belong to a non-cercopithecoid anthropoid by Fleagle and Simons (1978a). The later dispersal of the colobines and cercopithecines into Eurasia during the late Miocene was therefore their first occurrence outside Africa, and it still remains to explain what delayed their geographical and evolutionary diversification and whether this was the cause of the decline of the hominoids. To do this it is necessary to examine their dietary adaptations.

DIETARY ADAPTATIONS IN ANTHROPOIDS

Adaptations to herbivorous diets (as opposed to frugivorous) have been classified by Eisenberg (1978) into a five-point scale. He considers a number of adaptive strategies, but the two that I will consider here are modifications of the stomach and caecum and modifications of the dentition. Within broad limits these are correlated with each other and with change in diet (Kay and Hylander 1978), and they have been used to set up a simplified scheme in Table 3. Within this scheme the species with mainly frugivorous diets are classified in stage 1, and the developing herbivory of the monkeys can be seen as passing from stages 1 and 2 into stages 3 and 4, with the highly specialized colobines in stage 4 and the cercopithecines in stages 2 and 3 (Eisenberg 1978). Within this classification, and in all of the discussion in this section, it must be remembered that terms such as frugivorous and folivorous should not be rigidly applied so as to exclude other elements of the diet; for instance some of the most folivorous colobines eat a substantial proportion of fruit when it is available, and most of the frugivorous species of monkey and ape eat some leaves all year round to supplement their diet and to provide essential amino acids not present in their primarily frugivorous diet.

The original anthropoid diet appears likely to have been a generalized type of frugivory. Jolly (1970a) and Hylander (1975) correlated incisor size with degree of frugivory, and Cachel (1979) has linked the expansion of the incisive area with post-orbital closure as a means of providing the extra area of attachment for the temporal muscles needed for the larger incisors. Since the anthropoid primates are characterized by post-orbital closure, Cachel (1979) concludes that they were primitively frugivorous. This also was Kay's (1977a) conclusion on the basis of his studies on molar morphology: primate species with high projecting cusps having long shearing blades are more herbivorous (or folivorous) than species with low rounded cusps, and with one exception all the early and middle Miocene anthropoids that have been studied appear to be frugivorous (Kay 1977b,c). In the hominoids the species of *Proconsul* and *Limnopithecus* are all found to have the frugivorous pattern except for *P. gordoni,* and rather surprisingly the anthropoid *Dendropithecus* was also found to be frugivorous (Kay 1977b). It is impossible to be precise when dealing with fossils, but it seems likely that classifying them according to Eisenberg's scheme (table 3) would put them into stage 1 because of the complete absence of any modifications of the dentition. Possibly *P. gordoni* might belong in stage 2, for it has been shown to have longer shearing blades on its molar cusps than the other species (Kay 1977b), and in some ways this reinforces the stage 1 status of the rest of the species. The later specializations of the Cercopithecoidea, and to a lesser extent the Hominoidea, must therefore be seen in the perspective of a frugivorous ancestry.

Cercopithecoidea

There seems to be a general consensus that the original adaptive niche of the cercopithecoid monkeys was leaf-eating, at least to a certain degree. This was proposed by Jolly (1970a) and Napier (1970), although they further suggested that the colobine type of bilophodont molar was therefore primitive. The latter concept has been countered by Delson (1975a), who has convincingly shown the the ancestral cercopithecoid morphotype of the teeth is cercopithecine-like, with large shallow trigonids, low cusps connected by low transverse ridges, shallow notches between cusps and lophs, and well developed M[3] hypoconulids. Nevertheless, Delson (1975a) still accepted Napier's hypothesis (1970) that cercopithecoid monkeys might have evolved in seasonal forests

Table 3   *Classification of Herbivory (from Eisenberg 1978, Kay and Hylander 1978).*

	Modification of Digestive Tract	Modification of Dentition	Diet
Class 1	None	None	Mainly frugivorous
Class 2	None	Slight modification for grinding	Seed, fruit, buds, flowers, leaves, sap
Class 3	Caecum enlarged	Cusp modification for shearing	Buds, blossoms, leaves and shoots up to 30–40% of diet
Class 4	Stomach sacculated or caecum enlarged	Teeth high crowned with elongated shearing surfaces	Leaves and grass more than 40% of diet
Class 5	Specialized for breakdown of cellulose	As in 4	Leaves and grass predominate

where the ability to subsist on leaves during times of fruit scarcity would be selectively advantageous. Compared with most primates, cercopithecoids have molars with higher cusps, longer shearing blades, and larger crushing surfaces relative to body mass (Kay and Hylander 1978), and since other primates that eat a significant proportion of leaves approximate to this condition (for example *Propithecus* and *Alouatta),* it is reasonable to suppose that this group had a folivorous ancestry (Kay and Hylander 1978).

It is possible to synthesize these varying ideas into a single hypothesis. This is that molar bilophodonty, which is the characteristic attribute of all cercopithecoids, was developed as an adaptation towards increased leaf-eating in a still predominantly frugivorous diet, and be so doing the cercopithecoids were able to exist alongside the more numerous and adaptively successful early hominoids and primitive anthropoids by exploiting more marginal or seasonal habitats. The dental adaptation is therefore seen as the earliest acquisition of the cercopithecoids, and this initial cercopithecoid adaptation to increased folivory may be presumed to have given the ancestral cercopithecoids limited adaptive advantage over the other anthropoids. Later dietary specializations, acquired after the middle Miocene divergence of the colobines and cercopithecines, conferred greater advantage and led to their displacement of the hominoids over much of their ecological range. These adaptations will be discussed in the following sections.

Colobinae

The colobines will be considered only briefly. They constitute the most extreme element of cercopithecoid dietary adaptation, possessing an enlarged sacculated stomach in which bacteria and protozoa break down cellulose followed by absorption in the small intestines. Some colobines also seem to have a marked tolerance of vegetable alkaloids like strychnine (Kay 1978), and they may have the capacity to counteract the effects of other secondary compounds such as tannins that bind with proteins and prevent their digestion (Hladik 1978). The digestion of cellulose is aided by the modifications of the dentition, in which the shearing action of the high crowned bilophodont teeth, which are modified into thin enamelled blades passing across the blades of opposing teeth (Kay and Hylander 1978), cuts up the food into fine pieces in a sort of mincing action (Walker and Murray 1975) and so gives the gut microflora readier access to the food.

These adaptations are developed in varying degrees in the colobines, as some are more strictly folivorous than others, but the pattern stands for the group as a whole. There is variation in the proportions of young and old leaves that are eaten, and varying proportions of leaves as opposed to fruit (Hladik 1978).

In two species of *Presbytis* that Hladik compared with each other, *P. entellus* eats a higher proportion of shoots and young leaves (only 20% mature leaves), and when fruit is available it may concentrate on that; *P. senex* on the other hand eats up to 40% mature leaves and lower proportions of fruit even when it is freely available.

Colobine species generally attain higher biomass levels than do the cercopithecines. This appears to be related to the presence of common food trees which provide an abundant food source, particularly in times of scarcity, which is when most mature leaves are eaten. Mature leaves provide a widespread food source but do not allow for a variety of modes of exploitation, so that high biomass is linked with lower species diversity; for example, in Kibale forest, Uganda, two colobine species are present with a biomass of 18 kg per hectare, while in the same forest there are four species of cercopithecine with a biomass of only 3.6 kg per hectare (Struhsaker 1978). The same factors operate with the terrestrial ruminants, where the unselective grazers achieve high biomass but lower diversity, whereas more selective browsers living in woodland and forest exist at lower densities but higher species diversity (Jarman 1974). Like the grazing ruminants, the colobines dominate the mammalian arboreal folivore niche as a result of their specialization to an abundant food source.

It can be concluded that the colobines represent the present culmination of cercopithecoid evolution in that they carry the initial dietary adaptation to its most extreme development. In so doing they dominate the arboreal folivore niche in tropical Africa and Asia with their high population densities and biomass.

Cercopithecinae

The cercopithecines lack the dietary specializations of the colobines. If the premise is accepted that the original divergence of the cercopithecoids from other anthropoids was the result of a trend towards higher proportions of leaves in their diets, this being inferred from the presence of bilophodont molars in all Old World monkeys, then it can be said that in some ways the cercopithecines failed to establish the adaptation where the colobines succeeded. Certain of the cercopithecines appear in fact to have changed little from their Miocene ancestors, for example the macaques (Delson 1975a), and if they were slow to establish themselves in the Miocene it can reasonably be asked how it is they are so successful now. The reason clearly cannot lie with their folivorous adaptation, and it is suggested that they converged on the hominoid frugivorous niche as a result of competition from the folivorous colobines, and in so doing they became better adapted as frugivores than the hominoids. Most cercopithecines today are primarily frugivorous (Kay 1978), although a few species are more folivorous, with higher crowned molars having long shearing blades and with relatively small incisors (Kay 1977b, c, Hylander 1975). The instrument of their success does not appear to be the cheek pouch, although no doubt it was a useful specialization, but more probably it lies with their capacity to tolerate unripe fruit, enabling them to gain access to fruit before it is ripe enough for the hominoids.

Plants have evolved a variety of mechanisms to defend themselves against predators. Chemical defence involves the development of allelochemicals which render the plant parts toxic or unpalatable (Opler 1978). The effects of toxicity, however, have been found to depend partly on the nutrient content of the food item, so that the same absolute level of toxicity may be present in two items but one may be eaten and the other not if the one has a higher nutrient content than the other (Janzen 1978). In this case the energetic gain from the nutrient offsets the loss due to the toxicity (Janzen 1978), and it is the net gain that is significant. This has been illustrated most dramatically for the black colobus (*C. satanus*) which lives in an impoverished habitat and has switched from eating low-quality toxic leaves to more nutritious but equally toxic seeds (McKey 1978), and by this means it gains sufficient nourishment to tolerate the high level of toxicity of its food.

Evidence is now emerging that a major ecological distinction exists between hominoids and all cercopithecoids. On the basis of parallels in feeding behaviour, physiology and social organization of spider monkeys and chimpanzees on the one hand, and studies on sympatric monkey species on the other hand, Wrangham (1980b) has argued that a critical distinction between apes and monkeys is the greater tolerance of the latter for a variety of plant secondary compounds including tannins and alkaloids. One of the consequences of the colobine adaptations of the digestive tract is that they have the ability to detoxify many secondary plant compounds through the action of the bacteria and protozoa in the alkaline environment of their sacculated stomach (Freeland and Janzen 1974). In addition to this, however, data collected by Wrangham (personal communication) indicate that some cercopithecines eat food containing high levels of toxin and that they have higher rates of drug detoxification than does man. Conversely, hominoid species avoid food with high levels of toxicity, and in particular they avoid unripe fruit that is eaten by monkeys. There is some evidence linking the presence of astringent tannins with unripe fruit (Wrangham, personal communication) making them unpalatable and also partly indigestible because of the fixing action on the plant proteins. In developing a tolerance for tannins and other secondary compounds, therefore, the cercopithecines are able to tolerate less ripe fruit than the hominoids and thus gain access to fruiting trees before the hominoids. This gives them a selective advantage over the hominoids when ripe fruit is scarce, although of course it is reasonable to assume that other factors are also operating.

It is worth emphasizing that an adaptation such as this, which enables one group of animals to gain earlier access to a food source than other groups, is of enormous adaptive advantage. Fruit is a nonrenewable resource in any one year and the animals that reach it first are at a great advantage. It is not known at what stage the greater tolerance for plant secondary compounds was developed, but it is possible that it was already present in the stem cercopithecoids before the divergence of colobines and cercopithecines. This tolerance, which entails the detoxification of the toxic compounds, has an energetic price, and the price of detoxification has to be offset against the food value of the toxic food (Freeland and Janzen 1974). While most colobines have increased food value through their digestive specializations for breaking down cellulose, it may be that the cercopithecines have achieved the

same thing through changing their diet from the less nutritious leaves to more highly nutritious fruit. In so doing, however, they already had the advantage of the acquired toxicity tolerance which would enable them to tolerate less ripe fruit.

Hominoidea

The hominoids preserve a more primitive tooth pattern than the cercopithecoids, particularly in the structure of their molars. They retain a close approximation to the primitive anthropoid cusp pattern in these teeth, and it appears likely that they also have retained a greater degree of dietary similarity as well. The extant hominoids are mainly frugivorous, and Kay (1977b) has recently shown that most of the fossil ones were as well. The exceptions to this are gorillas and siamangs, which show some modification towards a more folivorous diet: both have a slightly enlarged caecum, relatively small incisors, and molars with pointed cusps and long shearing surfaces (Hylander 1975, Kay and Hylander 1978, Hladik 1978), all of which suggests a well established folivorous adaptation. It is interesting that the two most folivorous hominoids are also the two biggest in their respective families, but whether they became folivorous because of their large size or became larger because they were folivorous is uncertain. It has been suggested that their increase in size led to relatively larger muscles of mastication as a result of the positive allometry of muscle size with body size (Walker and Murray 1975), and this increased their ability to crush tough food items. This may be so, but it is not an adaptation that leads directly or necessarily to a more folivorous diet.

Another variation in hominoid diet is that seen in the species with thick enamel on their molars. It is not certain what the function of the thick enamel is, but since this feature first appears in the middle Miocene, it is possible that its evolution is related to competition from the monkeys. Thick enamel is known for three groups of hominoids: the orang-utan and man among the extant species and the middle Miocene fossil hominoids *Ramapithecus* and *Sivapithecus* (and *Gigantopithecus*). The fossil hominoids share with man and the fossil hominids an apparent habitat association of woodland savanna (Behrensmeyer 1975, Andrews and Evans 1979), and both have relatively small incisors, that is small relative to molar size and by inference to body size (Hylander 1975, Kay and Hylander 1978). They also have low crowned crushing teeth lacking almost completely the shearing surfaces present in herbivores, and this character they also share with the orang-utan. This combination of large posterior teeth with small anterior teeth suggests a method of feeding that combines a maximum of crushing effort from the posterior teeth with a minimum of preparation from the anterior teeth, and the best model for this is Jolly's hypothesis of small hard object feeding (Jolly 1970b). Rather misleadingly Jolly cites the seed-eating gelada as a typical example of this type of feeding behaviour, and while there certainly are similarities, it is unrealistic to assume so narrow an adaptation for so wide-ranging a group as the hominids (and the thick-enamelled Miocene hominoids). A great variety of food objects are available to a terrestrial forager; seeds are one of them, but more importantly there are the underground roots, rhizomes, bulbs and corms, fruits, buds and young leaves of herbaceous plants above ground, and insects, birds' eggs and even the occasional small animal for the carnivorously inclined. This is a more omnivorous diet than was envisaged by Jolly in his hypothesis, but I suggest that it is equally consistent with the morphological evidence and a lot more probable ecologically.

The hominoids possess one specialization of the gut that is lacking in other primates, and that is the appendix. This organ is variable in size, but it is usually smaller in the gorilla and man and larger in the chimpanzee and orang-utan (Sonntag 1923, 1924, Elftman and Atkinson 1950, Bourne and Bourne 1972). Developmentally it arises as part of the caecum, but its function appears to be immunological rather than digestive. In humans at least it is most strongly developed in childhood and atrophies with age, but there are a few data for the non-human hominoids. The predominance of IgA globulins in the mucous secretions in man may be of significance (Tomasi and Zigelbaum 1963), for it is the latest of the immunoglobulins to appear ontogenetically (Good and Gabrielson 1968), and might therefore also be expected to occur in the other hominoids but not necessarily in the monkeys. Its function is the neutralization of toxins, and this may have significance in the tolerance of vegetable toxins in hominoids, but the experimental work to test it has not yet been carried out.

With these exceptions, the hominoids have retained the primitive anthropoid dietary adaptation to frugivory, and lacking as they appear to do the tolerance to toxic or unpalatable substances that is present in cercopithecoids, they are less competitive in dietary terms than in cercopithecoids. The gorilla, siamang and the thick-enamelled hominoids have adapted their diet, but in addition to this, and exclusively in the case of the other hominoid species, they have evolved an extraordinary variety of postcranial and social adaptations that may also be considered to be the outcome from this competition with the monkeys. It has been seen how the early Miocene hominoids were monkey-like in their postcranial adaptations, but the extant species have quite different and almost bizarre adaptations: quadrupedal knuckle walking, bimanual brachiation and bipedal walking. No fossil evidence exists at present for the development of any of these postcranial adaptations. Middle Miocene hominoids from Hungary and France (Kretzoi 1975, Pilbeam and Simons 1971) and from Pakistan (Pilbeam et al. 1977a) have what might be called generalized orang-utan-like limb bones, particularly for instance the distal end of the humerus, so that they were clearly no longer monkey like but had not differentiated into any of the distinctive locomotor groups mentioned above. It is presumed, therefore that the adaptations for these may have developed since the middle Miocene and that they may be considered as part of the hominoid response to their partial dietary exclusion from the frugivorous niche, a response that has been partly successful in the hylobatids and hominids.

CONCLUSIONS

The species diversity of hominoids is high in the early Miocene and decreases at a fairly steady rate during the middle and late Miocene. At the same time the diversity of species within single habitats also decreases from four to six species in the early Miocene to maximum values of three in the middle Miocene and two in the late Miocene and present.

The species diversity of cercopithecoid monkeys is low in the early Miocene and increases at a steady rate (with an equal but opposite slope to the hominoid decrease) during the middle and late Miocene. Diversity within single habitats increases from one species in the early Miocene to two in the middle and late Miocene and up to 8 in the present.

These levels of species diversity for the Anthropoidea as a whole are comparable with, although slightly lower, than those present in living anthropoid faunas in equivalent geographical areas. The only major change that has taken place has been the relative contributions of the hominoids and cercopithecoids to the overall anthropoid diversity.

In evolutionary terms, the radiation of the hominoids took place in the early part of the Miocene and the radiation of the cercopithecoids in the later part of the Miocene and on to the present. Both groups are distinct in the fossil record by the early Miocene, so that during the time of the hominoid radiation the cercopithecoids were an uncommon and restricted group. This state persisted until near the end of the middle Miocene, a period of 10 million years at the very least.

The original anthropoid dietary adaptation is shown by available evidence to have been primarily frugivorous. It is unlikely that any of the dietary specializations of the extant anthropoids would have been present in their common ancestor because each is peculiar to its own group and not present in the others. The ancestral anthropoids would therefore have been restricted to a ripe fruit diet with some young leaves and shoots. All of the known species of Oligocene and early Miocene anthropoids, including hominoids and cercopithecoids where they were differentiated, were, with two exceptions, frugivorous and this is still the primary adaptation of the extant hominoids.

The divergence of the cercopithecoid monkeys from the anthropoid stock is associated with a tendency of dietary change towards arboreal folivory. More specifically, higher proportions of leaves were included in the diet. The evidence for this is provided by the bilophodont molar, which is more high crowned and has longer shear surfaces on its cusps than is the case in the molars of both the early anthropoids and the fossil and extant hominoids. These are specializations that in a number of other primate groups are associated with folivory, and it is reasonable to conclude, therefore, that the occurrence of the same specializations in the fossil monkeys indicates the beginning of a trend towards folivory in them. The fossil monkeys of the early Miocene (or of the Oligocene if the parapithecids are considered monkeys) show no evidence of further differentiation and thus do not appear to have carried the potential for leaf-eating very far, and it can be presumed that they were still largely frugivorous and that they existed alongside the more abundant hominoids by their ability to process additional quantities of leaves.

In the middle Miocene there is evidence for the divergence of the colobines and cercopithecines in Africa. The colobines show signs of increased folivory in their higher crowned molars, and it is presumed that the specializations of the digestive tract would have started to develop at about this time so that they were present by the beginning of the late Miocene when the colobines emigrated to Europe and Aisa. With the increased efficiency of their teeth and their digestive tract modifications, the colobines are better equipped for a folivorous diet than are the cercopithecines, and with the emergence of this group, the more conservative cercopithecines faced a new kind of competition. For this reason it is suggested that the cercopithecines adopted a new adaptive strategy to enable them to compete with the new folivores on the one hand and the old frugivores on the other: the colobines and hominoids respectively.

The adaptive response of the cercopithecines is presumed to have been the return to a more frugivorous diet with the added advantage of having developed the capacity to eat less ripe fruit. They did this by developing a tolerance for the secondary compounds such as tannins present in unripe fruit. In this way they could eat and digest fruit before it became ripe enough for the hominoids to eat and thus gain access to the main food resources in a wide range of habitats, effectively excluding the hominoids from all but a small part of these resources. This provides a mechanism by which the cercopithecines could have competed successfully with the hominoids, but it must be remembered that it is only a model of what could have taken place, not a known event.

The response of the hominoids to this (and possibly other) competition may be seen to have less to do with dietary than their physical and social adaptations. Their present locomotor diversity is unrivalled for such a limited taxonomic group, although it has been seen how the early hominoids were essentially monkey-like in their locomotor function. Gorillas and chimpanzees have adapted to an unusual quadrupedal form of locomotion, knuckle-walking, with complex and sophisticated social systems that may be seen as a means to optimizing foraging strategy, but while the chimpanzee has remained frugivorous, the gorilla converges on foliovorous monkeys in having an enlarged caecum and teeth with higher and more pointed cusps and smaller incisors. The siamang shows a similar convergence on the monkeys in dietary terms, but together with the gibbons has adapted to a highly mobile arboreal adaptation with bimanual brachiation as its locomotor function and a close-knit and territorial social system, possibly another strategy for optimizing foraging ability. The hominids and the thick-enamelled hominoids of the middle and late Miocene apparently occupied a more terrestrial niche in savanna and woodland habitats, and their jaws and thick-enamelled teeth became adapted for crushing food objects that needed little preparation.

Acknowledgments

I am very grateful to Dr. Richard Wrangham for permission to use unpublished data and for discussion and comments on various drafts of this paper. I am also grateful to Drs David Chivers, Libby Nesbit Evans, Alan Gentry and Chris Stringer, and to Stephen Dreyer, Terry Harrison and Lawrence Martin for valuable comments on the manuscript.

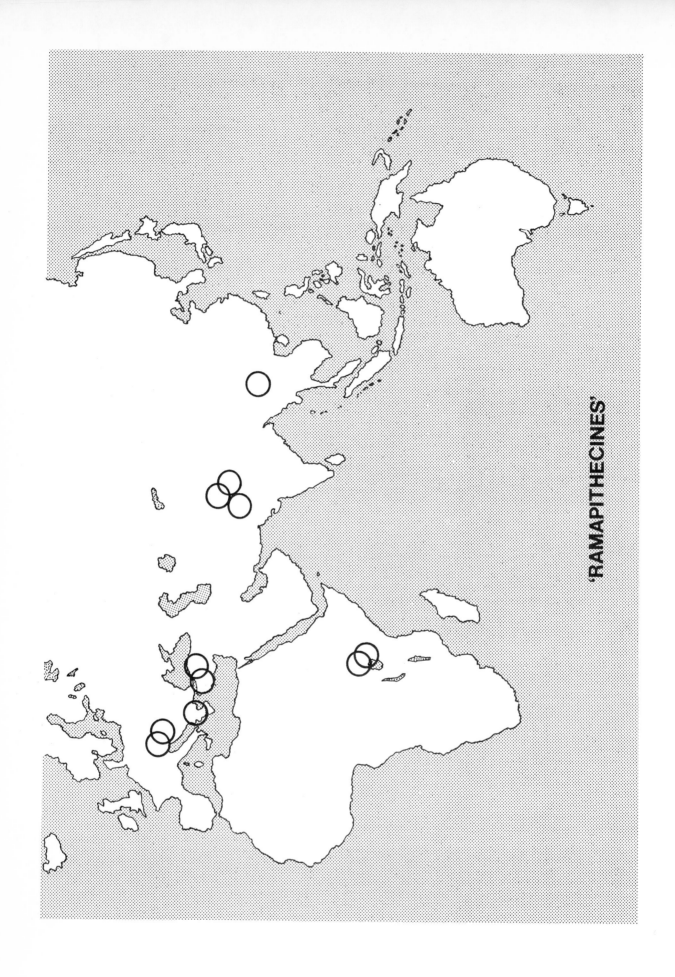

'RAMAPITHECINES'

Part V

Ramapithecus and Human Origins

Over the past two decades perhaps no other issue in primate evolution has sparked as much controversy and debate as the phyletic placement of *Ramapithecus*. The roots of this controversy can be traced to (1) the theoretical and philosophical differences about how one defines a human ancestor in the fossil record, (2) the fragmentary makeup of the actual remains of *Ramapithecus*, and (3) the conflicting implications of the biomolecular clock data that favor a late split of the ape/human lines. The chapters in this part survey the entire range of opinions concerning the position of *Ramapithecus* in hominoid evolution. The implications of the biomolecular data for ape/human ancestry discussed here receive further treatment in Part VII.

In 1910 Guy Pilgrim described the first fossils from the Siwalik Hills of India that would one day become the type species of *Ramapithecus*. More than two decades later, G. Edward Lewis (1934) named the genus *Ramapithecus* by describing a new upper-jaw fragment that he called *R. brevirostris* (meaning "Rama's short-faced ape"). He pointed out the apparent "human-like" affinities of *Ramapithecus* and argued for its placement in the Hominidae. Lewis' opinions remained largely overlooked until the early 1960s. In Chapter 27 Elwyn Simons (1961c) reformulates and strengthens Lewis' case for the hominid affinities of *Ramapithecus*. Simons discusses the dental and maxillary characters in which *Ramapithecus* appears to resemble the hominid pattern, concluding that many of Lewis' (1934) earlier observations were correct. He particularly emphasizes the parabolically shaped arcade of Lewis' type specimen. Unfortunately, as will be apparent, several of the key characters considered by Simons were based on reconstructions that were not supported by more complete fossils found subsequently.

In the next chapter David Pilbeam (1968b) discusses the problem of defining "the earliest hominid." He concludes that habitual bipedalism and a human dental apparatus are two major criteria of the Hominidae. Pilbeam supports Simons' (1961c, 1964c) observations on the human-like form of the *Ramapithecus* dentition, concluding that as a genus it is broadly ancestral to the Hominidae. According to Pilbeam, whether it should be placed in the hominid family is a matter of definition. Since *Ramapithecus* displays a hominid dental apparatus it apparently has crossed one of the two major adaptive boundaries and thus could be placed within the Hominidae. Based on a survey of the Miocene hominoids (see Simons and Pilbeam, 1965; Pilbeam, 1969a) Pilbeam concludes that hominids were recognizably distinct from pongids in the fossil record by 14 million years ago. He points out that this is in direct conflict with the biochemical data of Sarich and Wilson (1967a,b; see also Washburn, 1968b), which yields a 5-million-year-old date for the ape/human divergence.

A decade later Pilbeam (1978) wrote Chapter 29, a popularized account of his views on human origins at this time. Further research and fieldwork have brought about changes in Pilbeam's view of *Ramapithecus*. The simple dichotomy of pongids and hominids used in 1968 is replaced by a more complex four-family

division. A newly recovered complete jaw of *Ramapithecus* from the Potwar once and for all decides the issue of arcade shape: *Ramapithecus* has neither a parabolic arcade as argued by Simons (1961c) nor an ape-like U-shaped arcade but is instead rather V-shaped. Pilbeam observes, in retrospect, that arcade shape is not a particularly useful anatomical feature for defining hominoid groups. With regard to the nature of *Ramapithecus*, he speculates it was probably a small ape weighing no more than 40 pounds that was partially quadrupedal and partially bipedal, the latter due more to its small body size than to any other factor.

By the late 1970s a number of paleoanthropologists had begun to question both the phyletic position of *Ramapithecus* and its validity as a separate genus. In Chapter 30 Leonard Greenfield (1980) presents his view of hominoid relationships in a paper entitled "A Late Divergence Hypothesis." His hypothesis holds that the lineage leading to *Homo* did not become distinct from the lineage leading to the Great Apes until the middle-late Miocene (about 5–10 m.y.a.). This contrasts with the "early divergence hypothesis" as discussed in Chapters 27 and 28, where *Ramapithecus* is the stem hominid ancestor and the ape/human split predates 15 m.y.a. Following Greenfield's hypothesis, *Ramapithecus* can not be a hominid, and the acquisition of the dental and jaw features characteristic of the hominid line is a post-Miocene event. Instead, middle to late Miocene *Sivapithecus* is viewed as a plausible last common ancestor of both the living Great Apes *and* humans. Implicit in Greenfield's "late divergence hypothesis" is the combining (synonymy) of *Ramapithecus* with *Sivapithecus* (see Greenfield, 1979 and Chapter 30). The "late divergence hypothesis" is testable, and it is consistent with the biomolecular data set (see Greenfield, 1983). It has become the favored hypothesis of the 1980s.

Putting aside the question of evolutionary relationships, Richard F. Kay (1981) discusses the broader issue of the adaptations of the Ramapithecinae in Chapter 31. He reasons, by analogy with living primates, that the unusually thick enamel on the molar teeth of all ramapithecines indicates a diet of tough food stuffs such as hard nuts or seeds with tough pods. Kay characterizes the ramapithecines as "the nut-crackers." Since many thick-enamelled primates (and other mammals) living today are forest-dwellers, Kay suggests that the ramapithecines may have also been living in forested environments. There seems to be no compelling evidence that would associate thick molar enamel in ramapithecines with either open non-forested environments or with terrestrial adaptations as had been argued in the past. Thus Kay's (1981) study of the adaptive strategy of the Ramapithecinae indicates that this group does not appear to foreshadow the early hominid open woodland/savanna adaptation.

In Chapter 32 Peter Andrews and Jack Cronin (1982) review the molecular data that bear on the question of hominoid relationships and present new morphological data that link the *Sivapithecus-Ramapithecus* species group with the evolution of the orangutan. Specifically, they point out a number of derived dental and facial characters (see also Andrews and Tekkaya, 1980) that strongly suggest an evolutionary connection between *Sivapithecus* and the orangutan. Since *Ramapithecus* is judged synonomous with *Sivapithecus* by some researchers (see Greenfield, 1979, 1983, and Chapter 30, this volume) or a close relative by others (see Wolpoff, Chapter 33), the evidence marshalled by Andrews and Cronin (1982) is strong justification for rejecting its proposed hominid affinities. Further support for the linkage of *Sivapithecus* with the orangutan comes from the Potwar Plateau of Pakistan where a nearly complete face of *Sivapithecus indicus* was recovered in 1979 (see Pilbeam, 1982; Lipson and Pilbeam, 1982; Ward and Pilbeam, 1983) that has the same suite of derived characters that Andrews and Cronin discuss in Chapter 32.

In the final chapter, Milford Wolpoff (1982) approaches the analysis of *Ramapithecus* from historical and theoretical perspectives—quite different from the approach taken by Andrews and Cronin in the preceding chapter. He sees the question of whether or not to call *Ramapithecus* a hominid a far more complex issue than simply deciding on which side of the ape/human line it falls. Wolpoff also points out that there was not just one ape/human divergence but actually two separate divergences at different points in time (see also Greenfield, Chapter 30). This greatly complicates the problem of identifying the last common ancestor of apes and humans. Wolpoff agrees with Andrews and Cronin (1982) in their demonstration of a link between the *Ramapithecus-Sivapithecus* species group and the orangutan, but at the same time he argues that from within this "ramapithecine" group the ancestors of the earliest hominids will also be found. Thus Wolpoff would like to derive all the living Great Apes and humans from a single Asian-African "ramapithecine" group.

Wolpoff's arguments appear to bring the question of the phyletic placement of *Ramapithecus* full circle—from earliest hominid to primitive ape to ancestral orangutan and now as part of the ancestral stock for all of the above. New discoveries of an early Miocene form of *Sivapithecus* recovered by Alan Walker and Richard Leakey at the site of Buluk in northwestern Kenya (see news reports by Lewin, 1983 and Rensberger, 1984) appear to add a measure of support to Wolpoff's viewpoint. Alternatively, they could also indicate a much greater antiquity for the orangutan lineage. At the very least these new finds show us that simple solutions to the question of human origins may not always be the correct ones.

27

The Phyletic Position of *Ramapithecus*

E. L. SIMONS

Recent discoveries of early Pleistocene hominids at Olduvai gorge, Tanganyika, by expeditions under the direction of Dr. L.S.B. Leakey have pushed back certain knowledge of fossil man almost to the beginning of this epoch. To the extent that the K-A date suggested for these early men, 1.75 million years, (Leakey et al. 1961) is accurate, the beginning of the "Villafranchian" provincial age, and thus of the Pleistocene itself, is shown to be considerably earlier than most previous estimates. It therefore seems appropriate that renewed attention be drawn to the only Pliocene fossil primate specimen known to this writer, which can be defended as being within, or near, the population ancestral to Pleistocene and subsequent hominids, the type maxilla of *Ramapithecus brevirostris* at Yale Peabody Museum.

This maxilla, Peabody Museum No. 13799, was collected August 9, 1932 by the Yale North India Paleontological Expedition under Dr. G.E. Lewis (Fig. 1). The geologic occurrence of *R. brevirostris* was first given by Lewis (1934) as "Either latest Middle Siwalik [Dhok Pathan Zone] or basal upper Siwalik [Tatrot Zone]." However, Lewis (1937) later determined the horizon of Y.P.M. 13799 as being within the Nagri zone, which is of Pliocene early Middle Siwalik age. Gregory et al. (1938) also indicate the level of this specimen as Nagri.

Consequently, Hooijer and Colbert (1951) seem to have erred in listing *Ramapithecus* as occurring only in the Tatrot zone fauna which they suggest as being very close to the Plio-Pleistocene boundary. Regardless of these published differences in age determination the provenance of the specimen is known, so that, at least potentially, its temporal position can be verified. Faunal correlations indicate that, even in the unlikely event that *Ramapithecus* occurs as late as the Tatrot horizon, this primate is distinctly older than the "Villafranchian" hominids of Olduvai gorge.

In spite of the significance of Y.P.M. 13799, as being possibly the earliest known hominid, it has been largely overlooked, or briefly dealt with in the more recent summaries of hominid evolution, a common conclusion being that the type is too fragmentary to permit taxonomic assignment. Actually, such a conclusion is incorrect and misleading. This right maxilla provides at least some information as to shape, size or positioning of the entire upper dentition except for M³, in that alveolae of I¹⁻², C are preserved as well as the series P³ through M². Moreover the base of the nasal aperture can be seen above the incisors, and, contra Hrdlicka (1935), the dental arcade can be determined as parabolic and not U-shaped, as was correctly stated by Lewis (1934) in the original description of this form (see Fig. 2). Some may think (as Hrdlicka did) that extrapolating from the right maxilla alone, in order to determine that the disposition of the upper cheek teeth is in an arcuate line, instead of being arranged in the parallel series

seen in all pongids, is a rather uncertain procedure. However, at one point (see arrow 1, Fig. 2) the maxilla reaches nearly, if not entirely to the point of the palatal intermaxillary suture. Since we may safely assume that *Ramapithecus,* like other vertebrates, was bilaterally symmetrical, if the right maxilla and its mirror-image are pivoted around this point the amount of posterior divergence of the cheek tooth rows cannot be further decreased beyond the arrangement shown in Figure 2 without assuming an impossibly long basal diameter for the central incisor pair (Fig. 2, arrow 2). In

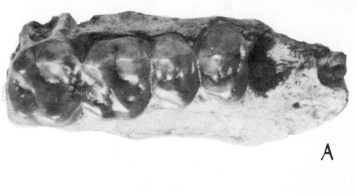

Figure 1 *Occlusal view (A) and lateral view (B) of right maxilla of type of* Ramapithecus brevirostris, *Y.P.M. 13799.*

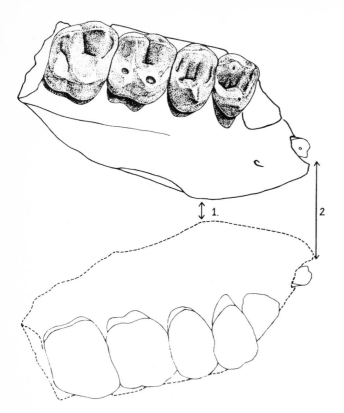

Figure 2 Ramapithecus brevirostris, *right maxilla, Y.P.M. 13799, and reverse of same, showing arcuate arrangement of teeth.*

fact, the space allowed for these teeth in Figure 2 (in order to be on the safe side) is intentionally made greater than it is likely to have been. Preservation of the entire length of the alveolar cavity of the right central incisor allows for comparative measurements as to its size. The central incisor root of *Ramapithecus* is only about half as long as it is in a series of chimpanzees examined in this connection and which had cheek teeth of the same absolute size as Y.P.M. 13799. In orangutans the central incisors have, comparatively, still longer roots than does *Pan*. As is well known, possession of large incisors relative to cheek teeth is a general feature distinguishing both fossil and living pongids from known hominids. In this feature of central incisor size, as in others, such as the highly arched palate, *Ramapithecus* agrees more closely with Hominidae than with Pongidae.

It is evident that most of the misapprehensions regarding *Ramapithecus* now current trace back to Hrdlicka's discussion of the specimen (1935) in which he insisted that the form could not be a hominid. Even a casual examination of this paper is sufficient to show that it bears every evidence of being a controversial and non-objective contribution. In contrast to this, all of the hominid resemblances cited for Y.P.M. 13799 by Lewis (1934) appear to this writer to have been correctly drawn, and these are reinforced by the additional hominid features called to attention here.

However, another possible source of uncertainty regarding the genus may derive from a mandible, Peabody Museum No. 13807, assigned by Lewis (1934) to *Ramapithecus*, but to a different species *R. hariensis*. This mandible shows heteromorphy in the lower premolars of the sort characteristic of pongids but

which is not known in undoubted Hominidae. In view of this heteromorphy, not indicated in P^{3-4} of *R. brevirostris* and inasmuch as the mandible of *R. hariensis* comes from a different locality, and from a horizon that may be considerably lower in the section, I see no convincing reason for associating generically the form it represents with that of the maxilla of *R. brevirostris*.

What then can be stated as fact regarding the type maxilla of *Ramapithecus*? As the species name implies, and as Lewis originally stated, this primate exhibits a reduction in prognathism, upper incisor size, and in length from the alveolar border above the incisors to the base of the nasal opening, when compared to pongids of its general size, whether living or fossil. This length from nasal aperture to I^2 in *Ramapithecus* is approximately 44 per cent of the length of $P^3 - M^2$ (see arrows, Figure 1) while corresponding percentages in a series of specimens of *Pan* range from 70 to 98. Specimens of *Pongo* and *Gorilla* examined fall within the range of *Pan*, in the proportion.

In addition to the foregoing differences, the upper incisors and canine, judging from their alveolae, cannot have been as large as they typically are in even the smallest Great Apes, a fact also pointed out by Lewis (1934), who remarked: "The face is very slightly prognathous, as contrasted with recent Simiidae. There are no diastemata in the dental series. The canine is small, not an antero-posteriorly elongated trenchant tusk but a hominid type with a transverse dimension exceeding the antero-posterior dimension." Lewis (1934: 163–166) fully discussed the dental characters of Y.P.M. 13799, consequently it is unnecessary to repeat this description here. In general, crown patterns resemble both *Dryopithecus* and *Australopithecus* about equally.

Without further extending the polemical atmosphere surrounding this specimen, so unfortunately initiated by Hrdlicka, this writer will simply call attention to his final statement regarding *Ramapithecus*, since he appears to be the only person to have studied the actual specimen who has published doubts as to its hominid status. The significance of this remark, in the light of modern understanding of the australopithecines as hominids, seems to have been overlooked. Hrdlicka (1935:36) observed: "The genus [*Ramapithecus*], although in the upper denture, in general, nearer to man than are any of the Dryopitheci or the *Australopithecus* cannot...be legitimately established as a hominid, that is a form within the direct human ancestry." This curious statement, indicates that Hrdlicka would now have to place the genus in the Hominidae since he regarded it as more man-like than *Australopithecus*, a genus universally accepted today by competent students as belonging to this family. Evidently if there are convincing reasons why *Ramapithecus brevirostris* should not be regarded as representing the earliest known hominid they have not been demonstrated to date.

To contend that the specimen is too inadequate for definite taxonomic assignment implies that pongids and hominids cannot be distinguished, even when reasonable information is available regarding the size, emplacement, structure and arrangement (whether arcuate or parabolic) of nearly all of the upper dentition, together with several characters of palate and face as well. Postcranial remains, if found, might make it easier to assign this primate taxonomically, but the six or seven distinct approximations to hominid morphology discussed here for Y.P.M. 13799 provide an adequate basis for associating it with the latter family. It seems illogical to choose the alternative of regarding this form as belonging to an otherwise unknown group of apes, parallelistic toward hominids but not closely related to them, when it occurs in the proper time and place to represent a forerunner of Pleistocene Hominidae.

28

The Earliest Hominids

D. R. PILBEAM

One of the principal interests of primate palaeontologists is the matter of human origins. This topic has generated a considerable amount of heat, much of it due to genuine disagreements but some to questions of definition. Before one can answer the question, "what is the earliest hominid?," agreement must obviously be reached on how it is to be recognized. The family Hominidae includes ourselves, our extinct ancestors and their extinct relatives. If the present hominids, *Homo sapiens*, and their closest living relatives, the Pongidae, are compared, there are many obvious contrasts between the two groups. Only some of the differences are of direct use to the palaeontologist. It is generally agreed (Simpson, 1966b) that there are two principal sets of criteria which must be satisfied if a fossil primate is to be considered a hominid.

The requirements are, first, evidence of habitual upright bipedalism as the chief method of locomotion, and, second, the presence of teeth which are essentially human in form. It has become almost a matter of anthropological dogma that the earliest hominids differed from their pongid ancestors because of subtle locomotor changes in the direction of bipedalism. Hominids are also characterized dentally by a particular type of occlusal pattern in the cheek teeth, vertically implanted incisors and small non-projecting canines morphologically resembling the incisors. The small canines are, at least in part, responsible for the fact that hominid teeth are arranged in a parabolic dental arcade. In apes the incisors are large and procumbent, and—particularly in males—the canines are projecting, pointed tusks. The cheek tooth rows are parallel. The basic differences between dentitions of men and apes are in canine size and in the fact that both male and female hominids have small canines. The second of anthropology's central dogmas is that reduction in hominid male canine size is correlated with the use of weapons (tools) in intra and inter-group feuding and in fighting between species. Clearly, then, these morphological complexes, associated with bipedalism and changes in tooth function, reflect marked shifts in behaviour between pongids and hominids (Washburn, 1968b).

Until quite recently, fossil hominids were thought to come only from deposits of Pleistocene age. The earliest forms, *Australopithecus africanus* from South Africa and *Homo habilis* (thought by some to be *A. africanus*) from East Africa, are 2 million or more years old (Campbell, 1967). Although in many features these creatures were primitive, they moved bipedally and possessed the dental adaptations typical of later men—small incisor-like canines and parabolic dental arcades. Only a few workers now hesitate to consider these forms unequivocally hominids.

The course of hominid evolution before the Pleistocene has been much less well understood. Many species of fossil great apes have been described (Simons and Pilbeam, 1965) since Lartet named *Dryopithecus fontani* from France in 1856. Large numbers of new *nomina* were coined for fossils from Europe, Asia and Africa, from deposits ranging in age between late Oligocene (about 30 million years ago) and middle Pliocene (about 6 million years ago). The whole group was classified in a subfamily of the Pongidae named Dryopithecinae after the type genus. More than fifty species, in numerous genera, have been described in the 112 years since 1856, and many workers came to believe that dryopithecines had undergone a very extensive adaptive radiation. It was also widely believed that because fossil apes were "so fragmentary" little could be said about their evolution. Recently more detailed studies by Simons and me (Simons and Pilbeam, 1965) have shown both beliefs to be incorrect. We concluded that the adaptive radiation consisted only of names, not of species, and that one genus, *Dryopithecus*, was adequate for those Tertiary species that were valid. The numbers of specimens of fossil apes now known run into hundreds, some of them relatively complete, and include enough post-cranial material for inferences on locomotor behaviour to be drawn for several African species (Le Gros Clark and Leakey, 1967).

The earliest *Dryopithecus* species (regarded by Leakey [1967a] as *Proconsul* species) are found in the early Miocene deposits of Kenya and Uganda. Some of the sites have been dated to about 20 million years (Bishop, 1967). One species, *D. africanus*, is known from cranial, dental and post-cranial material. Although this has been considered as a generalized hominoid "model" of an early hominid, *D. africanus* is now better regarded as a form ancestral to, or close to the ancestry of, the chimpanzee, *Pan troglodytes* (Pilbeam, 1969a). Related to *D. africanus*, and sharing with it a number of primitive features, is a second species, *D. major*. Its dentition and face have been described in detail; this species is a plausible ancestor for *Gorilla gorilla* (Pilbeam, 1969a). Although Leakey does not agree that all Ugandan and Kenyan specimens of *D. major* properly belong in that species, he does accept the relationship between the Moroto (Karamoja, Uganda) palate and gorilla (Leakey, 1963). Several other pongid species are also known in East Africa, none of which need concern us here.

Dryopithecines have also been recovered from Europe and Asia, from deposits of middle Miocene through middle Pliocene age (approximately 16 million to 6 million years) (Simons and Pilbeam, 1965). Several species are known in each of the major areas. In general, European forms were classified in *Dryopithecus*, Asian ones in *Sivapithecus*. Like *Proconsul* from Africa, all these species can be subsumed under one genus, *Dryopithecus*. Some of the dryopithecines from deposits in the Siwalik Hills, north-west India, are, in all probability, ancestral to the only living Asiatic pongid, *Pongo pygmaeus*, the orang-utan (Pilbeam, 1966, 1969a).

The Eurasian dryopithecines had their origins in Africa and

probably migrated thence some time during the middle Miocene, perhaps about 15 or 16 million years ago. Their ancestors are therefore to be sought in earlier Miocene deposits in Africa. Those species described previously as *Proconsul* almost certainly are not directly ancestral to Eurasian *Dryopithecus*, although they are probably very closely related to those ancestors.

In 1950, Le Gros Clark and Leakey (1950) named a new pongid species from Kenya, calling it *Sivapithecus africanus*. It consisted of one fragment of upper jaw with three teeth (British Museum number M.16649), together with two other isolated upper molars. This species differed from other African species of *Dryopithecus (Proconsul)* in several features, principally in that *S. africanus* lacked lingual cingula on the upper cheek teeth. Le Gros Clark and Leakey (1967) later discussed this palate in great detail, concluding that it was very similar to *Sivapithecus sivalensis* (as it was then known) from the Siwaliks of India. They thought it possible that *S. africanus* could be ancestral to *S. sivalensis*. As I have already noted, such a relationship, between African and Asiatic dryopithecines, should be expected. Simons and I (1965) transferred both species to *Dryopithecus*, and thought that there was enough temporal and morphological evidence to justify placing them in one species, *S. africanus*. I now think, however, that there is enough time separation between the two to make specific synonymy not particularly useful. For the moment I shall refer to the African type specimen (M.16649) as *"S. africanus"*; it cannot be transferred to *Dryopithecus* without being given a new species name, for there is already one *D. africanus*. In all features, this palatal fragment is ape-like.

It has come to be rather generally assumed, albeit in a rather vague fashion, that pre-Pleistocene hominid ancestry was rooted somewhere in the Dryopithecinae. Several species previously assigned to the sub-family have been considered to be rather more man-like than the others: for example, *Ramapithecus brevirostris* from the late Miocene of the Siwaliks, named by Lewis (1934) and described by him 2 years later in his unpublished PhD thesis as a hominid ancestral to *Australopithecus* and *Homo*. In 1961, Simons initiated a series (Simons and Pilbeam, 1965; Pilbeam, 1966, 1969a; Simons, 1961c, 1964c, 1968b; Pilbeam and Simons, 1965) of studies of *Ramapithecus*. The chief point to emerge from these studies, which still continue, is that remains of hominid-like forms from the Siwaliks of India are more plentiful than had been supposed. A new name also became necessary because of synonymies, and is *Ramapithecus punjabicus* (Simons, 1964c). The species includes upper and lower jaws and teeth, but unfortunately no post-cranial bones.

The chief contrasts between *Ramapithecus* and *Dryopithecus* are as follows (Simons and Pilbeam, 1965; Simons, 1964c; Pilbeam, 1969a). The cheek teeth of *Ramapithecus* resemble those of later hominids, being steep-sided and low-cusped, and the premolars are small. The canines are known only from their sockets, shaped and oriented like those of Pleistocene Hominidae. The crowns must have been small and non-projecting. The dental arcade was probably parabolic. In short, most of the features of hominid dentitions are found in *Ramapithecus,* and there are enough similarities between *Ramapithecus* and Pleistocene hominids to justify the view that they are part of one lineage or of closely related lineages.

In 1962, Leakey (1962) announced a new primate from the late Miocene of Fort Ternan in Kenya (absolute age [Bishop, 1967] for these deposits is 14 million years). This he called *Kenyapithecus wickeri*. Morphologically, it is extremely similar to *R. punjabicus* from India, as Leakey's first description shows (Lewis, 1934; Leakey, 1962). The specimen consists of two maxillae. The second premolar and first two molars are preserved and, fortu-

nately, a canine crown is present. The canine is relatively smaller than that of any living or extinct pongid and is as small as those of the later hominids (Pilbeam, 1969a). Morphologically the crown is a little ape-like (Pilbeam, 1969a, Leakey, 1962). Simons and I, separately and together, have suggested that these two species are so similar that they must be regarded as one (Simons and Pilbeam, 1965; Pilbeam, 1966). I now take the view that they are sufficiently similar to be placed in one genus, *Ramapithecus*, but the step towards complete specific synonymy is perhaps rather more difficult to take. The final result will have to wait until further material is recovered, both in Africa and India. Suffice to say that this is further evidence in favour of at least some faunal links between Africa and Eurasia in late Miocene times.

Should *Ramapithecus* be classified as a hominid, or is it a man-like pongid? I think it can be fairly argued that *Ramapithecus* is, broadly, ancestral to later Hominidae. Whether it should actually be classified in that family is as much a matter of semantics as anything else. I am inclined to the view that its hominid features so outweigh those that are primitive retentions as to indicate that at least one of those important and definitive hominid behavioural shifts had probably already occurred. Unfortunately, no post-cranial remains are known, so locomotion cannot be deduced. I would argue therefore that *Ramapithecus* is probably a hominid, and should perhaps remain such even if fossil evidence were produced which showed that it was not yet a habitual biped.

Recently, Leakey (1967) has stated at length his view that *R. punjabicus* and *"K. wickeri"* are not synonyms, that *"Kenyapithecus"* is a valid genus, and that early Miocene *"S. africanus"* is not a pongid but a hominid in the genus *"Kenyapithecus,"* ancestral to *"K. wickeri"* of the late Miocene. I find myself in agreement with Leakey on one point. Because we both accept late Miocene *Ramapithecus* as a hominid, earlier Hominidae should be expected in earlier Miocene African deposits. I am, however, in some disagreement with Leakey on a number of other points deriving from his article. Unfortunately there is only space here to mention a few of these. Lengthier comments can be found elsewhere (Pilbeam, 1969a)

First, it is necessary to restate my belief in the synonymy of *Ramapithecus* and *"Kenyapithecus wickeri,"* at least at the generic level. Recently, both Simons (1967b, 1968b) and I (Pilbeam, 1969a) have discussed this matter of synonymy in some detail, and I shall not dwell on it here. In my opinion then, *"Kenyapithecus"* is no longer available. This leaves several questions still to be discussed. First, what evidence is there that the hypodigm of *"S. africanus"* as listed by Leakey samples only one species? Second, what relationship does this "species" have to later Tertiary Pongidae or Hominidae? Third, if *"S. africanus"* can be shown to have any sort of ancestral relationship to late Miocene *Ramapithecus*, should it be classified in the same family? Fourth, if it can be ranked with Hominidae, is it similar enough to *Ramapithecus* to be congeneric?

"S. africanus" as described by Leakey (1967a, 1968b) consists of a small sample of rather fragmentary material. Maxillary specimens preserve evidence for the crowns of P^3, P^4 and M^1. Broken roots only are available for canines. The maxillae are preserved superiorly as far as the region of the canine fossa. Isolated upper incisors and what is thought to be M^2 are claimed to belong to the same species. In his first article, Leakey (1967a) included in *"S. africanus"* three mandibular fragments, all lacking tooth crowns. Subsequently, Leakey (1968b) added a fourth mandible to the hypodigm, this time with crowns of canines and all the cheek teeth preserved.

No positive evidence has been produced to support the claim

that this set of specimens belongs to a single species. Considering published dimensions alone, it is extremely unlikely that the later mandible (number 394) (Leakey, 1968b) and the type maxilla (M.16649) (Le Gros Clark and Leakey, 1967; Leakey, 1967a) are in fact sampled from the same species. However, this point is not so important here as a consideration of the features of "*S. africanus*" which might point to pongid or hominid ties.

First, the canine fossae of M.16649 and Sgr 111. These are not, as claimed in Leakey's (1967a) diagnosis, "somewhat of the general type seen in *Homo sapiens.*" They resemble much more closely that of the holotype of *D. nyanzae* (M.16647) (Le Gros Clark and Leakey, 1967) and of specimens of *Pan troglodytes* and *Pongo pygmaeus.* The forward position of the maxillary zygomatic process (above M[1]) is a resemblance to *Ramapithecus,* but is also to be found in *Oreopithecus* (Straus, 1963) as well as *Pongo pygmaeus* (Pilbeam, 1969a). It may be a primitive feature in some hominoid lineages, and is not necessarily indicative of ties with Hominidae.

As I have noted, no crowns of upper canines are preserved. Root dimensions, however (Leakey, 1963, 1967a) indicate that the crowns would have been large relative to cheek teeth. Simons and I (1965) noted (page 113) the presence above the P[3] roots in M.16649 of parts of a large canine alveolus. The evidence shows clearly that maxillary canine crowns would have been as large relatively as in any of the living apes.

In my preceding discussion of *Ramapithecus,* I mentioned the fact that the upper cheek teeth were steep-sided. In this feature they contrast markedly with those of fossil pongids (*Dryopithecus, Aegyptopithecus*) (Pilbeam, 1969a). The sides of the cheek teeth of these apes slope markedly, and the occlusal surface is therefore relatively constricted. Living pongids show trends away from this condition towards more vertically sided teeth. Sloping buccal and lingual surfaces are most marked in those species of *Dryopithecus* which lack lingual cingula on the upper cheek teeth (for example, *D. sivalensis*). In this feature the early Miocene "*S. africanus*" is closely similar to *D. sivalensis.* Table 1 lists ratios of

Table 1 *Ratio (Minimum B–L Breadth/Maximum B–L Breadth) × 100*

	R. punjabicus (n=3)	D. sivalensis	M.16649
P–	62·0–66·4	46·0–50·0 (*n*=2)	43·9
M–	61·8–63·5	48·8–57·0 (*n*=5)	46·6

minimum to maximum buccolingual breadth of P[4] and M[1], an index of the slope of the buccal and lingual borders or of the degree of constriction of the occlusal surface. (See also Fig. 8 of Simons [1967b], which shows clearly the contrasts between *Dryopithecus* (including M.16649) and *Ramapithecus.*)

In this feature of type of "*S. africanus*" is much closer to *D. sivalensis* than to *Ramapithecus.* No explanation for this contrast has been produced yet. "*S. africanus*" and *Dryopithecus* are similar, too, in detailed features of occlusal morphology. Leakey himself has pointed out the contrasts between the upper cheek teeth of M.16649 and East African *Ramapithecus* (Leakey, 1962)

Summing up the evidence from the upper jaws, "*S. africanus*" is in almost all features like other *Dryopithecus.* Basically, the maxillary characters are those of apes, not of hominids. Of particular interest here are the large canines. Although there is no definite information about the shape of the dental arcade, no complete palates being available, it is probable that the arcade would not have been parabolic. (Indeed, because this feature appears to be closely correlated with canine size and facial length,

it should be used only with circumspection as an additional character to "canine size.")

The mandibular fragments first described by Leakey (1967a) as belonging to this same species should, of course, have been grouped only tentatively with the maxillae. One specimen (Sgr 417) is small, smaller than *D. africanus,* while the largest of the three (CMH.142) is as big as a *D. major.* On grounds of size alone it is improbable that these specimens represent one species, even a highly dimorphic one. Sgr 417 may in fact be a *D. africanus.* CMH.142, however, is not, I think, a *D. (Proconsul)* at all. Its canine, to judge from the root, would have been large and probably projecting; the crown of P[3] too, known only from broken roots, was elongated and probably a sectorial, unicuspid tooth. The symphyseal cross-section and tooth root outlines are similar to those of the type specimen of *D. fontani* (Simons and Pilbeam, 1965) (Muséum National d'Histoire Naturelle in Paris, specimen No. AC.36). Whatever are the exact affinities of CMH.142 (and it is impossible to say anything definite about these because of the unsatisfactory nature of the remains), the specimen comes from an animal which was, dentally at least, a pongid. A third mandibular fragment (276) may belong to the same species as CMH.142. Little more can be said about it, except that probably canines projected and P[3] was sectorial (roots only are preserved for both teeth).

More recently, Leakey (1968b) has described a relatively complete mandible with cheek teeth and canines intact. As Leakey notes, this mandible contrasts in a number of ways with *D. major* and *D. nyanzae.* The symphyseal section is quite similar to that of 276, although a small simian shelf is present. The divergent tooth rows, relatively small M[1]s, triangular M[3]s, molar buccal cingula, and rounded anterior symphyseal contour are all similarities to *D. africanus.* Canines are relatively large and markedly pointed and projecting. The P[3] is unicuspid, elongated and sectorial. The overall similarities of the dentition are with *D. africanus* and other early Miocene African pongids. There are no features at all which point to links with known hominids, nor are there any reasons for not classifying this specimen (which may not, as I have noted, even be an "*S. africanus*") in the Pongidae.

The material listed by Leakey therefore is probably not sampled from one species, although the remains are too scrappy for a definite conclusion. All these specimens are ape-like as far as can be assessed from the rather limited remains. The post-cranial material of large hominoids from East Africa is also completely ape-like (Le Gros Clark and Leakey, 1967). There are no indications therefore of any trends towards the evolution of hominid dental or post-cranial characters.

Considering the holotype maxilla of "*S. africanus,*" M.16649, there are a number of features (canine size, and size, shape and morphology of the cheek teeth) which contrast with both Indian and East African *Ramapithecus* (see particularly Leakey [1962]). There are, on the other hand, a complex of characters linking M.16649, and any other specimens which might prove referable to the same species, with *D. sivalensis;* the mandibular evidence also points to ties with *D. fontani* and *D. sivalensis* (which are themselves closely similar forms). Symphyseal contour and cross-section, canine size and shape, premolar size and shape, and incisor disposition are all features linking these Eurasian and African groups. If the mandibles and maxillae from Kenya are conspecific, these resemblances would appear to strengthen the views that some, at least, of the early Miocene East African material is, broadly speaking, ancestral to Eurasian dryopithecines, and that it should be classified in *Dryopithecus.* I would still hesitate to give a species name to, or provide a differential diagnosis for, such inadequate material.

One specimen in the Kenya National Museum collections, CMH.133, is a right P⁴ classified until now as *D. nyanzae*. In this tooth the pattern of grooves and the shape and relative height of the cusps are about equally similar both to *Dryopithecus* and to *Ramapithecus*. The slope of the buccal and lingual surfaces resembles *Dryopithecus*. This specimen is almost certainly not a *D. nyanzae*; whether or not it is conspecific with M.16649 is uncertain but possible. Does this show that some specimens in the East African early (or middle) Miocene were part of a pre-*Ramapithecus* lineage? Unfortunately CMH.133 is but one tooth, has never been mentioned outside Le Gros Clark and Leakey's report (1967) and is from an unknown locality on Rusinga Island.

Concluding this discussion of the East African fossils, I believe it more probable than not that much of the material is sampled from a dryopithecine broadly ancestral to the *D. sivalensis/D. fontani* group. It is still possible, however, based albeit on slender and tentative evidence, that one or two specimens might be drawn from a pre-*Ramapithecus* species. If this is the case, should such a species be classified in Hominidae? The dental adaptations present inthis material are those of apes, as are the post-cranial ones.

I think that the division between Hominidae and Pongidae should be drawn somewhere along the lineage leading to the Pleistocene Hominidae, after at least one of the two major adaptive "boundaries" (dental or post-cranial) has been crossed. *Ramapithecus* can therefore be considered a hominid (provided that its "hominid features" are not parallelisms), whereas the early Miocene forms cannot. Leakey may have recovered an early hominid ancestor (although I doubt this), but he has not found an early hominid. It should be emphasized that this conclusion depends first on assigning fossils to their correct lineage, and second on semantics, on the definition of Hominidae.

Now, although Leakey and I do disagree on a number of points, we are in agreement on two things: that late Miocene *Ramapithecus* is a hominid and that still earlier hominids are to be expected. Recently several authors (Washburn, 1968b; Sarich and Wilson, 1967) have stated their view that the African apes shared a common ancestry with Hominidae as late as 5 million years ago. This common ancestor would have been, like the African apes, a knuckle-walker. If this theory is correct, *Ramapithecus* cannot be a hominid, as Leakey, Simons, I and others (Campbell, 1967, 1968) believe. None of the *Dryopithecus* species can be ancestral to any living pongids, nor can *Limnopithecus* or *Pliopithecus* be ancestral to Hylobatidae. Any characters of these (supposed) ancestral species suggesting ancestor/descendant relationships with living apes must therefore be parallelisms. I am more inclined to accept, however, the fossil evidence for the moment. This indicates that hominids were recognizably distinct from pongids 14 million years ago, that certain early Miocene species are ancestral to gibbons, chimpanzees, and gorillas, and that many of the "basic hominoid similarities" are likely therefore to be parallelisms.

To test these alternative hypotheses, discussed more fully elsewhere (Pilbeam, 1969a), we need more fossils, a closer look at the meaning of biochemical similarities and contrasts, and more complete comparative anatomical and biochemical studies. It is to be hoped that a wider consensus can then be achieved. Only when an agreed framework for hominoid evolution is produced can we hope to move on to the numerous questions of human morphological and behavioural evolution which still need to be answered.

29

Rethinking Human Origins

D. R. PILBEAM

I have spent a good deal of my professional career as a paleoanthropologist, searching for remains of our earliest ancestors. Since first coming to Yale as a graduate student in 1963, I have been particularly eager to find out more about an enigmatic hominoid (the term used to cover humans, their ancestors, their closest living relatives the apes, and ape ancestors) called *Ramapithecus*. The search has taken me to many exotic places, and to laboratories and museums all over the world. One wet, blustery, and rather unpleasant day in January 1976 it took me to a desolate and muddy hill in the middle of the barren badlands of the Potwar Plateau in Pakistan: Next to me stood Wendy Barry, wife of my postdoctoral colleague John Barry; we were both searching for fossils and complaining about the weather, in equal amounts. We were at Locality 182 because on the last day of our previous season, in March 1975, Martin Pickford, another coworker, had found a small piece of *Ramapithecus* lower jaw.

This had been very exciting for us, "us" being the members of a joint Yale-Geological Survey of Pakistan expedition which began field work under my direction in Pakistan in 1973. We had settled on Pakistan as a potentially important place because it had long been known as an area with extensive exposures of rocks between 4 and 14 million years old, the time during which the evolutionary lines leading to hominids (humans and their relatives) and pongids (apes and their relatives) first split. Also, many fossil mammals, including hominoids, had been recovered from those rocks (known collectively as the "Siwaliks" after a range of hills in India); among them were a few remains of *Ramapithecus,* thought by many of us to be the earliest recognizable hominid. The joint project has expanded since 1973 into a major effort involving scientists from several disciplines—anthropology, geology, paleontology, and botany. Our aim is to understand the changing climatic, geographic, vegetational, and faunal backdrop against which the human evolutionary story unfolded.

So, there Wendy and I stood, complaining on that depressing day. Suddenly, an excited shout: "I've got it, I've got it!" I looked down to where Wendy knelt in the mud, everything else forgotten; she cradled in her hand the complete left side of a *Ramapithecus* jaw, the best one found so far. We hugged one another, and shouted to the others, scattered about their tasks on the nearby hillsides. I recognized at once that we had found the opposite side of Martin's 1975 jaw fragment, which implied that the whole jaw might be there. We searched for the other parts, in vain.

An hour or so later Martin trudged up the hill after a rather gloomy day of mapping, anxious to see what all the excitement was about. Wendy and Martin had a good-natured competition going, to see who could find "the best," or "the most." Martin, both pleased and a little piqued about Wendy's discovery, began searching, going over the excitedly trampled ground: within min-utes two additional jaw fragments were uncovered. Incredulously, I held in my hands all four parts of the most complete *Ramapithecus* specimen ever found on the subcontinent. Many things ran through my mind, feelings of exhilaration and happiness for the discoverers, satisfaction for myself; but more than anything my mind was beginning to think along new lines, because I could see that our previous beliefs about *Ramapithecus* and the whole story of human origins needed rethinking.

Since that January day we have found over 80 hominoid fossils in Pakistan—many of them *Ramapithecus*—representing more than 40 individuals, which more than doubles the previously known collections from India and Pakistan. In addition, we have an enormous amount of geological and paleontological "back-up" data. At the least, we are ten times as knowledgeable about the area and the time period as we were earlier. In addition, several other expeditions, working in Kenya, Turkey, Greece, Hungary, India, and China have been finding hominoids of the same, 4 to 14 million year, age. The very tentative new syntheses involve radical changes in our thinking.

EARLIER IDEAS ON HUMAN EVOLUTION

I came to Yale in 1963 to study with Elwyn Simons, now of Duke University. For two years we worked together trying to sort out hominoid fossils of the Miocene, the geological time period between about 5 and 20 million years. Between 1963 and 1965, when I left Yale to go back to Cambridge, and after 1968 when I returned here, Simons and I developed our ideas. By the late 1960s we had articulated a view of hominoid evolution which, although by no means universally accepted, was widely shared by professional colleagues and was quite influential in shaping thinking. Fundamental to that view was the idea that the categories "hominids" and "pongids," represented today by the rather different humans and great apes, could be extended back into the past, into Middle Miocene times. They were, we believed, represented there respectively by *Ramapithecus*, the earliest hominid, and several species of *Dryopithecus,* an extinct genus of ape. Such forms all lived around 10 to 15 million years ago, and were found throughout the Old World. The most abundant sources of these species was India and Pakistan where the *Dryopithecus* species had previously been called *Sivapithecus*, a distinct name which Simons and I no longer believed was warranted. Another set of *Dryopithecus* species lived earlier in Africa, around 20 million years ago, and were thought to be ancestral both to the later *Dryopithecus* species and to *Ramapithecus.*

Several of the *Dryopithecus* species we saw as ancestral to living apes. Although they were represented mostly by jaws and teeth, a few bits and pieces of the skeleton were known, and we

Reprinted with permission from *Discovery* (Yale Peabody Museum of Natural History), Volume 13(1), pages 2–9. Copyright © 1978 by Yale Peabody Museum of Natural History, New Haven, CT.

believed that, reconstructed, they looked quite a lot like their supposed living descendants. The same applied to *Ramapithecus*. Actually, a good deal less was known about *Ramapithecus* than about some of the other creatures, and we only had a handful of jaws and isolated teeth. The concept of *Ramapithecus* we built up from these fragments was that of an animal quite similar to humans. It had, we believed, a parabolic dental arcade—its teeth were set in a row that was rounded at the front, gently broadening towards the rear—and in that feature resembled *Homo sapiens*. Also, it had small canines and incisors (the teeth at the front of the jaw), again like humans and in contrast to the large-fanged apes.

Furthermore we believed that *Ramapithecus* had evolved from a species of *Dryopithecus* around 15 million years ago, and had subsequently given rise to species of *Australopithecus*, the 2 million year old near-human fossil array thought to contain our direct ancestor.

The evolution of hominoids was thus seen by us as a rather simple story. At any one time there would have been, like today, relatively few species in existence. (Today there are only 5 species of hominids and pongids as opposed to dozens of monkeys, our next closes relatives.) Those fossils that did exist were thought to have been quite modern in aspect, to fall in clear ancestor-descendant lines, and to be divisible easily into two rather distinct clusters—hominids and pongids. In addition, we agreed with almost all other paleoanthropologists since the time of Charles Darwin that the prime driving force of human evolution had been the adoption of tool-use and the elaboration of symbolic, learned, "cultural" behavior, based on language. The progressive development of culture was linked in the emergent hominids to such fundamental traits as our small canines, upright walking and running abilities, dextrous hands, expanded brains, and, ultimately, prodigious mental feats. In fact, so firmly did we believe in the linkage together of tools/weapons, small canines (large ones superseded functionally by tools), and bipedalism (necessary for, and driven along by, handy tool use) that we took the small canines of *Ramapithecus* to imply very probably the existence of both tools and uprightness!

CHANGING PERSPECTIVES: THE FIRST HOMINIDS

In addition to many new finds of Middle Miocene hominoids, including that jaw from Locality 182 in Pakistan, expanded knowledge of geological, floral, and faunal contexts, a little more sophisticated thinking about the functional meaning of fossil morphology, and above all a realization of the pervasive influence of nonarticulated, implicit assumptions, have all combined to make me doubt that clearcut, logical, simple picture of human evolution. I don't claim to have now a "correct" answer. Indeed I now believe, although I didn't before, that one should not expect to make definitive statements about human evolution at the present time. But more on that later.

The new material and new thinking have generated a more complex, less clearcut picture of hominoid evolution. *Ramapithecus* still plays an important role, but it is no longer necessarily *the* star. Chronologically the story may go something like this. Around 20 million years ago, hominoids were confined to the African continent, which was cut off from the rest of the Old World by the Tethys Sea (the only remnant today of which is the Mediterranean). There the ancestors of what were to become human and ape lineages existed as a diverse array of primitive, rather monkeylike species. As judged from their skeletons and teeth they were browsers that lived a good part of their lives in the trees of the rich Miocene tropical forests, feeding on soft leaves, fruits, stems, flowers and, perhaps occasionally, insects. As

judged by numbers of species, they were more successful than the handful of present-day apes. Changes in the tropical forest brought about by declining temperatures, together with competition from the progressively more successful forest monkeys probably reduced the number of ape species, and also stimulated the evolution of modern types from the more primitive Miocene dryopithecids, as they are technically known (after an extinct French ape species, *Dryopithecus fontani*, first described in the 1850s).

We cannot specify which known dryopithecid, if any, gave rise to which modern ape. Indeed, unless the fossil record becomes many times more abundant, which is unlikely, we will probably never be able to name ancestors. Suffice it to say the dryopithecids were, as a group, ancestral to pongids.

The dryopithecids and pongids clung to the forests. But that was not true of another set of hominoids, a group I now call the ramapithecids, after their best known member. Beginning around 15 million years ago several important things happened. Africa and Eurasia became connected as Afro-Arabia drifted northwards to collide with the northern landmass; this precipitated a series of faunal migrations north and south resulting in mixing, competition, extinction, and evolution of many different groups of mammals, including new hominoids. At the same time, a cooling of earth climates began, opening up what had been forested areas and turning many previously closed habitats in the Old World into more open woodland and grassland.

The ramapithecids were the first group of monkeys or hominoids to venture into these new habitats. The ramapithecids, like other mammal groups such as bovids (cattle family), suids (pigs), elephantids, rodents, and many others, evolved adaptations of their teeth, jaws, and skeletons to living and feeding on the ground.

The ramapithecids were a diverse array of species, grouped into several genera (*Ramapithecus, Sivapithecus, Gigantopithecus*, and *Ouranopithecus* being just four of their technical Latin names) that exploited woodland habitats of Africa, Europe, and Asia around 10 million years ago. At the same time their dryopithecid cousins lived on in the Middle Miocene forests. We can now see that *Ramapithecus* was not markedly different from several of its contemporaries, and it was not uniquely similar to hominids. Instead it shared features with the other ramapithecids and is best considered part of that group, distinct from dryopithecids, as well as from pongids *and* hominids. Thus the simple dichotomy of hominoids into pongids and hominids is replaced by a more complex, more accurate, four-way split.

None of the ramapithecids is especially modern-looking. Contrary to Simons's and my original view, *Ramapithecus* itself does not have a parabolic dental arcade; we know this now from several complete specimens. And it turns out that arcade shape is not an especially important anatomical character anyway! Modern humans have parabolic arcades, living apes U-shaped ones, while dryopithecids, ramapithecids, and the *Australopithecus* group of hominids all show variations of a V-shape. Rather, *Ramapithecus* resembles other ramapithecids in what now seem more important features.

Thus, in contrast to forest dryopithecids and pongids, the woodland ramapithecids have enlarged chewing teeth (molars and premolars, or cheek teeth) with a thick coating of enamel, covering the softer dentine core of the tooth. The dryopithecids and pongids have smaller cheek teeth with thin enamel that wears quickly to expose a dentine pit at the tip of each cusp. Such an arrangement seems to facilitate the slicing and chewing of soft plant food. The ramapithecid thick-enamelled cheek teeth probably evolved for the powerful crushing and grinding of tougher kinds of vegetable foods typical of more open, nonforest settings.

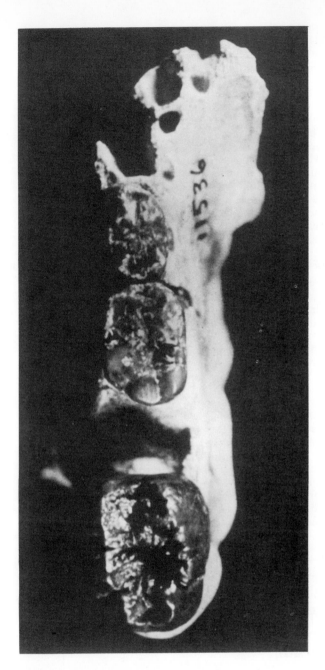

Figure 1   *An infant jaw of* Sivapithecus indicus *showing the milk molars and second permanent molar. This is the first infant* Sivapithecus *to be found and will yield much important information.*

Some of the ramapithecids had ape-sized canine teeth, others somewhat smaller canines, and a few species quite small ones. *Ramapithecus* and *Gigantopithecus* (as well as at least one other genus) apparently had quite diminutive, nonprojecting canines. However, even in those ramapithecid species with larger canines, heavy tooth wear truncates the crowns, and in fact they rarely project very much beyond the other teeth. It seems likely that both heavy tooth wear and reduced canines go together with enlarged and thick-enamelled cheek teeth as an adaptive response to chewing tough vegetable foods.

Our new finds from Pakistan comprise some of the very first skeletal fragments of ramapithecids. The medium and large forms—*Sivapithecus* and *Gigantopithecus* species—seem to have had skeletons like those of small chimpanzees and small gorillas, animals respectively around 70 and 150 pounds in weight. They were quadrupedal forms, moving on all four limbs. We know almost nothing of *Ramapithecus* except that it was probably tiny, weighing perhaps no more than 40 pounds, barely more than a large monkey. I have a hunch that when we know more about *Ramapithecus* we will find it to be part quadrupedal, part bipedal, its bipedalism made possible by small body size. In addition, I suspect that if it was a biped it was so, not because it needed to carry tools, but because it occasionally crouched or stood bipedally to feed, and that such postures gradually became used more and more in travelling and other movements.

Thus the ramapithecids, like their ancestors the dryopithecids, were a diverse array of species exploiting the newly available woodland habitats of the Old World Middle Miocene, spending a large amount of time on the ground away from the trees feeding on tough vegetable foods. All were megatoothed; all had teeth with thick enamel coatings; all were probably quadrupeds, although the smallest perhaps moved bipedally a good portion of the time. We known nothing of their brains, or of their social behavior. Probably they were like apes, both in social organization and in individual behavior, in "smartness." It seems unlikely indeed that they had cultural, symbolic behavior.

CHANGING PERSPECTIVES: LATER HOMINIDS

The ramapithecids almost disappear after about 8 million years, and are replaced in open habitats by the first true hominids, the australopithecines. These forms, found as yet only in Africa, are known from hundreds of specimens; the best material is 4 to 1 million years old. Unfortunately there is a great void in the hominoid fossil record between 8 and 4 million years, filled by only three specimens! The australopithecines gave rise to our own genus, *Homo,* around 2 million years ago. We ourselves are latecomers on the scene, arriving only 40,000 or 50,000 years ago.

The australopithecine and early *Homo* species share certain features: all were bipeds; all large-toothed, thick-enamelled forms with small canines; all had brains larger than those of apes, 50 to 100 percent larger; and some made simple stone tools and added at least some game to their diets. Whether or not any of them were

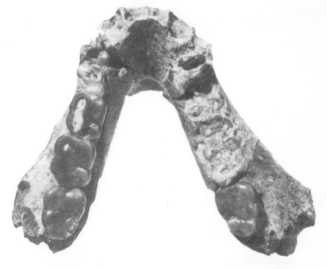

Figure 2   *The* Ramapithecus *jaw founded by Wendy Barry and Martin Pickford. This was once in four pieces.*

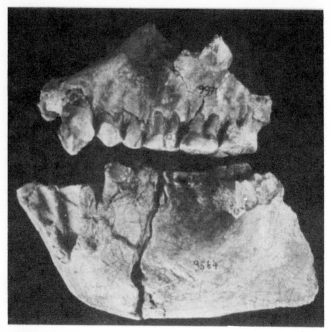

Figure 3 *Associated upper and lower jaws of* Sivapithecus indicus, *the most complete specimen of its kind found.*

capable of complex symbolic cultural behavior, mediated by language, is a hotly debated point. No evidence as yet adduced points unequivocally in that direction, and it is perhaps best at present to see them as complex-behaving animals, with much of their behavior learned, yet still falling far short of fully human behavioral achievements or capacities.

The hominids, first appearing some seven million years ago, could have evolved from the ramapithecids; alternatively they might derive from the dryopithecid-pongid cluster, in which case the ramapithecids would be an unsuccessful first attempt at open country living, with the independently derived hominid australo-

pithecines subsequently replacing them in woodland and grassland habitats. If ramapithecids are ancestral to australopithecines we still have several plausible choices for *the* ancestor. It is not necessarily *Ramapithecus,* nor necessarily any presently known form.

This is a much more complex scheme of hominoid evolution than our earlier version. The evolutionary picture resembles a bush rather than a ladder. The story is diverse, subtle, probabilistic; not simple, straightforward and inexorable. Connections through time are not clearcut. Earlier species are not pale copies of living ones, but unique creatures in their own right. Two groups are not enough to contain the diversity of extinct and extant forms. And "culture" looks less and less plausible as the driving force behind the early phases of human evolution; rather, food, feeding behavior, and habitat seem more likely evolutionary forces.

REFLECTIONS

The contrast between my earlier and current ideas is great, and has considerable implications for me. There are many lessons to be drawn from looking at the contrast, some perhaps of use in future thinking about human origins, others of a more personal nature.

First, and very important, is the replacement of "cultural" paradigms or organizing principles by those concentrating on what would be for nonhominoid mammal specialists the more normal attributes of food and feeding. This parallels the move of many paleoanthropologists towards a more general involvement in the study of mammals and their evolution, away from an obsession with the single species, *Homo sapiens.*

As a corollary, we can look at evolution now as an opportunistic process; it was not set in motion simply to end in us! Looking at the fossil record only to try to spot species ancestral to us or features that mark "key" stages on our evolutionary journey distorts our view of what really happened. Hominoid evolution was like that of many other mammal groups and not something special. Or at least, not until very late in the story.

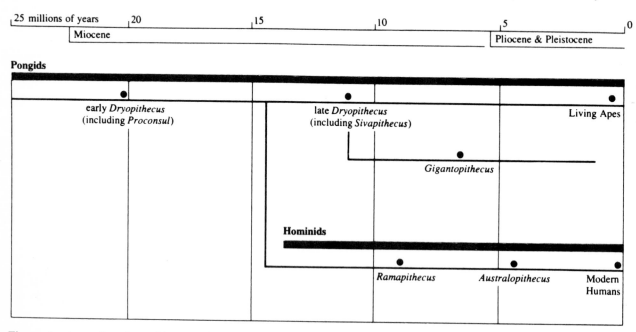

Figure 4 *An earlier view of hominoid evolution. Solid lines indicate what were considered highly probable relationships.*

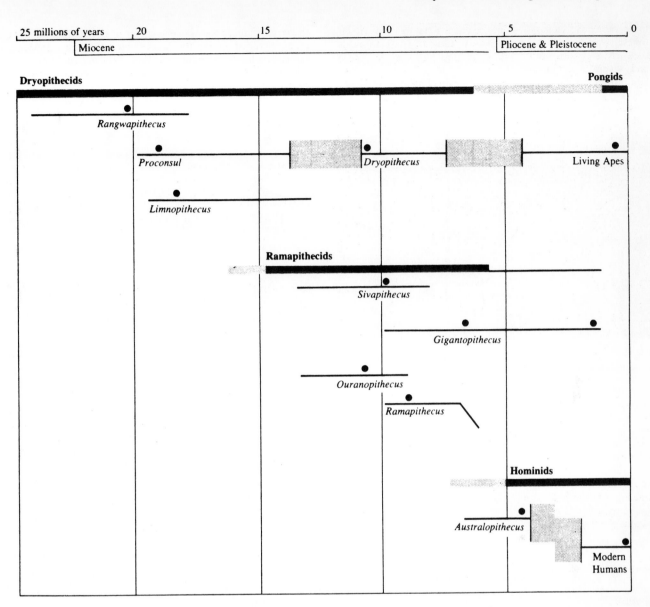

Figure 5 *A more complex view of hominid evolution. Gray areas indicate relationships that are now considered more tentative.*

Linked yet again to that point is the need to look at the past on its own terms; to try to understand past species as well-adapted forms, not as imperfect approximations to us. This would be what historians would call an historicist rather than a presentist approach. *Ramapithecus* did not evolve small canines and (perhaps) bipedal postures in order, eventually, to become human. It did so because both features were responses to a shift in habitat and diet. Their immediate significance was different from what, with the wisdom of hindsight, we can see as their retrospective significance. Incidentally, the apparently inevitable and logical interlocking of tooth, teeth, and limbs in a feedback model that can be traced back at least to Darwin may well be an illusion. Small canines and (perhaps) bipedalism are found in *Ramapithecus*—certainly in *Australopithecus*—not because they are inextricably linked through tools, but probably because they are each separate adaptations to a change in food and feeding patterns. *Gigantopithecus*, at least, had small canines, but was *not* a biped,

perhaps because it was too big. We can even conceive of bipeds with big canines.

What does this mean to me? Rather, how have I changed as a scientist so that I could postulate such a radically different scheme of human evolution? I am now aware of the existence of implicit guiding ideas. I try hard to detect them in my own thinking, to isolate those assumptions that are not articulated because they seem now so "obvious," yet will seem so silly a few years from now. I am also aware of the fact that, at least in my own subject of paleoanthropology, "theory"—heavily influenced by implicit ideas—almost always dominates "data." By that I mean that the fossils are rarely so abundant that hypotheses can be generated mainly from the data alone. Rather ideas that are totally unrelated to actual fossils have dominated theory building, which in turn strongly influences the way in which the fossils are interpreted.

Finally I'm now, I think, more philosophical about the inevitable subjective elements of my science. I try to keep myself honest, to be "objective" and "rigorous"; to state my assumptions as clearly as possible. But I realize that, until we have much more data, several plausible and different theories might fit the same data, rather than just one. I will never again cling quite so firmly to one particular evolutionary scheme. I have come to believe that many of the statements we make about the hows and whys of human evolution say as much about us, the paleoanthropologists and the larger society in which we live, as about anything that "really" happened.

30

A Late Divergence Hypothesis

L. O. GREENFIELD

One of the most interesting and unsettled topics in human evolution is the question of origins; where and when great ape and human lineages separated, initiating the course of events that led to the emergence of *Australopithecus*. At present, there is no consensus on this question. Due to varying methodological approaches (phenetic, stratophenetic, cladistic), the constantly changing data base (the reinterpretation of fossils and new discoveries), and large gaps in the fossil record, a variety of hypotheses have been proposed.

Among paleoanthropologists, two slightly different origin hypotheses have received the widest support. One of these is based on *"Ramapithecus"* (Simons, 1961c, 1964c, 1968b, 1972, 1976c, 1977; Simons and Pilbeam, 1965, 1972; Tattersall, 1975b; Conroy, 1972), a Middle Miocene hominoid considered to be the phyletic ancestor of *Australopithecus*. The other, and more recent, interpretation (Pilbeam et al., 1977a) is based on the presence of independent Middle Miocene hominoid lineages, distinguished from each other primarily by the thicknesses of their molar enamel caps.

While the respective interpretations of the available fossils differ somewhat, both hypotheses reach the same conclusions: that an independent human lineage is documentable by the Middle Miocene (16 m.y.a.); that the separation of the human lineage from the lineages leading to the extant great apes was an Early Miocene event; and that Early Miocene dryopithecines represent last common ancestors of man and the extant great apes.

While there is no available evidence which could be used to reject the *conclusions* of these early divergence hypotheses, phenetic and stratophenetic evidence also support an, at least, equally parsimonious and mutually exclusive third phylogenetic alternative. That alternative, a late divergence hypothesis, is presented here in detail. According to the late divergence hypothesis, the lineage leading to *Homo* did not become independent from the lineages leading to the extant great apes until the Middle-Late Miocene.

THE LATE DIVERGENCE HYPOTHESIS

Because of the paucity of the paleontological record, the late divergence hypothesis, like the early divergence hypotheses, begins with phylogenetic data derived from phenetic studies (comparative anatomy, ontogeny, and serology). However, beyond the shared phenetic premises (see below), the two sets of hypotheses differ. According to the late divergence hypothesis:

(1) No Middle Miocene hominoid can be singled out as the phyletic ancestor [1] of Plio/Pleistocene *Australopithecus/Homo*. The known Middle Miocene forms (including *"Ramapithecus"*) have dryopithecine-like [2] morphology in their known

parts and therefore cannot be ruled out as potential ancestors of both *Australopithecus* and extant great apes.

(2) Pliocene *Australopithecus afarensis* (Johanson and Taieb, 1976; Johanson et al., 1978; Johanson and White, 1979a; White, 1977a; Leakey et al., 1976) exhibits numerous dryopithecine characteristics in its dentition and gnathic region. These retentions suggest a post-Miocene (rather than a Middle Miocene) acquisition of most of the dental and gnathic features that might distinguish members of the human lineage from extant and fossil apes (and necessitate the adoption of new operational definitions for "ape" and "human"). The morphology of *A. afarensis* makes it more likely that a Middle Miocene forerunner of *Homo*, whether phyletically related or a common ancestor of extant great apes and *Australopithecus*, looked like a dryopithecine.

(3) The Early Miocene dryopithecines, particularly *D. (Proconsul)*, as far as is known, lack many of the shared characteristics of extant great apes and *Australopithecus*. If, as suggested by the early divergence hypothesis, the last common ancestor of extant great apes and man existed by the Early Miocene (20 m.y.a) then numerous, and unlikely, independent acqusitions of these shared characteristics must be cited in order to explain their presence. Instead of positing unlikely numerous independent acquisitions, the late divergence hypothesis holds that the "primitive" morphology of Early Miocene forms like *D. (Proconsul)*, indicates that the *Pongo, Pan, Gorilla,* and *Australopithecus/Homo* lineages had not yet diverged. Accordingly, *D.(Proconsul)* is considered to be a form which *predates*, rather than one which postdates, the separation of the human lineage from the lineages leading to the extant great apes.

While two species of *D. (Proconsul)* have been considered the phyletic ancestors of *Pan* and *Gorilla,* no special similarities to strongly suggest these phyletic relationships can be found by this author. It is primarily because of the presumed phyletic relationship between *"Ramapithecus"* and *Australopithecus* (and the collateral conclusion that the ancestors of the extant great apes were also present by the Middle Miocene) that these two *D. (Proconsul)* species have been considered the phyletic ancestors of the extant African apes.

(4) Middle-Late Miocene *Sivapithecus,* in all but a few of its known parts, represents a more likely last common ancestor of the extant great apes and man than the Early Miocene dryopithecines. If *Sivapithecus* is the last common ancestor, it would necessarily follow that some dental reversals have occurred in the extant great apes lineages.

According to the late divergence hypothesis, precise branching dates cannot be determined until undoubted phyletic ancestors of the extant great apes can be found. Present evidence suggests the earliest possible date for the separation of the *Pongo*

Reprinted with permission from *American Journal of Physical Anthropology,* Volume 52, pages 351–365. Copyright © 1980 by Alan R. Liss, Inc., Philadelphia.

lineage is about 15 m.y.a. The lineages leading to *Pan, Gorilla,* and *Homo* may have been separate from each other by 14 m.y.a. or as little as 6 m.y.a.

Both late and early divergence hypotheses are presently viable interpretations of human origins and all have strengths and weaknesses. However, it is felt that the late divergence hypothesis has not been given adequate attention by paleoanthropologists (see for example Pilbeam et al., 1977a). Thus, rather than presenting a point by point comparison of the various phylogenetic alternatives, the remainder of this paper is primarily devoted to a discussion of the phenetic premises and each of the major points of the late divergence hypothesis. It should be noted here that the late divergence hypothesis is not based upon the falsification of early divergence hypotheses. It is another hypothesis which needs to be tested; it could be used to set up a critical test between the competing hypotheses, and it almost certainly in part, or entirety, will be modified and/or rejected.

The Phenetic Evidence

The fossil record is the ultimate source of phylogenetic information. As Simpson has recently stated: "When relevant fossils are available and are well interpreted, their characteristics and their succession provide both the most direct and the most important data bearing on phylogeny." (1975:10). However, it is often the case that relevant paleontological evidence is lacking. No exception to this problem, the fossil record of early man, and especially that of the great apes, includes large gaps in time and space.

When fossils are unavailable, other sources of evidence have an especially important bearing on phylogenetic questions. Phenetic studies are directly relevant sources of phylogenetic information, and extensive comparative work on the living great apes and man has suggested overall relationships among these taxa (Fig. 1). Supportive evidence for this phylogeny derives from two levels of investigation. One is the comparative study of characters of complex inheritance: embryology, ontogeny, and gross anatomy (Schultz, 1926, 1930, 1937, 1951, 1956, 1969c, 1978; Washburn, 1950, 1963a, 1968b, 1978; Simpson, 1966b). The second level is the comparative study of serum proteins and DNA itself (Goodman, 1963, 1975, 1976a; Sarich and Wilson, 1967; Wilson and Sarich, 1969). This latter level provides the more convincing phenetic data bearing on phylogeny because genetic similarities and differences are more clearly defined (the underlying genetic bases for complex polygenic traits are unknown).

Phylogenetic conclusions derived from comparative studies of characters of complex and simple inheritance have, in general, been mutually reinforcing. Together, these phenetic studies suggest (Fig. 1) that *Pan, Gorilla,* and *Homo* are more closely related to each other (and about equally [3]) than any is to *Pongo.*

Severe tests of this phenetic phylogeny can only be based upon fossil evidence. However, because the relevant fossils necessary to set up a test are presently lacking, most paleoanthropologists, implicitly or explicitly, have considered the phenetic evidence of human/ape phylogeny (just the suggested general relationships and not the dating of the branching events) a relatively more important, or a more convincing source of phylogenetic information than the fossil record. Consequently, many workers have attempted to fit the fossil record into the phylogeny derived from phenetic studies.

A similar procedure is followed here. The less ambiguous phylogenetic data derived from phenetic studies is considered the first line of evidence [4] and paleontological materials are interpreted in light of the conclusions of pheneticists. Phenetics will continue to be of primary importance until the fossil record becomes complete enough to provide an unambiguous test of the phenetic phylogeny.

The Reinterpretation of Middle Miocene Hominoids (*Sivapithecus* and *"Ramapithecus"*)

Recent analyses of original specimens of *"Ramapithecus"* (Frayer, 1976; Greenfield, 1974, 1975, 1977–1979) have shown that the taxon is not unique; the dental and gnathic regions are like those of the contemporary dryopithecine, *Sivapithecus.* The differences between *"Ramapithecus"* and *Sivapithecus* were shown to be about as great as the differences between *Pan paniscus* and *Pan troglodytes* and certainly far less than the differences between *Pan* and *Gorilla* (which are often included in one genus, *Pan*). Consequently, *"Ramapithecus"* and *Sivapithecus* were formally synonymized (Greenfield, 1979), the latter having taxonomic priority. The hypodigm of *"Ramapithecus"* was transferred to two species of *Sivapithecus, S. brevirostris* (most of the *"R. punjabicus"* specimens) and *S. africanus* (*"R. wickeri"*). The *"Ramapithecus"* specimens were not found to be simply females of *Sivapithecus* as has often been suggested. Rather, male and female specimens were recognized in both Indian and African samples.

S. brevirostris and *S. africanus* were interpreted (Greenfield, 1979) as size variants of *Sivapithecus.* Some differences in relative proportions of front to back teeth, among *Sivapithecus* species, were also noted. These may be the result of allometric relationships between the various types of teeth and body mass. Dental and gnathic allometry in fossil and recent hominoids is now being investigated by this author to determine to what extent allometry affects functional, taxonomic, and phylogenetic analyses. Preliminary results indicate that the major differences between *S. brevirostris, S. sivalensis,* and *S. indicus* are related to scaling at different body sizes. For example, the interspecific positive canine allometry and cheek tooth isometry found by this author and others (Wood, 1979; Gingerich, 1977a; Kay, 1975) suggests that, all other factors being equal, smaller ape species (*S. brevirostris, Pan paniscus*) should have smaller canines relative to cheek tooth size than larger ape species (*S. sivalensis, S. indicus; Pan troglodytes*). This supports the congeneric status suggested (Greenfield, 1979) for the Middle Miocene forms because the differences among them appear to be merely the result of scaling that yields "functional equivalence at varying sizes" (Gould, 1975).

Although not evaluated systematically, other Eurasian materials (Pasalar *"Ramapithecus," "Ouranopithecus," "Bodvapithecus," "Rudapithecus," "Ankarapithecus,"* the Candir mandible, *"D. keiyuanensis,"* and *"Graecopithecus"*) were considered referrable to *Sivapithecus* because of their similar morphological patterns.

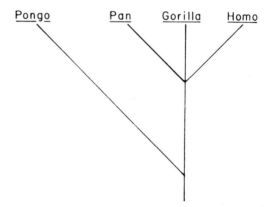

Figure 1 *A phylogeny of the extant great apes and man, based on phenetic evidence (comparative anatomy, ontogeny, serology), is pictured. These general relationships suggested by phenetics represent one of the premises of the late divergence hypothesis.*

Sivapithecus was defined (Greenfield, 1979) as a Middle to Late Miocene hominoid genus characterized by

highly heteromorphic maxillary incisors. Incisors comparable in size to those of *Dryopithecus*, relative to the cheek teeth. Slightly more vertically projecting I¹, and premaxillary prognathism greater than in *D. (Proconsul)*. Canines sexually dimorphic, but reduced relative to the cheek teeth as compared to *D. (Proconsul), Pan, Pongo,* and *Gorilla. Sivapithecus* retains C/P_3 interlock and associated shear facets, diastemata, and sectorial P_3. Unworn molars of *Sivapithecus* generally lack well developed cingula and other enamel surface complications. Molar enamel thick as in *Australopithecus.* Compared to *D. (Proconsul), Sivapithecus* has a slightly more robust mandibular corpus under the molars, more heavily buttressed symphysis, and deeper palate. *Sivapithecus* zygomatic arches are inserted more anteriorly and higher on the maxilla, and show greater lateral flare than in *D. (Proconsul).* A large masticatory apparatus relative to body size is indicated (tentatively) for *Sivapithecus.* The lower facia! region is narrow, abbreviated, and deep. Dental arcades vary from V-shaped to nearly parallel patterns.

The synonymizing of *"Ramapithecus"* and *Sivapithecus* represented more than just a complex taxonomic exercise. As a separate genus, *"Ramapithecus"* was perceived to be uniquely well on the road toward human dental form, so much in fact that it could no longer be considered a potential ancestor of extant great apes. That is, it exhibited a suite of derived, human-like characteristics which were considered to be irreversible. From this interpretation, it was proposed that the human lineage, represented by *"Ramapithecus,"* was independent from the lineages leading to the extant great apes by the Middle Miocene and that contemporary or earlier dryopithecines must be the phyletic ancestors of extant great apes. Because of the "primitive" morphology of Early Miocene forms like *D. (Proconsul)*, this interpretation of *"Ramapithecus"* turned into the paleontological argument which supported the "prebrachiationist" concept of human origins, rather than the "brachiation-knuckle walking" theory which is based largely upon phenetic studies (Washburn, 1590, 1963, 1978).

In contrast, inclusion of *"Ramapithecus"* in the dryopithecine genus *Sivapithecus* reflects the notion that it, as well as other species of *Sivapithecus,* predate the significant and unique events of hominization [5], and possess a morphology which makes them potential ancestors of *both* extant great apes and *Australopithecus.* According to this interpretation of *"Ramapithecus"* and *Sivapithecus,* there is no necessary expectation that *phyletic* ancestors of each of the extant great apes and *Australopithecus* existed by the beginning of the Middle Miocene, or if they did exist, that they were clearly distinguishable.

That *Sivapithecus* (unless otherwise stated, hereafter this genus also refers to all junior synonyms mentioned above) is not characterized, in its known parts, by a suite of features that could be considered both uniquely human and irreversible does not mean that it cannot be the phyletic ancestor of *Australopithecus.* Surprisingly, it is the absence of these characteristics which makes *Sivapithecus* a possible ancestor of *Australopithecus;* either a phyletic ancestor (i.e., an early member of an independent human lineage) or a last common ancestor of extant great apes and man. This is because comparisons among dryopithecines, extant great apes, and *A. afarensis,* the earliest undoubted member of the lineage leading to *Homo,* show that many unique and likely irreversible human dental adaptations did not evolve until after the Miocene, and that many dryopithecine-like characteristics persisted in the human lineage in the Pliocene. Ironically, most of

the "human characteristics traditionally used to support *"Ramapithecus"* as a forbear of *Australopithecus* are dryopithecine-like in *A. afarensis.* An excellent example of this is the morphology of the dental arcades and tooth rows (Greenfield, 1978). Given the morphology of *A. afarensis,* it is quite possible that a Middle Miocene forerunner of man, whether on an independent human lineage or a last common ancestor of extant great apes and man, was a dryopithecine and exhibited few, if any, of the characteristics that have traditionally been considered uniquely human and irreversible. Regardless of what the actual phylogenetic events were, present paleontological evidence suggests that the antiquity of many unique human dental traits appears to have been grossly exaggerated.

DRYOPITHECINE CHARACTERISTICS OF *A. AFARENSIS*

A comparison of Pliocene *A. afarensis* with geologically younger *Australopithecus* (and early *Homo*) and Middle Miocene *Sivapithecus* (Table 1) suggests that many dryopithecine characteristics in the dentition and gnathic region were retained in the lineage leading to *Homo* until post-Miocene times. *A. afarensis* is clearly a transitional form and closes, to a considerable extent, the morphological gap between the earliest undoubted member of the human lineage (now *A. afarensis*) and Middle Miocene dryopithecines. In this light, separate species status for the Hadar and Laetolil specimens (Johanson et al., 1978) seems appropriate. The morphology of *A. afarensis* also suggests that the traditional notion, that the earliest members of the human lineage were species with clear-cut human dental and gnathic traits (like a feminized male canine, nonhoning C/P_3 complex, parabolic dental arcades), may be false.

From a phylogenetic standpoint, the significance of the dental and gnathic morphology of *A. afarensis* is not yet clear. Johanson and White (1979a) suggested the possibility of a later divergence of human and great ape lineages than that estimated by the early divergence hypotheses. Another possibility is that an early divergence occurred, but members of the respective lineages were indistinguishable. In any case, the new evidence suggests it is far more likely that a Middle Miocene forerunner of *Homo,* whether a member of an independent human lineage or a common ancestor of extant great apes and *Homo,* was dryopithecine-like and predated the significant events of hominization.

Not all of the dental and gnathic characteristics of *A. afarensis* are dryopithecine-like or transitional. One outstanding feature which is unlikely to be found in an ancestor of *Pan, Pongo,* or *Gorilla* is the absence of two well defined canine morphs. The loss of canine sexual dimorphism, through the feminization of the male canine, is potentially one of the most important markers for direct ancestors of *A. afarensis.* Because of the known Middle Miocene dryopithecines *(Sivapithecus)* possess two canine morphs and a fully ape-like C/P_3 complex, it may be hypothesized that 1) a Middle Miocene phyletic ancestor is unknown; or 2) that it is known but is a dryopithecine and indistinguishable from contemporary species; or 3) that a phyletic ancestor did not exist until the Late Miocene.

THE MORPHOLOGY OF *D. (PROCONSUL)*

The extant great apes (Asian and African) and Plio/Pleistocene *Australopithecus/Homo* have many features in common. It is far more plausible that these shared anatomical structures (dental, gnathic, postcranial, ontogenetic) were derived from a last common ancestor rather than having been independently acquired in each of the lineages.

According to the early divergence hypotheses, lineage separation had already occurred by the Early Miocene (20 m.y.a.), at

Table 1 *Comparison between* Sivapithecus, A. afarensis, *and Later* Australopithecus *and Early* Homo

	Sivapithecus*	A. afarensis	Later Australopithecus and Early Homo
Incisors			
a. Relative size[1] $(\dfrac{I^1}{M^1 + M^2 \text{ area}}) \times 100$	$\overline{x} = 29$ OR 26-32 (N = 3)	$\overline{x} = 26$ — (N = 1)	$\overline{x} = 18$ OR 12-21 (N = 5)
b. Angle of I^1 projection	$\overline{x} = 57°$ OR 55-60° (N = 5)	$\overline{x} = 60°$ — (N =1)	$\overline{x} = 72°$ OR 69-77° (N = 4)
C/P_3 complex			
a. Relative size[1] $(\dfrac{C1 \text{ area}}{M1 \text{ area}}) \times 100$	Maxilla: males $\overline{x} = 104$ OR 81-131 (N = 6) Maxilla: females $\overline{x} = 68$ OR 57-76 (N = 3)	Maxilla: combined sexes $\overline{x} = 67$ OR 61-77 (N = 5)	Maxilla: combined sexes $\overline{x} = 48$ OR 33-73 (N = 14)
	Mandible: males $\overline{x} = 86$ OR 56-111 (N = 7) Mandible: females $x = 49$ OR 48-50 (N = 2)	Mandible: combined sexes $\overline{x} = 53$ OR 45-68 (N = 3)	Mandible: combined sexes $\overline{x} = 41$ OR 24-63 (N = 22)
b. Canine projection	Far beyond cheek tooth occlusal plane in males, slight projection in females	Both sexes, slight projection, reduction of crown tip to cheek tooth occlusal plane delayed	Both sexes, crown tip reduced to cheek tooth occlusal plane rapidly compared to A. afarensis
c. Canine interlock and wear	Commonly with P_3 and other canine, occasionally with I^2, shear, blunting in some old adults	Early interlock with P_3 and other canine, eventual loss of interlock with reduction of canine's height, shear, and blunting	Rare interlock, little or no shear, tip blunted

*Pooled data; see Greenfield (1979) for separate species data and measurement definitions.
[1]These are ambiguous indices. The relation of these measures to body mass is now being investigated by this author.

least the *Pongo* lineage and probably the others as well. Among the known Early Miocene hominoids postdating the proposed early divergence is *D. (Proconsul)*, a taxon often cited as containing the phyletic ancestors of *Pan* and *Gorilla*. *D. (Proconsul)* is the best known Early Miocene hominoid and is represented by a completely known dentition, two partial skulls, and a variety of postcrania.

The pertinent question, which has been discussed by several workers beginning with Le Gros Clark and Leakey (1951) is whether the known features of *D. (Proconsul)* suggest it to be a form which likely postdates the hypothetical last common ancestor of *Pongo, Pan, Gorilla,* and *Homo*—i.e., if it is a form which possessed the numerous shared characteristics of extant great apes and man.

Close examination (see below) of the known features of *D. (Proconsul)* shows that this Early Miocene form lacked many of the shared characteristcs. This is interpreted, within the context of the late divergence hypothesis, as an indication that the separation of the human lineage from the lineages leading to the extant great apes had not occurred by the beginning of the Middle Miocene (16 m.y.a.). Consequently, if *D. (Proconsul)* is an ancestor of extant great apes and man it is unlikely that it is (or post-dates) the *last common ancestor*. Some of the evidence which suggests that *D. (Proconsul)* predates the last common ancestor is presented below.

Dentition of *D. (Proconsul)*

It is difficult to find dental characteristics shared by extant great apes and *Australopithecus* that are not found in *D. (Proconsul)*. However, despite their widely varying dietary regimes and associated diverse dental adaptations, two dental characteristics shared by extant great apes and man cannot be found in *D. (Proconsul)*. The extant great apes and man possess relatively large first molars (expressed as a ratio of M_1 to M_2 areas in Table 2) and molars with an expanded phase II component. *D. (Proconsul)* retains the primitively small M_1 (like *Aegyptopithecus*) and molars with relatively short phase II (small trigone). In addition, the P^3 of *D. (Proconsul)* is more caniniform than might be expected (the buccal cusp, especially in male specimens, projects far beyond the P^4–M^3 chewing plane) in a hypothetical last common ancestor of extant great apes and *Australopithecus*.

Gnathic and Upper Facial Regions

The lower and upper facial regions of *D. (Proconsul)* (Fig. 2) are characterized by a suite of features which differ from the shared facial features of extant great apes and *Australopithecus*. In *D. (Proconsul)*, the subnasal region is short (the distance between the inferior margin of the nasal aperture and the alveoli of the maxillary incisors) and the maxillary incisors project at a very acute angle (Greenfield, 1977–1979). At the inferior margin of the nasal

Table 1 *(continued)*   *Comparison between* Sivapithecus, A. afarensis, *and later* Australopithecus *and early* Homo

	Sivapithecus*	A. afarensis	Later Australopithecus and Early Homo
d. Diastemata (crown)	Variable in size, between I^2 and C^1 and C_1 and P_3 (slight)	Variable in size between I^2 and C^1	Rarely seen—except in some early Homo
e. Canine dimorphism	Morphological and metric, two sizes and two morphs	Primarily metric, two sizes	Primarily metric, two sizes?
f. P_3 morphology and wear	Sectorial with occasional accessory metaconid, very elongate and oval, primarily shear and eventual blunting	Sectorial with major and minor cusps, elongate and less oval, early shear but followed rapidly by blunting	Multi-cusped, round blunting only
g. P_3 orientation	30–45° orientation with tooth row	45–60° orientation with tooth row	Not angled in most specimens
P_3/P_4 heteromorphy	Extreme	Intermediate-extreme	Slight-none
P^3 morphology	Triangular, buccal face much broader than lingual, mesial edge concave	Triangular or rectangular, buccal face broader than lingual, some mesial edge concavity	Rectangular
Tooth Rows	Straight, arcades V-shaped or parallel, general shape long and narrow	Primarily straight, some curve medially, V-shaped or elongate parabola, general shape long and narrow	Most curved medially, elongate parabola or parabola, some square anteriorly, general shape shorter and broader than A. afarensis
Zygomatic insertion (most anterior fibers of masseter)	Primarily M^1/M^2	Primarily M^1 and M^1/M^2	Primarily P^4/M^1, M^1, and M^1/M^2
DM_1 morphology	Sectorial, honed and premolar-like	Molarized premolar, some honing wear, raised trigonid	Molariform

*Pooled data; see Greenfield (1979) for separate species data and measurement definitions.
[1]These are ambiguous indices. The relation of these measures to body mass is now being investigated by this author.

aperture where the two premaxillae meet, their inner convex margins fold together to form a seam-like groove (as in cercopithecoids) which extends down the midline between the roots of

Table 2   *Size of M_1 Relative to M_2*

	N	x	OR	SD
Aegyptopithecus[1]	2	72.5	71–74	—
D. (Proconsul)[2]	12	70.2	58–80	6.1
Sivapithecus[2]	12	80.2	70–85	4.7
G. bilaspurensis[3]	1	79.0	—	—
A. afarensis[2]	2	79.0	78–80	—
later Australopithecus[4]	25	83.5	69–108	9.3
H. erectus[5]	17	96.1	84–115	7.8
Gorilla[2]	20	80.0	73–87	4.5
Pongo[2]	20	84.8	76–100	5.9
Pan troglodytes[2]	20	98.2	84–115	9.2

Sources of data: [1]Simons, 1965a; [2]data collected by this author; [3]Simons and Chopra, 1969; [4]Frayer, 1973; [5]Wolpoff, 1971a.

*(M_1 area/M_2 area) × 100, samples of D. (Proconsul) and Sivapithecus include data from several species *(D. africanus, D. nyanzae,* and *D. major; S. brevirostris, S. sivalensis,* and *S. indicus).* No significant differences were found between the pooled taxa in each group.

An index reflecting the size of M_1 relative to M_2 was calculated for a variety of extant and fossil hominoids. D. (Proconsul), like *Aegyptopithecus,* exhibits a small M_1 relative to M_2. Pleistocene and Recent taxa listed share an enlarged M_1 like *Sivapithecus,* their proposed last common ancestor.

the central incisors (Le Gros Clark and Leakey 1951). The zygomatic arch of *D. (Proconsul)* inserts low on the face, not far above the occlusal plane of the cheek teeth, and the palate is flat and very shallow (Greenfield, 1977, 1979). The orbits of *D. (Proconsul)* have a rectangular shape (Table 3 and Fig. 2) and their longer axes (orbital breadth) are oriented somewhat obliquely to the face.

In contrast, the extant great apes and *Australopithecus* possess long subnasal regions and more vertically projecting maxillary incisors (Greenfield, 1977, 1979). Their inferior nasal margins lack a groove or midline seam and are broad and guttered. Zygomatic arches insert high on the face, far above the level of the cheek tooth occlusal plane and all possess deep and arched palates. The orbits of extant great apes and *Australopithecus* are circular or squared (Table 3), not rectangular in shape, and their orientation is fully forward.

The Postcranial Skeleton

Interpretations of the locomotor behavior of *D. (Proconsul)* from the well known postcranial skeleton (Napier and Davis, 1959) and other isolated finds (Le Gros Clark and Leakey, 1951) have been anything but mutually reinforcing. Napier and Davis concluded that *D. (Proconsul)* was more like quadrupedal cercopithecoids than like any of the extant hominoids. Subsequent works (Lewis, 1971a, 1972a,b, 1974; Conroy and Fleagle, 1972, Preuschoft, 1973; Schön and Ziemer, 1973; Zwell and Conroy, 1973; Morbeck, 1975; O'Connor, 1976) have variably portrayed *D. (Proconsul)* as a brachiator, a part-time or potential knuckle walker, a quadrupedal form resembling some of the New World monkeys, and reiterating Napier and Davis' original conclusion, as a quadrupedal form resembling cercopithecoids.

These contradictory interpretations, in part, reflect the complex problems inherent in analyses of locomotor form and function. The subadult status of the most complete *D. (Proconsul)* specimen (KNM RU 2036) and incomplete documentation of postcranial characteristics, distinguishing extant hominoids from extant cercopithecoids, have also contributed to the problem. However, regardless of the functional interpretation, there clearly are postcranial characteristics shared by extant great apes and man which are not found in *D. (Proconsul)*.

While there has been debate over the morphology and functional implications of the *D. (Proconsul)* wrist, radius, and ulna, and general agreement that the distal humerus is very similar to that of extant hominoids, few have doubted the presence of two cercopithecoid-like (or "primitive") features, a retroflexed humerus and presence of an os centrale. The absence, or reduction of humeral retroflection among extant hominoids has been associated with habitual suspensory postures and it is more likely that their last common ancestor also lacked humeral retroflection. Similarly, fusion of the os centrale with the os scaphoid occurs early in ontogeny in *Pan, Gorilla,* and *Homo* (primarily in infancy in *Pan* and *Gorilla* and during the third intrauterine month in humans (Schultz, 1969c) and it is more likely that fusion occurred early in ontogeny in their last common ancestor. Again, *D. (Proconsul)* is more primitive than the likely hypothetical last common ancestor; specimen KNM RU 2036, with erupting M3s and C_1s already erupted (subadult), still lacks fusion. Because fusion occurs only in old adult *Pongo* specimens, it is not yet possible to tell, with respect to this characteristic, whether *D. (Proconsul)* had reached a developmental stage likely to have been present in the last common ancestor of *Pongo* and African apes and man.

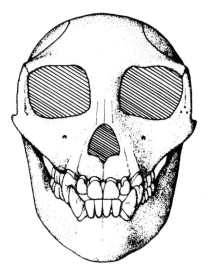

Figure 2   *This frontal view of the skull of* D. (Proconsul) africanus, *taken from Robinson (1952), illustrates some of the "primitive" characteristics of the taxon. Note the short subnasal region, the seam-like groove at the base of the nasal aperture, and the rectangular shape of the orbits.* D. (P.) africanus *lacks many of the shared dental, cranial, postcranial, and ontogenetic characteristics of the extant great apes and* Australopithecus. *This suggests that if* D. (Proconsul) *lies within their ancestry, it predates the branching of the lineages.*

Table 3   *Orbit Shape*

	N	x	OR	SD
D. (Proconsul)	1	80	—	—
Pan troglodytes	15	108	98–122	7.8
Pongo	7	116	111–123	3.8
Gorilla	15	112	96–125	8.0
Australopithecus[1]	6	101	94–109	3.2

[1]from casts

An index of orbit shape was calculated for the taxa listed by dividing inner orbital height by inner orbital breadth and multiplying this ratio by 100. The extant great apes and *Australopithecus* all possess round or squared orbits, while the one known *D. (Proconsul)* orbit is rectangular with orbital breadth being the long axis.

Problematical Characteristics

Additional shared features of extant great apes and *Australopithecus* have been found lacking or different in *D. (Proconsul)*, but these are more difficult to assess in terms of their relevance to phylogenetic problems. A few, however, are worth noting here.

Unlike extant great apes and *Australopithecus*, the frontal of *D. (P.) africanus* is inflated and lacks a supraorbital torus. This, however, may simply be due to the smaller body size of *D. (P.) africanus* and/or the sex of the specimen in question (a female, judging from the size and morphology of the canines). Similarly, the cranial capacity of 150 cc, estimated from the size of the one preserved foramen magnum (Radinsky, 1974), is less than might be expected for a last common ancestor. In order to resolve the problems inherent in comparing a small bodied animal like *D. (P.) africanus* with larger bodied extant great apes and man, Gingerich (1977a) utilized Jerison's encephalization quotient (EQ) and found the EQ of *D. (P.) africanus* was below average for living higher primates. However, parallel trends of increasing brain size/body size are not uncommon. Consequently, the utility of this criterion in suggesting branching events is diminished by the greater likelihood of EQs increasing independently in each of the lineages leading to the extant great apes and man.

Lastly, while Lewis has published on the wrist of *D. (P.) africanus* (1974) and concluded that many aspects are hominoid-like, a recent reanalysis (O'Connor, 1976), in which the differences between cercopithecoids and hominoids were clarified, has shown that *D. (Proconsul)* does not possess the wrist articular relationships shared by extant great apes and man. Instead, according to O'Connor, *D. (Proconsul)* exhibits cercopithecoid-like articular surfaces at the distal ulna, proximal pisiform, distal fourth and fifth metacarpals, and hamate. The dorsal surface of the radius was also found to be cercopithecoid-like. It is, however, the contradictory conclusions of the respective analyzers which makes the application of this and other postcranial evidence to questions of phylogeny problematical. If O'Connor's analysis is correct, this would provide additional evidence that *D. (Proconsul)* had not reached the structural grade expected in a last common ancestor of great apes and man.

Regardless of the assessment of these more problematical characteristics, there still remains the other "primitive" dental, gnathic, postcranial, and ontogenetic features of *D. (Proconsul)* discussed above. These suggest, that in order to accept *D. (Proconsul)* as a phyletic ancestor of extant African apes, or as a last common ancestor of all extant apes and man, numerous and unlikely independent acquisitions of shared characteristics must be cited. While independent acquisition is possible, it is far more likely that there was a last common ancestor that exhibited these shared features. Thus, phenetics and parsimony suggest that Early Miocene *D. (Proconsul)* predates the branching of the lineages. Below it is argued that the Middle Miocene dryopithecines, in particular *Sivapithecus,* come closer, in almost all of their known parts, to the morphology expected in a last common ancestor of extant great apes and man.

D. (Proconsul) is the only Early Miocene hominoid analyzed here in the context of early or late divergence hypotheses. This is because its anatomy is well known, and consequently it it possible to make numerous comparisons of its different anatomical systems with those of extant great apes and man. The late divergence hypothesis merely rejects the notion that *this* Early Miocene subgenus represents (or post-dates) the last common ancestor of extant great apes and man. The possibility that *D. (Proconsul)* may not be in the ancestry of later dryopithecines was not considered. Other Early Miocene dryopithecines like *D. (Rangwapithecus)* (Andrews, 1974), in some features, more closely resemble extant great apes and man and Middle Miocene hominoids than does *D. (Proconsul).* Thus, *D. (Rangwapithecus),* when it becomes better known (cranially and postcranially), could also be examined as a potential ancestor, or even a last common ancestor, of extant great apes and man.

SIVAPITHECUS AS A LAST COMMON ANCESTOR

Far less is known about the anatomy of *Sivapithecus* than is known about *D. (Proconsul).* The dentition and gnathic region of the Middle Miocene genus, however, are well known as a result of recent collecting (de Bonis et. al., 1974, 1975; de Bonis and Melentis, 1977b; Kretzoi, 1975; Andrews and Tekkaya, 1976; Andrews and Tobien, 1977; Pilbeam et. al., 1977a). In these preserved parts, *Sivapithecus* exhibits the characteristics which are shared by extant great apes and *Australopithecus* and which are absent in *D. (Proconsul).* Like the extant great apes and *Australopithecus, Sivapithecus* has large M1 relative to M2, molars with expanded trigones (lengthened phase II aspect of the chewing stroke), a long subnasal region, more vertically projecting maxillary incisors, a less caniniform P³, a broad and guttered inferior nasal margin lacking a midline seam, zygomatic arches which insert high on the face far above the level of the cheek tooth occlusal plane, and a deep and arched palate [6] (Greenfield, 1977, 1979). In addition, Morbeck (1979) has recently discussed the postcranial fragments of *"Rudapithecus"* and *"Bodvapithecus"* recovered by Kretzoi (1975). Her analysis of articular surfaces of the humero-ulnar, humero-radial, and proximal radio-ulner joints indicates that "at least one and perhaps both of the larger primates possessed morphological features and movement capabilities similar to that of extant Hominoidea." These features suggest that *Sivapithecus,* a genus which makes its first appearance just after the beginning of the Middle Miocene (Andrews and Tobien, 1977) is a better candidate for last common ancestor of extant great apes and man than *D. (Proconsul).*

The phylogenetic implications of the phenetic and paleontological arguments presented above are schematically represented in Figure 3. *D. (Proconsul), D. (Dryopithecus),* and *Sivapithecus* are shown as phyletically related (see Greenfield, 1979) and *Sivapithecus* as the last common ancestor of extant great apes and man.

If a late divergence did occur as suggested above, it necessarily follows that the present paleontological record is largely inadequate with respect to the determination of precise dates of important branching events. Late Miocene and Pliocene fossils, crucial to the identification of independent lineages leading respectively to each of the extant great apes and man, are rare, and the known Middle Miocene *Sivapithecus* species are not sufficiently distinct enough from each other to suggest, with much confidence, whether or not specific lineages can be identified. In addition to gaps in the fossil record, there is evidence of exchange

between Eurasian and African mammal faunas between 8 and 4 m.y.a. (Pilbeam, pers. comm.). This places less restriction on the potential locations of the significant branching events and increases the number of possible phylogenetic scenarios. Without additional evidence from the Late Miocene, the only statement that can presently be made with any confidence is that the lineage leading to *Homo* can be identified from the Pliocene to Recent; the separation of the human lineage from the lineages leading to the extant African apes could not have taken place later than about 6 m.y.a.

However, one feature, the cingulum, and its observed distribution among *Sivapithecus* species and the extant great apes and *Australopithecus,* might suggest when the *Pongo* lineage became independent. Within *Sivapithecus,* a widely distributed genus in time and space, the cingulum is a variable characteristic. Early Middle Miocene species from Pasalar and East Africa (Andrews and Tobien, 1977; Andrews and Walker, 1976) have much higher frequencies of cingulum than late Middle Miocene and early Late Miocene species (cingula are virtually absent on specimens from the Nagri of India and Pakistan) [7]. Among the extant great apes and *Australopithecus, Pan, Gorilla,* (Skaryd, 1971) and *Australopithecus* (Table 4) exhibit low-moderate frequencies of cingulum (with highly variable expression), while in *Pongo,* it is virtually absent. Because it is more likely that the last common ancestor of *Pan, Gorilla,* and *Australopithecus* also possessed moderate frequencies of cingulum, the temporal distribution of cingula among *Sivapithecus* species suggests that by the beginning of the Late Miocene (10 m.y.a.) there were taxa that were both likely phyletic ancestors of *Pongo* and unlikely ancestors of *Pan, Gorilla,* and *Australopithecus*—i.e., the *Pongo* lineage may have been independent from the lineage leading to *Pan, Gorilla,* and *Australopithecus.* This scenario, however, rests on the assumption that frequencies of cingula have decreased in all lineages since the Middle Miocene.

If *Sivapithecus* is assumed to be the last common ancestor of

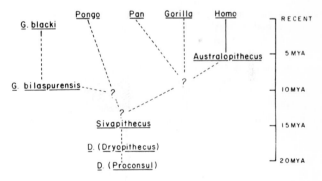

Figure 3 *This late divergence phylogeny is based on phenetic and paleontological evidence. Precise branching dates for the lineages leading respectively to the extant great apes and* Australopithecus *are unknown and therefore shown as question marks. If* Sivapithecus *was the last common ancestor of extant great apes and man,* maximum *dates of divergence are 15 m.y.a. for the* Pongo *lineage and at least a million years later for the lineages leading to* Pan, Gorilla, *and* Australopithecus. *One criterion, the temporal distribution of the cingulum in* Sivapithecus *(see text), suggests that the* Pongo *lineage was separate by 10 m.y.a.* Gigantopithecus *and* Pongo *are both thought to be descendants of Siwalik* Sivapithecus *because of the shared absence of molar cingula and numerous similarities in molar morphology. The last common ancestor of* Pan, Gorilla, *and* Australopithecus *probably had moderate frequencies of molar cingula.*

Table 4 *Distribution of the Cingulum in* A. afarensis

	Clear % (N)	Trace % (N)	None %(N)
M^1	0 (0)	50 (4)	50 (4)
M^2	17 (1)	17 (1)	66 (4)
M^3	0 (0)	0 (0)	100 (4)
M_1	0 (0)	31 (4)	69 (9)
M_2	0 (0)	24 (5)	76 (16)
M_3	0 (0)	18 (2)	82 (9)

extant great apes and man, it is possible to outline some of the important predicted dental evolutionary trends and to examine one of the major weaknesses of the late divergence hypothesis.

In decreasing order, *Sivapithecus* most resembles (dentition and gnathic region): *Australopithecus, Pongo, Pan/Gorilla* (about equally). Dental and gnathic resemblances between *Australopithecus* (especially *A. afarensis*) and *Sivapithecus* are numerous, and to date, few have questioned the likelihood of a relationship, phyletic or otherwise. The major differences between them are in the C/P_3 complex. In *A. afarensis*, the C_1 appears to have partially taken on an incisor's function. Some C_1s possess an anterior flange which has contributed to an incisor-like occlusal surface and P_3 was variably incorporated into the cheek tooth field, tending to be more molarized in larger specimens. In *Sivapithecus*, the canines and P_3 were part of the C/P_3 complex and their functions were not subsumed in either incisor or cheek tooth fields.

Sivapithecus and *Pongo* are almost as similar to each other. The major trends would have been some decrease in enamel thickness, a possible decrease in molar size/body mass proportions, an increase in the sizes of the anterior teeth relative to the cheek teeth, which may, in fact, be simply another way of expressing the decrease in molar size/body mass proportions, and the development of molar enamel wrinkling. If *Pan* and *Gorilla* are descendants of *Sivapithecus*, the major trends would have been a marked reduction in enamel thickness and a decrease in cheek tooth size/body mass proportions. In addition to these predicted dental trends there also would have been an increase in body mass in most of the proposed descendants.

The molar enamel thickness trends in the lineages leading respectively to each of the extant great apes, predicted by the late divergence hypothesis, could be interpreted as its major weakness. The comparatively thin molar enamel and lower cheek tooth size/body mass proportions shared by extant great apes are more likely to have been acquired from a last common ancestor than by parallel evolution. In these respects, the extant great apes (especially *Pan* and *Gorilla*) seem to be more similar to *D. (Proconsul)* than they are to *Sivapithecus*, most species of which are characterized by thicker molar enamel and (tentatively [8]) larger cheek teeth relative to body mass.

Pilbeam et. al. (1977a) considered thick and thin molar enamel important markers for ancestors of *Australopithecus* and extant great apes, respectively [9]. This early divergence hypothesis would be falsified, however, if a phyletic relationship between *Pongo* and *Sivapithecus* could be demonstrated. It is therefore important to note here some of the reasons why a relationship between these two genera is a viable possibility.

This is not the first time a phyletic relationship between *Pongo* and *Sivapithecus* has been proposed. More than half a century ago, Gregory and Hellman (1926) made detailed comparisons of their dental morphologies and concluded that *Sivapithecus* was a credible Miocene candidate for the ancestor of *Pongo*. In addition to the numerous similarities they found (see Fig. 4),

recent investigations have shown that the molar enamel of *Pongo* is thicker than that of *Pan* or *Gorilla* (Molnar and Gantt, 1977), that "*Pongo* molars are extremely large given its adult body size" (Kay and Hylander, 1978), and that *Pongo* zygomatic arches have a strong lateral flare (author's unpublished data). These resemblances to *Sivapithecus*, and those found by Gregory and Hellman, are sufficient enough to support the possibility of a phyletic relationship and to cast doubt on the early divergence phylogeny which has been based largely on a single and highly variable dental trait (molar enamel thickness).

CONCLUSIONS

The issue of human origins is far from resolved. As the preceding analysis has suggested, paleontological and phenetic evidence may be used to support a late divergence hypothesis in addition to the traditional early divergence hypotheses of paleoanthropologists.

The late divergence hypothesis was stated in general terms since, at present, this is all the available evidence will support. Its major points can be briefly summarized. First, there is yet no clear cut Middle Miocene candidate for phyletic ancestor of *Australopithecus*. Second, Middle Miocene forms *(Sivapithecus)* have dryopithecine-like morphology and cannot be ruled out as potential common ancestors of extant great apes *and Australopithecus*. Third, a Middle Miocene forerunner of Pliocene *A. afarensis*, whether phyletically related or a last common ancestor of extant great apes and *Australopithecus*, probably looked like a dryopithecine and exhibited few, if any, of the dental and gnathic characteristics that have been considered uniquely human and irreversible. Fourth, *D. (Proconsul)*, lacking many of the shared characteristics (dental, gnathic, cranial, postcranial, ontogenetic) of extant great apes and *Australopithecus*, probably predates the branching of the lineages. Fifth, Middle Miocene *Sivapithecus* is more likely last common ancestor of extant great apes and man than *D. (Proconsul)*. And sixth, if a late divergence did occur, then more precise branching dates cannot be determined until undoubted phyletic ancestors of the extant great apes are found or recognized.

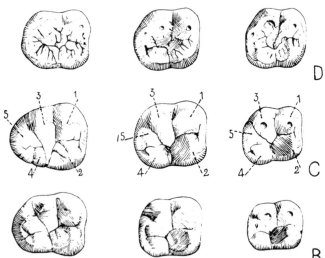

Figure 4 *Pictured are the* M_1, M_2, *and* M_3 *(right to left) of a modern orang (D) and two Siwalik* Sivapithecus *specimens, AMNH 19413 (C) and AMNH 19412 (B). These line drawings were taken from Gregory and Hellman (1926) who noted the numerous similarities in their dental morphologies and consequently supported a phyletic relationship between the two genera.*

The late divergence hypothesis is testable. It makes precise predictions about the course of dental evolutionary events in the extant great ape lineages (mutually exclusive predictions of the early divergence hypotheses represent part of the set of potential falsifiers). The late divergence hypothesis predicts that phyletic ancestors of *Pongo, Pan,* and *Gorilla,* will, with increasing antiquity, look more like *Sivapithecus* than *D. (Proconsul).* In particular, phyletic ancestors will, with increasing antiquity, exhibit thicker molar enamel and probably greater cheek tooth size/body size proportions.

Given that a relationship between *Sivapithecus* and *Australopithecus* is almost certain and that the phenetic evidence bearing on the phylogeny of the extant great apes and man is correct, the most relevant fossils to be collected in the future are those pertaining to the lineage leading to *Pongo.* If a phyletic relationship between *Pongo* and *Sivapithecus* can be firmly established then the late divergence hypothesis will not have been falsified by what could be considered a severe test. A failure to falsify will also mean that it is highly likely that *Pan* and *Gorilla* are also derived from *Sivapithecus* (this follows from the above assumptions) and that the early divergence hypotheses are incorrect. It will, however, be much more difficult to determine precise branching dates. One roadblock is the paucity of Late Miocene sites. Another may be that the lineages did not become distinctive enough to be identified with their respective descendants until long after the actual branching occurred.

ACKNOWLEDGMENTS

The work reported here was supported, in part, by a grant from the Horace H. Rackham School of Graduate Studies, The University of Michigan.

I gratefully acknowledge Mr. R.E.F. Leakey of the National Museums of Kenya, Mr. M.V.A. Sastry of the Geological Survey of India, Calcutta, Dr. D.R. Pilbeam of Yale Peabody Museum, Dr. E. L. Simons of Duke University, and Dr. D.C. Johanson of the Cleveland Museum of Natural History for permitting me to examine and measure fossil material in their care. I also thank Ms. Pat Helwig of the Cleveland Museum of Natural History and Dr. R.G. van Gelder of the American Museum of Natural History for the opportunity to examine and measure extant materials relevant to the present topic.

I further thank Dr. M.H. Wolpoff, who originally encouraged me to undertake the study of human origins.

Ms. Muriel Kirkpatrick drew the line drawing and did the photographic work and I thank her for her time and labor. Special thanks are due Ms. Patricia Russo, the rhetorician of this paper.

Notes

[1] The term, phyletic ancestor, denotes a species pertaining to an independent line of descent, as for example, an ancestor of *Australopithecus* which is *not* an ancestor of *Pan, Pongo,* or *Gorilla.*

[2] The term dryopithecine refers to the fossil ape genera, *Dryopithecus* and *Sivapithecus,* or their characteristics: small body size, nonbipedal, narrow faced, ape-like C/P$_3$ complex, V-shaped to parallel dental arcades, small incisors relative to the cheek teeth. The term, human lineage, refers to all species of the lineage leading to *Homo* that are not ancestors of *Pan* or *Gorilla.*

[3] There actually is no clear cut phenetic evidence suggesting whether (or which) two of the three might be more closely related.

[4] The author does not consider the protein clock a source for the dating of branching events. Because of the numerous questionable assumptions in the method, only the general relationships suggested by phenetics are assumed here.

[5] Loss of canine sexual dimorphism, adoption of bipedal locomotion, tool manufacture, delayed maturation, enlargement of the cerebral cortex, use of home bases, etc.

[6] Insertion of the zygomatic arch tends to be higher, and palates deeper, in geologically younger *Sivapithecus* species.

[7] Cingula or traces of them can be found on East African specimens FT 48, BMNH 16649, and two isolated molars, FT 34 and FT 40, which may belong to the hypodigm of *S. africanus.*

[8] Based upon tooth size/body size proportions suggested by provisional association (Pilbeam et al., 1977a) of postcrania with *Sivapithecus* species from Pakistan.

[9] Enamel thickness does not distribute Miocene hominoids neatly into two sets. Rather, it is a continuous variable. The enamel caps of Potwar Plateau specimens were reported (Pilbeam et al., 1977a) to vary in thickness between 2.5 and 3.0 mm, while those of Pasalar *"Ramapithecus"* and *Sivapithecus* varied between 1.1–1.5 mm and 1.7–2.5 mm, respectively (Andrews and Tobien, 1977). The enamel caps of *"Rudapithecus"* and *"Bodvapithecus,"* although not measured precisely, are probably thinner than those of the Potwar Pleateu specimens because they exhibit earlier onset of enamel perforations. All of the above specimens are considered to be part of the "thick enamel" group.

31

The Nut-Crackers—
A New Theory of the Adaptations
of the Ramapithecinae

R. F. KAY

Miocene Ramapithecinae (including *Ramapithecus, Sivapithecus,* and *Gigantopithecus)* have long figured prominently as a group which may be in the direct line of human ancestry (Simons, 1961c, 1972). This study provides new information about their way of life. Given the fragmentary nature of the ramapithecine fossil record most workers have projected the adaptive pattern of australopithecines, for which there is ample fossil evidence, backward to ramapithecines. Australopithecine terrestriality and feeding adaptations have been linked to a single, coevolving adaptive pattern. If this pattern is projected backward, it follows that ramapithecines, with a broadly australopithecine-like masticatory apparatus, must also have been terrestrial foragers. I hope to show in this paper, however, that ramapithecine feeding adaptations were not necessarily evolved in a terrestrial setting. It is proposed that species in this subfamily ate forest fruits with hard, tough rinds that require tremendous forces to open but, once opened, provide a rich source of nutrients previously available to only a few other mammals like peccaries and rodents. "Nut-cracking" epitomizes this adaptive strategy. Because a similar adaptive strategy is followed today by arboreal and terrestrial species, a determination of ramapithecine terrestriality awaits better postcranial evidence.

THEORIES OF RAMAPITHECINE ADAPTATIONS

Given the absence of informative postcranial material (Pilbeam, 1979, Pilbeam et al., 1977a), the evidence of ramapithecine adaptation is based first, on the vertebrate death assemblages and other evidence interred with the ramapithecine fossils and second, on the evidence of the dentition, face, and jaws.

Ramapithecine Paleoecology

Studies of the paleoecological setting in which ramapithecine fossils occur do not clearly indicate a preferred habitat. Middle Miocene African ramapithecine-bearing faunas from about 14-15MY ago have woodland affinities (Andrews and Evans, 1979) but we do not know whether ramapithecines lived also in rain forests. Similarly, these animals inhabited rain forests in the Indo-Pakistan region about 12 MY ago and are found in woodland forests about 9 MY (Pilbeam et al., 1977b), but we do not have evidence about their presence or absence in Indo-Pakistan woodland faunas of 12 MY or the rain forests of 9 MY. In any case, data about habitat preference per se will not indicate precise ecological or adaptational information; that is, what ramapithecines ate or whether or not they were developing terrestrial adaptations (unless, of course, it can be demonstrated that this group inhabited a steppe-type country where there are no trees).

Dental Evidence

Among other characteristics, ramapithecines had forward shifted jaw-closing muscles; shallow, transversely thickened mandibular corpora in the molar region; small lower incisors relative to molars; reduced canines (in some species); and heavily worn molars with low crown-relief and thick molar enamel (Greenfield, 1979; Simons, 1964c, 1976c; Simons and Pilbeam, 1972). This suite of dental and facial characteristics has served as the basis for a number of theories of ramapithecine diet and foraging pattern, the two most completely articulated of which are the seed-eating theory of Jolly (1970b) and the meat-eating, scavenging theory of Szalay (1972d). Each of these theories argues that the ramapithecine masticatory apparatus was adapted to generate and withstand comparatively great compressive forces on the teeth or to facilitate extensive and frequent food preparation. In this view, the ramapithecine jaw musculature was favorably situated to produce large molar biting forces and the heavily buttressed jaws and thick-enameled molars are well organized to withstand the stress of heavy mastication. Molar crowns with low relief, which flatten when worn, are supposed to be designed for crushing and milling. Canine reduction is thought to facilitate the free movement of the molars during grinding.

Jolly suggested that the masticatory apparatus of ramapithecines and australopithecines suited them for a diet which centered on cereal grains (grass seeds). In his view, the whole dental and facial structure is designed like a flour mill, where small hard spherical grains are pulverized by a grinding action. Although Jolly's theory may have a kernal of truth, there are several objections. First, although the theory would account for the massive jaws of these species, it cannot account for the characteristic molar pattern. Typically, grass-seed eating mammals, including the Old World monkey *Theropithecus,* have flat molar crowns made up of a series of cutting edges formed by complexly infolded enamel. This presumably allows thin-shelled cereal grains and grass blades to be milled efficiently. A number of rodent groups have evolved such a system convergently. By contrast, as Jolly recognized, ramapithecines and australopithecines have flat, relatively featureless molars on which shearing is deemphasized (see below) unlike mammalian grass-seed eaters. To account for this difference, Jolly argued that 1) *Theropithecus* and other grass-seed-eating mammals generally also eat grass blades, 2) it is the grass-blade constituent in the diet which is the agent selecting for a shearing design, and therefore, 3) early hominids must have eaten grass seeds but not grass blades. This argument is difficult to accept because it is hard to imagine how an early hominid could have separated successfully grass seeds from fibrous grass stems and leaves. Besides, even if this were possible, cereal grains would

Reprinted with permission from *American Journal of Physical Anthropology,* Volume 55, pages 141–151. Copyright © 1981 by Alan R. Liss, Inc., Philadelphia.

still require milling. The mechanical requirements of breaking apart thin-shelled grass hulls are an equally likely selective agent for the shearing designs of the teeth of graminivorous mammals. A second objection relates to the wear pattern of the teeth of ramapithecines. Species that eat grain ordinarily incorporate grit, seed hulls, and grass blades in their diet as well. Grasses produce a recognizable pattern of fine striations on the wear facets of mammalian teeth (Walker et al., 1978), as does fine grit (Covert and Kay, 1980). The wear surfaces on the molars of *Sivapithecus* which I have examined do not show such a striated pattern (Covert and Kay, 1980).

Third, no known seed-eating mammals reach the size of some ramapithecines. It seems doubtful that a species as large as the *Gorilla*-sized ramapithecine *Gigantopithecus bilaspurensis* could find enough grass seed the year around to survive on a diet of this sort.

Finally, Kinzey (1974) has demonstrated that a number of the features identified by Jolly as grain-feeding adaptations are seen in arboreal primates such as *Callicebus* and *Cebus apella,* which eat hard seeds and nuts, but not grain. Thus the similarities between the *Theropithecus* and early hominids were probably not due to both being grain eaters.

Szalay (1972d) has proposed a meat-eating, scavenging theory to explain the origins of ramapithecine facial and dental adaptations. As did Jolly, he notes that the characteristically thick enamel and robustly constructed jaws of early hominids might indicate heavy mastication. These adaptations he felt were most consistent with a meat-chewing or bone-cracking adaptation. However, the postcanine teeth of ramapithecines do not resemble those of meat-scavenging mammals. Modern scavengers like hyenas and certain dasyurids have pointed post-canine teeth for producing great force concentrations on large food objects. By contrast, flat occlusal surfaces would tend to distribute and disperse rather than concentrate bite forces on large objects, the reverse of the desired effect in bone munching. Flat crowned teeth are effective stress concentrators only on objects small enough to be rolled between the teeth.

Jolly and Szalay give contrasting reasons to explain the possible small size of the incisors of ramapithecines. According to Jolly, the presence of small incisors simply implies that these teeth were little used for food acquisition. Szalay suggests that the incisors were used for nipping off bites of food. He sees the smaller overall size of the incisors as a favorable situation for generating great forces per unit area. The work of Hylander (1975) on relative incisor size in extant Old World anthropoids favors Jolly's view that small incisors may imply less incisal preparation during ingestion. Whatever the case, patterns of incisor size and morphology among anthropoids tend to be better indicators of food object size than of dietary preference (Kay and Hylander, 1978).

Leopold and Ardrey (1972) argued that human ancestors were meat eaters but gave different reasons than did Szalay. They proposed that toxic substances commonly found in many plant foods would tend to exclude them from hominid diets until the invention of cooking. However, anthropoids are principally herbivorous and must have evolved detoxification mechanisms other than cooking for dealing with plant poisons (e.g., McCann, 1928).

Leopold and Ardrey's proposal stimulated Coursey (1973) to argue that early hominids may have eaten roots and tubers, which generally contain very few toxic compounds. If roots and tubers were routinely eaten, one would expect to observe heavily striated wear surfaces on ramapithecine molars from the incorporation of considerable amounts of grit in the diet. As noted above, this sort of wear is not observed.

Recently, Walker (1979) has proposed tentatively that aus-

tralopithecines may have been fruit eaters because their enamel microwear patterns are similar to those of the extant frugivorous anthropoids *Pan* and *Mandrillus*. Given that the gracile-jawed and thin-enameled early Miocene hominoids were probably frugivores as well (Kay, 1977b), this would imply either that thick molar enamel and mandibular strengthening evolved in ramapithecines without an accompanying dietary shift or that the term "frugivory," as used by Walker, embraces too much ground and inadequately characterizes the dietary shifts.

All the proposed feeding strategies discussed above (with the exception of that of Walker) have the common feature that a crucial element in the ramapithecine diet was found on the ground. Each would imply that the ramapithecines, although not necessarily bipedal, had already taken a critical step in that direction by being terrestrial. If the objections to these theories are valid, this would shed doubt on the assumption of ramapithecine terrestriality.

An additional piece of evidence for ramapithecine terrestriality that has yet to be examined critically concerns molar enamel thickness. Simons and Pilbeam note that both ramapithecines and australopithecines have very thick molar enamel (Simons and Pilbeam, 1972; Simons, 1976c). They conclude that because australopithecines were terrestrial (judging from their limb anatomy) and had thick enamel, perhaps ramapithecines also had a ground-feeding adaptive mode. In an extension of this argument Smith and Pilbeam (1980) suggest that as the arboreal orangutan (*Pongo*) has very thick enamel, it may have passed through a terrestrial stage in its evolution. The alternative possibility (as will be outlined below) is that thick enamel per se has nothing to do with terrestriality.

In summary, none of the existing ecological theories adequately accounts for the observed facial and dental morphology of ramapithecines. It appears unlikely that the group subsisted largely by eating cereal grains, roots, or scavenging carrion. Nor is there any persuasive evidence bearing on ramapithecine locomotion, apart from the argument that thick enamel implies terrestrial habits. If the available material is to yield ecological information it will be necessary to take a more detailed look at the adaptive significance of occlusal morphology and enamel thickness of ramapithecines in relation to those of Old World monkeys and apes.

COMPARISONS OF MOLAR ENAMEL THICKNESS

If the thick enamel of ramapithecines is to be taken as evidence that they were terrestrial foragers, then by analogy we should expect to find as a rule that extant ground-foraging anthropoids have relatively thicker enamel than their more arboreal close relatives.

The only published studies of enamel thickness among primates known to me are those of Molnar and Gantt (1977; Gantt, 1977, 1979), who used sectioned tooth material from eighteen extant nonhuman primates (12 Old World monkeys, four hominoids and two New World monkeys). Because of the difficulties of obtaining material for sectioning, they used very small samples. In order to gain much larger samples from a greater range of species, I measured enamel thickness on the slope of the M_2 oblique cristid, proximal to the hypoconid on museum specimens in which the tooth cusps had been naturally worn to the extent that the dentin is exposed (Fig. 1). Specimens were used only when the sum of the maximum mesiodistal dimensions of the windows in the enamel of the M_2 metaconid and hypoconid was between 5% and 15% of the mesiodistal dimension of M_2. This tends to eliminate specimens that have heavily worn crown slopes. It also tends to reduce the effect of interstitial wear on M_2 length.

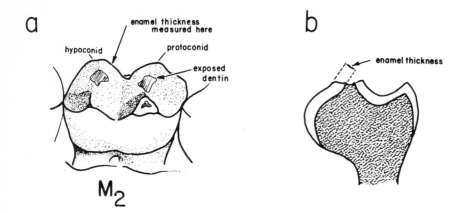

Figure 1 **(a)** *Dorsal-medial view of the left* M_2 *of* Cercopithecus aethiops *illustrating the measurement of enamel thickness used in this study. The tooth is rotated so that the line of sight is parallel to the slope of the oblique cristid at the point where the thickness is to be measured.* **(b)** *A schematic cross section of the tooth in 1A showing the enamel thickness dimension. The section is approximately in the coronal plane.*

Enamel thickness and tooth length measurements for 37 extant species are provided in Table 1. Enamel thickness was adjusted to account for differences in molar size by the following procedure. A regression equation was calculated by the least-squares method to express the relationship between the logarithm of mean M_2 mesiodistal length (independent variable) for each species and the logarithm of mean enamel thickness (dependent variable) for that species.

The regression equation relating mean enamel thickness (E_m) to mean M_2 length (M) is:

$$Ln\ (E_m) = -2.45 + 0.99\ Ln\ (M)$$

with a Pearson's r of 0.88 ($P < 0.00001$), the slope is 0.99, indicating that the mean enamel thickness changes roughly isometrically as a function of the mean M_2 length. From this equation can be derived the "expected" enamel thickness for any species with known M_2 length. The relative enamel thickness (E_r) is a measure of deviation from the expected estimated by the equation

$$E_r = \frac{100\ (E_m - E_p)}{E_p}$$

where E_m is the measured enamel thickness and E_p is the predicted enamel thickness from the regression equation mentioned above. My measure of buccal enamel thickness is very highly correlated with that of Gantt's (1979) "Buccal enamel width," which was measured on sectioned material. When Gantt's data are adjusted to express "relative thickness" in a fashion similar to that just outlined, the ranks of the relative enamel thickness between the 11 species common to the two studies (four apes and seven Old World monkeys) are significantly correlated using Kendall's coefficient of rank correlation ($P < 0.01$).

Comparison of E_r values of closely related cercopithecids (Table 1) shows no tendency for arboreal species to have relatively thinner, or terrestrial species relatively thicker, molar enamel (Fig. 2). Relative enamel thickness (E_r) ranges from –12 to +19 among species of the tribe Cercopithecini. *Erythrocebus*, the most terrestrial species (Hall, 1965), has thin enamel (–6.7), whereas *Cercopithecus aethiops*, which frequently feed on the ground (Struhsaker, 1967; Dunbar and Dunbar, 1974), has roughly average enamel thickness (+4.1). Conversely, predominantly arboreal species range from moderate to very thick enamel [compare *Cercopithecus mitis* (+4.9) with *Cercopithecus denti* (+19.0)]. A similar pattern of enamel thickness is seen among Papionini, in which species with thin molar enamel inhabit both arboreal and terrestrial niches and vice versa. A particularly revealing case is seen in species of *Cercocebus*. *Cercocebus albigena* is an arboreal species (Haddow et al., 1947; Waser, 1977), whereas *C. torquatus* fre-

quently feeds on the ground (Booth, 1956; Malbrant and Macclatchy, 1949); but the two do not differ significantly in enamel thickness (+37.4 compared with +38.6).

Just as with cercopithecids, there is no obvious relationship between habitat preference and enamel thickness among hominoids (Table 1, Fig. 2). Arboreal feeders (Chivers, 1972; Mackinnon, 1971, 1974, 1977) include species with relatively thick enamel (*Pongo pygmaeus*) and thin enamel *(Symphalangus syndactylus)*. *Gorilla gorilla,* the most terrestrial of all extant apes (Schaller, 1963), has the thinnest enamel of any ape (Table 1, Fig. 2).

Clearly, relative enamel thickness gives no indication of whether an animal forages primarily on the ground or in the trees. The hypothesis that enamel thickness in ramapithecines can tell us anything about whether or not they were terrestrial can be rejected. But are there other features of the feeding behavior of extant anthropoids that might explain why some species have very thick enamel? An apparent solution is that species with very thick enamel eat very hard foods.

Among extant primates, extremely thick enamel occurs in mangabeys *(Cercocebus),* which have enamel thickness coefficients of +37 and +39, and among orangutans *(Pongo pygmaeus),* with a coefficient of +34. Each of these thick enameled species eats very hard nuts, seeds, and fruits that other arboreal monkeys and apes cannot eat or cannot digest. The diet of *Cercocebus albigena* consists largely of fruit and berries (Haddow, 1952, Haddow et al., 1947; Waser, 1977). Its powerful specialized jaws allow it to consume tough fruits like palm nuts, which most other forest monkeys cannot handle. Haddow (1952) notes that the smaller redtail monkey swallows small hard seeds, which pass through the gut apparently undamaged, and avoids eating larger hard kernels and seeds. Mangabeys, however, crush and swallow the woody kernels of many of the same fruits. Kingdon (1971) reports that *C. albigena* teeth are frequently very heavily worn as a result of nut-crushing activities. Waser (personal communication) says that *Cercocebus* frequently can be located in the forest by the sounds it makes when crushing nuts.

MacKinnon (1977) reports that the orangutan's great size and strength enables it to feed on a variety of hard fruits and also very large fruits that gibbons are unable to tackle. In the area of his study, such fruits accounted for 35% of all fruit eaten by orangs. When *Pongo* feeds on acorns, loud crushing sounds can be heard well over 100 yards away. The hard stones of some green fruits are occasionally rolled between the teeth with the rind being rasped off, and the stone discarded or swallowed.

Of all the primate species examined, *Cebus apella*, the black-capped capuchin of South America, stands out as having by far the thickest enamel. A sample of five *C. apella* yield a mean enamel thickness coefficient of +118.9, with one individual having

Table 1   *Old World Anthropoidea: Second Lower Molar Enamel Thickness and Shearing Crest Development*

Species	N_e	E_m (S.E.)	M_2 length (S.E.)	E_r (%)	N_s	S_r
1. *Cercopithecus cephus*	10	0.54 (0.04)	5.73 (0.10)	−10.4	3	−15.9
2. *Cercopithecus aethiops*	25	0.58 (0.02)	6.53 (0.10)	+ 4.1	30	−12.3
3. *Cercopithecus denti*	10	0.57 (0.02)	5.61 (0.09)	+19.0	10	−18.2
4. *Cercopithecus neglectus*	10	0.52 (0.03)	6.60 (0.15)	− 7.7	26	−13.6
5. *Cercopithecus talapoin*	8	0.39 (0.01)	3.82 (0.05)	+19.3	23	−14.5
6. *Cercopithecus mitis*	13	0.58 (0.02)	6.48 (0.11)	+ 4.9	16	−13.1
7. *Cercopithecus lhoesti*	11	0.50 (0.04)	6.64 (0.14)	−11.7	18	− 5.4
8. *Cercopithecus ascanius*	29	0.58 (0.02)	5.89 (0.06)	+15.3	57	−17.3
9. *Erythrocebus patas*	7	0.60 (0.04)	7.54 (0.14)	− 6.7	16	− 6.5
10. *Cercocebus albigena*	44	0.80 (0.01)	7.34 (0.06)	+ 37.4	47	−20.1
11. *Cercocebus torquatus*	10	0.98 (0.04)	8.30 (0.18)	+38.6	35	−20.5
12. *Papio* species	41	1.19 (0.03)	12.47 (0.13)	+12.3	58	−11.6
13. *Mandrillus sphinx*	8	1.05 (0.05)	12.15 (0.22)	+ 1.7	9	−13.2
14. *Theropithecus gelada*	10	1.06 (0.03)	13.11 (0.24)	− 4.8	9	− 4.4
15. *Macaca speciosa*	7	0.84 (0.04)	9.40 (0.25)	+ 5.0	8	−10.5
16. *Macaca fuscata*	5	0.78 (0.03)	8.59 (0.12)	+ 6.6	4	−10.2
17. *Macaca sylvana*	6	0.90 (0.05)	9.07 (0.29)	+16.5	16	− 6.8
18. *Macaca nemestrina*	16	0.80 (0.02)	8.51 (0.13)	+10.3	15	−13.0
19. *Macaca fascicularis*	15	0.73 (0.04)	7.48 (0.11)	+15.7	30	−13.1
20. *Presbytis melalophos*	35	0.49 (0.01)	5.80 (0.06)	− 1.1	19	− 4.6
21. *Presbytis senex*	7	0.47 (0.08)	6.39 (0.11)	−13.7	20	+ 2.4
22. *Presbytis pileatus*	10	0.55 (0.03)	6.95 (0.08)	− 7.2	4	− 0.2
23. *Presbytis obscurus*	10	0.37 (0.02)	6.07 (0.08)	−20.8	16	− 0.9
24. *Presbytis frontatus*	10	0.52 (0.02)	5.66 (0.05)	+ 7.6	8	− 3.8
25. *Simias concolor*	10	0.49 (0.02)	7.05 (0.11)	−18.5	5	+ 0.5
26. *Nasalis larvatus*	10	0.58 (0.03)	8.02 (0.15)	−15.1	5	− 0.4
27. *Colobus verus*	8	0.34 (0.04)	5.40 (0.06)	−26.1	12	+ 5.9
28. *Colobus angolensis*	10	0.57 (0.03)	7.53 (0.11)	−11.2	34	− 0.4
29. *Pan troglodytes*	16	0.95 (0.05)	11.53 (0.17)	− 3.1	10	−27.3
30. *Gorilla gorilla*	23	1.14 (0.04)	17.53 (0.23)	−23.4	14	−19.5
31. *Pongo pygmaeus*	10	1.49 (0.05)	13.11 (0.40)	+33.8	5	−21.8
32. *Symphalangus syndactylus*	7	0.61 (0.06)	8.86 (0.14)	−19.2	6	−20.3
33. *Hylobates moloch*	24	0.52 (0.02)	6.62 (0.11)	− 7.9	5	−25.4
34. *Hylobates lar*	10	0.51 (0.03)	6.29 (0.13)	− 5.0	5	−23.1
35. *Hylobates concolor*	11	0.49 (0.03)	6.86 (0.14)	−16.3	—	—
36. *Hylobates klossi*	10	0.53 (0.02)	5.84 (0.08)	+ 6.3	5	−26.3
37. *Hylobates hoolock*	13	0.70 (0.04)	7.62 (0.07)	+ 7.8	5	−22.5

Columns from left to right: species identification number; N_e, sample size for enamel thickness estimates; E_m (S.E.), enamel thickness in millimeters with standard error of mean; M_2 length of sample on which the enamel thickness was measured with standard error of mean; E_r (%), relative enamel thickness expressed as a percentage; N_s, Sample size for shearing crest estimates; S_r, relative shearing crest development on M_2. Because estimates of E_r require worn specimens, whereas S_r requires unworn specimens, two samples of $N = 519$ and $N = 608$ were used. Eight shearing crests were measured to derive the total shearing estimate for M_2. These are named and defined by Kay (1977b).

an E_r of +150.4. The average E_r is 1.50 times thicker than that in *Pongo*. The high E_r of *C. apella* is particularly revealing because this species is perhaps the most behaviorly specialized in opening and eating the contents of palm nuts. The coats of these nuts are so hard that for a human to open them requires a hammer or heavy stone, but *Cebus* monkeys open them with their teeth and by pounding them against trees (Struhsaker and Leland, 1977; Izawa, 1979). These nuts in their fully ripe state are not eaten by other sympatric primate species (Izawa, 1979).

Other mammals that have comparatively thick molar enamel apparently also feed on soft foods covered with very hard rinds. For example, the sea otter, *Enhydra lutris,* which feeds on hard-shelled mollusks and crabs (Ewer, 1973) has lower-cusped "puffy" molars with thicker enamel than the fish-eating river otter *Lutra* (Ewer, 1973). Peccaries of the genus *Tayassu,* which have low-crowned teeth, are capable of cracking open hard nuts which constitute an important component of their diet. Kiltie reports that some of these nuts require sustained loads of more than 1,000 kg before cracking (Kiltie, 1979; Janzen and Higgens, 1979).

In summary, thick enamel is routinely seen among species that eat very hard foods and reaches its extreme development in the most behaviorly specialized hard nut eaters. In contrast, thick enamel has no correlation with a terrestrial habitus.

MOLAR SHEARING DEVELOPMENT IN OLD WORLD ANTHROPOIDS

Another feature of ramapithecine anatomy which might yield useful evidence about their dietary adaptations is the degree of development of the shearing crests on the molars in relation to molar length or enamel thickness. Recent studies have shown that within various taxonomic groups, the greater the proportion of structural carbohydrate a species eats, the longer will be its molar shearing crests in proportion to molar length (Sheine and Key, 1977). Thus fruit-eating species tend to have relatively shorter shearing crests than do leaf-eating species, because fruits generally contain less structural carbohydrate than do leaves.

In an earlier paper (Kay, 1977b) I developed a model to

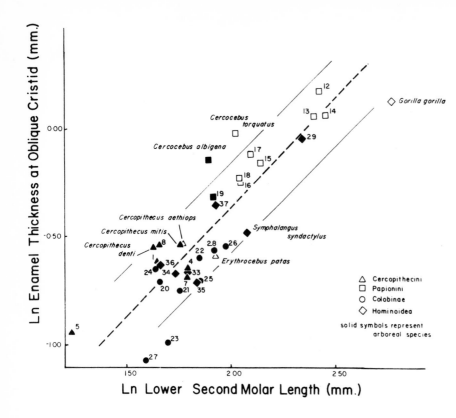

Figure 2   *Plot of Ln enamel thickness (in mm) against Ln second molar length (in mm). Each symbol represents a species. The species numbers are keyed to Table 1. Species mentioned in the text are named. The dashed line represents the regression equation relating enamel thickness to M_2 length (independent variable). The solid lines parallel to the solid line are fitted to the ±20% isolines.*

discriminate fruit-eating from leaf-eating hominoids on the basis of M_2 crest development relative to molar length. A regression equation was calculated to express the summed length of eight shearing crests as a function of the lower second molar length for the eight frugivorous hominoids: $S_E = 2.46 (M_2$ Length$)^{0.95}$, where S_E is the expected summed shearing blade length. From this equation, the relative amount of shearing development (SQ, or shearing quotient) may be expressed by the equation

$$SQ = 100 \left(\frac{S_O - S_E}{S_E} \right)$$

where S_O is the observed summed M_2 shearing blade length for a species. As expected from the way this equation is formulated, frugivorous hominoids have SQs clustering around zero (mean = +0.23, SD =3.40). Two extant folivorous hominoids *Gorilla gorilla* and *Hylobates (Symphalangus) syndactylus* have mean SQs of 7.06 and 9.04, respectively, exceeding the range based on a regression for leaf-eating hominoids yield similar clusterings. Thus it should be possible to derive some indication as to the dietary preferences of ramapithecines from a look at the amount of shearing on their molars relative to molar length.

To determine the degree of development of the cutting or shearing crests on M_2, a regression equation was developed between the natural logarithm of the mean of the summed lengths of eight M_2 shearing crests and the natural logarithm of mean M_2 mesiodistal dimension. From this equation, observed and expected shearing development are compared in a fashion analogous to that described above for enamel thickness to yield a value for relative shearing crest development. For extant colobines, cercopithecines, and hominoids, the slopes of these equations are virtually identical but the intercept values differ. As the slopes of the equations are similar, a calculation based on hominoids or cercopithecines yields virtually identical results. Relative shearing

crest development (S_r) in this case was expressed with relation to colobines alone (Table 1).

Additional insight into dental function may be gained from an examination of the relationship between molar shearing and enamel thickness. Relative enamel thickness (E_r) is highly negatively correlated with the relative shearing crest development (S_r) on the molars of the species studied (Fig. 3). Among 28 cercopithecids, the correlation between relative enamel thickness and relative shearing crest development is –0.85 (P < 0.00001). In general, species with thinner enamel have better developed shearing blades. A particularly striking example of this is illustrated by the very limited overlap of the ranges of cercopithecines and colobines (Fig. 3). The former characteristically have thick enamel and poorly developed shearing crests. The latter have thin enamel and well-developed shearing crests. This general trend is apparent also within the subfamilies of Old World monkeys. Among cercopithecines the correlation between relative enamel thickness and relative shearing crest development is –0.76 ($P <$ 0.0001); among colobines the correlation is –0.70 ($P <$ 0.002). Given the limited number of taxa that can be sampled, the situation is less clear-cut among hominoids (Fig. 3). If *Pongo* is removed from the sample, the correlation between these factors is significant ($P <$ 0.001); with *Pongo* included, the correlation between the factors is not significantly different from zero. This expresses dramatically that the relatively great enamel thickness of *Pongo* is not accompanied by reduced shearing potential by comparison with other hominoids.

The association of well-developed molar shearing with thin molar enamel has an apparent functional basis. Thin enamel is rapidly perforated by dental wear. Relatively soft, more rapidly wearing dentin appears in the windows in the enamel at the apices of the tall, sharp principal cusps. The enamel windows spread rapidly along the principal enamel crests with continued wear. The raised edge of the worn enamel on the crest margins forms a

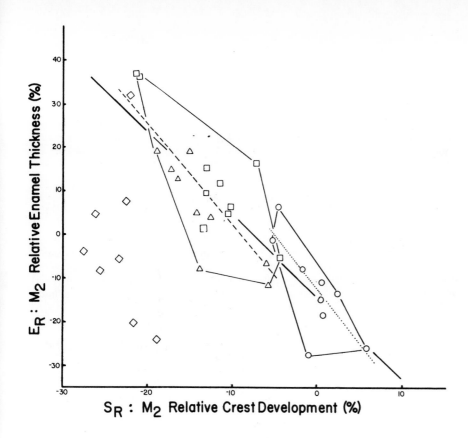

Figure 3   *Plot of relative enamel thickness against relative shearing crest development on M_2. Symbols are the same as Figure 2. Lines represent the regression equations derived for the colobine sample (dotted line), the cercopithecine sample (dashed line), and the combined cercopithecid sample (solid line). The polygons enclose cercopithecines and colobines. Pongo is represented by the uppermost diamond at the left of the graph.*

sharp edge, which enhances the shredding and slicing functions of the crests during mastication.

A molar system emphasizing high cusps (Kay, 1978) and thin enamel is associated with a leaf-eating diet. In such a diet, maximal food shredding is at a premium because leaves contain large quantities of structural carbohydrates, the digestion of which is greatly facilitated by reduced food particle size (Sheine, 1979a,b). High cusps with thin enamel would be expected to be particularly sensitive to heavy dental wear. However, heavy wear is not generally encountered among leaf-eaters, because the leaves of most plants other than grasses are easily shredded and contain very few opaline phytoliths (Rovner, 1971).

Species with poorly developed molar shearing crests have relatively thicker enamel than those with well-developed crests. This configuration produces relatively low crown-relief, particularly in advanced stages of wear. Enamel is perforated at the apices of the low cusps after considerable wear. By the time the apical enamel windows have expanded along the principal shearing crests, the molar basins are worn relatively flat. This sort of dental structure may be an adaptive response to selection for the more uniform distribution of very high occlusal forces engendered while masticating hard, tough food objects. Sharp, delicate cusps would tend to crack or splinter under such use. Relatively flat crowns may also facilitate crushing of food by the post-canine tooth surfaces. High crown-relief would interfere with the grinding action between cusps and basins. Furthermore, molar crowns covered with thick enamel should have better resistance to wear than those with thin enamel, other factors being equal. Animals with poorly developed shearing crests, however, chew their food less thoroughly, which results in a decrease in the digestibility of structural carbohydrates (Sheine, 1979a,b). Thus low-crowned teeth with relatively short shearing crests and thick enamel are

probably capable of withstanding high forces. Such teeth might be expected in species masticating hard foods but not utilizing structural carbohydrates as a primary source of energy.

ENAMEL THICKNESS AND SHEARING DEVELOPMENT IN RAMAPITHECINES

Table 2 provides some information on relative enamel thickness (E_r) and shearing quotient (SQ) for small samples of Indo-Pakistani ramapithecines. Considerable disagreement has arisen as to which species to attribute these specimens. Molar size suggests that more than one species is represented, but whether two or three taxa are present is uncertain (Pilbeam, 1979; Pilbeam et al., 1977a). I give one plausible taxonomic scheme in the table. In any case, the ranges of variation in relative enamel thickness and shearing coefficients do not exceed those typically found in single hominoid species. This suggests that these species do not have greatly different molar attributes and may be treated together. The molar enamel of ramapithecines is exceedingly thick: E_r averages about 1.75 times that of *Cercocebus* or *Pongo*, and about the same as that of *Cebus apella*. The shearing quotients of ramapithecines fall comfortably within the range seen in predominantly frugivorous extant hominoids and well outside the range of values for folivorous extant hominoids.

Extant species with thick-enameled molars have generally reduced molar shearing capabilities, an inefficient design for cutting up foods finely. Since dividing foods finely is at a premium among species that depend on structural carbohydrate (leaves, bark, buds, etc.) for a large proportion of their energy, we may assume that the thick-enameled ramapithecines were not efficient consumers of leaves or other high-fiber foods. A survey of the diets of various primate species, the teeth of which show thick

Table 2 *Indo-Pakistani Ramapithecinae*

Specimen	Taxon	SQ	E_r	M_2 length (mm)
GSP 4230	*Sivapithecus indicus*	−10.72	—	14.4
YPM 13828	*Sivapithecus indicus*	+ 1.50	—	13.9
AMNH 19412	*Sivapithecus sivalensis*	+ 4.39	—	11.5
YPM 13811	*Sivapithecus sivalensis*	+ 0.25	—	11.3
GSP 4635	*Ramapithecus punjabicus*	—	+175	10.9
GSP 6153	*Ramapithecus punjabicus*	—	+150	10.8
YPM 13814	*Ramapithecus punjabicus*	—	+101	10.6
GSP 11998	*Sivapithecus sivalensis*	—	+174	12.0
GSI 199	*Sivapithecus sivalensis*	—	+132	11.2
GSP 13566	*Sivapithecus sivalensis*	—	+177	12.6

Symbols: E_r, relative enamel thickness; SQ, shearing coefficient. See text for further description. Because estimates of SQ are upon unworn specimens and estimates of E_r can be made only on worn specimens, no single specimen can provide estimates of both. GSP, Geological Society of Pakistan; GSI, Geological Society of India; AMNH, American Museum of Natural History; YPM, Yale Peabody Museum.

enamel, reveals that they commonly masticate very hard foods such as seeds and hard nuts, which require great masticatory forces to open but whose contents are easily digested. Ramapithecines may have been doing the same. An adaptation for nut or hard fruit-eating would explain why thick enamel appears to have evolved in ramapithecines in combination with reduced molar crown relief, a forward shifting of the masseter muscle and a transverse thickening of the mandibular corpora. The lower crown relief seen in fossil hominoids may reduce the possibility of molar breakage; bringing the masseter origins forward over the molar row may have increased the amount of muscle force that could be brought to bear on the molar row (Jolly, 1970b); and altered mandibular shape (e.g., increased transverse thickness) was probably an adaptation for resisting torsional forces engendered from biting (Hylander, 1979).

RAMAPITHECINE ADAPTATIONS AND HOMINID ORIGINS

Ramapithecine molars have low crown-relief, poorly developed shearing, and thick enamel. The molars wear to a relatively flat surface and the shearing crests are soon worn away. Extant anthropoids that have a similar molar structure frequently eat nuts and seeds enclosed in tough pods. Once the hard rind is removed, these foods provide a rich source of readily assimilated nutrients. It is likely that ramapithecines had a similar diet. This reconstruction has some similarity to that of Jolly (1970b): Both theories emphasize the importance of small, hard food objects in the diet. However, Jolly's proposal that cereal grains were the ancestral hominid's central food item would imply that their molar adaptations evolved in open woodlands or savannas as a part of a terrestrial adaptation. By contrast, nut-cracking adaptations among primates are not necessarily (or even usually) associated with ground feeding. Many nut-eating mammals, including some primates, are arboreal.

The basic ramapithecine-like adaptation for nut feeding is seen today among primates that live in woodland as well as rain forest environments. Therefore there is no need to assume as has Pilbeam (1979) that these facial and dental adaptations had anything to do with the opening of new terrestrial adaptive zones during world-wide climatic cooling and increasing seasonality which may have reduced the area of tropical and subtropical rain forest, and expanded nonforest environments in the middle Miocene.

Available evidence about the ramapithecine dentition tends to refute the assumption of Jolly (1970b) and Simons and Pilbeam (1972) that the reduction of the canines in hominid evolu-

tion occurred as a direct result of selection for increased freedom of movement in the molar region required by the newly evolved grinding action between the molars. There is no linkage in ramapithecines between canine size or morphology and molar adaptations for grinding. Some ramapithecines like *Sivapithecus indicus* have large interlocking canines, whereas others like *Ramapithecus* may have comparatively reduced canines (Pilbeam, 1979). Irrespective of canine size, all species have thick-enameled molars which wear relatively flat. Therefore canine reduction cannot be viewed as a direct adaptation that allowed freer jaw movement for a grinding mode of molar occlusion. This is another instance of the mosaic pattern of the evolution of characteristically hominid features. If ramapithecines are hominid ancestors, we have evidence that a hominid dietary shift (inferred from the novel molar structure of ramapithecines) preceded canine size reduction and bipedalism, which in turn preceded the evolution of a large brain.

One of the most important unresolved conflicts about human origins concerns the date of the final separation of the human and ape lineages. The paleontological evidence (Simons, 1961c, 1972) suggests that the human lineage separated from that of the great apes at least 14 million years ago; the separation between orangutan and other great apes would have occurred much earlier. The biochemical evidence has been interpreted to mean that the human–African ape split occurred much later in time, no longer than 8 million years ago (Zihlman et al., 1978). A recent proposal to bridge this gap is that the ramapithecines are the ancestors of both great apes and humans; if so, the radiation of modern hominoids has occurred since 14 million years ago (Pilbeam, 1979, Zihlman et al., 1978). The latter interpretation would imply, among other things, that the last common ancestor of the great apes and hominids had evolved thick enamel from an early Miocene species with thin enamel. From this group would have arisen first the orangutan, which preserves the ancestral thick-enameled condition. More recently, the African ape stem would have split from the ancestral human stock. Hominids retain the primitive condition of thick enamel, whereas the thin enamel of African apes is a secondarily derived condition. The data presented here show that the evolution of thin enamel from a thick-enameled condition would be predicted in a lineage moving toward a diet of leaves or soft fruit.

SUMMARY

The morphology of the cheek teeth of the earliest ramapithecines, particularly the thick enamel and poorly developed shearing, suggests that they ate significant amounts of hard nuts or seeds enclosed in tough pods. Previous arguments that thick enamel implies terrestrial habits are unsound. Furthermore, canine reduction in early hominids should no longer be viewed as a direct adaptation to allow free movement of the molars during grinding.

Acknowledgments

I thank those who read and commented on this manuscript at various stages: M. Cartmill, R. MacPhee, W. Hylander, H. Covert, A. Walker, C. Jolly, R. Smith, D.R. Pilbeam, M. Wolpoff, and J. Fleagle. Measurements were made on specimens at the American Museum of Natural History; Field Museum of Natural History; British Museum (Natural History); Powell-Cotton Museum (Birchington, Kent); National Museum of Natural History (Smithsonian Institution); Congo Museum (Tervuren, Belgium); Comparative Anatomy Collection and Birds and Mammals Collection, Natural History Museum (Paris); and the private collection of N. Tappen, University of Wisconsin. I thank the keepers of those collections for their help. Dr. D.R. Pilbeam kindly allowed me to measure the ramapithecine fossils in his care. All measurements were made with a calibrated reticule mounted on a Wild M-5 microscope.

32

The Relationships of *Sivapithecus* and *Ramapithecus* and the Evolution of the Orang-utan

P. ANDREWS AND J. E. CRONIN

The Miocene epoch (25–5.5 Myr) was the period that heralded the appearance of mammalian faunas of modern appearance. It was also the period that saw the main radiations of the hominoid primates, culminating in forms that share many of the characteristics of the living apes and man (Pilbeam, 1979; Andrews, 1981b). The specific affinities within this complex of fossil and living hominoids are still, however, subjects of much controversy (Pilbeam, 1979; Andrews, 1981b; Simons and Pilbeam, 1965), and other than recent Pleistocene forms which are obviously related to the living orang-utan (Delson, 1977b) and gibbons (Szalay and Delson, 1979), and the Plio-Pleistocene forms of early hominids from East Africa (Howell, 1978; Johanson and White, 1979a), no fossil hominoid has been convincingly shown to be related uniquely to any of the extant hominoids.

Foremost in this controversy is the postulated hominid affinity of *Ramapithecus*, a form found now in middle to late Miocene deposits in East Africa (Pilbeam, 1979; Andrews and Walker, 1976), Turkey (Andrews and Tobien, 1977), China (Xu and Lu, 1976) and Indo-Pakistan (Simons and Pilbeam, 1965; Pilbeam, 1977; Simons, 1981). This phyletic linkage has been much debated and has been both rejected and supported on morphological (Pilbeam *et al.*, 1977a; Greenfield, 1979) and molecular grounds (Sarich and Cronin, 1976; Cronin and Meikle, 1982; Goodman, 1976b). Central to understanding this problem is: (1) the relationship of the complex of species within the *Sivapithecus-Ramapithecus* network, and (2) the relationships among the living hominoids and in particular the status of the orang-utan. To a large extent the adaptation and relationship of the orang-utan have remained substantial enigmas (Sarich and Cronin, 1976; Bruce and Ayala, 1979; Smith and Pilbeam, 1980), but they are clearly pivotal to our understanding of ape and hominoid evolution.

Recently, the relationships among the living species of hominoids have been greatly clarified by macromolecular comparisons (Sarich and Cronin, 1976; Cronin and Meikle, 1982; Bruce and Ayala, 1979; Ferris *et al.*, 1981). Although not yet definitive, these comparisons have provided a new insight into the branching sequence or cladistic relationships of the living apes and the timing of the divergence between apes and man. This evidence will be reviewed here, with particular reference to the relationships of the orang-utan with the other hominoids. Changes in morphology, particularly of the teeth and face, will be correlated with the pattern of relationships thus established.

As regards the fossils, a new fossil hominoid specimen has been recently described which sheds a substantial amount of light on the *Sivapithecus-Ramapithecus* complex. *S. meteai*, from the late Miocene of Turkey (Andrews and Tekkaya, 1980), shows similarities in morphology and a possible phyletic relationship to the living orang-utan. *S. meteai* clearly shares broad similarities with other Miocene *Sivapithecus* (Andrews and Tekkaya, 1980) forms, which in turn are closely allied to *Ramapithecus* (Pilbeam, 1977). If this linkage is correct then one segment of the hominoid evolutionary tree can be rooted through a purely paleontological approach.

MOLECULAR AFFINITIES AMONG THE HOMINOIDS

Each organism carries within itself a history that is encoded in the genome. Given the appropriate techniques we can decipher this codex and elucidate phylogenetic history. Such macromolecular comparisons may yield insights into both the sequence and timing of ancient speciation events. The molecular approach has long been critical to the resolution of several controversies concerning phyletic links over a wide taxonomic scale, for example, in the case of taxa such as Aves, Amphibia, Reptilia and Mammalia, particularly Primates. Such comparisons are not limited to living forms, as tissue from a frozen mammoth has supported its close molecular relationship to the modern elephant and skin from the recently extinct Tasmanian wolf has confirmed its affinities with the dasyuroid group of marsupials (Lowenstein *et al.*, 1981).

With respect to hominoid evolution, substantial genetic data are now available which shed light on the evolution of humans and apes. Data from immunodiffusion, microcomplement fixation, radioimmunoassay, amino acid sequencing, electrophoresis, nucleic acid hybridization, nucleotide sequence and restriction endonuclease mapping, and cytogenetics, have shown a fairly consistent pattern of the branching sequence or cladistics of the hominoids, and working within this framework one can attempt to analyse the rate and nature of morphological evolution along the specific lineages.

The biomolecular evidence (Fig. 1) argues for three clear major cladogenic events in hominoid evolution: first, the separation of the gibbon lineage from that leading to the great apes and man: second, the subsequent divergence of the orang-utan lineage from that linking the African apes and man: and thirdly, evidence from molecular comparisons that *Homo*, *Pan* (chimpanzee) and *Gorilla* share a substantial lineage before their subsequent divergence. Evidence as to which of the two extant lineages shares a period of common ancestry to the exclusion of the third is at present equivocal; evidence exists which favours pairing of *Pan-Gorilla*, *Homo-Pan* and *Homo-Gorilla* (in order of greatest preference). The difficulty in determining the sequence of divergence is primarily a result of extremely short lineages, in a temporal sense, and of the experimental error inherent in the techniques, and so far no data have conclusively demonstrated any unique shared derived molecular changes for any pair of the lineages. However, as interesting as this issue is, it is not central to the

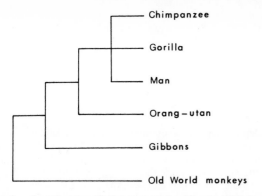

Figure 1 *A molecular cladogram of the Hominoidea incorporating data from immunological studies, electrophoretic analysis, protein sequencing, nucleic acid hybridization and restriction endonuclease analysis of mitochondrial DNA. The data are discussed in the test and in Goodman and Cronin (1982) and Cronin (1983) in more detail. Molecular clock calibration of the various sequences of divergence [as calibrated in the text and in Cronin and Meikle (1979, 1982) and Cronin (1983)] are (1)* Homo-Pan--Gorilla, *5 ± 15 Myr ago;* (2) Pongo-Homo, Pan, Gorilla, *10 ± 3 Myr ago; and* (3) Hylobates-Pongo, Homo, Pan, Gorilla, *12 ± 3 Myr ago.*

evidence for the common ancestry of the African apes and man and to the phyletic position of the orang-utan (Fig. 1).

Molecular data demonstrate convincingly the existence of the common lineage leading to the African apes and *Homo*. Along this common lineage (Fig. 1) numerous shared molecular changes have occurred. These involve derived amino acid sequence changes in the proteins albumin and transferrin, fibrinopeptides, carbonic anhydrase, haemoglobins, myoglobin, changes in mitochondrial DNA restriction sites, 1.5% base pair substitutions (involving $\sim 10^6-10^7$ base pair mutations) in nuclear DNA, and at least one locus with electrophoretic charge changes (Cronin, 1983; Goodman and Cronin, 1982; Table 1). Specifically determined events and inferred molecular change clearly document the existence of this lineage, and no molecular data reverse the position of *Pongo* and *Homo* shown in Fig. 1,

although some data provide more reliable results than others. The conclusion, therefore, is that, in so far as the molecular evidence reveals phylogenetic relationships, the orang-utan is removed from unique association with the African apes and is the sister group to the clade comprising the African apes and man.

In addition to the branching pattern, the molecular data may also elucidate the timing of evolutionary events. It has been hypothesized that due to their apparent approximate regularity of change, molecules may serve as an evolutionary clock to date past divergence events (Sarich and Cronin, 1976; Wilson *et al.*, 1977), and this has been used to suggest that the human and African ape lineage separated as recently as 4–6 Myr ago. Calculations based on rates of nucleic acid and serum protein evolution (Cronin and Meikle, 1982) give dates ranging from ~ 8.5–14 Myr for the origin of the orang-utan lineage ($x=9.9\pm$, $\gamma=1.7$). An estimate of 10 ± 3 Myr is thus quite in order, given the approximate nature of the molecular clock (Corruccini *et al.*, 1980), and this agrees with the fossil evidence described here for the origin of the orang-utan lineage. Previously, the fossil evidence has appeared to indicate dates of ~ 15–20 Myr for this event, but the data given here suggest that the date should be close to or slightly earlier than 10–11 Myr. Precluding naive interpretations of the molecular clock (Corruccini *et al.*, 1980), these results are quite compatible with each other (Zihlman *et al.*, 1978; Cronin and Meikle, 1979).

MORPHOLOGICAL AFFINITIES AMONG THE HOMINOIDS

The evidence from gross morphology generally agrees with the molecular evidence, except that it defines the relationships among the extant hominoids less precisely. The evidence must be viewed at various levels. First, there is the evidence linking the hominoids as a monophyletic group through such shared derived characters as the presence of a vermiform appendix and the loss of the tail. Second, there are the numerous characters of the postcranial and axial skeleton which link the great apes and man and exclude the gibbons (Andrews and Groves, 1975), particularly, for example, the morphology of the wrist (Lewis, 1969), and the structure and development of the shoulder musculature (Ashton and Oxnard, 1963) and the reduced lumbar region of the vertebral column (Schultz, 1961). Finally, there are the characters that provide evidence of pairing relationships within the great apes and man,

Table 1 *Molecular Comparisons among the Great Apes and* Homo

Molecular comparisons	Homo–Pongo	Pan/Gorilla–Pongo	Homo–Pan–Gorilla
Fibrinopeptide sequence (no. of amino acid differences) (Goodman and Cronin, 1982)	2	2	0
DNA hybridization (% base pair mismatch) (Beneveniste and Todaro, 1976)	2.4	2.0	1.0
DNA hybridization (% base pair mismatch) (Goodman and Cronin, 1982)	—	4.5	2.4
Immunological distance (ID units) (Sarich and Cronin, 1976)	37	39	14
Immunodiffusion distance (antigenic distance) (Goodman, 1976b; Goodman and Cronin, 1982)	1.6	1.6	0.8
Electrophoresis: plasma proteins (Sarich and Cronin, 1976; Cronin and Meikle, 1979)	≫2	≫2	~1.6
cladistic analysis based on number of electrophoretic charge changes at 23 loci (Cronin, 1983)	8	9	7
genetic distance value based on 23 loci (Bruce and Ayala, 1979)	0.349	0.300	0.367
Mitochondrial DNA comparisons using endonuclease restriction enzyme maps of site shared (Ferris *et al.*, 1981)	1*	2†	7‡

*Homo *shares a specific site with* Pongo *but others do not;* † Pan *and* Gorilla *share a site in common with* Pongo *but* Homo *does not;* ‡ Homo *shares this number of sites with at least one of the species of* Pan *or* Gorilla *but not with* Pongo *(or* Hylobates*).*

and as these are the principal concern of this paper these characters will be examined in more detail.

Evidence exists which previously led to the grouping of the great apes as a single group containing the chimpanzee, gorilla and orang-utan. These all have a diploid number of chromosomes of 48 compared with 46 in man. They have many superficial similarities in skull morphology, such as the prognathism of the face or the enlarged nuchal area of the occiput (Le Gros Clark, 1955), and in their postcranial morphology, particularly in their limb proportions and shoulder morphology which were thought to be functionally adapted to brachiation (Le Gros Clark, 1955). More recently, most of these characters have been shown to be either allometric consequences of increased body size (Andrews and Groves, 1975; Aiello, 1981; Biegert and Maier, 1972) or primitive characters retained by the living great apes (Delson and Andrews, 1975) and therefore having no relevance to the within-group relationships of the living great apes.

There is much stronger evidence linking the African apes, the chimpanzees and gorillas with each other, to the exclusion of the orang-utan, including the many anatomical specializations related to knuckle walking in chimpanzees and gorillas (Tuttle, 1967), and other characters such as the presence of the frontal sinus (Cave and Haines, 1940; Cave, 1961), a character which they also share with man. This evidence is in general agreement with the molecular data, and there is little morphological evidence that lends any support to alternative sets of relationships, for example, that the orang-utan is related more closely to gibbons (Romero-Herrara et al., 1976), that it forms the sister group to the group comprising the gibbons, African apes and man (Romero-Herrara et al., 1976), or that the orang-utan by contrast is more closely related to man than are the African apes (Benveniste and Todaro, 1976).

The lower face of chimpanzees and gorillas has many points of similarity with the human face despite the gross differences in facial height and the alveolar prognathism. Some of the characters shared between man, chimpanzees and gorillas are: the nasal aperture is broad; the subnasal plane is truncated and stepped down to the floor of the nasal cavity; the orbits are approximately square and often broader than high; the interorbital distance is broad; the infra-orbital foramina are usually three or less in number and are situated on or close to the zygomaticomaxillary suture; the zygomatic bone is usually curved and has a pronounced posterior slope; the zygomatic foramina (one or two) are small and are situated at or below the lower rim of the orbits; and the glabella is thickened (present in some fossil humans although not in modern man). The chimpanzee, gorilla and man also share the presence in the palate of small incisive foramina, large and oval-shaped greater palatine foramina, and large sphenopalatine fossae. Their teeth are basically similar in pattern except for the canines and premolars, and in particular the upper incisors lack a great size discrepancy between I^1 and I^2.

Many of the characters shared by man, chimpanzees and gorillas are present also in the gibbons. Where this is the case it is presumed that these characters are primitive for the Hominoidea, both because they are thus present in most widely separated members of the group. (In other words, they are not present in the orang-utan, which separates the clades of gibbons and the African apes and man in Fig. 1.) In three of these characters, however, gibbons differ from the African apes and man, and these must be examined briefly to determine whether the primitive condition can be identified for them. Gibbons have a narrower oval-shaped nose and from a consideration of other anthropoids, which generally have noses that are higher than broad, it seems likely that the primitive pattern for hominoids is a narrow oval-shaped nose as in gibbons. The infra-orbital foramina in gibbons are well removed from the zygomaticomaxillary suture, and again this is the usual condition in other anthropoids, thus it seems likely that this is the primitive condition in the hominoids. The condition of the glabella varies considerably in gibbons, from almost complete absence of thickening to fairly extensive thickening. This is true also of many other anthropoids, and it seems in fact to be size-related, therefore it is only possible to consider this character to be significant if glabellar thickening is either present on a small-bodied species or absent from a large one.

These characters have been summarized in a list (Table 2) which depicts the primitive condition in the Hominoidea. The two characters marked with an asterisk are present in gibbons, which thus retain the primitive condition, but not in the African apes and man, which thus share the derived conditions for these characters. All the other characters are shared between gibbons, African apes and man. One other character for which the polarity of evolutionary change is not yet understood concerns the presence of thick or thin enamel on the molars. The gibbons, chimpanzees and gorillas have thin enamel and man has thick enamel, and, on the face of it, thin enamel seems to be primitive for hominoids. Gibbons, chimpanzees and gorillas would thus have retained the primitive condition while man has developed the derived condition with thick enamel. Thick enamel, however, is a character that is shared with the orang-utan as well as with the fossil apes of the *Sivapithecus-Ramapithecus* group, and an understanding of the relationships between these two groups is essential to the understanding of human relationships.

RELATIONSHIPS OF THE ORANG-UTAN

The characters of the orang-utan differ in most respects from those listed in the previous section. Table 3 shows the characters of the orang-utan which it does and does not share with the fossil *S. meteai* (Andrews and Tekkaya, 1980) (Fig. 2). Other characters which it does not share with *S. meteai* could be added to the list, for example, the presence of wrinkling on the molar teeth in the orang-utan; but the absence of such clearly derived characters in the fossil hominoid need not be construed as evidence against an association between it and the orang-utan. Evidence for such an association comes from the list of shared characters, all of which (except where indicated) are interpreted as derived characters

Table 2 *Characters Considered likely to Be Primitive for the Hominidea*

*Nasal aperture higher than broad, oval-shaped
Subnasal plane truncated
Subnasal plane stepped down to floor of nasal cavity
Orbits as broad as or broader than high
Inter-orbital distance broad
Infra-orbital foramina few in number ($\leqslant$3)
*Infra-orbital foramina well removed from the zygomaticomaxillary suture
Zygomatic bone curved and with strong posterior slope
Zygomatic foramina small
Zygomatic foramina 1–2 in number
Zygomatic foramina situated at or below the lower rim of the orbits
Glabella thickening which may occur on large individuals/species
Small incisive foramina
Large, oval-shaped greater palatine foramina
Upper incisors lacking large size discrepancy
Thin enamel on molars (see text)

This list is not exhaustive but concentrates on those characters that are known from the fossil record.

*These characters are present in gibbons but not in the African apes and man.

Table 3   *Characters Shared between the Orang-utan and S. meteai*

---

*Nasal aperture higher than broad, oval-shaped

Subnasal plane smooth, not truncated

Subnasal plane continuous with floor of nasal cavity, not stepped

Orbits higher than broad—not known for *S. meteai* but known for *S. indicus* (Pilbeam, 1982)

Inter-orbital distance very narow

*Infra-orbital foramina few in number

*Infra-orbital foramina well removed from the zygomatico-maxillary suture

Zygomatic bone flattened, facing anteriorly

Zygomatic foramina relatively large

No glabellar thickening—not known for *S. meteai* but known for *S. indicus* (Pilbeam, 1982)

Extremely small incisive foramina

Slit-like greater palatine foramina

Great size discrepancy between I^1 and I^2

Thick enamel on molars

†Zygomatic foramen multiple (single in *S. meteai*)

†Zygomatic foramen situated above the level of the lower rim of the orbit (below the orbit in *S. meteai*)

---

*Characters considered to be shared primitive characters; all the rest are considered to be shared derived characters.

†Characters that are present in the orang-utan but for which *S. meteai* retains the primitive condition.

uniquely shared between *S. meteai* and the orang-utan. Where the fossil hominoid lacks the derived characters present in the orang-utan, this can be interpreted in terms of mosaic evolution, whereby some characters in an evolving lineage are developed before others. It would be more difficult to explain derived characters present in the fossil that are not present now in the orang-utan, but none have been observed.

No evidence exists as to whether the characters shared between *S. meteai* and the orang-utan are homologous or not. They may have arisen through convergent evolution rather than through common ancestry, but the latter is indicated by several sources. First, the characters shared between the orang-utan and *S. meteai* relate to separate functional complexes of the face and dentition. Some of the characters do inter-relate, such as the shape of the nose and the subnasal plane, or the nose and orbit shapes, or even possibly the shape of the nose and the narrow inter-orbital distance, but these characters relate directly neither to the characters of the palatine foramina, nor to two characters

of the dentition, that is, the large difference in size between the central and lateral incisors and the thickness of enamel on the molar teeth. These combinations of characters can be seen as separate functional complexes that show similarities between the orang-utan and *S. meteai*, and if they are all the results of convergence it is difficult to envisage the circumstances that would lead to similarities in such disparate sets of characters occurring together in the fossil and extant forms.

Homology rather than convergence is also indicated by the metrical study of *S. meteai* (McHenry *et al.* 1980), in which over a series of eight measurements of the palate and lower face, the extant hominoids showed a consistent and unified pattern differing markedly from that of other anthropoids. The pattern for *S. meteai* approximates fairly closely to the great ape pattern, demonstrating both that its affinities in most dimensions are with the great apes and that, seen as a pattern, each of these characters is consistent with the others and therefore more likely to be the result of shared ancestry rather than independently acquired convergences. As for the non-metrical data, the variation in some of the dimensions, for example, palate length and nasal width, are not signficantly inter-correlated and by inference are not functionaly related.

Even more tenuously, it can be observed that in certain other respects the face of *S. meteai* bears strong resemblances to that of the orang-utan. The size and conformation of the zygomatic region, for example, are similar in the two hominoids, although the degree of similarity is not sufficient to distinguish them absolutely from the other hominoids. The great bizygomatic breadth in *S. meteai* is most similar to the condition in the orang-utan, but is also falls within the range of variation for male gorillas (Andrews and Tekkaya, 1980). The same is true of the relatively great depth of the zygomatic region and the pronouonced premaxillary prognathism (Andrews and Tekkaya, 1980). The deep zygomatic region is combined in *S. meteai* with an estimated face length (nasal height measured from the estimated position of nasion, Andrews and Tekkaya, 1980) that is below the range for the gorillas but within the range for the orang-utan. Thus these characters combine to produce a face shape similar to that of the orang-utan and different from comparable-sized gorillas. However, as the fossil is incomplete, it is not yet possible to quantify these characters adequately and little reliance can be placed on them except as support for the evidence given earlier.

It can be concluded from the morphological evidence given here that the orang-utan possesses a number of derived characters not present in the other extant hominoids. This is consistent with the molecular evidence discussed in the previous section (Fig. 1). We have shown that the suite of characters that distinguishes the orang-utan from other hominoids is present also in a fossil ape, *S. meteai*, and as it is concluded that these similarities are probably homologous we have felt justified in adding the fossil hominoid to the original cladogram to produce Fig. 3.

Sivapitheucs indicus from India and Pakistan (Simons and Pilbeam, 1965; Pilbeam *et al.*, 1977a) can also be linked with *S. meteai* and the orang-utan on the basis of a newly described skull belonging to this species (Pilbeam, 1982). This skull, which was described after this review was written, shares with the orang-utan all the characters listed in Table 3, and it also provides evidence on two additional characters of the eye region which are included in Table 3. This strongly reinforces the case for the relationship of *Sivapithecus* with the orang-utan (Andrews, 1982).

DISCUSSION

S. meteai has long been identified as part of the *Sivapithecus-Ramapithecus* complex (Simons and Pilbeam, 1965; Andrews and Tekkaya, 1980). Simons and Pilbeam (1965) synonymized it with *S. indicus*, but it was later reinstated as a separate species

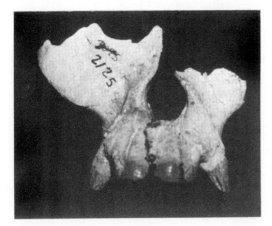

Figure 2   *The maxilla of* S. meteai *showing the lower part of the face.*

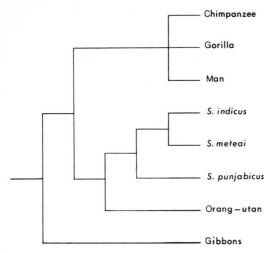

Figure 3 *Relationships of the extant and fossil Hominoidea. The relationships of the living species correspond to those in Fig. 1, and the fossils are shown as three species of* Sivapithecus *(including* Ramapithecus *as a junior synonym) which together constitute the sister group to the orang-utan.*

(Andrews and Tekkaya, 1980) on the basis of characters of the incisors, canines, premolars and posterior molars. In most of these characters it was presumed (Andrews and Tekkaya, 1980) that *S. meteai* possessed the derived condition with respect to the primitive hominoid morphotype and *S. indicus* the primitive condition. They both share, however, characters such as thickened enamel on the molar teeth, lower-crowned and more robust canines, and loss of molar cingula, which distinguish them from the primitive hominoid condition and indicate that the species presently assigned to *Sivapithecus* represent a clade, of which these are the distinguishing characters. As these characters are also shared by *Ramapithecus* it seems very probably that they are part of the same clade. This has been recognized implicitly by Pilbeam (1977) in his combination of 'ramapithecines' and 'sivapithecines' in the family Ramapithecidae, and more explicitly by Greenfield (1979) who dismissed even generic differences between them (although unfortunately he did so for the wrong reasons) and referred *Ramapithecus* to the genus *Sivapithecus*. This is certainly the simplest solution at present to a rather complicated situation, and although there is a risk that it is an oversimplification it does help to resolve many problems if the thick-enamelled Eurasian apes (excluding *Gigantopithecus*) are united in the genus *Sivapithecus*. This genus would then contain at least 3–4 species, which can be listed as follows with the geographical location and approximate ages:

S. indicus, India and Pakistan, middle to late Miocene, 11–8 Myr
S. punjabicus, India and Pakistan, middle to late Miocene, 11–8 Myr
Sivapithecus sp., China, late Miocene, 10–8 Myr
S. meteai, Turkey and Greece, late-middle Miocene, 11–10 Myr

Sivapithecus sivalensis (Simons and Pilbeam, 1965) is of uncertain status, especially the type specimen of the species, and thus is omitted from the present discussion. The species of middle Miocene age assigned to *Sivapithecus* and *Ramapithecus, Sivapithecus darwini* from Turkey and Czechoslovakia (Andrews and Tobien, 1977) and *Sivapithecus africanus* and *Ramapithecus wickeri* from Kenya (Andrews and Walker, 1976), are known only

from extremely fragmentary dental remains and they lack the body parts that have been used in the present analysis to determine relationships. Therefore, these species also have been omitted from the present discussion.

Two consequences ensue from the recognition of the *Sivapithecus* group as a clade. First, if one member of the clade, that is, a species such as *S. meteai,* is accepted as being closely related to the orang-utan, then the other fossil species are also linked in the same association. This would apply, for example, to species of *Ramapithecus* if they are congeneric with *Sivapithecus.* Only if characters are found to be present in more complete specimens that exclude them from this association (and also of course from the genus *Sivapithecus)* can it be accepted that they do not form part of this clade. Following this line of argument, in conjunction with the molecular and morphological evidence for the relationship of the extant apes, it is evident that '*Ramapithecus*' may not be related exclusively to man but may be related to the orang-utan.

The case for the hominid affinities of ramapithecines has most recently been reviewed by Kay (1982a). He also refers '*Ramapithecus*' to *Sivapithecus,* although not as a separate species but to the species *S. sivalensis.* He shows that the species of *Sivapithecus* share the following synapomorphies with man: (1) broad mandibular body; (2) low-crowned molars with thick enamel; (3) rather reduced canine size; (4) rather reduced canine sexual dimorphism; (5) buccolingually broad upper canines; and (6) tendency to enlarged P_3 metaconids. These are said to be characteristic of the whole *Sivapithecus* clade and that this is therefore the sister clade of man. We accept Kay's interpretation of the significance of characters 1 and 5, although mandibular robusticity varies considerably for the different species of *Sivapithecus,* but there are greater difficulties with the others. Two of the characters, the enlarged P_3 metaconids and the thick enamel of the molars, are present also in the orang-utan, and therefore provide equally good evidence for a *Sivapithecus-Pongo* relationship as for a *Sivapithecus-Homo* relationship. We do not accept Kay's conclusion that canine size is reduced in *Sivapithecus* for two reasons; first, his data show that the range of canine size relative to the first molar is equivalent to the middle of the range of extant great apes: in other words *Sivapithecus* canines are relatively bigger than female great ape canines as well as being relatively smaller than male canines. Secondly, it must be considered that *Sivapithecus* may have relatively large molars for its body size (Pilbeam *et al.,* 1977a), and if this is the case the ratio of canine/molar size may be misleading. Finally, we also disagree with the significance attached to reduced sexual dimorphism, as low sexual dimorphism has not been shown convincingly to be a valid characteristic of *Sivapithecus.*

The characters linking *Sivapithecus* with man are thus reduced to two: robust mandibles and buccolingually broad upper canines, two characters that are almost certainly functionally related. These characters certainly seem to be synapomorphies for *Homo* and *Sivapithecus,* but the likelihood of their being the result of parallel evolution must be given credence in view of the more numerous and functionally independent derived characters of the nose, the orbits, the palatal foramina and the incisors that are shown here to be shared between *Sivapithecus* and *Pongo.* We therefore consider it more likely that *Sivapithecus* species, particularly *S. meteai* and *S. indicus* but by inference also *Sivapithecus* (formerly *Ramapithecus) punjabicus,* are more closely related to the orang-utan than to man.

The other consequence arising from the proposed relationship of *S. meteai* and the orang-utan is that it provides an approximate minimum divergence data between the orang-utan and its extant sister group, the African apes and man. *S. meteai* itself comes from deposits about 10–11 Myr old, but the time span of *Sivapithecus* as recognized here is of the order 8–11 Myr. This age corresponds to the date of 10±3 Myr suggested by the molecular evidence and strongly indicates a minimium age of at least 10–11 Myr for the divergence between the orang-utan clade and the clade containing the African apes and man. The divergence between the members of the latter must be subsequent to this period, and the molecular evidence for a prolonged common ancestry between the African apes and man suggests that their eventual divergence was considerably later than the middle Miocene and perhaps as late as the end of the Miocene 5–6 Myr ago (Cronin and Meikle, 1982).

There is one other consequence of the relationships proposed here, which will be considered briefly. The classification of the cladistic groups proposed in Fig. 3 would differ greatly from many current schemes. The most recent one by Szalay and Delson (1979), for example, combines the great apes in the tribe Pongini with parallel tribes for *Dryopithecus* (Dryopithecini) and *Sivapithecus* (Sugrivapithecini). These are combined in the subfamily Ponginae, and *Ramapithecus, Australopithecus* and man are classified in the subfamily Homininae. These subfamilies are then grouped in the family Hominidae. Classifications based on Fig. 3 would have to recognize the chimpanzee, gorilla and man as one clade with the orang-utan and *Sivapithecus* (and '*Ramapithecus*') as another; a possible solution would be to classify these clades at subfamily level, as do Szalay and Delson (1979), but with Homininae for the African apes and man, and Ponginae for orang-utan and *Sivapithecus* (and '*Ramapithecus*'). The African

apes and man would then be divided at the level of tribe. An alternative classification at the family level is as follows:

Hominoidea (Gray, 1825)
Hylobatidae (Blyth, 1875)
Hylobates (Illiger, 1811)
Pongidae (Elliot, 1913)
Pongo (Lacepede, 1799)
Sivapithecus (Pilgrim, 1910)
Hominidae (Gray, 1825)
Gorillinae (Hurzeler, 1968)
Pan (Oken, 1816)
Gorilla (I. Geoffroy, 1852)
Homininae (Gray, 1825)
Homo (Linnaeus, 1758)
Australopithecus (Dart, 1925)

This change from traditional classifications has long been advocated by Goodman (see Goodman and Cronin, 1982), and it is one that is necessary with the recognition that the African apes are more closely related to man than they are to the orang-utan.

We thank N. Boaz, L. Brunker, E. Delson, S. Gould, T. Harrison, J. Lowenstein, L. Martin, E. Meikle and C. Stringer for helpful comments. V. Sarich provided helpful discussion on the molecular section of this paper, and much of the morphological evidence is the product of an ongoing collaboration with C.P. Groves. The Wenner-Gren Foundation, the L.S.B. Leakey Foundation and the Milton Fund of Harvard University provided grants to J.E.C. in support of part of this work.

33

Ramapithecus and Hominid Origins

M. H. WOLPOFF

According to the Baconian (i.e., inductivist) view of how science proceeds, the changing interpretations of *Ramapithecus* over the last two decades should reflect a shifting data base. As more data have been recovered, presumably the hypotheses regarding their interpretation have correspondingly changed to encompass a more complete (and more revealing) data set. Because it has been contended at various times that *Ramapithecus* is the earliest hominid [1], these changing interpretations should have resulted in different theories of homind origins.

In this paper, I argue that this does not describe what has actually happened. Instead, it has been shifting hypotheses that have altered the interpretation of the data base. The shifting hypotheses have been about hominid origins and not about *Ramapithecus* at all, because by and large the importance of *Ramapithecus* has historically been in the claim that it represents the earliest hominid. This argument is very Popperian in that it regards the interpretation of data as a consequence of theoretical framework and the function of new data recovery mainly as a potential refutation of current hypotheses (Wolpoff 1976b, 1978). Unlike Popper (1957), or at least the interpretations that have been given to his works (Halstead 1980, but see Popper 1980), I regard evolutionary theory as a falsifiable *scientific* theory, and specific evolutionary hypotheses such as those concerning hominid origins as *scientific* hypotheses because they *are* potentially refutable. The changing interpretations of *Ramapithecus* reflect just such a series of refutations.

Thus, whether or not *Ramapithecus* is a hominid, a hominid ancestor, or a collateral branch to the hominids, it is my contention that a complex interplay of theory, analysis and interpretation, and accumulating discoveries has wedded the interpretation of *Ramapithecus* to the problem of human origins.

RAMAPITHECUS AND DARWIN'S THEORY

The Darwinian theory of hominid origins has remained very powerful, and in one way or another it has influenced virtually every attempt to hypothesize about this event. Darwin posited what we would call a positive feedback relationship between what he viewed as the four critical elements that distinguish humans from the African apes: bipedalism, tool use, canine reduction, and the expansion of the brain. He hypothesized a fundamental adaptive shift associated with the origin of the human, in which an arboreal adaptation was exchanged for a terrestrial one and a primarily frugivorous diet was replaced by one emphasizing meat obtained by hunting. The tools attained their importance through their use in hunting and in defense. Bipedalism evolved as an adaptation for freeing the hands during locomotion so that tools and weapons could be carried and used easily. Canines diminished in size as tools replaced their functions in cutting, slashing, and social displays. Lastly, expanding brain size resulted from selection for more complex cooperative behavior and language, both of which were viewed as critical to the adaptations just discussed.

The initial interpretation of *Ramapithecus* as a hominid ancestor was completely within this Darwinian framework, just as were the initial interpretations of *Australopithecus africanus* (Dart 1925) and *Gigantopithecus blacki* (Wu 1962). Thus, the characteristics Lewis (1934) emphasized in his hominid interpretation of the *Ramapithecus* maxillary remains were the parabolic dental arcade, small canines that were transversely expanded, lack of a functional diastema, and a small degree of maxillary prognathism. Because a small canine and features related to it represent the only characteristic in a maxilla that can be related to the Darwinian theory, attention was focused on the interpretation of their morphology.

When Simons (1961c) resurrected the *Ramapithecus* argument, he emphasized the same canine-related features, even removing from the taxon one mandible that did not fit the functional model because of its sectorial P_3. In a subsequent publication (1964c) assigning a number of mandibles to the taxon, one of the primary criteria he used was an inward turning of the corpus at the M_1 position, indicating a parabolic arcade and a short snout. These interpretations were fully consistent with the canine-reduction aspect of Darwin's model.

Similarly, Leakey's (1962) publication of the *Kenyapithecus* maxilla emphasized the small canine seemingly associated with it (the tooth was found several feet away), although his discussion indicated that the tooth was probably used in cutting. While *Kenyapithecus* was not regarded as a hominid in this publication, its detailed resemblance to the *Ramapiihecus* palates was noted. In a later publication, Leakey (1967a) formally allocated *Kenyapithecus* to the hominids, emphasizing the vertically short canine crown and roots, the small shovel-shaped incisor crowns, and (what he termed) the arcuate shape of the dental arcade. Moreover, like Dart he sought evidence for the behavioral implications of the reduced canine, ultimately claiming evidence for artificially smashed bone at Fort Ternan (Leakey 1968a) as an indication that tools were being used.

The interpretive framework for these materials was probably most explicitly stated in a paper by Pilbeam (1966). *Ramapithecus* was claimed to be "completely hominid in known parts" (p. 3); the hominid characteristics were found mainly in the features associated with canine reduction described above. Moreover, Pilbeam argued that since the small canines were ineffective in agnostic behavior and group defense, "presumably, weapon use was established by this time" (p. 3). Finally, because "food must have been *prepared* for chewing by non-dental means; hands were probably used extensively and perhaps tools as well. . . . The evi-

Reprinted with permission from *Current Anthropology*, Volume 23, pages 501–510. Copyright © 1982 by the University of Chicago Press, Chicago.

dence, admittedly circumstantial at present, suggests a primate perhaps already bipedal and fully terrestrial" (p. 3).

With reduced canines, tool use, and (provisionally) bipedalism included in the *Ramapithecus* paradigm, there remained only increased intelligence to complete the predictions of Darwin's hypothesis. Indirect arguments suggesting improved intellectual capacities for the species were presented by Simons (1972) with an analysis of differential molar wear. He claimed that *Ramapithecus* differed from contemporary dryopithecines in showing a greater difference in wear between the adjacent molars (a steeper wear gradient). The steeper gradient was interpreted to indicate a longer period of time between successive molar eruptions and consequently delayed maturation of *Ramapithecus* offspring. The maturational delay presumably permitted the learning of more complex behaviors during childhood.

In sum, *Ramapithecus* was considered a hominid because the known remains fit the applicable aspects of Darwin's model. Thus, it was possible to use the model to speculate or interpret within the framework of interrelations that it provided. The arguments and interpretations indicating tool use, bipedalism, and more complex behavior clearly followed.

The inductivist interpretation would suggest that Darwinian-based arguments were ultimately dismissed because new data suggested a new theory about the phylogenetic status of *Ramapithecus*. Indeed, the period of the earlier 1970s in which this framework was effectively questioned was also a period of intensive fossil discovery. However, for several reasons I do not believe that the two phenomena are causally related.

The fit of the *Ramapithecus* data to the Darwinian model involved much more interpretation than actual analysis. The canine, for instance, was regarded as small and reduced long before a canine was found (the Fort Ternan specimen), and even when a canine was recovered and showed the morphology of a honing tooth, its "small size and reduction" were still emphasized. It did not require new materials to question these interpretations. Following the original paper by Lewis (1934), Hrdlicka (1935) systematically rejected the morphological arguments supporting the hominid interpretation, and other workers showed that the relative sizes of the canines and the incisors do not differ from those of some of the African pongids (Wolpoff 1971a, Yulish 1970). The reconstruction of the dental arcade as parabolic was questioned (Frayer 1976) and ultimately shown to be incorrect (Vogel 1975, Greenfield 1978). The evidence for a steep molar wear gradient in *Ramapithecus* was questioned (Greenfield 1974), and in any event the presence of steep gradients in primate species without delayed maturation had already been shown (Mann 1968, Wolpoff 1971b). The questions described above (as well as others from this period [see von Koenigswald 1972b, Robinson 1967]) were not raised because new materials were discovered; their basis was in the original specimens and their interpretations. However, historically it was too early for criticism to be effective, and these objections were not widely noted. First, the Darwinian hypothesis of *hominid* origins had to be replaced by another.

RAMAPITHECUS AND JOLLY'S THEORY

Publication of Jolly's "seed-eaters" hypothesis (1970b) received widespread attention (Jolly 1973) and was generally (although not universally) well accepted among paleoanthropologists. Although Robinson (1963b) had earlier argued that the masticatory apparatus of the *"Paranthropus"* remains from South Africa indicated a vegetarian adaptation in what he regarded as a little-modified descendent of the original australopithecine stock, it was Jolly who publicized a formal evolutionary hypothesis indicating mechanisms that connected hominid *origins* with a diet-oriented masticatory shift and the behaviors associated with it.

Indeed, while Jolly also focused on *"Paranthropus"* (and OH 5 from East Africa) in support of his ideas, it soon became clear that the earlier *Australopithecus africanus* evinced the same masticatory adaptation (Wolpoff 1973). The model was immediately applied to the interpretation of *Gigantopithecus bilaspurensis* (Simons and Ettel 1970), which was then still considered an "aberrant ape" (Pilbeam 1970). The application to *Ramapithecus* was somewhat slower.

Jolly's model linked canine reduction, upright posture and finally bipedalism, the development of thumb opposability, the appearance of language, the development of single-male groups, and evolution of a masticatory apparatus adapted for powerful grinding and crushing to a dietary and behavioral adaptation emphasizing the exploitation of seeds and other small objects. Jolly did not deal with brain-size (or, more generally, behavioral-complexity) increases, since, by the end of the 1960s, it was evident that Pliocene *Australopithecus* had an essentially ape-sized brain. Moreover, Jolly did not accept the arguments for delayed maturation (Mann 1968) or for neural reorganization (Holloway 1966a), so in essence he had nothing to explain. For these reasons, and because the early hominid dietary adaptation involved plant foods rather than meat (as Darwin had suggested), neither culture nor tool use played a role in his hominid-origins model (much to the apparent relief of some of the more paleontologically oriented paleoanthropologists).

Interpretation of an adaptation combining small-object feeding, powerful masticatory apparatus, and terrestriality found its way into the *Ramapithecus* discussions as new materials were discovered and older specimens reanalyzed. This interpretation was clear in the reconstruction and analysis of the Fort Ternan individual (Andrews and Walker 1976) and helped mitigate the effects of the narrow, parallel-sided dental arcade and associated mandible with a sectorial P_3 on the hominid interpretation of the specimen (although this process actually took some time, during which the premolar in question was "semi-sectorial" or "incipiently bicuspid" [Simons 1976c]).

By the earlier 1970s, the molars of already known *Ramapithecus* specimens had been reexamined and evidence found that was interpreted to show powerful mastication in this form (Simons and Pilbeam 1972), including thick enamel, interproximal attrition, and a steep molar wear gradient (the alternative to the delayed-maturation interpretation). Other features now brought into focus were the flat, deep face, vertical incisors, and heavily buttressed mandible (especially at the symphysis).

Newly discovered specimens were interpreted in a framework that itself was changing (Wolpoff 1975). Description of the Candir mandible (Andrews and Tekkaya 1976), for instance, emphasized the buttressing of the mandibular corpus and symphysis and shortening of the anterior face, while the shape of the dental arcade confirmed the Fort Ternan reconstruction. However, the sectorial P_3 was still being regarded as "molarized" (Simons 1976c). While the Pyrgos mandible was the only one known of the Greek specimens, it was described as extraordinarily australopithecine-like (Walker 1976), with an "arcurate" mandibular arcade (Simons 1978), although the (remains of) large molars and a thick mandibular body were also given some attention. The numerous more recent discoveries of much more complete Greek specimens have been interpreted somewhat differently (de Bonis and Melentis 1977a, 1978).

The Hungarian finds from Rudabánya (Kretzoi 1975) were viewed as *Ramapithecus*-like forms, although not actually allocated to *Ramapthiecus*. Indeed, Kretzoi (1976) found greater similarities between *Rudapithecus* and early *Homo* than between *Rudapithecus* and *Australopithecus*. Simons (1976c) disagreed on both points, allocating this form to *Ramapithecus* and noting the dental and gnathic adaptations for powerful mastication. At the same time, he compared the canine form and wear on the RUD 12 maxilla with the Hadar australopithecine palate AL 200 as well as with *Gigantopithecus*. The ground was shifting for the functional interpretation of the *Ramapithecus* canine; it no longer was incisiform (as it was in the middle of the decade [see Conroy and Pilbeam 1975]), but with the recovery of the Hadar australopithecines it could still be related to the hominids. In a similar manner, once analysis of australopithecine dental arcades showed that they were not parabolic (Genet-Varcin 1969), this condition was no longer claimed to characterize *Ramapithecus* (Simons 1977, Simons and Pilbeam 1978).

With the addition of the Candir and *Rudapithicus* specimens to the *Ramapithecus* sample, it became generally accepted that *Ramapithecus* combined a short face, thick molar enamel, a nonparabolic dental arcade (with varying degrees of posterior divergence), a relatively thick corpus and symphysis, and widely divergent zygomatic processes in a pattern that so clearly indicated powerful mastication (Wolpoff 1974, Hylander 1979) that at least one researcher described the complex as most resembling a miniature hyperrobust australopithecine (Walker 1976). At the same time, the canine function was seen to be pongid-like (although also like that of the earliest hominids). With the acceptance of Jolly's model, some of the criticisms of one decade were incorporated into the interpretations of the next.

Thus, by the closing years of the last decade, a firm case was being made that *Ramapithecus* showed the dental and gnathic adaptations of a powerful masticator and could be considered a hominid because it fit Jolly's model of hominid origins (Simons and Pilbeam 1978). These characteristics (in contrast to those emphasized by Darwin) related to an adaptive shift involving terrestrial small-object feeding in semiopen or open ecozones (Pilbeam 1979). This interpretation took most of the decade to develop fully because the anterior teeth were never really deemphasized until the discoveries at Rudabanya and Hadar allowed a rather different comparison to sustain the hominid interpretation.

In all, this was truly a case in which (to paraphrase Samuel Butler) the foundations were changed while the superstructure remained the same. The focus shifted from the front to the back of the jaw, and *Ramapithecus* remained a hominid.

Although Jolly's hypothesis was about hominid origins and the examples it drew upon were Pliocene *hominids*, it was the application of this model to *Ramapithecus* that carried the seeds of its destruction. Three factors have combined to set the stage for questioning whether the seed-eaters hypothesis can account for *hominid* origins.

First, a set of morphological characteristics indicating a powerful masticatory complex appears in a variety of mammals. Long before the hypotheses was published, the dental and gnathic characteristics associated with powerful grinding and crushing were recognized in a species whose diet was clearly not small objects. This complex, (perhaps mistakenly) referred to as the "T complex" because of Jolly's analogy using *Theropithecus*, was earlier described in the giant panda (Sicher 1944) in a comparison with bears that parallels Jolly's comparison of *Theropithecus* and *Papio*. Indeed, in his discussion of this bamboo-eating species, Davis (1964) found the closest analogy to be with *"Paranthropus robustus."* A similar morphological complex characterizes some of the leaf-eating ceboids (Kinzey 1974). The point is that, in living forms, the dental/gnathic complex described by Jolly does not necessarily indicate small-object feeding, let alone a terrestrial adaptation. It does not necessarily lead to hominization. Consequently, the same argument must apply to the interpretation of *Ramapithecus* as well as of other fossil primates. Thus, for instance, White (1975) indicates the possibility of a giant-panda-

like diet for *Gigantopithecus* and suggests that the pandas may have replaced this primate. Kay (1981) describes an alternative dietary interpretation for the *Ramapithecus* specimens themselves (nut eating) that involves neither a terrestrial adaptation nor any of the aspects of hominization hypothesized by Jolly. These arguments tend to remove some of the cause-and-effect aspects of the seed-eaters hypotheses by showing that a morphological adaptation to powerful grinding and crushing does not require a diet of seeds or grains. Moreover, there is an ample variety of other difficult-to-masticate food sources that early or pre-hominids might have utilized (Coursey 1973, Wolpoff 1973, Kay 1981).

Second, Jolly's hypotheses has always been weak in its explanation of how other basal hominid features might have followed from small-object feeding. For instance, whatever the validity of the argument that small canines remove the restriction of canine interlock and allow free lateral movement of the jaws, the argument can only account for canine reduction; it does not account for the change in canine form and function in the hominids. Bipedalism, too, has never been adequately explained by this hypothesis, and attempts to do so (Wrangham 1980a) have been less than convincing. Finally, not all workers are as willing as Jolly to regard culture, tool making, and expanded behavioral complexity as hominid attributes that evolved after hominids originated. It is not clear that these lack all causal relation to hominid origins. Thus, while small-object feeding could account for some of the features found in the earliest Pliocene hominids, it does not necessarily account for others.

Third, the focus on this dental/gnathic complex resulted in renewed support for the earlier claims that *Ramapithecus* was similar, or identical, to the other Asian hominoids (Frayer 1976, 1978; Greenfield 1974, 1975, 1977, 1979) because many of these earlier claims were based on the same diet-related characteristics but these features were not regarded as especially important under the Darwinian hypotheses. The special similarities (if not identity) of *Ramapithecus* and *Sivapithecus* (Andrews and Tekkaya 1980; Greenfield 1977, 1980; Frayer 1978; Pilbeam 1979), the combined mandibular morphology of both taxa in one species of *Ouranopithecus* (de Bonis and Melentis 1980), and the likelihood that these hominoids were markedly dimorphic (Frayer 1976, Greenfield 1979, Pilbeam 1979, Wolpoff 1980) suggest that at the very least the paradigm for the genus *Ramapithecus* must be expanded to include more specimens and a wider range of variation. The features that assumed importance in the newer interpretation, which was a consequence of the framework provided by the seed-eaters hypothesis, were widespread, and when sexual dimorphism was taken into account it became far from clear how many species were actually involved in what has increasingly come to be regarded as a single adaptive radiation (Wolpoff 1980) that might be referred to by the nontaxonomic term "ramapithecine" (see also Kay and Simons 1983). By many accounts, the ramapithecines would include specimens allocated to *Ramapithecus, Kenyapithecus, Sivapithecus, Ankarapithecus, Rudapithecus, Bodvapithecus, Ouranopithecus, "Hemianthropus,"* and *Gigantopithecus* (Pilbeam 1979, Kretzoi 1975, Wolpoff 1980).

An adaptive radiation can result from the appearance of a new adaptation, with consequences that allow a previously unutilized set of niches to be entered (Simpson 1953b). Rapid speciation almost invariably follows (Stanley 1979), and the resulting taxonomic group soon becomes highly diversified. In the ramapithecines, the common element in the remains now known for the group is the masticatory apparatus, adapted for a diet requiring powerful or prolonged grinding and crushing. It is likely that this reflects the novel adaptation that was the basis of the subsequent radiation. The exploitation of otherwise unusable dietary resources would allow adaptation to new ecozones because this dental/gnathic adaptation is not an adaptive specialization; it acts to expand the range of available resources.

Apart from the masticatory complex, other common elements are difficult to identify because few skeletal parts besides jaws and teeth have been found. Moreover, features unique to the ramapithecines cannot always be clearly distinguished from shared primitive features (such as the retention of marked sexual dimorphism) because little is known of the ancestral condition. Recent evidence suggests that the ramapithecines evolved from a *Proconsul* or *Proconsul*-like form of approximate *P. nyanzae* size (Pickford 1982).

Temporal and geographic considerations alone suggest the existence of a fairly large number of ramapithecine species. Of greater importance is the number of lineages in the adaptive radiation. Although workers such as Pilbeam (1979) have indicated five or more, the fact is that no ramapithecine-bearing locality has provided evidence of more than two contemporary lineages, and in many cases the data can be interpreted to show only one. This could be a consequence of the generalizing aspects of the masticatory morphology underlying the radiation. Instead of more finely subdividing the new niche, competition between the emerging ramapithecines seems to have promoted their rapid spread. The resulting pattern emphasizes allopatric more than sympatric species proliferation in a manner similar to the distribution of baboons and baboon-like forms such as *Theropithecus*, but the extent of allopatric species proliferation may have been markedly greater than in these cercopithecoids.

The evidence of variation within the ramapithecine adaptive radiation tends to be obscured by two circumstances. First gross size would appear to be the most dramatic variant in the recognized ramapithecine forms (molars, for instance, range from smaller than *Homo sapiens* to *Gigantopithecus* size). The problem this creates is one of separating variation due to scaling from that due to other factors. Even so, the adaptive importance of size variation over this range should not be understated. Second, because jaws and teeth are the most usual fossil remains, most comparisons are limited to the very adaptive complex that forms the basis of the radiation and consequently would not be expected to show dramatic adaptive differences within it.

In the context of this common adpative pattern there are, however, important morphological variants. For instance, enamel thickness varies continuously from the extremely thick condition in *Ouranopithecus*, *Gigantopithecus*, and *"Hemianthropus"* to a thinner expression in some of the Rudabánya forms. Even at individual sites this feature varies considerably. Thus at Rudabánya the larger specimens *(Bodvapithecus)* have quite thick enamel while the smaller ones range from this thickness to a thinner condition such as that seen at its extreme in RUD 12. Kay and Simons (1983) probably had this specimen in mind as the basis for their claim that molar enamel at Rudabánya is thin and consequently that *Rudapithecus* should be allocated to the European dryopithecines rather than to the ramapithecines. However, RUD 12 lies at the low end of a range of marked variation in enamel thickness, even within the *Rudapithecus* remains. Similarly, marked variation can be found in the deeply incised molar wrinkles that characterize some of the ramapithecine specimens. As in the case of enamel thickness, the variation is one of frequency between samples. Cingulum expression represents yet another varying feature. Variation can be seen in the development of the P_3 metaconid, which ranges from complete absence to the full development of an equal-sized cusp. Kay and Simons (1983) report that the frequency of P_3 metaconid enlargement observable in some of the ramapithecine specimens (exclud-

ing *Gigantopithecus*) is comparable to that observed in chimpanzees and orangs (although not gorillas). However, they further claim that the degree of development in some of the ramapithecine specimens exceeds that which they have observed in these living areas. This forms part of their argument for a special relationship between the ramapithecines and the hominids. Yet, the fact is that this feature is quite variable in the radiation, ranging from the gorilla condition (no metaconid) to the *Homo*-like bicuspid form of the *Gigantopithecus* premolars, with cusps of equal size.

Few nondental features allow comparison of the ramapithecine forms. The distal humeri from Hungary and Pakistan differ notably; the Fort Ternan humerus could represent a third variant, but its association with the African ramapithecine is uncertain. On the other hand, the cranial remains from Hungary, Pakistan, Greece, and the People's Republic of China are surprisingly similar.

The relation of the adaptive radiation of the ramapithecines to hominid origins will be discussed below. The acceptance of the notion of a ramapithecine adaptive radiation provides the third basis for questioning the seed-eaters model as a hypothesis about *hominid* origins. This is because the ramapithecine adaptive radiation was highly successful in terms of geographic range and survivorship of the taxonomic group. It has recently become evident that there were at least two nonhominid Pleistocene survivors.

RAMAPITHECUS AND THE LATE-DIVERGENCE HYPOTHESIS

The idea of a fairly recent divergence between humans and apes is hardly new. Early contentions of a late divergence were influenced by the very short estimates of the earth's age that preceded a full understanding of radioactive decay. The late-divergence hypothesis as presented by Greenfield (1980, 1983) specifically focuses on the divergence between humans and the *African* apes, emphasizing that there were two different divergence points in the evolution of the recent hominoids. This distinction is an important one.

Late divergence between humans and African apes and an earlier separation of *Pongo* have been supported morphologically, genetically, biochemically, and, most recently, paleontologically.

Morphologically, it has long been recognized that the chimpanzee is the most human-like of the pongids (Huxley 1861, 1863; Gregory 1930a; Simpson 1963; Washburn 1968b), and the idea of a very *Pan*-like ancestor for the hominid line has been maintained for decades (Schwalbe 1923, Coolidge 1933, Gregory 1934, Weinert 1944, Washburn 1968b, Zihlman et al. 1978, Zihlman 1979, among others [although not necessarily like *Pan paniscus* (McHenry and Corruccini 1981, Johnson 1981)]). This special relationship is one important aspect of the late-divergence hypothesis: the morphological data provide information about divergence sequence, although not absolute date.

These morphological comparisons are independently supported by genetic analysis. Chromosome banding studies show an especially close relation of *Homo, Pan,* and *Gorilla,* with *Pongo* somewhat divergent (Miller 1977, Yunis, Sawyer, and Dunham 1980). The comparison of protein coding sequences indicates an extraordinary genetic similarity between *Homo* and *Pan* (King and Wilson 1975). These genetic comparisons reveal less difference between the two genera than is common between sibling species. The extent of similarity far exceeds the minimum necessary to show that the relationship is genuine (Doolittle 1981). Taken at face value, these data would tend to indicate recent divergence as well as extreme closeness of relationship for the African hominoids and a more distant relation and earlier divergence for *Pongo*.

Biochemical analysis also supports the particular closeness of relationship of *Homo, Gorilla,* and *Pan.* In a recent series of summaries for numerous genetic systems (Goodman 1976b, Dene, Goodman, and Prychodko 1976), it was found that immunodiffusion studies, nucleotide replacements, and the analysis of various proteins consistently show humans, chimpanzees, and gorillas to be more closely related to each other than any of these are to orangutans. Thus, the divergence sequence suggested earlier on the basis of morphology is supported.

Biochemical studies have also been used to calibrate a divergence "clock" for this sequence (Sarich and Wilson 1967a, Sarich 1974, Sarich and Cronin 1976).This "clock" seems to provide direct evidence for a very late *Pan-Homo* divergence (estimates based on this procedure have ranged between 3,500,000 and 5,500,000 years ago). However, the techniques that result in these estimates give divergence times for other species that are at significant variance and virtually any interpretation of the fossil record (Uzzell and Pilbeam 1971, Jacobs and Pilbeam 1980, Read and Lestrel 1972, Radinsky 1978, Walker 1976 and references therein). Moreover, there have been an extraordinary number of criticisms of the molecular "clock" (Lovejoy, Burstein, and Heiple 1972, Lovejoy and Meindel 1973, Read 1975, Read and Lestrel 1970, Jukes and Holmquist 1972, Corruccini et al. 1980, Goodman 1974, Fitch and Langley 1976, Jukes 1980). Probably the best way to summarize the very disparate points raised is that the "clock" simply *should not* work. This conclusion supports the paleontological analyses that claim the clock *does not* work when applied to divergence times over broad time spans. Consequently although biochemical evidence seems to support a late *Pan-Homo* divergence, I believe that this is a red herring and the molecular "clock" does not support any divergence time, just as other independent evidence for late *Pan-Homo* divergence does not support the molecular "clock." However, while specific divergence *dates* may be rejected, I do not believe it is possible to dismiss the implications of the biochemical evidence for divergence *sequence* (Greenfield 1983).

Finally, paleontological evidence also supports the contention of an especially close relation between the hominid line and the ancestors of the African apes. The relationship has long been recognized for *A. africanus* (Le Gros Clark 1947), and the discovery of an even closer approach to the chimpanzee condition in *A. afarensis* (Johanson and White 1979) is not surprising. While the Pliocene fossil evidence can provide no more than a minimum divergence date (4,000,000 years), the number of primitive and specifically chimpanzee-like features in the known crania support the contention of a late divergence.

In sum, these data strongly support the notion that the divergence of the lineage leading to *Pongo* from the lineage leading to *Homo, Pan,* and *Gorilla* was earlier than the divergence of these three African hominoids from each other. The data further suggest that the later (African) divergence may have been fairly recent.

The effect of the late-divergence hypothesis on the interpretation of the ramapithecines is a consequence of the earlier divergence proposed for *Pongo.* This became evident when two of the Pleistocene survivors of the ramapithecine adaptive radiation were first recognized as such. One of these, *Gigantopithecus,* survived into the Middle Pleistocene, where contemporaneity with hominids has been established (Hsu, Wang, and Han 1975). The other survivor is *Pongo.*

Once again, the assertion that *Pongo* evolved from a ramapithecine was not initially established from recently discovered data, although the interpretations and implications of this contention are recent. Gregory and Hellman (1926) noted a number of morphological similarities between *Sivapithecus* and *Pongo* molars and suggested an ancestral relationship. Of course, they did not regard *Sivapithecus* as a ramapithecine (*Ramapithecus* had yet to be recognized). The dental similarities can now be shown to extend throughout the ramapithecine remains and include, in addition to those noted by Gregory and Hellman, the following variably expressed features: (1) enamel thickness; (2) deeply incised wrinkles that persist after crown morphology has been worn away (especially in the Lufeng ramapithecines and *"Hemianthropus"* but also sporadically in the remains from Pakistan and Hungary); (3) asymmetric heteromorphic lateral maxillary incisor size and form; and (4) central maxillary incisors that change in angulation during life so that in younger individuals a lingual wear plane typically extends from the tip to the base.

Newly discovered ramapithecine cranial remains also support the hypothesis of a ramapithecine ancestry for *Pongo.* In particular, the face and partial cranium from Pakistan (GSP 15000) reveal a suite of extraordinarily *Pongo*-like details (Pilbeam 1982). Specific resemblances include the facial profile, high hafting of the braincase on the face (a vertically oriented masticatory system), morphology of the zygomatic (especially size and position of the zygomatic foramen), orbit shape, palate shape and the form of the incisive and palatine foramina, relative I^2 size, and the form of the articular eminence of the glenoid fossa. The lower face and palate from Turkey (MTA 2125) have been similarly described: "the closest comparisons in most cases were with the orang-utan" (Andrews and Tekkaya 1980:94). The upper face from Rudabánya (RUD 44) is also characterized by *Pongo*-like features, including relative orbit heights, a very wide outer-orbital area, and marked converging anterior temporal ridges. Finally, *Pongo*-like features seem to characterize the newly discovered Lufeng crania (Wu 1981). As in the Rudabánya remains, the most marked resemblances are in the upper face and frontal, including the marked temporal ridges, the wide outer-orbital pillars, and the shape of the orbits and supraorbitals (Xu and Lu 1980).

Interestingly, Kay and Simons (1983) argue against a special affinity to *Pongo* in the ramapithecine facial remains, although their discussion omits the most complete, Lufeng and GSP 15000. Their contention is mainly based on the Turkish face, MTA 2125, and involves arguments that I regard as less than convincing. For instance, in claiming the specimen is not as prognathic as *Pongo,* they use as a measure of maxillary prognathism what is actually one of relative alveolar height (alveolar height/palate breadth at M2), bearing no relation to how *prognathic* the premaxilla might be, independent of its relative size. Similarly, they admit to the marked maxillary incisor heteromorphy in the Turkish specimen but feel it does not align it with *Pongo* because the incisor heteromorphy in this living ape was presumably attained by relative I^1 (this was calculated relative to molar size and therefore potentially confuses I^1 expansion with molar reduction). Besides positing an hoc explanation for what appears to be a simple relationship, this argument ignores the important morphological similarity of I^2 form in MTA 2125 and other specimens to that in *Pongo.* I would argue, in contrast, that the central and lower portions of the known ramapithecine faces do in fact show specific resemblances to *Pongo*'s. On the basis of the faces from Rudabánya, Ravin, Turkey, Pakistan, and Lufeng, these similarities would seem to include the incisor heteromorphy (metric *and* morphological), the changing incisor wear plane beginning with the very marked angulation characteristic of young individuals, the premaxillary prognathism as measured by the angulation of the premaxilla, and the strong superior-medial angulation of the canine roots. However, as I have said, I believe that an even stronger resemblance between these faces and *Pongo*'s is shown in the upper face and frontal as represented at Rudabánya, Pakistan, and Lufeng.

Kay and Simons conclude that the Turkish face shows some resemblances to those of all of the extant pongids. While I believe that there are more specific relations with *Pongo* than they admit, I would support this basic conclusion. To the general ancestral features they discuss might be added the variable characteristics of the ramapithecine dentitions described above, including cingulum development, molar wrinkling, enamel thickness, and P_3 metaconid expression. The fact is that ramapithecine variation in these facial and dental features would potentially allow the ancestry of every living hominoid group to be found.

However, in this context, the specific resemblances of *all* of the ramapithecine faces to *Pongo*'s have unavoidable implications for the hypothesis that the ancestry of *Pongo* is to be found within the ramapithecine adaptive radiation. Of course, only one of the ramapithecine lineages could be ancestral to this living ape; on the basis of the present evidence (especially the new material from China and Pakistan) and the argument of geographic proximity, this almost certainly would be an Asian lineage. However, if one of the ramapithecines is ancestral to *Pongo*, the implications are far-reaching for the entire adaptive radiation.

There are two hypotheses about how the ramapithecines may be related to hominids, depending on when the ramapithecines appeared relative to the split between the line leading to *Pongo* and the line leading to the African hominoids and subject to the constraints of the data discussed above..

If the split between the African hominoids and the line leading to *Pongo* occurred *before* the ramapithecines evolved, then the ramapithecines must be uniquely associated with either the *Pongo* line or the African hominoid line. The evidence for the ancestry of *Pongo* among the ramapithecines indicates that if this hypothesis were correct, no ramapithecines contributed to hominid ancestry because they were on the *Pongo* side of the split (Andrews 1982). It would then follow that the numerous dental and gnathic similarities of the earliest hominids (especially *A. afarensis*) with the ramapithecines would have to be interpreted as parallel independent acquisitions. This is far from impossible, since most of the similarities are in the dental/gnathic complex associated with powerful mastication and, as noted above, this complex has appeared again and again in unrelated forms as a common response to a similar adaptation. Thus, this interpretation cannot be easily dismissed. On the other hand, it was exactly this sort of evidence that led to the contention of a ramapithecine adaptive radiation. If the same criteria were applied to the jaws and teeth of *A. afarensis*, the functional interpretation of the morphology of this extraordinarily megadont early hominid form would suggest the same conclusion, namely, that *A. afarensis* is part of or closely related to the ramapithecine adaptive radiation. Indeed, this is how R. E. F. Leakey (1976a) interpreted the Hadar female AL 288-1. Thus, I consider this first hypothesis to be the far less likely one (Wolpoff 1981). Indeed, unless one is willing to postulate a fairly extensive amount of paralleled or even convergent evolution in the hominoids, it can probably be rejected.

If the split between the African hominoids and line leading to *Pongo* took place *within* the ramapithecines, some sort of ramapithecine ancestry for hominids as well as for *Pongo* is implied. Thus one could account for the fundamental aspects of both of the sets of interpretations that have developed in the last decade: the relationship of the ramapithecines to the earliest hominids. Indeed, an alternative way of stating this hypothesis is that the earliest hominids are part of the ramapithecine adaptive radiation.

Under this hypothesis there is a ramapithecine ancestry for the African apes (as discussed by Greenfield 1977), but these forms diverged markedly from the ancestral condition because of their specific dietary adaptations, with effects on their dental morphology and function described by Kay (1975, 1981) and Maier and Schneck (1981). At the same time, however, it would account for the specific resemblances of *Australopithecus* and *Pan* by presuming a late divergence between them.

This hypothesis thus necessarily posits a specific ramapithecine ancestor for the hominids, a species which is *also* ancestral to *Pan* and *Gorilla*. It is possible that one of the Eurasian ramapithecines represents this ancestor, but I suspect that the limited evidence now available does not allow a decision as to which (if any) of these lineages is the most likely (Wolpoff 198). Traditional comparisons have emphasized the hominid-like aspects of the smaller species (albeit as a hominid and not as a common ancestor of the hominids and African apes), such as *Ramapithecus* or *Rudapithecus*, but detailed dental morphology and size considerations could argue for one of the larger forms (*Sivapithecus*, *Ouranopithecus*). However, Middle Miocene geography would suggest the possibility that none of the Eurasian forms represent this ancestor.

If an African ramapithecine is a more likely candidate for this ancestor, the situation is little improved. There are virtually no hominoid fossils known between the Fort Ternan ramapithecines and *A. afarensis*; Lukeino, Ngorora, and Lothagam could be interpreted as either ramapithecines or early hominids, although there are substantial similarites between Lothagam and the mandibles from Laetoli. It is possible that the Fort Ternan ramapithecine represents or is closely related to the common ancestor of the African hominoids. On the other hand, it may be too early. The fact is that the date of the split between the African and Eurasian forms is unknown, and this split could be later that the Fort Ternan remains. Thus, in my view there is no ramapithecine ancestor for the African hominoids that can be identified unambiguously at this time. While it is possible that one of the Eurasian forms is this ancestor, I believe it more likely that, as the African fossil record spanning the later Miocene is better explored, a suitable ramapithecine lineage (perhaps beginning with Fort Ternan) will be found.

I predict that when found the Late Miocene African ramapithecine will be identifiably similar and fairly closely related to the known Eurasian remains. Craniodentally it will resemble *A. afarensis*, although with much more projecting and more sexually dimorphic canines. It will *not* resemble a chimpanzee to a significantly greater extent than *A. afarensis* does, and it will especially not resemble a pygmy chimpanzee. It is likely that at the time of the human–African-ape divergence, character displacement and other consequences of competition had initially greater effects on the apes than on the human line. Thus, I suggest that the last common ancestor of the African apes and humans was probably as different from living apes as it was from living humans. No living ape, even an enculturated one such as described by Kortlandt (1972), could form an adequate model for this ancestral form, any more than a living human group could.

Such an ancestral species would be expected to retain the dental/gnathic complex associated with powerful or prolonged chewing. At the same time, one might expect the more elongated cranial form also emphasizing anterior dental loading that characterizes the African apes in contrast to the shortened, more vertically oriented form of *Pongo* crania. If one were to describe this hypothetical ramapithecine cranium as a chimpanzee-like vault with a ramapithecine face and dentition (with the consequent molar-loading-related superstructures on the vault), this description would be fairly close to the known cranial remains of *A. afarensis* (Johanson 1980).

The locomotor complex of this ramapithecine might be expected to reflect an at least partly arboreal adaptation (climbing, arm hanging) without the brachiating specializations of the African apes or their associated knuckle-walking quadrupedalism. This is suggested by a number of comparisons, including the lumbar elongation of the australopithecines (STS 14 has six functional lumbars) combined with the arm-hanging abilities inherent in the human upper torso and even more markedly expressed in *Australopithecus*.

In sum, I believe the evidence best supports the late-divergence hypotheses presented by Greenfield (1980). I would suggest that adaptation for foods requiring powerful or prolonged chewing arose among one of the *Proconsuls* and that, because this provided the basis for utilizing a much wider range of dietary resources, a very successful adaptive radiation of hominoids resulted. With the geographic spread of the ramapithecines, the radiation was split into Eurasian and African branches (the *Pongo*-African hominoid split) and perhaps into European and Asian branches as well. I predict that the eventual discovery of more complete remains will demonstrate a much wider adaptive range within this radiation than the analysis of the dental/gnathic complex now indicates. It would appear that the western portion of the Eurasian ramapithecines became extinct while there were at least two Pleistocene survivors to the east : *Gigantopithecus* and *Pongo* (the so-called giant-orang teeth and the "*Hemianthropus*" specimens [2] from South China may represent a third Pleistocene survivor).

Relatively little is known of the African ramapithecine branch. Genetic evidence relating *Pan* and *Homo* and the morphological relations of *Pan* and *A. afarensis* indicate that during the late Miocene one of the African ramapithecine lineages further split into lines leading to the adaptively specialized African apes and a hominid line.

RAMAPITHECUS AND HOMINID ORIGINS

While I have argued that none of the ramapithecines are hominids, I propose that the earliest hominids were a special form of ramapithecine. I believe this is a distinction with a difference because it brings a different focus to the problem of hominid origins than is usually applied.

This difference is the result of two recent changes in interpretation. The first is a swing away from the earlier notion of australopithecines as fully human beings (albeit with somewhat diminished mental capacities) advanced by workers such as Le Gros Clark (1967, perhaps in reaction to Zuckerman) and Isaac (Isaac and Isaac 1975; see also Leakey and Lewin 1977). The second is the recognition, after a decade of behavioral studies, that the African apes, and by implication the latest common ancestor of these apes and humans, are far more hominized than was once believed. As a result, the ramapithecines can be viewed as more human-like while the earliest humans can be viewed as more ramapithecine-like without the contradictions that have historically been associated with this convergence of views.

An example of the first change is the way the importance of brain size has changed from the Darwinian expectation that dramatic improvement of mental capacity was a critical underlying factor in hominid origins. The evolution of modern brain size (and presumably the cultural/behavioral changes that came with it) has taken most of the Pleistocene. Evidently, only moderate or even minimal endocranial expansion can be associated with hominid origins. Holloway (1980) has published figures for Jerison's encephalization quotient for a sample of *Pan*, where $EQ =$ brain weight/$0.12*$ body weight[666]. The midsex value for *Pan* is 2.96. Using the midsex cranial capacity for *A. africanus* which I have determined (443 cc) and an approximate midsex body

weight average of 40 kg (Wolpoff 1973) produces an EQ of 3.20. These data, with the evidence for neural reorganization (Holloway 1966a, 1976), suggest that while some brain-related changes may have been associated with hominid origins, their magnitude was small compared with subsequent Pleistocene changes. The extent of actual change was probably even smaller and will be better estimated when data for the earlier species *A. afarensis* are available.

At the same time, the progressive hominization of the ramapithecines has brought the realization that hominid origins may not have involved the specific origins of hominid features. The changes surrounding this event appear to have been more concerned with emphasis and importance than with the appearance of evolutionary novelties. For example, the evidence for chimpanzee tool making that has accumulated over the past two decades increases the likelihood that the (African ramapithecine) common ancestor of hominids and the African apes was a tool maker. Yet, this contention seems to be contradicted by the numerous assertions that stone tools do not pre-date 2,500,000 years ago. The absence of any stone tools associated with *A. afarensis* has been taken to mean that tool making originated after hominid origins and thus did not play a role in the event (see, for instance, Lovejoy 1981). I believe this is an apparent contradiction stemming from a confusion between tool making and *stone* tool making that has existed over the last decade. Perhaps this confusion is a result of the difficulty of unambiguously identifying utilized or rudimentarily flaked stone (the ghost of the Kafuan is still with us), or perhaps it is a reaction to the enthusiastic interpretations of Dart or to the observed manufacture of nonlithic tools by chimpanzees. Whatever the case, it is probably a mistake to treat the two as synonymous in terms of their timing and their effects on the course of hominid evolution. Undoubtedly, stone tool making was critical in hominid evolution. Its development may well have been associated with the divergence of the two hominid lineages of the earliest Pleistocene and the marked and rapid development of features associated with further hominization in one of them. Nonetheless, it is likely that the manufacture of tools made of perishable materials and the use of unmodified stone long preceded the development of recognizable lithics. The potential adaptive importance of nonlithic tools could hardly be overstated (Wolpoff 1980). In view of the fact that this behavior is shared with *Pan* (see van Lawick-Goodall 1973, McGrew, Tutin, and Baldwin 1979, Teleki 1974, Harding and Teleki 1981), its origin likely pre-dates the hominid-African-pongid divergence, and changes in the importance of this behavior are a potentially critical aspect of hominid origins. If so, this does not at all mean that tool making was unimportant in the events leading to and following hominid origins, but rather that it did not originate then.

A similar example can be found in the changing views of diet and diet-related behavior in the course of hominid origins.The seed-eaters hypothesis, or something like it, would seem to pertain to *hominoid* rather than to *hominid* origins. Yet, this does not mean that powerful mastication was *unimportant* in the process of hominid origins. This adaptive complex simply did not originate then. Darwin emphasized the causal influence of hunting in his model of adaptive change in the earliest hominids. Applying Jolly's model to the earliest stages of hominoid evolution and taking the continued evolution of the powerful masticatory apparatus in subsequent hominid evolution (Wolpoff 1980) into account, it appears likely that organized hunting did not play a preeminent role in the earliest stages of hominization. Once again, however, this does not imply that organized hunting played *no* role in hominid origins, especially since most of the elements of organized hominid hunting appear in chimpanzee behavior (Galdikas and Teleki 1981). Like tool making, this behavior was probably characteristic of the ancestral African ramapithecine form.

One might say, in the end, that the gap between humans and their (presumably more apelike predivergence) ancestors has been successfully bridged from both directions.

In many respects, the model I am suggesting is very Darwinian in its scope, although not in its detail. Yet, in my view the main difference from Darwin's emphasis involves a distinction between *origins* and *importance*. Of the elements discussed by Darwin, it is possible that only bipedalism actually originated at the time when the hominids became a distinct lineage. Tool making and the beginnings of organized hunting may have preceded this event, while significant expansion of the brain and functional change in the canine probably followed it. Similarly, the development of a powerful masticatory apparatus seems to have preceded the event.

Thus, I contend the ramapithecine ancestor of the hominids and the African apes had already undergone a number of changes that are generally regarded as hominization. Shared characteristics of the living hominoids combined with the paleontological evidence discussed here indicate that this ramapithecine form was behaviorally complex, a tool user and a rudimentary tool maker, an omnivore utilizing a wide variety of dietary resources ranging from difficult-to-masticate foods to protein obtained through organized hunting and systematically shared by at least part of the social group, an incipient biped (the behavior was possible but not morphologically efficient), and just possibly a more complex communicator than is generally thought, utilizing a limited but symbolic-based open communication system. Hominid origins would seem to have involved as much a shift in the importance of these characteristics as the origin of uniquely hominid ones.

In sum, I propose that the continued reanalysis of *Ramapithecus* has ultimately affected the acceptable model of hominid origins, just as the reverse has been the case. In my view this continued reanalysis has resulted in the contention that the ramapithecines were not hominids, but the earliest hominid was a ramapithecine. Many of the interpretations and implications of Jolly's model have been focused on the problem of hominoid, rather than hominid, origins. Finally, this reanalysis, along with the last decade's advances in pongid behavioral studies and fossil hominid recovery, renews the focus on a modified Darwinian model for the origin of our lineage.

AN ECOLOGICAL MODEL OF HOMINID ORIGINS

There clearly were changing interrelated ecological adaptations associated with hominid origins. These changes are best discussed in the context of the fact that the effects on the African *pongid* [3] lineage (or lineages) were at least as great as on the hominids. It is critical to remember that *hominid* origins are also *African ape* origins.

The actual speciation event separating hominid and (African) ape lineages will probably always remain unknown, since it need not be directly associated with any of the adaptive changes that followed. The genetic isolation and ultimate speciation of ramapithecine populations was not necessarily a consequence of adaptive divergence (although this is always a possibility). However, following this event the adaptive shifts in the lineages could be interpreted as the result of subsequent competition between them.

This suggested model for divergence makes no assumptions about the niche of the ancestral ramapithecine species. The group may have been primarily arboreal or largely terrestrial prior to the speciation event. I believe that actual finds, ecological associations, and analysis of the locomotor skeleton (when discovered)

will ultimately provide the data needed to resolve this question. However, given the consequences of competition following speciation, one credible hypothesis is that the initial niche was a broad one, involving dense to fairly open parklands with some utilization (perhaps seasonal) of even more densely forested localities. Besides best fitting a dietary regime indicated by the masticatory complex, such a hypothesis would allow one to view the initial results of competition as dividing the niche occupied by the parent ramapithecine species into two narrower and less overlapping adaptive zones. These would be the more open parklands grading into savanna (hominids) and the denser woodlands grading into forest (African apes). The effects of subsequent competition on the differing adaptations outlined by this initial division would presumably be continued niche divergence that proceeded until the adaptive zones were sufficiently separate significantly to reduce competition over limiting resources. This model ties the appearance of a terrestrial, open-grasslands adaptation in the hominids to a *combination* of competitive exclusion *and* opportunism allowed by the expanded dietary resource base available to a hominoid species with both a powerful masticatory apparatus and rudimentary tool and weapon use. The model implies that there is no necessary link between the specific development of a powerful masticatory apparatus and any terrestrial adaptation (*contra* Pilbeam 1979). Moreover, it suggests a ramapithecine adaptation to ecological circumstances that would support any of the several models of preadaptations for bipedalism that have been recently proposed (Tuttle 1975b, Post 1980, Stern and Susman 1981).

As I reconstruct the divergence process, apes presumably reduced competition through dietary (and eventually dental/gnathic) specialization and locomotor changes (true brachiation, knuckle walking) allowing an effective woodland/forest adaptation (Andrews 1981b). Precluded from these ecozones by competition, the hominid adaptation was to more open regions. Building on their ramapithecine inheritance, a combination of powerful masticatory apparatus, the probably rapid development of efficient bipedalism, the use of rudimentary tools and weapons (digging sticks, clubs), and a series of social changes possibly related to the recognition of extended kinship relations (Wolpoff 1980, Allen et al. 1982) allowed a wide range of difficult-to-gather and difficult-to-masticate foods to help form the basis of an effective adaptation to a unique open-country niche. One would suspect that dietary items included seeds and grains, nuts, roots, and hunted, gathered, and scavenged protein during the dry season (Coursey 1973, Peters 1979). Wet-season food resources cover a much wider potential range and do not necessarily involve difficulties in acquisiton or mastication.

The primary hominid adaptation was, of course, behavioral, with the morphological changes coming as a consequence. Some of the elements of this behavioral change probably involved shifts in the emphasis and adaptive importance of preexisting behaviors such as the manufacture and use of tools. Others, however, were fundamental and perhaps novel. These involve what may have been a dramatic restructuring of socially defined roles and the consequent social expectations. For instance, Mann (1972) has emphasized the shifts in regulatory behavior that must have followed from the replacement of canines by tools in social displays and the consequent loss of physiological control (i.e., late canine eruption in males) over the initiation of adult behavior. Recent studies of labor division in chimpanzees (Galdikas and Teleki 1981) help call attention to the hominid shift in provisioning strategy and behavior already posited by Isaac (1978). Isaac's contention is that the use of a home base reflects reciprocal provisioning, and the recent studies of chimpanzee provisioning suggest that such reciprocity must have resulted from the appearance of two new elements for this behavior. Female chimpanzees are known to share foods, and regular male provisioning of females (although not necessarily along lines of biological relationship) usually occurs when there is an episode of hunting. The important elements introduced to this pattern by the emerging hominids almost certainly were the female provisioning of males and the orientation of provisioning networks along lines of biological relationships. If it were posited that early hominid females regularly gathered difficult-to-masticate foods as part of the developing grasslands adaptation, such a shift in provisioning strategy would fit the morphological evidence of increasing masticatory power in the early hominid evolutionary sequence. Interestingly, recent analysis of microscopic cut marks on animal bones from Olduvai (Lewin 1981) suggests that the animal skins were salvaged and possibly used as rudimentay containers (to carry gathered plant foods to a home base?).

The social basis for such a shift in provisioning has recently been discussed by Lovejoy (1981), who posits the appearance of monogamy as the binding mechanism for reciprocal provisioning and suggests that shorter birth spacing resulting from the improved diet for monogamous pairs underlay the population expansion of the emerging grasslands-adapted hominids. While monogamy is probably neither necessary nor perhaps even sufficient to account for this critical change in provisioning strategy (Allen et al. 1982), some form of social role definition based on kinship relations almost certainly evolved as its basis. Moreover, the potential complexity of relations in such a social pattern combined with the requirements of reciprocal food sharing would render the development of an open communication system very advantageous (cf. Isaac 1978, Wolpoff 1980, Holloway 1969, and others). The comparative paleoneurology of australopithecine endocasts provides some direct (although limited) supportive evidence for the contention of an early appearance for hominid language (Holloway 1976).

Thus, in sum, there are reasons to suggest that a number of recognizably cultural elements were associated with hominid origins and/or the subsequent adaptive changes that occurred early in the evolution of our lineage. While these behavioral changes must remain the most contentious element of any theory of human origins, their importance as primary causal factors in the sequence of observable morphological, demographic, and adaptive changes underlies the continued attempts to use the available comparative and interpretive data to delimit the conditions surrounding their appearance and evolution.

Judging from its expression in *A. afarensis,* bipedalism would also appear to have been an early critical aspect of the developing hominid adaptation. The main advantages of bipedal locomotion—freeing the hands, carrying, and long-distance energy-efficient stride—probably all played a role in what was very likely a rapid locomotor shift. Arguments about whether carrying babies was more important than carrying clubs (e.g., Lovejoy 1981) miss the entire point of this locomotor change; a *group* of early hominids could carry *all* of the items that have been deemed important in the various arguments about the origin of bipedalism.

The *A. afarensis* dentitions show that, unlike the appearance of bipedalism, the change in the canine cutting complex was more gradual. Individuals in this earliest hominid species show canine and premolar wear indicating a range including chimpanzee-like honing (White 1981), occlusal chiseling (Wolpoff 1979a, Wolpoff and Russell 1981), and flat grinding (Taieb, Johanson, and Coppens 1975). There is a corresponding polymorphism in the form of the P_3, ranging from single-cusped forms (sectorial [Coppens 1977]) to a bicuspid morphology with equal-sized cusps

(Johanson 1980) and including all of the variants between. Thus, while the functional change in the cutting complex may have begun with hominid origins, it required continued selection to attain the modern condition, in which the canine is morphologically and functionally incisorlike and the P_3 is incorporated into the grinding dentition.

It is tempting to suggest that the gradualness of this change reflects the gradualness with which tools replaced the cutting functions of the canine, but this argument is essentially circular and requires independent confirmation. Moreover, I have hypothesized (1979a) that the development of the bicuspid P_3 crown, with a ridge connecting the two cusps, may have provided means of retaining a form of the cutting function of the canine while reducing its projection and overlap. This would presumably be an intermediate step in the process of functional change in the anterior cutting complex. Whatever the case, the association of this change with tool use remains an unsettled issue.

Finally, just as brain-size expansion associated with complex behavior and cooperation were important in Darwin's model (we would term these developments "cultural" today), I believe that the origin of the cultural adaptation was probably critical in this modification of it. Behavioral interpretations without obvious morphological correlates are very difficult to assess, and the relevant morphological data (relative brain size, endocast analysis) have not been published for the earliest hominid species. Nevertheless, the later species *A. africanus* shows evidence of both limited brain-size expansion and neural reorganization, as well as delayed maturation, and it is surely short of a wild leap of faith to hypothesize that these correlates of cultural behavior had their origin in the social changes associated with the hominid adaptations to open country. Indeed, the specific elements suggested by Darwin, cooperation and language, may well have played a critical role in the group adaptations to this ecozone. Two decades of baboon studies have shown the importance of structured cooperative behavior in the savanna adaptations of this species, and the ramapithecine ancestor of the hominids probably brought a much more complex repertoire to the behavioral basis of the hominid adaptation, judging from the behavioral complexity of the African apes.

Whether or not culture (meaning structured learned behavior) actually originated with the hominids or developed as part of their successful open-country adaptation, its effects are demonstrable in the morphology of *A. africanus.* This early hominid species had already embarked on an evolutionary pathway that was and has remained unique.

CONCLUSIONS

I remarked at the beginning that the complex historical interplay between theories of hominid origins and the interpretations of *Ramapithecus* has affected the development of both. In many respects, what I have described is a full circle in which modified versions of the original hominid-origins theory and the original interpretation of *Ramapithecus* can be sustained, but not quite in the way they were first presented, while incorporating the bulk of the discoveries, interpretations, and criticisms that have appeared along the way. In the circumscribing of this circle, it is clear that the development of one could not have preceded without the development of the other, which is to say that there are neither factless theories nor theoryless facts.

If *Ramapithecus* itself was not a hominid, there is a great likelihood that the earliest hominid was a ramapithecine. If all of the details of Darwin's theory of hominid origins are not fully correct, virtually every one of them must still be accounted for by any current origins hypothesis. If Jolly's model cannot be applied to *hominid* origins, a modification of it may have critical importance in the interpretation of *hominoid* origins, and in any event the dental/gnathic complex he described for hominids was there at their beginning and played an important role in their earliest adaptive changes.

Of the participants in the intertwined developments of the last few decades, if it can be said that none were completely right it is also true that few were completely wrong, at least in the context of the model I have presented. Moreover, the interpretations that can be sustained in one form or another far outnumber those that must be completely rejected. In all, the historical development of human-origins theories and ramapithecine interpretations presents a satisfying contrast to the Piltdown fiasco and reflects the scientific aspect of paleoanthropological studies in a most positive manner [4].

Acknowledgments

For permission to examine the specimens in their care, I am very grateful to P. Andrews, L. de Bonis, C. K. Brain, D. J. Johanson, G. H. R. von Koenigswald, M. Kretzoi, M. Leakey, R. E. F. Leakey, J. Melentis, M. Pickford, D. Pilbeam, Wu Rukang, P. V. Tobias, and A. C. Walker. I thank R. Ciochon, R. Corruccini, D. W. Frayer, L. O. Greenfield, W. Jungers, R. Kay, F. B. Livingstone, M. Russell, and L. Shepartz for help in preparing the manuscript. This research was supported by NSF grant BNS 76-82729.

Notes

[1] By "hominid" I specifically mean those taxa in the lineage leading to *Homo sapiens,* and any collateral branches of this lineage, after the divergence of this lineage from the one leading to the African apes.

[2] The confusion of *"Hemianthropus"* teeth with *Australopithecus* teeth (von Koenigswald 1957b) is a consequence of their similar size and shared primitive characteristics, related to the basal ramapithecine masticatory adaptation. These teeth are actually worn versions of the thick-enameled dental remains that have been attributed to "giant orangs." Even in Asian Middle Pleistocene deposits, it is often difficult to distinguish worn orang postcanine teeth from worn hominid teeth. Since there are no remains other than dental known for this third form (giant orangs, *"Hemianthropus"*), exactly what is represented is unclear. However, size extends completely into the *Gigantopithecus* range, and there also appears to be considerable morphological overlap. Thus, at least the dental evidence suggests that it may be a third lineage, contemporary with and intermediate between *Pongo* and *Gigantopithecus.*

[3] If this model is correct, "pongid" is no longer an appropriate name for this African group, since the Pongidae are named after *Pongo* and it is my contention that *Pongo, Pan,* and *Gorilla* no longer form a natural group by themselves (they would only if *Homo* were included). Their relationship is one of *grade,* and therefore they should be referred to together by a nontaxonomic term such as "great apes."

[4] When this article was originally reviewed for *Current Anthropology* it was sent out to a number of scientists for comment. The following responded: L. de Bonis, J. G. Fleagle, D. W. Frayer, L. O. Greenfield, K. H. Jacobs, R. Protsch, G. P. Rightmire, V. Sarich, J. H. Schwartz, I. Tattersall, M. J. Walker and A. L. Zihlman & J. M. Lowenstein. This CA* treatment was published along with the original text. Because of space limitations these comments can not be included here (The Editors).

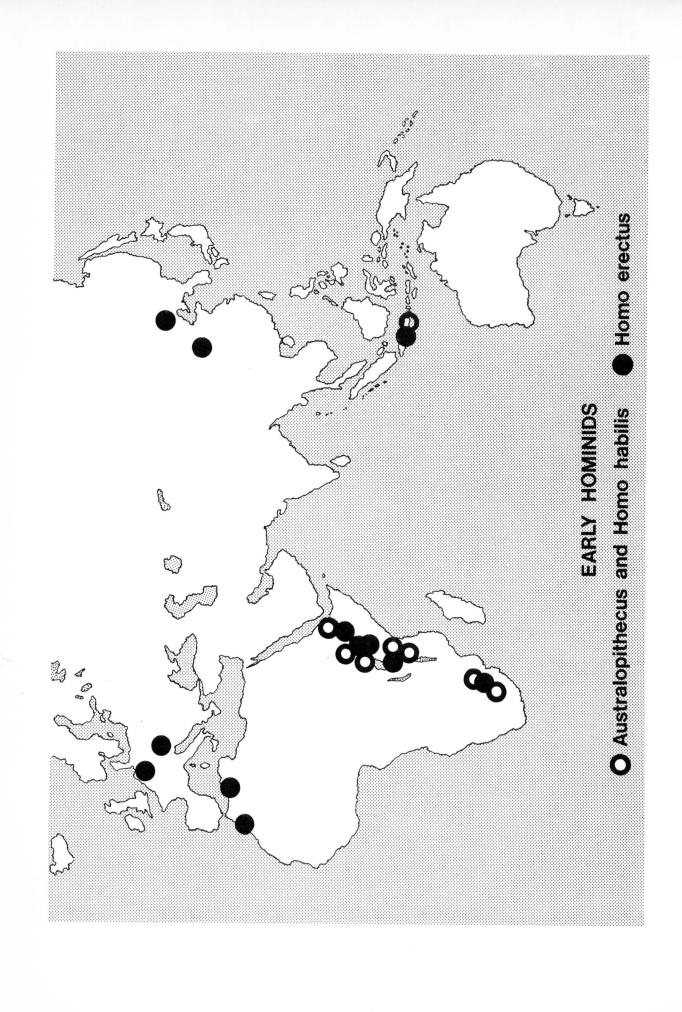

EARLY HOMINIDS

○ Australopithecus and Homo habilis   ● Homo erectus

Part VI

Early Hominids

Although there was considerable often bitter debate about the hominid status of *Australopithecus* when it was first discovered (compare Dart, 1925 with Keith, 1925 and Smith, 1925), for the past three decades there has been nearly universal agreement that australopithecines represent the earliest fossil evidence of the hominid line. As the chapters in this part reflect, the major questions in early hominid evolution now concern the relationships among species of *Australopithecus* with regard to their habits, adaptations, and phyletic relationships to later hominids.

John T. Robinson (1963b) in Chapter 34 summarizes a viewpoint common in the 1950s and 1960s regarding the diversity and relationships within australopithecines. He argues that all of these early hominids can be placed in two adaptive groups (gracile and robust) that are quite distinct in morphology, ecological relationships, and behavior. The gracile form, which Robinson calls *Australopithecus*, is considered more "human-like," whereas the robust form, *Paranthropus*, is viewed as divergent, a sort of slightly hominized pongid. Phyletically, *Australopithecus* is placed in the ancestry of the genus *Homo* whereas *Paranthropus* represents the overspecialized ancestor of *Australopithecus*, which became extinct after giving rise to that genus. Robinson sees a major dichotomy in cranio-dental form separating his two australopithecine groups, the robust form exhibiting an enlargement of the cheek teeth, development of a massive facial skeleton and chewing musculature, and a size reduction and crowding of the front teeth resulting in a dietary adaptation emphasizing the crushing and grinding of tough vegetable material. In contrast, the gracile form shows none of these cranio-dental specializations and was instead characterized by a much more eclectic, omnivorous diet. This morphological dichotomy, according to Robinson, is correlated with behavioral and ecological differences that have become well known as the "Dietary Hypothesis." The proposals put forth in this insightful chapter form a solid foundation on which to base much of our current views of early hominid evolution.

In striking contrast to Robinson's view of early hominid diversity, Milford Wolpoff (1971c) argues (Chapter 35) that robust and gracile australopithecines represent but a single species of early hominid. Evoking the principle of competitive exclusion (Gauss, 1934) and stressing his own version of the "Single Species Hypothesis" (see also Brace, 1967a), Wolpoff argues that since culture is the *primary* hominid adaptation, all early hominid species would have inhabited the same broad adaptive niche and thus would have been forced to compete with one another for scarce resources. Two ecologically similar, culture-bearing hominid species *could not* co-exist without entering into a struggle for existence resulting in the extinction of one species. Thus a strict theoretical application of the "Single Species Hypothesis" would predict the existence of a single hominid species at any one point in time. Such is the case for modern *Homo sapiens* and for *Homo erectus*, Wolpoff observes, and the same principle should apply to the earliest hominids. In support of his view Wolpoff (see also

1968, 1970, 1971a, 1974) presents a battery of metrical data that demonstrate substantial overlap in the morphologies of robust and gracile australopithecines. In Wolpoff's closing sentence he states that his model of early hominid diversity fits the available fossil evidence, and he challenges non-believers, to replace it and come up with a model that better fits the evidence.

Apparently Richard Leakey and Alan Walker (1976) accepted Wolpoff's challenge, but rather than coming up with a new model they instead present unequivocal fossil evidence for the contemporaneous existence of two hominid species in the lower Pleistocene beds of the Koobi Fora Formation of northwestern Kenya. In their brief contribution presented in Chapter 36 they illustrate and describe (see also R. E. F. Leakey, 1976b) a new skull of *Homo erectus* and directly compare it with a previously recovered skull of a robust australopithecine from the same geological formation. Leakey and Walker observe that the "Single Species Hypothesis" might have been the simplest solution to the problem of early hominid diversity, but it was apparently not the correct one. They conclude that more complex models are needed to understand early hominid diversity (see also Walker and Leakey, 1978). Further challenges to the "Single Species Hypothesis" on both theoretical grounds (Washburn and Ciochon, 1976) and from the fossil record (Clarke, 1977) have been made, and Wolpoff (1980), to some extent, has recanted his earlier position. The lesson to be learned from this exchange is a point previously emphasized several times: new fossil discoveries are very often the major arbitrators of theoretical disputes in the fossil record.

In Chapter 37, Donald Johanson and Tim White (1979) present one of the most recent views of the systematic and phyletic relationships of the australopithecines and introduce the earliest member of this group, *Australopithecus afarensis*. Johanson and White argue that *A. afarensis,* known from the middle Pliocene localities of Hadar, Ethiopia, and Laetoli, Tanzania, is demonstrably the earliest and most primitive australopithecine yet recovered and occupies a position ancestral to both East African *Homo habilis* and South African *A. africanus.* Following their scenario, *A. afarensis* assumes the central position as the common ancestor of all later hominids and pushes back the origin of the australopithecine group at least one million years. The views of Johanson and White have elicited a number of comments critical of their evolutionary scheme (see Tobias, 1980a,b; Olson, 1981; Boaz, 1983). A recent contribution by White et al. (1981) answers many of these comments and does much to establish this view of australopithecine evolution as the major hypothesis of the 1980s.

In Chapter 38, the final contribution in this part, Owen Lovejoy (1981) presents a model of hominid origins that challenges the traditional notion that the origin of humans was a direct consequence of material culture and the expansion of the brain such as was argued by Robinson in Chapter 34 and Wolpoff in Chapter 35. Instead, Lovejoy provides a scenario that attempts to integrate all that is known about the biology of the early hominids into a model consistent with living higher-primate behavior patterns as well as evidence from the fossil record. His model accounts for the exclusively ground-dwelling nature of hominids, the origin of bipedalism, pair-bonding and the development of nuclear families, establishment of home bases, food sharing, beginnings of tool use, brain expansion, development of human sexual epigamic features, continuous sexual receptivity, and multiple infant care. Lovejoy stresses that a significant hominid adaptation lies in our ability to produce offspring more frequently than our Great Ape forebears. According to Lovejoy, this was accomplished by the development of bipedalism and its integration with an entirely unique social-behavioral adaptation involving the features listed above. Lovejoy's model has aroused considerable controversy (see replies in *Science,* Vol. 217, pp. 295–306, 1982). Yet, as far as scenarios of human origins go, this one integrates uniquely human patterns more than any model yet proposed.

The chapters in this part provide but a brief overview of early hominid evolution. The major focus of each chapter has been the evolution and diversification of *Australopithecus*. It is comforting to note how much we know about *Australopithecus* today compared with what Raymond Dart knew about human origins when he described the Taung skull. Although tomorrow's discoveries and theories will almost certainly take us in new and unimagined directions, the basic framework laid out in these chapters should remain valid.

34

Adaptive Radiation in the Australopithecines and the Origin of Man

J. T. ROBINSON

The chief purpose of this paper is to speculate about the selective forces which brought the australopithecine group into existence, caused adaptive radiation within the group, and resulted in the origin of man.

AUSTRALOPITHECINE TAXONOMY

Although our prime concern in this discussion is with ecology, behavior, and functional anatomy, it seems advisable to discuss taxonomy first. Adaptive radiation cannot be discussed fruitfully if it is not clear what categories are involved in the group concerned; indeed whether adaptive radiation has actually occurred must first be decided by taxonomic analysis. From the literature it would appear that most workers regard the Australopithecine group as being virtually uniform taxonomically; some apparently go so far as to regard the known forms as members of the same phyletic sequence and therefore as exhibiting no adaptive radiation whatever. The literature also contains many statements that such and such are Australopithecine characteristics as though all known Australopithecines have the same characteristics.

Long and intimate acquaintance with almost all the known material has convinced me that the specimens fit into two quite distinct groups which are well differentiated in morphology and apparently also in ecology and behavior. Not only this, but it seems to me that the split between the two Australopithecine subgroups is far more fundamental than is that between one of these *(Australopithecus)* and the hominines. The other *(Paranthropus)* is quite aberrant as a hominid, whether it is compared with the contemporary *Australopithecus* or with the more advanced hominines. In order, therefore, to leave no doubt about the background, and the reasons for it, against which the later discussion occurs, taxonomy will be discussed here briefly, though it has been discussed elsewhere (Robinson 1954a, 1954b, 1956, 1961, 1962b).

South African Australopithecines

South African Australopithecines are divided into two taxa, *Australopithecus africanus* Dart and *Paranthropus robustus* Broom (Robinson 1954a).

The practicing taxonomist is normally primarily concerned with identification and uses for this purpose good diagnostic "characters" which allow him to distinguish between closely related forms. The characters chosen must be selected and used with due regard to their ranges of variation. In this practical taxonomic sense there are a number of very good characters distinguishing *Australopithecus* and *Paranthropus*.

The first lower deciduous molar, for example, not only allows instantaneous recognition (even when considerably worn) of which group is being dealt with, but also serves to distinguish

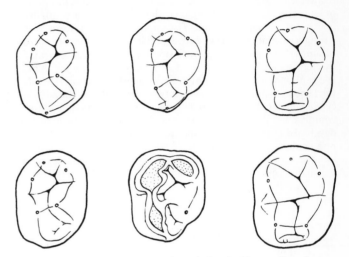

Figure 1 *Upper row: examples of the deciduous first lower molar in* Paranthropus, Australopithecus, *and* Homo *(modern Bushman). Lower row: diagrammatic representations of the cusp and fissure patterns of the teeth in upper row.*

Paranthropus from all other hominids in which the tooth is known. On the other hand, the *Australopithecus* form of dm_1 is also found in all living and fossil hominines in which its nature is known, including Pekin and Neanderthal man. The morphology of dm_1 thus serves to emphasize not only the distinction between *Paranthropus* and *Australopithecus,* but also the similarity between the latter and hominines. The permanent lower canine is another good diagnostic feature. The two Australopithecines can be separated without the slightest difficulty by means of this tooth; in *Australopithecus* the crown is large and highly asymmetric, while in *Paranthropus* the crown is small, more symmetric and with little relief on the lingual surface but the root is substantial. The large difference in proportion between the anterior and posterior teeth in the two forms is also a good diagnostic feature. In *Australopithecus* the canines and incisors are fairly large for a hominid and the postcanine teeth are of proportionate size. On the other hand in *Paranthropus* the postcanine teeth are larger than those of *Australopithecus*, as could be expected in a larger and more robust animal, but the incisors and canines are distinctly smaller than those of the latter. The condition seen in *Australopithecus* fits very well with that found in hominines; that in *Paranthropus* is quite aberrant and unlike that seen in any other known hominid. There are many other diagnostic features: the nasal cavity floor and its relation to the subnasal maxillary surface, the nature and shape of the palate, the shape and structure of the face and of the braincase, and others which cannot be discussed here because of lack of space.

In contrast to this practical, workaday taxonomic approach, there is a more inclusive and satisfying view which sees the animal as a member of a population in its natural environment rather than as a series of taxonomic characters. In such a view the isolated characters of the other approach are seen as parts of an integrated pattern. Viewed from this standpoint the differences between the two Australopithecine types are brought out even more clearly.

In *Paranthropus* it seems clear that the architecture of the skull and head in general is closely related to specializations of the dentition. The small anterior teeth, which in the maxilla are set in relatively lightly constructed bone and in the mandible in a more or less vertical symphysial region with no trace of chin, result in a relatively orthognathous face. The massive postcanine teeth, with strongly developed root systems, are set in massive bone. The area of support and the channels of dissipation of the forces generated by chewing are well developed. Examples of these are the thickened columns up either side of the nasal aperture, the enormously thickened palate anteriorly (over a centimeter thick in one adolescent where it can be measured opposite M^1), the pterygopalatine complex and the zygomatic process of the maxilla. The strongly developed musculature required to operate this massive postcanine dental battery has also affected the architecture of the skull in an obvious manner. Since in all known adults of both sexes in which the appropriate part of the skull is preserved a sagittal crest is present, the temporal muscles were clearly large in relation to brain-case size. The origin of the masseter, especially the superficial portion, is very clearly marked and extensive, as is the insertion on the broad and high ramus. The masseter must thus have been large and powerful. This was evidently true also of the pterygoid muscles in view of the relatively great development of the lateral pterygoid plate.

Maxillary prognathism is reduced by the relatively poor development of the anterior teeth. The support needed for the relatively massive postcanine dentition has resulted in a strongly stressed, hence completely flat, nasal area. The massive chewing muscles are associated, among other things, with a strongly developed zygomatic region. These factors result in the typically wide and massive, but either flat or actually dished, face. The total lack of a true forehead and the comparatively great postorbital constriction make the brow ridges seem massive and projecting, though in actual fact they are not especially massive. The well-developed postorbital constriction, which in part at least is associated with the great development of the temporal muscle; the sagittal crest, which is directly due to the relatively great size of the temporal muscles in relation to brain-case size; and the absence of a true forehead, result in a brain-case shape which is unique among hominids. The robustness of the jugal arch and the large attachment area required by robust nuchal muscles cause the mastoid region to project laterally appreciably more than does the braincase above this region.

In *Paranthropus* the effect of the unusual dental specializations on the architecture of the skull—of the whole head, in fact—has been far-reaching. The result is a skull which bears considerable superficial resemblance to that of some pongids. That it is not a pongid skull is clearly evident, however, as a result of the effect of another important factor affecting skull architecture: erect posture. This has resulted in a very significant lowering of the relative height of the nuchal area of the occiput, which is quite differently oriented in the erectly bipedal hominids than it is in the quadrupedal pongids and all other terrestrial vertebrates. The altered orientation of the nuchal plane clearly distinguishes the skulls of both types of Australopithecine from those of pon-

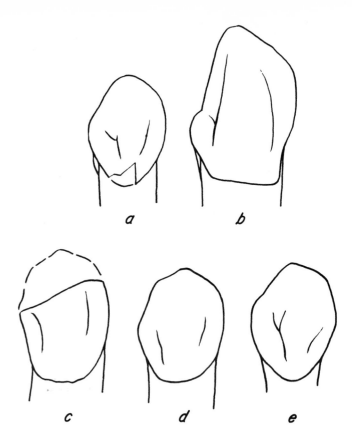

Figure 2   *Mandibular canines of (a)* Paranthropus, *(b)* Australopithecus, *(c) and (d)* Homo *(Pekin man), and* Homo *(modern Bantu). It will be recalled that (a) is from a very robust form while (b) is from a small and lightly built form; (c) and (d) after Weidenreich.*

gids, though not from each other, as has been shown by the use of Le Gros Clark's nuchal-area height index (Le Gros Clark 1950a, 1955; Ashton and Zuckerman 1951; Robinson, 1958).

In *Australopithecus* the dental picture is quite different from that in *Paranthropus*. The anterior teeth are relatively larger and the postcanine teeth relatively smaller than in the latter; a condition which very closely resembles that found in early hominines. Because of the large anterior teeth the face is more prognathous. Owing to the smaller postcanine dentition and the chewing forces are weaker and the musculature less strongly developed. This is shown by such features as the much weaker root system of the postcanine dentition, less robust bone in which the teeth are set, slenderer zygomatic bone and zygomatic processes of maxilla and temporal, as well as the lateral pterygoid plate. Furthermore muscular attachments are far less obvious than they are in *Paranthropus*. Besides these points, there is normally no trace of a sagittal crest since the temporal muscles do not normally approach the dorsal midline of the calvaria at all closely. However, while the temporal muscles in *Australopithecus* were clearly smaller than those of *Paranthropus*, the lack of sagittal crest is not due to this fact alone since another factor is yet operative in this case: the braincase is relatively higher. The index devised by Le Gros Clark and called by him the supraorbital height index, shows clearly (Robinson 1961) that calvaria height above the superior margin of the orbits is very near the hominine condition in *Australopithecus* but of approximately average pongid condition in *Paranthropus*. The usual absence of a sagittal crest in the

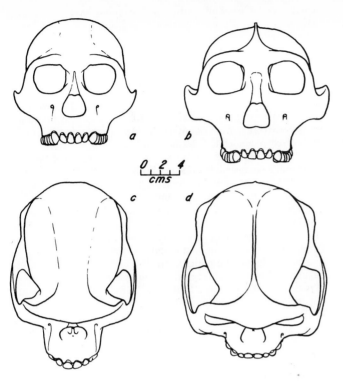

Figure 3   *Facial views of skulls of (a)* Australopithecus *and (b)* Paranthropus. *Top views of skulls of (c)* Australopithecus *and (d)* Paranthropus. *Both skulls are of females.*

former is thus due both to reduced temporal muscle size and increase in relative height of the brain-case as compared to the latter (see also Robinson 1958).

Dietary Specialization in the Australopithecines

Both types of Australopithecine are hominids, hence the basic similarity of their skulls derived from that of a common ancestor. Since both were also erectly bipedal, the modifications of the occiput resulting from this locomotor specialization are also found in both. Beyond this the two skull types differ sharply. The differences, as I have tried to show, appear to belong in each case to a pattern controlled chiefly by the specializations of the dentition.

Considerable significance therefore attaches to the reason for the differences in dental specialization; hence it is important to try to discover that reason. For there must be a reason: it is not acceptable to say that all hominids have one sort of dentition except *Paranthropus*, which has quite a different type, but that there is no adaptive significance in this fact—it just happened that way without cause. The difference between the two dental types is very clearcut: on the one hand *Australopithecus* and all known hominines have a balanced pattern of tooth size and on the other *Paranthropus* alone has an unbalanced pattern, as it were, in which the size relation of anterior and postcanine teeth is quite different. It would therefore seem likely that there is also a fairly clearcut reason for the difference.

In *Paranthropus* the postcanine dentition is clearly very important. The tooth crowns are large, the enamel is thick, the occlusal surfaces large and of low relief and the root systems very well developed. The relatively great and flat occlusal surfaces and the massiveness of the postcanine teeth clearly point to a prime dietary function of crushing and grinding. The massiveness of the

entire masticatory apparatus and the relatively rapid rate of wear of the teeth indicates a diet of tough material and one that probably needed much chewing and had a relatively low nutritive value. The anterior teeth either were of much less importance in relation to the diet than the postcanine ones, or the diet *required* small anterior teeth and that this feature thus represents positive adaptation to dietary needs. The latter seems doubtful since it is not obvious what manner of diet it could be that once placed a premium on small anterior and massive posterior teeth. It seems more probable that reduction of the incisors and canines resulted from reduced need for them in respect not only of diet but also of other things such as behavior affecting other members of the same group or those of other groups in its environment. A fairly obvious and old explanation seems to be that in an erect biped the hands, either alone or in conjunction with objects used as tools, take over much of the function of the anterior teeth in obtaining food and in offense and defense. If this is the reason for the small anterior teeth of *Paranthropus*, as seems to me likely, then the cheek teeth were either being maintained at, or increased to, a large size by natural selection while reduction was occurring in the anterior teeth. In spite of their small size, the anterior teeth do not wear down rapidly. This again suggests reduced usefulness.

In *Australopithecus* there is much less stress on crushing and grinding and obviously the anterior teeth were considerably more important than those of *Paranthropus*. *Australopithecus* was at least as well adapted to erectly bipedal posture as was *Paranthropus*; consequently one must conclude that the tendency to reduction of the anterior teeth, brought about by bipedalism, was

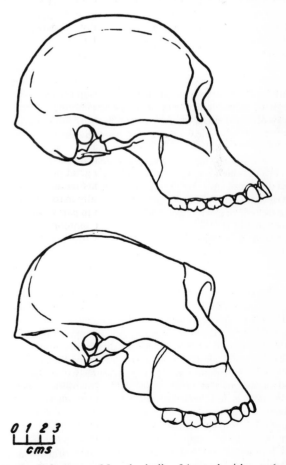

Figure 4   *Side views of female skulls of* Australopithecus *(top) and* Paranthropus.

counteracted by some selective advantage for keeping them large. This could either have been diet or needs of defense and offense.

The implication of the absolute and relative size difference between the anterior teeth of the two forms clearly is that there was either a difference of behavior of considerable magnitude between the two, or one of diet, perhaps both. The fact that the cheek teeth of *Paranthropus* are larger than those of *Australopithecus* need not be due to anything more than the bodily size difference between them since the former is appreciably more robust than the latter.

The very great similarity in dentition and general skull structure between *Australopithecus* and the hominines suggests that they were basically similar in diet and behavior. That is to say that they were omnivores, eating both vegetable food and meat. This provides a logical reason for the relatively large anterior teeth since the need for these in meat-eating would be substantial until tool-making had reached a fairly advanced level. Since the teeth are unlikely to have been important for fighting in an erect biped with a relatively flat face, it seems likely that the canines of *Australopithecus* were large, like those of early hominines, because their diet included meat. This is perhaps supported also by the fact that this form is known mainly from periods of appreciable aridity when vegetable food will not have been plentiful, other than for grazers, hence the probability that animal food formed part of their diet. Finally, as appears later in this paper (also Robinson 1962b), the strong development of tool-using and of tool-making was very probably associated with the tendency for early members of the phyletic line leading to man to take to a certain amount of meat-eating. *Australopithecus* could very easily be representative of this stage. The hypothesis that *Australopithecus* was an omnivore that ate at least a moderate amount of meat does not therefore seem unreasonable.

The emphasis on crushing and grinding and lack of emphasis on the canine in the *Paranthropus* dentition seems inconsistent with a diet such as that suggested for *Australopithecus*. But these features are entirely consistent with a vegetarian diet, as is the massiveness of the whole masticatory mechanism and the much greater body size as a whole. This conclusion is supported also by the presence of grit in the diet, as suggested by damage to the teeth in the form of flakes of enamel detached from the edge of the occlusal margin. The flaking was clearly caused by the application of considerable pressure over a very small area—such as is produced, for example, by the stiletto heel of a lady's shoe as compared to the heel of a man's shoe—as would be the case in biting on a particle of grit but which would not be produced by biting on bone. Grit in the diet would seem to suggest roots and bulbs as part of the diet. The available evidence also points to the fact that *Paranthropus* is known only from wet climatic periods, both in Africa and the Far East. This again is consistent with a vegetarian diet.

The objection has been raised that it is not possible to tell, for example, what the diets are of various sorts of monkey by looking at the teeth, and therefore that reasoning such as the above is unsound. It would seem to me clear that distinguishing accurately between different sorts of vegetarian diet from dentitions which all are basically adapted to vegetarianism is a difficult task, especially as this example concerns the notoriously stable cercopithecoid dentition. But in the case of the Australopithecines the distinction is not between two forms of vegetarian diet but between a vegetarian and an omnivorous diet. That basic dietary adaptation *can* be recognized in many cases from teeth is a commonplace of mammalogy and vertebrate paleontology. In this instance one is faced with the fact that one form of Australopithecine fits so well in all relevant aspects into the picture presented by the hominines that it is not necessary to seek for it other

explanations of basic adaptations of diet and behavior than apply to the hominines. But the other Australopithecine does *not* fit into this pattern, in respect of either the dentition or the skull morphology. So there must be some definite explanation of this difference. The dietary explanation seems entirely logical and, so far as I am aware, no evidence exists which is clearly inconsistent with such an explanation. On the contrary, independant confirmatory evidence seems to exist in the fact that *Paranthropus* appears to have shared the same territory with an early hominine. In the Swartkrans site "Telanthropus" and *Paranthropus* remains were directly associated. In Java the Sangiran site has yielded both "Pithecanthropus" and "Meganthropus" remains. According to von Koenigswald "Pithecanthropus" IV and the type mandible of "Meganthropus" came from the Black Clay (Putjangan beds) not far from each other, but "Pithecanthropus" II and III came from the later Kabuh conglomerate. According to Marks (1953) the "Meganthropus" mandible which he found in 1952 was not *in situ* but was lying on a slope of hard conglomerate and had enough matrix adhering to it to indicate satisfactorily that it had weathered out of the conglomerate. This conglomerate is one which von Koenigswald regards as a boundary bed (basal Kabuh) between the Kabuh and Putjangan beds. Therefore it would appear that "Pithecanthropus" and "Meganthropus" occurred synchronously at Sangiran, not merely at one time horizon, but over a significant period of time. As indicated elsewhere (Robinson 1953, 1955, 1962b) "Meganthropus" is fairly clearly a *Paranthropus*. The implication is clear that *Paranthropus* differed sufficiently in ecological adaptation from early hominines for them to be able to coexist in at least two places at opposite ends of the Old World. Continued coexistence is unlikely if the ecological requirements of the two were virtually identical but if their ecology was as different as is here suggested such coexistence is entirely logical and in no way remarkable.

The resistance to accounting for the differences between the Australopithecines by the dietary hypothesis seems to spring from the *assumption* that they represent a single phyletic line, hence that all known Australopithecines must, willy-nilly, be very closely related. The reason for such an assumption is not clear. Paleontology has provided much evidence to show that adaptive radiation usually follows the achievement of a new grade of organization. The emergence of the Australopithecine group represents the achievement of an important new grade of organization: that of the first known erect biped in evolutionary history. The new group must have arisen from a higher primate. Vegetarianism is the basic dietary adaptation of non-hominid higher primates, but meat-eating is an important part of hominine diet; hence it seems clear that dietary changes must have occurred. Some adaptive radiation in early Australopithecine history, particularly if it involved difference of diet, should consequently occasion no surprise. In later phases of hominid evolution adaptive radiation of any magnitude is very unlikely. This is because of the new adaptive mechanism, additional to the genetic one, which became available with the effective development of culture (*vide* Dobzhansky 1961). This more rapid, artificial adaptation results in a slowing down of natural adaptation, hence of speciation, owing to the fact that man can so easily adapt himself to different environments long before the genetic mechanism has the opportunity to do so. This does not mean that natural selection does not operate on man, but artificial adaptation modifies its action to a considerable extent. The Australopithecines in the initial phases of their evolution will have had no more than a rudimentary level of culture, hence the slowing down of the rate of speciation will not have applied. In the later phases of their evolution cultural development may have been sufficient to make its effect felt to some extent.

Some workers who disagree with the notion of significant adaptive radiation within the Australopithecines wish to regard *Australopithecus* and *Paranthropus* as successive members of the same phyletic sequence. If by this is meant that the Australopithecines were ancestors of the hominines, but through a sequence *Australopithecus*—*Paranthropus*—early hominine, then one is faced with the fact that the first member of this sequence is morphologically, and evidently ecologically too, much more like the last member of the sequence than it is like the middle one. Retrogressive steps *do* occur in evolution, but to have so many reversals in what would presumably have been a fairly rapid evolutionary sequence does not correspond with what is known of phyletic lines in other vertebrates where the evidence is more complete. It is worth considering a few of the reversals involved. *Australopithecus*, as already mentioned earlier in this paper, has the vertex of the skull significantly higher above the upper orbital margin than has *Paranthropus*. The supraorbital height index of the latter falls right in the normal range for pongids whereas that of the former agrees closely with early hominines. This is not a simple feature as it not only involves considerable alterations in skull architecture, but is doubtless also a reflection of expansion in the brain. The latter, it is now well recognized, is evolutionarily a very conservative organ. A very conservative organ thus developed to essentially the hominine condition, retrogressed to a pongid condition and then rapidly advanced to essentially the same hominine condition that it previously had, all in the space of a few hundred thousand years at most. Similarly dm$_1$ in *Australopithecus* has a structure also found in all hominines but not in *Paranthropus*. So a typically hominine condition has to give way to one which is unique among hominids and then return to identically the condition it previously had. At the same time the canines reduced from an early hominine size to a modern hominine size but then had to revert again to their previous size a short time later. The reduction in canine size would have occurred in a stage when meat-eating was being developed but cooking was not known and tools were not well developed. Furthermore, if my interpretation of the tool situation in the Sterkfontein Valley is correct, then a hominine was there present for a short while with *Australopithecus* before *Paranthropus* appeared.

In my opinion the known evidence indicates that the possibility of hominines having arisen from *Australopithecus* via *Paranthropus* is remote in the extreme and need not be considered seriously. If the latter, as known, is regarded as a direct lineal descendant of the known specimens of *Australopithecus*, then it means that Australopithecines were not ancestors of hominines, or if so, the *Paranthropus* group represents a side branch. It now appears that these two genera were contemporaries in Africa south of the Sahara since both are known from East Africa and apparently a little later in time both occur in South Africa with *Australopithecus* occurring in the latter area at about the time that *Paranthropus* was living in the former. If the latter is a lineal descendant of the known specimens of the former, then the transition must have occurred more than once in different places. However, the fact that all the known forms, wherever they occur in time or space, fall easily into one or the other of two very different groups, suggests that it is quite improbable that *Paranthropus* represents simply a later evolutionary phase of the known *Australopithecus* specimens.

Theoretical considerations thus indicate that adaptive radiation in the early stages of hominid evolution is by no means improbable. The available facts indicate considerable morphological difference between the two well-defined groups, more than can be found among living pongids for example, and imply that considerable differences in ecological adaptation were also involved. The ecological differences appear rather clearly to have involved differences of diet. At least differences in diet may logically be inferred from the morphological differences present; such a conclusion adequately explains the different morphology of the two groups and no alternative explanation has yet been advanced which does so. Furthermore, the dietary hypothesis is very fruitful in throwing light on the entire question of Australopithecine evolution and the origin from them of the hominines (Robinson 1962b, and later in this paper).

Non-South African Australopithecines

Australopithecines are at present known from two areas outside of South Africa: Java and East Africa. Coppens (1961) has reported a new Australopithecine find from another region, the Lake Chad basin. This specimen comes from Koro-Toro and apparently, was coeval with an early Villafranchian fauna. The specimen has not yet been described in detail; if the dating and identification are correct, then this is the oldest of the known Australopithecines.

The Javanese form was first designated *Meganthropus palaeojavanicus* (Weidenreich 1945), but detailed analysis of the available information resulted in its being placed in the genus *Paranthropus* (Robinson 1953, 1955). The reason for this is that, with only the most trivial exceptions, the features of the known specimens fall within the observed range of the known *Paranthropus* material. Among these features are the massive mandible and the combination of small canines and enormously robust postcanine teeth. Although most of the crowns are missing from the 1952 mandible (Marks 1953), the roots are present in the sockets, including those of the left incisors and canines; along with those of the cheek teeth they reflect the characteristically *Paranthropus* condition. The conclusion that "Meganthropus" is a *Paranthropus* has been contested by von Koenigswald who has, however, produced no cogent evidence to refute it. Some of his main objections have been considered recently (Robinson 1962b). Not only is there so far no valid evidence differentiating the two groups of specimens, but "Meganthropus" exhibits some features which are diagnostic of *Paranthropus*. It is therefore reasonable to regard the former as a member of the latter genus.

Leakey (1959) reported the discovery of a good skull of a late adolescent Australopithecine from Olduvai. This form was regarded as new and named *Zinjanthropus boisei*. It has been shown, however, that the skull and dental characters, and their pattern of specialization, are typically those of *Paranthropus* (Robinson 1960). As in the case of "Meganthropus," the morphological differences held to validate generic separation from *Paranthropus* either disappear or become very slight if the observed range of variation of these features is taken into account.

In 1939 Kohl-Larsen discovered in the Laetolil beds, near Lake Eyassi in East Africa and in the same general region as Olduvai, a fragment of maxilla containing P^3 and P^4 and also an isolated upper molar. They were named *Präanthropus* (a *nomen nudum* since no species name was given) by Hennig (1948) and *Meganthropus africanus* by Weinert (1950, 1951), a view supported by Remane (1951). This matter has been considered at some length (Robinson 1953, 1955) and the conclusion drawn that (1) since one form is known only by mandibular and the other only by maxillary material, no evidence exists for placing them in the same genus; (2) since the East African specimen exhibits characters which all fall within the observed range of variation of the corresponding features of *Australopithecus*, the logical course is to refer the material to the latter genus. This is also the opinion of von Koenigswald (1957a).

Recently Leakey (1961a,b) has announced the discovery of further material at Olduvai, including a juvenile mandible and two parietals from a Bed I horizon a little lower than that from which "Zinjanthropus" came. The mandible appears to have the characteristics of *Australopithecus* but the size of the parietals suggests that perhaps they belonged to a larger-brained creature than either of the Australopithecine types. Additional material is therefore needed before reliable conclusions can be drawn about whether the material from this horizon belongs to one or more forms and what its or their identity is.

We may conclude that

1. *Paranthropus* is a well defined genus consisting of a somewhat aberrant type of hominid whose morphological, ecological, and behavioral adaptations are quite distinct from those of all other known hominids. It is known from East and South Africa as well as Java; in the latter two places it is known to have coexisted at the same site with an early hominine.

2. *Australopithecus* differs clearly in morphological, ecological, and behavioral adaptations from *Paranthropus* but exhibits very considerable similarity in these respects to hominines. It occurs in South and East Africa but is not at present known from the Far East.

CULTURAL STATUS OF THE AUSTRALOPITHECINES

The level of culture achieved by the Australopithecines also concerns the subject of this paper. The relationship between Australopithecines and the stone industries found with them in the Sterkfontein Valley and at Olduvai has been discussed elsewhere (Robinson & Mason 1957; Robinson 1959, 1961, 1962a) and will not be dealt with here. The conclusion was reached that, despite commonly held opinion to the contrary, there is as yet no proof that either form of Australopithecine possessed a settled stone culture.

The evidence in fact favors the conclusion that the Australopithecines were no more than tool-users, employing whatever came to hand in the form of sticks, stones, bones, etc. This aspect of *Australopithecus* behavior has been dealt with at considerable length by Dart (e.g., 1957a, 1957b, 1958b, 1960). In my opinion the evidence provided is enough to establish that this form was a tool-user, though it would appear that its osteodontokeratic prowess has been overrated considerably. This tool-using ability has been disputed by some authors. For example Mason (1961) holds that since a bone culture (due presumably to *Homo sapiens*) has been found in a Middle Stone Age (end-Pleistocene) deposit at Kalkbank and since early hominines were already in existence in Australopithecine times, therefore the Makapan Limeworks bone culture should be attributed to a hominine that preyed on *Australopithecus* there. Washburn (1957) has also argued against the latter form having had a bone culture. His argument turns on whether the associated bones represent bone accumulation by the latter or by carnivorous animals such as hyaenids. Washburn and Howell (1960) accept the bone associated with the Olduvai *Paranthropus* as food remains of this vegetarian form and therefore as proof of predatory activity by it. However, in the same paragraph they state, "It is very unlikely that the earlier and small-bodied Australopithecines (i.e., *Australopithecus*) did much killing," without explaining why associated faunal remains are to be accepted as food remains of an Australopithecine in the one case but not in the other.

On theoretical grounds the probability that Australopithecines used tools would seem to be high. As is well known, tool-using of an indisputable character occurs among non-primate mammals, birds and even invertebrates. Australopithecines would appear in general to be no less well endowed than these other tool-users and in addition were possessed of the very great advantage of being erect bipeds with emancipated fore limbs.

Furthermore, Australopithecines are very closely related to hominines, the supreme users and makers of tools. The probability that they used tools would thus seem very high. That bone was used would seem equally clear. Many hominines are known to have used bone for tools (indeed some still do) and incontestable evidence of the use of bone in Australopithecine times exists in the form of the bone tool from Sterkfontein (Robinson 1959) and the less perfect example from Olduvai (Leakey 1960), both of which were evidently used for working leather. Whether or not these were products of Australopithecine activity is not clear; but they do represent examples from two widely separated areas of the use of bone in Australopithecine times. That Australopithecines used bone is therefore a distinct possibility. It is interesting to note that some authors who do not accept tool-using for *Australopithecus*—in bone at any rate—nevertheless firmly believe that they were stone tool makers.

Acceptance of *Australopithecus* as a tool-user, in bone or anything else, therefore seems entirely reasonable. On general grounds it would seem probable that *Paranthropus* also used tools, though such activity may have been much more poorly developed in this vegetarian and may well not have included the use of bone.

THE ORIGIN OF THE AUSTRALOPITHECINAE

The Subfamily Homininae includes forms characterized morphologically by erect posture and large brain and behaviorally by relatively complex cultural activity. The latter feature appears to be dependant on the large brain since it seems that intelligence of the hominine caliber is not associated with brains smaller than an ill-defined lower limit in volume of the general order of 800 cubic centimeters.

The Subfamily Australopithecinae includes forms which have the erect posture, but not the large brain, of the hominines. Erect posture is more than adequately proven by the morphology of one virtually complete pelvis with most of the spinal column and a proximal portion of femur; three other adult innominate bones and two juvenile specimens; two proximal ends of femora and two distal ends, as well as a number of skulls showing the structure and orientation of the occiput. The pelvic morphology closely resembles that of hominines and differs sharply from that of pongids. A short, broad innominate is present which has the posterior part of the ilium expanded, hence there is a deep and well-developed greater sciatic notch; the iliac crest is in the form of a sinusoidal curve when seen from the top; the sacrum is broad; distinct lumbar lordosis is present and a femur with a strong lateral lean of the shaft from the vertical when the distal articular surfaces are placed on a flat horizontal surface with the shaft as nearly vertical as possible. The nuchal plane of the occiput has the near-horizontal disposition found in erect biped. For example the arrangement of the origin and insertion of *gluteus maximus* are such that this muscle must have acted as an extensor of the thigh. *Gluteus medius* was evidently an abductor. A well-developed anterior inferior iliac spine suggests a powerful *rectus femoris*, and therefore probably *quadriceps* as a whole. This is a very important muscle in erect bipedal locomotion and in standing without additional support. A well-defined attachment area just below that for the direct head of *rectus femoris*, and a pronounced femoral tubercle, indicate a powerful ilio-femoral ligament strengthened and functioning in the manner of that in hominines. There is also evidence for the "locking" of the knee joint with the leg straight. The best available evidence for erect posture is the *Australopithecus*; that for *Paranthropus* indicates a basically similar condition, though perhaps not quite as well adapted to erect bipedalism as in *Australopithecus*.

The Australopithecines must have originated from some

more primitive primate group, but it is not our aim here to inquire closely into what that group might be. The ancestral form may have been a member of the same early hominoid stock to which *Proconsul* belongs, as is commonly believed, or it may have been part of an independent line already quite distinct at the time the early Miocene East African pongids lived. *Amphipithecus* and *Oreopithecus* suggest that the hominids may have resulted from a line which was slow-rate for most of its history and which has been independent since the prosimian stage. The evidence is as yet too scanty for definite conclusions to be drawn.

Australopithecines appear to differ from pongids primarily in having erect bipedal posture, a primitive culture, and in the nature of the dentition. The differences between pongid and Australopithecine dentitions are most striking in the anterior teeth, especially the canines, incisors and P_3 as well as dm_1. The reduction in canine size, as was suggested already by Charles Darwin, probably resulted from the use of tools. Effective tool-using could only have become possible after erect posture had been acquired. The altered character of the incisors and canines in the early hominids may therefore have been a consequence chiefly of changed posture and locomotion. The differences between the pongid and hominid types of P_3 cannot primarily have been due to these changes, however, as the evidence clearly shows.

Since the improved cultural level and reduced canines of Australopithecines followed upon the change to erect bipedalism, the latter would appear to have been the key feature in the origin of the Australopithecines. This locomotor change represents a major adaptive shift which opened up entirely new evolutionary opportunities in this primate line as compared to all known previous ones.

The Origin of Erect Posture

The manner of origin of erect posture is, however, not clear. A critical part of the change centers around the shift in function of *gluteus maximus* from being primarily an abductor of the thigh to an extensor. The rest of the pelvic and thigh musculature of pongids and similar quadrupeds is very similar in function; but *gluteus maximus* functions very differently in the two groups and this difference is of profound importance in locomotion. The power provided by this muscle, particularly in the second half of a stride, is largely responsible for the efficacy of upright locomotion and is mainly responsible for the difference between the efficient erect locomotion of man and the far less efficient erect walking of a pongid, although the whole story is more complicated. It is readily apparent that a short, broad innominate, in which the breadth increase is mainly in the posterior part of the ilium, is a major cause of the change in function of *gluteus maximus,* since in such a case the origin of the muscle is placed well behind the acetabulum. This, coupled with the fact that the thigh is normally in at least a fairly extended position in erect bipeds, places the main line of action of the muscle behind the hip joint, hence contraction causes extension of the thigh, not abduction.

Higher primates commonly rear up on their hind legs under various circumstances normal to their way of life. The gibbon often does this in trees as part of locomotion, but apart from this the upright posture is probably mainly used for improving visibility, getting food or during play. The former is probably by far the most significant function, especially in the case of a mainly or entirely terrestrial animal living in the forest verge or in tree savanna where vision may be obstructed by shrubs and grass. It seems reasonable to suppose that members of a population in which the point had been reached where, in the erect position, *gluteus maximus* functions chiefly as an extensor would find it easier to use this posture or mode of locomotion.

Under these circumstances such animals would have the advantage of being able to keep alert about what was happening in their neighborhood rather more easily than in the case of pure quadrupeds. But this would certainly not be the whole of the advantage gained; the advantage conferred by having the hands free for manipulations such as tool-using is very obvious. It is now well recognized that even a small advantage is sufficient to allow selection to operate effectively. In this case the advantages would not be small. To start with, the advantage of freed hands would perhaps be comparatively unimportant, but the more erect posture was used, the more important it would become.

If erect posture and locomotion came to be used frequently in such a population, the nature of selection on the locomotor apparatus would alter considerably. Relatively minor changes only would at that stage be required to adapt fully to erect posture as the normal habitat. Rapid adaptation to erect bipedalism could thus be expected. It does not, therefore, seem especially difficult to see how natural selection would bring about a rapid readaptation, in respect of posture and locomotion, in a group in which the innominate had become sufficiently broad and short for the change of function of *gluteus maximus* to occur.

The difficult part to explain, as far as I am concerned, is the process which led to the changes in the innominate. Starting from the general pelvic type found in the prosimians and the arboreal monkeys, it is very difficult to see what manner of locomotory specializations could have brought about the required pelvic changes. Forms specializing in the direction of brachiating seem to acquire a pelvis which is long and narrow. This is the case in the living pongids, even though some are no longer primarily brachiators, and also in *Ateles,* a New World monkey which brachiates to an appreciable extent. The gorilla is now essentially a ground-dweller and when in the trees is mainly a quadrupedal climber (see Schaller & Emlen, 1963) and it also has a broadened ilium. The descent to the ground of an erstwhile brachiator might thus be regarded as suitable means for bringing about the required changes in the pelvis. This is, unfortunately, not what the case of the gorilla demonstrates. Not only is the pelvis quite clearly still of the pongid sort, but even the broadened ilium is not, as simple statistics alone might suggest, more nearly like that of hominids. All the increase in breadth is in the anterior part of the ilium, not the posterior, and appears to be related to the stout trunk of this animal. The greater iliac breadth therefore does not affect the function of *gluteus maximus*. Brachiators, whether modified for ground-dwelling or not, do not appear to offer any suggestion of tendencies in the required direction. Postulating that an arboreal form without brachiating specializations descended to the ground does not appear to help either. This experiment has also been performed, as it were, and the chacma baboon is an example of the result. Here again there is no evidence of a tendency for the innominate bone to shorten significantly or for the ilium to broaden in the direction of the sacroiliac articulation.

The known non-hominid primate locomotory specializations therefore do not appear to afford any help in explaining how an arboreal primate pelvis could have become modified to the point where changed muscular function could provide a basis for altered selection pressures causing adaptation to erect posture.

It might be felt that looking for the origin of the change in locomotor specialization is wrong; that behavioral changes might have led to erect posture. For example it might be supposed that the development of a proclivity for tool-using altered selection so that the pelvic changes were brought about. This does not seem likely. In the first place, primates which have enough manual dexterity and grasping ability to use tools and which in captivity have been shown to be capable of a certain amount of tool-using, do not appear to make use of these abilities in their natural state. In some cases it would seem that even the most primitive use of tools would be very beneficial, but is not resorted to. An example is the baboon; a simple digging stick or other object would help considerably in recovering grass rhizomes, which are eaten to a considerable extent (see DeVore & Washburn, 1963). Such tool-using is used to very good effect by one of Darwin's finches of the Galapagos Islands and by the sea otter of the California coast with far less suitable equipment anatomically than the baboon has. In the second place, it is difficult to see how use of tools could bring about the required anatomical changes. If tool-using of the primitive sort under consideration was adopted by a quadrupedal apelike animal and in the use of tools it stood erect quite often, would this cause a selective situation favoring a short and broad pelvis? After all monkeys and apes commonly do stand erect, or even move in this position, and have presumably done so for millions of years without the pelvis changing in the required direction. Many other quite different animals also do this, bears and mongooses, for example. Only when the innominate had reached a state in which when the animal stood erect *gluteus maximus* functioned as an extensor of the thigh, would a situation exist where use of this posture would bring positive selection to bear in the direction of better adaptation to the bipedal locomotor habit. It is therefore very easy to see tool-using following adoption of erect bipedalism, but the latter being a consequence of tool-using seems very improbable.

The pelvic modifications were associated with changes which were not primarily concerned with locomotion, but which rendered the pelvis preadaptive for erect posture. That is to say that the changes responsible for making the change in function of *gluteus maximus* possible were being controlled by selection which was not concerned with erect posture. But when they had proceeded to a point where the change in muscular function was a possibility, a new situation existed in which selection for bipedality became a reality. A distinct adaptive shift would then occur and only after it had occurred was it possible to refer to the previous adaptive situation as being preadaptive for erect posture. However, I am unable to offer any explanation of what was causing the changes before the adaptive shift occurred.

Whatever the reason for the pelvic changes, it is a fact that they did occur, and once they had, a new adaptive trend came into being. According to the view presented here the process occurred in two phases: the first, during which it is difficult to see how selection for erect posture as such could have been operating, can in retrospect be regarded as the preadaptive phase; this was followed by the adaptive phase during which selection pressures were concerned directly with erect posture. This is, of course, typical of instances where a sharp adaptive shift occurs. In this case the threshold involved the changed function of *gluteus maximus*. Before this the pelvic changes represented a prospective adaptation; after the threshold had been crossed, adaptation to the new adaptive zone was rapid under the direct control of selection.

In respect of the preadaptive phase, it is of great significance that *Oreopithecus* had a somewhat shortened innominate with a relatively broad ilium (Schultz 1960). Not only is the ilium broad, but the increased breadth is posteriad, in the region of the sacroiliac articulation. Precisely the required changes required to bring about the changed function of *gluteus maximus* were therefore in progress in *Oreopithecus* and the early Pliocene horizon at which it occurred (Hürzeler 1958) would place it at about the right period in time. From the point of view of the pelvis, therefore, *Oreopithecus* is remarkably suitable as an Australopithecine ancestor in the stage before the adaptive shift to erect posture occurred. Some other skeletal features are also quite consistent with this view: the short face, canines of moderate size (substantial in males), compact tooth row, occlusal pattern of upper

molars and fairly vertical chin region. If P_3 is indeed bicuspid in all cases, with the lingual cusp well developed, then this is additional and good evidence for this view since hominids are the only higher primate group so far known in which well developed bicuspedness in this tooth is normal. The cusp pattern of the lower molars is somewhat unusual: this may indicate that *Oreopithecus* was off the main line leading to the others in which the dryopithecus pattern developed and therefore that the latter pattern was not yet fully in evidence. More information is needed about the nature of the upper limbs, which are known in the most complete skeleton. It is not inconceivable, therefore, that this form is, as Hürzeler (1958) has suggested, an early member of the ancestral line of the hominids.

The Nature of the First Australopithecines

The consequence of the adaptive shift involving the locomotor apparatus was to give rise to an erectly bipedal primate which was the first hominid and also the first Australopithecine. This form was probably a vegetarian. The reason for this conclusion is that vegetarianism, in its broadest sense, is characteristic of non-hominid higher primates and the hominid ancestors are therefore likely to have shared this characteristic. There is no reason to suppose that diet could have been an important factor in the locomotory changes. Furthermore, if the hominid ancestors had been meat-eaters from long before the Australopithecine stage, and therefore for a very considerable period of time, then it is likely that the dentition would have reflected this fact in adaptations to such a diet. This is not the case; hominid dentitions have no definite adaptations to a carnivorous diet. It thus seems fair to conclude that the first Australopithecine was an essentially vegetarian biped.

Since the conclusion has already been reached that *Paranthropus* is not as advanced in the hominine direction as is *Australopithecus* and had an essentially vegetarian diet, the possibility exists that the former has diverged less from the ancestral Australopithecine than has *Australopithecus*. The skull of *Paranthropus* is primitive for a hominid in some respects. There is no true forehead, the brow ridges are rendered prominent by a well-developed postorbital constriction and the vertex rises very little above the level of the upper orbital margins. This latter point is very well demonstrated by the supraorbital height index of Le Gros Clark (1950a). The value of this index for *Australopithecus* (Sts. 5) is 61 (68 according to Le Gros Clark, 74 according to Ashton and Zuckerman, 1951). This approaches the figure for the index in *modern* hominines which, according to Ashton and Zuckerman, averages about 70 and ranges from about 63 to about 77. The value for a few specimens of Pekin man, determined from illustrations, appears to range from about 63 to 67 for the small sample. On the other hand the three great apes have mean values for this index which range from 49 for the orang to 54 for the gorilla, according to Ashton and Zuckerman. The figure for *Paranthropus* from Swartkrans is 50 and that for *Paranthropus* from Olduvai, determined from photographs, appears to be just over 50. This feature reflects a significant feature of cranial morphology, which in turn almost certainly reflects some aspects of brain morphology. It is significant, therefore, that in these respects *Paranthropus* agrees with the pongids while *Australopithecus* exhibits a condition closely resembling that of the hominines. The vegetarianism of the former may also be interpreted most easily as a relatively primitive feature. If in fact the Australopithecine ancestors were vegetarians, then the simplest explanation of this condition in *Paranthropus* is that it was retained from ancestors. It is possible that it was a secondary condition; in the absence of evidence suggesting that this form was originally a vegetarian, then became an omnivore and then once more became

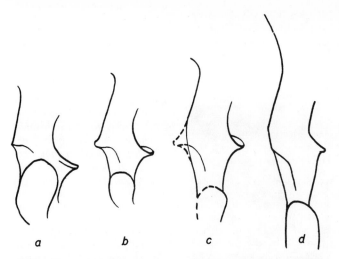

Figure 5   *The ischial regions of the innominate bones of (a)* Homo *(Bantu female), (b)* Australopithecus *female, (c)* Paranthropus *(probably female), and (d) female Orang. The distance from the acetabular margin to the nearer edge of the ischial tuberosity in* Paranthropus *is half as much again than it is in* Australopithecus, *whereas the difference in size between the innominate bones of these two is much less than this.*

a vegetarian, much the most economical hypothesis is to assume that it always had been vegetarian. There are still other significant features in which *Paranthropus* is more primitive than *Australopithecus*. In the former the ischium is well developed and the "bare area" between the acetabular margin and the area of muscular attachment on the tuberosity relatively much longer than in *Australopithecus*. Also the total lack of separation between the anterior portion of the nasal cavity floor and the subnasal surface of the maxilla presents an almost ultrapongid appearance in *Paranthropus* while *Australopithecus* has a considerably more hominine condition (see Robinson 1953, Figs. 4, 5, 6). Another feature seems at first glance also to fit into this pattern; the robust root systems of the premolars in which the fusion process has not gone as far as it has in *Australopithecus*. However it seems to me that perhaps this is partly due simply to absolute size and thus maybe to a heterogonic effect. The large size is, in any event, due to the dietary demands, and thus the less advanced root fusion really is part of the difference in diet and is not a separate difference.

There do not appear to be any features in which *Paranthropus* is clearly more advanced in the hominine direction than is *Australopithecus*. The small, hominine-like canines may appear to be an obvious exception to this statement. There is no doubt that the canines of *Paranthropus* match those of modern man better, in both size and structure, than do those of *Australopithecus*. However the canines of the latter form are very much more like those of *early* hominines and therefore are in fact better antecedents for the hominine canines than are those of *Paranthropus*. But the latter form cannot be held to be more primitive than *Australopithecus* in this respect. Here the combination of vegetarian diet and erect posture has allowed the canines to reduce and has thus produced what can only be regarded as a specialized condition. A somewhat similar condition holds for dm_1; the tendency in the deciduous molars and the permanent premolars of hominids as compared to the other higher primates is toward greater molarisation. In *Paranthropus* dm_1 is clearly more molarised than is that of *Australopithecus*; hence it may be argued that in this respect the former is more advanced than the

latter. This is undoubtedly a valid conclusion—in regard to degree of molarisation. But no hominines have the highly molarised *Paranthropus* type of tooth—they all have precisely the type of tooth that is found in *Australopithecus*. That is to say that where this tooth is known among hominines it is of the *Australopithecus* type. Again *Paranthropus* cannot be said to be more primitive in this respect. As in the case of the canines it is merely aberrant as far as the hominine line is concerned.

The position thus is that in some important respects *Paranthropus* demonstrably is more primitive than *Australopithecus*, judged from the viewpoint of hominines as the end-forms with which we are concerned. In some respects the former is merely aberrant, but in none does it appear to be more advanced than *Australopithecus*. There seems to be no alternative therefore to the conclusion that *Paranthropus* is the more primitive of the two Australopithecines. In view of this conclusion and the evidence which suggests that the forms coexisted in Africa but that *Paranthropus* probably survived longer in the Old World than did *Australopithecus*, it would seem clear that they belonged to separate, divergent phyletic lines. It would also seem clear that these lines had a common ancestor.

If the above conclusions are valid, then *Paranthropus* must be a less modified descendant of the ancestral Australopithecine than is *Australopithecus*. Not that the ancestral form will have had the exaggerated characters seen in the known specimens of *Paranthropus*. The canines, for example, will have been larger and therefore probably the skull will have been somewht more gracile. This early *Paranthropus* will therefore have differed less from the known *Australopithecus* material than does the known, later, *Paranthropus* material. But it will nevertheless have been more nearly *Paranthropus* than *Australopithecus* because of diet, absence of forehead, pongid-like ischium, primitive nasal area, and probably many other things of which we are as yet unaware.

ECOLOGY AND ADAPTIVE RADIATION

If *Paranthropus* represents basically the original Australopithecine stock and *Australopithecus* represents an adaptively different line evolving in a different direction, how did the latter line arise?

It seems unlikely that the earliest Australopithecines can have been as recent in age as the Pleistocene since the two phyletic lines were already well differentiated early in that period. On the other hand it seems logical to suppose that tool-using, tool-making, and increased brainsize are virtually inevitable consequences of erect posture and that they will have followed the origin of the latter fairly rapidly in terms of the geological time scale. Consequently it is more likely that Australopithecines originated in the latter half of the Tertiary than in the earlier half; probably in the Pliocene, just possibly in the Miocene.

There is reason to believe that most of the Miocene was a period of expanding forests in Africa, but that the late Miocene and Pliocene was a time of desiccation and shrinking forests. The Kalahari sands of central and southern Africa throw some light on this matter. The original Kalahari sands overlie unconformably the Kalahari Limestone plain, which resulted from the African erosion cycle of early to mid-Tertiary times. However they predate the cutting of the Kalahari rivers into the Limestone in the lower Pleistocene. It would therefore seem that between the wetter period of the earlier Miocene and that of the early Pleistocene, considerable desiccation occurred during which the extensive deposits of Kalahari sand formed. These extend from fairly far south in South Africa right up into the Congo basin. The studies of botanists and of entomologists studying humicolous faunas support these conclusions in demonstrating marked forest expansion in the Miocene and equally marked recession in the Pliocene,

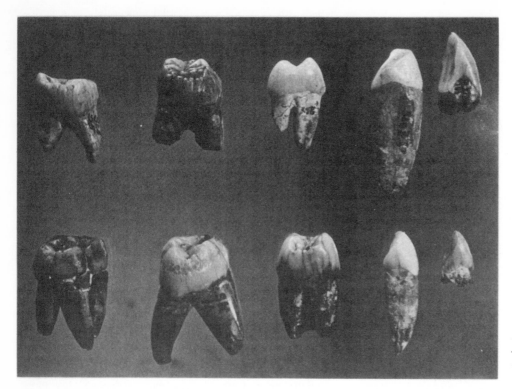

Figure 6 *Teeth in the upper row belong to* Australopithecus *those in the lower, to* Paranthropus. *The cheek teeth in both cases are maxillary, while the canines are mandibular. Note the greater development of the root systems in* Paranthropus: *these are average-sized teeth, while the second from the left in the upper row is a very large tooth for* Australopithecus. *The canines are in marked contrast to the cheek teeth—the canines of the small and lightly built* Australopithecus *are robust, while those of the large* Paranthropus *are small. Note hypoplastic enamel of the second tooth from left, lower row. This was presumably the result of a disease in childhood.*

leaving residual forests in a ring round the central Congo basin and in East Africa, and with a certain amount of expansion again in the Pleistocene. (See, for example, Mabbutt 1955, 1957; Cahen & Lepersonne 1952; also private communication from Leleup on humicolous faunas.)

One may conclude from this that suitable habitats for the vegetarian, original Australopithecines (*Paranthropus*) line will have become increasingly scarce through the later Tertiary. This will have been as true for other forms requiring forest or broken forest habitat and reasonably moist conditions, hence it could be expected that competition for such environments may have been more severe than usual. On the other hand grass savanna and other more arid environments will have expanded at this time, thus providing increased opportunity for animals adapted to, or capable of adapting to, such conditions.

The climatic changes in the desiccation process will not have been sudden. Australopithecines living in areas which subsequently became semiarid will have found that the dry season

gradually became longer and drier. The critical time of the year, the latter part of the dry season, will gradually have become more difficult to cope with. It is reasonable to suppose that in these times of hardship insects, reptiles, small mammals, the eggs and nestlings of birds, etc., will have been eaten to supplement their diet. It is known that purely vegetarian primates will readily eat meat in captivity and that baboons, for example, will occasionally do so in the wild. Taking to a certain amount of meat-eating under environmental pressure could therefore occur fairly easily. As desiccation proceeded, such a deme will have found that it had to rely on the seasonal supplement to its normal vegetarian diet more frequently and to a greater degree. Under these circumstances it could be expected that population density will have dropped, probably to vanishing point in the most heavily affected areas. But is is probable that in at least some areas the creatures will have adapted satisfactorily to the altering circumstances and adopted a certain amount of carnivorousness as a normal part of their way of life. That is to say, the originally vegetarian diet will

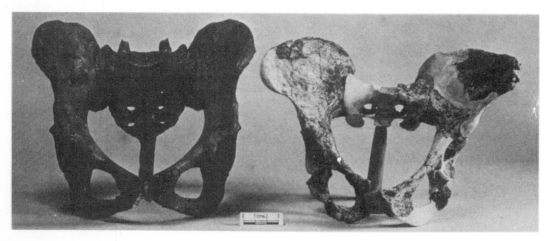

Figure 7 *The pelvis of a female Bantu (left) and a female* Australopithecus *(right). In the latter, virtually nothing is missing except the lower portion of the sacrum. Owing to distortion during fossilization, the pelvis is no longer quite symmetric.*

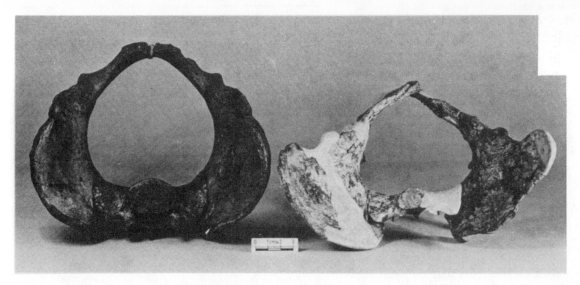

Figure 8   *The pelvis of a female Bantu (left) and a female* Australopithecus *(right). The pelvic opening in the latter specimen is slightly smaller than it originally was and not quite symmetrical, owing to post-mortem distortion. This individual, of which a good deal of the skeleton is known, was about 4 feet in height and very lightly built. The pelvis is therefore relatively large.*

have become altered by the addition of a certain amount of meat-eating to an omnivorous diet.

It is quite clear, however, that such modifications to the environment will have altered the nature of selection acting on the group. Even an elementary level of tool-using will have had obvious advantages in the changing food situation. For the vegetarian part of their food, implements for digging will have made possible greater exploitation of the larger number of bulbs found in drier areas. Implements for bashing, hitting, or throwing, as well as digging, will have made capture and consumption of small animals much easier. Improved tool-using will thus have been favored by selection and any improvements in this respect will have improved adaptation, especially in respect of the carnivorous aspect of their diet. It is also obvious that improved intelligence will have been of great benefit in improving tool-using ability and dealing generally with the stresses of a somewhat hostile environment. Improved intelligence will consequently also have been favored by selection. Since there appears to be some relationship between intelligence and brain volume with regard to that portion of the range between the brain size of the larger pongids and that of the early hominines, it is probable that this part of the process of selection for improved intelligence will have been accompanied by an increase in brain volume. This would probably have shown up first as an increase in the size of the cerebral hemispheres; this would in turn have affected the braincase by expanding the frontal region laterally and especially vertically. This is precisely what is seen in the brain-case of *Australopithecus* as compared to *Paranthropus*. Improved intelligence at this stage will have improved tool-using ability so that progress in the direction of improved intelligence and improved use of tools will have reinforced each other and led to increasingly improved adaptation to the changing environment. This will have been especially true as far as meat-eating is concerned since improved tool-using will have made it increasingly easy to deal with the mechanics of capturing prey, penetrating the skin and removing the meat from larger animals, etc. Improved intelligence will have led also to improved hunting methods.

The changed environmental circumstances resulting from the known desiccation of a substantial part of Africa during the later Tertiary could therefore very easily have led to a second adaptive shift and the establishment of a second phyletic line in the Australopithecines. In this the introduction of a carnivorous element in the diet and an enhanced level of cultural activity were important features. *Australopithecus* is evidently precisely such a line. It is of interest that this form is present in the Sterkfontein Valley in the more arid periods, while *Paranthropus* is present only in the wetter periods (for climatic data see Brain 1958). The canines of *Australopithecus* are appreciably less reduced than those of *Paranthropus*, which suggests that the former genus arose from the latter well before the reduction of the anterior teeth in the latter had reached the stage seen in the known forms. Adaptation to an omnivorous diet which included an appreciable amount of meat-eating will have kept the canines as large as they originally were or perhaps even increased their size slightly.

THE ORIGIN OF HOMININES

Once the line adapting to drier conditions and altered ecology had become established, thus producing *Australopithecus* as we know it, its evolution would not have stopped there. The selection pressures operating, and entirely different from those controlling the direction of the *Paranthropus* line, would not cease to operate and it was virtually inevitable that adaptation would be carried well past the *Australopithecus* stage. The cultural situation would by then have become the vital factor. The need for tool-using in successfully adapting to the different way of life would, as indicated, place a high premium on intelligence. As this improved, presumably by an increase in size of the cerebral cortex so as to provide increased correlation and association areas, cultural facility also improved. When the modification of the brain had reached a point where hominine levels of intellectual ability began to appear, apparently when brain volume reached the general order of about 800 to 1,000 cu. cm., facility with tools reached the point where a characteristic hominine phenomenon appeared: the deliberate manufacture of tools for particular purposes. This provided still further scope for development and the increase in brain size occurred rapidly to about the modern volume. At this point it seems that correlation between brain size and intelligence is not especially close since all manner of other factors are involved. Cultural ability did not improve at the same rapid rate that applied to brain volume, presumably because there were many problems of communication and organization. But momentum gradually built up and the rate of change seems still to be increasing. "Telanthropus," from Swartkrans, was apparently an early member of this hominine stage. This form has now been included in the genus *Homo* (Robinson 1961) in which, it seems to me, all hominines should go. Definitions of the genera *Australopithecus, Paranthropus,* and *Homo* are given in Robinson (1962b). From the Sterkfontein Valley have come, therefore, members of both the major lines of Australopithecine evolution as well as members of both stages of the *Australopithecus-Homo* stream. This is also true of the Olduvai region of Tanganyika.

CONCLUSION

The evolution of hominids thus seems to have involved two critical points or thresholds where adaptive shifts occurred. The first was an essentially anatomical change, the shift from quadrupedal to erectly bipedal locomotor habit. This freed the hands and opened up the possibility of becoming an efficient tool-user. However, with a vegetarian diet and suitable environmental conditions to provide the necessities for the way of life of the first Australopithecines, there would probably be comparatively little stimulus for any major change in the adaptive situation then existing.

The second critical point was the inclusion of meat as a normal part of the diet. This was a direct response to altered and altering environmental conditions. Increasing aridity over much of Africa, from which the vast majority of Australopithecine specimens are known, brought about the change in diet. The second adaptive shift was therefore primarily ecological. But the change in diet placed a premium on tool-using, that is on cultural activity, and on improved intelligence. The change in climatic conditions thus provided precisely the stimulus needed for rapid development of the potentialities present in the prospective adaptation represented by erect bipedal posture and freed hands. An inevitable consequence was the appearance of hominines with their relatively very high intellectual and cultural ability and the new means of adaptation, artificial adaptation, with the consequent effect on the pattern of speciation.

Once *Paranthropus* had given rise to the *Australopithecus* line it remained, as it were, in an evolutionary backwater. In the long run it was unable to hold its own and became extinct. *Australopithecus* evolved fairly rapidly, compared to *Paranthropus,* and soon the first hominines came into existence. This did not happen by the transformation of all demes of *Australopithecus,* since this form was still in existence after hominines had already appeared. But the competition between the two was probably such that, as the hominines spread, *Australopithecus*

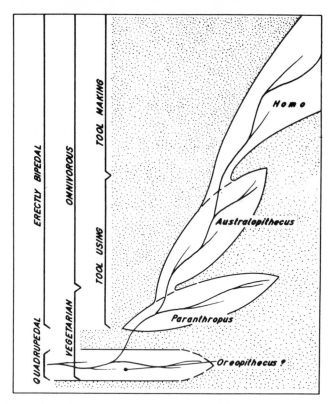

Figure 9   *Diagrammatic representation of the more important adaptive zones occupied by the hominid evolutionary stream. The threshold between the quadrupedal and bipedal stages is a major one between essentially discontinuous zones. The second and third thresholds—change to omnivorous diet and tool manufacture—are of great importance but did not involve clearly discontinuous zones. It should be emphasized that this is not a family tree, but an adaptive grid.*

lost ground and became extinct. The whole process is graphically summarized in Figure 9.

The second adaptive shift, involving the incorporation of meat-eating in the diet, seems to me to have been an evolutionary change of enormous importance which opened up a vast new evolutionary field. This change in my opinion ranks in evolutionary importance with the origin of mammals—perhaps more appropriately with the origin of tetrapods. With the relatively great expansion of intelligence and culture it introduced a new dimension and a new evolutionary mechanism into the evolutionary picture, which at best are only very palely foreshadowed in other animals. Furthermore it brought into existence the first member of the animal kingdom able to colonize other worlds—or at the very least, able physically to leave this one permanently.

35

Competitive Exclusion among Lower Pleistocene Hominids: The Single Species Hypothesis

M. H. WOLPOFF

As early as 1950, Ernst Mayr suggested applying the competitive exclusion principle to early hominid evolution. Other authors, such as Schwalbe (1899; 1913), Weinert (1951), and Weidenreich (1943a), interpreted hominid evolution within the framework of (what we would now describe as) a single evolving polytypic lineage, based purely on morphological considerations. Mayr, however, was the first to give this interpretation a theoretical basis in the synthetic theory.

The single species hypothesis rests on the nature of the primary hominid adaptation: culture (structured learned behavior). Because of cultural adaptation, all hominid species occupy the same, extremely broad, adaptive niche. For this reason, allopatric hominid species would become sympatric. Thus the competitive exclusion principle can be legitimately applied. The most likely outcome is the continued survival of only one hominid lineage.

THE ORIGIN OF BIPEDAL LOCOMOTION

As Mayr originally claimed and as Washburn has often stressed (1951; 1960; 1963a), one of the primary hominid morphological adaptations centres about bipedal locomotion. Other distinctive hominid characteristics not related to diet either arise from this adaptation, or from secondary adaptations.

What selective pressures lead to bipedalism? What selective advantages did bipedalism confer on very early hominids? Many answers to these questions have been proposed. For instance Hewes (1961) suggests food transport across the savanna as the primary adaptive advantage of bipedal locomotion, and L. S. B. Leakey points to the ability to see over tall grass (personal communication). While these suggestions obviously form part of an adaptive explanation, by themselves they simply do not explain the presence of an adaptive advantage strong enough to compensate for the loss of quadrupedal mobility (Washburn 1951: 69; Pfeiffer 1969: 43–53; Oakley 1959: 443); nor do they take sufficient account of the concomitant dangers due to predators in a savanna existence (Bramblett 1967). As Brace (1962: 607) put it: 'It would seem that a weaponless biped trudging over the savanna with a load of ripe meat would be an exceedingly poor bet for survival'.

Thus the use of tools as a means of defence appears to be critical. A dependence upon tools both in offensive and defensive behaviour explains the selective advantage of bipedal locomotion: the hands are freed during locomotion so that a tool or weapon is available at all times. The question of availability at all times is crucial, for the great apes can both produce and carry tools (Goodall 1964). However, tool use in chimpanzees differs from human tool use in one important respect: chimpanzees do not regularly use their tools as weapons, nor do they depend upon

tools as a means of defence (Goodall 1964). Still, the established use of tools by these pongids as part of their normal way of life makes it likely that the ancestors of the earliest hominids were also capable of this behaviour.

Since the ability to make and use tools as a learned and ecologically important type of behaviour is not restricted to hominids, the unique hominid *dependence* upon tools for defence is all the more revealing. Faced with a predator, a hominid who knew how to use a club for defence but did not have one available was just as dead as one to whom the notion never occurred. The advantage of carrying weapons continuously probably provided the greatest impetus to the morphological changes transforming bipedal locomotion from a possibility to an efficient form of movement.

All the relevant australopithecine skeletal material known indicates a completely bipedal stance (Lovejoy et al. 1973). Of most importance is the presence of a greater sciatic notch in pelvises from Swartkrans, Sterkfontein, Makapansgat and Kromdraai. This brings the action of gluteus maximus posterior to the acetabulum, causing it to act as an extensor. It has been suggested that the femoral attachment for gluteus maximus appears somewhat more lateral than modern man in the Swartkrans and Olduvai femora. However, in the only complete and relatively undistorted australopithecine *os innominatum*, STS 14 from Sterkfontein, the orientation of the acetabulum is also somewhat more lateral than the more anterior-lateral orientation often found in modern man. Articulation of either Swartkrans femur in the Sterkfontein acetabulum shows a *posterior* attachment of gluteus maximus on the femur, when the femur shaft is placed in the plane of the iliac pillar. Thus the 'human' condition is achieved in the total morphological pattern. Actually, the position of this muscle attachment and the lesser trochanter in australopithecine femora is consistent with moderate anteversion (Heiple & Lovejoy 1971). There is no reason to believe the morphology of the (undiscovered) pelvises actually associated with the Swartkrans femora would be any different. In addition, Day (1969) infers a lateral acetabulum for robust australopithecine pelvises from the posterior position of the lesser trochanter of the femur. Moreover, the acetabulum orientation is not outside the range of *Homo sapiens* variation (Lovejoy & Heiple 1972).

Other pelvic features indicate erect posture. The acetabulum in both Swartkrans and Sterkfontein specimens is very large, compared with the total size of the pelvis (Schultz 1969b). This indicates that the full weight of the upper part of the body passes through the joint. Schultz showed that the australopithecine ratios of acetabulum to pelvis size fall within the human and outside the ape range of variation. All the specimens have an iliopsoas groove. Indeed, all the australopithecine pelvises are similar, except for size, in every feature not obviously altered by post mortem distortion. It is particularly significant that this similarity includes features in which both gracile and robust pelvises differ from what is normal for *Homo sapiens* (although within the range of variation), such as the horizontal orientation and more marked distance between margins of the ischial tuberosity relative to the position of the acetabulum, the presence of an *anterior* iliac pillar, the more lateral orientation of the acetabulum, and so on.

A number of authors have pointed to details of femoral morphology indicating erect posture in the australopithecines (Le Gros Clark 1947; Straus 1962; Day 1969; Lovejoy & Heiple 1970). All femora have an obturator externus groove (Day 1969). The heads of australopithecine femora from Swartkrans, Sterkfontein (reconstructed), and East Rudolf are large, relative to the size of the bone.

The Olduvai foot is completely human in form, indicating both effective weight transmission and a striding gait (Day & Napier 1964). The Olduvai talus, the object of a separate study, also shows features suggesting bipedalism as does the talus from Kromdraai (Day & Wood 1968). Even the single terminal phalanx from Olduvai, hominid 10, shows evidence of bipedality (Day & Napier 1966). The Sterkfontein vertebral column suggests the presence of curvature in the lumbar area (Robinson 1956). The low position of the nuchal line in all specimens (Le Gros Clark 1967) indicates a long lever arm for the nuchal musculature, and a vertical orientation of the vertebral column.

Since the living African apes are capable of prolonged bipedal locomotion *without* all these morphological adaptations, the presence of such a total morphological pattern in the australopithecines most likely indicates habitual bipedalism in the same sense that modern man is habitually bipedal, although not necessarily in the same detailed way.

The pre-australopithecine hominids could be described as primates with the morphological and behavioural capabilities of living apes, using these capabilities in order to adapt to a way of life similar to that of baboons. The australopithecines themselves show the morphological consequences of this adaptation. The special fracturing of the battered baboon skeletal material at australopithecine-bearing sites such as Taung and Makapansgat suggests competition between these ecologically similar primate species.

THE MEANING OF REDUCED CANINES

According to this hypothesis, the reduced canines found in even the earliest australopithecines indicate a replacement of the canine defensive function (Washburn & Avis 1958: 425) by regularly employing implements as a means of defence, as Darwin suggested a century ago.

In this regard, it is important to establish that both gracile and robust australopithecines have reduced canines. Specifically, the canines of both australopithecine types are not significantly different in size from each other, or from those of *Homo erectus*. Table 1 gives mandibular and maxillary canine area (L∗B) data for the three groups. Area is used as a measure of canine size, combining maximum length and breadth. Because maximum breadth occurs at the base, and maximum length is usually not too far above it, only the most extreme wear invalidates area as a size measure. Sources for the *Homo erectus* material are given in another publication (Wolpoff 1971a). The australopithecine data were measured by C. L. Brace in South Africa and at Olduvai. Additional specimens from east Africa were published by Howell (1969), Walker (1971), Coppens (1970), and R. E. F. Leakey (1971). The three Java mandibles were described by Marks (1953) and Tobias & von Koenigswald (1964). Table 1 shows an extensive similarity among canine areas. Living groups of man show far

greater differences in average areas (Wolpoff 1970a: tables 22 and 27). Extant population averages for mandibular canines range 52 mm² to 71 mm². The corresponding range for maxillary canine averages is from 68 mm² to 90 mm².

A *t* test was used to test for significance of difference, using the data in table 1. In the mandible as well as the maxilla, both gracile and robust australopithecines are *not significantly different* from *Homo erectus* even at the 5 per cent level. In fact, they are not significantly different from each other at the 5 per cent level! Apparently, australopithecine canines were already reduced to Mid-Pleistocene hominid size.

Australopithecines were small to moderate-sized creatures. The most recent body size estimation for STS 14 (one of the smallest specimens) is 42–43 inches, weighing between 40 and 50 pounds (Lovejoy & Heiple 1970). The Olduvai tibia suggests a height of 56 inches (Coon 1963: 285). No estimates for the robust australopithecines have exceeded 200 pounds, and the small size of the femur heads of SK 82 and 97 (Napier 1964) suggests considerably less for at least some. Others were larger, as a comparison of Swartkrans mandibles shows (Wolpoff 1971d, fig. 1); the size range at this site alone was extensive. A similar size range occurs for postcranial material from East Rudolf (R. E. F. Leakey 1971).

Graciles and robusts overlapped considerably. Robinson indicates that male gracile and female robust australopithecines were approximately equal in robustness and stature (1970: 1219). A small body weight for many of the specimens is suggested by the size of the bones and joint surfaces. All the joints through which the weight of the body passes are very small: often smaller, in fact, than the joints in very small *Homo sapiens* (Broek 1938). The Swartkrans femur heads are as small as those of chimpanzees, although the full weight of the upper body does not usually pass through this joint in chimpanzees. The size of the neck, shaft, and reconstructed head in the Sterkfontein femur is even smaller. The sacral articular area of the Sterkfontein pelvis is considerable smaller than that of a comparably sized bushman pelvis.

There is every indication that early Pleistocene savanna primates of this size range would not lack natural predators, and would thus require a means of defence. The robust australopithecines have been compared to gorillas. Gorilla males, however, weight between 300 and 400 pounds (Napier & Napier 1967) and do not live on savanna. The necessity of maintaining an adequate means of defence for a savanna-dwelling lower Pleistocene primate in the australopithecine size range is suggested by the fact that savanna-dwelling baboons of this time period were characterised by large projecting canines coupled with a body size sometimes exceeding that of male gorillas (R. E. F. Leakey 1969; Freedman 1957; von Koenigswald 1969; L. S. B. Leakey & Whitworth 1958).

THE EVIDENCE FOR TOOL USE

Chimpanzees have been observed both walking erect and using tools. This indeed is part of their normal repertoire. The common ancestors of pongids and hominids were probably also capable of such behaviour. In view of the complex morphological and neurological prerequisites of primate tool use, and its transmission, and the very different adaptive patterns of the living hominid and pongid species now capable of such activity, it is unlikely that tool use arose in parallel after the lineages separated.

Given this reconstruction of pre-divergence behavioural capability, we must ask why the lineage separation occurred. The single species hypothesis rests on the *assumption* that the *dependence* upon tools as a means of defence allowed the savanna-forest niche divergence to occur, and thus formed the basis for this split. The resulting adaptation allowed effective hominid utilisation of

Table 1 *Canine Area Averages (mm²)*

	Mandibular				Maxillary			
	$\overline{X}$	*(Range)*	*CV*	*N*	$\overline{X}$	*(Range)*	*CV*	*N*
Australopithecines								
Gracile	87	(73–104)	12	13	93	(73–127)	18	8
Robust	81	(66–96)	11	18	88	(86–111)	14	19
Homo erectus	77	(62–98)	15	16	97	(83–113)	11	10
Homo sapiens	58	(37–81)	15	311	72	(41–120)	15	283
Australian aborigines	71	(64–80)	8	15	90	(58–120)	20	15
New Britain	59	(50–74)	10	44	73	(59–92)	11	43
European	56	(37–77)	17	42	68	(53–98)	16	42
Malay	52	(44–72)	14	12	71	(64–85)	17	3

savanna resources: first seeds and roots (Jolly 1970b), and later scavenged game (Dart 1957a; 1964; M.D. Leakey 1967). Pongids, on the other hand, became better adapted to a forest niche than their dryopithecine ancestors.

What were australopithecine tools like? The earliest dated implements, at this time, come from deposits on the east side of Lake Rudolf (M. D. Leakey 1970b), Kenya. These deposits are associated with a date of at least 2.6 million years (Fitch & Miller 1970). On the other hand, I believe, the earliest direct evidence of hunting activity, as opposed to scavenging or single kill sites (of possibly already incapacitated animals) derives from sites near the top of Bed I in Olduvai Gorge, *at least 1.5 million years later*. If tools were invented by 'killer ape' ancestors as a means for gathering animal protein (Ardrey 1961), archaeology indicates that these early hominids were not good at it, because most early tools are simple cutting edges and digging implements (L. S. B. Leakey 1960): there are no hunting tools. Such implements could be interpreted as part of a dietary adaptation based on scavenging and gathering roots (unobtainable for baboons). Thus, Jolly's seed-eater hypotheses seems supported. *Homo erectus* is apparently the earliest hominid to show the dental reduction commensurate with significant meat-eating.

Unfortunately, the implements first used for defence were probably simple clubs of wood and bone. Their use is only occasionally, and often indirectly, shown (Dart 1957a). To ask which came first, defensive implements or tools for cutting while scavenging and digging while gathering, is a 'chicken and egg' type of question. They are both part of the same adaptive complex: the hominid ecological equivalent of savanna-adapted baboons.

The early dependence on implements as a means of defence allowed an effective savanna adaptation, and consequently led to the differentiation of the hominid stock, necessitating bipedal locomotion and consequently providing its selective advantage. Culture, in this context, can be viewed as an adaptation to insure the effective transmission of tool use from generation to generation. Selection acted to modify the hominid morphology in the direction of producing a more efficient culture-bearing animal, allowing both the structuring and the transmission of survival-oriented kinds of learned behaviour.

Any bipedal small-canined hominid population should not only have been culture-bearing, but indeed should have been dependent upon culture for its survival. African archaeology offers support for the first part of this contention, for tools have been associated with both gracile and robust australopithecines on a series of living floors in Bed I (M. D. Leakey 1966; 1967). In the Koobi Fora area of Lake Rudolf, an Oldowan industry and numerous australopithecines derive from the same deposits (M. D. Leakey 1970b; R. E. F. Leakey 1970a; 1970b; 1971). At Makapansgat, australopithecines are found both within and between breccia strata with pebble tools (Dart 1955b; 1962a; 1962b). The well-described industry at Sterkfontein (Robinson 1957; Mason

1962: 472–5) derives from the middle breccia. Recent investigations indicate that STS 5 also comes from this breccia (Tobias & Hughes 1969). In fact it has been suggested that all Sterkfontein australopithecines come from the middle breccia (Tobias & Hughes 1969: 164) rather than from the lower breccia as originally claimed. An industry characterised as Acheulean derives from the deposit which yielded the Natron mandible (Isaac 1965; 1967). Definite stone tools occur at Swartkrans (Brain 1958; 1967; 1970; M. D. Leakey 1970a), and at least one convincing artefact came from Kromdraii (Brain 1958). Indeed, some of the australopithecine industries are surprisingly advanced. In addition to Natron, the Sterkfontein assemblage has been called early Acheulean (Mason 1962: 472–5), and the Swartkrans collection likened to Olduvai Bed II sites such as BK II (M. D. Leakey 1970a).

In numerous cases the evidence directly associates *both* gracile and robust australopithecines with the use and manufacture of stone tools. More often than not, most of these seem associated with scavenging activity. Australopithecine scavenging has been adequately demonstrated from the body part distribution at both Makapansgat (Dart 1957a) and Olduvai Bed I (M. D. Leakey 1967).

CULTURE AND COMPETITIVE EXCLUSION

Man has adapted *culturally* to the physical environment, and has adapted *morphologically* to effectively bearing culture. Thus culture, rather than any particular morphological configuration, is man's primary means of adaptation. His morphological evolution has been consistently directed by selection for a more effective culture-bearing creature. Culture plays the dual role of man's primary means of adaptation, as well as the niche to which man has morphologically adapted. In this sense, all hominids occupy the same adaptive niche.

The fact that culture is an integral part of man's adaptive pattern suggests that cultural evidence is as important as osteological evidence in reconstructing hominid evolutionary history.

Although culture may have arisen as a defensive survival mechanism, once present, it opened up a whole new range of environmental resources. Culture acts to multiply, rather than to restrict, the number of usuable environmental resources. Because of this hominid adaptive characteristic implemented by culture it is unlikely that different hominid species could have been maintained. Mayr (1950) originally applied Gauss's principle (1934) of competitive exclusion to the understanding of hominid evolution. As he interprets the principle:

the logical consequence of competition is that the potential coexistence of two ecologically similar species allows three alternatives: (1) the two species are sufficiently similar in their needs and their ability to fulfill these needs so that one of the two species becomes extinct, either (a) because it is 'competitively inferior' or has a smaller capacity to increase or (b) because it has an initial numerical disadvantage; (2) there is a sufficiently large zone of ecological nonoverlap (area of reduced or absent competition) to permit the two species to coexist indefinitely (1950: 68).

There are two conditions that must be met for closely related species to coexist sympatrically: (1) they must be able to tolerate the hazards occurring in the area of overlap; (2) they must differ from each other in such a manner that they do not enter into a 'struggle for existence' in which one succeeds at the expense of the other.

In culture-bearing hominids, it is particularly difficult to meet these conditions. For the separation of two species, a fortuitous isolation of part of the parent species must occur over sufficient time for genetic isolating mechanisms to become established. If this separation were to have occurred before the hominid differentiation then culture presumably arose independently in each lineage, as shown by both the archaeological evidence and by the same morphological evidence for bipedalism found in both gracile and robust australopithecines. On the other hand, a separation after the hominid differentiation is questionable for exactly the same reasons that sympatry itself is questionable (Mayr 1963a: 66).

Even if two separate hominid lineages could have arisen, how could they have been maintained for an appreciable length of time? One of the advantages afforded by culture is the great ecological diversity in the utilisation of a broad ecological base which it allows. In consequence, hominids tend to spread over a broad range, occupying areas where only *some* resources are available at a given time. That is, hominids can utilise more resources than are ever available at one place. Thus, the australopithecines spread over the Old World tropics and semi-tropics from South Africa to Java, occupying a large variety of climatic habitats and living sites. Synchronic culture-bearing hominid species could not help but become sympatric (Cain 1953) in a number of different areas. Related species are more likely to be found in similar habitats than are unrelated ones (Williams 1947; Bagenal 1959). The sympatric hominid species would then be in competition for *different* resources in *different* areas of overlap. For the total range of each species, the overlapping resource base would necessarily be extensive.

With competition occurring in different areas for different resources between species each able to utilise a broad ecological base, subsequent adaptation *could not reduce competition*. New adaptations would have to be learned. Rather than narrowing the range of utilised environmental resources for each species, such further adaptation would probably broaden this range by increasing the capacity to learn how to utilise additional resources, and thus *increase* the amount of real competition for the whole species. That is, competition would most likely cause each hominid species to develop the ability to utilise a wider range of resources and thus increase the amount of competition. One surely must succeed at the expense of the other.

APPLICATION TO THE LOWER PLEISTOCENE

Most authors now apply the results of competitive exclusion interpreting Mid-Pleistocene and more recent hominids, recognising only one synchronic hominid species. The single species hypothesis is primarily applied here to hominid origins, predicting the valid application of competitive exclusion in interpreting earlier hominid remains.

There are excellent reasons to believe that culture played a crucial role in australopithecine survival, in and apart from the logic dictated by the single species hypothesis. Mann (1968), for instance, has been able to demonstrate that the rate of australo-

pithecine development and maturation was delayed, as in modern man, rather than accelerated as in modern chimpanzees. Based on molar eruption timing, Mann showed that australopithecine children took as long to mature as do our own. If selection for increased learning capacity, associated with cultural behaviour, resulted in delayed maturation in Lower Pleistocene hominids, it must have been operative *before* the Pleistocene.

Similarly, McKinley (1971) demonstrated that australopithecines (graciles and robusts) followed a 'human' model of short birth spacing. In baboons, gorillas and chimpanzees, successive births are spaced apart by the length of child dependency. Thus in slow maturing chimpanzees, he calculated an average birth interval of 4.6 years from data given in Goodall (1967), the only available source. In man the maturation timing is about half as fast, so the corresponding period of child dependency is close to eight years. Human births, however, are not spaced by this period, but rather can be as close together as one to two years. This seems primarily due to the influence of complex social factors on the effect of child dependency. The result is highly adaptive to rapid population expansion. Australopithecines follow the human model of delayed maturation timing. The corresponding birth spacing, following the chimpanzee model but based on the delayed maturation rate, is close to eight years. However, the average time between births calculated by McKinley (1971), three to four years in the robusts and four to five years in the graciles, is less. Social behaviour, far more complex than that evinced by baboons, acted to *shorten* the effects of child dependence in australopithecines, although Mann's work shows that this period of dependence was *longer* for them than it is in modern pongids. In both cases, the evidence clearly indicates extensive australopithecine adaptation to social-cultural behaviour. This suggests that such behaviour was adaptively important prior to the Lower Pleistocene.

Is the conclusion that gracile and robust australopithecines are members of the same hominid lineage really so unlikely? Authors other than myself have demonstrated greater differences among groups of anatomically modern *Homo sapiens* (Bielicki 1966; Brace 1963a; 1963b; 1967a; 1967b; Buettner-Janusch 1966; Dart 1955a; Le Gros Clark 1967; Oppenheimer 1964). Variation among all australopithecines appears on a par with variation in both gorillas and chimpanzees (Wolpoff 1970; Remane 1959; Schultz 1937; 1954; 1963; 1968; 1969a). As Campbell has pointed out (1969), this interpretation fits the established pattern of both gracile and robust groups of people in Africa from the Lower Pleistocene to the present.

The synchronic occurrence of both gracile and robust australopithecines has been demonstrated in east Africa from terminal Pliocene well into the Lower Pleistocene (Arambourg & Coppens 1967; 1968; Coppens 1970; Arambourg et al. 1967; 1969; Howell 1968a; 1968b; 1969; Patterson & Howells 1967; R. E. F. Leakey 1970a; 1970b; 1971; L. S. B. Leakey 1960; Martyn Tobias 1967). The available evidence indicates that social and cultural behaviours acted as evolution-orienting factors over this time span.

If the graciles and the robusts truly are separate lineages, there can by definition have been no gene flow between them. Given the facts that 1) they were supposed to be adapting quite differently, and that (2) they were synchronic for at least two million years (Howell 1969), one would expect non-overlapping differences in adaptively differentiated features to have occurred. Conversely, if this expectation were not permissible it would be impossible to test the hypothesis suggesting that the gracile and robust australopithecines were two different lineages. At that point, the question of separate australopithecine species would become unanswerable, and hence phylogenetically meaningless.

The interpretation of separate australopithecine lineages

yields two testable predictions: (1) One expects non-overlapping sets of differences between the two lineages which indicate different adaptations, and consequently separate total morphological patterns; (2) One expects these differences to become greater through time.

The first prediction is best approached by directly testing the dietary hypothesis (Robinson 1956; 1963a; 1963b). Do graciles and robusts evince different dietary adaptations? The robusts are supposed to be adapted to a far more vegetarian diet than the graciles. Consequently, the grinding area of their cheek teeth should be considerably greater. The predicted difference in grinding area is fundamental to the dietary hypothesis. Without it, the demonstrable differences between gracile and robusts can only be related to size. Actually, the size difference complicates comparison of the cheek teeth. The robusts should have larger teeth corresponding to their larger size (Robinson 1963b), and at the same time they should also have larger teeth because of their more vegetarian diet. In sum, they should have *much* larger cheek teeth than do the graciles.

Table 2 presents data to test this prediction. Sources for the australopithecine measurements have already been given. Sources for the remaining hominid data are given in another publication (Wolpoff 1971a). Data are given for the summed areas (length times breadth) of the posterior tooth rows (PM_3—M_3 of individual specimens. In both mandible and maxilla, the graciles are about 88 per cent the size of robusts. Averages representing the range of modern populations are given for comparison. In the posterior dentition, Lapps are only 83 per cent of the size of other Europeans and in the extreme case they are 73 per cent the size of Australian aborigines in the mandible and 67 per cent in the maxilla. In both absolute and percentage differences, gracile and robust australopithecines are considerably closer together than numerous modern populations, in some cases living side by side. While these data do not mean that the graciles and the robusts are identical to each other, they clearly indicate extensive similarity in an adaptive complex which is supposed to show significant difference.

The summed posterior areas for 318 gorilla maxillae are included: I measured some of these specimens at the American Museum of Natural History. Specimens from the Yale Peabody Museum and other sources were published by Pilbeam (1969a), and those at the University of Wisconsin by Booth (1971). Most of the material was measured by Mr. P. Mahler, of the University of Michigan, and myself, from the Hamann-Todd collection. Mandibles could not be used because of the lack of P_3 analogy. The averages for gorillas and robust australopithecines are within 5 per cent of each other, although the smallest gorillas have at least double the maximum body mass estimated for robust australopithecines, and the larger specimens have three to four times the

mass. The presence of a robust australopithecine-sized posterior grinding area in a species considerably larger, restricted to a primarily vegetarian diet, gives indication of the diet of the robusts. Jolly (1970b) has argued that hominids arose from a savanna-based primate adapted to the intensive mastication involved in small object feeding. Given this tooth size comparison with the much larger gorillas, adapted to a forest foliage diet, the dietary part of Jolly's 'seed eaters' hypothesis seems substantiated.

According to Robinson's dietary hypothesis, gracile australopithecines are supposed to have reduced posterior dentitions because of their presumed more omnivorous diet. As table 2 indicates, the graciles have posterior summed grinding areas completely within the range of variation of gorillas, in spite of the likelihood that their body size was between one-quarter and one-eighth gorilla size. Most estimates give robusts about double the average weight of graciles. If even approximately true, the graciles have relatively *larger* posterior teeth. Is it really so likely that the graciles had a more omnivorous diet than the robusts? Or rather, in keeping with Jolly's hypothesis that early hominids appear to be dentally adapted to a heavily masticated diet, does it not appear more probable that *both* the smaller graciles and the larger robusts subsisted on a diet of scavenged game, small objects and roots.

The second prediction is also questionable. Robust specimens from Omo (for instance L7-125) and Olduvai are considerably *more* extreme than robusts from apparently younger sites such as Swartkrans and Kromdraai. I do not conclude that there is a trend from greater to lesser robustness in the robusts (although this is possible), both because the earlier sample size is too small and because the range of variation in the later specimens is almost great enough to include the earlier ones. Certainly, however, there is no indication of the reverse tendency, as is predicted from the interpretation of separate lineages. This observation also directly contradicts the character displacement hypothesis suggested by Schaffer (1968).

With the increasing amount of data accumulated, numerous workers have come to recognise the intensive intergradation and overlap of gracile and robust australopithecines both in terms of individual specimens and in terms of entire sites such as Kromdraai and Makapansgat where 'intermediate' populations could be represented (Le Gros Clark 1967; Brace 1963a; 1963b; 1967a; Campbell 1969; Dart 1955a; 1964; Buettner-Janusch 1966; Simons 1968a; Mann 1970; Wolpoff 1968; 1971a; 1970; 1971b). Indeed, one of those most familiar with the actual specimens has recently concluded:

> The distinction between these two lines should not be over-stated: it is suggested that gene exchange between the two

Table 2   *Averaged Summed Posterior Areas (mm²)*

	Mandibular				Maxillary			
	$\bar{X}$	(Range)	CV	N	$\bar{X}$	(Range)	CV	N
Australopithecines								
Gracile	901	(770–1006)	10	8	849	(788–946)	7	6
Robust	1010	(666–1883)	17	16	969	(781–1354)	16	12
Homo erectus	656	(557–731)	8	14	643	(529–858)	18	8
Homo sapiens	496	(353–649)	11	216	501	(301–710)	14	238
Australian aborigines	539				581			
American aborigines	498				493			
Europeans	464				464			
Japanese	449				449			
Lapps	391				388			
Pan gorilla					1011	(744–1481)	14	318

lines might have been possible. Such hybridizing effects might have aided the 'toning down' of the extremely robust earlier *A. boisei* into the later less robust *A. robustus*; and secondly, might account for a number of *A. robustus* features in the otherwise *A. africanus* forms of Makapansgat (Tobias 1969b: 311–12).

This description can only refer to subspecies within the same species.

There are certainly differences between what I have referred to as gracile and robust australopithecines, just as there is a difference between Bushmen and Bantu, or Norwegians and Lapps, or Lowland and Mountain gorillas. These differences can be extensive. One would never confuse crania of Bushmen and Bantu with each other, whether a single cranium or a much larger sample was involved. Different *types*, however, are not necessarily different *species*. If a sampling of australopithecine mendelian populations were available, I believe that we would characterise some as gracile, some as robust, and still others as intermediate. Thus a plausible interpretation of australopithecine variation refers the differences between them to sub-species level. This interpretation fits the predictions of the single species hypothesis.

The single species hypothesis is concerned with consequences of the pongid-hominid lineage separation. As such, it presents a way of approaching early hominid interpretations which would otherwise be untestable. While Jolly's 'seed eaters' hypothesis deals with the consequences of how early hominids *subsisted* on the savanna, this hypothesis considers the consequences of how they *defended* themselves. The two views are not contradictory.

The single species hypothesis is grounded in the synthetic theory, and fits the available evidence. Those who believe it to be invalid must not only demonstrate the unequivocal value of their proposed refutations, but must also replace the hypothesis with one which fits the evidence better.

36

Australopithecus, Homo erectus, and the Single Species Hypothesis

R. E. F. LEAKEY AND A. C. WALKER

An enormous wealth of early hominid remains has been discovered over the past few years by expeditions within eastern Africa. Evidence has been presented for the existence over a considerable period of time of at least two contemporaneous hominid species (Leakey, 1971, 1972, 1973, 1974). Some of this evidence is compelling, but some less so for a variety of reasons such as the lack of association, fragmentary specimens, geological uncertainties, equivocal anatomical differences and suchlike. Many of these new specimens are of great antiquity and have led to suggestions that an early form of the genus *Homo* was contemporary with at least one species of *Australopithecus*. The evidence presented here deals not with the early states of human evolution, but with the unequivocal occurrence of *H. erectus* from the Koobi Fora Formation, east of Lake Turkana (formerly Lake Rudolf).

Among the variety of hypotheses put forward to accommodate the evidence in an evolutionary framework, the most explicit and directly simple is the single species hypothesis (Brace, 1967). This hypothesis rests on the assumption that dependence upon tools was the primary hominid adaptation that enabled expansion into open country environments. In the clearest exposition of the hypothesis (Wolpoff, 1971), basic hominid characteristics, including bipedal locomotion, reduced canines and delayed physical maturity are seen to have come about in response to more effective and greater dependence on culture. This assumption of a basic hominid cultural adaptation allied with the principle of competitive exclusion (Gauss, 1934; Mayr, 1950) leads to the conclusion that two or more hominid species would be extremely unlikely to exist sympatrically. Here we present decisive evidence that shows the existence of two contemporaneous hominid species in the Koobi Fora area.

The adult cranium KNM-ER 3733 (Leakey, 1976) was found *in situ* in the upper member of the Koobi Fora Formation (Bowen and Vondra, 1973). The sediments lie stratigraphically between the KBS and the Koobi Fora/BBS tuff complexes. The cranium consists of a complete calvaria and a great deal of the facial skeleton, including the nasal and zygomatic bones. The nearly complete alveoli of the anterior teeth and examples of the premolars and molars are preserved. The third molars were lost before fossilisation. Preparation has been limited so far to the external cranial surfaces and part of the brain case. An endocranial capacity is unknown, but by comparison is likely to be of the order of 800–900 ml. In all its features the cranium is strikingly like that of *H. erectus* from Peking (Weidenreich, 1943). Such orthodox anthropometric comparisons that can be made at present fall well within the range of the Peking specimens.

The cranium is large (glabella to inion/opisthocranion is 183 mm) with large projecting supraorbital tori and little postorbital constriction (minimum postorbital breadth 91 mm). There is a

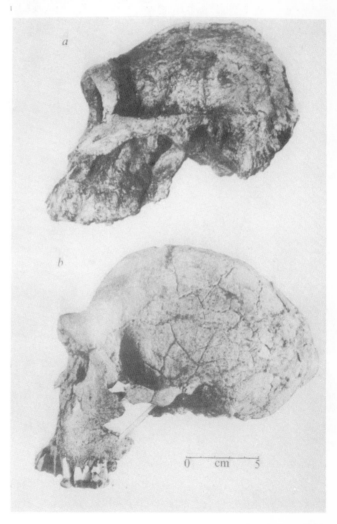

Figure 1 *Lateral aspect of KNM-ER 406 (a) and KNM-ER 3733 (b).*

marked postglabellar sulcus and the frontal squama rises steeply from behind it to reach vertex at bregma. The skull decreases in height from bregma and the occipital bears a pronounced torus where the occipital and nuchal planes are sharply angled. The greatest breadth is low at the angular torus (biauricular breadth 132 mm). There are strong temporal lines that are 60 mm apart at

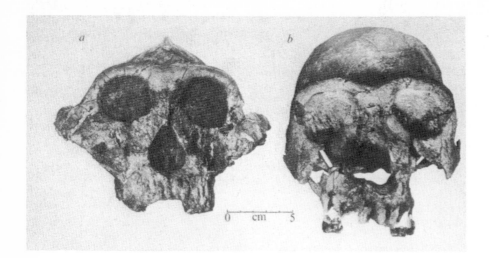

Figure 2 *Frontal aspect of KNM-ER 406 (a) and KNM-ER 3733 (b).*

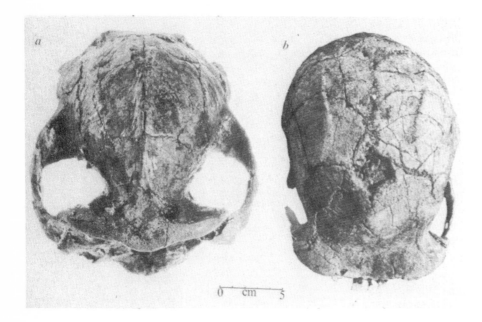

Figure 3 *Superior aspect of KNM-ER 406 (a) and KNM-ER 3733 (b).*

their nearest point. As far as can be judged at present, the vault bones are thick (about 10.0 mm in mid-parietal). The temporal fossae are small. The facial skeleton is partly preserved and shows deep and wide zygomatic portions, longitudinally concave and laterally convex projecting nasals and wide, low piriform aperture. The incisive alveolar plane is short and wide and the large incisive alveoli are set almost in a straight line. There are strong canine juga. The palate is high and roughly square in outline. The facial skeleton is flexed under the calvaria and in the preliminary reconstruction is set at about the same angle as that reconstructed

for a female *H. erectus* by Weidenreich (1943). The prosthion to nasion length is 87 mm. This is the best preserved single *H. erectus* cranium known, the facial skeleton being more complete and less distorted than that of 'Pithecanthropus VIII' (Sartono, 1971).

Figures 1-3 show KNM-ER 3733 together with KNM-ER 406, another cranium discovered *in situ* in the Upper Member of the Koobi Fora Formation. This latter specimen has been discribed (Leakey et al., 1971) in some detail and is clearly that of a robust *Australopithecus*. Other specimens of robust *Australopithecus* have been found in deposits of both the Upper and

Lower members of the Koobi Fora Formation. Two *in situ* mandibles that must have been associated with this type of cranium are the most massive representatives of this species. KNM-ER 729 (Leakey, 1972) was excavated in 1971 from the base of the Middle Tuff and KNM-ER 3230 (Leakey, 1976) from the upper part of the BBS complex. Other, more incomplete, specimens of robust *Australopithecus* mandibles have come from the higher levels of the Formation between the Karari/Chari and Koobi Fora/BBS tuffs. The radioisotopic dating of these tuffs, in spite of a continuing controversy over one of them (Curtis et al., 1975), is not an issue here. The contemporaneity of *Homo erectus* and a robust *Australopithecus* is now clearly established over the period during which the Upper Member of the Koobi Fora Formation was deposited. Using the time scale given by Fitch et al. (1974), this would be from earlier than about 1.3 to earlier than 1.6 Myr ago.

The new data show that the simplest hypothesis concerning early human evolution is incorrect and that more complex models must be devised. The single species hypothesis has served a useful purpose in focusing attention on variability among the early hominids and also on the ecological consequences of hominid adaptations. Alternative concepts, especially those concerning niche divergence and sympatry, should now be formulated. We think that populations antecedent to *H. erectus* ones have been sampled in the Koobi Fora Formation (specimens include KNM-ER 1470, 1590 and 3732). The clear demonstration of at least two hominid species earlier in time should enable us to reconsider our approaches to the problems of earlier hominid evolution. We also think that there is no good evidence for the presence of *Australopithecus* outside Africa, and the finding of an apparently advanced *H. erectus* cranium at Koobi Fora provokes issues such as why *Australopithecus* is only an African form, the nature of its extinction and the apparent stability of some hominid morphologies over the long periods of time.

We thank the Museum Trustees of Kenya and the National Museums of Kenya for access to material and facilities and the National Geographic Society for support. The NSF supported A.C.W. We also thank colleagues who helped with suggestions and observations on this paper.

37

A Systematic Assessment of Early African Hominids

D. C. JOHANSON AND T. D. WHITE

Paleoanthropological research in eastern and southern Africa has provided an extensive fossil record documenting human evolution over the last 2.5 million years. The accumulated fossil remains from sites such as Koobi Fora, Olduvai Gorge, Omo, Sterkfontein, and Swartkrans (Fig. 1) have been studied, described, and afforded diverse phylogenetic and taxonomic interpretations (Walker and Leakey, 1978; Coppens et al., Eds., 1976; Jolly, Ed., 1978; Tobias, 1973a; Howell, 1978; Robinson, 1972).

The sites of Laetolil in Tanzania and Hadar in Ethiopia (Fig. 1) have yielded abundant remains of human ancestors that have been dated firmly between 3 and 4 million years ago. These new hominid fossils, recovered since 1973, constitute the earliest definitive evidence of the family Hominidae [1]. The morphology and attributes of these remains are demonstrably more primitive than those of hominid specimens from other sites. Because of their great age, abundance, state of preservation, and distinctive morphology, the Laetolil and Hadar fossils provide a new perspective on human phylogeny during Pliocene and Pleistocene times.

It is not our aim in this article to review the extensive literature that deals with hominid origins, phylogeny, and taxonomy. Our first intention is to describe some of the most salient morphological features of the newly recovered Pliocene hominids from Laetolil and Hadar. We will then assess the phylogenetic position of the new specimens within the Hominidae in light of their distinctive skeletal anatomy. Finally, we will express the implications of these findings in a taxonomic evaluation.

BACKGROUND

The major hominid collections from Laetolil were made by Mary D. Leakey's expedition. Laetolil lies about 50 kilometers (30 miles) south of Olduvai Gorge in northern Tanzania (Fig. 1). The ongoing fieldwork was initiated at Laetolil in 1974, and the geology, paleontology, and history of the site have been described by M. D. Leakey et al. (1976). The fossil hominids consist primarily of dental and gnathic remains derived from the Laetolil Beds and are radiometrically placed between 3.6 and 3.8 million years ago (M. D. Leakey et al, 1976). Laetolil hominids (L. H.) 1 through 14 have been described (White, 1977b) and nine additional specimens have been recovered.

Hadar is located in the Afar triangle of Ethiopia (Fig. 1). Intensive paleoanthropological fieldwork was conducted at the site between 1972 and 1977 by the International Afar Research Expedition (Taieb et al., 1976; Aronson et al., 1977; Johanson and Taieb, 1976; Johanson, Gray, and Coppens, 1978). Abundant, diverse, well-preserved fossils were recovered from the Hadar Formation. On the basis of geochronologic and biostratigraphic evidence, this formation has been dated between 2.6 and 3.3 million years ago (Aronson et al., 1977). A remarkable collection

Figure 1 *Geographic location of the major fossil hominid sites discussed in the text: 1, Hadar; 2, Omo; 3, Koobi Fora; 4, Olduvai Gorge; 5, Laetolil; 6, Makapansgat; 7, Sterkfontein, Swartkrans, and Kromdraai, and 8, Taung.*

of hominid specimens representing a minimum of 35 and a maximum of more than 65 individuals was recovered. Preservation is outstanding and some Hadar specimens are exceptionally complete. In several cases there are associated skeletal parts of the same individual (Fig. 5; Johanson and Taieb, 1976; Johanson et al., 1978; Taieb et al., 1978). Nearly all anatomical regions of the body are represented in the collections from Hadar. This situation is unprecedented for the earlier portion of the fossil hominid record. For example, we have nearly 40 percent of a skeleton known as "Lucy" from Afar Locality (A.L.) 288 and more than 200 specimens representing an absolute minimum of 13 individuals from A.L. 333 and 333w. Some of the material has been presented (Johanson and Taieb, 1976; Johanson, Gray, and Coppens, 1978; Taieb et al., 1978; Johanson, Taieb, Coppens, and Roche, 1978; Johanson and Coppens, 1976; Taieb, 1975; Johanson, 1976; Taieb, et al., 1974; Taieb, Johanson, and Coppens,

1975; Taieb, Johanson, Coppens, and Bonnefille, 1975), but a large portion of the sample is currently under investigation and will be fully described in the near future.

A comparative study of the Hadar and Laetolil hominids has clarified the relationship between the two collections. The strong morphological and chronological continuity seen between the Hadar and Laetolil fossil hominid samples strongly suggests that these collections are most conveniently and effectively considered together in the following systematic assessment.

ANATOMICAL EVIDENCE

The Laetolil and Hadar fossil hominid remains have a distinctive suite of primitive cranial and postcranial characteristics. Some of these have been mentioned in earlier publications but this is the first report on the combined sample as of September 1978. It is not possible in an article of this length to describe them in detail; instead, some of the major anatomical features of the material are outlined below.

Dentition

As with other paleontological materials from these sites, the dental elements comprise the largest portion of the Pliocene hominid sample from Hadar and Laetolil.

Incisors   The upper centrals are characterized by their great mesiodistal dimension, which contrasts with the diminutive mesiodistal diameter of the lateral incisors (A.L. 200-1a; L.H.-3).

Canines   The large, asymmetric, pointed lowers project slightly above the tooth row and usually have a pronounced lingual ridge (A.L. 400-1a, 128-23; L.H.-3). The uppers also are large and project slightly. When worn, they often bear an exposed strip of dentine along the distal occlusal edge (A.L. 200-1a; L.H.-5). Apical wear is often present as well. Both upper and lower canine roots are massive and long.

Premolars   The lower third premolars (P_3) are characterized by a dominant, mesiodistally elongate buccal cusp. The extensive buccal face often shows vertical wear striae produced by occlusion with the overlapping upper canine. A smaller lingual cusp is usually present, but some specimens (A.L. 288-1, 128-23) display only an inflated lingual ridge. The P_3 often possesses two distinct roots with the anterior one angulated mesiobucally (A.L. 333w-60; L.H.-4). In occlusal view, P_3 crown shape is normally an elongated oval, the long axis of which is oriented mesiobuccal to distolingual at 45° to 60° to the mediodistal axis of the tooth row. The upper third premolar (P^3) is sometimes three-rooted, with a pointed buccal cusp and an extensive, asymetric buccal face (A.L. 200-1a; L.H.-6). The buccal cervicoenamel line projects toward the mesiobuccal root, and in occlusal view the mesial placement of the lingual cusp gives the crown an asymmetric appearance. The P^3 tends to be slightly larger than the upper fourth premolar (P^4), and the latter does not show mesiodistal elongation of the buccal crown portion.

Molars   The lower molars, particularly the first and second, tend to be square in outline. The cusps are usually arranged in a simple Y-5 pattern, surrounding wide occlusal foveae. The third molars are generally larger and their distal outlines are rounded. The molar size sequence is normally $M_3 > M_2 > M_1$. The upper molars usually follow the same size sequence, their occlusal foveae are wide, and their hypocones are fully developed.

Deciduous dentition   The deciduous canines are morphologically similar to their adult counterparts in relative size, morphology, and occlusal projection (A.L. 333-99, 104; L.H.-2). The deciduous first molars conform to the molarized human pattern and show deep buccal grooves (A.L. 333-43, -86; L.H.-2).

Overall, the adult and deciduous dentitions are variably intermediate between Hominidae and Pongidae in most of the features enumerated by Le Gros Clark (1950b). Neither metric data (Table 1) [2] nor morphological considerations [3] suggest to us that more than one evolving hominid lineage is represented in the dental samples from Hadar and Laetolil.

Cranium

Portions of several adult and juvenile faces are available from Hadar and Laetolil. The adults show strong alveolar prognathism associated with somewhat procumbent incisors, the curved roots of which promote a convex clivus. The lower margin of the pyriform aperture is marked laterally by a raised border (A.L. 200-1a, 333-1). The large canine roots are reflected in strong canine jugae, which contribute to the formation of pillars lateral to the pyriform aperture. These pillars act to set this region apart from the zygomatic processes of the maxillae. The anterior margins of these large processes are located above the junction of P^4 and M^1 and are oriented nearly perpendicular to the tooth rows. The inferior margins of the zygomatic arches are flared anteriorly and laterally. The palates are shallow anteriorly and their lateral margins tend to converge posteriorly (Fig. 2). The dental arcades are long, narrow, and straight-sided instead of parabolic. The tooth row is sometimes interrupted by diastemata between the lateral incisors and canines (A.L. 200-1a).

Preserved portions of the adult crania A.L. 333-45 (Fig. 3) and A.L. 288-1 show a host of primitive features. There are strong muscle markings including a compound temporal-nuchal crest on both sides of A.L. 333-45. The temporal lines converge anteriorly and closely approximate the midline. An anteriorly placed sagittal crest is possible, but the relevant portions are not preserved. The smaller specimen, A.L. 288-1, is less robust but is morphologically similar in its preserved portions [4]. Specimen A.L. 333-45 is heavily pneumatized in lateral portions of the cranial base. The nuchal plane is concave and is longer than the occipital plane. The mastoid region is flattened posteriorly and the mastoid tips point anteroinferiorly. The external auditory meatus takes on a tubular appearance when viewed basally, strongly resembling the pongid condition. The mandibular fossae are broad, have little relief, and are placed only partially beneath the braincase. There is a strong entoglenoid process. A very weak articular eminence results in a mandibular fossa that is open anteriorly. The preserved occipital condyle is located below the external auditory meatus in lateral view and bears a strong angulation across its articular surface. It has not yet been possible to make satisfactory estimates of cranial capacity on the basis of preserved portions of crania, although preliminary observations suggest that it is small, probably within the known range of other *Australopithecus* species (*sensu stricto*). Studies of the cranial remains from Hadar and Laetolil have shown the distinctiveness of this anatomical region and promise to provide additional information concerning the ontogeny and functional anatomy of these early hominids [5].

Mandible

A combined sample of at least 25 adult and juvenile individuals represented by mandibular remains is available from Hadar and Laetolil. The mandibles from the two sites are strikingly similar (Fig. 4). Some major parts of the complex of features distinguishing this collection from other fossil hominid mandibles are described here.

Although ascending rami are poorly represented, available adult mandible specimens (A.L. 333-108) indicate large but not

Table 1 *Combined metric data for the Laetolil and Hadar hominid dentitions. Only measurements on intact teeth are provided. The measurement technique is described elsewhere (White, 1977b). Mesiodistal diameters for postcanine teeth are corrected for interproximal attrition except in cases where that was impossible. For anterior teeth, (w) indicates worn teeth representing range values. Other abbreviations: MD, mediodistal; BL, buccolingual; N, number; R, range; X, mean; and S.D., standard deviation.*

Dentition	Lower		Upper	
	MD	BL	MD	BL
Permanent				
First incisor (I1)				
N	1	3	4	5
R		7.3-7.7	90w-11.8	7.1-8.6
X	5.6w	7.50	10.36	8.16
S.D.		0.20	1.17	0.60
Second incisor (I2)				
N	4	3	6	8
R	5.72-7.1w	6.7-7.8	6.7w-8.2	6.2-8.1
X	6.28	7.37	7.65	7.18
S.D.	0.59	0.59	0.59	0.65
Canine (C)				
N	5	9	10	10
R	7.9-11.7	8.8-12.0	8.9-11.6	9.3-12.5
X	9.16	10.17	9.92	10.94
S.D.	1.54	1.15	0.74	1.11
Third premolar (P3)				
N	14	14	8	7
R	8.2-12.6	9.5-12.6	7.2-9.3	9.8-13.4
X	9.51	10.60	8.50	12.03
S.D.	1.09	0.98	0.74	1.19
Fourth premolar (P4)				
N	13	12	8	5
R	7.7-10.9	9.8-12.8	7.6-9.7	11.1-12.6
X	9.58	10.93	8.95	12.00
S.D.	0.95	0.92	0.68	0.60
First molar (M1)				
N	18	16	9	9
R	10.1-14.6	11.0-13.9	10.8-13.7	11.2-15.0
X	12.85	12.62	12.22	13.23
S.D.	1.05	0.90	0.92	1.24
Second molar (M2)				
N	17	17	3	3
R	12.1-15.4	12.1-15.2	12.1-13.5	13.4-15.0
X	14.02	13.44	12.83	14.40
S.D.	1.08	1.06	0.70	0.87
Third molar (M3)				
N	11	12	5	5
R	13.3-16.3	11.7-14.9	11.4-14.3	13.1-15.5
X	14.55	13.23	12.54	14.22
S.D.	0.8	1.02	1.32	1.05

necessarily high mandibular rami. The condyles (A.L. 333w-1e, -16) are large and concordant with the broad articular surfaces of the preserved crania. The A.L. 288-1 mandibular ramus slopes somewhat posteriorly. The ramus usually joins the corpus at a high position, defining a narrow, restricted extramolar sulcus (A.L. 266-1; L.H.-4).

The mandibular corpora are variable in size, and larger specimens are relatively deep in their anterior portions. The lateral contours in the region of the mental foramen are usually hollowed (A.L. 333w-60; L.H.-4). The mental foramina tend to occupy positions low on the corpus and open anterosuperiorly (A.L. 277-1, 288-1; L.H.-4). The mandibular canal passes immediately below the distal root of the third lower molar (M_3). The base of the corpus is everted, and the anterior portion of the corpus is rounded and bulbous. The symphyseal section usually shows a moderate superior transverse torus. The inferior transverse torus is low and rounded rather than shelf-like. There is strong posterior angulation of the symphyseal axis (A.L. 400-1a; L.H.-4). In occlusal aspect, the molars and premolars form straight rows and the anterior portion of the dental arcade tends to be narrow, especially in the smaller specimens. Some specimens show slight postcanine diastemata (A.L. 266-1; L.H.-4). The dramatic size differences seen between such morphologically similar mandibular specimens as A.L. 333w-60 and A.L. 333w-12 suggest a high level of sexual dimorphism within a single hominid lineage [6].

Postcranium

Comparison of the Hadar and Laetolil postcranial material with other Plio-Pleistocene remains is hampered at this time by diffi-

Table 1   *(continued)*

Dentition	Lower		Upper	
	MD	BL	MD	BL
Deciduous				
First deciduous incisor (di1)				
N	1	1		
R				
X	4.2	3.6		
S.D.				
Second deciduous incisor (di2)				
N	2	3	1	1
R	4.8-5.7	4.2-5.0		
X	5.25	4.63	5.7	4.5
S.D.		0.40		
Deciduous canine (dc)				
N	3	2	3	4
R	6.2-6.6	5.8	6.8-7.7	5.3-6.5
X	6.43	5.8	7.37	5.95
S.D.	0.21		0.49	0.49
First deciduous molar (dm1)				
N	4	4	4	3
R	8.5-9.6	7.6-8.4	8.1-9.4	8.9-9.3
X	9.15	7.93	8.68	9.17
S.D.	0.48	0.36	0.54	0.23
Second deciduous molar (dm2)				
N	2	2	4	4
R	11.6-12.6	9.7-10.6	9.9-10.8	10.5-12.6
X	12.1	10.15	10.23	11.20
S.D.			0.40	0.95

culties in associating cranial and postcranial material found at other sites. In addition, a number of skeletal elements found at Hadar (particularly some of the hand and foot bones) are either absent or poorly represented at other sites, which makes meaningful comparisons impossible. However, some anatomical features of the postcranium are already obvious and deserve mention.

The postcranial skeleton is well represented, and all analyses so far indicate that the hominids were adapted to bipedal locomo-tion. This is especially evident from the analysis of the knee joint anatomy (Johanson et al., 1976).

The most complete adult skeleton is that of A.L. 288-1 ("Lucy," Fig. 5). The small body size of this evidently female individual (about 3.5 to 4.0 feet in height) is matched by some other postcranial remains (A.L. 128, 129) and these smaller spec-imens can be contrasted with other larger but morphologically identical individuals from Hadar (A.L. 333 and 333w, Fig. 6). We

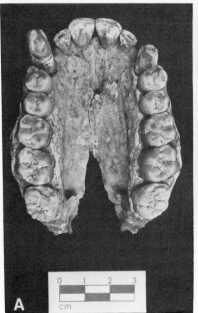
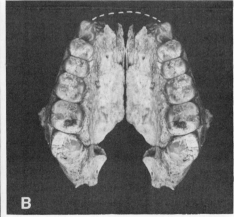

Figure 2   *Comparison of the (A) A.L. 200-1a and (B) A.L. 199-1 palates found at Hadar. The palate in (B) consists of a right half, but the intact midline permits photo-graphic mirror-imaging. Note the morpho-logical identity but different size of the specimens.*

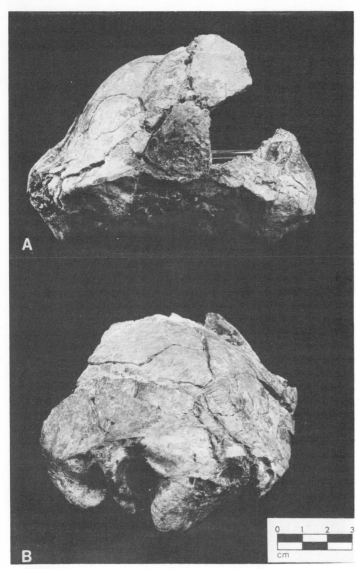

Figure 3 *(A) Occipital and (B) left lateral views of the A.L. 333-45 partial cranium from Hadar. The specimen suffered postmortem distortion, but many important anatomical details are discernible.*

consider that much of this body size difference reflects sexual dimorphism [7]. All of the postcranial elements indicate high levels of skeletal robustness with regard to muscular and tendinous insertions.

The humerofemoral index (ratio of the length of the humerus to the length of the femur) of the A.L. 288-1 specimen is approximately 83.9 (Johanson and Taieb, 1976). This value is high relative to modern humans. The hand bones from Hadar also differ from those of modern humans—for instance, in the "waisted" appearance of the capitate (A.L. 288-1, 333-40), the lack of a styloid process on the third metacarpal (A.L. 333-16, -65), and the longitudinal curvature of the phalanges (A.L. 333-19, -63). A cervical vertebra with a long spinous process (A.L. 333-106) is quite distinct. Two pedal navicular bones (A.L. 333-47, -36) exhibit extensive cuboideonavicular facets and the pedal phalanges are highly curved. One of the potentially most significant bones, the A.L. 288-1 innominate, is currently being recon-

structed. Its morphology is commensurate with a bipedal mode of locomotion. The specimen displays a straight anterior margin between the anterior superior and inferior spines, lending a heightened appearance to the ilium. These and additional postcranial features will be elucidated by biomechanical and anatomical studies [8].

In summary, the Hadar and Laetolil remains seem to represent a distinctive early hominid form characterized by substantial sexual dimorphism and a host of primitive dental and cranial characteristics. We interpret this material as representing a single hominid lineage [9]. An alternative interpretation would be that some smaller individuals, particularly the partial "Lucy" skeleton, represent a distinct lineage contemporary with the majority of the Hadar and Laetolil fossil hominids (Leakey, 1976a; Leakey and Lewin, 1977, 1978). For the reasons discussed above, we consider that the available evidence cannot be used to convincingly argue for the presence of two distinct hominid species at either site. The Hadar and Laetolil hominids are most parsimoniously interpreted as representing one sexually dimorphic hominid taxon.

PHYLOGENETIC CONSIDERATIONS

The overview of the Laetolil and Hadar remains presented above indicates that these forms represent the most primitive group of demonstrable hominids yet recovered from the fossil record. Although clearly hominid in their dentition, mandibles, cranium, and post-cranium, these forms retain hints of a still poorly known Miocene ancestor.

The Laetolil and Hadar fossil hominids are important primarily because of their bearing on questions of early hominid phylogeny. They allow a perception of human evolution that was hitherto impossible. However, before we deal specifically with hominid phylogeny, it is necessary to view hominoid evolution in broader perspective.

Miocene relations

The ancestry of the Laetolil and Hadar hominids is not well understood. It must lie within the Miocene hominoid radiation of Africa and Eurasia, and *Ramapithecus* is the candidate most often considered to fulfill this role (Simons, 1977). Pilbeam et al. (1977a) suggested that characters typical of extant Pongidae are not necessarily useful in understanding or classifying Miocene hominoid radiation. They proposed instead that the more advanced members of this radiation be divided into two families, the Dryopithecidae and the Ramapithecidae. We concur with the observation that *Ramapithecus* shares numerous adaptive similarities in its dental and gnathic composition with other Miocene forms such as *Sivapithecus* and *Gigantopithecus*. Many of these features were once thought to be distinctive of the family Hominidae (Simons, 1976c).

Some interpretations of the postcranial anatomy (Washburn, 1971; Lewis, 1973) and biochemical affinities (Sarich, 1971; Sarich and Wilson, 1967) of modern humans and extant African apes suggest that the pongid-hominid divergence was late in time. Some paleontologists, anatomists, and biochemists, however, place the divergence earlier—in the middle Miocene or even the Oligocene (Simons, 1977; Straus, 1949; Lovejoy and Meindl, 1973). Of course, genetic divergence (lineage separation) does not necessarily coincide with morphological divergence. The lack of a consistent definition of *Ramapithecus* and its detailed similarity to other Miocene hominoid genera combine with the primitive appearance of the Laetolil and Hadar material to suggest that a late divergence must remain a possibility. Ultimate resolution of the question will come only with the collection and analysis of

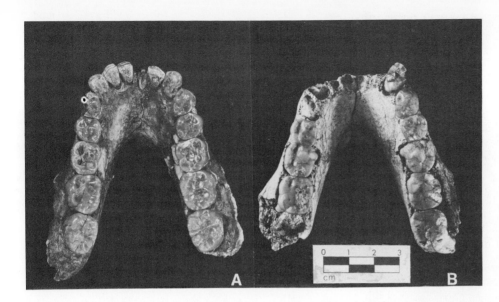

Figure 4 *Occlusal views of the mandibles from (A) Hadar (A.L. 400-1a) and (B) Laetolil (L.H.-4). Note the similarities in dentition, dental arcade shape, and mandibular morphology.*

further hominoid remains dating between 5 and 15 million years ago. Critical to this resolution will be the recovery of specimens representing lineages of the extant pongids.

Plio-Pleistocene Relations

Bipedalism appears to have been the dominant form of terrestrial locomotion employed by the Hadar and the Laetolil hominids. Morphological features associated with this locomotor mode are clearly manifested in these hominids, and for this reason the Laetolil and Hadar hominoid remains are unequivocally assigned to the family Hominidae. Representing, as they do, the earliest well-known hominids, what are their relationships with previously discovered Plio-Pleistocene hominids dating later in time? Our interpretations of hominid phylogeny during this period are given in Fig. 7, which indicates some of the more important sites and specimens along with their chronological placement.

The interpretation of hominid phylogeny presented in Fig. 7 relies heavily on the remains recovered since 1960 in eastern Africa. To fully appreciate this new resolution of early hominid phylogeny, it is necessary to consider the historical framework of fossil hominid discoveries. This is particularly true because the recent discoveries from eastern Africa have usually been interpreted in terms of a framework formulated on the basis of the South African discoveries.

South African Discoveries

The description and naming of the Taung skull from South Africa as the holotype of *Australopithecus africanus* by Dart (1925) represented a milestone in human evolutionary studies. Until the discovery of the Olduvai Hominid 5 (O.H. 5) cranium in 1959 (Leakey, 1959), Plio-Pleistocene hominids from the South African cave breccias at Taung, Sterkfontein, Makapansgat, Kromdraai, and Swartkrans dominated thinking on the earlier phases of human evolution. The distinctive character of the Kromdraai find led Broom (1938), to propose a different type of hominid, which he called *Paranthropus robustus*. Additional discoveries at Swartkrans reinforced Broom's recognition of a distinct, robust hominid lineage. However, hints of a second hominid type in the deposit at Swartkrans prompted Broom and Robinson (1949) to name the species *Telanthropus capensis*, which they considered to be ancestral to later forms of humans. Differences between fossil hominids from Taung, the Sterkfontein Type Site, and Makapansgat (collectively known as gracile australopithecines) and those from Kromdraai and Swartkrans (collectively known as robust australopithecines, with the exception of *Telanthropus*) were detailed by Broom (1950) and Robinson (1972). Doubts concerning the dating of these hominids have obscured their phylogenetic relationships, and some authors have suggested that the gracile and robust hominids represent nothing more than large and small forms of the same hominid species (Maier, 1963b).

East African Discoveries

The 1959 discovery of a very large and robust cranium at Olduvai Gorge demonstrated the presence of the robust hominid form in East Africa and focused attention on this part of the world. Soon thereafter, a smaller-toothed and apparently larger-brained hominid (O.H. 7) was recovered from equivalent levels and named *Homo habilis* (Leakey, et al., 1964). Debate concerning the differences between *H. habilis* and the gracile australopithecines from South Africa ensued (Robinson, 1965, 1966; Leakey, 1966; Tobias, 1966; Brace et al., 1973). The debate illustrates the difficulties encountered when interpreting the East African collections in a framework devised for the South African fossil hominids. While taxonomic considerations received paramount attention, phylogenetic aspects tended to be obscured. More recently, Brace (1973) and Wolpoff (1974) have claimed that only one lineage of Plio-Pleistocene hominid could be demonstrated in southern or eastern Africa at any point in the past.

In 1975, fieldwork at Koobi Fora in northern Kenya resulted in the demonstration of contemporaneity between KNM-ER 3733, an unequivocal *Homo erectus* cranium, and KNM-ER 406, an obvious robust australopithecine (Leakey and Walker, 1976) [10]. This was dramatic confirmation of earlier interpretations that had suggested the existence of two distinct hominid lineages in the African early Pleistocene. One lineage, commonly referred to as robust australopithecine, is represented by specimens exhibiting craniofacial and dental features that apparently reflect an adaptation involving a very heavily masticated diet [11]. Members

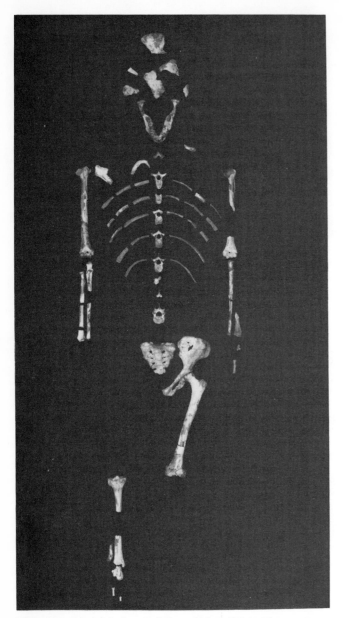

Figure 5   *Partial skeleton of "Lucy" (A.L. 288-1). This specimen is the most complete Pliocene hominid thus far discovered. The total length of the femur is 28 cm.*

Figure 6   *Comparison of large (A.L. 333-4) and small (A.L. 129-1a) distal femora from Hadar. Note the size difference but morphological identity.*

maximizes vertical occlusal force and spreads this force across an enlarged postcanine dentition (White, 1977b). This lineage displays no substantial tendency to expand cranial capacity.

Members of the second lineage are characterized by a contrasting suite of dental and cranial features and have been referred to the genus *Homo.* This lineage lacks the specializations related to a heavily masticated diet, but exhibits a definite tendency toward expansion of the brain. Among hominid populations comprising this second lineage there were undoubtedly substantial ranges of variation in cranial capacity, and to sort single specimens into either lineage solely on the basis of this criterion could be misleading. Mandibles, dentitions, and other cranial characteristics, aside from overall cranial capacity, serve to distinguish this from the other, more specialized lineage. Ultimately, the tendencies for brain expansion and gracilization of the masticatory apparatus characteristic of the earliest portions of this lineage culminated in the species *Homo sapiens* [12] (Lovejoy, 1974; Walker, 1973). Some investigators (Walker and Leakey, 1978; Leakey, 1976; Leakey and Lewin, 1977, 1978) have alluded to the existence of a third lineage in eastern Africa between 1 and 2 million years ago. The evidence for this third species, usually regarded as northern gracile *Australopithecus,* consists of three or four fragmentary crania. The morphology and dimensions of these specimens suggest to us that they are better considered as representatives of a variable, sexually dimorphic *Homo* lineage sampled through time.

Gracile Australopithecine Affinities

With the demonstration of two evolving lineages in the early Pleistocene (1.5 million years ago) of eastern Africa, it is necessary to reassess the phylogenetic affinities of the South African fossil hominids. Many students of early hominid evolution consider the gracile australopithecines to most closely approximate the ancestral hominid stock (Tobias, 1967). Both robust australopithecines and the earliest representatives of the genus *Homo* are thought to have arisen either from the gracile species represented at Taung, the Sterkfontein Type Site, and Makapansgat or from a closely related form. Before the recovery and analysis of the

of this lineage have been recovered from both eastern and southern African deposits. Important derived characteristics that differentiate more evolved members of this lineage have been recognized by numerous authors (Robinson, 1972; Broom and Robinson, 1949; Broom, 1950; Robinson, 1956; Howell, 1972; Tobias, 1967). These include extremely molarized deciduous and adult premolars, a relatively expanded postcanine dentition, and development of mandibular and cranial features related to a large masticatory apparatus. The latter are seen especially well in such specimens as O.H. 5 and KNM-ER 406, which have large, anteriorly placed zygomatics, large temporal fossae, and anteriorly placed sagittal crests. The mandibles have broad, deep rami and heavy buttressing of the corpus. Most if not all of these anatomical specializations are related to a craniofacial adaptation that

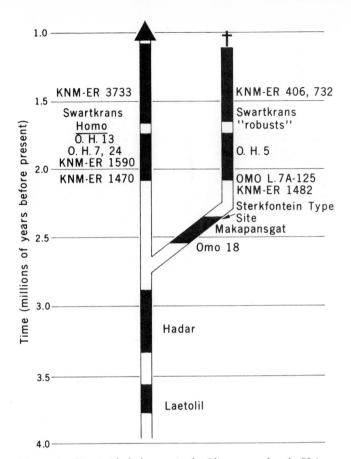

Figure 7 *Hominid phylogeny in the Pliocene and early Pleistocene based on the available fossil record. Some of the more important fossil samples and specimens are shown in their approximate chronological positions. The dark portions indicate periods from which hominid fossils are well known.*

Pliocene fossils from Hadar and Laetolil, such an evolutionary model best fit the available evidence. We presently enjoy a unique perspective afforded by the Hadar and Laetolil material. Study of these new fossils has prompted us to reexamine earlier hypotheses concerning affinities of the South African gracile australopithecines.

Of primary consideration in the phylogenetic interpretation of the South African gracile australopithecines is their chronological placement. The South African cave breccias have not been radiometrically dated. Consideration of the fauna from these sites relative to dated fauna in eastern Africa leads to the placement indicated in Fig. 7 (White and Harris, 1977; Vrba, 1975). It should be noted that faunal data place the Sterkfontein Type Site and Makapansgat deposits earlier than Bed I Olduvai, and postdating the Hadar and Laetolil remains. The third site yielding a gracile australopithecine, in fact the holotype of *A. africanus* (Dart, 1925), is Taung. Despite the recent claims of Partridge (1973) and Butzer (1974), Taung must be considered undated (Bishop, 1978).

It is significant that some of the gracile australopithecine specimens from Makapansgat have been considered robust by various workers (Aguirre, 1970; Wallace, 1975). Even 48 years after its description, the Taung specimen was hypothesized to represent a late surviving *A. robustus* (Tobias, 1973). Many workers have pointed out the similarities between gracile and

robust australopithecines from South Africa in dietary adaptation (Wolpoff, 1974; Tobias, 1967; Wallace, 1975) as well as locomotion (Lovejoy, 1974; Walker, 1973). Others have consistently maintained a generic distinction between the forms (Robinson, 1956, 1972; Clark, 1978). Our own examination of the relevant fossils suggests an alternative to these opposing interpretations.

Detailed morphological analysis of the gracile australopithecine sample from South Africa indicates an evolutionary status consistent with its relative chronological placement. The sample differs from the Hadar and Laetolil material in the direction of robust australopithecines. The South African gracile australopithecine group lacks elements in the suite of primitive characteristics described above for the Hadar and Laetolil hominids. It seems to share several distinctive, derived characters with later robust australopithecines. These include stronger molarization of the premolars, increased relative size of the postcanine dentition, increased buttressing of the mandibular corpus in the symphyseal region, and increased robustness of the corpus itself. Dental metrics reinforce the hypothesis that the Sterkfontein Type Site and Makapansgat gracile australopithecines represent a link between the basal, undifferentiated hominids at Hadar and Laetolil and the later robust australopithecines (Fig. 8).

Of course, morphological and metrical comparisons should not be expected to unerringly place every single individual along an evolving lineage. Our interpretation of the South African gracile australopithecines is based on a consideration of the available sample characteristics for the fossil hominids. We are fully aware that individual traits and even single specimens can be matched in samples that we consider to represent different evolutionary entities and ultimately taxa. For example, the matching of individual specimens and demonstration of overlap between the samples from Sterkfontein and Swartkrans serve to point out the general similarities of these groups, but at the same time conceal real and biologically meaningful differences which we consider to have phylogenetic significance.

Likewise, it is possible to emphasize the similarities between the Laetolil and Hadar fossils and the gracile australopithecines from South Africa. To include the more archaic material from eastern Africa in an already established gracile australopithecine phylogenetic or taxonomic category would obscure the evolutionary relationships and significance of the new material. We propose below a taxonomy consistent with these observations.

TAXONOMIC CONSIDERATIONS

The ultimate goal of human evolutionary studies is to understand phylogenetic relationships and adaptive patterns among the hominids. Such understanding has sometimes been hampered by an emphasis on naming the hominid specimens. We recognize the usefulness of classifying fossil materials, and we agree with Simpson (1963) that "classification is not intended to be an adequate expression of phylogeny but only to be consistent with conclusions as to evolutionary affinities." The evolutionary affinities of the Hadar and Laetolil material are discussed above. Our interpretation of hominid phylogeny during the Pliocene and Pleistocene is presented in Fig. 7.

Taxonomic debate often stems from the inability of Linnean nomenclature to cope with an evolutionary progression of paleontological remains. This becomes particularly evident when the members of an evolving lineage are represented by a fairly complete fossil record. To us, this appears to be the case for Plio-Pleistocene hominids, and this situation is not unique among vertebrates (White and Harris, 1977; Maglio, 1973).

Several alterntive taxonomic schemes may be generated on

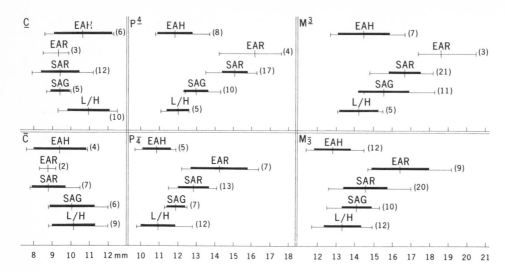

Figure 8   *Buccolingual tooth crown dimensions for the fossil hominids discussed in the text. The observed sample ranges are indicated by light horizontal lines, the arithmetic means by light vertical lines, and 1 standard deviation from the mean by darker horizontal bars. The number of specimens is shown in parentheses. Mesiodistal crown lengths and crown areas show the same tendencies, but buccolingual dimensions are used in this graphic treatment because they are not affected by interproximal attrition during the life of an individual. Only tooth crowns that are complete or that can be estimated within 0.2 mm are included. All specimens were measured and assigned to a sample set by one of the authors. Abbreviations: L/H, Laetolil and Hadar; SAG, South African gracile (Taung, Makapansgat, Sterkfontein Type Site); SAR, South African robust (Kromdraai, Swartkrans non-Homo); EAR, East African robust (Olduvai Beds I and II, East Turkana Lower and Upper Members, Natron, Chesowanja); and EAH, East African Homo (Olduvai Beds I and II, East Turkana Lower and Upper Members). Note the relative placement of the gracile australopithecine sample between the earlier Laetolil and Hadar sample and the later robust australopithecines of South Africa. Only C, P4, and M3 are displayed graphically, but the mediodistal and buccolingual means of the SAG sample are intermediate between the means for the L/H and SAR samples for every postcanine tooth, upper and lower (except the P_3 buccolingual dimension, which is larger in SAG than in SAR). These diagrams lend graphic support to morphological considerations described in the text. They are presented merely as supplementary evidence for the arguments presented there.*

the basis of our phylogenetic reconstruction (Fig. 7). A number of examples are shown in Fig. 9. Alternatives a to c would adopt generic distinction for the new material based on Senyürek's study (1955) of the original Garusi maxillary fragment recovered from Laetolil in 1939. He used the genus *Praeanthropus* of Henning (1948) and the species named *africanus* suggested by Weinert (1950), producing the binomen *Praeanthropus africanus*. Among other problems, adoption of such a distinction would imply that the Hadar and Laetolil fossil hominids were significantly different in their adaptation from later hominids. Our examination of the material suggests that such distinction is inconsistent with its observed phylogenetic and adaptive affinities.

A scheme that places the Laetolil and Hadar remains in the genus *Homo* (Fig. 9, d to f) will undoubtedly be favored by some. Such a scheme, as shown in Fig. 9d, follows Mayr's (1951) suggestion that all hominid fossils be placed in species of the genus *Homo*. He later withdrew this suggestion (Mayr, 1963), stating that "The extraordinary brain evolution between *Australopithecus* and *Homo* justifies the generic separation of the two taxa, no matter how similar they might be in many other morphological characters." We concur with this contention that the unique adaptive and evolutionary trends seen in the lineage leading to *H.*

sapiens merit generic distinction. This trend is not yet evident in the Laetolil and Hadar hominids. For this reason, we favor the schemes shown in Fig. 9, g to i.

The alternatives shown in Fig. 9, h and i, would tend to obscure phylogenetic continuity by unnecessary generic splitting. The taxonomic scheme we consider most useful in expressing our phylogenetic findings is shown in Fig. 9g. We follow Mayr (1951) in his perception of the genus *Homo* as being characterized by progressive brain enlargement associated with increasing cultural elaboration. The first species for which these trends can be discerned is *Homo habilis* (Leakey, et al., 1964).

The juvenile status of the Taung holotype specimen of *A. africanus* precludes its precise phylogenetic placement. We agree with the traditional and widely accepted approach in which the specimen is considered to be indistinguishable from the Sterkfontein Type Site fossils (Tobias, 1965, 1967, 1973a; Robinson, 1956, 1972; Le Gros Clark, 1950b; Mayr, 1951; Howell, 1972; Wallace, 1975; Pilbeam, 1972). Since the later sample is significantly less primitive than the Hadar and Laetolil material, a new species of the genus *Australopithecus* has been created (Johanson, White and Coppens, 1978). This most primitive *Australopithecus* species is *A. afarensis* and is based on the holotype specimen L.H.-4

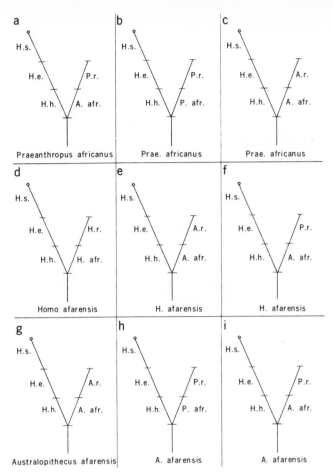

Figure 9 *Alternative taxonomic schemes available for representing Plio-Pleistocene human evolution. Abbreviations: H., Homo; A., Australopithecus; Prae., Praeanthropus; P., Paranthropus; s., sapiens; e., erectus; h., habilis; r., robustus; and afr., africanus. Australopithecus boisei is considered conspecific with A. robustus.*

and Hadar (Fig. 10). The recovery of the well-dated Hadar and Laetolil hominids extends our understanding of human origins well into the Pliocene. The implications of the new material for understanding the mode and tempo of hominid evolution are great. The apparent lack of morphological differences between fossils separated by at least 0.5 million years at Laetolil and Hadar suggests relative stasis in the earliest documented portions of hominid evolution. The dramatic morphological changes initiated between 2 and 3 million years ago suggest that this relative stasis was upset. Although the precise reasons for the phyletic divergence that led to *A. robustus* through the earlier, intermediate *A. africanus* are not well understood, a South African origin for this stock is plausible. Whatever the case, the clear niche divergence between *H. erectus* and *A. robustus* about 1.5 million years ago indicated by the eastern African fossil record indicates that an increased evolutionary rate for the period between 2 and 3 million years ago may ultimately be shown by larger fossil samples.

Another implication of the new fossil hominid material concerns sexual dimorphism. The extent of size and morphological variation in the Pliocene hominids from Hadar and Laetolil comes as no surprise, since later portions of the hominid fossil record also show greater sexual dimorphism than exists among modern humans (Brace, 1973; Weidenreich, 1943; de Lumley and de Lumley, 1974). However, although the Laetolil and Hadar fossil hominids show marked body size dimorphism, the metric and morphological dimorphism of the canine teeth is not as pronounced as in most other extant, ground-dwelling primates. This implies a functional pattern different from that seen in other primates and may have significant behavioral implications.

In this article we have avoided placing emphasis on taxonomic problems inherent in paleontological material. Instead, we have tried to provide a phylogenetic framework for the early Hominidae that will allow anatomical, biomechanical, and behavioral studies of fossil humans to proceed constructively.

Acknowledgments

We express our appreciation to the Provisional Military Government of Socialist Ethiopia and the United Republic of Tanzania for providing encouragement and cooperation during fieldwork at Hadar and Laetolil. We thank the following for financial support: National Science Foundation, National Geographic Society, L. S. B. Leakey Foundation, Wenner-Gren Foundation, Cleveland Museum of Natural History, and Centre National de la Recherche Scientifique (CNRS). M. D. Leakey kindly made the Laetolil fossils available for our study. Thanks are due F. C. Howell for critically reviewing the manuscript. Y. Rak, O. Lovejoy, and G. Eck provided helpful comments and observations. Dr. Lovejoy kindly provided the Libben Amerindian data. Special gratitude is expressed to W. H. Kimbel, who contributed valuable assistance and comments during the preparation of this article. This study was based on hominid specimens from Hadar and Laetolil listed in Johanson, White and Coppens (1978).

as well as a series of paratypes from both Laetolil and Hadar. It obtains its name from the Afar region of Ethiopia, which has produced the most abundant evidence.

DISCUSSION

We have presented the phylogenetic hypothesis that most parsimoniously accommodates the new fossil hominids from Laetolil

Notes

[1]   We are aware of the hominid specimens from Lothogam (5.5 million years), and Kanapoi (about 4.0 million years). However, these remains are so fragmentary that accurate assessment of their taxonomic affinities is difficult.

[2]   Figure 8 clearly indicates that the observed ranges of buccolingual tooth diameters in the *A. afarensis* teeth are no larger than in the other fossil hominid samples. The coefficients of variation of the buccolingual diameters for the lower molars of *A. afarensis* are as follows: M1, 7.1 (N=16, M2, 7.9 (N=17), M3, 7.7 (N=12). Comparable figures for *Pan troglodytes* (Johanson, 1974a) are: M1, 5.2 (N=409), M2, 5.8 (N=333), M3, 6.0 (N=252); for *Pan paniscus* (Johanson, 1974b): M1, 6.8 (N=91), M2, 6.6 (N=50), M3, 5.9 (N=28); *Gorilla gorilla* (Mahler, 1973): M1, 6.4 (N=374), M2, 7.0 (N=370), M3, 7.9 (N=335); *Pongo pygmaeus* (Mahler, 1973): M1, 8.1 (N=129), M2, 9.6 (N=133), M3, 9.6 (N=110); *Homo sapiens* (Wolpoff, 1971a): M1, 7.2 (N=558), M2, 8.7 (N=529), M3, 10.5 (N=448). Hence, the coefficients of variation for the *A. afarensis* molars are well within the ranges derived for extant hominoids.

[3]   The lack of a distinct lower P3 metaconid in some Hadar specimens may be interpreted by some as taxonomically diagnostic. However, our own investigation of the variability in this feature in both extant and extinct hominoids suggests that the presence or absence of a lower P3 metaconid has minimal phyletic valence.

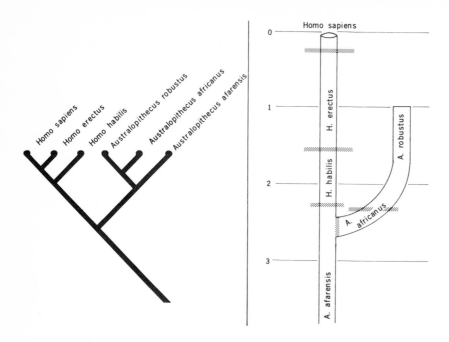

Figure 10 *(A) Cladogram of the family Hominidae. (B) Phylogenetic tree of the family Hominidae. See text for a discussion of the views represented by these diagrams.*

[4]   Not enough of the A.L. 288-1 cranium is preserved to provide meaningful metric comparisons between it and the A.L. 333-45 specimen. However, differences in vault thickness, temporal line expression, and nuchal crest development are indicative of a degree of sexual dimorphism similar to that found in the common chimpanzee.

[5]   W. H. Kimbel of the Cleveland Museum of Natural History, together with the authors, is currently undertaking an in-depth study of the cranial remains from Hadar.

[6]   The variation in Hadar and Laetolil adult mandible height and breadth measurements at the junction of lower M1 and M2 is large [$N=12$, coefficient of variation (CV) for height = 11.40; CV for breadth = 11.19], but not excessive when compared to the variation observed in samples of modern *Homo sapiens* ($N = 20$; CV for height = 8.85; CV for breadth = 11.59), *Pan troglodytes* ($N = 21$; CV for height = 10.79; CV for breadth = 9.42), and *Gorilla gorilla* ($N = 22$; CV for height = 11.07; CV for breadth = 7.54). The high degree of variation in mandibular breadth among hominids reflects the relatively forward position of the corpus-ramus junction.

[7]   Using an average of biepicondylar breadth and medial and lateral condylar height measurements, we find that the difference between the largest (A.L. 333-4) and smallest (A.L. 129-1a) Hadar femora is 80.8 percent. Comparable figures for Libben Amerindians (X, 89.6 percent; maximum, 73.5 percent, $N = 50$), *Pan troglodytes* (X, 95.1 percent; maximum, 80.6 percent, $N = 20$), and *Gorilla gorilla* (X, 76.0 percent; maximum, 61.9 percent, $N = 30$) indicate that the percentage difference between Hadar femora does not exceed ranges for other hominoids.

[8]   C. O. Lovejoy of Kent State University, together with the authors, is currently engaged in intensive descriptive and analytic study of the Hadar postcranial remains.

[9]   Some of our previously published interpretations of the Laetolil and Hadar fossil hominids differ from those presented here. This is because a more thorough study of the material has prompted us to elaborate and clarify some of these earlier preliminary statements.

[10] KNM-ER is Kenya National Museum's designation for East Turkana fossils.

[11] Robinson (1958 and 1972) pioneered the interpretation that the cranial and dental specializations of the robust hominid species reflect a diet demanding extensive mastication. This diet was thought to consist almost exclusively of vegetal materials.

[12] The emphasis on distinctive cranial and dental anatomy of members of these two lineages is not intended to overshadow the possibility that differences in postcranial anatomy may be found. No essential differences in basic locomotor mode of Plio-Pleistocene hominids have yet been convincingly delineated [Lovejoy (1974) versus Robinson (1972)], but minor morphological differences may ultimately allow distinctions to be made between or within the two lineages on the basis of the postcranial skeleton (Walker, 1973).

38

The Origin of Man
C. O. LOVEJOY

During the last quarter-century, the study of human origins has proved remarkably successful. Crucial fossils and primate behavioral data are now available from which to reconstruct man's evolution during the last 15 million years. Equally important is the recognition of a close genetic relationship between man and the other extant hominoids (especially *Pan* and *Gorilla*) (Goodman, 1976b; King and Wilson, 1975; Sarich, 1968a, 1971; Kohne, 1972; Kohne et al., 1970; Lovejoy and Meindl, 1973). Experiments on DNA hybridization indicate at least 98 percent identity in nonrepeated DNA in man and chimpanzee, sufficient similarity to suggest the possibility of a viable hybrid. These data confirm studies by comparative anatomists who have emphasized the striking anatomical similarities of apes and man (Huxley, 1863; Gregory, 1930c; Keith, 1923, 1929; Le Gros Clark, 1962). As a consequence of this physical similarity, models of human origin must directly address the few primary differences separating humans from apes. Clearly, the rate of acquisition of these differences, the fossil evidence bearing on their first appearance, and their underlying selection are crucial to an understanding of human evolution.

MATERIAL CULTURE

The most commonly cited distinction between man and apes is the former's reliance on material culture. The belief that tools were pivotal to the divergence of hominids was initiated by Darwin (1871) and has remained the most popular view (Bartholomew and Birdsell, 1953; Mann, 1972; Washburn, 1960, 1963b, 1968b; DeVore, 1965; Tobias, 1971; Washburn and Ciochon, 1974). Darwin was impressed by the absence of large canines in man and attributed their reduction to tool use. As Holloway (1967) and Jolly (1970) have cogently argued, however, tool use is not an explanation of canine reduction since there is no behavioral contradiction in having both functional canines and tools. There is little doubt that material culture has played a role in the evolution of *Homo sapiens* and *H. erectus*, but this does not require it to have been a significant factor in the origin of hominids. In fact, the earliest recognizable tools are only about 2 million years old [1], but there is considerable evidence placing the phyletic origin of hominids in the middle to late Miocene (12 to 6 million years ago) (Simons, 1972, 1978; Pilbeam, 1972; Jolly, Ed., 1978; Greenfield, 1979). Although the earliest tools will have left no record because of the use of perishable materials, there is still the necessary presumption of a 6- to 10-million-year period dominated by reliance on material culture—a view with numerous shortcomings.

The use of primitive tools by extant pongids (Kortlandt, 1967; Goodall, 1965; Wright, 1978) supports the contention of comparable abilities in early hominids, but it also demonstrates that tool use is a general capacity of pongids, none of which exhibit the unique characters of hominids [2]. If tools were the primary determinant of early hominization, why should their first appearance be so late in the hominid record? More importantly, what activity requiring tools was critical to early hominid survival and phyletic origin [3]? It is now clear that hunting does not qualify as such an activity [4]. From the first recognizable tools to the industrial revolution required only 2 million years, whereas if tools played a part in the origin of hominids, they must have remained primitive and unchanged for at least 5 million years. It is likely that either the earliest hominids made no use of tools at all, or that such use was comparable to that in other extant hominoids and was not critical to their survival or pivotal to their origin.

EXPANSION OF THE NEOCORTEX

It is now clear that the marked expansion of the hominid cerebral cortex took place during the last 2 to 3 million years (Johanson and White, 1979; Radinsky, 1975). Detailed study of the Hadar crania from Ethiopia, recently attributed to *Australopithecus afarensis* (Johanson, White and Coppens, 1978), has revealed that they were strikingly primitive [5]. Preliminary estimates of cranial capacity indicate a brain size well within the range of extant pongids (Holloway, personal communication). The pelvis of the skeleton known as "Lucy" from Afar Locality (A.L.) 288 has been fully reconstructed (Lovejoy, 1979). One of its most salient features is a birth canal whose shape and dimensions show little or no effects of selection for passage of enlarged fetal crania, adaptations that so clearly dominate the form of the modern human pelvis (Lovejoy et al., 1973; Lovejoy, 1974, 1975).

BIPEDALITY

Bipedality is an unusual mode of mammalian locomotion. Contrary to the so called efficiency argument, energy expenditure for bipedal walking is probably not significantly different from that during quadrupedal locomotion (Lovejoy, 1978; Taylor and Rowntree, 1973; Cavagna et al., 1964). Yet the adoption of non-saltatory bipedal progression is disadvantageous because both speed and agility are markedly reduced (Lovejoy, 1974, 1975, 1978; Lovejoy et al., 1973) [6]. All present evidence, especially that made available by the postcranium of *A. afarensis*, confirms an essentially complete adaptation to bipedal locomotion by at least 4 million years ago (Lovejoy, 1979; Johanson et al., 1976). This conclusion is provided unequivocal support by the hominid footprints discovered at Laetoli in Tanzania (Johanson and White, 1979).

DENTITION

Additional distinctions between hominids and pongids are found in their respective jaws and teeth. In fact, these differences have allowed the identification of possible hominids in the Miocene—

there are no distinctive postcranial or cranial remains of un-doubted hominid affinities before about 4 million years ago. As a result of recent field work in Mio-Pliocene deposits (Johanson and Taieb, 1976; Johanson and Coppens, 1976; M. D. Leakey et al., 1976; White, 1977; Coppens et al., Eds., 1976; Tobias, 1973b; R. E. Leakey et al., 1971, 1972; Day and Leakey, 1973, 1974; Leakey and Wood, 1973; Day et al., 1974, 1976; Howell, 1969; Howell and Wood, 1974; Pilbeam, et al., 1977a), it is now possible to suggest a broad schedule of phases in the evolution of the hominoid dentition that can serve as an outline of hominoid phyletic events during the last 23 million years.

Phase I

This phase has a generalized dryopithecine dentition including a distinct Y-5 lower molar cusp pattern with bunodont crowns, thin enamel, and cheek teeth small relative to body size; incisors are broad with canine-premolar shear. This phase is associated with forest faunas and floras (Pilbeam, 1969a, 1976; Simons and Pil-beam, 1965, 1972; Simons, 1976a) and is shared by all hominoids before 15 million years ago (range, 23 to 15 million years) (Pil-beam, et al., 1977a; Pilbeam, 1969a, 1976; Simons and Pilbeam, 1965, 1972; Simons, 1976a; Tobias and Coppens, Eds., 1976).

Phase II

This phase shows a shift toward greater molar dominance. About 14 million years ago, hominoids fall into two groups. The first retained phase I characters and may constitute ancestral popula-tions of extant apes (*Proconsul*, "*Rangwapithecus*," and *Limno-pithecus, Dryopithecus* (Pilbeam et al., 1977a). A second group exhibits enamel thickening, increased molar wear gradient, and moderate anterior dental reduction or increased relative molar size, or both. Mandibles are more robust and prognathism is reduced. The shift toward greater molar dominance has partially been attributed to greater reliance on terrestrial food sources. This group includes genera (*Ramapithecus* and *Sivapithecus*) probably related to hominids, an extinct ape (*Gigantopithecus*), and possibly the modern orang-utan (range, 14 to 8 million years) (Greenfield, 1979; Pilbeam et al., 1977a; Pilbeam, 1969a, 1976; Simons and Pilbeam, 1965, 1972; Simons, 1976a, Tobias and Coppens, Eds., 1976; Greenfield, 1977) [7].

Phase III

This phase represents a conservative period. The dentition of *A. afarensis* appears only moderately changed in morphology and proportions from phase II; the features include comparatively large incisors, frequently a unicuspid lower first premolar, canines of moderate size, molars of moderate size (relative to body size and later hominids), and loss of canine-premolar shear (range, 7 to 2½ million years) (Greenfield, 1979; Johanson and White, 1979).

Phase IV

This phase represents Plio-Pleistocene specialization. The sample in this time range is divisible into two clades or phyletic lines (Johanson and White, 1979). The first was possibly restricted to savannah and grassland. It displays extreme anterior tooth reduc-tion and excessive molar dominance and became extinct by mid-Pleistocene (*A. africanus* → *A. robustus* → "*A. boisei*"). A second clade, ancestral to *H. erectus*, retained a more generalized denti-tion in the early Pleistocene but underwent dentognathic reduc-tion in the middle and upper Pleistocene as a consequence of reliance on material culture [for example, reduced dental manipu-lation and greater preoral food preparation (Brace and Montagu,

1977)]. My view is that this clade occupied more varied habitats. Both groups are probably directly descendant from *A. afarensis* (Johanson and White, 1979).

MODELS OF HUMAN ORIGIN

A model of hominid origin proposed by Jolly (1970) uses analogy to anatomical and behavioral characters shared by *Theropithecus gelada* and some early hominids. He suggests that early hominid populations relied on small-object feeding, that this dietary spe-cialization led to a suite of adaptations to the grassland savannah, and that bipedality developed in response to feeding posture. Yet geladas, which do rely on small-object feeding, are not bipedal and show no significant adaptations to bipedality. Bipedal loco-motion is clearly not required for extensive small-object feeding especially on grasslands where speed and agility are of great value in animals who also lack wide visual fields and sensitive olfaction [8]. Furthermore, the dental morphology of *A. afarensis* is con-siderably more generalized than that of later hominids. The die-tary specialization seen in *A. robustus* is possibly accountable by Jolly's model, but the more generalized dentition of *A. afarensis* is not [9]. It is more likely that hominids venturing into open habi-tats were already bipedal and that their regular occupation of savannahs was not possible until intensified social behavior was well developed.

Other theorists have viewed hominization as the direct result of savannah occupation by prehominids. Proponents of this view believe that the selective pressures of life on grassland savannahs directly produced the human character complex. Bipedal loco-motion is posited as sentinel behavior and as an adaptation allowing weapons to be used against predators. Intelligence is said to be favored because highly integrated troop behavior is neces-sary for predator repulsion. Differences in some behaviors of chimpanze populations now living in woodland savannahs versus those inhabiting more forested areas are cited as evidence (Kor-tlandt, 1962, 1972).

There are many problems with this view. Bipedality is useless for avoidance or escape from predators. Occasional bipedality, as seen in many primates, is sufficient for the use of weapons. Most importantly, brain expansion and cultural development remotely postdate hominid divergence.

Furthermore, Miocene ecology is inconsistent with the savannah selection theory. While cooling, aridity, and increased seasonality had pronounced effects on Old World floras, the predominant effect of these climatic trends, in areas where homin-ids are known to have been present, appears to have been the development of diversified mosaics, rather than broad-scale forest reduction (Moreau, 1951; Butzer, 1977; Campbell, 1972; Campbell, Ed., 1972; Andrews and Van Couvering, 1975). It would be more correct to say that hominids of the middle and late Miocene were presented with a greater variety of possible habitats than to view them as having suffered an imposed "terrestrializa-tion." It is also clear that some Miocene sites at which possible hominids have been recovered had canopy forest conditions (Greenfield, 1979; Kretzoi, 1975; Kennedy, 1978). While in-creased seasonality would have imposed a need for larger feeding ranges, occasional use of woodlands and edaphic grasslands would not necessarily impose elevated carnivore pressure. Nor, as was pointed out above, would early hominids be required to abandon quadrupedality in order to use more orthograde posi-tional behavior during feeding. Quite the contrary, it would appear that late Miocene habitat mosaics would allow adoption of bipedality (in forests and transition mosaics) rather than directly select for it. All present evidence therefore indicates that hominid clade evolved in forest or mosaic conditions, or both

[10], rather than only on grassland or savannahs, and that bipedal locmotion was not a response to feeding posture, material culture, or predator avoidance.

In summary, four major character complexes are usually cited as distinguishing hominids from pongids. Hominids have remarkable brain expansion, a complex material culture, anterior dental reduction and molar dominance, and bipedal locomotion. Only bipedal locomotion and partial dental modifications can be shown to have an antiquity even approximating the earliest appearance of unquestioned, developed hominids (*A. afarensis*).

DEMOGRAPHIC STRATEGY AND THE EVOLUTION OF HOMINIDS

The order Primates has long been recognized to display a *scala naturae* consisting of "intercalary types"—extant forms that represent earlier stages in the development of major adaptive trends. Figure 1 is a well-known diagram of the chronology of life phases in living primates. There is an obvious trend toward prolonged life-span, which has both physiological and demographic correlates bearing directly on the phyletic origin of hominids.

The physiological correlates (Fig. 1) include a longer period of infant dependency, prolonged gestation, single births, and successively greater periods between pregnancies. Cutler (1976) has demonstrated that such developmental parameters are "qualitatively and sequentially similar in different mammalian species" but proceed "at different characteristic rates defined by the reciprocal of the MLP" (maximum life potential). The progressive slowing of life phases can in turn be accounted for by an increasingly K-type demographic strategy [11]. With each step in the *scala naturae*, populations devote a greater proportion of their reproductive energy to subadult care, with increased investment in the survival of fewer offspring. Among chimpanzee populations, this trend appears to have resulted in marginal demographic conditions. Field studies at Gombe in Tanzania show the

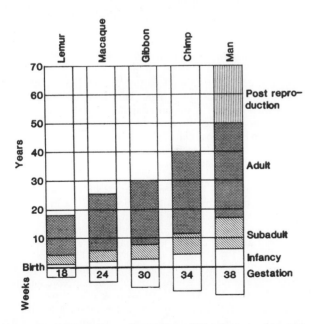

Figure 1   *Progressive prolongation of life phases and gestation in primates. Note the proportionality of the four indicated phases. The post-reproductive phase is restricted to man and is probably a recent development (Lovejoy et al., 1977; after Schultz, 1969).*

Figure 2   *Mechanical model of demographic variables in hominoids. The R is the intrinsic rate of population increase (1 = static population size). An increase in the lengths of the four periods on the bar to the right (birth space, gestation, infant dependency, and sexual maturity) is accompanied by a comparable shift of longevity to the left, but without realization of that longevity, prolonged maturation reduces R and leads to extinction or replacement by populations in which life phases are chronologically shorter. Of the four variables on the right, only birth space can be significantly shortened (shifted to the left) without alteration of primate aging physiology.*

average period between successful births to be 5.6 years (Teleki et al., 1976). This can be attributed in part to a greatly prolonged period of subadult dependency. Van Lawick-Goodall's (1976) description of the chimpanzee life phases is instructive:

> The infant does not start to walk until he is six months old, and he seldom ventures more than a few yards from his mother until he is over nine months old. He may ingest a few scraps of solid food when he is six months, but solids do not become a significant part of his diet until he is about two years of age and he continues to nurse until he is between four-and-a-half and six years old. Moreover, while he may travel short distances...when he is about four years old, he continues to make long journeys riding on his mother's back until he is five or six....

This extreme degree of parental investment has profound demographic consequences. A chimpanzee female does not reach sexual maturity until she is about 10 years old. If she is to reproduce herself and her mate, that is maintain a stable population, she must survive to an age of 21 years [12]. Whereas in rhesus macaques, the age is only about 9 years [13].

Figure 2 shows a balance depicting the reciprocal relation between longevity and the primary demographic elements of parental investment. The two sides of this hypothetical balance are physiologically interdependent; as longevity is increased, each of the developmental stages is proportionately prolonged. The relationships between these variables, in fact, are not exactly linear, but they do have remarkably high correlations in most mammals (Cutler, 1976). As the scale indicates, greater longevity is accompanied by a proportionate delay in reproductive rate and therefore requires a female to survive to an older age in order to maintain the same reproductive value (measured at birth) (Fisher, 1930). Put another way, the total reproductive rate of a primate species can remain constant with progressive increases in longevity only if the crude mortality rate is correspondingly reduced. Actual mortality rate is dependent on both maximum life potential, a genetic factor, and environmental interaction. Deaths caused by predation, accident, parasitism, infection, failure of food supply, and so forth, are at least partially stochastic events beyond the complete control of the organism. Only if mechanisms are developed to increase an organism's resistance to such factors, can the effects of increased longevity be reproductively accommodated. Strong social bonds, high levels of intelligence, intense parenting, and long periods of learning are among factors used by higher primates to depress environmentally induced mortality. It is of some interest that such factors also require greater longevity (for brain development, learning, acquisition of social and parenting skills) and that they constitute reciprocal links leading to greater longevity. This positive feedback system, however, has an absolute limit; environmentally induced mortality can never be completely under organism control, no matter how effective the mechanisms developed to resist it.

Suppose that late Miocene hominoids were approaching the effective limit of this feedback system or at least were sufficiently near the limit not to thrive in novel environments [14]. Two demographic variables could be altered to improve reproductive success—survivorship (the probability of surviving) and the time period between successive births (the birth space). All other factors are direct linear functions of mammalian developmental physiology and could not be altered. The argument is subject to the following simple quantification

$$RV = l(s) \int_{s}^{MLP} l(x)b(x)dx$$

where RV is reproductive value of a cohort measured at birth, that is, the expected number of offspring produced by a unit radix; $b(x)$ is fertility at age x; $l(x)$ is survivorship at age x; s is age at sexual maturation; and MLP = maximum life potential. Assuming that a female gives birth at age s years and subsequently every β (birth space) years until reaching MLP, her total offspring would be given by

$$\frac{MLP - s}{\beta}$$

Fertility is then seen to be dependent on birth space β according to

$$\int_{s}^{MLP} b(x)dx = \frac{MLP - s}{\beta}$$

A simple solution (but one which is fully acceptable because of the proportionate relation between MLP and s) is $b(x) = 1/\beta$. The expression for RV then becomes

$$RV = \frac{1}{\beta} \left[l(s) \int_{s}^{MLP} l(x)dx \right]$$

Because the term in brackets is independent of β, RV is inversely proportional to β, and RV is increased by a shorter birth space, by greater values of $l(x)$ for any age, or by both. Table 1 provides reproductive values for chimpanzees, Old World monkeys, and man from estimated values of β, s, and MLP under the simplifying assumption of $l(x) = l^x$. It can be seen from this table that both chimpanzees and humans have considerably lower reproductive values than Old World monkeys for low values of $l(x)$. As the values used for calculation are conservative, the existence of successful hominid clades in Pliocene mosaics suggests that both birth space reduction and elevation of survivorship had probably been accomplished. This is without explanation unless a major change in reproductive strategy accompanied occupation of novel environments by these hominids. Yet neither brain expansion nor significant material culture appear at this time level and were therefore not responsible for this shift.

A BEHAVIORAL MODEL FOR EARLY HOMINID EVOLUTION

Any behavioral change that increases reproductive rate, survivorship, or both, is under selection of maximum intensity. Higher primates rely on social behavioral mechanisms to promote survivorship during all phases of the life cycle, and one could cite numerous methods by which it theoretically could be increased. Avoidance of dietary toxins, use of more reliable food sources, and increased competence in arboreal locomotion are obvious examples. Yet these are among many that have remained under strong selection throughout much of the course of primate evolution, and it is therefore unlikely that early hominid adaptation was a product of intensified selection for adaptations almost universal to anthropoid primates. For early hominids we must look beyond such common variables to novel forms of behavioral change. The tendency has been to concentrate on singular, extraordinary traits of later human evolution such as intense technology, organized hunting, and the massive human brain. Yet these adaptations were not likely to have arisen de novo from elemental behaviors seen in extant nonhuman primates, such as the primitive tool using of the chimpanzee, in the absence of a broad selective milieu. It is more probable that significant preadaptations were present in early hominids that served as a behavioral base from which the "breakthrough" adaptations [15] of later hominids could progressively develop. We are therefore in search of a novel behavioral pattern in Miocene hominoids that could evolve from typical primate survival strategies, but that might also include important elements of other mammalian strategies, that is, a behavioral pattern that arose by recombination of common mammalian behavioral elements and that increased survivorship and birthrate.

In her essay on mother-infant relationships among chimpanzees, Van Lawick-Goodall (1969) [16] noted two primary causes of mortality among infants: "inadequacy" of the mother-infant relationship and "injuries caused by falling from the mother." An intensification of both the quality and quantity of parenting would unquestionably improve survivorship of the altricial chimpanzee infant. The feeding and reproductive strategies of higher primates, however, largely prevent such an advancement. The mother must both care for the infant and forage for herself. A common method of altricial infant care in other mammals is sequestration of offspring at locations of maximum safety. Nests, lodges, setts, warrens, dreys, dens, lairs, and burrows are examples of this strategy. A similar adaptation in primates is usually not possible, however, because the need to forage requires both mother and infant to remain mobile. The requirement of mother-infant mobility is a significant cause of mortality and is at the same time the most important restriction on primate birth spacing.

Many primates display significant sex differences in foraging. Diet composition, selection of food items, feeding time, and canopy levels and sites differ in some species (Clutton-Brock, 1977). In at least *Pongo pygmaeus* and *Colobus badius*, males often feed at lower canopy levels than females (Clutton-Brock and Harvey, 1977; Sussman, Ed., 1979) [17]. In the gelada baboon, all-male groups "tended not to exploit quite the same areas as the reproductive units thus reducing indirect competition for food" (Crook, 1972). Clutton-Brock (1977) notes that an increased separation of males from female-offspring foraging sites is advantageous where (i) animals feed outward from a fixed base, (ii) the adult sex ratio is close to parity, and (iii) feeding rate is limited by search time rather than by handling time, which is the time spent both preparing and consuming food. Similar feeding differences by sex are found in birds and other mammals. (Selander, 1966, 1972; Hutchinson, 1965).

It is reasonable to assume that Miocene hominoids traveled between food sources on the ground and that these primates would be best characterized as omnivores (Greenfield, 1979). These are ecologically sound assumptions. Increased seasonality coupled with already occurring local biotic variation (edaphic grasslands, savannah, woodland, forest) (Jolly, 1970b; Greenfield, 1979; Moreau, 1951; Butzer, 1977; Campbell, 1972; Campbell, Ed., 1972; Andrews and Van Couvering, 1975) would have presented variable and mosaic conditions. Occupation of heterogeneous ("patchy") environments and use of variable food sources favors a generalist strategy, whereas reliance on a homogeneous diet requires high food concentrations [18, 19]. The time spent searching for food is greatest among generalists who live in food-sparse environments (Pianka, 1974). In short, Miocene ecological conditions support the view that feeding rate would have been more dependent on search time than handling time.

Greater seasonality and the need to increase both birthrate and survivorship would also favor at least partial separation of male and female day ranges since this strategy would increase carrying capacity and improve the protein and calorie supply of females and their offspring. Terrestriality, however, would require a centrifugal or linear displacement of males, as opposed to vertical stratification in canopy feeding. Given the Miocene conditions described above, such separation could become marked especially in the dry season. If such separation were primarily due only to an increase in the male day range, moreover, the range of the female-offspring group could be proportionately reduced by progressive elimination of male competition for local resources. This separation would be under strong positive selection. Lowered mobility of females would reduce accident rate during travel, maximize familiarity with the core area, reduce exposure to predators, and allow intensification of parenting behavior, thus elevating survivorship [20]. Such a division of feeding areas, however, would not genetically favor males unless it specifically reduced competition with their own biological offspring and did not reduce their opportunities for consort relationships. Polygynous mating would not be favored by this adaptive strategy because the advantage of feeding divergence is reduced as the number of males is reduced. Conversely, a sex ratio close to parity would select for the proposed feeding strategy. Such a ratio would obtain if the mating pattern were monogamous pair bonding. In this case, males would avoid competition with their bonded mates and biological offspring (by using alternative feeding sites) and not be disadvantaged by physical separation, that is, there would

be no loss of consort opportunity. In short, monogamous pair bonding would favor feeding divergence by "assuring" males of biological paternity and by reducing feeding competition with their own offspring and mates.

Such a system would increase survivorship and would also favor any increase in the reproductive rate of a monogamous pair so long as feeding strategy was sufficient to meet the increased load on the sources of protein and calories. One element of feeding among forest chimpanzees is the "food call" sometimes made by males upon discovery of a new food source (Reynolds and Reynolds, 1965; Van Lawick-Goodall, 1968; Sugiyama, 1968, 1969) [21]. In the proposed system, however, selection would not favor this behavior; instead, selection would favor a behavior that would benefit only the male's own reproductive unit. The simple alternative to the food call would involve collecting the available food item or items and returning them to the mate and offspring. Contrary to the opinion that such behavior would be altruistic, it would not be so in the proposed system, because it would only benefit the biological offspring of the male carrying out the provisioning and thus would be under powerful, direct selection. If this behavior were to become a regular component of the male's behavioral repertoire, it would directly increase his reproductive rate by correspondingly improving the protein and calorie supply of the female who could then accommodate greater gestational and lactation loads and intensify parenting [22]. The behavior would thus achieve both an increase in survivorship and a reduction in birth space. It would allow a progressive increase in the number of dependent offspring because their nutritional and supervisory requirements could be met more adequately.

Behaviors associated with similar reproductive strategies are in fact present in other primates. In both the Callitrichidae and Aotinae, extensive paternal care of the young constitutes a critical part of reproductive strategy in some species (Mitchell and Brandt, 1972; Kummer, 1971; Hershkovitz, 1977). Among callitrichids, the social unit is usually an adult male and female, plus one to several subadults. Maternal care is largely restricted to suckling and grooming, the male being responsible for subadults at all other times. The modal birth is dizygotic twins (Clutton-Brock, 1977; Hershkovitz, 1977). It is likely that this system is a partitioning of care in response to the high protein and calorie requirements of these small species. Male care during foraging tends to equilibrate the high caloric load imposed on females by lactation and gestation of two (and sometimes three) offspring— the process of twinning being an obvious demographic adaptation of elevated birthrate. As Hershkovitz (1977) notes: "survival of a population [of callitrichids] in the wild depends on close synchronization between cyclical nutritional requirements for young and old and the seasonal changes in the quality and quantity of available food." This same statement could be as well applied to early hominids, especially given increased Miocene seasonality and the need for a decrease in birth spacing. The altricial infants of Miocene hominoids, however, would have required reduced mobility and therefore prevented a callitrichid strategy of male care, with the simplest solution being the male provisional model proposed above.

THE ORIGIN OF BIPEDALITY

Provisioning is, of course, the primary parental care strategy of most canids and birds (Grzimek, Ed., 1975; Lack, 1968; Eisenberg, 1966) [23]. Both groups exhibit direct male involvement similar to that described for callitrichids. Their offspring are normally immature at birth, immobile, and require constant pro-

Table 1   *Relative reproductive values of Old World Primates calculated from Eq. 4 (see text) and multiplied by 10 for clarity.*

Annual Survivorship	Reproductive Values		
	Old World Monkeys*	Chimpanzees†	Man‡
.90	17	4	2
.92	23	7	4
.94	31	13	9
.96	42	25	24
.98	58	50	64

*Maximum life potential = 20; sexual maturity = 4; birth space = 2 (Drickamer, 1974; Napier and Napier, 1967) [13]. †Maximum life potential = 40; sexual maturity = 10; birth space = 3 (Napier and Napier, 1967; Teleki et al., 1976; Van Lawick-Goodall, 1969) [16]. ‡Maximum life potential = 60; sexual maturity = 15; birth space = 2.5.

visioning and parenting. In some species, a sexual division of labor, like that posited here for early hominids, is observed. Female hornbills (Bucerotidae), for example, depend totally on male provisioning for their survival and that of their offspring. Monogamous pair bonding is characteristic of 90 percent of bird species (Lack, 1968; Emlen and Oring, 1977) [23] and is the most common mating system in provisioning canids (Grzimek, Ed., 1975). Both groups, as a fundamental feature of reproductive strategy, commonly sequester their offspring at home bases [24].

One critical difference separates provisioning in birds and canids from that suggested for early hominids. Birds and canids can carry in their mouths or regurgitate (or both) a significant proportion of their body weight. Oral carrying would have been inadequate for early hominids, however, and a strong selection for bipedality, which would allow provisions to be carried "by hand," would thus accompany provisioning behavior [25].

Chimpanzees are fully capable of short-range bipedal walking and a variety of hindlimb stances (Bauer, 1977), but because they lack the pelvic and lower limb adaptations characteristic of hominids, bipedal walking leads to rapid fatigue (Lovejoy, 1978). It appears likely that the skeletal alterations for bipedality would be under strong selection only by consistent, extended periods of upright walking and not by either occasional bipedality or upright posture. While primitive material culture does not impose this kind of selection, carrying behavior of the type suggested above, does. It is likely that the need to carry significant amounts of food was a strong selection factor in favor of primitive material culture (Leakey and Lewin, 1978; Hewes, 1961). Although it is not a significant shift from primitive tools of the type used by chimpanzees today, such as "termite sticks" and "leaf sponges," to simple and readily available natural articles that could be used to enhance carrying ability, it is a significant shift from such primitive and occasional tool use to the stone tools of the basal Pleistocene. Development of such tools is most likely to have followed an extended period of more primitive material culture, which was not critical to survival. It has been suggested frequently that the earliest tools were weapons. However, the progressive development of more advanced stone tools from rudimentary weapons is unlikely. A prolonged and extensive period of regular and habitual use of simple (primitive) carrying devices could eventually allow the coordination and pattern recognition necessary for a more advanced reliance on material culture.

The sequential evolution of behavior proposed in this article has a high probability of mirroring actual behavioral events during the Miocene. In most higher primates, male fitness is largely determined by consort success of one sort or another (Trivers, 1972). Male enhancement of offspring survival is for the most part indirect and is expressed more in terms of demic or kin selection by general behaviors such as territory defense or predator recognition and repulsion (Hrdy, 1976). Females are solely responsible for true parenting and their ability in this is under strong selection. However, progressive intensification of higher primate K strategy elevates parenting requirements and lowers reproductive rate. The most obvious, and perhaps only, additional mechanism available with which to meet this "demographic dilemma" is an increase in the direct and continuous participation by males in the reproductive process. Whatever the actual sequence of events, whether as posed above or by some alternative order, such additional investment would improve survivorship and favor a mating structure that intensified energy apportionment to the male's biological offspring. Two mating patterns satisfy this latter requirement: polygyny (one male and several females) or monogamy. The former, however, requires male energy to continue to be devoted to maintaining consorts, and a pool of competing males

is ensured by polygynous structure itself, thereby directing it away from direct enhancement of survivorship.

In their synthesis of the evolution of mating systems, Emlen and Oring (1972) stress three factors common to polygynous mating structure. (i) One sex is predisposed to assume most, or all, of the parental care. (ii) Parental care requirements are minimal. (iii) A superabundant food resource enables a single parent to provide full parental care. As has been noted above, however, survivorship of offspring must have been critical to Miocene hominoids; further female parenting is negated by the mobile feeding strategy; hominoid males may be considered an "untapped pool of reproductive energy; and Miocene ecological conditions required a generalist feeding strategy. Conditions were prime for the establishment of male parental investment and a monogamous mating structure. Finally, it should be pointed out that only among primates in which the male is clearly and directly involved in the parenting process should monogamy be found. This is exactly the case, as this mating structure is found only in gibbons, siamangs, and the New World taxa discussed above (Clutton-Brock, 1977).

HUMAN SEXUAL BEHAVIOR AND ANATOMY

The highly unusual sexual behavior of man may now be brought into focus. Human females are continually sexually receptive [26] and have essentially no externally recognizable estrous cycle; male approach may be considered equally stable. Copulation shows little or no synchronization with ovulation (McChance et al., 1937) [27]. As was pointed out above, the selective emergence of a monogamous mating structure and male provisioning would require that males not be disadvantaged in obtaining consorts. Provisioning in birds and canids is normally made possible by highly restricted breeding seasons and discrete generations—the female normally is impregnable for only brief periods during which parental care is not required. The menstrual cycle of higher primates (Beach, 1947; Chance, 1962), however, requires regular male proximity for reproductive success. The progressive elimination of external manifestations of ovulation and the establishment of continual receptivity would require copulatory vigilance in both sexes in order to ensure fertilization. Moreover, copulation would increase pairbond adhesion and serve as a social display asserting that bond. Indeed, any sequestration of ovulation (J. Lancaster, personal communication) would seem to directly imply both regular copulatory behavior and monogamous mating structure. It establishes mathematical parity between males restricted to a single mate and those practicing complete promiscuity, and the balance of selection falls to the offspring of pair-bonded males, since their energetic capacity for provisioning (and improved survivorship and reproductive rate) is maximized.

Man displays a greater elaboration of epigamic characters than any other primate (Jolly, 1970b; Crook, 1972; Morris, 1967; Beach, 1978). Frequently, our sexual dimorphism is tacitly accepted as evidence for a polygynous mating structure because marked sexual dimorphism is most often a product of elaboration of characters of attraction, display, and agonistic behavior in males of polygynous species. Among primates, the degree of sexual dimorphism corresponds closely to the degree of male competition for mates (Clutton-Brock and Harvey, 1977; Crook, 1972; Morris, 1967). Yet human sexual dimorphism is clearly not typical as is even made clear by the fossil record. In their discussion of *A. afarensis*, Johanson and White (1979) [28] note that although this species shows "marked body size dimorphism, the metric and morphological dimorphism of the canine teeth is not as pronounced as in other extant, ground-dwelling primates. This implies a functional pattern different from that seen in other primates and may have significant behavioral implication." There can be no doubt that large male canines are part of the "whole anatomy of bluff, threat, and fighting" (Washburn and Ciochon, 1974). The reduction and effective loss of canine dimorphism in early hominids therefore serves as primary evidence in favor of the proposed behavioral model [29]. But it is important to stress that while canine dimorphism was undergoing reduction, other forms of dimorphism were apparently being accentuated, as judged from their expression in modern man, who remains the most epigamically adorned primate.

Since man displays a highly unusual mating structure, it is perhaps not surprising that his epigamic, or perhaps parasexual, anatomy is equally unusual and fully explicable by that mating structure. If pair bonding was fundamental and crucial to early hominid reproductive strategy, the anatomical characters that could reinforce pair bonds would also be under strong positive selection. Thus the body and facial hair, distinctive somatotype, the conspicuous penis of human males, and the prominent and permanently enlarged mammae of human females are not surprising in light of Mayr's (1972) observation that in "monogamous species such as herons (egrets) in which the pair bond is continuously tested and strengthened by mutual displays, there has been a 'transference' of the display characters from the males to the females with the result that both sexes have elaborate display plumes." In man, however, marked epigamic dimorphism is achieved by elaboration of parasexual characters in both males and females, rather than in males alone. Their display value is clearly cross-sexual and not intrasexual as in other primates. It should be stressed that these epigamic characters are highly variable and can thus be viewed as a mechanism for establishing and displaying individual sexual uniqueness, and that such uniqueness would play a major role in the maintenance of pair bonds (Crook, 1972). This is especially important when other epigamic features of man (pubic, axillary, and scalp hair), which have been elaborated in both sexes, are considered. Such characters may also contribute to individual sexual uniqueness [30]. Redolent individuality is clearly the most probable role of axillary and urogenital scent "organs" (eccrine and apocrine glands plus hair), which are unique among mammals (Montagna, 1975). An objection that might be voiced in response to these suggestions is that such auxiliary pair-bond "enhancers" are eclipsed by the paramount role of culture in the mating practices of nontechnological societies. Quite the contrary, the more that culture can be shown to dominate the mating structure and process of recent man, the more ancient must be the anatomical-physiological mechanisms involved in the formation and maintainence of pair bonds [31]

HIGHER PRIMATE PALEOGEOGRAPHY

The present-day geographic distributions of Old World monkeys and apes are shown in Fig. 3. The great apes are markedly restricted and occupy only minor areas where minimal environmental changes have taken place since the early Miocene. Yet the fossil record shows that their lineal ancestors (dryopithecines, *sensu lato*) spread throughout the Old World following the establishment of a land bridge and forest corridor between Africa and Eurasia about 16 to 17 million years ago, and that they enjoyed considerable success after their colonization of Europe and Asia (Andrews and Van Couvering, 1975; Delson, 1975a). Old World monkeys, on the other hand, were much less abundant during this period (Delson and Andrews, 1975). After the middle and late Miocene, however, a marked reduction in dryopithecine numbers occurred. While this cannot be deducted from the sparse fossil

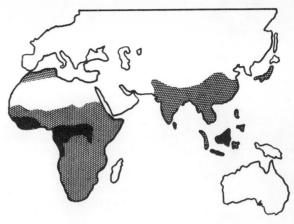

Figure 3   *Approximate distribution of extant Old World monkeys (hatched) and pongids (gorilla, chimpanzee, orang-utan) (solid) (Kortlandt, 1972; Sussman, Ed., 1979; Schultz, 1969; Napier and Napier, 1967).*

record of the late Miocene and early Pliocene, the distribution of extant descendants of the dryopithecines is ample evidence of their relict status. Today, Old World monkeys are clearly the dominant and successful group, having replaced the dryopithecines and their descendants during the last 12 million years (Delson, 1975a). One hominoid group did survive and remain relatively abundant—the Hominidae. It is probable that the hominoid trends of prolonged longevity and increased parental investment are the key to the replacement of most pongid taxa by Old World monkeys, which are reproductively more prosperous. If only a portion of Miocene hominoids made the adaptations described above, two distinct groups would subsequently result. One group might counter the "demographic dilemma" according to the model suggested in this article; a second group could survive by occupying habitats with minimal environmental hazards. Hominids, being more demographically resistant to environmentally induced mortality, would be more capable of expanding into novel and varied habitats, especially mosaics, and of competing with the radiating Old World monkeys. Conversely, the extant pongids are by implication descendant of populations progressively more restricted to highly favorable forest conditions, where minimal seasonality in food supply, low predation pressure, and limited size of the home range would be in effect. These differences in habitat preference would result in a more extensive fossil record for hominids than pongids, both by virtue of the geographic expansion of hominids and as a consequence of the occupation of habitats with more favorable conditions of fossilization. It is therefore quite possible that the sivapithecines (*sensu lato*) of the middle and late Miocene, which already evince dental modifications adumbrating those of late Pliocene hominids (Greenfield, 1979), may have contained primitive emergent hominids, at least behaviorally, if not phylogenetically.

THE NUCLEAR FAMILY

Man's most unique character is without question his enormous intelligence, and its evolutionary pathway has fascinated all who have attempted to explain the human career. Hunting and toolmaking are most frequently cited as "primal causes" for the Pleistocene acceleration in hominid brain development. Yet have these not figured so prominently because they leave ubiquitous evidence—the archeological record? Other human behaviors at least as critical to survival (especially reproductive behavior) are not "fossilized." It is now clear that man probably remained an omnivore throughout the Pleistocene and that hunting may have always been an auxiliary food source [32].

As Reynolds (1976) stressed, intense social behavior would seem the most likely single cause of the origin of human intelligence if one origin must be isolated. Tools are used to manipulate the environment and are thus a vehicle of intelligence, not necessarily a cause. Chimpanzees occasionally use tools (a behavior that has fascinated many early hominid theorists), but tools are not critical to their survival. Primates, which are the most intelligent mammals, have achieved evolutionary success primarily by their social and reproductive behavior, which is their most developed original character. It seems reasonable therefore to propose that a further elaboration of this adaptive strategy is the most likely "cause" of early hominid success and the further development of intelligence.

It is of interest to explore one further effect of the proposed model on early hominid social structure. The strong maternal and sibling ties of higher primates are now well documented (Goodall, 1976; Van Lawick-Goodall, 1969; Yamanda, 1963; Koford, 1963; Reynolds, 1976; Kaufmann, 1965a). The matrifocal unit of chimpanzees continues throughout the life of the mother, as do sibling ties. In the proposed hominid reproductive strategy, the process of pair bonding would not only lead to the direct involvement of males in the survivorship of offspring, in primates as intelligent as extant hominoids, it would establish paternity, and thus lead to a gradual replacement of the matrifocal group by a "bifocal" one— the primitive nuclear family (Beach, 1978) [33]. The effects of such a social unit on survivorship and species success could be profound. It could lead to a further shortening of birth space, which would accelerate the reproductive rate and amplify sibling bonds. Reduction of birth space would allow coincident protraction of the subadult (learning) period [34]. Behaviors that in other primates are common causes of infant death (for example, agonistic buffering) (Hrdy, 1975) would be largely eliminated, while those that might improve survivorship (for example, adoption) (Teleki et al., 1976; Kummer, 1967b; Sade, 1967) would be facilitated. The age until which an orphaned chimpanzee does not survive the death of its mother is "around 5 years of age, but may stretch another 3 to 4 in special circumstances" (Teleki et al., 1976). Survival of a second parent may have been a crucial reproductive advance in early hominids (Wolpoff, 1979b, cited in *Mosaic*). Primiparous females are much less adept than multiparous mothers. Drickamer (1974) found that in free-ranging *Macaca mulatta* "between 40 and 50% of the infants born first or second to a female did not survive their first year, but by the fourth infant born to the same female only 9% died during the first 12 months." Lancaster (1972) notes that: "Recent field and laboratory workers have shown that in many species of mammals, and especially in monkeys and apes, learning and experience play vital roles in the development of the behavior patterns used in mating and maternal care." The effect of intensified parenting, protracted learning, and enhanced sibling relationships would have a markedly beneficial effect upon survivorship. Such projections of the behavior of developing hominids are certainly not new, but they have not received their due emphasis. Can the nuclear family not be viewed

as a prodigious adaptation central to the success of early hominids? It may certainly be considered as being within the behavior repertoire of hominoid primates, provided that the reproductive and feeding strategies commensurate to its development were themselves under strong selection. This brief review of the fossil record and some primate behavioral and ecological adaptations would seem to strongly favor the correctness of this view.

CONCLUSION

It is a truism to say that even late Pliocene hominids must have been unusual mammals, both behaviorally and anatomically. As was pointed out above, emphasis in models of human origin has traditionally been on singular, extraordinary traits of later human evolution. The model proposed in this article has placed greater emphasis on a fundamental behavioral base from which these unusual adaptations could be directionally selected.

The proposed model accounts for the early origin of bipedality as a locomotor behavior directly enhancing reproductive fitness, not as a behavior resulting from occasional upright feeding posture. It accounts for the origin of the home base in the same fashion as it has been acquired by numerous other mammals. It accounts for the human nuclear family, for the distinctive human sexual epigamic features, and the species' unique sexual behavior. It accounts for a functional, rudimentary material culture of long-standing, and it accounts for the greater proportion of r-selected [11] characters in hominids relative to other hominoids. It accounts for these characters with simple behavioral changes common to both primates and other mammals and in relatively favorable environments, rather than by rapid or forced occupation of habitats for which early hominoids were clearly not adaptively or demographically equipped. It is fully consistent with primate paleogeography, present knowledge of higher primate behavior patterns (as well as those of other mammals), and the hominid fossil record.

If the model is correct, the conventional concept that material culture is pivotal to the differentiation and origin of the primary characters of the Hominidae is probably incorrect. Rather, both advanced material culture and the Pleistocene acceleration in brain development are sequelae to an already established hominid character system, which included intensified parenting and social relationships, monogamous pair bonding, specialized sexual-reproductive behavior, and bipedality. It implies that the nuclear family and human sexual behavior may have their ultimate origin long before the dawn of the Pleistocene.

Acknowledgments

I thank G. J. Armelagos, T. Barton, B. Campbell, T. Gray, F. C. Howell, K. Jacobs, D. C. Johanson, B. Kimbel, A. E. Mann, R. S. Meindl, R. P. Mensorth, M. H. Wolpoff, P. Shipman, A. C. Walker, T. D. White, and S. Ward, who read earlier versions of this paper and provided valuable comments. I thank D. C. Johanson, C. J. Jolly, J. Lancaster, R. S. Meindl, and T. D. White for valuable discussions about its content. I thank T. Barton for discussions and advice with respect to the quantitative approach used, L. don Carlos and R. P. Mensforth for research assistance, and R. S. Meindl for listening to endless anecdotes about the behavior of canids, rodents, and birds.

Notes

[1]   Artifacts have been found in situ in the Gona region of the Hadar formation by the International Afar Research Expedition. Stratification of the Gona region is at present under investigation and correlations with the KH member are as yet uncertain. At present the artifacts are thought to overlie deposits equivalent to BKT_2 tuff of the KH member, which has a potassium-argon age determination of about 2.6 million years. The artifacts may therefore be older than 2 million years (D. C. Johanson, personal communication).

[2]   A convincing and more detailed argument is provided by Jolly (1970b).

[3]   Whether or not the early evolution of material culture did in fact proceed in a gradualistic manner is difficult to establish. A punctuated equilibrium model is equally applicable to the early artifact record, and it is not unlikely that material culture proceeded at variable rates.

[4]   Contrary to popular opinion, there is no evidence whatsoever that early hominids hunted. Bipedality is probably the mode of locomotion least adapted to hunting, unless sophisticated technology is available (or unusually high levels of intelligence, or both). The evidence made available by *A. afarensis* is particularly striking. Further australopithecine evolution from that species is documented by a reduction of the anterior dentition and further enlargement of the grinding teeth. Artifacts do not appear until 2 million years ago, and when they do appear it is difficult to interpret them as hunting implements. In short, if the evidence made available by the fossil record is to be used in reconstructing early hominid evolution, one of its clearest implications is that hunting was not a dietarily significant behavior. See also Jolly (1973), Andrews and Van Couvering (1975), Mann (1981), Delson (1975a) and Schultz (1969c).

[5]   Among the salient features of the *A. afarensis* cranium are a convex nasal clivis, a uniformly shallow palate anterior to the incisal foramen, an independent juga with individual sharp lateral margins of the nasal aperture, a true canine fossa, a rounded tympanic plate with an inferiorly directed surface, a compound temporal-nuchal crest in large and small individuals, an occipital plane that is short relative to the anterior nuchal plane, and a relatively vertical nuchal plane and posterior positioned foramen magnum (W. H. Kimbel, in preparation). See also Johanson and White (1979).

[6]   I have elsewhere pointed out that the resting lengths of the major propulsive muscles about the hip and knee in quadrupedal primates are so substantially altered by the adoption of erect posture that the regular effective use of both quadrupedal and bipedal locomotion is not possible. Thus the transition to bipedality as a habitual mode of locomotion must have been relatively rapid [Lovejoy (1978); and Lovejoy, Heiple, and Burstein (1973)].

[7]   While there is general agreement as to those morphological features directly associated with molar dominance in the group of hominids referred to as phase II, there is some disagreement as to the distinctiveness of other dentognathic characters. See P. Andrews and I. Tekkaya (1976).

[8]   This is especially true of *Erythrocebus patas* which is both a small-object feeder and the fastest ground-living primate (Hall, 1965).

[9]   It should also be pointed out that changes in the masticatory apparatus of hominids are a reflection of changes in habitat and are not necessarily the initial cause of clade differentiation. Care must be taken not to view those characters that allow identification of early hominids as synonomous with actual forces of divergence.

[10]   Greenfield (1979) concludes that *"Sivapithecus* utilized a broad range of zones including tropical rain forests (Chinji of India), subtropics (Rudabanya and Chinji-Nagri of India and Pakistan), and woodland and bush habitats (Late Nagri and Early Dhok Pathan of India and Pakistan, Fort Ternan)."

[11]   The K and r are opposite ends of the continuum of reproductive strategy. In the r strategy, the number of offspring is maximized at the expense of parental care; at the K end (the effective limit of which is 1), parental care is maximized.

[12]   A female chimpanzee reaches sexual maturity at about age 10 years. Using the average span of 5.6 years between successful births (Teleki, Hunt and Pfifferling, 1976) gives a required life expectancy of about 21 years (a chimpanzee infant usually dies after his mother's death if it is not at least 4 years old). The authors (Teleki, et al., 1976, p. 577) conclude similarly that "The mean generation span, or elapsed time between birth of a female and birth of her median offspring, is about 19.6 years for a sample of ten Gombe females with three or more recorded births." Their most realistic estimate of achieved reproduction is three to four "offspring that are successfully raised to sexual maturity" (Teleki, et al., 1976).

[13]   Demographic studies of Old World monkey populations are at present insufficient to provide accurate data for unprotected and undisturbed populations (D. S. Sade et al., 1976), but approximations can be made adequately from observations of protected or introduced populations. Ninety percent of adult females in the Chhatari population studied by Southwick and Siddiqi (1976) gave birth yearly during the 14-year study period. Infant mortality averaged 16.3 percent, and juvenile mortality was judged very low (the actual figure was 33 percent but most of this loss was attributed to trapping). A reasonable figure with this data is 7 to 9 years, from methods similar to those used in Teleki, Hunt and Pfifferling, (1976). L. C. Drickamer (1974) found first birth to occur in females of 4 years on the average in free-ranging rhesus at La Parguera, Puerto Rico, and in animals of this age 68 to 77 percent produced infants each year. Although first-year mortality was high for first and second born offspring (40 to 50 percent), it was only 9 percent for third born. These data support the above conclusion as do those of C. B. Koford (1965) from rhesus on Cayo Santiago.

[14]   It is in fact more likely that they were not, or at least not in the extreme forms seen in extant hominoids. The hominid adaptations proposed in this article are more likely to have been developed to prevent the "demographic dilemma." Modern pongids probably represent terminal phases of extreme parental investment only because of long-term occupation of particularly favorable environments, hence their very restricted present-day distribution.

[15]   By breakthrough adaption is meant that which allows or precedes an adaptive radiation. While the Old World monkeys replaced all other hominoids, the hominids were successful in many of the same environments occupied by monkeys, including the Pleistocene savannahs.

[16]   It should also be pointed out that falls as a consequence of mother-infant travel would be a more critical selection factor in early hominids than other primates. Van Lawick-Goodall (1969, p. 424) points out that "in striking contrast to most other primate species, the small chimpanzee infant often appears unable to remain securely attached to its mother if she makes a sudden movement. For several months after birth of her infant the mother may have to support it, thus hindering her movements . . . " Mortality as a consequence of falling may thus be viewed as one selection factor in favor of terrestrial care, but more importantly, the adoption of bipedality would mean the loss of most or all of the prehensility of the infant foot, which is an important grasping organ in the chimpanzee infant. This, in turn, selects for a more secure infant carrying ability in the mother and thus bipedality. This form of selection can clearly not be viewed as the initial selective force for bipedal locomotion, but in conjunction with others, would certainly contribute to the total selective pattern.

[17]   Clutton-Brock (1977) notes that "although sex differences in feeding behavior are common among primates there is little evidence to suggest that they have evolved to minimize feeding competition between the sexes." Yet for the two species just cited this may indeed be the case. *Colobus badius* uses a more generalist feeding strategy than its sympatric congenerics, has a larger number of adult males within the troop [T. T. Struhsaker and J. F. Oates, in Sussman (1979, pp. 165–1860)] and has the highest population density of any Old World monkey [T. H. Clutton-Brock, in Sussman (1979, pp. 503–512)]. It is clearly true of the orang [Horr, in Sussman, (1979)].

[18]   This forms the basis for an additional criticism of the hominid model proposed by Jolly (1970b). Gelada baboons are highly specialized feeders whose feeding rate is largely limited by handling time.

[19]   "Although nutritional factors alone would not preclude the possibility that early hominids were dietarily quite specialized, (from) the available archaeological evidence and (from) what is known of the dietary patterns of living gatherer-hunters and chimpanzees, it appears unlikely that all early hominids were almost exclusively carnivorous or herbivorous. It is more reasonable to suggest that the diet fell within the broad range of today's gatherer-hunter diets, but that within the wide spectrum of this adaptation, local environment resources and seasonal scarcity may have forced some individual populations to become more dependent on vegetable or animal tissue foods than others" (A. E. Mann, personal communication; 1981).

[20]   D. A. Horr (in Sussman, 1979, p. 320) comments with respect to the orang-utan: ". . . orang social organization might easily be explained as follows: In order not to overload the food supply, orangs disperse themselves in the jungles. Females carrying infants or tending young juveniles can best

survive if they don't have to move far. Young orangs could also best learn the jungle in a restricted, familar area. . . . Adult males are unencumbered by young and can more easily over wider areas. This means that they compete with females for food only for short periods of time, and thus do not overload her food supply and force her to move over wide areas."

[21] See R. W. Wrangham (1977) for futher discussion on the possible functions of the "food call."

[22] Such loads can become intense (up to 1.5 times normal resting basal metabolic rate in females) (O. W. Portman, 1970). The modern human preparation for lactation is an average accumulation of 9 pounds subcutaneous fat (D. B. Jelliffe and E. F. P. Jelliffe, 1978). A major birth interval limitation in hominoid primates may well be the lactational loads placed on the mother (in contradistinction to the needs of the infant) and any improvement in feeding strategy could "support" a reduction in birth space on this basis.

[23] Monogamy is especially characteristic of long-lived (K-selected) birds (Davis, 1976; Mills, 1973; Coulson, 1966).

[24] Such sequestration is common among rodents as well with perhaps its most classic expression in castorids, which are comparatively K-selected (requiring 3 to 4 years to mature sexually) and live in stable family groups.

[25] A second, also important, element of food-handling behavior may have been premastication. Reduction of birth space would have required an earlier reinitiation of ovulation, which would in turn have required a reduction of mechanical stimuli to the mechanoreceptors of the nipple and areola and thereby a reduction in prolactin levels (R. C. Kolodny, L. S. Jacobs, W. H. Daughaday, 1972); hence, an earlier age of weaning. Parental premastication would have facilitated such behavior, at the same time enhancing parental bonds (see discussion of the nuclear family). This could have also increased the rate of dental wear and have been an auxiliary selection component of the dentognathic changes characteristics of early hominids.

[26] D. C. Johanson, personal communication.

[27] J. R. Udry and N. M. Morris (1968), did find such a relationship, but it is a moot point since with sequestration of ovulation and its external manifestations, copulation would reqture female initiation.

[28] In other primates that are monogamous, there is little dimorphism. All of these, however, live in territorial family groups (A. Jolly, 1972) and there is therefore no intragroup competition for mates. Strong sexual dimorphism is usually a consequence of either differential competition for mates or differential exploitation of resources (Clutton-Brock, 1977; Selander, 1972; Hutchinson, 1965). In *A. afarensis,* according to the proposed model, dimorphism would be favored on the latter basis. Small female body size would reduce caloric-protein requirements, while large body size would increase male mobility and predator resistance.

[29] This is not meant to imply that it was a cause of canine reduction, but only that the process could occur by a combination of relaxation of selection on large male canine size and a positive selective mechanism for reduced canines. The latter is most likely to be found in the concurrent dentognathic changes of greater molar dominance and general anterior tooth reduction (Jolly, 1970b; Johanson and White, 1979).

[30] Modern man displays a remarkable number of morphological traits that may be considered epigamic (hair color and type, lip size and form, corporal hair patterning, eyebrows, facial countenance, and so forth). Attempts have been made to correlate some of these with geographic variables, but they have been largely unsuccessful. An alternative explanation is that disruptive selection acts to maximize the variability of these features within populations, thereby enhancing the distinctiveness of potential and actual mates in establishing and maintaining pair bonds. The subsequent geographic isolation, whether partial or complete, of a population could then have resulted in a truncation of expression and apparent uniqueness of some features that maintained their epigamic significance in the population (for example, the epicanthic eye fold). The obvious polygenic basis of such traits and their reappearance in unrelated populations (Bushman, Lapps, infant Euroamericans) indicate that their expression is a consequence of elevated frequencies of genes that may be universal in *H. sapiens,* but below an expressive threshold in some populations.

[31] Further evidence of the age of pair bonding is provided by the absence of strong canine dimorphism in *A. afarensis* (Johanson and White, 1979). The only other Old World higher primates without canine dimorphism are the gibbon and siamang, which are monogamous (Washburn and Ciochon, 1974).

[32] The provisioning model proposed here effectively accounts for the origin of hunting by means of a progressive elaboration of provisioning behavior (that is, collecting → scavenging + collecting → hunting + scavenging + collecting) without the requirement that hunting be critical to human evolution at any point. The similarity in social behavior between canids and early humans has often been cited and attributed to hunting. It is more likely that such similarities take origin in reproductive strategy (pair bonding, intratroop cooperation, provisioning, male involvement in subadult care, and so forth) and that hunting merely represents one food procurement method that satisfies the economic requirements of the social system. There are numerous carnivores that do not display this form of reproductive strategy, and there are some rodents that do but, of course, do not hunt (see footnote [24]).

[33] The term bifocal is preferable to nuclear family because the latter carries manifest connotations from its application to Western and non-Western modern human cultures, none of which are implied by its use here.

[34] A. E. Mann's (1975) extensive studies of dental development and wear in australopithecines indicate a prolonged period of development was established by about 2.5 million years ago. K. R. McKinley's survivorship calculations based upon Mann's (1971) data led him to conclude that australopithecines show a hominid rather than a nonhuman primate "birth spacing pattern." While these calculations require a number of assumptions about the origin and nature of the death assemblages at Swartkrans and Sterkfontein, they are strong evidence that a major demographic shift was fully developed by 2.0 to 2.5 million years ago which included an extended period of subadult dependency.

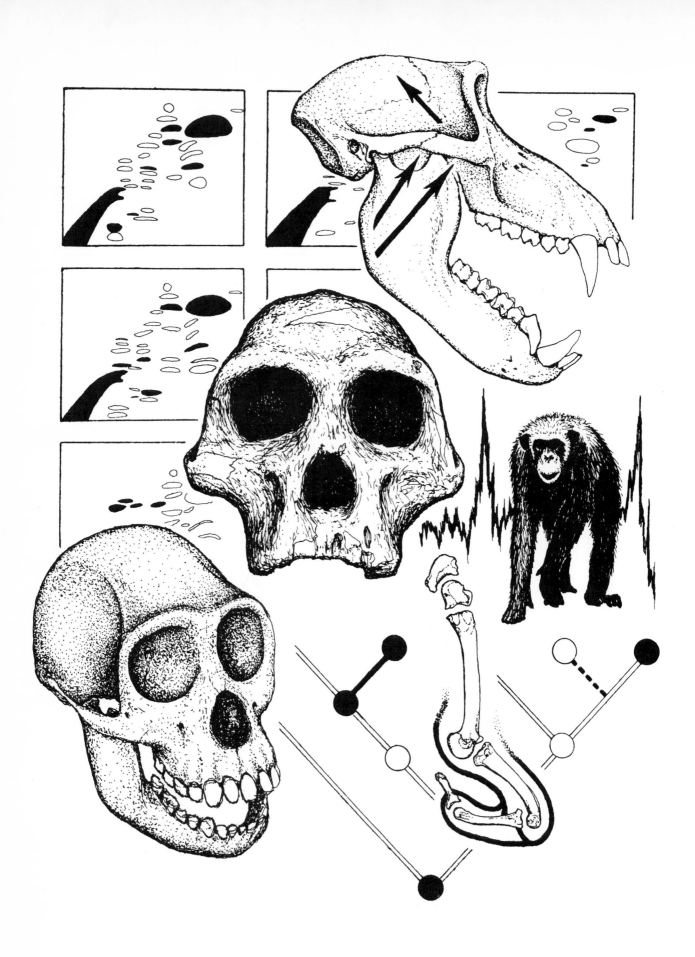

Part VII
Diverse Approaches in Human Evolution

As many of the earlier chapters showed, the fossil record is only one of the many sources of data that contribute to our understanding of primate and human evolution. The articles in this part illustrate some of the diverse approaches to human evolution.

Chapter 39, by Morris Goodman (1963), is one of the first papers in which immunological studies were used to investigate the evolutionary relationships among living primates. Goodman's studies demonstrated quite clearly that the human lineage was not distinct from that of all Great Apes. Rather, humans, chimps and gorillas are more closely related to one another than any of the three is to the orangutan or to any other primates. Furthermore, the immunological data indicate that the biomolecular differences separating humans and apes are extremely small compared with the magnitude of differences that seem to separate apes and humans as a group from other primates. Goodman interpreted this serological similarity among humans and African apes as evidence for a slow rate of biomolecular evolution among hominoids rather than as evidence for a very recent divergence. He argues that the unique form of placentation that characterizes Great Apes and humans, in which maternal and fetal blood is intimately apposed rather than being separated by membranous layers as in many other primates, has led to a slowdown in the rate of molecular evolution in ape and human lineages.

Vince Sarich, the author of the next chapter, and Allan Wilson (see Sarich and Wilson, 1967a,b) agreed with Goodman that humans and African apes were extremely similar immunologically but disagreed with Goodman's interpretation of the data. They argued that rates of biomolecular evolution have been roughly constant and that immunological differences between living animals can be used to construct a molecular clock that could tell the amount of time any two species had been separated from a common ancestor. Thus, they argued that the immunological similarity between African apes and humans was evidence of a very recent common ancestor. Although Sarich's results concerning the times of divergence between living primate groups and the implications for interpreting the fossil record were initially rejected by most anthropologists, and especially paleontologists (see Simons, 1969b, 1979b; Walker 1976), they now enjoy very wide support (see Chapter 44).

Chapter 41, by Clifford Jolly (1970), has had an important effect on the study of human evolution because it approaches the problems of human origins from a nonhuman primate perspective. Jolly's paper deals not with the issue of when the human lineage appeared, but why. Most discussions of human origins, from Darwin through the present, have considered the group of features that distinguish living humans from living apes (bipedal locomotion, small canines, large brains, and tool use) as an integrated adaptive package. Jolly, however, attempted to separate these features and analyse them one at a time, as they seemed to have appeared in the fossil record. He deals first with the dental features of small canines and large molars that characterized the initial divergence of humans and apes, which he attempts to interpret as primate rather than human adaptations. Only later

does he incorporate the uniquely human features of bipedalism, tools, and brain enlargement in his scenario of human evolution.

Chapter 42, by Russell H. Tuttle (1974), deals with the other side of the ape-human divergence: Which living ape most closely resembles the ape-human ancestor? Drawing heavily from comparative anatomy and his own experimental studies of ape locomotion, Tuttle reviews the various ape models of human origins. Although the anatomical evidence for a common ancestor between living apes and humans is unassailable in his opinion, he thinks that the specific nature of the ape that gave rise to early hominids remains an open question.

The authors of Chapter 43 are less indecisive than Tuttle regarding the type of living ape most similar to the earliest hominids. In their view, the pygmy chimpanzee, *Pan paniscus*, is a good model for the ape that gave rise to the other African apes and humans. Although their views have raised considerable controversy (see Johnson, 1982 and comments; also Susman, 1984), the similarities between pygmy chimps and early hominids have been supported by other workers (e.g., Stern and Susman, 1982).

One of the most persuasive methodological changes in the study of primate and human evolution over the past two decades has been the widespread influence of a "cladistic" methodology derived from the work of Willi Hennig (1966). Although few students of human evolution have followed the methods of phylogenetic systematics rigorously (Tattersall and Eldridge, 1975), much of the basic philosophy of cladistic methodology has become commonplace. Thus, over the past decade students of primate and human evolution have become far more conscious of the distinctions between primitive (or ancestral) characters and derived characters—the latter being the most important for reconstructing phylogeny.

The final chapter, by R. L. Ciochon (1983), illustrates a phylogenetic methodology that has become increasingly common in attempts to reconstruct the pathways of primate evolution. By analysis of dozens of anatomical features, he has attempted to reconstruct the phylogenetic history of hominoid primates by interpolating fossil primates into a phylogenetic framework based on living species (see Delson and Rosenberger, 1980; for a similar analysis of early anthropoid evolution). Ciochon's approach is in direct contrast with that of some authors (e.g., Gingerich and Schoeninger, 1977) who would argue that the fossil record takes priority in determining phylogenetic relationships, or others who would argue that fossils have little to offer in understanding evolution.

Although the chapters in Part VII are more eclectic than those which made up earlier sections of the book, they emphasize an important feature of paleoanthropology—its many facets. Many, but not all, of the differing interpretations of primate and human evolution can be related to the different types of data and theoretical bases that underlie those interpretations. Ultimately, however, all these approaches can provide us with a broader understanding of the evolutionary events that occurred during the evolution of the primates and the origin of the human lineage.

39

Man's Place in the Phylogeny of the Primates as Reflected in Serum Proteins

M. GOODMAN

The evolutionary development of the Primates has resulted from the sum of the processes of cladogenesis and anagenesis. Cladogenesis, or phylogenetic branching (Rensch, 1959), is the splitting of a species into two or more species and the ensuing divergence of these species from each other. Anagenesis, or progressive evolution (Rensch, 1959), is the process by which the more highly organized forms of life evolved from the simpler. In the order Primates, cladogenesis is exemplified by the prosimian phylogenies, and anagenesis is exemplified by hominine phylogeny.

The living primate fauna can be arranged according to grades of morphological organization into a remarkable sequence which goes from the tree shrew to man and simulates the successive anagenetic stages of the history of the Primates. The placement of the organisms in this *scala naturae* has guided systematists in classifying the members of the order. For example, although man is taxonomically grouped with the apes in the superfamily Hominoidea, he is invariably separated from the latter by being the sole contemporary member of the family Hominidae. In turn the apes are usually grouped together in the family Pongidae. However it is quite possible that certain of the living apes are more closely related to man than to the other apes. In other words certain of the apes may show a more recent common ancestry and more genetical affinities with man than with other apes. There are a number of such questions concerning the phylogenetic relationships of various living Primates. Indeed even the boundaries of the order are in dispute with some systematists placing the tree shrews in the Insectivora.

To solve these problems of primate systematics we must use very discriminating criteria to evaluate the genetic similarities and differences of organisms. The comparative analysis of proteins is potentially capable of providing such criteria, for recent developments in protein chemistry and in genetics have demonstrated that there is a close correspondence between the structural specificity of proteins and the code of information in genetic material (Anfinsen, 1959; Ingram, 1961; Crick, 1961; Chantrenne, 1961). Hence by utilizing some of the newer biochemical and serological methods for studying the structural specificity of proteins it is possible to ascertain the genetical affinities of contemporary organisms with a reasonable degree of objectivity.

Furthermore from our growing knowledge of the structure of proteins and the functional and immunological properties of proteins we can draw theoretical conclusions about the nature of the evolutionary process which led to the emergence of *Homo sapiens*. This article, following previous efforts of the author (Goodman, 1960a, 1961), will attempt to state some of the conclusions. An immunological theory of primate evolution will be presented and the results of a comparative serological study of the serum proteins of the Primates will be reviewed [1]. The data gathered in this study provide experimental evidence for the taxonomic position of man and the course of his evolution predicated by the theory.

PROTEINS AND THE EVOLUTIONARY PROCESS

A higher organism differs from a lower by the greater informational content of its genome. This is revealed by the greater biochemical complexity and more extensive molecular differentiation of the higher organism. If we compare microorganisms to lower metazoans and the latter to higher metazoans, a rough correlation is found between increases in the mass of DNA per cell and increases in the histological and morphological complexity of the organisms (Sneath, 1962). However, an increase in the size or number of the DNA polymers can only provide more raw material for the coding of hereditary information. The actual development of a more highly organized living system must await the evolution of a genome whose sections or genes show an increase in the integration of their informational contents. In other words a more highly organized living system has a more highly organized code of information in its genome.

The validity of this view will become apparent after we consider some of the relationships between gene mutation, protein structure and function, and natural selection. Our inquiry will show that the process of speciation must have caused a much deeper divergence of genetic codes and a much more rapid evolution of the structural specificity of proteins among lower organisms at the primitive stages of phylogenesis than among higher organisms at advanced stages of phylogenesis. Although morphological evolution appeared to accelerate with the emergence of higher organisms, molecular evolution decelerated. The central argument we shall develop is that the possibilities for the specificity of any particular protein to vary among the members of a species decreased as the biochemical complexity and molecular differentiation of organisms increased.

A protein is composed of one or more long polypeptide chains whose complete sequence of amino acid residues (the primary structure of the protein) is under strict genetic control. The nature of this control is such that the substitution of one amino acid for another in a chain can be related to a single gene mutation (Ingram, 1961). The three dimensional architecture of a protein molecule (the secondary and tertiary structure of a protein) is determined by the coiling and specific foldings of the polypeptide chains. A typical chain may contain from 100 to 500 amino acids of up to 20 different kinds. Thus mutations can cause an exceedingly large number of permutations in amino acid sequence. A change in amino acid sequence could so alter the conformations of a polypeptide chain as to create a protein with a new tertiary structure and a new set of surface configurations. Alternatively the amino acid substitution need not alter the conformations of the chain but could nevertheless occur at a position on the chain which would alter the topography of a small portion of the surface of the protein.

It is convenient to consider the biological functioning of globular proteins in terms of three categories of surface configurations: (a) the configurations which are called the active site because they are responsible for the primary functional activity of a protein, for example an enzymatic, hormonal, or antibody activity, (b) the ancillary configurations which enable a protein to be transported preferentially across certain membrane barriers and to be taken up preferentially by certain cell types, and (c) the neutral configurations with no apparent functional activity. Obviously any mutation of a primary structure which affected the active site of a protein would also affect the survival chances of the organism. However, it is clear from the structural analysis of hormones, enzymes, and antibodies that the active sites usually occupy a small proportion of the total surfaces of the proteins (e.g., Li, 1957; Smith, 1957). Also studies of the species specificity of proteins (e.g., Tristram, 1953; Porter, 1953) establish that a protein type can vary markedly in amino acid composition and antigenic structure and still execute its characteristic physiological role. Furthermore from the x-ray analysis and other structural studies of sperm whale myoglobin (Kendrew et al., 1960) and horse hemoglobin (Perutz et al., 1960) the principle has been established that proteins and major subunits of proteins can be dissimilar in primary structure and remarkably similar in three dimensional arrangement. Thus more than one primary structure is compatible with a single tertiary structure and many of the informational bits of a genetic code can be varied without affecting a protein's active site. We may deduce that in the beginning stages of phylogenesis many permutations in the primary structure of any one protein were permissible and that a vast amount of selectively neutral genetic variability could accumulate in any stable species.

It may be argued that due to the reducing nature of the environment during the early stages of the evolution of life (Oparin, 1957) a primordial protein would have had very few if any disulfide bonds and a large proportion of its backbone polypeptide chain would have had the disordered arrangement of the random coil. Such a protein would have had a minimum of tertiary structure and for this reason alone would have lacked surface configurations which could acquire functions ancillary to the active site. Looked at in this way metazoan evolution could not begin until conditions had materialized for the synthesis of proteins with well organized tertiary structures.

A typical soluble protein in our unicellular ancestors who preceded the start of metazoan evolution probably had a well organized tertiary structure which exposed two kinds of surface configurations: a small constellation of organic groups (the active site) executing the primary functional activity of the protein, and the large proportion of remaining configurations without functional activity. The latter, in contrast to the former, must have showed extensive structural heterogeneity among the organisms of the species as the result of an accumulation of neutral genic alleles in the species. Let us refer to this protein as protein A and trace it to a descendent species in our line of ancestry much later in evolution.

The organisms in this metazoan species still produce a protein A whose tertiary structure has hardly changed since the dawn of metazoan evolution. Nor has there been any change in the active site, which may be labeled a. In addition to a the protein has other surface configurations which are uniform throughout the population. These are the configurations with ancillary effects on the functional activity of protein A. For example, the b configuration of protein A interacts with complementary configuration x of the characteristic protein of tissue X, and thus A is taken up preferentially by tissue X where it executes its primary functional activity. Any mutations which altered either the b or the x configuration would be detrimental unless a simultaneous mutation occurred which so altered the other that b' and x' were also complementary (as b and x had been). Since this would be a highly improbable event, the evolutionary stability of both b and x configurations would be relatively high. In contrast to this state of stability, configuration c of protein A interacts weakly with configuration y of plasma protein Y, which tends to divert protein A from tissue X. Consequently, mutations would be favored which altered c to either of several alternative configurations (c', c'', or c''') since none of these react with y or other accessible configurations of the plasma proteins.

Arbitrarily we shall assign to the therapsid reptilian stage of phylogenesis the species which tolerates equally well configurations c', c'', an c'''. Now we shall consider a descendent species later in anagenesis after eutherian evolution had begun. The evolution of molecular complexity in the lineage leading to this species provided the basis for such improvements as a finer regulation of physiological processes by the central nervous system, the development of a superconcentrating kidney for the conservation of water, and the establishment of a placenta which grafted fetal tissues to maternal tissues and protected the fetus from many of the vagaries of the external environment. In directing the molecular evolution underlying these improvements, natural selection had to stabilize additional protein configurations with primary and ancillary functions. Also with the increasing specificity of the stereochemical interactions of protein molecules, the third category of protein configurations, those without apparent function, had to be selected more carefully to insure their state of non-interference with physiological processes.

The intensity of the selection for functional and for non-interfering configurations is indicated by the large number of serologically distinct proteins found in plasma and by the high specificity of their molecular associations. For example, if red blood cells rupture, the released hemoglobin is preferentially bound to haptogobin (a distinct species of the plasma proteins). Hormones are transported in plasma by specific proteins. Thyroxine, for instance, is preferentially bound to a particular species of the alpha proteins, but also to the tryptophane rich prealbumin (Winzler, 1960; Antoniades, 1960). The selective transport of gamma globulin in preference to other maternal serum proteins across the placenta into the fetus exemplifies the high specificity of molecular associations which involve ancillary configurations of the proteins. Not only is gamma globulin transported in preference to other types of plasma proteins, but homologous gamma globulin is more readily transported into the fetus than is heterologous gamma globulin (Hemmings and Brambell, 1961). For example, when rabbit and bovine immune gamma globulins were administered to a pregnant rabbit, the rabbit protein crossed the placenta in preference to the bovine protein. Other examples including some of transport across the placenta of man, are also cited by Hemmings and Brambell (1961). Such studies show that, in addition to sites for antibody activity, immune gamma globulin has surface structures of high specificity for the selective transfer of the protein to the fetus.

In this example of the selective transport of gamma globulin across the placenta, the structural specifications of gamma globulin by the genetic code are coadapted to the structural specifications of the other plasma proteins and the placental membrane proteins. We can see from examples of this type that the process of anagenesis decreased the chances that any single gene mutation would be neutral to natural selection since any permutation in a polypeptide chain would with greater probability affect the physiological activity of a protein.

In addition to altering the structural specifications of proteins, mutations can alter the epigenetic histories of proteins. At any particular time only a portion of the DNA polymers of a cell act (through the relay system of "messenger RNA") in protein synthesis, and the portions which act depend on the environmental conditions prevailing in the cell. For instance, the presence in the cell of a substrate can activate the mechanisms for channeling specific genetic information into the synthesis of certain enzymes. Then the accumulation of the terminal product of the enzymatic reactions can repress this mechanism and activate another. Although such induction and repression can be initiated by relatively small organic molecules, the mechanisms for channeling "messages" to and from the DNA polymers must involve chains of specific reactions among macromolecules.

Ontogenetic development is in essence an epigenetic process and depends at each stage on a cascade of interactions between a changing cellular environment and genetic material [2]. At the beginning of embryogenesis only a small fraction of the total genetic information is channeled into protein synthesis. Then with each cell division a new set of environmental conditions in the descendent cells induces different portions of the genetic information to be expressed in the protein synthesis of these cells. At the completion of ontogenetic development the total amount of genetic information channeled into protein synthesis is much greater than it was in the beginning stages, but this greater sum of genetic information in the structural specificities of proteins results from addition of the activities of the various differentiated cells of the mature organism.

Natural selection has so acted on the genome that the expression of its structural information during ontogenetic development is programmed against a fixed cycle of environmental changes, one repeated from generation to generation. If the cycle is disturbed by the introduction into the fetal environment of a new factor (such as a new chemical substance) from the external environment, developmental anomalies are likely to ensue. Similarly any gene mutation which caused the early fetus to synthesize a protein with a mutated structure would introduce a new factor into the fetal environment and would tend to distabilize the subsequent course of ontogenetic development. It is not surprising that a gene mutation which acts on development from the early stages of ontogeny is more likely to be detrimental to the organism than a gene mutation which only acts at later stages of ontogeny.

A gene mutation will be selectively advantageous if it increases the capacity of the organism to withstand the disintegrating impact of the external environment and selectively disadvantageous if it decreases this capacity. In the case of selectively advantageous gene mutations the new configurations on the surfaces of the mutated proteins can be looked upon as molecular adaptations to the conditions of the environment. It will help us to pursue our inquiry if we distinguish in the higher metazoans between two kinds of molecular adaptation, the kind directed "inward" to integrate the stereochemical interactions of the proteins (and thus perfect the metabolic machinery of an organism) and the kind directed "outward" to meet the exigencies of the surroundings of an organism. With dichotomies of this type we can characterize a certain aspect of the mechanism underlying the process of anagenesis.

Let us do so by considering a paradoxical situation. It concerns the species at the forefront of the main stream of anagenesis. Compared to all other organisms the members of this species have the most elaborate endogenous makeup with the greatest differ-

entiation of parts and the largest number of "inward-directed" molecular adaptations. As a result of the superior metabolic efficiency of its members, the advanced species is supplanting competitors and colonizing the largest range of exogenous habitats. Hence the new conditions of the external environment select for new gene mutations and genetic heterozygosity since only this state can give the species a variety of new "outward-directed" adaptations. However such a state tends to disorganize the established network of highly specific protein interactions. Indeed the probability of interference with physiological processes by any mutated protein specificity is in direct proportion to the number of "inward-directed" molecular adaptations of an organism. Thus two opposing selective pressures (one favoring genetic heterozygosity, the other favoring genetic homozygosity) operate with special force on the anagenetically advancing species.

A new equilibrium is established between these two opposing selective pressures by a selection of genetic codes which increase the complexity of ontogenetic development. Such codes dissociate the rates of maturation of different somatic systems. Thus the systems which mature early in ontogeny establish the basic features of the endogenous environment, and the genes which control these early maturing systems become more uniform throughout the population. It is the later maturing systems which develop the capacity for adaptive responses to exogenous conditions and which serve as a reservoir for gene determined polymorphisms.

The ontogeny of the immunological system illustrates how the conflicting needs of an organism can to some extent be met by the late maturation of somatic systems with "outward-directed" molecular adaptations. The immunological system is one of the major achievements of vertebrate evolution. It gives the individual the capacity to respond in a highly selective manner to a fantastically large number of molecular configurations. Without the capacity to produce antibodies the vertebrate organism could not survive the invasion of microorganisms and animal parasites. Yet with this capacity the vertebrate organism can produce autoantibodies which interfere with endogenous physiological processes and thereby lessen the chances of survival of the organism. Vertebrate evolution, however, has minimized this dilemma by selecting genetic codes which delay maturation of the immunological system and thus allow a state of immunological tolerance [3] to be acquired to endogenous proteins. As an acquired state, immunological tolerance is limited in scope. It only pertains to the proteins which contact the cellular progenitors of the immunological system during a critical stage of fetal development. Thus endogenous proteins may be autoantigenic if they appear late in ontogeny or are confined to tissues and rarely enter the blood circulation. Furthermore, many proteins and other macromolecules which are non-antigenic within the individual synthesizing them show some degree of isoantigenicity within the species.

ROLE OF THE IMMUNOLOGICAL SYSTEM IN THE EMERGENCE OF *HOMO SAPIENS*

The development of placentation in the mammals initiated a new stage of progressive evolution. It provided opportunity for an intimate apposition of fetal and maternal blood vessels which could secure for the developing tissues of the embryo an abundant supply of nutrients and oxygen and an effective channel for waste removal. It also permitted fetal proteins to come in contact with maternal lymphoid cells and maternal antibodies to cross into the fetus. In this situation, a fetal protein altered by gene mutation was a potential isoantigen, injurious to the fetus by its capacity to immunize the mother. Thus in the mammalian stage of phylogenesis maternal isoimmunizations further decreased the heterozygosity of prenatally acting genes. There would also be in some lineages a selective advantage to genetic codes which delayed the appearance of proteins involved in adaptions to the external environment until after birth, since changes in the specificities of such proteins would not immunize the mother.

It was only after the transition from oviparity to viviparity in the early mammals that maternal immunizations to fetal antigens could become a factor in mammalian evolution. At first this factor was minimal. The primitive placenta of the basal stock of the Eutheria can be depicted (Hamilton, et al., 1952) as having a simple non-deciduate epithelichorial arrangement in which several avascular layers separated the fetal and maternal blood vessels. This placenta functioned with low efficiency in the transport of metabolites and would hence have blocked the entrance of fetal isoantigens into the maternal circulation. We can assume that just as small modern mammals have a relatively brief period of intra-uterine existence (Rensch, 1959) so did the small insectivore-like mammals of the Cretaceous epoch. If these animals were the size of *Sorex*, gestation may have lasted no longer than three weeks. Moreover, during this time much of embryonic development probably occurred in the lumen of the uterus before the placenta was fully elaborated. Since in mammals of any size it takes from one to two weeks for homografts to elicit immune reactions, the time during which maternal immunizations could injure the fetus was relatively short in the Cretaceous eutherian mammals. Thus isoantigens could exist in these mammals not only among the molecular structures which differentiate late in fetal development (as the gamma globulin isoantigens do in modern mammals; see Oudin, 1960; Dray and Young, 1960), but also (though to a lesser extent) among the molecular structures which differentiate early. Because there was still considerable genetic plasticity in the average populations of these basal eutherian mammals, speciation could cause extensive genetic divergencies among the radiating lineages and a rapid evolution of the structural specificities of the proteins.

The lineages which retained an epitheliochorial placenta would not be as subject to the selective action of the maternal immunological system as the lineages which evolved the hemochorial placenta with its intimate apposition of fetal and maternal blood streams. As lineages of the former type radiated into new ecological habitats they could in time acquire specializations for these habitats through the evolution of all sections of the ancestral genetic code, not just the ontogenetically late acting genes. The various molecular adaptations ("inward-directed" ones as well as "outward-directed" ones) would be selected in relation to the special features of the external environment of each lineage. A different picture would be presented by the lineages with the hemochorial placenta. In these lineages fetal proteins would have become more accessible to the maternal immunological system through the more intimate placental arrangement. If there was then a prolongation of gestation, maternal immunizations would markedly slow down the rate of divergence from the ancestral genetic code, particularly in the case of the ontogenetically early acting genes.

This immunological view of the evolution of placentation in the mammals supports the standpoint of Hill (1932) and Le Gros Clark (1959) that the hemochorial placenta is an advanced eutherian character. However, it has also been argued (Wislocki, 1929) that the hemochorial placenta is a primitive eutherian character. This argument is based on the distribution of the hemochorial placenta in contemporary eutherian orders. It is found more frequently in the orders of small placental mammals such as the insectivores (the "primitive" or lower representatives of the Eutheria) than in the orders of large placental mammals such as the ungulates. (The major exception to this pattern of distribution

is the uniform presence of highly developed hemochorial arrangements in the catarrhine Primates.) There is no a priori reason to consider the presence of hemochorial placentas in many small eutherian mammals as a primitive feature of these mammals. Rather the presence of such placentas could conceivably be an example of an advanced feature which was acquired independently in many separately evolving phyletic groups at later stages of evolution. There is no rule which requires the contemporary small mammals to be primitive in all respects. The crucial question is whether the hemochorial placenta had developed in the basal stock of the Eutheria before the extensive radiation of this stock into the various orders of the placental mammals.

The immunological considerations stated in previous paragraphs provide an a priori basis for arguing that the hemochorial placenta was not present in the early stages of eutherian cladogenesis. The reduction of genetic plasticity in the basal eutherian-population by maternal immunizations would have been minimal in the absence of an intimate connection of fetal tissues to the maternal circulation. Conversely the genetic plasticity of this population would have been reduced by changes in the conditions of gestation. Thus the development of a placenta of the hemochorial type which increases the opportunity for immune responses to the fetus would have reduced genetic heterogeneity in the population. It is reasonable to assume that the formation of diverse orders of placental mammals resulted from the cladogenesis of a basal population with a relatively high degree of genetic plasticity rather than from one with a low degree. Therefore, a placenta of the epitheliochorial type which minimizes maternal immune responses to the fetus would have best fitted the conditions required for the first major adaptive radiations of the class Mammalia.

The immunological considerations also offer an explanation for why the hemochorial placenta is more frequently found in small rather than large placental mammals. Let us assume that in mammalian evolution there was a general trend toward an elaboration of placental mechanisms which secure optimal metabolic conditions for the developing fetus and that in the absence of maternal isoimmunizations the logical outcome of the trend was the elaboration of the hemochorial placenta. However, in the presence of maternal isoimmunizations, the trend toward the hemochorial placenta could progress more rapidly in small mammals with short gestation periods than in large mammals with long gestation periods, for the immune response of a small mammal to an antigenic challenge is no more rapid than that of a large mammal. Thus the hemochorial placenta in a small mammal with a brief gestation period would provide less opportunity for maternal immune attacks on the fetus than this same type of placenta in a large mammal with a lengthy gestation period.

These immunological considerations suggest that significant strides toward the development of a hemochorial placenta occurred in the line of human ancestry during the early stages of primate phylogeny when man's prosimian ancestors still had a relatively short period of gestation. It is not necessary to assume that these strides were equally strong in all the prosimian lineages, particularly if the adaptive radiations of the early Primates commenced shortly after the emergence of the basal members of the order. Thus the epitheliochorial placenta of the lemurs can be considered a retained primitive character rather than a regression or specialization away from a supposed ancient type of hemochorial placenta. The lemurs, however, do show advances in the mechanisms of placentation which foreshadow the advances found in higher Primates. For example, the chorion of the lemur placenta becomes vascularized relatively early in fetal development (Le Gros Clark, 1959).

Another important trend in primate evolution has been the

progressive expansion of the brain. There is reason to suspect that this trend depended on the one which secured the fetus optimal conditions for its development. In contrast to the paleoencephalon, the neocortex is easily injured by oxygen deprivation (Himwich, 1952) and a rich supply of nutrients and oxygen is required during fetal life for the differentiation of this phylogenetically new part of the brain. There is perhaps a causal relationship between the rapidity by which the early embryo of man establishes an intimate connection with the maternal circulation and the remarkably expanded brain which develops in man. Although the evolution of the neocortex depended on an evolution of more efficient placental mechanisms, the acquisition of these mechanisms would not by itself guarantee that cerebral evolution would occur. Such evolution would then depend on whether the organisms in their particular environmental habitat would gain some selective advantage by acquiring bigger brains.

The arboreal environment of the primitive Primates, while favoring the preservation of such generalized anatomical features as pentadactyl limbs, selected for visual acuity and for an expansion of the higher cerebral centers controlling the motor responses to visual and tactile impressions. However, as already noted, the evolution of a more complex brain could not take place until the functional efficiency of placentation had increased. Thus we can postulate that the lineages responsible for the phylogenetic advances of the Primates were those in which an elaboration of the hemochorial placenta with its intimate apposition of fetal and maternal blood streams occurred comparatively early in the evolution of the order. The evolution of our ancestors must have then been shaped by the following immunological conditions.

Since fetal proteins could more readily contact the maternal immunological system through the more intimate placental arrangement, isoimmunizations could decrease the incidence of genetic heterozygosity and increase the homozygosity of the most commonly occurring genes. This would preserve (along with the selective action of the arboreal environment) the primitive generalized features of the Primates. However, mutant genes could alter the specificities of proteins which were adequately isolated from the maternal lymphoid cells, for such mutant genes would have escaped the selective pressure of maternal isoimmunizations. The proteins of the central nervous system depend on genes which fall in this category, as do also all proteins which do not appear before birth. Therefore, the mutant genes which act only in the postnatal organism have the selective advantage of not provoking the maternal immunological system. On the other hand a delay in the appearance of adult proteins would increase the helplessness of the offspring. Genetic codes which retarded the maturation of the new born could only be selected if there were corresponding increases in the protective care of the young by the mother or other adults. The Primates with their expanding cerebral cortex and their increasing ability for adaptive psychological responses were uniquely suited for this development.

The phylogenetic advances of the Primates which culminated in man can be considered progressive because at each stage they increased the autonomy of the individual from the external environment and gave the individual the capacity to master a larger number of external challenges. This growing capacity to master new conditions depended of course on the progressive expansion of the higher centers of the brain. The basal Primates emerging from the Cretaceous eutherian radiation appear to have had in their arboreal environment more stimulus than the basal members of the other eutherian orders for the evolutionary development of the cerebral cortex. The early members of the Hominoidea were probably still at the prosimian grade of phylogenetic development in their skeletal structure. Thus at this stage of

primate evolution an arboreal environment continued to offer the hominoids the best stimulus for the elaboration of the brain. Somewhat later in the phylogenesis of the Hominoidea, after larger sized organisms had evolved, an ecological domain which was both arboreal and terrestrial offered a greater stimulus than a strictly arboreal environment for further cerebral development. Those hominoids who entered and successfully occupied this broadened ecological domain can be considered the basal members of the Hominidae. Later yet in phylogenesis, the most progressive of the Primates were the bipedal hominids who inhabited a predominantly terrestrial domain and moved over vast geographical territories. Again the broadening of the environment favored the further elaboration of the brain.

We can now recapitulate and stress the dynamics of the process depicted by our immunological theory of primate evolution. Early in the anagenesis of the Primates the hemochorial placenta developed, providing the metabolic basis for a continuing progressive elaboration of the brain. In turn this cerebral evolution, which correlated with increasing body size, allowed the Primates to invade more varied ecological domains. This led to two opposing tendencies. On the one hand the more varied conditions of a broader environmental zone selected for mutant genes and the state of genetic heterozygosity. On the other hand, as the barrier between embryonic tissues and maternal blood decreased and as the period of gestation lengthened, the maternal immunological system selected for the state of genetic homozygosity. An equilibrium was established between these contradictory tendencies by a selection of genetic codes which delayed maturation until the postnatal phase of ontogeny. The somatic systems which establish the basic features of the endogenous environment still matured before birth and were increasingly controlled by genes with a high degree of homozygosity. The somatic systems immature at birth were those with the largest concentration of "outward-directed" molecular adaptations. The expression of genetic heterozygosity then began to center in these late maturing structures. Furthermore, this process necessitated the evolution of family and social life by increasing the immaturity and helplessness of the offspring. It is not by chance that man who shows the greatest cerebral development and occupies the broadest environmental range is also the most retarded in his maturation.

Another aspect of the process of primate evolution which can be characterized by our immunological theory concerns the rate of molecular evolution in different primate lineages. The theory requires the rate of diversification of genetic codes by phylogenetic branching to decrease after each anagenetic advance. The prosimians would show the greatest rate of diversification and the most extensive divergencies in the protein specificities of the radiating lineages. There would also be marked differences among these lineages in their continuing rates of evolution. Certain groups such as the Lemuroidea who retained the epitheliochorial placenta would continue to show rapid rates of evolution and in the sanctuary of Madagascar a striking diversification of genetic codes at a fairly late stage of the evolutionary history of the Primates. In contrast, the Tupaioidea who were restricted to a narrow ecological niche, resembling the one most frequented by the Cretaceous proto-primates, must have had their genetic plasticity drastically reduced after they acquired the hemochorial placenta. This then placed a brake on the subsequent evolution of their genetic codes.

It is consistent with our theory to seek the origins of the Anthropoidea in that group of generalized tarsioids of the Paleocene in which the evolutionary advances towards the hemochorial placenta were progressing rapidly. The Platyrrhini and Catarrhini could then have independently evolved from prosimian ancestors along roughly parallel lines. They would retain despite their ancient separation a fair amount of correspondence in many sections of their genetic codes, since divergence in fetal proteins would have encountered adverse selective pressure from maternal immunizations. Due perhaps to the isolation of the South American continent during most of the Tertiary epoch, the process of anagensis was less marked in the platyrrhine Primates or ceboids than in the Old World or catarrhine members of the Anthropoidea, the cercopithecoids and hominoids. According to Hill (1932) the platyrrhine placenta only attains structural efficiency at a late period of gestation. Thus, compared to the cercopithecoids and hominoids, the ceboids show the least fetal-maternal vascular intimacy during gestation. In this situation the ceboid radiations would have diversified genetic codes at a somewhat faster rate than either the cercopithecoid or hominoid radiations.

It would accord with the process of primate evolution depicted by the theory if the basal group of the Hominoidea, as distinct from that of the Cercopithecoidea, was that catarrhine lineage which, compared to the others, showed the largest degree of fetal-maternal vascular intimacy during gestation. This phyletic group would then have shown a relatively low degree of genetic plasticity. Consequently, during the radiations of the Hominoidea only small sections of the genetic codes would be diversified, and the serological specificities of the proteins of the separately evolving lineages would retain a high degree of similarity. A stage of the process would be reached with the emergence of the bipedal hominids where genetic plasticity was so reduced that the most advanced lineage (the one with the largest brain) would retain its reproductive cohesiveness and fail to speciate even though its population expanded over the surface of the earth and occupied more geographical territory than had the members of any previous primate radiation. This last and greatest expansion of the Primates can be attributed to the basal members of our own species *Homo sapiens*.

The process depicted by our immunological theory must also be affecting the current biological evolution of our species. Erythroblastosis fetalis due to the Rh and other blood group incompatibilities is perhaps the classic example that the destructive side of the maternal immunological system is very much in evidence in the human population. It is also clear that each individual has in addition to the blood group substances a unique assortment of histocompatibility antigens in his tissues. Maternal isoimmunizations may prove to be a key factor in abortions and in the birth of premature and defective children. The process, then, of the weeding out of isoantigens is by no means over. Our thesis is also supported by the fact that some blood group substances and histocompatibility antigens do not show final maturation until the stage when the possibility of transplacental immunization has ended. We might expect even more pronounced shifts in the maturation of isoantigens to the postnatal phase of ontogeny if evolution can continue under a protective cover of an advanced technology in a society which protects its young.

However there are selective pressures apposed to further delays in maturation. These pressures emanate from the individual's own immunological capabilities. In the current stage of human evolution, the neonatal organism cannot long survive microbial invasion unless its own immunological system rapidly matures. However, if the mature immunological system is suddenly faced with "strange" endogenous molecular specificities for which it lacks tolerance, it will tend to respond to these specificities with destructive autoimmunizations [4]. Thus the threat from autoimmunizations as well as isoimmunizations has tended to further limit the evolutionary development of new molecular specificities to intracellular proteins so immobilized in stationary tissues as to normally be inaccessible to antibody producing cells. Nevertheless, the immunological threat still remains, for such immobilization cannot be absolute and tends to break down with a release of autoantigens when the tissue is under stress.

A good example is provided by the high incidence of autoimmunizations to human thyroid proteins in thyroid diseases (Roitt et al., 1958; Roitt and Doniach, 1960) and in old age (Goudie et al., 1959; Goodman et al., 1963). The autoantibodies to human thyroid proteins cross react with thyroid extracts from catarrhine Primates (reacting more strongly with chimpanzee than with rhesus monkey) but not significantly with the extracts from non-primate mammals. In other words, the antoantigenicity of human thyroid proteins resides primarily in structural configurations which were acquired during the more recent stages of evolution. Any protein is a potential autoantigen if it does not make contact with the cellular progenitors of the immunological system in the fetus. Thus the process of anagenesis in the Primates must have caused human tissues to become rich in autoantigens [5], for the evolution of molecular complexity created membranes such as those of the "blood-brain barrier" which effectively block the entrance of intracellular proteins into the blood stream.

SEROLOGICAL ANALYSIS OF PRIMATE PHYLOGENY

A comparative study of the serum proteins of the Primates is in progress (Goodman, 1960b, 1961, 1962a,b,c, Goodman et al., 1960; Goodman and Poulik, 1961a,b). The principal techniques are two-dimensional starch-gel electrophoresis (Poulik and Smithies, 1958) and agar-gel precipitin testing with a variety of antisera to proteins of different Primates. These methods provide data on the phylogenetic relationships of man and other Primates. In addition, the data of the precipitin method (Goodman, 1960b, 1962a) demonstrate that certain types of proteins have evolved more slowly than other types during the radiations of the Primates. Furthermore it can be shown, though only in an oblique way, that the rate of diversification of genetic codes has been greater in certain taxa than in others. Thus it is possible to test experimentally whether or not the trends in the molecular evolution of the Primates, predicated by the theory set forth in this chapter, have occurred.

Two-dimensional starch-gel electrophoresis (filter-paper electrophoresis in one dimension followed by starch-gel electrophoresis in the other) separates the proteins of serum into 19 to 25 components. The arrangement of these components in starch-gel provides a pattern which is characteristic of the species of the organism whose serum is analyzed. Such patterns are especially useful for determining if organisms show a phylogenetic relationship at a subfamily or generic level (Goodman et al., 1960; Goodman, 1961, 1962b,c). For example, *Perodicticus* and *Galago* (who are grouped together in the Lorisidae, but placed in separate subfamilies) have quite divergent patterns, whereas *Galago crassicaudates* and *Galago senegalensis* (two species of the same genus) have almost identical patterns.

When the patterns of members of the Hominoidea (man, gorilla, chimpanzee, gibbon and orangutan) are compared with each other (Figure 1), the differences among the various hominoid types are much larger than those found between macaques and baboons or between macaques and vervets (Figure 3). Indeed the various species of the subfamily Cercopithecinae examined by two-dimensional starch-gel electrophoresis (five species of *Macaca*, one of *Papio*, and one of *Cercopithecus*) show a high degree of similarity. Although the starch-gel pattern of each hominoid type diverges sharply from the others, there is a constellation of

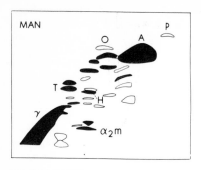

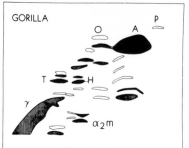

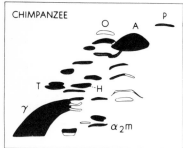

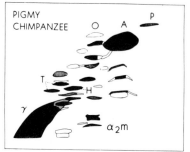

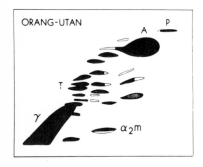

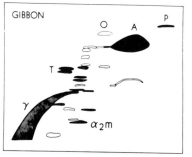

Figure 1   *Diagrams of two-dimensional starch-gel electrophoresis patterns of hominoid sera. P = prealbumin; A = albumin; O = orosomucoid; T = transferrin; H = haptoglobin; and γ = gamma globulin. Type 1 haptoglobin is represented in the human pattern. (Inherited variations of haptoglobin occur in the human population, but appear to be minimal in nonhuman primate populations. Transferrin shows an extensive polymorphism among chimpanzees and perhaps as indicated by our first gorilla results, among gorillas.)*

about ten faster migrating proteins in man, chimpanzee, and gorilla (Figure 2), but not in gibbon or orangutan. The faster migrating proteins of the gibbon and orangutan present patterns in each case which are quite dissimilar from those of the other hominoids.

Such data allow us to make educated guesses about some of the questions concerning the phylogenetic history of the Hominoidea. For example, the known hominoid fossils trace a gibbonoid line of apes back to the Oligocene epoch, but otherwise fail to establish the times of phyletic separation within the Hominoidea. However, there are fossil baboons and macaques dating from the Middle Pliocene (Le Gros Clark, 1959). Thus when the divergencies of the hominoid patterns are contrasted to the similarities of the cercopithecine patterns it can be deduced that the phyletic separations of the five major groups of living hominoids are at least as ancient as the Lower Pliocene (if not more ancient). Furthermore there is some suggestion from the divergencies of the hominoid patterns that the phyletic line which branched to give rise to gorilla, man, and chimpanzee did so after it had separated from more ancient lines leading to gibbon and orangutan.

The suggestion that *Gorilla*, *Homo*, and *Pan* form a phyletic unit within the Hominoidea is strongly supported by the results of the second set of experiments in which an agar-gel precipitin technique (Goodman, 1962a) was employed. This technique determines the degrees of correspondence in antigenic structure existing among the protein homologues of different species. To do so an antiserum is produced to a protein isolated from the

organisms of a particular species. Then cross reactions are developed in immunodiffusion plates between the antiserum and the counterparts (or homologues) of this protein in other species. In serological terminology the immunizing protein is called the homologous antigen and the counterparts of this protein in other species are called the heterologous antigens. Since antigenic specificity corresponds closely to genetic specificity, it turns out that the cross reaction of the heterologous antigen decreases as the evolutionary separation of the heterologous species from the homologous species increases.

In the studies reviewed here a variety of antisera were used for each particular protein or antigen. This ensured that the antigenic and species specific properties of many of the protein's surface configurations would affect the measurements. For example, if a rhesus monkey is immunized with a human protein, only a small proportion of the surface configurations of the protein will be antigenic, and these will tend to be the more species specific configurations. On the other hand, if a chicken is immunized, a large proportion of the surface configurations of the protein may be antigenic, and some of these (the configurations which have been stable over long periods of evolution) will show a low degree of species specificity. Chicken, monkey, and rabbit antisera to purified human serum proteins were used to develop cross reactions with heterologous antigens from a number of different Primates (Goodman, 1962a). These cross reactions consistently demonstrated in accord with the precipitin studies of other investigators (Nuttall, 1904; Mollison, references in Kramp,

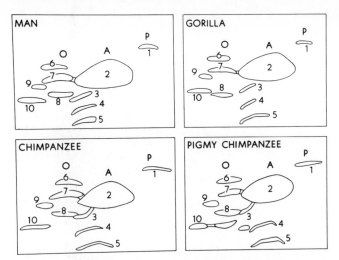

Figure 2 *Outline diagram of ten faster migrating components in two-dimensional starch-gel electrophoresis which show similar but not identical positions in man, gorilla, chimpanzee, and pigmy chimpanzee. Note that in gorilla component 8 is to the right and beneath component 10; whereas in man component 8 is to the right and above component 10, but the separation between 9 and 10 seems to be greater in chimpanzee than in man or gorilla. Note the extra unnumbered components between 10 and 4 in the pigmy chimpanzee.*

1956; Wolfe, 1933, 1939; Boyden, 1958; Paluska and Korinek, 1960; Williams and Wemyss, 1961) the close phylogenetic relationship of man to anthropoid apes, his decreasing relationship first to cercopithecoids, then to ceboids, and his distant relationship to prosimians. Chimpanzee and gibbon were the first anthropoid apes to be tested (Goodman, 1962a), and chimpanzee developed the larger cross reactions.

During the past year Drs. Arthur J. Riopelle and Harold P. Klinger generously contributed for the study serum and plasma samples from several gorillas and orangutans. Many more precipitin tests have now been carried out. A large body of data (Goodman, 1962b,c) demonstrate that man is more closely related to the African apes (chimpanzee and gorilla) than to the Asiatic apes (orangutan and gibbon). For example, hominoid albumins were compared with rhesus monkey antisera to human albumin. Gorilla albumin was quite similar to human albumin; chimpanzee albumin diverged slightly from the human protein, and gibbon and orangutan albumins clearly diverged from human albumin as well as from each other. The orangutan albumin appears to be the most divergent of the hominoid albumins. The rhesus monkey antisera also reacted with another protein (Goodman, 1962a,b) in addition to albumin. This additional reaction was afforded by the serum of man, chimpanzee, and gorilla, but not by that of gibbon and orangutan. With a capuchin monkey antiserum to human serum, orangutan and gibbon showed moderate divergencies from man, whereas gorilla and chimpanzee showed only slight divergencies from man, that of the gorilla being the smallest.

With chicken antisera to human ceruloplasmin (the copper binding protein of plasma) and rabbit and chicken antisera to human transferrin (the iron binding protein of plasma), chimpanzee and gorilla appeared identical with man, whereas orangutan and gibbon diverged from man. The gibbon transferrin appears to be the most divergent of the hominoid transferrins. With a rabbit antiserum to human alpha$_2$ macroglobulin, only the gorilla

appeared identical with man; chimpanzee, gibbon, and orangutan showed small divergencies. On the other hand, with rabbit and chicken antisera to human gamma globulin, only the chimpanzee was very similar to man; gorilla, orangutan, and gibbon, respectively, were increasingly divergent. These data, therefore, not only show that the Asiatic apes are more distant from man than are the African apes, but also that all the anthropoid apes have diverged from each other.

Evidence that the chimpanzee has more recent common ancestry with man and gorilla than with orangutan or gibbon is furnished by the cross reactions of antisera to chimpanzee serum. The antisera were produced in chickens, a spider monkey, and a woolly monkey. With these antisera orangutan and gibbon showed greater divergencies from chimpanzee than did gorilla or man. Examples of the cross reactions developed by the woolly monkey antiserum are presented in Figure 4. Chicken antisera have also been produced to gorilla serum. The cross reactions of these antisera do not show any divergence of chimpanzee or man from gorilla but do show a divergence of orangutan and gibbon, the gibbon in this case showing the most divergence. It may be mentioned that Zuckerkandl et al. (1960) found that the primary structure of adult hemoglobin is very similar in man, chimpanzee, and gorilla, but somewhat divergent in orangutan. This finding provides additional evidence for the close phyletic relationship of man, chimpanzee, and gorilla within the Hominoidea.

All these serological findings argue for a revision of the taxonomy of the Hominoidea since in the classifications now in use *Gorilla* and *Pan* are invariably placed with *Pongo* rather than with *Homo*. A broadening of the Hominidae to include *Gorilla* and *Pan* as well as *Homo* would reflect more closely the cladistic and genetic relationships suggested by the serological data. *Hylobates* would be in the Hylobatidae and *Pongo* would remain in the Pongidae.

Placing Hylobatidae, Pongidae, and Hominidae together in the Hominoidea is supported by the cross reactions of a chicken antiserum to gibbon serum and another to orangutan serum. With these antisera, man and each of the apes yielded larger cross reactions than cercopithecoids (stump tail macaque—a member of the Cercopithecinae, and langur—a member of the Colobinae, were tested). Using chicken antisera to macaque serum, cross reactions were stronger with the langur than with the various apes and man. This supports the established primate taxonomies which separate the cercopithecines from hominoids and group them with colobines in the Cercopithecoidea. With anti-macaque sera (see also Paluska and Korinek, 1960) ceboids do not show as much relationship to cercopithecoids as do hominoids. However, cercopithecoids share some antigenic configurations with ceboids that hominoids lack. Similarly, hominoids share configurations with ceboids that cercopithecoids lack (Goodman, 1962a,c). This type of data suggests that shortly after the dual branches of Platyrrhini and Catarrhini had formed within the Anthropoidea the catarrhine Primates branched into the Hominoidea and Cercopithecoidea. Finally, the cross reactions of various chicken anti-ceboid sera (Goodman 1962b,c) show that the suborder Anthropoidea is a valid phylogenetic taxon. These cross reactions demonstrate that the ceboids are more closely related to the catarrhine Primates (cercopithecoids and hominoids) than to the lorisiform or lemuriform lower Primates. Thus a systematics of the higher Primates derived solely from serological data would not be much different from the established systematics which was largely derived from morphological data. A correspondence of the serological and morphological approaches is to be expected since the building blocks of morphological structures are proteins. Even the revision of hominoid taxonomy suggested by the serological data, that of grouping *Pan* and *Gorilla* with *Homo*

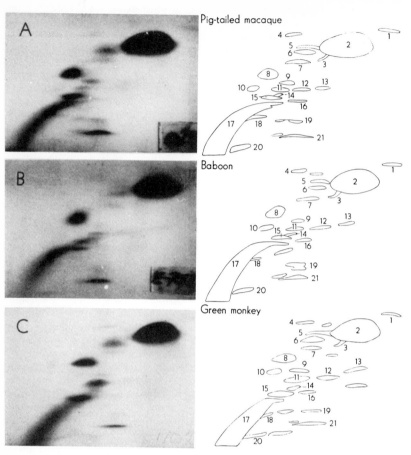

Figure 3 *Photographs and matching outline diagrams of twodimensional starch-gel electrophoresis patterns of Cercopithecine sera. A pig-tailed macaque (*Macaca nemestrina*), a baboon (*Papio*) and a green monkey (*Cercopithecus aethiops*) are compared. Note that a constellation of 21 components can be identified in each pattern.*

rather than with *Pongo*, is perhaps anticipated by morphological data. Indeed, Darwin (1871) only had the evidence of morphology when he suggested that our closest relatives were the gorilla and chimpanzee and that "it is somewhat more probable that our early progenitors lived on the African continent than elsewhere."

The close phyletic relationship of gorilla and chimpanzee to man that the serological data demonstrate agrees with the concept developed in a preceding section of this chapter, that the basal members of the Hominidae were the progressive hominoids who entered and successfully occupied a widening ecological domain, one which was terrestrial as well as arboreal. The gorilla and chimpanzee, who continue to occupy a terrestrial-arboreal domain, could readily have descended along with man from such a basal hominid. In turn the gibbon and orangutan, who are almost solely arboreal in their mode of life, probably trace back as separate branches (as may be deduced from the observed difference in their serum proteins) to earlier and more primitive members of the Hominoidea.

The fact that the gibbons can be traced as a separate branch of the Hominoidea back into the Oligocene, 30–40 million years ago, highlights the significance of the finding (Goodman, 1962a,b,c) of an extensive correspondence in antigenic structure among hominoid albumins. Using chicken and rabbit anti-human albumin sera, human, gorilla, chimpanzee, and gibbon albumins appeared to be identical in the immunodiffusion plates and orangutan albumin was only slightly divergent from the others. Thus the antigenic structure of albumin has been remarkably stable during the long course of hominoid evolution. Furthermore the chicken and rabbit anti-human albumin sera developed sizeable cross reactions with cercopithecoid and ceboid albumins. (Among the Primates only lemurid, lorisid, and tupaiid albumins developed small cross reactions.) In contrast to the albumin results, the cross reactions of rabbit and chicken anti-human gamma globulin sera (Goodman 1962a,b,c) revealed distinct antigenic differences among hominoid gamma globulins and very little correspondence between the catarrhine and platyrrhine Primates, with respect to gamma globulin. We can conclude that during the development of the suborder Anthropoidea albumin evolution has not been nearly as marked as gamma globulin evolution. The data also demonstrate that within this suborder the hominoids show the least albumin evolution and the ceboids the most (Goodman, 1962a,b,c). Although anti-human albumin sera detected almost no diversification of albumin within the Hominoidea, the anti-ceboid sera detected considerable diversification of albumin within the Cebidae.

These findings can be explained by the immunological theory of primate evolution developed in this chapter. The theory proposes that during the progressive evolution of the Primates, as the placental structure came to permit more interchange between fetal and maternal circulations, maternal isoimmunizations decreased the heterozygosity of prenatally acting genes and thereby increasing the evolutionary stability of proteins such as albumin which are synthesized early in fetal life. These isoimmunizations also increased the selective advantage of genetic codes which delayed maturation in ontogeny. Thus proteins which are not synthesized until after birth such as gamma globulin could evolve at a relatively rapid rate, since divergence in such proteins would not be selected against by maternal immunizations. The evolutionary stability of albumin would be greater in the Hominoidea than in the Ceboidea, since fetal-maternal vascular intimacy (and thus opportunity for selection against mutant genes by maternal immunizations) is greater in the hominoids than in the ceboids.

We can not prove that the greater diversification of albumin in the Ceboidea as compared to the Hominoidea resulted from a more rapid evolution of ceboid albumins, since the times of phyletic branching within the Ceboidea are unknown. Possibly many of the ceboid branches are more ancient than the hominoid

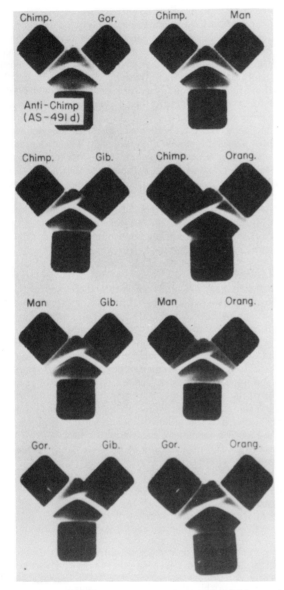

Figure 4 *Photographs of cross reactions in type IV immunodiffusion plates developed by a woolly monkey antiserum (AS-491d) to chimpanzee serum. The antiserum was added in 0.1 ml amounts and the antigens (undilute sera) were added in either 0.2 ml or 0.1 ml amounts. Chimpanzee developed a weak spur against gorilla, a very weak spur against man, and strong spurs against gibbon and orangutan. Man and gorilla each developed moderate spurs against gibbon and orangutan. A protein develops a spur against its counterpart (or homologue) in another serum if it contains antigenic configurations to which antibodies are directed that its counterpart lacks. Thus the human and gorilla homologues of the principle protein of chimpanzee serum to which the antibodies were directed developed larger cross reactions than the gibbon and orangutan counterparts of this protein. (For an explanation of cross reactions in immunodiffusion plates see Goodman, 1962a.)*

branches. However we can test our theory by examining other taxa. For example, the fossil record suggests that the family Bovidae first emerged during the Miocene (Simpson, 1949) after extensive branching of the Hominoidea had already occurred. Yet by the immunodiffusion plate technique chicken anti-beef albumin sera could detect divergencies even within the subfamily Bovinae, kuda albumin diverging from beef albumin (Goodman, 1962b), whereas the anti-human albumin sera failed to show any comparable divergencies among hominoid albumins. Clearly, then, albumin evolution has been more rapid in the subfamily Bovinae then in the superfamily Hominoidea. This type of finding could be predicted by our theory, for the artiodactyls have an epitheliochorial placenta which, as previously noted, minimizes the possibility of transplacental immunizations.

The serological reactions of lower Primates and insectivores also provide data upon which a theory of human evolution can be built. The data pertain to the antigenic (and thus genetic) correspondence of these organisms to man as judged by the cross reactions of antisera to purified human serum proteins. It is best to consider these data after briefly reviewing some unpublished results obtained with chicken antisera produced to the serum of lemur, potto, hedgehog, and tree shrew. It was found with anti-lemur and anti-potto sera that marked antigenic differences separate the lorisids (galago, loris, and potto) from the lemurids (lemur). However each group showed more antigenic correspondence to the other than to either the tupaiids or to members of the Anthropoidea. Thus the lorisiform prosimians and the Malagasy prosimians appear not to have separated from each other until sometime after their common ancestor had separated from the forerunners of the Anthropoidea. Nor could any special phylogenetic relationship between the Tupaioidea and Lemuroidea be discerned. With anti-hedgehog and anti-tree shrew sera, hedgehog and tree shrew appeared to show more correspondence to each other than to any of the other mammals whose sera were tested (tenrec, rat, rabbit, dog, horse, beef, elephant, man, lemur, galago, and kangaroo). But even this apparent relationship between tree shrew and hedgehog must be a very distant one, since in each case the heterologous cross reaction was many times less than the homologous. Possibly the separation of the Tupaioidea from other mammalian groups dates back to the first great radiations of the primitive eutherian mammals in the late Cretaceous epoch.

We can now consider the cross reactions of the lower Primates and non-primate mammals with the antisera to human serum proteins. When the antisera were to gamma globulin and to alpha$_2$ macroglobulin, the Strepsirhini (lemur and galago) developed larger cross reactions than the tree shrew and the non-primate mammals. However when the antisera were to albumin (Goodman, 1962a,b,c) the tree shrew cross reactions were comparable (perhaps slightly larger) than the lorisid (galago and potto) cross reactions and clearly larger than the cross reactions of lemuroids (*Lemur* and *Propithecus*) and the other mammals tested (hedgehog, tenrec, rat, rabbit, elephant, [6] dog, horse, beef, and kangaroo). These data in confirming morphological evidence that the tree shrew has strong affinities with the Primates again suggest that maternal isoimmunizations played a role in the molecular evolution of the Primates. Such immunization could account for human albumin corresponding more to tree shrew albumin than to lemur albumin even though man appears to have a somewhat longer period of common ancestry with lemur than with tree shrew. Albumin evolution would not have been slowed in the Lemuroidea as much as in the Tupaioidea by maternal isoimmunizations, since the lemurs retained during their phylogenesis an epitheliochorial placenta while the tree shrews acquired a fairly advanced hemochorial placenta (Le Gros Clark, 1959). The serological correspondence between man and the tree shrew would seem to be due to the retention in each of very ancient or primitive genetic patterns. However the full significance of the relationship between man and tree shrew will not emerge until serological comparisons are also carried out with such groups as elephant shrews and common shrews.

SUMMARY AND CONCLUSIONS

1. Proteins play a central role in the evolutionary process. Each change in the amino acid sequence of a protein can be related to a genic mutation. In turn a change in amino acid sequence may affect the biological functioning of a protein and thereby influence the survival chances of the organism carrying the mutant gene. The biological functioning of a protein depends on the stereochemical properties of its surface configurations. Only a few surface configurations were responsible for the biological functioning of a typical soluble protein in an undifferentiated unicellular organism at primitive stages of evolution.

In contrast, many surface configurations affected the biological functioning of a typical soluble protein in a higher metazoan at advanced stages of evolution. As in the primitive organism there were the surface configurations which constituted the active site responsible for the primary functional activity of the protein. In addition there were the ancillary configurations which enabled the protein to preferentially cross particular membrane barriers, and there were also the remaining surface configurations, so contoured as to not specifically interact with any of the surface configurations of the large number of other serologically distinct proteins. As progressive evolution increased the biochemical complexity and molecular differentiation of organisms, the possibility also increased that any single mutant gene would be detrimental to the organism carrying it. This line of reasoning suggests that the process of speciation caused a much deeper divergence of genetic codes and a much more rapid evolution of the structural specificity of proteins among lower organisms at the primitive stages of phylogenesis than among higher organisms at advanced stages of phylogenesis.

2. At the phylogenetic stage of the placental mammals a new factor, maternal immunizations to fetal antigens, further restricted the divergence of genetic codes during the process of speciation. Proteins synthesized early in fetal life such as albumin showed a decreasing rate of evolution. But proteins not synthesized until after birth such as gamma globulin still evolved at a relatively rapid rate, since in this case divergence was not selected against by maternal immunizations. It is proposed that such an immunological process operated with particular force in the phyletic line leading to man. It was in this line that cerebral evolution progressed most rapidly. But the progressive expansion of the neocortex could only occur on the basis of a placental evolution which increased fetal-maternal vascular intimacy, allowing the developing fetus to obtain a rich supply of nutrients and oxygen.

As cerebral evolution progressed and body size increased, the ancestors of man invaded broader ecological domains. The new and more varied conditions of the external environment selected for a state of genetic heterozygosity. As the barrier between embryonic tissues and maternal blood decreased and as the period of gestation lengthened, maternal immunizations selected for a state of genetic homozygosity. This contradiction was resolved by a selection of genetic codes which increasingly shifted to the postnatal phase of ontogeny the maturation of somatic systems concerned with adaptive responses to the external environment, since maternal immunizations would not select against gene-determined polymorphisms in such late maturing systems. But the delay in the appearance of adult proteins increased the helplessness of the young and thereby spurred the evolution of family and social life.

3. A comparative study of the serum proteins of the Primates provides evidence for the theory of man's evolution set forth in this paper. The data show that there was a marked deceleration of albumin evolution, but not of gamma globulin evolution, during the phylogenetic advances of the Primates. This deceleration can be correlated with a placental evolution that increased the opportunity for maternal immune attacks on the fetus. Man's distant but nevertheless striking serological relationship to the tree shrew is further evidence that sections of the human genotype are extremely ancient, dating back to primitive members of the Eutheria.

The serological data provide information on the taxonomic relationships of man and other Primates. The dendrogram presented in Figure 5 summarizes the author's judgment of these relationships. A full presentation of the data concerning hominoid relationships appears in *Human Biology*, 35:377–436.

Notes

[1] This study is being supported by grant G14152 of the National Science Foundation. Certain phases of it have also received support from grant MY-2476 of the National Institute of Mental Health and grant 226 of the National Multiple Sclerosis Society.

[2] See, e.g., Chantrenne (1961) and Platt (1962) who describe the gene-environmental interactions of ontogenetic development in terms of newer concepts and analogies of biochemical genetics.

[3] See Hasek et al. (1961) and Smith (1961) for a review of facts and theories on the mechanisms of immunological tolerance.

[4] Even if immunological unresponsiveness (or paralysis) developed the danger would exist that the paralysis would impair the general functioning of the immunological system and thus make the organism more susceptible to infections. In this connection Liacopoulos et al. (1962) found that when large doses of bovine albumin were administered to guinea pigs the immune unresponsiveness which developed to bovine albumin spread to affect unrelated protein antigens.

[5] If our thesis is correct that the tissues of man are rich in autoantigens, it follows that the production of autoantibodies may play a key role in the degenerative diseases and in the process of senescence. The very extensive morphological and molecular differentiation of the mammalian organism prevents development of immunological tolerance to many intracellular molecular specificities. Thus the suggestion (Goodman et al., 1963) that autoimmune reactions cause a gradual destruction of the viable cells of the organism amounts to a version of the morphogenetic theories of mammalian senescence (Comfort, 1956). Our particular version is quite sweeping in the role that it attributes to immunological phenomena, the increase in tissue autoantigens being related to the threat to the mammalian fetus from maternal isoimmunizations.

Conceivably the science of immunobiology will be able in the not too distant future to offer a program for slowing down the rate of senescence in man. Immunobiology will first have to assess the possibilities for a molecular fetalization of our species, one which either reduces the level of autoantigens in tissues or instead decreases sensitivity to such substances by further delaying the maturation (or permanently retarding the ontogenetic development) of the immunological system. It will then be necessary to determine a means of directing such a fetalization to ensure a progressive rather than degenerate evolution of our species.

[6] It is also of interest that of the non-primate mammals tested with chicken anti-human albumin sera to the elephant developed the largest cross reactions (Goodman, 1962a,b). This suggests that a marked deceleration of albumin evolution occurred in elephant phylogeny as in human phylogeny. In view of the elephant's exceptionally long period of gestation and the hemochorial arrangement to the zonary portion of his placenta (Mossman, 1937), maternal immune attacks on the fetus could have operated in elephant phylogeny (as in human phylogeny) to stabilize sections of the genetic code.

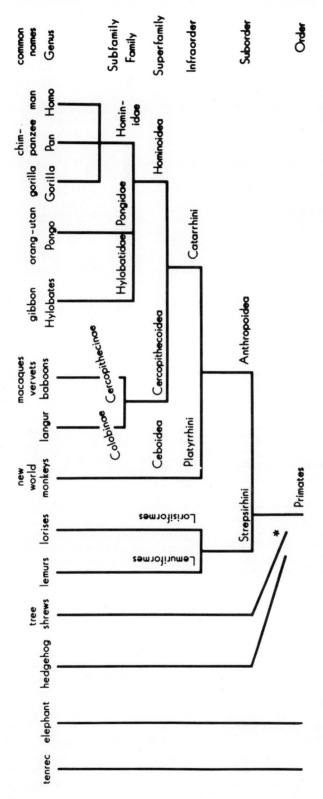

Figure 5   *A classification from serological reactions. The line leading to elephant is interposed between hedgehog and tenrec to demonstrate that even though hedgehog and tenrec are grouped in the Insectivora by the established taxonomies of mammals no special serological relationship exists between them. (*Although serologically the tree shrews show some affinities with the Primates, they also show some with hedgehogs. Additional serological data may help identify an assemblage in which the tree shrew can be placed.)*

40

A Molecular Approach to the Question of Human Origins

V. SARICH

A little more than 100 years ago Thomas Henry Huxley wrote that

> whatever system of organs be studied, the comparison of their modifications in the ape series leads to one and the same result—that the structural differences which separate Man from Gorilla and Chimpanzee are not so great as those which separate the Gorilla from the lower apes. [Ed. note: by "lower apes" Huxley meant Old World monkeys.]

Time has served to vindicate this early view, even among the general public, for it is the African apes (chimpanzee and gorilla) with whom man most closely identifies upon his visits to zoos. Their basic anatomy and behavior are so strikingly reminiscent of our own that it is difficult to avoid a feeling of close kinship. The translation of this empathy into an evolutionary context explains the similarities as being characters retained from a common ancestor, and the differences as having developed since the time that the lines leading to ourselves and the African apes began their independent development. The nature of this gap, however, Huxley saw as a very important issue:

> It would be no less wrong than absurd to deny the existence of this chasm, but it is at least equally wrong and absurd to exaggerate its magnitude, and, resting on the admitted fact of its existence, refuse to inquire whether it is wide or narrow.

Now, in spite of the vast effort that has been devoted to the study of human evolution since Huxley's time, one can still find no measure of agreement on the size of this chasm. Although most anthropologists generally regard the gorilla and chimpanzee as man's closest living relatives, estimates of the time of divergence of the human lineage from that (or those) leading to the African apes vary widely. In recent years this time has been given to be as short as two to four million years by Washburn in 1965 and as long as twenty-five to thirty million years independently by Leakey, Simons, and Pilbeam in 1967. These disagreements among people looking at the same data are due in part to the fragmentary nature of the fossil record that consists mainly of teeth and jaws, and in part to the failure of traditional comparative anatomy to develop methods that would lead to agreement, even among anatomists, as to the evolutionary meaning of such data. Even more extreme opinions have been ventured, but this tenfold range of time has to be narrowed appreciably before one can begin any serious investigation of human origins and evolution. A beginning for the human line some three million years ago would almost necessitate an ancestor not unlike the modern pygmy chimpanzee (*Pan paniscus*), whereas a beginning thirty million years ago would require a derivation from an ancestor that had barely reached the monkey grade of primate evolution. The australopithecines do take the hominid line back to about three million years but the hominid and pongid fossil record beyond this point is a most unsuitable one. The basic question then becomes one of whether the gap between data points is to be no more than one or two million years or about twenty-five million years.

Looking at this situation in early 1965, Allan Wilson (of the Biochemistry Department, University of California, Berkeley) and I decided that the continued analysis of anatomical characters did not appear likely to soon resolve the problem convincingly. As its solution is basic to the whole study of human evolution, we undertook to look more closely at the proteins of man and his primate relatives to see if perhaps biochemistry might succeed in answering Huxley's 100 year old question. Though, as will be seen later, our answer (that man and the African apes derive from a common ancestral species still living in the late Pliocene five million years ago) has not yet proved acceptable to major segments of the anthropological and paleontological community, it is difficult to fault in terms of its own logic. In addition, as will be seen, it has had the far more important virtue of being consistent with the whole body of other molecular evidence obtained subsequent to our original proposal (see Sarich and Wilson, 1967a). Finally, this recent date in the late Pliocene can enormously simplify the study of our own evolution.

One should not, however, get the impression that the idea of obtaining evolutionary information from molecules rather than muscles, bones, or teeth was the creation of Sarich and Wilson—it goes back nearly seventy years to an English parasitologist named Nuttall, who wrote in 1902 after a pioneering study in the then nascent science of immunology:

> I do not wish these numbers to be taken as final, nevertheless they show the essential correctness of the previous crude results. To obtain a constant it will be necessary to make repeated tests with the blood of each species with different dilutions and different proportions of antiserum. I am inclined to believe that with care we shall perhaps be able to "measure species" with this method, for it appears that there are measurable differences in the reactions obtained with related blood, in other words, determine degrees of relationship which we may be able to formulate (Nuttall, 1902, p. 825–27).

The basis of the molecular approach is, of course, the fact that as organisms evolve, so does their constituent genetic material (DNA) and its functional product (proteins). The peculiar advantage lies in the commonality of DNA and proteins structures in all organisms; i.e., the units (four organic bases for DNA and twenty amino acids for proteins) are identical and it is only the particular sequences in which they are arranged that vary from species to species. Since the units are identical, then, the extent of evolutionary change between a pair of related species is readily quantifiable in a manner common to all extant forms of life.

This evolutionary process is illustrated in Table 1 by a single DNA codon, say GCT (guanine-cytosine-thymine), that instructs the protein synthesizing machinery of the cell to incorporate the amino acid arginine into a developing protein that may have several hundred amino acids in its complete structure, each one of which will be coded for by a sequence of three bases in the DNA. Evolutionary change develops when a mutation occurs and survives the crucible of the selective process (that is, when it is found to be either neutral or advantageous). For example, our GCT codon might mutate to GAT (A for adenine) and we would obtain leucine instead of arginine in the protein. In about 20 per cent of all mutations, because of the redundancy (degeneracy) of the code, a mutation might have no effect on protein structure—GCA, GCT, GCC, and GCG all code for arginine. Thus a mutation from GCT to GCA would affect only the DNA and might well have no functional effects—no matter what the third base in the GC codon, we always obtain arginine in the protein. The evolutionary process, then, at its most basic level, involves a change (mutation) in the DNA which is incorporated into the ongoing gene pool of the evolving species and which can be reflected by a corresponding change in the amino acid sequence of the particular protein coded for by that gene. Such a process, multiplied millions of times in millions of different populations over hundreds of millions of years has given rise to the bewildering diversity of molecular and morphological structures that we see in today's world.

We might state as a basic rule that such a process will generally have to produce divergence when any two populations become isolated from one another, as a relative rarity of mutations and the finite size of populations make it statistically improbable that identical changes will be available for either natural selection or drift to incorporate into the ongoing gene pools. As long as no interbreeding takes place, then, our two populations will become increasingly different from one another and we may then compare protein structures in the modern representatives as a measure of the differences between the two lines.

Thus, to measure species, in Nuttall's terms, we need only to measure the extent of DNA and/or protein sequence difference between them. Several techniques of measurement are available, none of which is technically simple, but three—DNA hybridization, protein sequencing, and immunology—have been widely used with the results showing a comforting concordance to the student of human evolution. To summarize at the outset, all of the presently available data are inconsistent with any date outside the range of about three–eight million years for the most recent time that a common ancestor of man, chimpanzee, and gorilla could have existed; they indicate a most probable figure in the area of four–five million years, i.e., in the latest Pliocene.

To date by molecules rather than by fossils, a process we have termed an evolutionary or molecular clock, would appear to have been first proposed by Emile Zuckerkandl and Linus Pauling in the early 1960's. Though a number of scientists have discussed this concept since that time, to our knowledge no one had seriously applied it to a specific evolutionary problem prior to the Sarich-Wilson *Science* article already referred to.

That work was based on the study of a single molecule, serum albumin. Questions have been raised about the value of a single locus for probing evolutionary relationships among species containing thousands of loci. The justification lies in the fact that this locus has evolved as a part of the organisms that make up the lineages culminating in those species we see and compare today. Its history is, therefore, necessarily the history of those lineages. In other words, the times of divergence between the various albumin lineages are the same as the times of divergence between the species themselves.

Serum albumins are single polypeptide chains consisting of ca. 570 amino acids and are found in the serum of all tetrapod vertebrates (and probably also in fish). The order of the 570 amino acids (of twenty different kinds) of course varies from species to species and it is this variation that formed the basis for

Table 1   *The DNA Codons as Presently Known. ATT, ATC, and ACT Do Not Code for Any Amino Acid.*

AAA AAG	Phenylalanine	AGA AGG AGT AGC	Serine	ATA ATG	Tyrosine	ACA ACG	Cysteine
AAT AAC	Leucine			ATT ATC	Stop	ACT	Stop
		GGA GGG GGT GGC	Proline			ACC	Tryptophan
GAA GAG GAT GAC	Leucine			GTA GTG	Histidine	GCA GCG GCT GCC	Arginine
		TGA TGG TGT TGC	Threonine	GTT GTC	Glutamine		
TAA TAG TAT	Isoleucine			TTA TTG	Asparagine	TCA TCG	Serine
TAC	Methionine	CGA CGG CGT CGC	Alanine	TTT TTC	Lysine	TCT TCC	Arginine
CAA CAG CAT CAC	Valine			CTA CTG	Aspartic acid	CCA CCG CCT CCC	Glycine
				CTC CTT	Glutamic acid		

our study. Ideally a comparative study of protein evolution would be carried out by working out the exact sequences of amino acids in a series of primate albumins (or some other appropriate protein). This ideal, however, was not technically feasible for our study because amino acid sequences are not easy to generate; we therefore took a somewhat less precise but far more efficient approach to the evolutionary information contained in these sequences. Instead of working out the sequences, we measured the differences between them by the use of immunological techniques.

Our procedure began by purifying the desired serum albumin from a single primate—for the human sample, I was the source. After purifying my albumin and those of several other primate species, they were injected into a large series of rabbits. When a vertebrate is exposed to a foreign substance, it responds by making antibodies to it. This reaction is the basis of immunizations against disease (e.g., smallpox or polio), rejections in transplants, and allergies. A rabbit does of course contain albumin, but since rabbits and primates shared a common ancestor their respective albumins have been evolving in different directions and so are today very different molecules. Thus a primate albumin is very much a foreign substance to a rabbit. The antibodies the rabbit produces are themselves proteins and possess reactive ends that "recognize" specific areas (called antigenic sites) on the albumin with which they are injected. These antigenic sites are in no sense well-defined areas intrinsic to the albumin; they are sets of five to eight amino acids which are different from those in the corresponding area on rabbit albumin and which are therefore potentially capable of triggering antibody production in the rabbit. No single rabbit will recognize all possible antigenic sites, nor will each rabbit recognize the same sites. The use of a number of rabbits per primate albumin species, then, provides immunological coverage for a large part of the albumin molecule. This mixture of sera from injected rabbits was our test reagent that was to be reacted with a series of primate albumins. The degree of recognition or cross reaction would then depend on the number of amino acid differences between the homologous albumin (the albumin injected) and the heterologous test albumin. Simplifying the actual situation only slightly the total reaction would represent the sum of all individual antigenic site-antibody reactions where, if the sites are identical in the homologous and heterologous albumins, one gets a 100 per cent reaction; if the heterologous site has one amino acid difference, one obtains about 50 per cent reaction; and if the heterologous site has two substitutions, there is no reaction. The difference between the homologous and heterologous albumins is calculated on a scale that states that if the homologous and heterologous albumins are immunologically identical, the ID is zero, whereas the weakest cross reactions represent immunological distances of about 200 units. For serum albumins one ID unit is equal to slightly less than one amino acid substitution; i.e., the weakest cross reactions represent about a 30 per cent difference in two sequences.

These sera from rabbits injected with my own serum albumin and with those of many other primate species were then used to measure differences between many pairs of primate albumins. This process has been repeated for many carnivore, artiodactyl (odd-toed hoofed mammals), perissodactyl (horses, rhinos, tapirs), and other mammalian albumins. The data obtained are almost without exception consistent with the hypothesis that albumin evolution has been the result of a situation where the probability of an amino acid substitution occurring in a given length of time has been the same for *each* albumin lineage throughout its existence. Recent work in our laboratory would suggest that the *same* statistical rate of change applies to amphibian and reptilian albumins as well.

The results of such an evolutionary process can be glimpsed by considering a series of primate albumins: man, chimpanzee, rhesus monkey, and spider monkey. Antisera to all four are available and the following data were obtained (in immunological distances):

Man-chimp	7	Chimp-rhesus monkey	30
Man-rhesus monkey	32	Chimp-spider monkey	56
Man-spider monkey	58	Rhesus monkey-spider monkey	56

If one goes beyond the higher primates (man, apes, and New and Old World monkeys) to the prosimians, the distances jump sharply to 110–140 units, and for those non-primate mammals that show measurable reactions they are greater than 160 units. Even from this limited survey several things are apparent. Clearly the human-chimp differences, relative to the total evolutionary differentiation at the mammalian albumin locus, is a very small one. I might point out here that it is also quite comparable to the distances between the albumins of the following pairs of familiar and closely related species: domestic cat-lion, 7; sheep-goat, 6; dog-fox, 9; horse-donkey, 5; horse-zebra, 8. In addition, we can readily deduce the meaning of this miniscule human-chimp difference by asking whether it might be part of a pattern of less rapid evolution of ape and human albumins relative to the rates seen in other primate lineages. If this were the case, then ape and human albumins would be seen as less changed than, for example, those of the Old World monkeys because of their increased reactivity with antisera to the albumins of the New World monkeys or prosimians. That this is not the case is evident from the fact that human, chimp, and rhesus monkey albumins are almost equally different from that of the spider monkey indicating that since man, chimp, and rhesus monkey last shared a common ancestor, their albumins have undergone very similar amounts of albumin evolution.

One can readily extend this type of analysis into the development of an actual phylogenetic tree for man, chimp, and the two monkey species by using two prosimian albumins, those of lemur and slow loris, as reference points. The following mean distances have been obtained: to man, 123; to chimp, 120; to rhesus monkey, 120; to spider monkey, 121 (see Figure 1). The albumin differences between man and spider monkey is 58 units, therefore $a + c + e + f = 58$. By the same logic $a + c + e + g + h = 123$ and $f + g + h = 121$. Now:

$$a + c + e + g + h = 123$$
$$f + g + h = 121$$

subtracting, however

$$a + c + e - f = 2$$
$$a + c + e + f = 58$$
$$2a + 2c + 2e = 60$$
$$\therefore a + c + e = 30 \text{ and } f = 28$$

Reasoning along similar lines: $a = 4$, $b = 3$, $c = 13$, $d = 15$, $e = 13$. Clearly g and h cannot be calculated unless a non-primate reference point is used. Such a calculation has been done, but it is not necessary to discuss it here.

It is always possible, given the appropriate reference species, to reconstruct the phylogeny of any group of modern species and, of particular importance, calculate the amounts of evolutionary change that took place along those lineages leading to the modern species. To reconstruct the phylogeny of Figure 1, for example, we need only assume that lineage g existed; that is, that man, chimp, rhesus monkey, and spider monkey shared a common ancestor that lived subsequent to the divergence of the lines leading to lemur and loris. Given that assumption, verifiable by using non-primate albumins as reference points, the internal

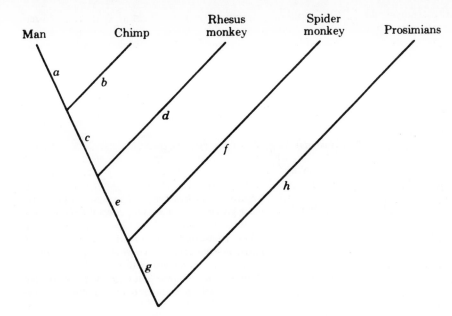

Figure 1   *A phylogeny of several primate taxa. Each letter indicates the amount of albumin evolution that took place along that lineage. The albumin distance separating any two species is thus the sum of the changes that occurred along the lineages that connect the two species.*

phylogeny of the higher primates can be developed using only molecular data (Figure 2).

The picture of evolutionary regularity shown by the four primate albumin does not change when the number is increased— the pattern of regularity always remains. This regularity, as discussed above, would appear to have a large random component whose evolutionary outcome is best described in terms of a Poisson distribution. An appropriate homology is the process of radioactive decay. Let us, for example, posit a C14 sample that over a period of thirty days provided 1500 decay events. This total of 1500 would not, however, be made up of a set of thirty 50-count days, but of a set of daily counts whose distribution would approximate that given in Figure 3 where the Poisson mean is fifty. The Poisson distribution describes the expected results of a process where the probability of a given individual event remains constant over time and is thus appropriate to the radioactive decay situation—where each C14 atom has an equal probability of decaying in a given interval of time. It is also appropriately applied to the process of protein evolution, not so much because we know a priori the mechanisms of protein evolution, but because evolutionary rate tests (such as that discussed above for human, chimp, rhesus monkey, and spider monkey albumins) consistently give a distribution approximating the Poisson.

One need not explain this phenomenon of regularity to use it, but it does perhaps require some consideration in view of the well-known haphazard nature of the evolutionary process when dealing with anatomical features. A currently feasible explanation has been suggested by Kimura (1968) and King and Jukes (1969) who have pointed out that this regularity could be due to the inexorable fixation of neutral mutations at a low but constant probability in the reality of evolutionary time. Neutral mutations are those "unseen" by the selective process—unseen because they are functionally and therefore selectively equivalent to the original amino acid. As King and Jukes have so elegantly put it:

> Natural selection is the editor, rather than the composer, of the genetic message. One thing the editor does *not* do is to remove changes which it is unable to perceive.

Beyond this quota of surviving neutral mutations, whose evolutionary accretion will follow a statistically determined pattern, will be those substitutions conferring a change in fitness. One might have amino acid changes that increase the fitness in a particular environment; conversely the relative proportion of maladaptive to neutral mutations might be increased in some evolutionarily conservative context. Such positive or negative selection will introduce a nonrandom element that should then be seen as "irregularity." On the basis of the presently available data, both immunological and direct sequence, one suspects that this nonrandom element must be rather minor (though hardly unimportant). Whether or not the neutral mutation model is correct has no bearing on the observed regularity of mammalian albumin evolution. The regularity is there (and not assumed as some of our critics would have it) and its presence allows the use of the primate albumin locus as an evolutionary clock subject to statistical considerations.

A clock must be calibrated in the form of an equation ID = kT where T is the time of divergence between two species and ID is the difference between their albumins in immunological distance units. k must be calculated using a known T corresponding to a measured ID. Once k is set, other times of divergence can be calculated. Since these calculated dates will be relative dates dependent on the accuracy of the assumed calibrating date, one desires a calibrating date subject to as little uncertainty as possible. Looking at the available mammalian data, it appears that the most reasonable equation would be an albumin immunological distance of 100 units set equal to a most probable (in the Poisson sense) time of separation of sixty million years. That is 100 = k (60 x 10^6 years) and thus, k = 1.67 units per million years of separation or 0.83 units per million years per lineage. Unless the paleontologists are completely wrong in their assessment of all early mammalian history (and whatever our disagreements on the question of human origins this seems very unlikely), this figure is unlikely to require serious adjustment in the future. The picture for primate evolution in general and ape and human evolution in particular derived from this albumin study is given in Figures 2 and 4.

On the specific point of the human-African ape relationship, the albumins of man, chimpanzee, and gorilla stand equidistant from one another with seven units separating any pair of the trio. As there is no indication that the albumins of the apes and man are conservative in their pattern of evolutionary change (that is, they are just as evolved as those of the New and Old World monkeys) a direct application of the albumin clock is indicated. At seven units of albumin immunological distance and 1.67 units

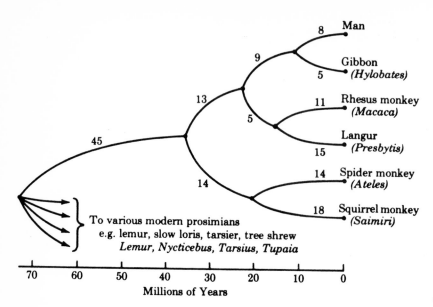

Figure 2   *A phylogeny of the major higher primate groups. The number on each lineage indicates the amount of albumin evolution (measured immunologically) that took place along it. It is of particular importance to note the large amount of albumin evolutionary history that the higher primates share subsequent to their divergence from any lineage leading to a modern prosimian.*

of evolution per million years of separation this works out to slightly more than four million years ago as the last time man, chimp and gorilla shared a common ancestor. The albumin clock also indicates that the Asiatic apes diverged at an appreciably earlier time—the orang at around seven–eight million years and the gibbon at around ten–twelve months.

These dates, particularly for the evolution of the modern apes and man, are, in terms of most current anthropological and paleontological opinion, far too recent. The idea that all the modern apes (chimpanzee, gorilla, orang, gibbon, siamang) share with man a late Miocene to early Pliocene ancestor and that the chimpanzee, gorilla, and man derive from a single species still living in late Pliocene times is disturbing to most students in the field. In view of this and in view of the fact that some degree of statistical and experimental uncertainty exists in our calculations, it is interesting to analyze other genetic data concerning human evolution along the lines given above. Table 2 lists some of these data where comparisons among ape, human, and Old World

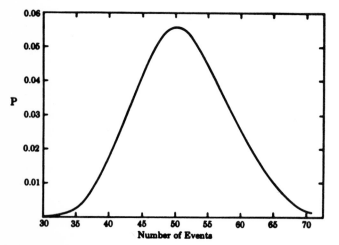

Figure 3   *A Poisson distribution representing the probability of obtaining a specific number of decay events when the mean is fifty events per day. Note that quite appreciable deviations from the mean are to be expected—and found, as some of the data in Figure 2 indicate.*

monkey molecules are available. One notes the interesting fact that the albumin difference seems in no way unique—a result we should have expected from our earlier conclusion that there is a single phylogeny for any group of species and that the phylogeny of a protein locus (or the whole DNA) must be coincident with the phylogeny of the species themselves. Nevertheless, in view of the relative novelty of the molecular approach to evolutionary history, the relative concordance among these various comparisons is comforting.

It is possible to analyze the hemoglobin data in somewhat greater detail and provide some probability values for our temporal conclusions. As mentioned in Table 2, man and chimpanzee have identical hemoglobins (the gorilla is different from both by two substitutions), whereas the recently sequenced hemoglobin of the rhesus monkey differs from them by about fifteen mutational events. It could be argued that human and chimpanzee hemoglobins are identical because hemoglobin evolution has been slow among the African apes and man. However, if this were the case, then human and chimpanzee hemoglobins should be less different than the hemoglobin of rhesus monkeys is from that of some species outside the catarrhines. One such reference point is the horse, from which human, chimpanzee, and rhesus monkey hemoglobins are equally different. Again this result should not be surprising in view of what has already been said concerning the regularity of the protein evolutionary process. Some calculations are also in order. The estimated time of divergence of the apes and Old World monkeys is about twenty-three million years using the albumin data. The average rate of hemoglobin evolution among mammals is about one amino acid substitution per 3.5 million years. Human and rhesus monkey hemoglobins differ by fifteen amino acid substitutions that would correspond to 15 × 3.5 or 52 million years of evolution apportioned along the two lineages indicating twenty-six million years for the most recent common ancestor of man and rhesus monkey. In view of the small number of changes involved, the concordance of the albumin and hemoglobin dates is heartening.

Now clearly the calculation of an exact time of divergence from the human-chimp-gorilla hemoglobin data is unwarranted for statistical reasons, but these data are obviously compatible with the albumin date of four-five million years. More important, the results are incompatible with the divergence time of twenty--thirty million years that is the general view of most paleontolo-

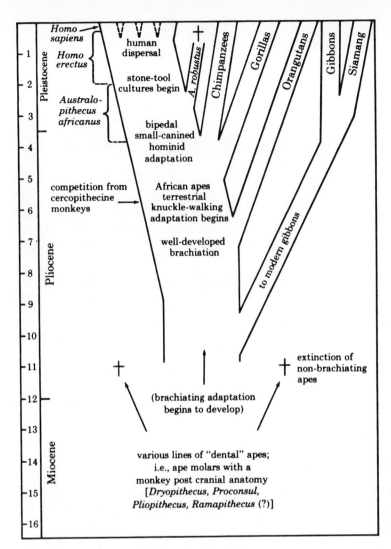

Figure 4   *A detailed history of recent hominoid evolution. The indicated time of divergence are calculated from the available molecular evidence (albumin, hemoglobin, DNA). The descriptive captions are interpretive attempts to synthesize the biochemical, anatomical, and paleontological evidence.*

gists and anthropologists today. If we apply the probability model for protein evolution discussed in this article to this case, then we can calculate using the Poisson distribution that there is less than one chance in 10^5 that a sequence difference of zero-two residues could result from a divergence time of thirty million years, and from one chance in 100 to one chance in 1,000 if the date were fifteen million years. Obviously the same probability considerations can be applied to the independently derived albumin date, thus making the combined probability that any date for the hominid-pongid divergence could approach even ten million years vanishingly low. The other protein data given in Table 2, though not subject to as detailed analysis because of the lack of data from the more primitive primates, are entirely consistent with the albumin-hemoglobin date and thoroughly inconsistent with the postulation of any appreciable antiquity for the origin of our lineage.

Added note: After this article was prepared, Dr. David Kohne of the Carnegie Institution kindly provided us with a manuscript on primate DNA evolution (see Kohne, et al., 1970, 1972). Their technique depends on the fact that the separated strands of the classic DNA double helix can, under proper conditions, be caused to renature and form the original double helix.

Interestingly enough two single stranded DNAs from different species (provided their relationship is sufficiently close) can also form a "hybrid" DNA double helix. Because the DNAs of the two species have been evolving (changing) over the time period since the two species last shared a common ancestor, they will no longer be of perfect complementary fit; e.g., instead of the expected thymine, an adenine on one strand might find itself opposite a cytosine with no resultant bonding. This lack of perfect base pairing in the hybrid DNA will lower the heat stability of the hybrid molecules so that they will fall apart at a temperature lower than that found for the native DNA. For example, whereas the human DNA double helix is 50 per cent separated into its constituent single strands at $82.9°C$, the temperature for the human-chimp hybrid is $81.2°C$. This decrease of $1.7°C$. translates into approximately 2.5 per cent of the nucleotide pairs being mismatched (e.g., adenine-cytosine instead of adenine-thymine). The precise data are shown in Table 3.

Considering the statistical and experimental uncertainties implicit in the albumin approach, and the relatively untested nature of the DNA approach, I see this agreement as remarkable and, in its way, supportive of both methods and interpretive deductions based on their data.

It is also interesting that the available albumin and fibrino-

Table 2

Molecule	Human-Chimp Difference	Human-Old World Monkey Difference (Mean)	Type of Comparison
Albumin	7	35	immunological
Hemoglobin	0	15	amino acid sequences
Transferrin	3	30	immunological
Fibrinopeptides	0	7	amino acid sequences
DNA	2.6	10.7	hybridization

peptide data make it possible to scale the human-chimpanzee differences against those found in other groups. The human-chimp difference for both these proteins is already pointed out for the albumins very similar to that found in other undoubtedly closely related pairs of mammalian species: dog-fox, domestic cat-lion, and horse-zebra. All in all, what appeared to be a position out on the end of a long limb taken by Allan Wilson and myself when we first published the albumin conclusions concerning the recency of the hominid-African ape split in 1967 is now beginning to look more and more reasonable as the molecular data accumulate.

It might be instructive to reflect upon what the molecular data and the paleontological times of divergence would lead us to conclude. In the albumin case, using a human-chimp divergence time of twenty million years (Simons and Ettel, 1970) we should expect about thirty-three ID units of difference. We find seven. In the case of the fibrinopeptides, where the average rate of change is about one substitution per lineage per six million years, we might expect six differences between man and chimpanzee. We find none. For the hemoglobins (α and β chains combined) the average is about one substitution per vertebrate lineage per 3.5 million years (Kimura, 1969). Between man and chimp we might then expect ten differences. We find none. We have already shown that this lack of evolutionary change in the albumins and hemoglobins (see Wilson and Sarich, 1969) cannot be due to retarded evolution in the chimp and/or human lineages. In view of the generally stochastic nature of protein evolution one might consider that much of the argument here is, in large part, an accident of history. It is entirely possible to imagine a situation where the molecular evidence might have come first and the paleontological and comparative anatomical data at some later time. Suggestions of any appreciable antiquity, then, for the African ape-hominid divergence might be greeted with as much skepticism as have been our protein conclusions. A couple of recent quotes might be appropriate here:

A scale of relationship can be calculated from the magnitudes of the reactions between antihuman albumin and albumins from many primates. This scale can then be converted into units of time. Using this method, Sarich and Wilson have suggested that hominids and pongids were part of a common population about five million years ago. If Sarich and Wilson had looked more carefully at paleontological investigations, they would have found their suggestion is unwarranted. Their suggestion implies that some as yet undiscovered fossils, representative of a population ancestral to modern man, gorillas, and chimpanzees, will provide the evidence to support their interpretation...there are some things that cannot be done with molecular data and some things that cannot be done with fossils, and I object to careless assumptions and thoughtless statements about evolutionary processes in some of the conclusions drawn from the immunological data mentioned.

Unfortunately there is a growing tendency, which I would like to suppress if possible, to view the molecular approach to primate evolutionary studies as a kind of instant phylogeny. No hard work, no tough intellectual arguments. No fuss, no muss, no dishpan hands. Just throw some proteins into a laboratory apparatus, shake them up, and bingo!—we have an answer to questions that have puzzled us for at least three generations (Buettner-Janusch, 1969, p. 132–33).

If the immunological dates of divergence devised by Sarich are correct, then paleontologists have not yet found a single fossil related to the ancestry of any living primate and the whole host of species which they have found are all parallelistic imitations of modern higher primates. I find this impossible to believe. Some fossil primates do exhibit evolution parallel to other forms, as is particularly well demonstrated in the case of the subfossil Malagasy lemurs, but it is not presently acceptable to assume that all the fossil primates resembling modern forms are only parallelisms, that highly arboreal apes wandered hundreds of miles out of Africa across the Pontian steppes of Eurasia in search of tropical rain forests, or that *Australopithecus* sprang full-blown five million years ago, as Minerva did from Jupiter, from the head of a chimpanzee or gorilla (Simons, 1969b, p. 330).

Table 3

Species Compared	DNA[a] Difference	DNA Time of[b] Divergence	Albumin[c] Time of Divergence
Human-chimp	2.5	5	4
Human-gibbon	6.1	13	11
Human-rhesus monkey	10.3	21	23
Human-capucin monkey	17.4	36	36
Rhesus monkey-green monkey	2.9	6	5

[a]These numbers represent the calculated percentage of base pair mismatches.
[b]Calculated using 36 million years for the human-capucin divergence.
[c]Calculated as discussed above using $k = 1.67$ Immunological distance units/-million years and adjusting for any deviations from average amounts of change.

Rhetoric aside, all this says is that our conclusions somehow don't agree with the presently more or less accepted interpretations of the higher primate fossil record.

The point is that one has to be very careful in distinguishing between data and interpretation. There are molecular and paleontological and anatomical data, but the data often mean different things to different people and thus there are molecular and paleontological and anatomical interpretations of those data. There is, however, but a single evolutionary history or phylogeny for a group of species—therefore all interpretations of what happened and why and how it happened in the history of that group must use this single phyletic framework. One cannot legitimately use one phylogeny to explain the molecular data, another for the paleontological, and still a third for the anatomical. Conversely, one cannot legitimately derive three or more phylogenies from three such sets of data. We require that the various lines of evidence be used to place more and more marked constraints upon interpretations derived from other areas, and in this cybernetic fashion to more closely approach evolutionary reality. No single way leads to perfection; at present the protein picture lacks something in resolution, the fossil record is necessarily rather incomplete, and the anatomical data are often difficult to interpret. I have already discussed the particular advantages of the protein approach in developing the phyletic picture. As might be expected, then, it is not particularly difficult to interpret the available paleontological and anatomical evidence in terms of the short time scale protein phylogeny [see also O. J. Lewis, (1969)]. I have yet to see any suggestion as to how a twenty million year date for the origin of the hominid line can possibly be used to explain the molecular evidence. To put it as bluntly as possible, I now feel that the body of molecular evidence on the *Homo-Pan* relationship is sufficiently extensive so that one no longer has the option of considering a fossil specimen older than about eight million years as a hominid *no matter what it looks like.*

If the living apes and man shared a common early Pliocene ancestor (ten–twelve million years ago), and if the African apes and man shared one in the late Pliocene (four–five million years ago), then the use of these dates allows the construction of a picture of ape evolution in general and human evolution in particular that has a certain elegance in its simplicity and utility. Washburn points out in the following article that the evolutionary unity of the modern apes and man is exemplified in certain details of anatomy in the upper part of the body that can all be viewed as parts of a locomotor-feeding adaptation, usually termed brachiation. Some specific aspects of this adaptation that serve to differentiate the living apes and man from their monkey relatives include the wide sternum, long clavicle, the large acromion process, and large deltoid muscle, and the fact that the flexor muscles of the upper limb (those facing forward when one stands with palms facing forward) are large relative to the extensors. The structure of the elbow and forearm allows a full 180° of pronation and supination (i.e., with upper arm immobile, the forearm can rotate through a full 180°). Full extension at the elbow is possible. The forearm bone in line with the little finger (ulna) does not articulate with the wrist bones, thus permitting man (and the apes, of course) to freely abduct and adduct in swinging from a limb and to flip a football fifty yards with a so-called "flick of the wrist." The upper limb and hand are very long in relation to the trunk (man, chimpanzees, and gorillas overlap appreciably in upper-limb/trunk length ratios). The lumbar vertebrae are few, and the lumbar region is short. The trunk is short, wide, and shallow. The preceding is condensed and annotated from another Washburn publication. He then went on to state:

It seems most improbable that this detailed structural-functional similarity could be due to parallelism...parallelism means that animals resemble each other because similar groups have adapted in similar ways but that the lines are

genetically independent. To suggest that man evolved the structure of a brachiator by parallelism and not brachiating is to misunderstand the nature of parallel evolution (Washburn, 1963b, p. 195).

In addition to these details of post-cranial anatomy is the diagnostic difference between the specialized bilophodont molars of the Old World monkeys and the more primitive ape and human molars. The ape molar cusp pattern is a paleontologically ancient one going back at least to the latter part of the Oligocene period (twenty-five–thirty million years ago), but the brachiating adaptation would appear not to be demonstrable in the primate fossil record. The Miocene apes, of which a fair number exist in the fossil record, I have therefore termed "dental apes." Unless we are being misled by the fossil record, then, the brachiating adaptation evolved after the time represented in the ape fossil record and before the beginning of the modern ape radiation. The molecular data discussed above place this beginning at ten–twelve million years in the past.

The numerous branches of the widespread group of Miocene apes (dryopithecines) must have therefore left but a single surviving lineage—a lineage that owes its unique evolutionary success to the development of the brachiating adaptation. The modern apes and man, as the living products of the adaptive radiation following this development, show similarities in basic pattern (reflecting the attainment of this new grade of organization), but differences in detail (representing the specific adaptations made along the various line comprising the adaptive radiation). The unity of the modern Hominoidea (apes and man) is thus based on the relatively short period of time during which a major adaptation was being evolved; their diversity is based on the relatively long period of time in which each line has been evolving independently of the others (Figure 4).

Our ancestors were, then, functional monkeys until about ten–fifteen million years ago, and brachiating apes until about 6 million years ago. Then apparently a terrestrial adaptation began. The chimpanzee and gorilla are both significantly terrestrial forms—using knuckle-walking to get along on the ground. In this context it is interesting to note the usual lack of hair on the back of our middle phalanges and speculate that this is an indication of a recent phase of knuckle-walking in our own ancestry. It would certainly be a perfect transition state from full brachiation (of which we are of course still perfectly capable) to a bipedal adaptation—the first indication of which comes with the earliest of the South African australopithecines who probably date to some three million years. One is thus left with from one to three million years of our history that remain undocumented by the fossil record and to which the molecular approach can contribute little. It is during this period that the hominid grade, characterized by the bipedal adaptation and reduction in size of the canine complex, was reached.

As indicated in Figure 4, the suggestion is made that the adaptive radiation leading to the African apes and man began as a result of competition in the trees by the evolving cercopithecines (today the macaques, baboons, mangabeys, and guenons). I doubt that it is coincidental that the maximum immunological distances among the albumins of the modern cercopithecines (about eight units) and the differences in their DNAs are practically identical to those within the man-chimp-gorilla triad. This doubt is furthered when the absence of small and truly arboreal apes in Africa is noted. The modern cercopithecines are perhaps the most successful of the modern nonhuman primates and comprise about thirty-five species. The reasons for their recent evolutionary success are not yet clear but there is every reason to think that it was this success that, in effect, chased the African apes out of the trees into varying degrees of terrestrial adaptation made by the common ancestor of man, chimp, and gorilla; and that knuckle-walking represents the "transitional" phase only for man since the chimp and gorilla still operate in the terrestrial locomotor mode abandoned by our ancestors several millions of years ago.

This knuckle-walking transition phase (supported mainly by Washburn) has recently been most strongly contested by Tuttle (1969a):

> In so far as I can discern from dissections of numerous ape and human hands there are no features in the bones, ligaments, or muscles of the latter that give evidence for a history of knuckle-walking in man.

Accepting this assessment, how are we to incorporate it into the model being discussed? The point Tuttle misses, by managing to completely avoid mention of the molecular evidence, is that of time in grade. If knuckle-walking developed in the common ancestor of man and the African apes, then the protein data discussed here suggest that this phase could hardly have begun earlier than about seven million years ago. In view of the australopithecine evidence, the hominid knuckle-walking phase presumably ended no more than two or three million years later. Thus the chimpanzee and gorilla would have been knuckle-walkers some three to five times as long as our ancestors, and in addition, during the time that human ancestors have not been knuckle-walking (and presumably losing the adaptation) the gorilla and chimpanzee have been perfecting theirs. If we ignore these factors—suggested by the molecular data as well as evolutionary logic—then basic disagreement of a pattern familiar to any student of the history of anthropology appears. If, on the other hand, we first develop a phylogenetic framework and then use it, many problems, of which the knuckle-walking contretemps is but one, are resolved.

The reason for knuckle-walking would appear to be a mechanical one representing the most efficient way of transmitting the stresses of forelimb support in an animal where the articulation between the forearm and wrist bones is incomplete, as in the modern apes and ourselves. If we or the apes were to attempt quadrupedal locomotion palms down and wrist hyperextended (palm horizontal and forearm vertical) we would have a distinct tendency to drive the forearm bones right through the wrist—mechanically, a most inefficient situation.

Going back to the cercopithecine-pongid competition in the middle Pliocene tropical forests of Africa, however, it is interesting and more than a little ironic that our potential for eventual success was probably in no small part due to the failure of our ancestors in this competition. They, losing in one niche, exploited a new and open one, and in so doing made man possible.

If one were to ask for the best description that can be made of this species most recently ancestral to the gorilla, chimpanzee, and ourselves, then the most logical answer is that it must have most closely resembled the least specialized of the three. Thus we begin the reconstruction and understanding of our recent history with a form not unlike a small chimpanzee; we move through a phase where the basic hominid grade was reached; and then we draw a fuller picture from the detailed data that the australopithecine fossil record allows. To give an answer to Huxley's 100 year old query it can be concluded that the temporal gap separating man from the African apes is indeed narrow.

In conclusion, it might be appropriate to consider the disagreements between the picture developed in this article and the more traditional ones concerning human origins, and to then suggest how a rapprochement might be effected. What I have attempted here is to provide the DNA and protein data, derive evolutionary conclusions from them, and then, in the light of those conclusions, develop an intepretation of some of the more pertinent anatomical and paleontological data. To the extent that the paleontologists and other physical anthropologists disagree—to the extend that they continue to see discordance where I see concordance—it is up to them to provide some alterntive interpretation of the protein data. We are not unaware of this discord but we have also been utterly unable to interpret the protein data in other than the fashion presented here. We are, though, beginning to focus on specific areas of disagreement and as those are ironed out, a closer and closer approach to general agreement concerning the course of early hominid evolution will be made. One trusts that this will not be too long in coming.

Added note: A recent paper by Read and Lestrel (1970) has questioned the linearity of the immunological distance-sequence difference function used in this paper. A fuller defense of the particular scale we use is in preparation but involves issues too complex to be properly and lucidly treated here. Basically the problem with the Read-Lestrel accommodation of the albumin immunological and paleontological interpretations of hominid evolution lies in the fact that their proposed model has no predictive value. It would predict that in other protein and DNA comparisons man and chimpanzee would show large differences in the two species and this is simply not the case—as has been pointed out in this chapter. It is also disturbing to note that the Read-Lestrel article, though purporting to be a "critical appraisal" (the title is *Hominid Phylogeny and Immunology: A Critical Appraisal*), does not even mention the Wilson and Sarich (1969) article in which the predictive aspects of the albumin model are tested and found not wanting. The reader, having digested this chapter, should now be capable of providing his own critical appraisal of the Read-Lestrel paper.

41

The Seed-Eaters:
A New Model of Hominid Differentiation
Based on a Baboon Analogy

C. J. JOLLY

Despite years of theorising, and a rapidly accumulating body of fossil evidence, physical anthropology still lacks a convincing causal model of hominid origins. Diverse lines of evidence point to a later common ancestry with the African pongids than with any other living primate, and studies of hominid fossils of the Basal and Early Lower Pleistocene (Howell 1967) have elucidated the complex of characters which at that time distinguished the family from African and other Pongidae (Le Gros Clark 1964). It is also possible to argue that the elements of the complex form a mutually reinforcing positive feedback system. Bipedalism frees the forelimb to make and use artefacts; regular use of tools and weapons permits (or causes) reduction of the anterior teeth by taking over their functions; the elaboration of material culture and associated learning is correlated with a cerebral reorganisation of which increase in relative cranial capacity is one aspect. Bipedalism is needed to permit handling of the relatively helpless young through the long period of cultural conditioning, and so on.

Preoccupied with the apparent elegance of the feedback model, we tend to forget that to demonstrate the mutual relationship between the elements is not to account for their origin, and hence does not explain *why* the hominids became differentiated from the pongids, or why this was achieved in the hominid way. From their very circularity, feedback models cannot explain their own beginnings, except by tautology, which is no explanation at all. In fact, the more closely the elements of the hominid complex are shown to interlock the more difficult it becomes to say what was responsible for setting the feedback spiral in motion, and for accumulating the elements of the cycle in the first place. Most authors seem either to avoid the problem of origins and causes altogether (beyond vague references to "open country" life), or to fall back upon reasoning that tends to be teleological and often also illogical. This article is an attempt to reopen the problem of origins by examining critically some of the existing models of hominid differentiation, and to suggest a new one based on a fresh approach.

PREVIOUS MODELS OF HOMINID DIFFERENTIATION

Direct fossil evidence for the use of "raw" tools or weapons is necessarily tenuous, and that for the use of fabricated stone artefacts appears relatively late (Howell 1967). Nevertheless, as Holloway has pointed out (1967), the currently orthodox theory regards these elements as pivotal in the evolution of the hominid adaptive complex, probably antedating and determining the evolution of upright posture, and certainly in some way determining the reduction of the anterior teeth, the loss of sexual dimorphism in the canines, and the expansion of the cerebral cortex (Bartholomew & Birdsell 1953; Washburn 1963a; DeVore 1964). A variant of this theory, proposed by Robinson (1962b), sees bipedialism as the primary adaptation (of unknown origin), from which tool-using developed and hence anterior dental reduction.

Holloway (1967) rejects the orthodox, tool-and-weapon-determinant model, partly on the grounds that it postulates no genetic or selectional mechanism for anterior dental reduction, and thus implies Darwin's "Lamarckian" notion of the gradual loss of structures through the inherited effects of disuse. It seems a little carping to accuse Washburn and his colleagues of Lamarckism because they omit to make explicit their view of the selective factors involved. These are in fact stated by Washburn in his reply to Holloway (Washburn 1968c): natural selection favours the reduction of canines after their function has been subsumed by artificial weapons since this reduces the chance of accidental injury in intra-specific altercations. Since, as we shall see, orthodox natural selection is adequate to explain the reduction of teeth to an appropriate size following a change of function, it is hard to see why Washburn should avoid the Scylla of Lamarckism only to fall into the Charybdis of altruistic selection. Why should natural selection favour the evolution of a structure that is of no benefit to its bearer, for the benefit of other, unrelated conspecifics? Even if we swallow altruistic selection, a basic illogicality remains. If the males use artificial weapons to fight other species, why should they bite one another in intra-specific combat? If they do not, then the size of their canines is irrelevant to the infliction of any injury, accidental or otherwise. In any case, the best way to avoid accidental and unnecessary intra-specific injury, of any kind, is to evolve unambiguous signals expressing threat and appeasement without resort to violence. The ability to make and recognise such signals is of advantage to *both* parties to the dispute, and therefore can be favoured by orthodox natural selection, is independent of the nature of the weapons used, and is found in the majority of social species, including both artefact-using and non-artefact using higher primates.

We may now consider the underlying proposition that it was artefact use, which, by making the canine redundant as a weapon, and the incisors as tools, led to their reduction. It is known that hominoids with front teeth smaller than those of living or fossil Pongidae were widespread at the close of the Miocene period: *Oreopithecus* in southern Europe and Africa (Hürzeler 1954b, etc.; Leakey 1967b), and *Ramapithecus* (probably including *Kenyapithecus*) in India, Africa, and perhaps southern Europe and China (Simons & Pilbeam 1965). If the theory of artefactual determinism is to be applied consistently, regular tool- and weapon-making has to be extended back into the Miocene, and also attributed to Hominoidea other than the direct ancestor of the Hominidae, whether one considers this to be *Ramapithecus*, *Oreopithecus*, or neither. Simons (1965b) regards *Ramapithecus* as too early to be a tool-*maker*, but he and Pilbeam (1965) suggest that it was a regular tool-*user*, like the savannah chimpanzee

Reprinted with permission from *Man*, Volume 5, pages 5–26. Copyright © 1970 by the Royal Anthropological Institute of Great Britain and Ireland.

(Goodall 1964; Kortlandt 1967). This is eminently likely, but is no explanation for anterior dental reduction since the chimpanzee has relatively the largest canines and incisors of any pongid, much larger than those of the gorilla, which has never been observed to use artefacts in the wild. To explain hominid dental reduction on these grounds, therefore, we presumably have to postulate that the basal hominids were much more dependent upon artefacts than the chimpanzee, without any obvious explanation of why this should be so. One would also expect signs of regular toolmaking to appear in the fossil record at least as early as the first signs of dental reduction, rather than twelve million years later. The more artefactually sophisticated the wild chimpanzee is shown to be, of course, the weaker the logic of the tool/weapon determinant theory becomes, rather than the other way about, as its proponents seem to feel.

Clearly, some other explanation is needed for anterior tooth reduction, at least at its inception. Recognising this, Pilbeam and Simons (1965) and Simons (1965b) regard tool-use by *Ramapithecus* as compensation rather than cause for anterior tooth reduction, adopting as a causal factor Mills's (1963) suggestion that upright posture leads to facial shortening, and that canine reduction would then follow to avoid "locking" when the jaw is rotated in chewing. The main objection to this scheme (Holloway 1967) is that there is no logical reason why facial shortening should follow upright posture. Indeed, if brachiation is counted as upright posture, then it clearly does not. (Among extinct Madagascan lemurs, for instance the long-faced *Palaeopropithecus* was a brachiator, the very short-faced *Hadropithecus* a terrestrial quadruped (Walker 1967b).) Nor does a reduced canine accompany a short face in, for instance *Hylobates* or *Presbytis*. Furthermore, the explanation extends only to the canines, and does not account for the fact that incisal rather than canine reduction distinguishes the known species of *Ramapithecus* from small female Pongidae.

The same criticism applies to the model proposed by Holloway (1967), who finds an explanation of canine reduction in hormonal factors associated with the adoption of a hominid way of life:

> Natural selection favoured an intragroup organisation based on social cooperation, a higher threshold to intragroup aggression, and a reduction of dominance displays...a shift in endocrine function took place so that natural selection for reduced secondary sexual characteristics (such as canines) meant a concomitant selection for reduced aggressiveness within the group [1967: 65].

Thus, reduced canine dimorphism is apparently attributed to a pleiotropic effect of genetically-controlled reduction in hormonal dimorphism, itself favoured by the "co-operative life" of hunting.

This argument is vulnerable on several counts. First, there is no obvious reason why even *Homo sapiens* should be thought less hormonally dimorphic than other catarrhines; in structural dimorphism the "feminised" canine of the male is a human peculiarity, but humans are rather more dimorphic in body-mass than chimpanzees, and much more dimorphic than any other hominoid in the development of epigamic characters, especially on the breast and about the head and neck, which can only be paralleled, in Primates, in some baboons. Equally, there seems little to suggest that human males are any less competitive and aggressive among themselves than those of other species; the difference rather lies in the fact that these attributes are expressed in culturally-determined channels (such as vituperative correspondence in the *American Anthropologist*) rather than by species-specific threat gestures or physical assault, so that expression of rage is postponed and channelled, not abolished at source. It

seems unlikely that the basal hominids had departed further than modern man from the catarrhine norm. In fact, an elaboration of dominance/subordination behaviour, and thus an intensification of the social bond between males, is often attributed to a shift to "open-country" life (Chance 1955; 1967).

Second, the hypothesis that the canines which are disclosed when a male primate yawns are functioning as "organs of threat" is not unchallenged; Hall (1962) found that in Chacma baboons yawning appeared in ambivalent situations where it could more plausibly be interpreted as displacement. The size of the canines "displayed" by a male in a displacement yawn would be of no consequence to his social relations or his Darwinian fitness.

Third, and most trenchant, we must critically examine the assumptions, accepted by "orthodox" opinions as well as by Holloway, that an increase in meat-eating beyond that usual in primates would follow "open-country" adaptation, and that the peculiarities of the hominids ultimately represent adaptations to hunting. The first of these assumptions is perhaps supported by the fact that chimpanzees living in savannah woodland have been seen catching and eating mammals (Goodall 1965), while those living in rain-forest have not. The flaw lies in the second part of the argument, and is like that in the artefact-determinant theory; the more proficient a hunter the non-bipedal, large-canined, large-incisored chimpanzee is found to be, the less plausible it becomes to attribute the origin of converse hominid traits to hunting. Moreover, the hunting and meat-eating behaviour of the chimpanzee does not, to the unbiased eye, suggest the selective forces that could lead to the evolution of hominid characters. Neither weapon-use nor bipedialism is prominent. Prey is captured and killed with the bare hands, and is dismembered, like other fleshy foods, with the incisors. Thus, if a population of chimpanzee-like apes becomes adapted to a hunting life in savannah, there is absolutely no reason to predict incisal reduction, weapon-use, or bipedalism. On the contrary, it is most difficult to interpret the hominid characters of the australopithecines functionally as adaptations to life as a carnivorous chimpanzee. Incisal reduction would make for less efficient processing of all fleshy foods, including meat. A change from knuckle-walking, which can be a speedy and efficient form of terrestrial locomotion, to a mechanically imperfect bipedalism (Washburn 1950; Napier 1964) would scarcely improve hunting ability, especially since a knuckle-walking animal can, if it wishes, carry an artefact in its fist while running (cf. illustration in Reynolds & Reynolds 1965: 382). Once these characters existed as preadaptations in the basal hominids, they may well have determined that when hunting was adopted as a regular activity, it was hunting of the type that we now recognise as distinctively human, but to use this as an explanation of their first appearance is inadmissibly teleological.

This view is supported by the absence of fossil evidence for efficient hunting before the latter part of the Lower Pleistocene. It seems most unlikely that the hominid line would become partially and inefficiently adapted to hunting in the Miocene, only to persist in this transitional phase until the Lower Pleistocene (becoming, meanwhile, very specialised dentally, but no better at hunting or tool-making!), when a period of rapid adaptation to hunting efficiency took place. Perhaps recognising this, adherents of the "predatory chimpanzee" model tend to situate the hominid-pongid divergence in the late Pliocene, and regard all known fossils of basal Pleistocene Hominidae as representative of a short-lived transitional phase of imperfect hunting adaptation (Washburn 1963a). This is a view that is intrinsically unlikely, and difficult to reconcile with the fossil evidence of Tertiary hominids. The obvious way out of the dilemma is to set aside the current obsession with hunting and carnivorousness, and to look for an alternative activity which is associated with "open-country" life but which is functionally consistent with the anatomy of basal hominids.

Impressed by the bipedal charge of the mountain gorilla, and his tendency to toss foliage around when excited, two authors (Livingston 1962; Wescott 1967) have suggested that therein might lie the origin of human bipedalism and other elements of the hominid complex. The objection to this notion is again that it is illogical to invoke the behaviour of living apes to explain the origin of something that they themselves have not developed; if upright display leads to habitual bipedalism, why are gorillas still walking on their knuckles? Conversely, if hominid bipedalism were initially used solely in display, why should they have taken to standing erect between episodes? Even if we grant that the savannah is more predator-ridden than the forest (a view often stated but seldom substantiated, even for the recent, let alone the Tertiary), it is difficult to believe that attacks were so frequent as to make defensive display a way of life.

The occasional bipedalism, tool- and weapon-use, and meat-eating of the pongids are useful indicators of the elements that were probably part of th hominid repertoire, ready for elaboration under particular circumstances. To explain this elaboration, however, we must look *outside* the normal behaviour of apes for a factor which agrees functionally with the known attributes of early hominids. As we have seen, "hunting" is singularly implausible as such a factor. The object of this article is to suggest an alternative, based initially on the observation that many of the characters distinctive of basal hominids, as opposed to pongids, also distinguish the grassland baboon *Theropithecus* from its woodland-savannah and forest relatives *Papio* and *Mandrillus*, and are functionally correlated with different, but no less vegetarian, dietary habits.

THEROPITHECUS—HOMINID PARALLELISMS

The assumption is made here that both hominids and living African pongids are descended from Dryopithecinae, a group intermediate between the two in most of its known characters (most of which are dental), though rather closer to Pongidae that to Hominidae. The chimpanzee can then be seen as manifesting evolutionary trends away from the ancestral condition more or less opposite to those of the Hominidae, while the gorillas retain a more conservative condition, at least dentally. This model would work as well on the less likely assumption that the chimpanzee represents the primitive condition. It is also assumed that the Cercopithecine genera *Theropithecus* and *Papio* either diverged from an intermediate common ancestor, or, more probably, that *Theropithecus* and *Mandrillus* have become differentiated in opposite directions from a *Papio*-like form (Jolly 1970a). This process can be documented for *Theropithecus* during the course of the Pleistocene (Jolly 1965; 1972).

Table 1 summarises characters by which *either* early Pleistocene Hominidae differ from *Pan*, or *Theropithecus* from *Papio* and *Mandrillus*, listed without regard to their functional interrelationships or significance. Those which distinguish early Hominidae from Pongidae constitute the "Hominid adaptive complex," and are indicated in column A, while those which form part of the *"Theropithecus* adaptive complex" are indicated in column B. Rectangles show those features common to the two complexes, of which there are twenty-two out of forty-eight, reasonable *prima facie* evidence for parallelism between them. This hypotheses can be tested by checking the elements of the complexes for cross-occurrence in *Papio* and *Pan*. If the high number of common characters were simply due to chance, rather than to parallelism, we should not expect significantly fewer of the Hominid characters to appear in *Papio* (as opposed to *Theropithecus*), or significantly fewer of the *Theropithecus* complex

characters to occur in *Pan*. In fact, none of these cross-correspondences occurs. There are some grounds, therefore, for assuming the existence of evolutionary parallelism, and perhaps some degree of functional equivalence between the differentiation of *Theropithecus* and that of the basal hominids, and the common features may be used to construct a model of hominid divergence from pongids. To do this, we must examine the functional implications of the "AB" characters.

Of these, only one certain one appears in the "behaviour" category, largely because of the impossibility of observing the behaviour of fossil forms. Inferences of behaviour from structure are, of course, not permissible at this stage of analysis. The single common character is the basic one of true "open-country" habitat, inferred largely from the death-assemblages in which early *Theropithecus* and Hominidae are found, as well as the habitat of *T. gelada*.

Three skeletal "AB" characters are postcranial. The abbreviated fingers and unreduced thumb makes a pollex-index grip possible for the terrestrial monkeys (Napier & Napier 1967). Bishop (1964) showed that *Erythrocebus patas* made more consistent use of such a grip than its more arboreal relatives, the guenons. Recent work by Crook (e.g. 1966), including filmed close-ups of hand-use in the wild, has made it clear that the gelada (in contrast to , for instance, *Papio*) uses a precision-grip for most of its food-collecting. Food consists mainly of grass-blades, seeds and rhizomes which are picked up singly between thumb and index, and collected in the fist until a mouthful is accumulated. The index is thus continually used independently of the other digits. This feeding method is facilitated by the well-developed pollex and the very short index finger (Pocock 1925; Jolly 1965), a combination giving the gelada the highest "opposability index" (Napier & Napier 1967) of any catarrhine, not excluding *Homo sapiens* (J. R. Napier, personal communication). It is significant that the precision-grip of the gelada, which like other Cercopithecinae has not been seen making or using artefacts in the wild, should far outclass that of the tool- and weapon-using chimpanzee (Napier 1960).

The two common features of the foot are attributable to terrestrial adaptation which requires pedal compactness rather than hallucal gripping-power (Pocock 1925; Jolly 1965); the rest of the foot structure is different in the two forms and reflects the fact that their move to terrestrialism was quite independent and analogous. Most of the postcranial elements of the hominid complex are absent in *Theropithecus*, being related to upright bipedalism (Clark 1964). The post-cranial features of the *Theropithecus* complex are much fewer, expressing the fact that apart from the absence of tree-climbing its locomoter repertoire scarcely differs qualitatively from that seen in its woodland and forest relatives, although the frequency of the elements differs considerably (Crook & Aldrich-Blake 1968).

The mastoid process of the large Pleistocene *Theropithecus* is perhaps unexpected, since in Hominidae it can be related to erect posture (Krantz 1963). However, unlike the hindlimb characters which *Theropithecus* does not share, the mastoid is related to poising the head on the erect trunk, not the trunk upon a hyperextended hindlimb. The gelada spends most of the day in an upright *sitting* position, as, probably, did its Pleistocene relatives, and, when foraging, even moves in the truncally erect position, shuffling slowly on its haunches, hindlimbs flexed under it. Thus, *truncal* erectness is more habitual than in any non-bipedal catarrhine, and the mastoid process becomes explicable. Also, the forelimb is more "liberated" from locomoter function in *Theropithecus* than in any other non-biped, simply because the animal rarely locomotes. Sitting upright allows both hands to be used simultaneously for rapid gathering of small food-objects, a pat-

Table 1 *Adaptive Characters of the Villafranchian Hominidae and* Theropithecus. *(Column A. Characters distinguishing early Hominidae from* Pan *and other Pongidae. B. Characters distinguishing* Theropithecus *from* Papio *and* Mandrillus. *C. Features of the Hominid complex not seen in* Theropithecus. *D. Features of the* Theropithecus *complex not seen in Hominidae.)*

	A	B	C	D	Note no.
1. Behaviour					
a. Open-country habitat, not forest or woodland	X	X	—	—	
b. Trees rarely or never climbed when feeding	(X)	(X)	—	—	1
c. One-male breeding unit	(X)	(X)	—	—	1, 2
d. Foraging mainly in sitting position	?	(X)	—	—	1
e. Small daily range	?	(X)	—	—	1
f. More regular use of artefacts in agonistic situations	X	—	X	—	3
g. Regular use of stone cutting-tools	X	—	X	—	4
h. Most food collected by index-pollex precision grip	?	(X)	—	—	1
2. Postcranial structure					
a. Hand more adept, Opposability Index higher	X	X	—	—	5, 6
b. Index finger abbreviated	?	X	—	—	7
c. Hallux short and weak	—	X	—	X	7
d. Hallux relatively non-abductible	X	X	—	—	7, 8
e. Foot double-arched	X	—	X	—	8
f. Phalanges of pedal digits 2–5 shorter	(X)	X	—	—	7
g. Ilium short and reflexed	X	—	X	—	9
h. Sacroiliac articulation extensive	X	—	X	—	9
i. Anterior-inferior iliac spine strong	X	—	X	—	9
j. Ischium without flaring tuberosities	X	—	X	—	9
k. Accessory sitting pads (fat deposits on buttocks) present	(X)	(X)	—	—	7
l. Femur short compared with humerus	?	X	—	—	7
m. Distal end femur indicates straight-knee 'locking'	X	—	X	—	9
n. Epigamic hair about face and neck strongly dimorphic	(X)	(X)	—	—	1, 7
o. Female epigamic features pectoral as well as perineal	(X)	(X)	—	—	1, 10
3. Cranium and mandible					
a. Foramen magnum basally displaced	X	—	X	—	11
b. Articular fossa deep, articular eminence present	X	—	X	—	9
c. Fossa narrow, post-glenoid process appressed to tympanic	X	X	—	—	9, 7
d. Post-glenoid process often absent, superseded by tympanic	X	—	X	—	9
e. Post-glenoid process long and stout	—	X	—	X	7, 12
f. Basi-occipital short and broad	X	X	—	—	9, 7
g. Mastoid process regularly present	X	X	—	—	9, 7, 13
h. Temporal origins set forward on cranium	X	X	—	—	9, 7
i. Ascending ramus vertical, even in largest forms	X	X	—	—	9, 7, 12
j. Mandibular corpus very robust in molar region	X	X	—	—	9, 7, 12
k. Premaxilla reduced	X	X	—	—	9, 7

Table 1   *(continued)*

	A	B	C	D	Note no.
l. Dental arcade narrows anteriorly	X	X	—	—	9, 7
m. Dental arcade of mandible parabolic, 'simian' shelf absent	X	—	X	—	9
n. Dental arcade (especially in larger forms) V-shaped; shelf massive	—	X	—	X	7
4. *Teeth*					
a. Incisors relatively small and allometrically reducing	X	X	—	—	9, 7
b. Canine relatively small, especially in larger forms	X	X	—	—	9, 7
c. Canine incisiform	X	—	X	—	9
d. Male canine 'feminised', little sexual dimorphism in canines	X	—	X	—	9
e. Third lower premolar bicuspid	X	—	X	—	9
f. Sectorial face of male P_3 relatively small and allometrically decreasing	—	X	—	X	7
g. Molar crowns more parallel-sided, cusps set towards edge	X	X	—	—	14, 7
h. Cheek-teeth markedly crowded mesiodistally	X	X	—	—	14, 7
i. Cheek-teeth with deep and complex enamel invagination	—	X	—	X	7
j. Cheek-teeth with thick enamel	X	—	X	—	
k. Canine eruption early relative to that of molars	X	X	—	—	7, 9
l. Wear-plane on cheek-teeth flat, not inclined bucco-lingually	X	—	X	—	9
m. Wear on cheek-teeth rapid, producing steep M_1–M_3 'wear-gradient'	X	X	—	—	14, 7

1. Crook 1966; 1967; Crook & Aldrich-Blake 1968. Parentheses indicate behavioural and soft-part features which are present in the living representative of the group (*Theropithecus gelada* or *Homo sapiens*), but which cannot be demonstrated on fossil material.
2. In all but a very few human societies, where polyandry sometimes occurs (Murdock 1949).
3. The 'bashed baboons' of the South African cave sites (Barbour 1949; Dart 1949b, etc.) are the most direct evidence for this.
4. Not, apparently, in the earliest Pleistocene hominid sites (Howell 1967).
5. Napier 1962.
6. J. R. Napier, personal communication.
7. Pocock 1925; Jolly 1965; 1970a; 1972.
8. Day & Napier 1964.
9. Le Gros Clark 1964.
10. Matthews 1956.
11. The functional interpretation of this character is disputed (cf. Le Gros Clark 1964; Biegert 1963).
12. Leakey & Whitworth 1958.
13. Not in *T. gelada*, but regular in larger Pleistocene forms.
14. Simons & Pilbeam 1965.

tern seen more rarely in *Papio* where a tri-pedal stance leaving one hand free is associated with a diet mainly of large items (Crook & Aldrich-Black 1968).

The majority of "AB" characters are in the jaws and teeth. The temporal muscles (which are large in *Theropithecus* and some, at least, of the early Hominidae) are set well forward, so that their line of action lies almost parallel to that of the masseters, and their moment-arm about the temporo-mandibular joint is relatively long, as compared to that of the resistance of food-objects between the teeth, thus exerting a high grinding or crushing force per unit of muscular exertion. On the other hand, the gap between the opposing occlusal surfaces is small per unit of muscular extension, limiting the size of objects that can be tackled. Also, the horizontal component of temporal action is reduced, lessening its effectiveness in bilateral retraction of the mandible against resistance ("nibbling"), and in resisting forces tending to displace the mandible forwards, as when objects are held in the hand and stripped through the front teeth. Thus efficiency of incisal action, which is used by catarrhines in fruit-

peeling, nibbling flesh of fruits from rinds, stripping cortex from esculent vines and tubers, and, occasionally but very significantly, tearing mammal-meat from bones or skin (DeVore & Washburn 1963; Goodall 1965), is sacrificed to adding the power of the temporals to that of the masseters and pterygoids, used mainly for cheek-tooth chewing.

In the larger forms (of both taxa), with their allometrically longer faces, the forward position of the temporals is preserved by making the ascending ramus of the mandible high but vertical, deepening the posterior maxilla. The tooth row scarcely lengthens, although the face is long from prosthion to nasion, and the cheek-teeth become mesio-distally crowded. In the fruit-eaters the horizontal component of temporal action is maintained by keeping the ascending ramus low as the face lengthens; the corpus becomes elongated and marked diastemata tend to appear in the tooth row. The short basi-occiput, and anterio-posteriorly narrowed articular fossa of both *Theropithecus* and Hominidae can be seen as part of the same functionally-determined developmental pattern. The proportions of the molars and incisors fit the same functional complex. Both *Theropithecus* and the hominids have narrower, smaller incisors than their woodland and forest relatives. Molar area is greater per unit of body-mass, and incisal width less, in *Theropithecus* than *Papio* or *Mandrillus*, and absolute incisal breadth is no greater in *T. oswaldi mariae*, as big as a female gorilla, than it is in *T. gelada*. In *Papio* both incisors and molars increase proportionately to body-mass in larger forms, and in the forest-dwelling *Mandrillus* it is the molars which in males are scarcely larger than those of females half their weight (Jolly 1969; 1972). Among the Hominoidea, the Villafranchian Hominidae are *Theropithecus*-like in their dental proportions, and perhaps in their allometric ratio; the large form "Zinjanthropus" has the most extreme relative incisal reduction, while *Pan* is *Mandrillus*-like in proportions and ratios (Jolly & Chimene in preparation).

In the monkeys, the evidence for molar dominance in *Theropithecus* agrees well with data on diet in the natural habitat. In the few areas where the Pleistocene sympatry of *Theropithecus* and *Papio* still exists in the Ethiopian highlands, *Theropithecus* eats small food objects requiring little incisal preparation, but prolonged chewing, while *Papio* (which elsewhere in its wide range is a most catholic feeder) here concentrates on flesh fruits and other tree products, most of which require peeling or nibbling with incisors (Crook & Aldrich-Blake 1968). There seems no good reason against attributing the *Theropithecus*-like incisal proportions and jaw characters of the early hominids to a similar adaptation to a diet of small, tough objects. There is no need to postulate a compensatory use of cutting-tools for food preparation, until it can be shown archaeologically that such tools were being made (cf. Pilbeam & Simons 1965).

To avoid the charge of Lamarckism, I should perhaps suggest some selectional mechanisms leading to incisal reduction in molar-dominant forms. One is a general explanation of the reduction of structures to a size related to their function. Every structure is at once a liability, in that it can become the site of an injury or infection and requires energy and raw materials for its formation and maintenance, and an asset in so far as it performs a homeostatic function. Natural selection will favour the genotype producing a structure of such size and complexity as to confer the greatest *net* advantage. In the monkey or hominoid adapting to a gelada-like diet, each unit of tooth-material allotted genetically to a molar will bring a greater return in food processed than a unit allotted to an incisor. Thus selection should favour the genotype which determines the incisors at the smallest size consistent with their residual function. This "somatic budget effect" differs from Brace's (1963c) "random mutation effect" (criticised by Holloway (1966b) among others) chiefly in that it proposes a positive advantage in reduction.

A second mechanism is specific to teeth. While dental size is genetically (or at least antenatally) determined, the development of the alveolus depends partly upon the stresses placed upon it during its working life (Oppenheimer 1964). An underexercised jaw may thus be too small to accommodate its dental series, which tends to become disadvantageously crowded and maloccluded. Natural selection will then favour the genotype which reduces the teeth to a size fitting the reduced alveolus. The "Oppenheimer effect," originally proposed to explain the reduction of complete dentitions (as in the case of *Homo sapiens* after the introduction of cooking and food-preparation), could equally operate on particular dental regions, as in the case of *Theropithecus* and the early hominids, where the incisors were reduced but the molars were, if anything, larger than those of their forest- and woodland-dwelling relatives.

One of the most surprising findings is that canine reduction is one of the shared characters. This is contrary to the weapon-determinant hypothesis which states that "open-country," terrestrial primates, being more exposed to predation, should have *larger* canine teeth, unless they use artificial weapons. The situation in *Theropithecus* is somewhat complicated by the existence of both allometric and evolutionary trends toward canine reduction (Jolly 1972). The canines decrease in relative size from the Basal to the Upper Pleistocene, when compared between forms of approximately equal body-size. And within each palaeospecies, there is evidence that the canine is smallest, relative to the molars and to the general size of the animal, in the largest forms, where the males were about the same size as a female gorilla. The allometric trend is exactly opposite to that seen in present-day *Papio*, in which males of the largest forms have relatively and absolutely the largest canines. (*Theropithecus gelada*, a very small form of highly-evolved *Theropithecus*, has a male canine size that is relatively large for the genus, but entirely predictable from the allometric ratios between molars and canines characteristic of the whole genus.)

Both the allometric and the evolutionary trends are incompatible with the theory that canine size in males is positively correlated with terrestrial life in non artefactual primates, but at least two alternative explanations for canine reduction are possible. It may be favoured as an adaptation to increased efficiency in rotary chewing, by avoiding canine "locking" and producing more even molar wear; this would be consistent with the evidence for "molar dominance." An early stage of adaptation might involve the canines being worn flat as they erupted. This would obviously be a wasteful situation which might be expected to be corrected by the "somatic budget" effect, and natural selection.

The other possible explanation relates canine reduction to reduction of the incisors. The fact that the two trends parallel each other so closely in both *Theropithecus* and the Hominidae, both evolutionarily and allometrically, suggests, *a priori*, a relationship between these processes. Since incisal reduction can be plausibly interpreted as an adaptation to small-object feeding, it seems reasonable to create canine reduction as the secondary, dependent, character. The dependence can be attributed to either, or both, of two mechanisms. First, the direct effects of incisal disuse upon the anterior alveolar region might produce an extended "Oppenheimer effect," acting primarily on the incisors, and secondarily and less intensely, on the neighbouring canines. This might explain the fact that while incisal reduction seems fully evolved already in the rather primitive, probably early, Makapan *Theropithecus*, canine reduction proceeds through the Pleistocene. Alternatively, the dependence might be at the genetic level, with canine reduction being a simple pleiotrophic effect of a genotype which primarily determined incisal reduction. There is some evidence for a canine-incisal genetic "field" in both Cercopithecoidea (Swindler *et al.* 1967) and Hominoidea (Jolly & Chimene, unpublished data). It may well be that both selective factors are operative in canine reduction; adaptation to rotary chewing favouring crown height reduction, and effects stemming from incisal reduction acting upon crown-area dimensions. Since the genetic factors determining these two parameters of canine size are most unlikely to be independent of each other, the two processes would be mutually reinforcing.

This scheme for canine reduction in *Theropithecus* is distinct from that of Simons and Pilbeam (1965) who also attempted to explain both incisal and canine reduction (in *Ramapithecus*) in terms of diet. They proposed that canine size was related to its own function in food preparation, thus exposing themselves to Washburn's objection that if this were so canine sexual dimorphism would imply a sexual dietary difference in non-human catarrhines which is not borne out by field observations. The scheme proposed here recognises the essentially agonistic function of the canine but suggests that its reduction in *Theropithecus* is unrelated to this function, and is a secondary effect of dietary influences on incisors and molars. If as a consequence of a dietary aspect of terrestrial adaptation, and in the complete absence of either bipedalism or use of artificial weapons, a trend towards canine reduction can be initiated in *Theropithecus* (a highly terrestrial catarrhine), then there is no need to postulate that these characters either preceded or accompanied the earliest stages (at least) of canine reduction in Hominidae, which could similarly be attributed to dietary factors.

Having, I hope, established that adaptation to terrestrial life and small-object feeding constitutes at least a reasonable working model for the initial hominid divergence from Pongidae, I should now like to speculate about the characters of soft tissues and social organisation, although admittedly these can never be tested against the fossil record.

Unlike the savannah-woodland species, all three truly 'open-country' Cercopithecinae (*Theropithecus, Erythrocebus* and *Papio hamadryas*) have a social organisation involving exclusive mating-groups with only one adult male. Of the three, the patas is peculiar in that a female-holding male maintains his exclusive sexual rights by vigilant and agonistic behaviour directed against other adult males, whose presence he will not tolerate (Hall *et al.* 1965). Patas 'harems' do not, therefore, co-exist as parts of a higher-order organisation. In the hamadryas and gelada, however, the male maintains cohesion of his group by threatening and chastising his own females when necessary, and many one-male groups co-exist within semi-permanent troops or bands (Kummer 1967a; Crook 1967). In both species, cooperation between males is not excluded, and is frequently seen in situations of extra-troop threat. It would not be unreasonable to expect a similar social organisation, with permanent, monogamous or polygamous one-male groups set within the matrix of a larger society, to be developed by a hominoid adapting to a gelada-like way of life. This pattern is also, one might add, still distinctive of the vast majority of *Homo sapiens* (Murdock 1949).

In both hamadryas and gelada the attention-binding quality of the adult male is enhanced by the conspicuous cape of fur about his shoulders, which is groomed by his harem. It is likely that this feature has been favoured by Darwinian sexual selection (Jolly 1963). Only one other living catarrhine has such striking sexual dimorphism in epigamic hair about the face and neck: *Homo sapiens.* Similarly, bearing the female epigamic features pectorally and ventrally, rather than perineally, is a feature which can be correlated, in geladas, with a way of life in which the majority of the foraging time is spent sitting down (Crook & Gartlan 1966). It is unique to *Theropithecus* among non-human primates, but also occurs in *Homo sapiens.*

Fatty pads on the buttocks, adjacent to the true ischial callosities, are another *Theropithecus* peculiarity (Pocock 1925) which can be plausibly related to the habit of sitting while feeding, and also occur uniquely in *Homo sapiens* among the Hominoidea.

DIVERGENCES BETWEEN THE THEROPITHECUS AND HOMINID ADAPTATIONS

So far I have concentrated on the parallelisms between the *Theropithecus* and hominid adaptive complexes. Their divergences are, however, equally instructive. Columns C and D of table I have been added to extract those characters which occur as part of one of the complexes, but not of the other. Since characters were initially selected because they fitted either A or B (that is, they discriminate *within* superfamilies), I omit the large number of attributes which simply indicate that *Theropithecus* is a monkey but the Hominidae are Hominoidea, a point that is not at issue. The last two columns therefore indicate features adaptive to different aspects of the 'open-country' habitat, and analogous

adaptations to the same aspect. Again, we must examine their functional implications.

The 'AB' characters of the jaws and teeth seem to comprise a functional complex related to a diet of small, tough objects. In *Theropithecus,* these are known to be mostly grass blades and rhizomes, accounting for the tendency towards cheek-tooth crown complexity which is convergent upon similar structures in grass-eating animals as diverse as voles, warthogs and elephants. This character is not part of the hominid complex, which instead includes cheek-teeth with relatively low cusps but thick enamel which wear to an even, flat, and uniform surface. Also, the temporo-mandibular joint of the hominids is unique among catarrhines in the possession of an articular eminence upon which the mandibular condyle rides in rotary chewing. Thus, the molar surfaces do not simply grind across one another in chewing, they also swing towards and away from one another. Such molars and jaw action are clearly not adapted to mincing grass blades, but rather to breaking up small, hard, solid objects of more or less spherical shape, by a combination of crushing and rolling such as is employed in milling machines. The efficiency of the combined action lies in its seeking out, by continuous internal deformation, the weaknesses in the structure of the object to be crushed.

The possession of a parabolic dental arcade (rather than the V-shape seen in large *Theropithecus*), and the absence of the 'simian' symphyseal shelf, are also explicable on the basis of such a diet. The features cannot be related to tooth-function in any obvious way, nor can they be attributed simply to a Hominoid rather than Cercopithecoid inheritance, since the Hominoid *Gigantopithecus bilaspurensis* has a *Theropithecus*-like arcade and symphysis. A possible functional explanation involves the tongue. Experiments with grain-chewing in *Homo sapiens* suggest that objects not crushed by one masticatory stroke tend to be pushed by the rolling action into the oral cavity, whence they are guided back to the teeth by the tongue. This demands much more constant, agile, lingual motion than is needed in masticating a fibrous bolus of fruit (or turf). Thus, a chewing apparatus of the hominid type might be expected to include a thick, muscular, mobile tongue, accommodated in a large oral cavity. The highly-arched palate, capacious interramal space, and absence of symphyseal shelf may all be interpreted as elements of the large oral cavity, also, incidentally, providing preadaptations to articulate speech.

We can thus distinguish a sub-complex of unique hominid characters which suggests that the 'small objects' of the basal hominid diet were solid, spherical, and hard. Many potential foods fit the description, but only one is widespread enough in open country to be a likely staple. This is the seeds of grasses and annual herbs, which still provide the bulk of the calories of most hominids. This is not to say that other resources were not exploited when available, but that the diet of basal hominids was probably centred upon cereal grains as that of the chimpanzee is upon fruit and that of the gorilla on herbage.

Although canine reduction is one of the AB features, the incisiform shape of the canine, and its 'feminisation' in the male are unique, among catarrhines, to hominids of australopithecine and later grades. Presumably these features represent an extreme stage of reduction in adaptation to rotary chewing, reflected in the flat wear-plane of the cheek-teeth. One might therefore ask why a similar degree of reduction has not appeared in the *Theropithecus* line. Several explanations are possible. First, the chewing motion involved in grass-mincing might not be as demanding as that used for seed-milling. Second, a relatively high canine/incisor ratio and high sexual dimorphism are probably Cercopithecoid heritage charaters, which would tend to blunt the effect of 'genetic fall-out' from incisal reduction. A reduction in anterior tooth size which leaves the cercopithecoid *Theropithecus* with a canine reduced in size, but still a useful weapon, might reduce the canine of a hominoid beyond the point of usefulness. Perhaps because the canines of all Pongidae are shorter and blunter than those of Cercopithecoidea, a tendency to use artefacts as weapons was probably another hominoid heritage character, thus permitting the Hominidae to compensate for the loss of biting canines favoured by rotary chewing, and allowing the extreme degree of reduction represented by incisification and feminisation. Alternatively, it may be that *Theropithecus,* a much more recent lineage than the hominids, has not yet had time to achieve full canine reduction.

A second group of C and D characters is postcranial and reflects the fact that while *Theropithecus* is a quadruped, at least when moving more than a few paces, the Villafrancian hominids were evidently bipeds of a sort. Again, we can evoke a combination of adaptational and heritage features in explanation. A gelada-like foraging pattern leads to constant truncal erectness in the sitting position, with the trunk 'balanced' on the pelvis, and the forelimbs free. In *Theropithecus,* this behavioural trait (and its associated adaptive features) are superimposed upon a thoroughgoing, cercopithecoid quadrupedalism, producing a locomotor repertoire in which the animal abandons 'bipedal' bottom-shuffling for quadrupedal locomotion when it moves fast or for more than a few paces. The hominoid ancestor of the Hominidae, on the other hand, is most unlikely to have been postcranially baboonlike. The smallest, and best known, dryopithecine (*D. (Proconsul) africanus),* described as a 'semibrachiator' (Napier & Davis 1959), shows limb proportions and other features recalling generalised arboreal climbers like some of the Cebidae, but no distinctively cercopithecoid features. Its larger relatives had probably moved in the direction of truncal erectness, forelimb independence and abductibility, and the other characters distinctive of the Pongidae as a group, which are generally attributed to brachiation and can be seen as the consequences of large size in an arboreal habitat (Napier 1967a).

Recently Walker and Rose (1968) and Walker (personal communication) have detected signs of locomotor adaptations very like those of the living African pongids in the fragmentary (and largely undescribed) postcranial remains of the larger African Dryopithecinae. In the basal hominid, therefore, the 'gelada' specialisations would be superimposed upon a behavioural repertoire and post-cranial structure already attuned to some degree of truncal erectness. This combination of heritage and adaptation may have been the elusive determinant of terrestrial bipedalism, a gait that is inherently 'unlikely', and which would thus have begun as a gelada-like shuffle. Locomotion of any kind is infrequent during gelada-like foraging, so that (unlike hunting!) it is an ideal apprenticeship for an adapting biped. Furthermore, as a final bonus, if the hominids were derived from a 'brachiator' or knuckle-walker stock, they would have carried, in preadaptation, the high intermembral index which *Theropithecus* has had to acquire as part of his adaptation.

A NEW MODEL OF HOMINID DIFFERENTIATION: PHASE I, THE SEED-EATERS

The anatomical evidence seems to suggest that at some time during the Tertiary, the populations of Dryopithecinae destined to become hominids began to exploit more and more exclusively a habitat in which grass and other seeds constituted most of the available resources, while trees were scarce or absent. However, the great majority of contemporary tropical grasslands and open savannahs (especially those immediately surrounding patches of evergreen rainforest in all-year rainfall areas), are believed to be recent artefacts of burning and clearance by agricultural man (Rattray 1960; Richards 1952; Hopkins 1965; White *et al.* 1954). Under climatic climax conditions the vegetation of the seasonal

rainfall tropics would almost always include at least one well-developed tree stratum, ranging from semi-deciduous forests through woodlands to *sahel* where paucity of rainfall inhibits both herb and tree strata (Hopkins 1965).

What, then, would have been the biotope of the grain-eating, basal hominids? The obvious answer is provided by the areas of treeless edaphic grassland which exist, even under natural conditions, within woodland or seasonal forest zones, wherever local drainage conditions cause periodic flooding, and hence lead to perpetual sub-climax conditions by inhibiting the growth of trees and shrubs (Richards 1952; Sillans 1958). These areas range in extent from hundreds of acres, like the bed of the seasonal Lake Amboseli, to a network of strips interlaced with woodlands (*dambos;* Ansell 1960; Michelmore 1939; Sillians 1958). Edaphic grasslands produce no tree-foods, but support a rich, all-year growth of grasses and other herbs, and are the feeding-grounds for many grass-eating animals (Ansell 1960). The remains of Villafrancian hominids are often found in deposits formed in such seasonal waters, as is Pleistocene *Theropithecus* (Jolly 1972), adding some circumstantial evidence that this was their preferred habitat. Seasonality in rainfall, producing a fluctuating water-level, is important to the development of edaphic grasslands, since perennial flooding leads to a swamp-forest climax. While there is little evidence for catastrophic desiccation in the tropics of the kind demanded by some models of hominid differentiation, there are indications that a trend towards seasonality persisted through the Tertiary, especially in Africa (Moreau 1951).

The first stages of grain-feeding adaptation probably took place in a *dambo*-like environment, later shifting to wider floodplains. The change from a fruit (or herbage)-centered diet to one based upon cereals would lead, by the evolutionary processes discussed, to the complex of small-object-feeding, seed-eating, terrestrial adaptations (see fig. I). Other grassland resources obtainable by individual foraging or simple, *ad hoc* co-operation like that seen in the woodland chimpanzee, would also be utilised. Such items as small animals, vertebrate and invertebrate, leafy parts of herbs and shrubs, and occasional fruits and tubers would be qualitatively vital, if only to supply vitamins (especially ascorbic acid and B_{12}), and minerals, and could easily be accommodated by jaws adapted to grain-milling.

The ability to exploit grass-seeds as a staple is not seen in other mammals of comparable size, though it is seen in birds and rodents, presumably because the agile hand and hand-eye co-ordination of a higher primate is a necessary preadaptation to picking up such small objects fast enough to support a large animal. With these preadaptations, and the adaptive characters of jaws, teeth and limbs, the basal hominids would have faced little competition in the exploitation of a concentrated, high-energy food (a situation which would hardly have existed had they, as the 'hunting' model demands, started to eat the meat of ungulates in direct competition with the Felidae, Canidae, Viverridae, and Hyaenidae). They would thus have attained a stable, adaptive plateau upon which they could have persisted for millions of years, peacefully accumulating the physiological adaptations of a terrestrial, 'open-country' species. There is no reason to suppose that they would show radical advances in intellect, social organisations, material and non-material culture, or communication, beyond that seen in one or other of the extant higher primates. The 'third ape' in Simons's phrase, remained an ape, albeit a hominid ape.

PHASE 2, "HUMAN" HOMINIDS

We do not therefore need to invoke late, 'human' characteristics in teleological explanation of initial hominid divergence. However, Phase I hominids would be uniquely preadapted to develop such features following a further, comparatively minor, ecologi-

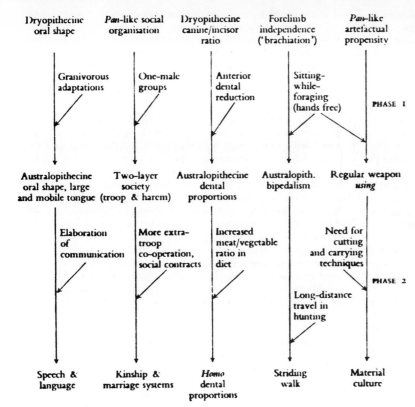

Figure I *A model of the development of some of the major hominid characters. During Phase 1 a series of Dryopithecine heritage characters (top line) is modified by the functionally determined require- ments of the sedentary seed-eating complex (second line), producing the characters of evolved Phase 1 hominids (line three). These are the preadaptive her- itage characters of Phase 2, which determine the fact that adaptation to the demands of a hunting way of life (fourth line) takes the form of the human traits listed in the bottom line. The illogicalities of previous models tend to arise from omitting the vital second line, inserting the elements of the hunting complex in its place, and invoking feedback.*

cal shift. The latter may have involved the increasing assumption by the adult males of the role of providers of mammal meat, with the equally important (but often neglected) corollary that the females and juveniles thereby became responsible for collecting enough vegetable food for themselves *and* the hunters. The adult males would perhaps be behaviourally predisposed to hunting by an existing role as 'scouts' (Reynolds 1966). This is as inherently likely in a species adapted to exploiting patches of seeding grasses as it is in the fruit-eating chimpanzee which Reynolds uses as a model of a pre-hominid hominoid. The environmental change prompting the inception of hunting need have been only slight; perhaps an intensification of seasonality in a marginal tropical area which would put a premium on exploiting meat as an addi- tional staple instead of an occasional treat. The dietary change would be small; an increase in the ratio of one high-energy, concentrated food (meat) to another (grain), which would be reflected dentally in a moderate reversal of the Phase I back-tooth dominance in favour of the incisal breadth needed to tear meat. The major impact of the change would be upon culture and society rather than upon diet itself. To a female, collecting vegeta- ble food for herself and her mate, there would be great advantage in developing techniques for more rapid harvesting, for carrying the day's booty, and for preparing it by a less laborious means than chewing. On the male side, there would be a premium on the development of cutting-tools for preparing the kill for transport back to the band. The skillful hands, upright posture, and reduced anterior dentition acquired as part of the Phase I complex would predispose the hominids to solve these problems of adaptation by the development of their hominoid artefactual propensity into true material culture, a solution which could not be predicted on the basis of the 'hunting chimp' model (see fig. I). In both sexes the division of labour would involve constantly postponing feeding for the sake of contributing to the communal bag, and, in males, the impulse to dominate would likewise have to be controlled in the interests of the hunt. The need for co-operation between local

bands may have led to the elaboration of truly human kinship systems, in which rights to females are exchanged.

All these factors, and others, were probably related to the evolution of complex forms of symbolic communication (largely, thanks to the seed-eating mouth, in the form of speech rather than gesture), language, ritual, and intellect. Thus the beginning of true hunting and the division of labour would initiate a second period of quantum evolution in hominid history, which we are still experiencing. The effects of this step upon human physical and behavioural evolution have been examined at length by others (e.g. Washburn & Lancaster, 1968), and are beyond the scope of this article. The point to be emphasised here is that this second, distinctively human phase is most comprehensible when it is built upon a firm base of preadaptations which had their initial signifi- cance in the seed-eating complex, not upon a chimpanzee-like or semi-human condition. By distinguishing the elements of the Phase 1 adaptive plateau, those of the Phase 2 'hominisation process' are thrown into relief.

After some populations had shifted into the Phase 2 cycle, there is no reason why they should not have existed sympatrically with other hominid species which continued to specialise in the Phase 1 niche.

THE FOSSIL RECORD IN THE LIGHT OF THE NEW MODEL

The new model must now be compared with the fossil record, both to test its compatibility and to relate its events to a timescale.

A medium-sized *Dryopithecus* of the Miocene is a reasona- ble starting-point for hominid differentiation, and increased sea- sonality in the Middle-to-Upper Miocene and Lower Pliocene makes it likely that Phase I differentiation began at that time. The fragmentary Upper Miocene specimens referred to *Ramapithecus* (Simons 1965b), though representing only jaws and teeth, are of precisely the form to be expected in an early Phase I hominid:

narrow, uprightly-placed and weak incisors, broad, large and mesio-distally crowded cheek-teeth, set in a short but very robust mandibular corpus. Recent work has shown (D. Pilbeam, personal communication) that the molar enamel is thicker than that of contemporary Dryopithecinae, though thinner than that of later hominids, and that the wear-gradient from anterior to posterior molars is steeper. In the Fort Ternan specimen the canine crown is small, but still conical, like that of a Mid-Pleistocene *Theropithecus* female of comparable size. The material is as yet insufficient to show whether or not the male canine was yet 'feminised'; this is immaterial to the argument that by Fort Ternan times the Hominidae had entered a granivorous niche in edaphic grasslands. Tattersall's recent (1969a) appraisal of the habitat of the Siwalik *Ramapithecus* is consistent with this hypothesis. He describes forested country crossed by watercourses which at the latitude of the Siwaliks must have fluctuated with a seasonal rainfall resulting in *dambo*-like conditions. The very incompleteness of the *Ramapithecus* material enables predictions to be made to test the seed-eating model. We may predict that the mandibular ascending ramus of *Ramapithecus* will be found to be relatively vertical, its postorbital constriction narrow, supraorbital radge projecting and face concave in profile; the postcranial skeleton should show a short ilium and short, stout phalanges, but also rather long arms and short, stout legs.

Among the Early Pleistocene hominids, the 'robust' australopithecines show exactly the combination of characters to be expected from long-term Phase I adaptation. Indeed, a major advantage of the two-phase model is that it makes sense of the apparent paradox of these hominids. Their specialisations (such as 'superhuman' incisal and canine reduction) are related to the seed-eating complex, while their apparent primitiveness (represented by characters like relatively small cranial capacity and comparativey inefficient bipedalism (Napier 1964; Day 1969; Tobias 1969b) is simply *absence* of Phase 2 specialisations.

Robinson (1962b; 1963b) is one of the few to see the robust australopithecines as representative of a primitive stage of hominid evolution, rather than a late and 'aberrant' line, and to recognise that basal hominids are unlikely to have been more carnivorous than pongids. He therefore comes closest to the present scheme, but does not solve the paradox of *robustus* by recognising the significance of *small-object* vegetarianism to the characters of Phase I adaptation. Recent discoveries suggest that robust australopithecines have a time-span in African running into millions of years; this would be compatible with our model of Phase I differentiations leading to an adaptive plateau, but not with schemes which see all australopithecines as incompetent and transient hunters, not those that see the robust group as a late 'offshoot'.

The population represented by the specimens called *Homo habilis* (Leakey *et al.* 1964) fit the model as early, but clearly differentiated Phase 2 hominids, as their describers contend, with a dentition in which the trend to back-tooth dominance has been partially reversed, a cutting-tool culture, and increased cranial capacity. At several African sites there is evidence of contemporary and sympatric Phase I *(robustus)* hominids, as predicted by the model.

If this interpretation is correct, there would seem ample justification for referring *habilis* to the genus *Homo,* on the grounds of its departure along the path of Phase 2 adaptation.

The 'gracile' australopithecines (excluding *habilis*) might fit into one of three places on the model (and, conceiveably, different populations referred to this species do in fact fit in different places). Possibly they are evolved Phase I hominids whose apparently unspecialised dental proportions might be attributed to an allometric effect of smaller size, as in *Theropithecus*. This interpretation, however, is unlikely, mainly because the size-difference between the robust and gracile groups seems insufficient to account for their divergences in shape by allomorphosis alone. Alternatively, they might be a truly primitive (Phase I) stock from which both robust australopithecines and Phase 2 hominids evolved. This view has been widely espoused, and if dental proportions were all, it would be most plausible. However, the known *africanus* specimens are probably too late and too cerebrally advanced to be primitive. Most likely, they are at an early stage of Phase 2 evolution, with their osteodontokeratic culture, perhaps improved bipedalism, and some cerebral expansion beyond that seen in the *robustus* group (Robinson 1963b). In this case their anterior dentition could be seen as secondarily somewhat enlarged from the primitive condition. This interpretation would favour sinking *Australopithecus* in *Homo,* while retaining *Paranthropus* for the evolved Phase I forms, as Robinson suggested.

The nature of an evolutionary model, concerned with unique events, is such that it cannot be tested experimentally. Its major test lies in its plausibility, especially in its ability to account for the data of comparative anatomy, behaviour, and the fossil record inclusively, comprehensively, and with a minimum of subhypotheses. It should also provide predictions which are in theory testable, as with a more complete fossil record, thus enabling discussion to move forward from mere assertion and counter-assertion. An evolutionary model which is designed to account for nothing beyond the data from which it is derived, may be entertaining, but has about as much scientific value as the *Just so stories*.

While none of the previous models of hominid differentiation is without plausibility, none is very convincing. Too few of the elements of the hominid complex are accounted for, and too often the end-products of hominid evolution have to be invoked in teleological 'explanation'. On the other hand, the nature of the causal factors invoked, especially behavioral ones, is often such as to make the hypothesis untestable.

The model presented here is based upon the nearest approach to an experimental situation that can be found inevolutionary studies, the parallel adaptation to a closely similar niche by a related organism. While based initially upon diet, and dental characters, it also accounts for hominid features as diverse as manual dexterity, shelfless mandible and epigamic hair, and for features of the fossil record such as the apparent paradox of *Paranthropus,* and the fact that hominid (or pseudohominid) dentitions apparently preceded tools by several million years. On the other hand, there seem to be no major departures from logic or from the data. It is therefore suggested that the two-phase model, with a seed-eating econiche for the first hominids, should at least be considered as an alternative working hypothesis against which to set new facts and fossils.

Note

This article is a revised and expanded version of a paper read in the Department of Vertebrate Paleontology, Yale Peabody Museum, on February 14, 1969. The helpful comments of Drs. Colin Groves, David Pilbeam, Elwyn Simons, and Alan Walker on this and other occasions are gratefully acknowledged.

42

Darwin's Apes, Dental Apes, and the Descent of Man: Normal Science in Evolutionary Anthropology
R. H. TUTTLE

The appearance of Darwin's evolutionary thesis (1859) created a commotion, and its centenary generated a spate of commemorative volumes (Bennett 1958, Huxley 1958, Huxley et al. 1958, Darlington 1959, Loewenberg 1959, Tax 1960). By contrast, Darwin's major anthropological essay, *The Descent of Man* (1871), received little commemoration (Campbell 1972, Day 1973). Partly as an apology for this apparent pretermission by the anthropological community, I should like to recall briefly Darwin's model on the early phases of hominid evolution, summarize durable models of several latter-day evolutionists, and comment on select aspects of contemporary knowledge on hominoid evolutionary biology. Next I will venture an appeal for a renewed quest for arboreal imperatives in hominid phylogeny. I will conclude with a brief commentary on normal science in evolutionary anthropology.

DARWIN'S MODEL

Darwin considered three basic aspects of human evolutionary biology in *The Descent of Man: (a)* whether man had evolved from preexisting forms, *(b)* the manner of human development, and *(c)* the relative importance of differences between the "so-called races" of man. The bulk of the book is devoted to the first and the third of these subjects.

Darwin was properly brief and cautious concerning the course of ante-hominid phylogeny. While admitting that some form of "anthropomorphous ape" was antecedent to man, he warned that "we must not fall into the error of supposing that the early progenitor of man was identical with, or even closely resembled, any existing ape or monkey" (1871: 520). Like Huxley (1863) before him, Darwin concluded that man's nearest relatives among extant forms are the African apes. He clearly favored Africa, the contemporary homeland of gorillas and chimpanzees, as the continent where the hominids had diverged from the pongids. He noted that Africa had not been explored by geologists; hence, the nonavailability of fossil forms that might reveal the connections between man and ancient apes.

Darwin ventured very little by way of a model on what the common ancestor of the pongids and man may have looked like or what selective factors were operant during the early stages of the hominoid adaptive radiation. He did speculate in some detail on the appearance of penultimate progenitors of man. These he described (p. 524) as hairy, bearded, tailed creatures that could move their pointed ears freely. They possessed prehensile feet with opposable great toes. The males had large canine teeth that were used as weapons. They were arboreal inhabitants of tropical or subtropical forests. In brief, Darwin's apes are fuzzy apes indeed.

Huxley (1863), Haeckel (1866, 1868, 1874), and other contemporaries of Darwin also neglected to describe in a clear-cut fashion the appearance of hypothetical hominoid ancestors of man or to speculate in much greater detail than Darwin did on the mechanisms whereby the hominid functional-morphological complex evolved. Instead, they were of necessity preoccupied with the demonstration that man had special affinities with the anthropoid apes and that he was indeed a natural product of evolution.

THE KEITH-GREGORY MODEL

Evolutionary theorists did not begin to deal substantively with the mechanisms whereby apes of some sort evolved into upstanding bipedal striding hominids until the turn of the century. In 1903, Keith published a preliminary rendering of his four-stage hypothesis on the evolution of man based upon his behavioral observations and dissections of langurs, macaques, and gibbons in Thailand and a wider acquaintance with primate morphology from anatomical studies in England. He stressed that specific arboreal locomotive postures of apes contain important clues to the manner by which man obtained upright posture. According to Keith's thesis, pronograde catarrhine-monkey-like primates gave rise to relatively small-bodied, orthograde, "brachiating" gibbon-like primates—the *hylobatians*. From the hylobatians evolved large-bodied orthograde apes, which Keith initially termed "giant primates" (1903) and later *troglodytians* (1923:453). He believed that the fundamental functional-morphological complex of the brachiators was virtually full-fledged in the hylobatian stage (1923). The troglodytian stage was achieved primarily by an increase in body size (fig. 1). The hominid lineage was established when a stock of arboreal giant apes became adapted "by what means we know not" to plantigrade progression on the ground (1903:19). During the transition from the troglodytian to the human stage, morphological changes were confined almost entirely to the lower limbs.

During the period between 1903 and the mid-1930's, Keith (1912–34) reiterated and refined his brachiationist theory and defended the ape model of human ancestry against a series of influential pongidophobiacs including Boule (1912: 165–166; 1913:59–61), Klaatsch (1923), Osborn (1926–30), and his own student Wood-Jones (1915–48). He was ably joined in this cause by the American paleontologist Gregory (1916–49), who argued eloquently and vigorously that "the habit of brachiating" must have played an imperative role in the development of upright posture in man.

Among the shared postcranial anatomical features of man and apes, the following may be cited to support the brachiationist theory of common heritage from a stock of ancient orthograde apes that engaged in suspensory behavior (Tuttle 1969a):

> long upper limbs relative to trunk length and long forearms and hands in particular

Reprinted with permission from *Current Anthropology*, Volume 15, pages 389–398. Copyright © 1974 by the University of Chicago Press, Chicago.

Figure 1 *Suspensory postures of* Pan troglodytes *and* Hylobates lar. a, *juvenile male chimpanzee hanging from wire of cage roof; note full extension of wrist, elbow, and shoulder joints, relatively long forelimbs, prominent thumbs, and broad chest; b, juvenile male chimpanzee arm-swinging across the roof of wire cage; note full extension of right forelimb joints and rotation of right shoulder joint; c, adult male gibbon in single-forelimb suspensory posture; note full extension of right wrist, elbow, and shoulder joints and remarkable rotation of right shoulder joint; d, two-forelimb suspensory posture (same subject as in c); note broad chest and relatively long limbs compared with trunk; e, two-forelimb suspension (same subject as in c) with notable augmentary support supplied by feet resting on a lower horizontal branch; note that thumbs do not participate in the grip; f, suspension from right forelimb and right hindlimb (same subject as in c) while engaged in manipulatory behavior with left hand; g, vertical climbing on three juxtaposed trees (same subject as in c); the halluces are opposed to the other pedal digits; on the smaller tree, the long right pollex is opposed to the other right manual digits to effect a grip while on the larger tree the left pollex is aligned with the other left manual digits.*

thumbs of notable absolute size

broad, anteroposteriorly flattened chests

strongly angled ribs and protrusion of the vertebral column into the chest cavity

a broad sternum, with progressive fusion of the sternal elements in adults

long collar bones

shoulder blades located on the posterior aspects of the chest wall

large acromial processes on the shoulder blades

laterally directed shoulder joints

remarkable mobility of the shoulder and elbow joint complexes

relatively large muscles that raise the arm and rotate the shoulder blade

a characteristic arrangement of muscle fibers in the diaphragm

characteristic positions of the heart, lungs, and other organs in the body cavity

close attachment of some abdominal organs to the diaphragm and posterior abdominal wall

a relatively short truncal segment of the vertebral column

lumbar back muscles that are smaller and show less fasciculation than those of monkeys

no tail

reduced coccyx and more sacral vertebrae than monkeys have

a muscular pelvic diaphragm in the pelvic outlet

MORTON'S HYLOBATIAN MODEL

Next to Keith and Gregory, the most sophisticated early advocate of the brachiationist argument was Morton, an orthopedic surgeon who worked with Gregory and others at the American Museum of Natural History. Although he agreed wholeheartedly with Keith and Gregory that the hominid lineage derived most of its postcranial resemblances with extant apes during a common stage of arboreal orthograde habits (Morton 1922–35), he proposed that hominids evolved from hypothetical gibbon-sized apes instead of large-bodied troglodytians. Unfortunately, the intriguing potential theoretic tensions thereby created between his model and the Keith-Gregory model were never openly debated by the principals, perhaps because they were all preoccupied with defending the general brachiationist perspective against frequent assaults by others.

Unlike Keith and Gregory, who, except for occasional minor digressions, left the reader to imagine forms remarkably like extant apes, Morton (1924a:39) provided a rather detailed description of the hypothetical prehominid ape:

1. About the size of the modern gibbon but stockier in build.
2. Arms and legs of equal length or nearly so.
3. Active and agile in its movements, freely brachiating after the manner of the chimpanzee, but not as the modern gibbon.
4. Completely quadrumanous with well-developed pollex and hallux.
5. Erect posture habitual, bipedal movements in the trees and on the ground, as in the modern gibbon. (The extended legs and the adaptation of the nervous system had unfitted it for quadrupedal locomotion.) [See my Fig. 2]
6. Cranium gibbon-like.
7. Tailless; the ischial pads had probably disappeared.

Morton believed that predecessors of modern hylobatid apes diverged from the hylobatian stock before the divergence of man and the radiation of the pongid apes. The small, agile protohominids separated from the great-ape stock by adopting terrestrial habits. Morton (1924b:88) postulated that the transition to terrestrial habitation progressed rapidly and that, from the outset, the protohominids were upstanding bipeds.

In designating Morton's model "hylobatian," I do not wish to imply any greater resemblance between his hypothetical protohominid and extant hylobatid apes than Morton himself did. If read fully, his theory is clearly distinct from the early paleontologically based ideas of Pilgrim (1915), Werth (1928:874) and Winge (Jensen, Sparck, and Volsφe 1941:320) [1], who posited especially close phylogenetic relationships between hylobatids and hominids.

Like Morton, Schultz (1927a, b, 1930, 1936, 1969c), the Swiss physical anthropologist who provided an unparalleled bulk of information on comparative primate morphology, favored early divergence of the hominid lineage from the common hominoid stock at a point near, but somewhat later than, furcation of the hylobatid apes and prior to radiation of the great apes. Schultz (1927a:40) initially expressed his predilection for a generalized protohylobatian model as follows: "For reasons not yet worked out I incline to the view that gibbons have retained a greater number of primitive characters which may have been the property of the 'missing link' than any other of the living apes." He closed his classic paper of 1936 on "characters common to higher primates and characters specific for man" with the comment (p. 452): "The Hylobatidae and man have evolved in opposite directions from a common, early stock, yet in a number of important features they have retained a closer similarity than

exists in regard to the same features between man and the great apes." That Schultz still abides by this thesis is evidenced by the diagrammatic family tree of the Hominoidea in his recent book *The Life of Primates* (1969c:251).

WASHBURN'S TROGLODYTIAN MODEL

The 1940s were the nonbrachiationists' heyday. In 1940, English anatomist-anthropologist Le Gros Clark formulated the hypothesis that although man and apes evolved from a common hominoid stock, the hominid furcation occurred relatively early and at a time when the characteristic postcranial features of modern apes were only incipiently developed in the basal hominoids. In the same year, Keith (1940) abandoned the brachiationist model, perhaps partly because of Le Gros Clark's influence. Wood-Jones (1940) reveled in Keith's recantation and redoubled his efforts to dissassemble the brachiationist models. Straus (1940, 1942a, 1949) emerged as an ardent opponent of brachiationist models, bringing to bear on the question of man's origins considerable knowledge derived from his extensive dissections of primates (1941–49). Straus, like the contemporary Keith, cited evidence from the comparative morphology of catarrhine hands as especially compelling against a brachiating stage in human evolution. Like Boule, he favored a generalized pronograde catarrhine-monkey-like progenitor for man.

Schultz (1951) also implied that brachiation had been relatively unimportant in human phylogeny. He derived the hominid lineage from an early hominoid stock in which incipient upright posturing had developed. He did not, however, delimit the hypothetical alternative mechanism whereby hominid uprightness had been achieved. Thus, insofar as one can identify Schultz's theoretical position on this matter, he seems to occupy the narrow border region between the more strictly catarrhine-monkey modelers, like Boule and Straus, and the prebrachiationists, represented by Le Gros Clark and the latter-day Keith.

Despite the new challenge to his theory from colleagues with remarkable comparative-anatomical expertise, Gregory experienced no change of mind on the subject. Thus the foundations of the American school of troglodytian brachiationists remained steadfast. But though Gregory (1949) reaffirmed his confidence in the orthodox brachiationist model, he never engaged Straus, Le Gros Clark, or Schultz in lively and prolonged exchanges like his debates with Osborn and Wood-Jones.

Morton also remained faithful to the brachiationist cause. In 1952, he stated that he was not convinced that available fossil evidence merited rejection of the hypothesis (Morton and Fuller 1952).

Hooton (1946) basically upheld Gregory's view that hominid phylogeny sequentially included small-ape and large-ape stages. He argued that the earliest ape acquired some of its distinguishing features by selection for habitually sitting upright and climbing tree trunks; it brachiated in moderation, sometimes walked bipedally on branches, and "perhaps even on occasion went on all fours along the boughs" (1946:134). He rejected small apes in favor of large apes as the immediate precursors of man, stating (p. 135) that "approximately modern human size...[was] probably antecedent to, and prerequisite for, man's career as a terrestrial biped." Hooton, like Keith, thus believed that large size and great weight might have forced the protohominids from the trees.

From 1950 onward, Washburn, a student of Hooton's who had the good fortune to work also with Gregory, Schultz, Straus, and Carpenter, picked up the gage cast down by the nonbrachiationists and became the new champion of the troglodytian brachiationist cause. Washburn (1950) first publicly expressed his strong support for a troglodytian precursor of man in the same

Cold Spring Harbor Symposium at which Schultz presented his prebrachiationist view. Thereafter Washburn and several inspired students and associates set out to refine the concept of brachiation and to further substantiate the troglodytian model with some new as well as traditional biological techniques, experimentation, and natural scientific perspectives (Washburn 1959-73; Washburn and Avis 1958; Avis 1962; Washburn and Hamburg 1965). By the 1960s, Washburn was so impressed by the resemblance between chimpanzee and man and the potency of "quantum evolution" that he surmised a remarkably late time of divergence for African apes and man—"perhaps some 2 to 4 million years ago"(Washburn and Hamburg 1965:11). This aspect of Washburn's model is reminiscent of Weinert's (1932) troglodytian theory.

Washburn believed that available Dryopithecinae did not possess the characteristic postcranial functional-morphological complex of the brachiators, inclusive of man. And since a form of advanced pongid brachiator was probably antecedent to extant African apes and man, many of the available dryopithecine fossils must represent collateral extinct Neogene lineages of "dental apes."

THE KNUCKLE-WALKING HYPOTHESIS

During the 1960s a renaissance and refinement of biomolecular perspectives on hominoid evolution produced overwhelming evidence that man, chimpanzee, and gorilla have remarkably close genetic affinities (Goodman 1962-67, Chiarelli 1962, Klinger et al. 1963, Williams 1964, Sarich and Wilson 1967, Wiener, Gordon, and Moor-Jankowski 1964). Concurrently, my thesis research and augmentary postdoctoral studies on catarrhine primate hands showed that the African apes are unique among Primates in possessing a functional-morphological complex termed knuckle-walking (Tuttle 1965-1975; Fig. 3). I suggested that the following anatomical features are especially related to knuckle-walking versus other aspects of African pongid locomotive repertoires (Tuttle 1965-1975, Tuttle et al. 1972, Tuttle and Basmajian 1974c):

> knuckle-pads of friction skin on manual digits II–V (Fig. 4)
> shortening of the long digital flexor muscles of manual digits II–V (Fig. 5)
> osseo-ligamentous structures and relationships that produce close-packed positions in metacarpophalangeal joints II–V and in the wrist joint (Figs. 6–8)

Washburn (1967–73), the champion of the troglodytian model, was quick to account for this mass of novel data in a manner minimally perturbing to its classic form. Thus he interpolated a stage of knuckle-walking between the stages of brachiation and bipedalism. He believed that knuckle-walking provided an answer to the long-standing puzzle of how a relatively large-bodied brachiator could survive during the transition from quasi-terrestrial, tool-less, herbivorous pongid to bipedal, weapon-bearing, carnivorous hominid. He proposed that knuckle-walking would enable troglodytian prehominids to move rapidly away from predators and to carry tools during locomotion (Washburn 1967, 1968a,b).

Since Washburn advanced the knuckle-walking hypothesis, several followers have variously entertained knuckle-walkers on man's family tree. Two of Washburn's colleagues at Berkeley—Clark (1970) and Sarich (1971)— have endorsed the new model. Pilbeam (1972:41) states that the Miocene protohominids may have been "at least incipiently knuckle-walkers," Napier (1970), Leakey (1971), Kortlandt (1972), and Robinson (1972:196) subscribe to the idea that at least *Paranthropus*, the robust australo-

pith, may have been a knuckle-walker. However, Robinson (1972:245) explicitly denies that *Homo africanus,* the gracile australopith, was anything but a highly advanced biped. Following my suggestion of 1967, Robinson argues (p. 206) that the hand bones from Olduvai F.L.K. NN I (Napier 1962) are as reasonably attributed to *Paranthropus* as to *Homo africanus* and cites them as possible evidence for knuckle-walking in early hominids; but he cites the absence of features related to knuckle-walking on a second metacarpal from Kromdraai as evidence that "knuckle-walking was at best a very insignificant part of the locomotor activity of this form." Perhaps the term "knuckling" would be more appropriate here for what would appear to be no more than a facultative hand posture with no clearly identifiable morphological correlates.

As evidence for a knuckle-walking stage in hominid phylogeny, Washburn (1968b:26) cited "the general ape anatomy" of the human hand, sparseness of mid-phalangeal hair on human fingers, and the hominoid hand bones from Olduvai F.L.K. NN 1. He did not, however, fully describe what he meant by "general ape anatomy." Although I have dissected more than 200 anthropoid hands, I cannot perceive a general ape hand. Though the human hand is clearly derived in common with those of the extant apes, certain particularities of gibbon, orangutan, and African pongid hands are so striking that I found it impossible to divine a general ape pattern to which the human hand could be compared or one that clearly associates man especially with the African apes. The condition of the much discussed "flexor pollicis longus" portion of the deep digital flexor complex is more similar in hylobatid apes (Fig. 9) and man than either is to pongid apes (Fig. 5), but the development of the contrahentes complex and the interosseous muscles places man closer to the great apes (Tuttle 1969b). Again, fusion of the os centrale to the scaphoid bone in the wrist (Schultz 1936) might place man with the African apes, but certain mensural studies on the thumb (Tuttle 1972b) and the configuration of the metacarpal heads (Fig. 7) might place him with the hylobatids.

The problem of glabrousness of the mid-phalangeal regions of the fingers is just as hairy. Although many persons possess depilated mid-phalangeal regions on the fingers, there is no histological evidence for a prehistory of knucklepads. Further, depilation of the toes is more common than depilation of the fingers in man, but only a lampooner would portray the protohominids as full-fledged pedal knuckle-walkers (Tuttle 1969a).

While the hand bones from Olduvai F.L.K. NN 1 strongly suggest a recent heritage or habitus or both of arboreal climbing and perhaps suspensory behavior in the hominoid to which they belonged, the morphological bits requisite to infer habitual knuckle-walking comparable to that of extant African apes are not available (Tuttle 1967:198; 1969a). Consequently, contrary to Washburn's commentaries on the subject, there is no unequivocal direct evidence to support a knuckle-walking stage in hominid evolution. Further, if Robinson (1972) is correct in assigning the metacarpal from Kromdraai to *Paranthropus*, the troglodyto-philes must now resolve the theoretic tensions created by at least one bit of direct evidence against a significant knuckle-walking stage in hominid phylogeny.

OTHER RECENT PERSPECTIVES

The brachiationist cause is now receiving a variety of new supporters, many of whom deny, sidestep, or simply pass over the problem of knuckle-walking.

Oxnard's (1968a,b, 1969) statistical studies on mammalian shoulder girdles, including available fossil hominid scapular and clavicular fragments, have led him to conclude that hominids at Sterkfontein and Olduvai Gorge Bed I "possessed structural modifications of the shoulder girdle towards considerable ability for

Figure 2 *(above left) Terrestrial bipedal walking (a–d, f) and running (e) postures in an adult male* Hylobates lar *(a–e) and an adult female* Symphalangus syndactylus *(f). The trunk tilts forward throughout the locomotive cycle, thereby displacing the center of gravity forward and perhaps reducing the amount of muscular force required to propel the body (as in running man). a, mid-stance phase of left hindlimb and swing phase of right hindlimb; note flexion of knees and curling of pedal digits II-V; b, early contact part of stance phase in left hindlimb and release part of swing phase in right hindlimb; note nearly full extension of left knee; c, early support part of stance phase in left hindlimb and early propulsive part of stance in right hindlimb; d, approximately full support part of stance phase in left hindlimb and precontact part of swing phase in right hindlimb; note remarkable flexion of left hip and knee joints and lateral displacement of the trunk to the left.*

Figure 3 *(upper right) Knuckle-walking by an adult female* Pan troglodytes *carrying an infant ventrally. The left hand demonstrates a remarkable degree of hyperextension at the metacarpophalangeal joints II-V. In the left forelimb, note the alignment of arm, forearm, and metacarpus, full extension of the elbow joint, and slight flexion (cf. volarflexion) of the wrist.*

Figure 4 *(lower right) a, left hand of a juvenile gorilla (*Pan gorilla*), showing knuckle-pads over the dorsal surfaces of middle phalanges II-V; b, left hand of a facultative knuckle-walking adult male* Pongo pygmaeus, *showing hairy skin over dorsal aspects of the middle phalanges of digits II-V.*

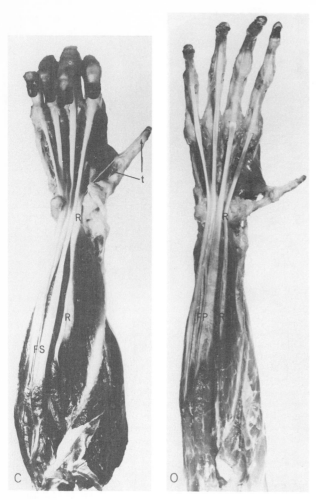

Figure 5 *Fresh preparations of the ventral aspect of the right forearm and hand of an adult male* Pan troglodytes *(C) and a juvenile* Pongo pygmaeus *(O). In C, the proper flexors of the wrist have been removed to expose the flexor digitorum superficialis muscle (FS) and the radial component of the flexor digitorum profundus muscle (R). In O, the proper flexors of the wrist and the flexor digitorum superficialis muscle have been removed to expose the ulnar and humeral (FP) as well as the radial (R) components of the flexor digitorum profundus muscle. Note that C possesses a vestigial tendon (t) to the pollex and O does not.*

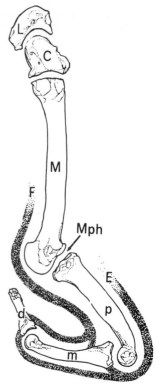

Figure 6 *"Exploded" third digital ray and associated carpal bones and tendons in the hand of an adult female chimpanzee in knuckle-walking posture; note the close-packed hyperextended position of the metacarpophalangeal joint (Mph) due to the backward extension of the articular surface on the metacarpal head (M). L, lunate; C, capitate; F, long digital flexor tendons; E, long digital extensor tendons; p, proximal phalanx; m, middle phalanx; d, distal phalanx. (Drawing by P. Murray; Copyright 1969 by the American Association for the Advancement of Science.)*

suspension of the body by the arms" (1968b:431). He does not subscribe to the knuckle-walking hypothesis.

I have expressed support (1969a) for the hypothesis that man evolved from a large-bodied arboreal ape that was adapted for some degree of suspensory posturing and locomotion. Contrary to Washburn's (1973:475) assertion, I did not state that man's ancestors *could not* have been knuckle-walkers. I simply elected not to subscribe to the thesis of a knuckle-walking stage in human evolution until its advocates knuckle down to the task of verifying it with detailed empirical evidence.

McHenry's (1973a) recent statistical studies on the distal humerus of hominoids, as I read them, show that *Paranthropus* did not possess those features that set the extant knuckle-walking apes apart from other hominoids. Whether the forelimbs of *Paranthropus* were otherwise employed in locomotion promises to be a subject of continued debate (Kay 1973, McHenry 1973b).

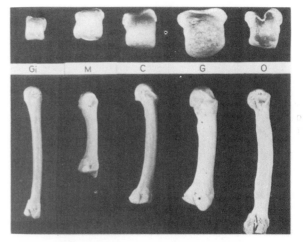

Figure 7 *Distal end (above) and medial aspect (below) of left metacarpal III in* Hylobates concolor *(Gi),* Homo sapiens *(M),* Pan troglodytes *(C),* Pan gorilla *(G), and* Pongo pygmaeus *(O). Note the considerable extension of articular surface onto the dorsal aspect of the metacarpal head and the well-developed ridge on the dorsodistal end of the metacarpal in C and G.*

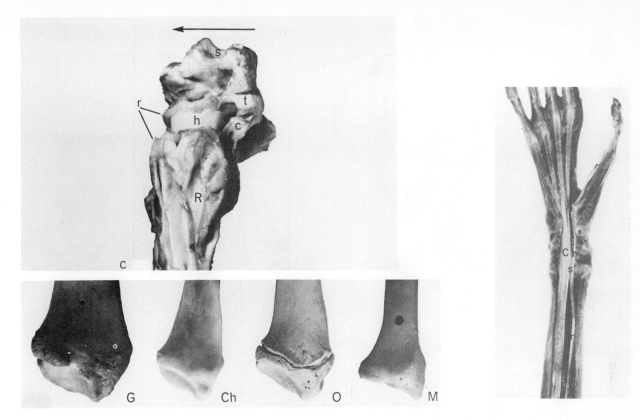

Figure 8 *Structure of the proximal wrist joint. Above, fresh preparation of the lateral aspect of the left wrist in an adult* Pan troglodytes. *A well-developed radiocarpal ligament (c) attaches proximally to the styloid process of the radius (R) and distally to carpal bones. As the hand moves into dorsiflexion (arrow), the radiocarpal ligaments become taut and the head (h) of the scaphoid bone rotates ventrally so that a close-packed position is effected at the proximal wrist joint. t, robust tubercle of the scaphoid bone; s, saddle-shaped facet on the trapezium bone which articulates with the base of the pollical metacarpal bone; r, ridges on the scaphoid bone and distal radius (see below) that are implicated in the close-packed position. Below, right distal radii of* Pan gorilla *(G),* Pan troglodytes *(Ch),* Pongo pygmaeus *(O), and* Homo sapiens *(M). In G and Ch, the carpal articular surface is deeply concave and oriented so as to limit extension (cf. dorsiflexion) of the wrist.*

Figure 9 *Fresh preparation of the ventral aspect of the right forearm and hand of an adult male* Hylobates lar. *The proper flexors of the wrist and the flexor digitorum superficialis muscle have been removed to expose the flexor digitorum profundus muscle. Note that in contrast to that of the pongid apes (Fig. 5) the radial component of the flexor digitorum profundus muscle here sends a robust tendon (r) to the distal phalanx of the thumb. Except for a "flexor shunt" (s) over the carpus, the pollical segment is separate from the ulnar and humeral segments of this muscle (C) and closely resembles the flexor pollicis longus muscle of man.*

In accordance with the classic troglodytian brachiationist model, Lewis (1969–73), a British anatomist, argues from hominoid wrists that man, chimpanzee, and gorilla share an especially close common ancestry characterized by advanced brachiation. He finds no striking features in the wrist bones of the African apes that might be especially related to knuckle-walking versus a heritage of brachiation. Although Lewis may continue to ignore certain facts in publications by Schreiber (1936) and me (Tuttle 1965–1975) on the subject, it will be interesting to note his reaction to the superb experimental radiographic and comparative morphological studies on ape and macaque wrists of Jenkins and Fleagle (1975). The latter's observations fundamentally confirm that special features related to knuckle-walking are present in the wrists of African apes. Further, Jenkins and Fleagle provide more reasonable mechanical explanations for these features in the context of a knuckle-walking hypothesis than Lewis does within the brachiationist model he has adopted from Washburn and Avis (1962).

Following vintage ideas of Straus (1940), Lewis (1972a) simplistically explains the knuckle-walking of African apes as the direct result of brachiating habits. Thus he suggests that especially limited wrist extension in chimpanzees and gorillas is "a logical consequence of aligning hand and forearm" (p. 211) for brachiation. By this conceptual scheme, orangutans and hylobatid apes, brachiators par excellence, which I have documented (1965–1975) as possessing highly extensible wrists, become illogical apes indeed. Lewis (p. 211) cites the relative shortness of the manual long digital flexor muscles as "a potent factor in limitation of wrist extension of great apes." This too is denervated by readily available facts on *Pongo*—a great ape which, contrary to Lewis, often exhibits no shortening of the long digital flexor tendons, at least into early adulthood (Fig. 10). Freely extensible wrists and fingers are probably advantageous, if not requisite, for important aspects of the arboreal locomotive repertoires of the Asian apes (Tuttle 1969b). By contrast, shortened manual digital flexor muscles and osseo-ligamentous limitations of wrist extension are

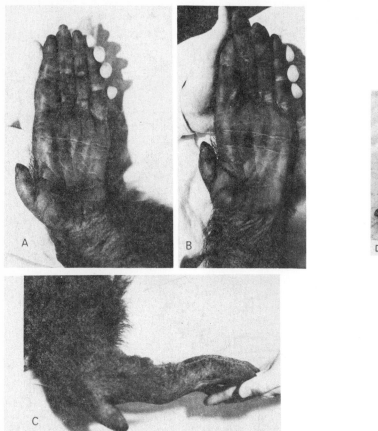

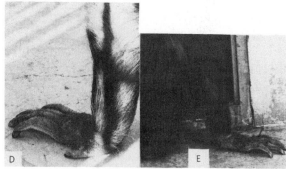

Figure 10   *Mobility of the wrist and finger joints in* Pongo pygmaeus. *A–C, left hand of an anesthetized subadult male orangutan in maximum passive adduction (cf. ulnar deviation), neutral position, and maximum passive extension (cf. dorsiflexion); note that the fingers are readily extended even when the wrist is maximally dorsiflexed. D, anterior view of a load-bearing palmigrade posture in the right hand of an adult orangutan; note that the load appears to be borne mainly by the posterior aspect of the palm; the curvature of the phalanges and the permanent flexion set of the metacarpophalangeal and interphalangeal joints prevent the fingers from lying flat against the substrate. E, posterior view of a palmigrade posture in the right hand of an adult orangutan; the hand was being raised from the substrate as the picture was taken.*

important components in the knuckle-walking complex of the African apes. These features probably evolved de novo as selection for efficient terrestrial hand posturing acted on the relatively long-fingered, flexible hands of the protroglodytians (Tuttle 1969a).

As might be anticipated, while the brachiationist perspective acquires fresh impetus, new criticisms and alternative models are also proliferating. Thus, consequent upon the new perspective on knuckle-walking in extant African apes and renewed paleontological research on the Neogene dental apes, a scheme which, following Hooton (1946), may be designated a "ground-ape" model has recently surfaced. While it has had several spokesmen, its authorship is difficult to pinpoint precisely. The major center for the ground-ape model is Yale University (Simons and Pilbeam 1972, Conroy and Fleagle 1972), though its subscribers there have cited a generous supply of personal communications on the topic from Alan Walker (Simons and Pilbeam 1972, Pilbeam 1972). If the ground-ape model truly gets off the ground, historians of the subject might wish to consider Kortlandt's (1968) influence on its formulation.

The ground-ape hypothesis would have all great apes (implicitly or explicitly including the orangutan), and in most cases man also, derived from terrestrial quadrupedal ancestors. Many of the features traditionally associated with "brachiation" in extant large-bodied hominoids are attributed primarily to terrestrial knuckle-walking in Miocene apes. Accordingly, knuckle-walking was precedent and prospectively adaptive for the arboreal climbing and suspensory behavior of orangutans, chimpanzees, and gorillas. Simons and Pilbeam (1972; Simons 1972; Pilbeam 1972) have posited an extensive period of parallel evolution of knuckle-walking in lineages that culminated in gorillas and chimpanzees because they recognize dentally distinct ancestors for the two lines among Miocene dryopithecine apes some 15 to 20 million years ago.

The ground-ape model contrasts markedly, with the brachiationist/knuckle-walking model and is not concordant with Washburn's latest biomolecular surmise on the probable time of African pongid-hominid divergence, between 5 and 10 million years ago (Washburn 1973:472–73).

Concepts of brachiators and knuckle-walkers have undergone further revision quite recently. Washburn (1973) has confessed that he had previously exaggerated the importance of arm-swinging under branches as a method of locomotion in the Hominoidea. Instead he would now emphasize reaching in many directions while climbing and feeding in trees as responsible for the similarities in the arm and trunk of apes and man. In some ways, Washburn's current perspective on the adaptive complex of brachiators recalls Hooton's (1946), but the latter did not stress feeding behavior so much as climbing.

I have also been rudely propelled to the confessional on overstatement by the documentation of knuckling and knuckle-walking (Fig. 11) in several captive orangutans (Tuttle and Beck 1972, Tuttle and Basmajian 1974c, Susman 1974, Tuttle 1975). Clearly, I should not have stated categorically that orangutans cannot knuckle-walk (Tuttle 1967:171). I would now submit that if a highly advanced arboreal climber and arm-swinger like the orangutan can develop ontogenetically into a knuckle-walker (albeit in special circumstances of captivity), ancestors of the African apes may have been somewhat similarly predisposed to knuckle-walking by their own arboreal heritage. It also seems reasonable to argue that if the common ancestor of chimpanzee and gorilla possessed somewhat advanced features of the hand for suspensory behavior and climbing, it could have developed into a knuckle-walker quite rapidly after adopting terrestrial habits. On the basis of what can now be inferred about the mechanics of knuckle-walking in extant African apes, I surmise that selection would act rather strongly and rapidly to develop structures which facilitate close-packed positions of extension in the wrist and metacarpophalangeal joints of digits II–V of novice knuckle-walkers. Accordingly, the imprint of knuckle-walking should appear in the fossil record rather soon after this behavior became habitual in the terrestrial activities of the prototroglodytians and any other hominoids that have had knuckle-walkers in their lineage. Because this evidence is not apparent in available fossils, my theoretic sympathies on hominoid phylogeny incline much closer to a conventional brachiationist model than to the ground-ape theory.

RENEWED QUEST FOR ARBOREAL IMPERATIVES

My experimental (Tuttle et al. 1972; Tuttle and Basmajian 1974a,b,c; Tuttle, Basmajian, and Ishida 1975), behavioral, and functional-morphological (Tuttle 1965–1975) studies of apes and a review of classic and contemporary views of hominoid evolution lead me to conclude that there is still much to be gained by searching for arboreal imperatives in the selective complex that produced the protohominids. Perhaps the dramatic discoveries of bipedal South African australopithecines in relatively dry, open paleohabitats have led evolutionary anthropologists too far into the savanna for them to uncover the deeper roots of man's family tree. Washburn (1967, 1968a) recognized this recently and promptly sent his protohominid knuckle-walking to the forest fringe and forest floor. My hypothetical protohominid would step bipedally forward and call a welcome to his model, but from a steady perch in the forest canopy. Thus I see no compelling reason to revise my suggestions (1969a) that the initial divergence of ancestral man and troglodytians occurred fundamentally in arboreal contexts and that, from the outsets of their terrestrial careers, the African apes were predisposed to knuckle-walking and the hominids to bipedalism because of somewhat different arboreal heritages. Perhaps even the pelvic tilt mechanism which uniquely equips man for a highly efficient bipedal striding gait was at least incipiently developed during a phase of bipedal branch-walking. Among all previous brachiationist models, Morton's strikes me as the most reasonable. However, there are many morphological,

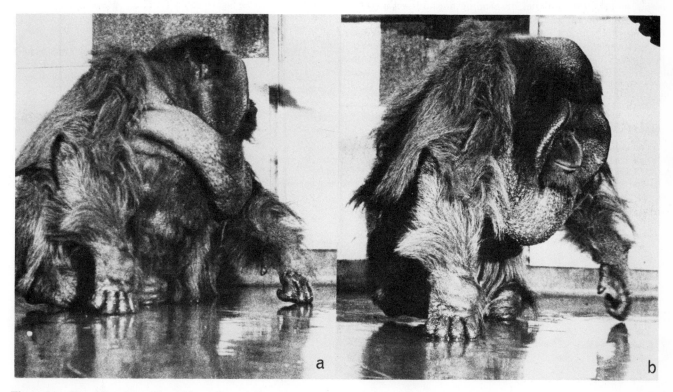

Figure 11 *Knuckling (a) and knuckle-walking (b) hand postures in an adult male* Pongo pygmaeus *(same subject as in Fig. 4b). a, outset of squatting progression, in which the knuckled hands rest lightly on the substrate as the right foot swings forward; b, left forelimb and right hindlimb swinging forward while right forelimb and left hindlimb are supportive. This behavior has been observed only on wet substrates.*

biomechanical, and evolutionary puzzles to be solved and experiments to be performed before subscription to a detailed modernized rendering of Morton's model can be wholeheartedly encouraged. A new thoroughgoing study of the 1951 skeleton of *Dryopithecus (Proconsul) africanus* (Napier and Davis 1959) in modern perspective would probably also be rewarding to theoretical evolutionary anthropology.

NORMAL SCIENCE IN EVOLUTIONARY ANTHROPOLOGY

In his remarkable essay *The Structure of Scientific Revolutions* (1970), Kuhn describes the establishment of "normal science" in a particular scientific community on the basis of a foundation paradigm that shares two characteristics: (1) that it was sufficiently unprecedented to attract an enduring group of adherents away from competing modes of scientific activity and (2) that it was sufficiently openended to leave all sorts of problems for the redefined group of practitioners to resolve. The thesis of Darwin, Huxley, and Haeckel, based on comparative morphological evidence, that man had evolved by natural selection from anthropoid apes seems to fulfill the first of Kuhn's criteria. The second criterion is met in that subsequent to their statements there has been no end to research and controversy over the precise mechanisms of hominid emergence and the appearance of the earliest hominids. Hence it seems reasonable to speak of normal science in evolutionary anthropology.

We are now in a vigorous phase of "puzzle solving" or "mopping up," as Kuhn (1970:35–42, 24) terms it. Getting a handle on an evolutionary anthropological problem often means employing sophisticated technical devices like computers and cineradiographic and electromyographic equipment in experimental morphological studies; or spending long hours excavating those plots of the earth that are not covered by concrete; or trying to glimpse a free-ranging ape before it and its habitat enter the roster of vanished species; or decoding the amino-acid sequences in complex protein molecules. With that done, the fortunate evolutionary anthropologist will be able to provide a small piece of the puzzle of hominoid evolution and to begin work on the next one.

ABSTRACT

This paper sketches successive refinements, optional renderings, and disclaimers of the ape model of hominid evolution from its initial statement by Darwin, Huxley, and Haeckel until 1973. I suggest that of the four principal ape models that have been advanced—brachiating troglodytian; brachiating, bipedal hylobatian; knuckle-walking, brachiating troglodytian; and pristine ground ape—the hylobatian model of Morton is the most convincing. Thus, future biomechanical, experimental, and theoretic studies might be profitably focused on determination of the arboreal adaptive complexes that were fundamental to the initial divergence of the hominids and the pongid apes. The 19th-century ape model and the burgeoning research provoked by it during the past century seem to satisfy Kuhn's criteria for the establishment of a normal science. Hence we might refer to the ape paradigm as the foundation of a normal science of evolutionary anthropology, especially in the aspect which deals with the emergence and other early phases of the hominid career [2].

Acknowledgments

Research upon which this essay is based was supported by NSF Grants Nos. GS-834, GS-1888, and GS-3209; a Public Health Service Research Career Development Award No. 1-KO4-GM16347-01 from the National Institutes of Health; the Wenner-Gren Foundation for Anthropological Research; the Marian and Adolph Lichtstern Fund of the University of Chicago; NIH Grant Nos. FR-00165 and RR-00165 to Yerkes Regional Primate Research Center; NIH Grant No. FR-00164 to the Delta Regional Primate Research Center; NIH Grant No. FR-00169-04 to the National Center for Primate Biology; and NIH Grant No. 1-P06-FR00278-02 to the Southwest Foundation for Research and Education. I thank Les Siemens, University of Chicago, for assisting with figures 1a and b, 7, and 8b; Frank Kiernan and Tim Gill of Yerkes Regional Primate Research center for Figures 1c-g and 2; Frank Kiernan for Figures 4a and 10; Irwin Bernstein of Yerkes Regional Primate Research Center for Figure 3; and Leland LaFrance of the Chicago Zoological Park for Figures 4b and 11.

This paper was read at the Harvard Anthropology Colloquium on October 30, 1973. I should like to thank David Maybury-Lewis for his invitation and comments and Sol Tax for encouraging aspects of the project. Sherwood Washburn's influence on the piece should be apparent. As ever, I am indebted to him for provocation and perspective.

Notes

[1]   However, aspects of Winge's model rather interestingly foreshadow Morton's.

[2]   When this article was originally reviewed for *Current Anthropology* it was sent out to a number of scientists for comment. The following responded: B. Blumenberg, E. Delson, W. W. Howells, J. H. Kress, S. C. Malik, H. McHenry, F. E. Poirer, A. Sharma, and N. B. Todd. This CA* treatment was published along with the original text. Because of space limitations these comments can not be included here.

43

Pygmy Chimpanzee as a Possible Prototype for the Common Ancestor of Humans Chimpanzees and Gorillas

A. L. ZIHLMAN, J. E. CRONIN, D. L. CRAMER, AND V. M. SARICH

A convincing theory of human origins must clarify man's relationships with living primates and with the ancestral forms known only through fossils. Phylogenetic relationships have previously been determined mainly by anatomical similarities, but now, biochemical similarities provide independent criteria for evolutionary relationships. Albumin and transferrin immunology, immunodiffusion, DNA annealing and amino acid analysis all indicate that chimpanzees, gorillas and humans share a substantial common ancestry, and that the Asiatic apes (gibbons and orangutans) diverged earlier from this lineage (Goodman, 1976b; Benveniste and Todaro, 1976; Sarich and Cronin, 1976). These findings directly conflict with the more widely held view that all the great apes diverged from a common ancestor long after the 'origin' of the evolutionary line leading to modern humans (Delson and Andrews, 1975). The molecular data consistently suggest a much more recent origin of the man-chimpanzee-gorilla separation than was previously imagined, namely, in the range of 4–6 M yr ago (Sarich and Cronin, 1976, 1977). These data show that, although the two chimpanzee species (*Pan paniscus* and *P. troglodytes*) are biochemically distinct, they are more closely related to each other than either is to humans or gorillas (Goodman et al., 1970; Cronin, 1977). The chimpanzees speciated, then, after the initial three-way split. We therefore, here contend that, among living species, the pygmy chimpanzee (*P. paniscus*) offers us the best prototype of the prehominid ancestor. Biochemical, morphological, behavioural and palaeontological data support this proposition and argue for a relatively recent and accelerated divergence of the hominid from the pongid line.

If this cladogram is correct, then an ancestor for the earliest hominids must also be a suitable ancestor for chimpanzees and gorillas. The morphological similarities between the three groups should logically be derived from the common ancestor, rather than by evolutionary convergence, or by parallelism from distinct primate lineages. In fact, the modern African apes may be viewed as size variants in a single morphotypic series, going stepwise from the smaller *P. paniscus* to the large male gorilla. The scant hominid fossil record during the past 4–8 M yr, and the absence of any fossil chimpanzees or gorillas, forces us to work backwards from the living hominoids to reconstruct a prototype of the common ancestor.

Chimpanzees have generally been considered closer to this prototypic ape than gorillas, because the latter are more specialised morphologically and behaviourally. They are larger and males are twice the size of females; they have a more specialised diet, more restricted habitat and less flexible social behaviour than chimpanzees. Morphologically, pygmy chimpanzees are more 'generalized' in body size, body build and sexual dimorphism than the common species and may provide a better proto-

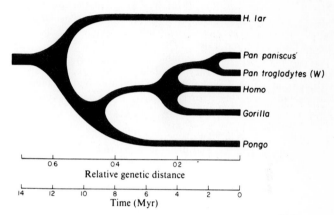

Figure 1 *Molecular cladogram showing genetic relationships of the ape and human lineages.*

type for studying the derivation of the earliest hominids and African apes.

Pygmy chimpanzees on the average weigh less than the average common chimpanzee, but with much overlap: 25–48 kg, compared to 25–60 kg or somewhat more (Zihlman and Cramer, 1978). Pygmy chimpanzees have smaller facial and canine dimensions, smaller average cranial capacities (350 cm³ compared with 390 cm³), and the two species can be discriminated on mandibular length alone (Cramer, 1977).

Pygmy chimpanzees have a narrower trunk and more gracile upper body, as reflected in significantly smaller clavicular length, scapular dimensions and iliac and sacral breadths (Zihlman and Cramer, 1978). Their upper and lower limbs are of about equal length, whereas arms of common chimpanzees are relatively longer.

As to sexual dimorphism, pygmy chimpanzees are moderately sexually dimorphic in body weight, but no dimorphism can be detected in cranial capacity, limb bone lengths or robusticity, or in anterior and posterior dentition (Cramer and Zihlman, 1978). They have relatively small and only slightly dimorphic canine teeth (Johanson, 1974). Overall, this is less dimorphism than occurs in common chimpanzees and no more than occurs in modern humans.

Observations on the behaviour of pygmy chimpanzees, either in their natural habitat in the Zaire River Basin or in captivity, have not been extensive. They feed high in trees, move and feed on the ground (Badrian and Badrian, 1977; Kano, 1979), and in captivity walk bipedally more than do common chimpanzees (S. Savage-Rumbaugh, personal communication). They therefore encompass the spectrum of locomotor behaviour of living African hominoids: arboreality, terrestriality and bipedality. Laboratory and field studies reveal pygmy chimpanzees as highly intelligent and flexible in feeding, locomotor, sexual and communicatory behaviour (Savage and Bakeman, 1978; Savage-Rumbaugh et al., 1977; Jordan and Jordan, 1977).

Given these morphological and behavioural data, we maintain that pygmy chimpanzees present a general pattern from which other African hominoids could have developed. This contention is further supported by comparing pygmy chimpanzees with the earliest hominids.

The fossil record, based on dentition, shows the occurrence of *Australopithecus* by 3.5 M yr ago in the mosaic savannas of Eastern Africa (Leakey et al, 1976). By 3 M yr ago, the hominids were distinct from apes in the broader and shorter pelvis, larger cranial capacity, larger postcanine teeth and somewhat smaller canines (Johanson and Taieb, 1976; Aronson et al., 1977).

P. paniscus provides a suitable comparison for *Australopithecus* (Zihlman, 1977); they are similar in body size, postcranial dimensions and as Figs 2 and 3 show, even in cranial and facial features, although this *Australopithecus* has a larger cranial capacity (485 cm³) (Holloway, 1972) than *P. paniscus* (350 cm³).

Postcranial dimensions are given in Table 1. Femoral length, femoral head and acetabular diameters are very similar. Humerus and innominate lengths, and breadths of sacrum and ilium are different. These findings demonstrate early hominid affinities with apes. Probably the initial acquisition of hominid upright locomotion entailed reduction of upper limb length and broadening of the ilium and sacrum.

Homo and *Pan* (both species) differ in their genomes by approximately 1%, a value characteristic of sibling species (King and Wilson, 1975). The genetic changes accounting for the morphological differences between man and chimpanzee may be only a small fraction of that amount. A few regulatory genes may have large somatic effects, whereas the remaining structural genes may be time dependent in their rate of fixation, and of little selective significance.

The ultimate strength of any hypothesis rests on its compatibility with all lines of evidence, on its ability to withstand falsification and on its power of prediction. We hypothesise (1) that as our knowledge of the fossil record improves, hominids before 3.0 M-yr ago should continue to converge on the pygmy chimpanzee-like condition, cranially and postcranially, as they seem to do dentally, (2) that Pliocene pongids ancestral to chimpanzees and gorillas should do likewise, and (3) that at least one lineage of late Miocene ape (perhaps *Ramapithecus*) should continue to develop evolutionary trends begun in the early Miocene and evolve into a form similar to the extant pygmy chimpanzee.

We thank L. Brunker, T. Grand, F. Jenkins, J. Lowenstein and K. Wcislo for comments, and A. C. Wilson, University of California, for laboratory facilities. For research support, we acknowledge the Wenner—Gren Foundation for Anthropological Research and the Faculty Research Committee, University of California (to A.L.Z.).

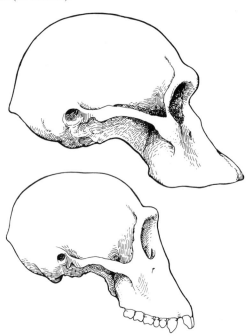

Figure 2   *Side view, drawn to scale, of Sterkfontein 5 (above) and adult female* P. paniscus *(below).*

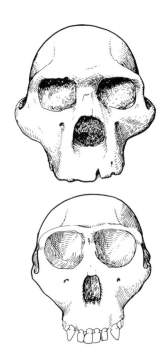

Figure 3   *Front view, drawn to scale, of Sterkfontein 5 and adult female* P. paniscus.

Table 1   *Comparison of Pygmy Chimpanzee and Early Hominid Pelvis, Femur, and Humerus (in mm).*

Measurement	Early Hominids	Pygmy Chimpanzees
Femur length	280*,280†	293 (264–315)
Femoral head diameter	31†	30.5 (28–34)
Acetabular diameter	37†	36 (32–38)
Innominate length	170†	253 (232–272)
Iliac breadth	113†	97 (80–118)
Sacral breadth	76†	63 (51–70)
Humerus length	235*	285 (250–307)

*AL 288–1 from Hadar, Ethiopia (Johanson and Taieb, 1976)
†Sts 14 from Sterkfontein, South Africa. Measurements from A.L.Z. (unpublished data); femur length from Lovejoy (1975).

44

Hominoid Cladistics and the Ancestry of Modern Apes and Humans

R. L. CIOCHON

INTRODUCTION

The major purpose for assembling the volume *New Interpretations of Ape and Human Ancestry* was to provide a forum for the presentation of alternative viewpoints on the subject of ape and human ancestry. In this chapter I will reduce those viewpoints to a summary statement reflecting the contributors' intent through an evaluation of the volume's major themes and by the presentation of a series of cladistic models and character analyses. Naturally it will be impossible to cover all of the points dealt with in the preceding 29 chapters. Therefore, only the salient points of the major themes of hominoid phylogeny will be considered. In particular I will consider (1) the branching order (cladogenesis) of fossil and recent hominoid primates, (2) the structural components (morphotype) of the last common ancestor of humans and the living apes as well as the morphotypes of earlier hypothetical ancestors in the diversification of the Hominoidea, (3) the timing and geographical placement of the ape-human divergence and the origin of the extant ape and human lineages, and (4) the adaptive nature and probable scenario of the Miocene hominoid cladogenesis with specific focus on the initial differentiation of hominids from their ape forebears.

In my view, the chapters in *New Interpretations of Ape and Human Ancestry* represent a major step toward reconceptualization of the phyletic position and relationships of the Miocene hominoids *vis à vis* modern apes and humans. As was earlier suggested in a report of the Florence conference on which this volume is based (see Ciochon and Corruccini, 1982), this reconceptualization could foster new evolutionary pardigms of ape and human ancestry, melding together a wide variety of neontological, paleontological, and biomolecular data. This final chapter should help to document this reconceptualization.

CLADISTIC MODELS OF HOMINOID EVOLUTION

In the following sections a series of cladograms is presented that address the issue of relationships of modern hominoid groups to fossil groups from which they are possibly derived. Several of these cladograms deal with issues touched on previously but are presented here to provide a more sequential view of the major events in hominoid evolution. It is naturally not possible to present every cladistic model considered by contributors to this volume. However, the models selected acknowledge all of the major issues and in the process provide a series of testable cases that can be modified as future evidence dictates.

Evidence For Initial Differentiation Of Hominoids

The origins of the Hominoidea can be traced back to the Fayum catarrhine radiation (Fleagle and Kay, 1983; Fleagle, 1983). In the cladistic sense, however, this is difficult to justify, since, as Fleagle and Kay (1983) demonstrate, the Fayum "hominoid" Propliopiths (*Aegyptopithecus* and *Propliopithecus*) share no derived features that link them with any Miocene taxa or with any extant catarrhine group. Falk's (1983) comments on the only known endocast of a Propliopith (*Aegyptopithecus*) also support this view. Thus, the Propliopiths occupy the position of primitive basal catarrhines and are suitable phyletic ancestors of the Hominoidea *sensu stricto* and Cercopithecoidea (Fig. 1).

Because of their uniformly primitive nature, the classification of Propliopiths within the Catarrhini presents a problem. Fleagle and Kay (1983) argue for retaining the group within the Hominoidea in a separate subfamily of the family Pliopithecidae. This implies a broader definition of the Hominoidea and from a strict cladistic perspective is not wholly defendable since the Fayum Propliopiths are viewed as giving rise to *both* extant catarrhine superfamilies yet are ranked within the Hominoidea. One cladistically correct alternative would be to set up a separate superfamily for the Propliopiths; another would be to rank the Old World monkeys as a family within the Hominoidea. Since neither of these is an adequate arrangement in my opinion, I will support Fleagle and Kay's choice as a temporary albeit, "incorrect" solution.

Following Fleagle and Kay's (1983) lead, the realization that Old World monkeys are derived from a primitive hominoid ancestor might force reconsideration of the nature of the cercopithecoid adaptation. For instance, Andrews (1983) points out that the ancestral cercopithecoid morphotype represents the derived condition *vis à vis* hominoids with respect to their specialized bilophodont dentition, diet, habitat preference, and posture. Fleagle (1983) and Rose (1983) add that the postcranial skeleton of cercopithecoids also represents the derived condition. Therefore, living cercopithecoids can be regarded, in many ways, as a wholly derived group, whereas living hominoids may approximate the primitive ancestral catarrhine condition to a much greater extent than has been previously considered.

Finally, the consideration that Propliopiths are ancestral to both cercopithecoids and hominoids is a delight to advocates of the biomolecular clock evidence (Cronin, 1983; Zihlman and Lowenstein, 1983), who have insisted for many years that the cercopithecoid-hominoid split postdated the Fayum catarrhine

Editor's Note: This chapter presents an overview and summary of the 30-chapter, edited volume *New Interpretations of Ape and Human Ancestry.* Due to space limitations the section of this article entitled "Evaluation of Major Themes" could not be reprinted here. This section presents a summary of the major themes of that volume. Please refer to Ciochon (1983, pp. 784–804) for this discussion.

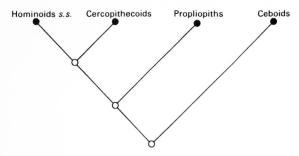

Figure 1   *Cladogram depicting relationship of the Hominoidea to other major groups in the Anthropoidea. Hominoids are here defined in the strict sense to include the six extant genera* (Hylobates, Symphalangus, Pongo, Gorilla, Pan, Homo) *and their hypothetical ancestors and collaterals, all of which share at least some of the suite of derived characters definable in the living apes and humans (see Table 1). The sister group of the Hominoidea sensu stricto is the Cercopithecoidea (Old World Monkeys), as diagrammed here. The Oligocene Fayum Propliopiths are, in turn, the sister group of the hominoid–cercopithecoid clade and may very well prove to be the ancestral stock from which the clade is derived (see text). Finally, the Ceboidea (New World Monkeys) join the others as the sister group of the Catarrhini. Note that the nonhominoid, noncercopithecoid Parapithecidae from the Fayum have not been diagrammed here since they have little bearing on the higher category relationships discussed herein. The solid circles in this cladogram represent existing phyletic groups, whereas the open circles define hypothetical ancestors.*

radiation. Cronin (1983) dates this divergence at 20 ± 2 m.y.a., while Boaz (1983) offers a paleontological-biomolecular averaged estimate of 26 m.y.a.

Evidence for Gibbon Lineage

The ancestry of the Hylobatidae is perhaps the most unresolved issue in the evolution of the Hominoidea. It is not so much a lack of fossil evidence that has caused this situation but rather the way that evidence is viewed. Fleagle (1984) in a recent review of gibbon ancestry observes there is a notable lack of consensus of what a primitive gibbon should look like. Gibbons are phyletically the most primitive hominoids. Their clade represents the first branch of the hominoid cladogenesis that survived to the present. Yet, primitive as they are in some features, they are extraordinarily specialized in others (see morphotype 4 in Table 1). It is the muscular and skeletal specializations that make gibbons so unique but at the same time these specializations almost certainly mask much of the morphological evidence of gibbon heritage (Fleagle, 1984).

Gibbon evolution was not widely discussed by contributors to this volume. Ward and Pilbeam (1983) do point out that extant hylobatids evince the primitive "dryomorph" subnasal pattern. Gantt (1983) notes that *Hylobates* exhibits enamel prism Pattern 3A, like the great majority of the Hominoidea, and also has relatively thin molar enamel caps. Kluge (1983) observes that gibbons display a number of primitive hominoid soft anatomical features (see also morphotype 3 in Table 1). Fleagle and Kay (1983), Prasad (1983), and Chopra (1983) provide background discussion of some potential fossil gibbons. Finally, Cronin (1983) suggests the gibbon lineage diverged from the hominoid stem at 12 ± 3 m.y.a. based on molecular clock evidence and Boaz (1983) estimates a divergence date of 16.5 m.y.a. on the basis of averaged data sets.

Figure 2 illustrates three cladistic relationships of gibbons to other hominoid groups that have been proposed at various times. In a similar diagram Fleagle (1984) cites the existence of four different relationships. These are both particularly apt representations of the unresolved nature of hylobatid ancestry. I will now briefly review some fossil evidence of the Lesser Apes in the hope of delimiting the choices available in Fig. 2.

Gibbons are today restricted to the rainforests of Southeast Asia. In the Pleistocene their range was considerably larger, since numerous sites in China, Indonesia, and Southeast Asia have yielded primate teeth indistinguishable from those of living hylobatids (Szalay and Delson, 1979). To my knowledge there is no corroborated evidence of Pliocene gibbons. From the late Miocene of North India Chopra and Kaul (1979) and Chopra (1983) describe the existence of a potential hylobatid left M^3 from the Nagri level beds of Quarry D at Haritalyangar. The Quarry D site has recently been dated by paleomagnetism at an age of 7.4 m.y.a. (G. Johnson, personal communication). In size and morphology, the left M^3 is remarkably similar to that of extant gibbons (Chopra, 1983, p. 543, Fig.1). A significant derived character uniting this specimen with modern gibbons, according to Chopra (1983), is the relatively weak development of its lingual cingulum. Just as important in my estimation are both its phenetic similarity to living gibbons *and* its occurrence in late Miocene exposures of southern Asia, the correct temporal and zoogeographical provenience for an early gibbon. The specimen was named *Pliopithecus krishnaii* by Chopra and Kaul (1979) and more recently has been assigned to a new genus, *Krishnapithecus* Ginsburg and Mein 1980. I endorse this opinion since it is unlikely that the Quarry D M^3 was related to the primitive European catarrhine, *Pliopithecus*.

Prasad (1983) notes the occurrence of a mandibular fragment containing a partial M_3 from a Nagri level horizon near Haritalyangar. This specimen was originally described as *Hylopithecus* by Pilgrim (1927) and judged by him as a potential early hylobatid, a view favored by Simons and Pilbeam (1965). Prasad (1983) hints that the isolated M^3 from Quarry D at Haritalyangar now assigned to *Krishnapithecus krishnaii* may represent the same primate taxon as *Hylopithecus*. Recently, I had the opportunity to examine the type specimens of both of these species in India in 1982 and can confirm that they are quite different in size and morphology. The *Hylopithecus* type is now best viewed as further evidence of a middle-sized and thin-enameled Dryopith in Asia, while the *K. krishnaii* specimen is very possibly an early gibbon.

The scant but tantalizing evidence of a fossil gibbon from Quarry D at Haritalyangar [1] in North India takes on additional significance in light of new discoveries at the Lufeng hominoid locality in southern China. This locality, recently dated to 8 m.y.a. (Flynn and Qi, 1982), has yielded a partial face, about a dozen jaws, and many isolated teeth of a small hominoid with possible gibbon affinities (E. Delson, personal communication). An even earlier find in China, *Dionysopithecus* (Li, 1978), from the late middle Miocene (?12 m.y.a.), also has possible gibbon affinities. Though only known from a partial maxilla, it is morphologically, temporally, and paleozoogeographically the most likely candidate for the title of earliest true fossil gibbon (see also Fleagle, 1984). However, its uncanny resemblance to *Micropithecus* (Fleagle, 1978a) from the early Miocene of East Africa forces consideration of that taxon as a possible precursor of *Dionysopithecus*. *Micropithecus* is known from a larger sample of specimens and is decidedly part of the early Miocene African Dryopith radiation. Yet its maxilla differs in virtually no significant way from the type of *Dionysopithecus* (Fleagle, 1984). Both taxa were described within a few months of each other, but unfortunately neither had associated postcrania. Fleagle (1984) points out that should future finds of *Dionysopithecus* establish its identity with *Micropithecus*, then the former nomen would assume priority. This would then extend the temporal and paleozoogeographic range of the ancestral gibbon lineage to between 17–19 m.y.a. in East Africa. This would conflict with the biomolecular data and would also usher in the return of an "African fossil gibbon," a concept with which I have never been comfortable. This concept seems especially unlikely when one considers the newly documented Asian *Sivapithecus*-orangutan link (see following section).

Reviewing the evidence for a still more distant ancestry of the Lesser Apes necessitates comparisons with the Pliopithecidae. *Propliopithecus* is a particularly unlikely candidate for the title of earliest gibbon, due to its complete lack of extant catarrhine synapomorphies (see also comments in preceding section). *Pliopithecus* also has neither the derived cranial features (Fleagle and Kay, 1983) nor postcranial specializations (Fleagle 1983) that would link it with the ancestry of gibbons. As concluded by earlier researchers (e.g., Remane, 1965, Ciochon and Corruccini, 1977a; Szalay and Delson, 1979) and supported in this volume by Fleagle and Kay (1983), *Pliopithecus* is best regarded as a phyletically primitive catarrhine related most closely to the Fayum Propliopiths. Though slightly more derived in a few features, many of these same arguments can also be applied to the early Miocene African *Dendropithecus* in rejecting its ancestral gibbon status (Ciochon and Corruccini, 1977a; Fleagle, 1984).

Fleagle (1984) summarizes and reflects on the way past interpretations have been drawn from the fossil record of gibbon ancestry. He states:

fossil gibbons have in virtually all cases been identified on the basis of either primitive hominoid features or 'trends,' rarely on unique 'shared-derived' features linking the fossils with the extant taxa. Therefore, at present, our understanding of extant hominoid phylogeny contributes more to our interpretation of the fossil record of gibbon evolution than *vice versa*.

From Fleagle's comments and my discussion of the fossil evidence for gibbon ancestry I feel it is now possible to reject the relationships (see Fig. 2) of the gibbon (3) clade and the gibbon (2) clade without further discussion. The gibbon (1) clade relationship has no fossil branches indicated, yet I do feel there is now evidence in the 7–12 m.y.a. time range in southern Asia for early gibbons. Perhaps the cladogram should be redrawn to include an Asian Dryopith (?) branch to represent such forms as *Krishnapithecus*, Lufeng fossil "gibbons," and possibly *Dionysopithecus* (see Fig. 8 in Appendix). Even if future discoveries prove that some or all of these taxa are not early gibbons, I suggest that the search for evidence of fossil hylobatids should be concentrated in the later Miocene and Pliocene of southern Asia among the small-bodied Asian Dryopiths.

Evidence For Orangutan Lineage

Until quite recently the fossil record of the orangutan lineage was limited to a few isolated teeth of questionable affinity from the Siwaliks, such as "*Paleosimia*" Pilgrim 1915 (see also Prasad, 1983), and to hundreds of isolated teeth from the early Pleistocene to Recent deposits of southern China, Thailand, Malaysia, Sumatra, Java, and Borneo (Szalay and Delson, 1979). With the

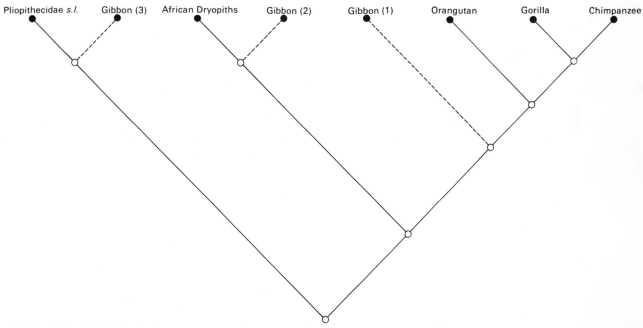

Figure 2 *Cladogram depicting three potential cladistic relationships of gibbons (Hylobatidae) to other members of the Hominoidea. In the gibbon (3) clade, hylobatids are represented as the sister group of the Pliopithecidae based on their long-noted resemblance to* Propliopithecus *and* Pliopithecus. *In the gibbon (2) clade, hylobatids are viewed as the sister group of the African Dryopiths because of their resemblance to* Dendropithecus *and* Micropithecus. *In the gibbon (1) clade hylobatids are presented as the sister group of the extant Great Apes without further identifying their relationship to any particular fossil group. This implies an ancestry more closely linked with the cladogenesis of the Great Apes. The solid circles in this cladogram represent existing phyletic groups, whereas the open circles define hypothetical ancestors. Note that in succeeding cladograms the position of the hylobatid clade is always indicated by a dashed line to indicate that its position is conjectural.*

announcement and interpretation of a new partial skull of *Siva-pithecus* (Pilbeam, 1982; Andrews, 1982; Ward and Pilbeam, 1983) from the U level (~8 m.y.a.) of the Potwar Plateau of Pakistan our understanding of the fossil record of orangutans has increased dramatically. This unique discovery not only documented the earliest evidence of the orangutan lineage, it also altered many researchers' perceptions of the phyletic relationships of Ramapiths* as a group.

The change in viewpoints regarding the affinities of Ramapiths in general and *Sivapithecus* in particular can be related directly to the identification of a suite of derived craniofacial features in several new fossil skulls (see Table 1). From this perspective it is useful to review the opinions expressed over the last decade regarding the position of *Pongo* in hominoid evolution (compare Figs. 3–5). In my opinion the current evidence from craniofacial anatomy makes Fig. 4 the most likely choice. This view is also supported by a plurality of contributors. Specifically, Boaz (1983, p. 708) suggests that *Sivapithecus* "is ancestral at least at a generic level to *Pongo*." Ward and Pilbeam (1983, p. 220) observe that the subnasal patterns of *Sivapithecus* and *Pongo* "indicates a close phylogenetic relationship." Finally, Andrews (1983, p. 457) concludes that *Sivapithecus* "may be considered to be broadly ancestral to the orangutan." Wolpoff (1983) takes a slightly different view by suggesting that *Pongo* as well as the African ape—human clade are derived from within the Ramapith radiation, indicating his support of a Fig. 5 arrangement. Kay and Simons (1983, p. 616), on the other hand, support the alternative in Fig. 3 by cautiously stating, "African Great Apes and *Pongo* derive from stocks of thin-enameled apes represented perhaps collaterally by western and northern European later Miocene *Dryopithecus*." They suggest that *Dryopithecus* in Europe may represent a similar grade of development in postcranial, dental, and gnathic structure relative to that seen in Ramapiths and the living Great Apes. Kay and Simons (1983) also strongly question the validity of the *Sivapithecus-Pongo* link (see disputed charac-

ters in Table 1), although their own scenario for the origin of *Pongo* is equally open to challenge.

Putting aside the question of later Miocene *Sivapithecus*, it is worth mentioning that Chopra (1983) describes and names a new genus and species of "orang-like" hominoid based on the occurrence of an isolated left M^1 from the upper middle Miocene (Chinji level) deposits of Ramnagar in North India. According to Chopra (1983, p. 544), "*Sivasimia chinjiensis* is significantly different from all other Chinji level hominoids in possession of strongly wrinkled enamel on its occlusal surface." Chopra feels this character and other aspects of the cusp morphology indicate potential phyletic ties to the orangutan. Von Koenigswald (1981) has also recently described a new genus and species of hominoid based on an isolated molar (right M_2) from Chinji level beds near Chinji, Pakistan (von Koenigswald, 1983, p. 520, Fig. 1). This M_2 exhibits wrinkling of its occlusal enamel surface like the Ramnagar specimen. However, von Koenigswald (1981, 1983) does not posit phyletic ties to *Pongo*, but rather, he proposes that *Chinjipithecus atavus* von Koenigswald was ancestral to the late Miocene *Gigantopithecus giganteus*. Both of these isolated molars appear to be unerupted tooth germs. This may account, in part for the high degree of occlusal enamel wrinkling that characterizes each specimen. On the other hand, it is possible that one or both of these specimens could indicate earlier evidence of the orangutan lineage. However, observations made on the molars of the Potwar *Sivapithecus* skull (GSP 15000) show no significant degree of occlusal enamel wrinkling (Pilbeam, 1982, and personal communication). Therefore it is unlikely that these Chinji level molars would be exhibiting derived occlusal features shared with extant orangutans when the more recent Nagri level GSP 15000 does not appear to show these features. To complicate matters, the Chinese Lufeng *Sivapithecus* material (of Nagri age equivalence) is characterized by strong enamel wrinkling (Pilbeam, personal communication; Wolpoff, personal communication).

It should be noted here that there is one paleoecological

*Editor's note: Ramapiths are an informal taxonomic grouping of Miocene hominoids comprising *Sivapithecus*, "*Ramapithecus*," and related forms.

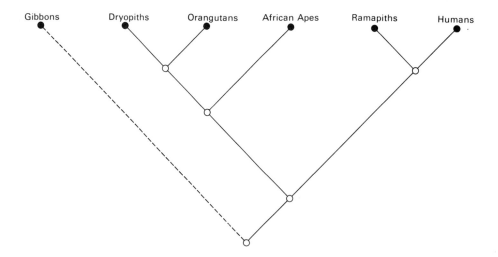

Figure 3 *Cladogram depicting the traditional view of hominoid relationships supported by many researchers during the 1960s and early 1970s. Ramapiths are viewed as the sister group of humans and were considered by many to be the earliest representatives of the Hominidae. Orangutans and the two African apes were thought to be independently derived from different groups of Dryopiths and are here represented as sister groups of the Dryopiths. Gibbons were considered a distant outgroup not closely related to the Great Ape–human radiation. This view of hominoid relationships has maintained a group of adherents and recently resurfaced in a much more sophisticated formulation. Solid circles and open circles defined as in Figs. 1 and 2.*

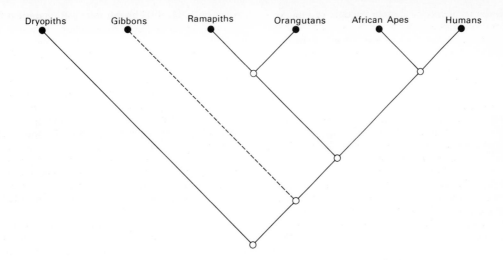

Figure 4 *Cladogram depicting a recent view of hominoid relationships supported by a growing number of researchers. The African apes are viewed as the sister group of humans, reflecting a close phyletic relationship between the two groups. Orangutans and Ramapiths are considered sister groups based on a suite of newly recognised derived characters. Dryopiths are considered a distant out-group defined mainly on the basis of primitive characters (see Table 1). The gibbon clade is placed in a position more closely related to the ancestry of the Great Apes but its exact position vis a vis Ramapiths and Dryopiths is still viewed as conjectural. Solid circles and open circles defined as in Figs. 1 and 2.*

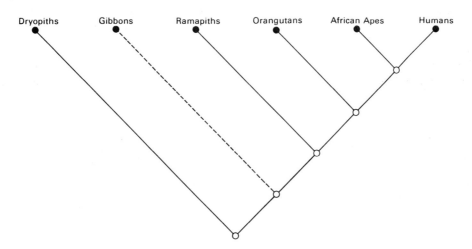

Figure 5 *Cladogram depicting relationships within the Hominoidea supported by morphologists and molecular biologists who favor a temporally recent cladogenesis of the extant apes and humans. The Ramapiths are here considered the sister group of the extant Great Apes and humans, which indicates that the derivation of those groups would postdate the earliest known occurrence of Ramapiths. Note also that humans and the African apes share a sister group relationship which is next joined by the orangutan clade. The position of the gibbon clade is once again indicated as conjectural; some researchers who favor the biomolecular evidence have even suggested its positioning as the sister group of the Great Ape–human clade. Following this view, researchers would have to look no further than the Ramapiths to derive all the extant Hominoidea. Dryopiths occupy the extreme out-group position based on their suite of shared primitive characters (see Table 1). Solid circles and open circles defined as in Figs. 1 and 2.*

obstacle which does not lend support to the derivation of *Pongo* from a *Sivapithecus*-like ancestor. Andrews (1983) acknowledges that if the temperate woodland-forest-adapted omnivore *Sivapithecus* is the closest known relative of the tropical-forest-adapted frugivorus orangutan, then (1) the ancestral orangutan could have been more terrestrial and omnivorous in the recent past and/or (2) a major adaptive shift could have taken place along the *Sivapithecus–Pongo* lineage, resulting in the development of an arboreal frugivore from a partly terrestrial omnivore. Smith and Pilbeam (1980) and Ward and Pilbeam (1983) provide elements of support for both of these views. However, this is one issue where parsimony is not apparent and adequate resolution may have to await the recovery of a middle Pliocene orangutan with associated floral remains.

Finally, to close on a more positive note, it is worthwhile once again emphasizing that the newly proposed *"Sivapithecus* as orangutan ancestor model" is basically consistent with the molecular clock estimate of 10 ± 3 m.y.a. for the divergence of *Pongo* (Cronin, 1983; Andrews and Cronin, 1982). Boaz (1983) arrives at a similar "averaged age" estimate of 11.0 m.y.a. for the orangutan split with the African ape-hominid clade. Zihlman and Lowenstein (1983) also offer an estimate of 9-10 m.y.a. for this same event.

Evidence for African Great Ape Lineage

Fossil evidence for the initial differentiation and later evolution of the African Great Apes is lacking. Except for the brief mention of an incisor from the lower Pleistocene of Uganda described by Von Bartheld et al. (1970) as an early gorilla [2], not one single piece of fossil evidence is available to document the lineage leading to the African apes. Therefore many contributors to this volume have chosen to combine their discussions of the early evolution of the African apes with the early evolution of the hominid line. This view has been adopted by those supporting African apes and humans as sister groups (as in Figs. 4 and 5). Especially prominent among this group are contributors suggesting a relatively recent divergence of the African Great Ape and hominid clades.

Wolpoff (1983) believes that a suitable Ramapith ancestor of the African ape-human clade will be found in later Miocene exposures of the African continent. This hypothetical Ramapith, according to Wolpoff, would have a chimpanzee-like cranial vault with a Ramapith-like face and dentition equipped for handling anterior dental loading stresses. Wolpoff (1983) sees this hypothetical ancestor fairly closely resembling the known cranial parts of *Australopithecus afarensis* [see figures in White et al. (1983) for examples]. Zihlman and Lowenstein (1983) approach the African ape-human common ancestor concept from a different perspective based on analogs to living forms. They argue that "pygmy chimpanzees (*Pan paniscus*) are more generalized than the common species and make a reasonable beginning point for the evolution of three different adaptations: the slender, bipedal hominids, the large dimorphic gorillas, and the moderately dimorphic fruit-eating common chimpanzees" (p. 685). The term "generalized" to Zihlman and Lowenstein (1983) implies "approaching the ancestral state" and thus they consider *P. paniscus* has undergone the least amount of morphological change *vis a vis* the last common ancestor of the African ape-hominid clade, which makes it a suitable candidate to represent that ancestor.

Wolpoff (1983) comments further on the common ancestor of chimpanzees, gorillas, and humans, suggesting this ancestor will *not* resemble a pygmy chimpanzee to a significantly greater degree than *A. afarensis* already does. Now, since Zihlman and Lowenstein (1983, p. 687-688, Fig. 1) draw attention to a large number of features shared by both *Pan paniscus* and *Australopithecus*, it appears on the surface that Wolpoff's views and their own are not so different. However, Wolpoff (1983, p. 666) also suggests that "the last common ancestor of the African apes and humans was probably as different from living apes as it was from living humans. No living ape…could form an adequate model for this ancestral form, any more than a living human group could." This interchange leads me to suggest that Wolpoff and Zihlman-Lowenstein are arguing at cross purposes. In my estimation they actually have both defined elements of the same African ape-human ancestral morphotype (see morphotype 9 in Table 1) but are in disagreement over its semantic expression. The ancestor probably was a generalized chimplike creature with a primitive hominoid "dryomorph" subnasal morphology (Ward and

Pilbeam, 1983). It probably was a Ramapith as Wolpoff (1983) suggests, but a uniquely African variety, which might better be termed a "dryomorph Ramapith."[3] When fossil evidence of the African ape/human clade is one day recovered, these observations will be ultimately falsifiable.

Those contributors supporting a more ancient origin of the African apes do not engage in the same sort of scenario-building as that presented in the discussion above. Rather, they emphasize the Dryopith-Ramapith dichotomy linking the origin of the African apes (and usually *Pongo*) to the Dryopiths and the origin of the Hominidae to the Ramapiths (see Fig. 3). As outlined in the preceding section on the orangutan lineage, Kay and Simons (1983) suggest deriving the African apes from later Miocene thin-enameled Dryopiths, perhaps represented, in a collateral sense, by European *Dryopithecus*. Pickford (1982, 1983) suggests a similar scheme, focusing on the East African late lower Miocene Dryopiths, from which he derives the "Pongidae." Pickford (1982) specifically excludes *Proconsul* from the "Pongidae" since it does not possess the apelike facial anatomy that he suggests later Dryopiths share. De Bonis (1983) also looks to the Dryopiths for the ancestral stock of the African apes. Based on dental and postcranial features, he suggests that European *Dryopithecus* and *Pan* are sister groups, with *Gorilla* joining the clade next, followed by *Proconsul*. From this arrangement he dates the *Pan-Gorilla* divergence to between 14 and 22 m.y.a.

One interesting outgrowth of de Bonis' (1983) cladistic analysis is that he regards knuckle-walking as a shared derived behavioral mode specific to the African apes. It has long been known that the African apes and humans are very similar in their upper arm anatomy but quite divergent in most aspects of lower limb anatomy. These differences naturally relate to human bipedality vs. African ape knuckle-walking quadrupedality. For many years it was assumed that knuckle-walking was part of the ancestral "pongid" morphotype and therefore that the earliest hominids had passed through a knuckle-walking stage in human ancestry. Once it became possible to isolate the anatomical correlates of knuckle-walking in the wrists and hands of the African Great Apes (Tuttle, 1969a, 1975b; Corruccini, 1978), this information could be applied directly to the hominid fossil record. To date, none of the African ape knuckle-walking features have been found in the wrist or hand bones of any fossil hominid, including the recently described *A. afarensis* (Tuttle, 1981; Bush et al., 1982). It may therefore be assumed that knuckle-walking is a shared derived character complex that arose in the African Great Apes (see morphotype 10 in Table 1) after the separation of the hominid clade. Regarding the uniqueness of knuckle-walking in the African apes, it should be noted that the existence of a parallel behavior has been documented in a captive orangutan (Tuttle and Beck, 1973; Susman, 1974). This observation prompted Tuttle (1974, p. 397) to remark:

> if a highly advanced arboreal climber and arm-swinger like the orangutan can develop ontogenetically into a knuckle-walker (albeit in special circumstances of captivity), ancestors of the African apes may have been somewhat similarly predisposed to knuckle-walking by their own arboreal heritage. It also seems reasonable to argue that if the common ancestor of chimpanzee and gorilla possessed somewhat advanced features of the hand for suspensory behavior and climbing, it could have developed into a knuckle-walker quite rapidly after adopting terrestrial habits.

Thus Tuttle (1974) offers a scenario for the unique development of knuckle-walking in the African apes based on "preadaptive" features of the common hominoid arboreal heritage.

Evidence for Differentiation of the Hominid Lineage

The origin of the Hominidae has been the central theme of paleoanthropological studies for more than a century. It is therefore not surprising, then, that it is also the central theme of many chapters in this volume. The issues raised here by contributors focus specifically on the early differentiation of the hominid clade with regard to its geographical and temporal placement and its adaptive nature. Later aspects of the hominid cladogenesis were not the primary focus of this volume on Miocene Hominoidea, but nevertheless have also received a well-documented coverage (e.g., White et. al., 1983; Boaz, 1983).

As long ago as 1871 Charles Darwin observed that Africa was the probable continent of origin of our early progenitors, based on human resemblances to the African Apes. The first archaic human discoveries in Java and China appeared to shift the balance of opinion on continental origins to Asia. Thus the Africa/Asia issue in hominid origins became a major point of contention as paleoanthropological discoveries began to document the human fossil record. The issue of continent of origin of the hominid clade has been discussed by a variety of contributors in this volume. Basically, those who support Ramapiths as exclusive hominid ancestors (as in Fig. 3) favor the Asian continent, where most of the fossil evidence has been found. Those contributors viewing Ramapiths as specially related to the orangutan (as in Fig. 4) or who suggest Ramapiths as the common ancestor of all the Great Ape and human lines (as in Fig. 5) tend to favor Africa. It should be noted that my use of the word "favor" in reference to a contributors' position is a rather reductionist view. The basic issue is that few contributors actually discuss the paleozoogeographic implications of their viewpoints, with the notable exception of Bernor (1983).

Bernor presents an in-depth overview of the zoogeographic relationships of the Miocene Hominoidea. In the later Miocene he documents the spread of truly open-country habitats (Bernor,

1983, p. 51, Fig. 6) that developed in parts of the Old World in the Turolian (10–5 m.y.a.). Bernor argues that though Ramapiths were the most open-habitat-adapted of the Miocene hominoids, they could not make the shift to these new habitats. Thus the presence of a seasonally-adapted Turolian open-country biotope interposed between the Siwalik Province and the East African Province provided an effective ecological barrier to hominoid migrations between the two regions by the earliest Turolian (~10 m.y.a.). Bernor (1983) also notes that throughout the Miocene the predominant mammalian dispersal pattern appears to have been into, not out of, the Siwalik Province, making it a sort of provincial biogeographic *cul-de-sac*. Combining these points with the earliest occurrence of bipedal hominids in Africa at the 3–4 m.y.a. time range, Bernor (1983, p. 59) concludes that "the evolution of *Australopithecus* would most likely have been an African event," probably confined to sub-Saharan Africa. Thus Bernor presents specific zoogeographic evidence excluding Siwalik Ramapiths (and other Eurasian hominoids?) from participation in the evolution of the earliest documented hominids. This does not prove an African origin of the Hominidae, but it strongly supports one.

The temporal placement (dating) of the origin of the Hominidae has always been a thorny issue. This probably can be traced back to the Victorian Age, when Thomas Henry Huxley first documented our close affinity with the African apes and Charles Darwin expounded on the "missing link" concept in human evolution. Simply stated, if the common link between ape and human could be demonstrated as ancient, then it was surmised that humans would be less closely related to the living apes. Naturally, the converse would apply if the "link" was more recent. Among contributors to this volume, the actual dating of the hominid lineage was closely dependent upon one's cladistic views of hominoid relationships (compared in Figs. 3–5). If it can be assumed that views favoring the African apes and humans as sister groups among living hominoids are the more correct interpretations [*contra* Kluge (1983), Kay and Simons (1983)], then the biomolec-

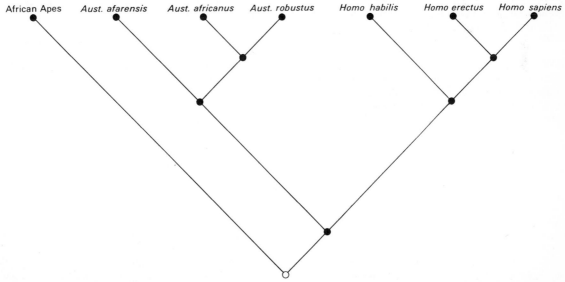

Figure 6 *Cladogram depicting relationships among currently recognized taxa of the Hominidae based on the most recent synthesis of the data (see White et al., 1981). Note that the genus* Australopithecus *shares a sister group relationship with the genus* Homo. *Also note that the African apes are viewed as the sister group of Hominidae, which would indicate that these groups shared a common ancestry subsequent to the Ramapith radiation. Within the Hominidae,* A. afarensis *occupies a unique position as the sister group of all later hominids.* Homo habilis *occupies a similar position within the genus* Homo. *Solid circles in this cladogram represent existing phyletic groups; the single open circle represents the hypothetical common ancestor of the African apes and the earliest known hominid,* Australopithecus.

ular data set can be brought to bear directly on this issue. Zihlman and Lowenstein (1983) estimate the human–African ape divergence at 5–6 m.y.a. based on their interpretation of the molecular clock. Cronin (1983) also discusses a similar age estimate, but later states "I am increasingly confident that the hominid–pongid divergence is in the range of 4.0–8.0 m.y.a." (p. 134). I support this more liberal phrasing of the dating issue and also submit that Goodman et al. (1983) and others will probably tone down their criticisms of clock dates based on this interpretation.

If the origin of the Hominidae is in the 4.0–8.0 m.y.a. time range, how early can we reasonably expect to document this lineage in the fossil record (based on shared derived characters like those in morphotype 11 of Table 1)? Greenfield (1983, p. 700) indirectly answers this question by concluding that divergences "may or may not have immediately given rise to clearly distinguishable members of independent lineages leading to humans and the extant Great Apes." Therefore, I feel that isolating the exact point of divergence of earliest Hominidae in the fossil record may be an unresolvable issue. This is precisely why students of cladistic analysis emphasize the need to document sister group relationships rather than search for perhaps unknowable ancestors. This is also why the biomolecular evidence may prove more useful than the fossil evidence for defining the precise divergence date of two extant lineages. However, the biomolecular data cannot at present resolve the relationship of fossil groups to one another or to modern groups (but see Lowenstein, 1983) so both sources of evidence *must* be used in tandem to develop an "alpha phylogeny."

Cronin (1983) presents a genetically based, time-oriented scenario for the origin of the hominid lineage. He suggests that the actual morphological shift from an apelike forebear to a distinguishable early hominid occurred relatively rapidly, perhaps in only 1 or 2 million years (approximately 50,000 to 100,000 generations). Cronin (1983, p. 131) states:

> A rapid shift in locomotor adaptations was the first major derived change, followed by unique changes in the brain and dentition, etc. One need not posit millions of years for the initial hominid adaptations to evolve from the ape ancestral conditions. Rates of evolution vary dramatically and traits may be responding to either intense selection or random fixation. Small population size, coupled with inbreeding, may facilitate rapid change during the process of speciation. This type of small population bottleneck, or small population mode of allopatric speciation, may have predominated in ancestral hominid populations.

Zihlman and Lowenstein (1983) also observe that origin of the Hominidae from an apelike (pygmy chimp-like) precursor may have taken place in 1 million years or perhaps less time. They speculate on whether 1 million years of transformation can be viewed as a gradual or punctuated event in the context of evolutionary processes. I think it probably comes as no surprise that these biomolecular-related views of hominid origins are characterized by a temporally rapid shift for the origin of the hominid lineage. Since these contributors both favor a late ape–human divergence, the transitions they propose are concordantly rapid. Those who favored an earlier divergence, such as in the now-dated *"Ramapithecus"* as early hominid model, usually supported a more gradual transition in morphology. Because I consider that the origin of higher categories occurs primarily by rapid, steplike, punctuated events, I find myself in full agreement with the views offered by Cronin and by Zihlman and Lowenstein.

The adaptive nature of the earliest hominids, or rather the delineation of the behavioral and morphological feedback systems that resulted in the acquisition of our distinctly human characteristics, has been the subject of much recent discussion (e.g., Lovejoy, 1981). Wolpoff (1983) among contributors to this volume touches on this issue more than any other contributor. He suggests that the common ancestor of hominids and the African apes was probably a special form of Ramapith. Based on shared similarities of African apes and humans, together with evidence from the hominid fossil record, Wolpoff (1983, p. 667–668) characterizes this proto-African ape, proto-human as:

> behaviorally complex, a tool user, and a rudimentary tool maker, an omnivore utilizing a wide variety of dietary resources ranging from foods difficult to masticate to protein obtained through organized hunting and systematically shared by at least part of the social group, an incipient biped (the behavior was possible but not morphologically efficient), and just possibly a more complex communicator than generally thought, utilizing an albeit limited but symbol-based open communication system.

Wolpoff observes that a shift in importance of these features was as much responsible for the origin of the hominid clade as the acquisition of additional hominid apomophous features. Therefore, it may not have been development of new characteristics that marked the adaptive shift of the earliest hominids, but rather an emphasis on specialization of an already existing suite of features. This view of the hominization process differs from the conventional one in that many of the previously considered uniquely hominid adaptations would have been at least incipiently present in the earliest African ape. As the African ape lineage developed, these features would have been subject to an alternate set of selective forces and modified still further after the separation of the *Pan* and *Gorilla* lineages. If Wolpoff (1983) is correct in this view, then the early members of the African ape clade would have been more "human-like" in their behavior and morphology than either of the living African Great Apes. Of course, the need then arises to explain how many of the traditional, once sacred, hominid features arose in the common proto-African ape proto-human ancestor. So the question of adaptive significance has simply been shifted to a different level [4].

One aspect of Wolpoff's (1983) hominid origins model that deserves further discussion is his characterization of the common African ape-human ancestor as an "incipient biped." I take this to mean that the common ancestor was more predisposed to bipedalism than are the extant knuckle-walking chimpanzee and gorilla. If so, then Wolpoff's model supports aspects of Tuttle's (1974, 1975b, 1981) hylobatian model. Specifically, Tuttle (1981) argues that arboreal bipedalism on horizontal boughs and vertical climbing on trunks and vines were aspects of a locomotor repertoire that evolved *before* the emergence of the Hominidae. Tuttle's pre-hominids utilized bipedal running and hindlimb-propelled leaps to forage in the arboreal milieu. In this sense their arboreal behavior was preadaptive for terrestrial bipedalism. Stern and Susman (1981) also relate the arboreal activity of vertical climbing to the emergence of terrestrial bipedalism, emphasizing its preadaptive basis. They conclude that the earliest hominids were at least partially arboreal. These points all focus attention, in my opinion, on the adaptive mosaic that must have characterized the very early Hominidae.

This early phase of hominid evolution is not well documented in the fossil record. However, by the 3–4 m.y.a. time range in East Africa the record becomes considerably more complete. It is from this time period that the earliest evidence of bipedalism is documented in taxon *Australopithecus afarensis* Johanson et al.

1978. With regard to *A. afarensis*, the critical issue lies not in the nature of its bipedal adaptation, but rather in its phylogenetic affinities. Boaz (1983, p. 712) observes that "The primary question in regard to *A. afarensis* is not whether it represents a separate (though perhaps earlier and closely related) species distinct from *A. africanus*, but is the assessment of its phylogenetic position." Once raising this issue, Boaz (1983) addresses two questions: (1) Whether the *A. afarensis-Homo* relationship is valid and (2) whether *A. africanus* is the ancestor of robust australopithecines. Boaz (1983) argues the evidence for a link between the hominid populations of Laetoli and Hadar (*A. afarensis vide* Johanson et al. 1978) and the first species of *Homo* is tenuous. He observes there is a temporal and morphological gap spanning this phylogenetic connection that is better bridged by the gracile australopithecines from South Africa. In this regard he would place the southern African (Sterkfontein and Makapansgat) and eastern African (Laetoli and Hadar) gracile australopithecines in the same species, *A. africanus*. Boaz (1983, p. 713) concludes that a more parsimonious hypothesis is one where the "South African *A. africanus* represents the population descendent from earlier, slightly more primitive *A. africanus* and ancestral to early *Homo*."

Regarding the second question concerning Johanson and White's (1979a) views on the origin of the *A. robustus/boisei* lineage, Boaz (1983) critiques the evidence for *in situ* evolution of robust australopithecines in South Africa. Since they appear rather suddenly in the fossil record of East and South Africa (about 2.0–2.1 m.y.a.), Boaz suggests the robust forms may have evolved elsewhere on the African continent (by allopatric or parapatric speciation) and dispersed into East and South Africa in the latest Pliocene. He then briefly discusses the South African australopithecine record, commenting on the lack of appropriate "robust-like" ancestors.

Boaz (1983) does focus on two valid points of contention concerning the phylogeny and scenario proposed by Johanson and White (1979a). However, a more recent contribution on the subject (White et al., 1983) in my opinion firmly establishes *A. africanus* as the sister group of the robust australopithecines (see Fig. 6) and thus very probably the ancestral stock from which that group is derived. Furthermore, White et al. (1983) demonstrate beyond any reasonable doubt that *A. afarensis* is a species distinct from *A. africanus* and in those characters in which they differ, the former represents a better ancestral morphotype for *Homo* than the latter. This was all accomplished by carefully circumscribing and then minutely defining the variation present in the taxon *A. africanus*. White et al. (1983) observe:

> Although *A. afarensis* and *A. africanus* share many common primitive features, the latter taxon exhibits a morphological composite of the skull and dentition derived toward the *A. robustus* + *A. boisei* character state. This pattern and all its components can be related to a functional trend toward maximizing and spreading vertical occlusal forces along the postcanine tooth row in response to dietary specializations. Members of the *Homo* clade do not show this specialization but exhibit a pattern indicative of encephalization and masticatory apparatus reduction (p. 769). Our hypothesis predicts that structural and temporal intermediates between *A. afarensis* and *H. habilis* will not exhibit the *A. africanus* morphological pattern (p. 771).

I view Fig. 6 as defined by White et al. (1983, p. 729, Fig. 4) as the most parsimonious arrangement of the known paleontological evidence for Plio-Pleistocene hominid evolution. As is evident from the cladogram in Fig. 6, I feel there is reasonable evidence

that all the ancestors for the various clades presented here have been identified, with the single exception of the first node at the bottom, which links the African apes with the Plio-Pleistocene Hominidae. This in no way signifies that all of Plio-Pleistocene hominid evolution is now understood. What it does show is that, insofar as the diagramming of relationships is a rough approximation of past events, we now have enough evidence down to the species level to draw an "alpha" cladogram of hominid relationships that is falsifiable. In my opinion if we possessed as much knowledge concerning Miocene hominoid evolution as we do concerning Plio-Pleistocene hominid evolution this volume would be considerably shorter and much more succinct! But knowing the relationships of taxa in a cladogram does not always signify a concomitant understanding of the evolutionary process and factors that brought about the neatly diagrammed speciation events. More evidence from the fossil record and studies based on living representatives will always be necessary to further reconstruct the scenario of events responsible for the origin of each clade.

SUMMARY CLADISTIC MODEL

In order to provide a final overview of the many facts and opinions expressed throughout this volume I have decided to put forth a summary cladistic model and combine this with a tabular presentation of hominoid characters that support this model. Figure 7 represents the views of a plurality of contributors, including those of the editors. I have added a series of numbered boxes between the nodes of this cladogram, which represent the morphotypes of each clade. The characters of each morphotype appear in Table 1 under the appropriate numerical heading. The analysis of character states for each morphotype was conducted following the principles of the cladistic method as outlined in this volume by Kluge (1983) and de Bonis (1983) and elsewhere. Each morphotype can be viewed as a list of features likely to be diagnostic of the ancestor of one or more taxa in its clade. It is unlikely that any single fossil specimen will possess all of the characters of any one morphotype. On the other hand, if it can be shown that a fossil possesses none of the derived characters of a morphotype, then it can be rejected as a member of that clade. Four different categories of characters are defined in the footnote to Table 1. Naturally, only the derived (apomorphic) characters are of use in establishing the sister group relationships diagrammed in Fig. 7. However, some morphotypes are defined primarily on the basis of primitive characters, in many cases due to the lack of appropriate fossil material. Fig. 7 therefore is probably best viewed at this stage as a heuristic device capable of enhancing our understanding of hominoid evolution rather than providing any definitive answers.

In the reconstruction of phylogeny no evolutionary scenario can be completely parsimonious. The fact that reversals and parallelisms are common features in mammalian evolution is now considered a truism. Therefore a certain number of characters used to establish any cladogram or phylogeny can be expected to show homoplasy. In some cases the occurrence of reversals, parallelisms, or convergence can involve as many as half of all characters (Cartmill, 1982). Parsimony at this point becomes a relative term. The cladogram presented in Fig. 7 and documented in Table 1 is, in my opinion, the most parsimonious arrangement of the data assembled in this volume and gleaned from other recent sources. Some instances of parallelism or convergence have been noted; others almost certainly exist but have not been recognized. With the acknowledgment of this caveat, I feel it is possible to view Fig. 7 and Table 1 as a series of falsifiable steps or stages that quite possibly occurred in the evolution of the Hominoidea.

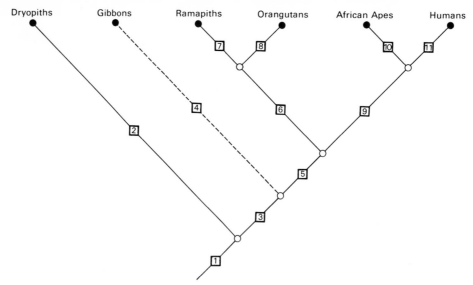

Figure 7   *Summary cladogram of hominoid relationships supported by a plurality of contributors to this volume and reflecting the views of the editors and a growing number of researchers in the field of paleoanthropology. To this cladogram, which was first presented in Fig. 4, I have added a series of numbered boxes that correspond to the 11 hominoid morphotypes described in Table 1. These boxes have been centered between nodes in this cladogram so that the morphotypes might not be confused with ancestral stages in hominoid evolution but rather used to define the character states of features likely to be diagnostic of those ancestors. In reality, no hypothetical or actual ancestor would be likely to possess all of the characters of any one morphotype described in Table 1. In this sense the morphotypes might best be viewed as a summary of mosaic evolutionary events along a clade (see text for further explanation). It is not possible to label this cladogram the "consensus" view of the volume, since contributor support was split between Figs. 3, 4, and 5 (see text). However, a plurality (including the editors) did specifically favor this cladogram, along with a number of other contributors who described the anatomical characters listed in Table 1 that support this cladogram. The 11 morphotypes are labeled as follows: (1) Proto-Hominoid, (2) Dryopith, (3) Stem extant Hominoid, (4) Proto-gibbon, (5) Pre-Great Ape/human, (6) Stem Ramapith, (7) Ramapith, (8) Proto-orangutan, (9) Pre-African ape/human, (10) Proto-African ape, (11), Proto-human.*

Many of these steps are testable through the recovery of additional fossil remains. New fossils also allow more precise determination of homoplasy. Though no ancestor is expected to completely resemble the reconstructed morphotypes presented in Table 1, the demonstration of one or more of the derived character states in newly discovered fossil forms provides an independent assessment of the morphotype's validity, and thus the sequence of events that occurred in hominoid evolution. Conversely, if enough new fossil discoveries constantly lacked one set of derived character states and yet preserved others, then one might question the validity of the morphotype.

It should be noted that some characters that cannot be assessed in the fossil record have also been included in Table 1. These characters, such as soft anatomical features or chromosome number, can be significant additions since they often provide an independent check on the "hard" osteological data and may one day also be used to corroborate evolutionary scenarios based on new neontological analyses. Finally, I want to emphasize, once again, the preliminary nature of the relationships and character determinations presented in Fig. 7 and Table 1, respectively. A more complete set of data is now being assembled and will be presented in a future publication (Ciochon and Myers, in preparation). That publication will also contain further elaboration of the hypothetical nodal ancestors identified in Fig. 7.

CONCLUSION

It is my hope that *New Interpretations of Ape and Human Ancestry* will demonstrate that the field of paleoanthropology is progressing toward the development of new, fully testable para-

digms of ape and human ancestry. The summary model of hominoid cladogenesis outlined above certainly represents progress toward the attainment of an enduring consensus. Concrete evidence that progress has occurred during the last two decades can be found by comparing this volume's contents with what is now regarded as a classic in paleoanthropology, Washburn's (1963b) *Classification and Human Evolution*. Both volumes deal with a remarkable similar range of topics (comparative anatomy, biochemistry, hominid phylogeny), although Washburn's volume emphasized the new field of primate behavior to a much greater extent than can be found here.

Classification and Human Evolution achieved a major milestone when it provided the first legitimate forum for the presentation of the biomolecular viewpoint of human origins. I suggest that that viewpoint has assumed a preeminent position in the minds of many contributors to *New Interpretations of Ape and Human Ancestry*, as the preceding chapters illustrate. This can be legitimately labeled as progress. In further comparisons of the two volumes it is possible to see that progress has also occurred in our understanding of the fossil record. In several of the chapters on hominid phylogeny in Washburn's volume, much discussion and debate took place over the position of the enigmatic primate from Italy, *Oreopithecus*, in human phylogeny. Twenty years later in this volume on ape and human ancestry *Oreopithecus* receives hardly a mention through the 29 preceding chapters, and rightly so, since this "swamp ape" has little bearing on current theories of ape and human origins. Perhaps a decade from now "*Ramapithecus*" will be relegated to a similar position within human phylogeny. This illustrates that progress in paleoanthro-

Table 1 *Selected Characters of the Eleven Hominoid Morphotypes[a] Diagrammed in Fig. 7*

Polarity	Character	Source
(1) Proto-Hominoid morphotype		
D	Reduction of snout together with development of short, nonprojecting nasals	Fleagle (1984), Fleagle and Kay (1983)
D	Choanal shape wide and low	Delson and Andrews (1975)
D	Alveolar recess of maxillary sinus invades the alveolar process	Cave and Haines (1940)
D	Presence of tubular ectotympanic bone	Fleagle and Kay (1983)
D	Dorsal reduction of premaxillae	Fleagle and Kay (1983)
D	Reduction of postorbital constriction	Fleagle and Kay (1983)
D	Appearance of enamel prism Pattern 3A	Gantt (1983)
D	Loss of entepicondylar foramen of humerus	Fleagle and Kay (1983)
P	Humerus with large projecting medial epicondyle and with spherical capitulum separated from trochlea by shallow groove	Fleagle (1983)
P	Ontogenetically early appearance of ischial callosities	Delson and Andrews (1975)
P	Diploid ($2n$) number of chromosomes equal to 44	Chiarelli (1975), Kluge (1983)
P	Moderate sexual dimorphism	Fleagle et al. (1980)
(2) Dryopith morphotype		
P	Broad interorbital septum	Andrews and Cronin (1982), Kay (1982a)
P	Nasal aperture higher than broad, oval-shaped	Andrews and Cronin (1982), Kay (1982a)
P	No fronto-ethmoidal sinus	Martin and Andrews (1982)
P	Subnasal alveolar process appears as a flattened oval in sagittal section	Ward and Pilbeam (1983)
P	Incisive fossa forms transversely broad basin that opens directly into oral cavity (subnasal plane stepped down to floor of nasal cavity)	Ward and Pilbeam (1983), Andrews and Cronin (1982)
P	Nasoalveolar clivis abbreviated, terminating in incisive fossa (subnasal plane truncated)	Ward and Pilbeam (1983), Andrews and Cronin (1982)
P	Narrow troughlike maxillary sinus that extends to the dental alveoli	Ward and Pilbeam (1983)
P	Occasional thickening of glabellar region	Andrews and Cronin (1982)
P	Flaring zygomata with strong posterior slope	Kay (1982a), Andrews and Cronin (1982)
P	Zygomatic foramina situated at or below the lower rim of the orbits	Andrews and Cronin (1982)
P	Zygomatic foramina small	Andrews and Cronin (1982)
P	Large, oval-shaped greater palatine foramina	Andrews and Cronin (1982)
P	Infra-orbital foramina well removed from zygomaticomaxillary suture	Andrews and Cronin (1982)
P	Uniformly small lower incisors	Kay (1982a)
P	Upper incisors lacking large size discrepancy	Andrews and Cronin (1982), Kay (1982a)
P	Single cusped and sectorial P_3	Martin and Andrews (1982), Kay (1982a)
P	Slender canines	Martin and Andrews (1982)
P	High degree of canine sexual dimorphism	Kay (1982a, 1982b)
P	Canine root aligned along tooth row axis or rotated internally	Ward and Pilbeam (1983)
P	Canine roots elliptical in transverse section	Ward and Pilbeam (1983)

[a]The polarity or status of each character appears in the left-hand column preceding the character description according to the following conventions: D, derived unique character by comparison to sister taxon; P, primitive, ancestral character shared by sister taxon; C, convergent (or parallel) nonhomologous character indicating phenetic similarity only; V, variable character of doubtful phyletic significance due to its random expression in sister taxa under comparison or in other out-groups. The source or sources which provided the character and in most cases the interpretation of polarity appears in the right-hand column. In some cases where the source made no explicit statement of polarity, I made the determination. I did not always choose the author who first mentioned the existence of a feature, but rather chose as the source the most cladistically phrased view of that character's representation. In many cases, the characters and sources are derived from this volume. When a difference of opinion existed over a character the alternative polarity and author can be found listed together in the source column. An attempt has been made to assess character correlation throughout this tabulation to avoid duplication of characters. Some repetition of anatomical features, however, was unavoidable. The exact positioning of a few characters within particular morphotypes is open to question due to the lack of appropriate fossil evidence. Therefore, even though a character's primitive or derived state may be firmly established its position within a particular morphotype is sometimes conjectural. Data from biomolecular studies, such as amino acid sequencing of proteins and DNA endonuclease restriction mapping of base sequences, can easily be added to this tabulation in the future to further augment the designated morphotypes.

Table 1 *(continued)*

Polarity	Character	Source
P	Molars with moderate to large cingula	Von Koenigswald (1973), Martin and Andrews (1982), Kay (1982a)
P	M_2^2 larger than M_1^1	Corruccini and Henderson (1978)
P	Intermediate thickness of enamel caps on molars	Gantt (1983), Ward and Pilbeam (1983)
P	Gracile mandibles with symphyses buttressed by large inferior and superior transverse tori	Martin and Andrews (1982), Kay (1982a)
P	Presence of superior parallel sulcus (a^1) and anterior occipital sulcus (a^3) on frontal lobe of cortex	Falk (1983)
P	Prominent to moderate expression of deltopectoral crest of humerus	Morbeck (1983)
D?	Presence of relatively broad, spool-shaped trochlea on humerus	Rose (1983)
D?	Reduction of olecranon beak of ulna	Rose (1983)
D?	Presence of spherical femoral head with articular surface extending onto neck anteriorly	Rose (1983)

(3) Stem extant Hominoid morphotype

Polarity	Character	Source
D	Acquisition of fronto-orbital sulcus (*fo*), opercular sulcus (*io*), and frontalis superior sulcus (*fs*) on frontal lobe of cortex	Falk (1983)
D	Lengthening and torsioning of clavicle	Napier and Napier (1967), Oxnard (1968c)
D	Broadened sternum with progressive fusion of the sternal elements in adults	Tuttle (1974)
D	Dorsal positioning of scapula with glenoid fossa directed more cranially	Le Gros Clark (1959), Washburn (1963a)
D	Glenoid fossa widened, rounded, and flat with nonprojecting supraglenoid tubercle	Corruccini and Ciochon (1976)
D	Spinoglenoid notch of scapula deepened	Corruccini and Ciochon (1976)
D	Presence of coraco-acromial ligament spanning the laterally projecting coracoid process and acromion	Ciochon and Corruccini (1977b)
D	Lengthening of humerus with humeral head becoming large and globular	Le Gros Clark (1959), Corruccini and Ciochon (1976)
D	Intertubercular sulcus of humerus narrow and deep with lateral extent of head articular surface restricted from extending into the sulcus	Corruccini and Ciochon (1976)
D	Distal humerus with broad trochlea relative to capitulum with midportion constricted; capitulum is bulbous, coranoid fossa is deep, and radial fossa is shallow	Morbeck (1983)
D	Lateral edge of trochlea that separates trochlea and capitulum anteriorly distinctly wraps around distally to meet with olecranon fossa	Morbeck (1983)
D	Proximal ulna with segmented trochlear notch and U-shaped deep radial notch	Morbeck (1983)
D	Some reduction of olecranon process of ulna	Tuttle (1975b)
D	Moderate reduction of ulnar styloid process with development of interarticular meniscus and occasional bony lunula but maintaining some direct ulna—triquetral contact	Lewis (1972b), Corruccini (1978)
D	Radial head approaches circularity and is not tilted	Rose (1983)
D	Loss of sesamoid bones from the tendons of the hand	Washburn (1963a)
D	Shortening of lumbar region of vertebral column by 1–2 vertebrae, yielding a mode of five	Schultz (1961, 1963)
D	Reduction of the tail not by atrophy but by transformation into the shelflike coccyx	Andrews and Grove (1975)
D	Development of suspensory foraging and feeding adaptations including some underbranch leaping and hauling actions	Tuttle (1975b), Washburn (1968a)
D	Development of pelvic diaphragm	Tuttle (1975b)
D	Presence of vermiform appendix	Andrews and Cronin (1982)
D	Fetal membranes with interstitial implantation, decidua capsularis but no uterine symplasma	Luckett (1975)
P	Urethrovaginal septum distinct with labia majora forming relatively defined narrow, cutaneous folds	Kluge (1983)

Table 1   *(continued)*

Polarity	Character	Source
P	Well-developed glans penis with prominent corona glandis	Kluge (1983)
P	Flexor pollicus longus muscle and tendon well developed	Kluge (1983)
P	Tendons of flexor hallucis longus and flexor digitorum longus are fused and evenly distributed to five digits	Kluge (1983)

(4) Proto-gibbon morphotype

D	Substantial elongation of all forelimb elements with emphasis on ricochetal arm-swinging and vertical climbing	Tuttle (1975b)
D	Interarticular meniscus in wrist joint strengthened by full development of bony lunala, the os Daubentonii	Lewis (1969), Delson and Andrews (1975)
D	Proximal capitate and hamate form distinctive narrow, knoblike process	Jenkins and Fleagle (1975)
D	Carpal and pedal digits II–V elongated	Biegert (1963), Tuttle (1972b)
P	Long, fully opposable thumb	Fleagle (1984)
D	Carpal digit I forming ball-and-socket joint at the trapezium and metacarpal base	Van Horn (1972), Andrews and Groves (1976)
D	Ventral curvature (bowing) of digital rays II–V of hand notable in proximal and middle phalanges	Tuttle (1972b), Andrews and Groves (1975)
D	Phalanges of pedal digits II–V curved ventrally and act together with the widely divergent, heavily muscled hallux to aid in the development of above-branch bipedal running	Tuttle (1972b, 1975b)
D	Lateral flexion and rotation of spine with lowered center of gravity	Tuttle (1981)
D	Reduction of sexual dimorphism	Delson and Andrews (1975)
D	Loss of superior parallel sulcus (a^1) on frontal lobe of cortex	Falk (1983)
D	Thin enamel cap on molars	Gantt (1983), Frisch (1973)

(5) Pre-Great Ape/human morphotype

D	Choanal shape, narrow and high	Delson and Andrews (1975)
P or D	Intermediate or thick enamel caps on molars	Ward and Pilbeam (1983)
D	Cartilaginous meniscus fully interposed between greatly reduced ulnar styloid process and pisiform, resulting in total exclusion of ulnar-carpal articulations	Lewis (1969, 1972) McHenry and Corruccini (1983)
D	Reduction of pisiform size and loss of its articular facet for the ulnar styloid process	McHenry and Corruccini (1983), Morbeck (1972)
D	Presence of fossa on distal ulna for attachment of interarticular meniscus	McHenry and Corruccini (1983)
D	Development of a pronounced hook on the hamate	O'Conner (1975), Corruccini et al. (1975), McHenry and Corruccini (1983)
D	Dorsal tubercle (Lister's) of radius shifts laterally to overlie the scaphoid facet	O'Conner (1975), Corruccini et al. (1975), McHenry and Corruccini (1983)
D	Sickle-shaped development of scaphoid bone	O'Conner (1975), Corruccini et al. (1975)
D	Lack of extension of articular cartilage onto dorsum of metacarpal V	O'Conner (1975), Corruccini et al. (1975)
D	Progressive increase in mobility of wrist, elbow, and shoulder joints until full range of forelimb movement capabilities is achieved, resulting in various suspensory postures and hanging/feeding adaptations	Rose (1983), Morbeck (1983), Corruccini et al. (1976), Fleagle (1983)
D	Ontogenetically late appearance of ischial callosities	Delson and Andrews (1975)
D	Acquisition of arcurate (*arc*) sulcus on frontal lobe of cortex	Falk (1983)
D	Diploid ($2n$) number of chromosomes equal to 48	Yunis and Prakash (1982), Mai (1983), Kluge (1983)

(6) Stem-Ramapith morphotype

D	Narrow interorbital septum	Andrews (1983), Kay and Simons (1983, ?V)
D	Short, broad nasal aperture	Andrews (1983), Kay and Simons (1983, ?P)
D	Subnasal alveolar process that appears elliptical in sagittal section	Ward and Pilbeam (1983)
D	Incisive fossa forms narrow depression just behind nasospinale (subnasal plane continuous with floor of nasal cavity, not stepped)	Ward and Pilbeam (1983), Andrews and Cronin (1982), Lipson and Pilbeam (1982), Kay and Simons (1983, ?V)

Table 1   *(continued)*

Polarity	Character	Source
D	Nasoalveolar clivus extends posteriorly into nasal cavity (subnasal plane smooth, not truncated)	Ward and Pilbeam (1983), Andrews and Cronin (1982), Lipson and Pilbeam (1982), Preuss (1982)
D	High alveolar or premaxillary prognathism	Andrews and Tekkaya (1980), Wolpoff (1983), Kay and Simons (1983, ?P)
D	Wide maxillary sinus that rapidly diminishes anteriorly	Ward and Pilbeam (1983)
D	Orbits higher than broad, rather ovoid in shape	Andrews and Cronin (1982), Wolpoff (1983), Kay and Simons (1983), Preuss (1982)
D	Temporal edge of orbit marked by a well-defined marginal process	Preuss (1982)
D	Face is concave in lateral view, which can be termed "simognathic"	Preuss (1982), Napier and Napier (1967)
D	No glabellar thickening	Andrews and Cronin (1982)
D	Deep zygomatic process with strong lateral flare	Andrews (1983), Kay and Simons (1983, ?P)
D	Zygomatic bone flattened and facing anteriorly	Andrews and Cronin (1982), Preuss (1982), Kay and Simons (1983, ?V)
D	Presence of pronounced malar notch (or incisura malaris) on inferolateral surface of zygomatic body	Preuss (1982)
D	Zygomatic foramina relatively large	Andrews and Cronin (1982), Kay and Simons (1983, ?V)
D	Slitlike greater palatine foramina	Andrews and Cronin (1982), Kay and Simons (1983, ?V)
D	Extremely small incisive foramina	Andrews and Cronin (1982), Kay and Simons (1983, ?V)
D	Broad I^1 that is much larger than I^2	Andrews (1983), Wolpoff (1983), Preuss (1982), Kay and Simons (1983, ?C)
D	Central maxillary incisors that change angulation during life	Wolpoff (1983)
D	Some external rotation of upper canine roots	Ward and Pilbeam (1983)
D	Canine roots quadrangular in transverse section	Ward and Pilbeam (1983)
D	Well-developed canine fossa	Ward and Pilbeam (1983)
D	Strong superior–medial angulation of canine roots	Wolpoff (1983)
P or D	Intermediate thickness of enamel caps on molars	Ward and Pilbeam (1983), Gantt (1983)
D	Deeply incised wrinkles that persist after crown morphology has been worn away	Wolpoff (1983), Gantt (1983), Andrews and Cronin (1982, ?V)

(7) Ramapith morphotype

Polarity	Character	Source
P	Poor development of supra-orbital torus	Andrews (1982), Kay and Simons (1983)
P	Arched palate	Kay and Simons (1983), Kay (1982a)
P	Zygomatic process set well forward	Kay and Simons (1983), Kay (1982a)
P	Infra-orbital foramina few in number	Andrews and Cronin (1982)
D	Shallow, broad mandibular corpora	Kay and Simons (1983), Kay (1982a), Andrews and Cronin (1982, ?C)
P	V-shaped lower dental arcade	Kay and Simons (1983), Kay (1982a)
P	Small upper and lower incisors relative to cheek tooth size	Kay and Simons (1983), Kay (1982a)
D	Small upper and lower canines relative to cheek tooth size	Kay and Simons (1983), Kay (1982a), Andrews and Cronin (1982, ?V)
D	Long axis of upper canine cross section set more or less buccolingually	Kay and Simons (1983), Kay (1982a), Ward and Pilbeam (1983, ?C), Andrews and Cronin (1982, ?C)
D?	P_3 with well-developed metaconid	Kay and Simons (1983), Kay (1982a), Wolpoff (1983, ?V), Andrews and Cronin (1982, ?V)
P	Lower molar buccal cingula poorly developed or absent	Kay and Simons (1983), Kay (1982a)
D	Thick enamel caps on molars	Kay and Simons (1983), de Bonis (1983), Wolpoff (1983), Ward and Pilbeam (1983), Andrews (1983), Preuss (1982)
D	Low canine sexual dimorphism	Kay (1982b), Kay and Simons (1983), Andrews and Cronin (1982, ?V)

(8) Proto-orangutan morphotype

Polarity	Character	Source
D	Zygomatic foramen situated above the level of the lower rim of the orbit	Andrews and Cronin (1982)
P	Poor development of supra-orbital torus	Andrews (1982), Kay and Simons (1983)

Table 1 *(continued)*

Polarity	Character	Source
D	Increase in size of upper premolars relative to molars	Preuss (1982)
P	Intermediate thickness of enamel on molar caps with prominent wrinkling	Ward and Pilbeam (1983)
D?	Shortening of lumbar region of vertebral column by one vertebra, yielding a mode and mean of four	Schultz (1961, 1963), Kluge (1983)
V	Occasional ossification of the os centrale and scaphoid of the wrist occurring only in aged or arthritic individuals	Schultz (1936, 1963), Corruccini (1978)
C	Proximal capitate and hamate form distinctive narrow, knoblike process	Jenkins and Fleagle (1975)
C?	Further lengthening of forelimb elements with emphasis on vertical climbing, hanging, and terminal branch foraging and feeding	Tuttle (1975b)
D	Hallux reduced dramatically, often resulting in absence of distal phalanx	Tuttle and Rogers (1966)
D	Pedal digits II–V elongate and the extrinsic pedal digital flexor musculature is well developed	Tuttle (1975b)
D	Metatarsal and proximal and middle phalangeal bones of digits II–V possess marked degree of plantar curvature important in powerful grasping	Tuttle (1970)
D	Tensor fasciae lata muscle and iliotibial tract are lost	Tuttle (1975b)
D	Refinement of suspensory adaptations to specialized canopy dwelling takes place	Tuttle (1975b)
D	Urethrovaginal septum indistinct together with absence of well-defined labia majora	Kluge (1983)
D	Glans penis either not differentiated or only moderately developed with little or no corona glandis	Kluge (1983)
D	Flexor pollicis longus muscle and tendon greatly reduced	Kluge (1983)
D	Tendons of flexor hallucis longus and flexor digitorum longus are markedly discrete and unevenly distributed to five digits	Kluge (1983)

(9) Pre-African ape/human morphotype

Polarity	Character	Source
P	Broad interorbital septum	Andrews and Cronin (1982), Kay (1982a)
D?	Nasoalveolar clivis projects well back into nasal cavity and drops sharply into incisive fossa, which is transversely broad	Ward and Pilbeam (1983)
D	Incisive fossa is divided into two chambers by the vomeronasal contact with the hard palate being deflected beneath nasospinale, resulting in formation of true incisive canal	Ward and Pilbeam (1983)
D	Large sphenopalatine fossae	Andrews and Cronin (1982)
D	Presence of fronto-ethmoidal sinus	Cave and Haines (1940), Andrews and Cronin (1982)
P	P_3 sectorial (C^1 honing) and bilaterally compressed	Delson and Andrews (1975), Kay (1982a)
D	Marked reduction of trigonid in lower molars	Delson et al. (1977)
P	Canine roots aligned along tooth row axis or rotated internally and are elliptical in transverse section	Ward and Pilbeam (1983)
P or D	Intermediate thickness of enamel caps on molars	Ward and Pilbeam (1983)
D	Nearly universal and complete fusion of the os centrale to the scaphoid at a very early age (usually prenatal)	Schultz (1936), Corruccini (1978), Kluge (1983)
D?	Foot is fully plantigrade with prominent plantar flexing mechanism	Tuttle (1975b)
D	Iliac blades exhibit some shortening, widening, and lateral projection	Tuttle (1981)
D?	Vertical climbing (hoisting, hauling, and transferring) emphasized together with suspensory posturing and incipient bipedal behavior	Tuttle (1975b)
D	Incipient bipedalism is performed with knee joints flexed and femora abducted, flexed, and laterally rotated about the hip joint	Tuttle (1975b), Wolpoff (1983)
P	Diploid ($2n$) number of chromosomes equal to 48	Yunis and Prakash (1982), Mai (1983), Kluge (1983)
D	Reduction or loss of dorsal hair from middle segments of fingers and toes	Washburn (1968b)

Table 1   *(continued)*

Polarity	Character	Source
	(10) Proto-African ape morphotype	
D	Thin enamel caps on molars	Ward and Pilbeam (1983), Gantt (1983), Kay and Simons (1983, ?P)
D	Volar and ulnar inclination of concave articular surface of the distal radius	Jenkins and Fleagle (1975), Tuttle (1974, 1975b)
D	Prominent bony ridge on dorsodistal aspect of radial articular surface and on distal surface of scaphoid	Tuttle (1975b)
D?	Concavoconvex facet on the capitate for the os centrale	Jenkins and Fleagle (1975)
D	Prominent transverse ridge at base of dorsal articular surface of metacarpal heads	Tuttle (1967, 1969a)
D	Pronounced extension of articular surface onto the dorsal aspect of metacarpal heads II–V	Tuttle (1967)
D	Presence of friction skin pads (knuckle pads) over the dorsal aspects of the middle phalanges and their associated osteoligamentous support mechanisms	Schultz (1936), Tuttle (1969a)
D	Remarkably strong development of the flexor digitorum superficialis which guards against excessive stress on the metacarpophalangeal weight-bearing joints	Tuttle (1975b)
D	Development of steepness on the lateral aspect of the olecranon fossa forming a ridgelike structure superior to fossa together with a deepening of the fossa to accommodate overextension, which stabilizes the humeroulnar joint	McHenry (1975), Tuttle (1975b)
C	Shortening of lumbar region of vertebral column by 1–2 vertebrae to yield a mode of four and mean of 3.5	Schultz (1961, 1963), Kluge (1983, ?D)
D	Occurrence of knuckle-walking as a distinctive mode of locomotion	De Bonis (1983)
C	Urethovaginal septum indistinct together with absence of well-defined labia majora	Kluge (1983, ?D)
C	Glans penis either not differentiated or only moderately developed with little or no corona glandis	Kluge (1983, ?D)
C	Flexor pollicus longus muscle and tendon greatly reduced	Kluge (1983, ?D)
C	Tendons of flexor hallucis longus and flexor digitorum longus are markedly discrete and unevenly distributed to the five digits	Kluge (1983, ?D)
D	Set V, Y, Y′, Y″, and Z events involving unique changes in chromosomes 1, 2, 4, 5, 7, 8, 9, 10, 12, 13, 14, 16, 17, and 22	Mai (1983)
	(11) Proto-human morphotype	
D	Progressive increase in cranial capacity	Tobias (1971)
D	Lack of external evidence of the fronto-orbital sulcus (*fo*) on the cortical surface of the frontal lobe	Connolly (1950), Falk (personal communication)
D	Progressive anterior shift in position of foramen magnum and occipital condyles	Le Gros Clark (1964), Tobias (1971)
D	Development of a less flattened and straight nasoalveolar clivis	Boaz (1983)
D	Reduction in overall size of canine	Le Gros Clark (1964), Washburn and Ciochon (1974), White et al. (1983)
D	Reduction in degree of canine sexual dimorphism	Kay (1982b), Kay and Simons (1983)
D	Progressive loss of canine—premolar honing with incipient development of metaconid on P_3 and enlargement of protocone or paracone on P_3, P_4, resulting in bicuspid upper premolars	Delson et al. (1977), de Bonis (1983), Kay and Simons (1983)
D	Progressive reduction in size and frequency of I^2–C^1 and C_1–P_3 diastemata	Boaz (1983)
C	Long axis of upper canine cross section set more or less buccolingually (externally rotated canine roots)	Ward and Pilbeam (1983), Andrews and Cronin (1982), Kay and Simons (1983, ?D)
D	Decrease in molar length relative to molar width	Corruccini and McHenry (1980), Boaz (1983)
D	Increase in molar crown height	Corruccini (1977), Corruccini and McHenry (1980)
D	Reduction in frequency of Carabelli cusp complexes in M^1	Boaz (1983)
P	Intermediate thickness of enamel caps on molars	Ward and Pilbeam (1983)
D	Appearance of enamel prism Pattern 3B	Gantt (1983)

Table 1   *(continued)*

Polarity	Character	Source
C	Shallow and broad mandibular corpora	Andrews and Cronin (1982), Kay and Simons (1983, ?D)
D	Progressive reduction in forelimb length relative to trunk height	Jungers (1982)
D	Shortening of ischium and splaying and broadening of illia	Washburn (1963b), Zihlman and Lowenstein (1983)
D	Pronounced curvature in lumbosacral region, where the broadened sacrum is abruptly bent back (dorsoflexed) so that it forms a striking promontory with the lumbar region	Schultz (1968), Washburn (1963b), Zihlman and Lowenstein (1983)
D	Well-developed anterior inferior iliac spine	Lovejoy (1975, 1978)
D	Clear definition of ilio-psoas groove	Lovejoy (1978)
D	Increase in bicondylar angle of femur	Lovejoy (1978)
D	Lengthening of femoral neck	Lovejoy (1978)
D	Deepening of patellar groove of femur	Lovejoy (1975, 1978)
D	Development of rotational astragalocalcaneal joint	Delson and Andrews (1975)
D	Permanent convergence of the hallux	Biegert (1963), Tuttle (1975b)
D	Cranial retreat of the lower portion of gluteus maximus	Tuttle et al. (1975)
D	Occurrence of bipedalism as a distinctive locomotor pattern	White et al. (1983), de Bonis (1983)
D	Complete loss of ischial callosities	Delson and Andrews (1975)
D	Diploid (2*n*) number of chromosomes reduced to 46 via event 2b, which changed the number of acrocentrics in the diploid karyotype from 14 to ten, the number of metacentrics by two, and the diploid number by two, and also set X events involving unique changes in chromosomes 1, 2, 4, 9, and 18	Delson and Andrews (1975) Mai (1983)

pology is often slow to come, but once a major shift occurs, the implications can be suddenly far-reaching.

APPENDIX

As stated in the Introduction, the purpose of this chapter was to provide a summary statement reflecting the various points of view concerning hominoid phylogeny expressed by contributors to this volume. While compiling the data for presentation in Table 1 and Figs. 1–7 it became apparent that no cladogram presented or implied in the pages of this volume accurately reflected my own developing viewpoint. Hewing to the principle that this chapter should primarily summarize, rather than engage in speculative clade-building, an attempt was made to limit my own opinions to a minimum. However, this volume would not be complete unless I had an opportunity to air my viewpoint like so many have done in the preceding chapters. Fig. 8 therefore represents my view of hominoid relationships.

As is apparent, this cladogram is based on the summary cladogram of Fig. 7. To this a number of new clades have been added which are based only on fossil groups. The groups are defined as follows: Prop, Propliopiths (*Aegyptopithecus, Propliopithecus, Pliopithecus*) Proc, Proconsulmorphs (*Proconsul, Rangawapithecus, Limnopithecus, Micropithecus*); L Dry, larger Dryopiths (*Dryopithecus, "Hispanopithecus", "Rudapithecus," Hylopithecus,* Domeli jaw, some isolated teeth from Ramnagar); S Dry, small Dryopiths (*Dionysopithecus, Krishnapithecus*); Gib, Gibbons (*Hylobates, Symphalangus*); Ken, Kenyapiths [= dryomorph Ramapiths (*Kenyapithecus*)]; Ram, Ramapiths [= ramamorph Ramapiths (*Sivapithecus, "Ramapithecus," Gigantopithecus, "Ouranopithecus"*)]; Orang, orangutan (*Pongo*); C+G, chimpanzee and gorilla (*Pan* and *Gorilla*); Hom, Hominids

(*Australopithecus, Homo*). It is easy to see why I did not present this cladogram in the summary cladistic model section. The addition of all these new groups would have necessitated the expansion of Table 1 to include several new morphotypes for which there is not adequate fossil evidence. Hence, I consider this a speculative view which in some instances is not supported by much evidence. My reasoning for these steps is nevertheless presented throughout the text (see also footnote [3]).

An abbreviated classification of the Hominoidea that is at least partially consistent with this cladogram is as follows: Hominoidea—Pliopithecidae: Propliopithecinae (*Aegyptopithecus*), Pliopithecinae (*Pliopithecus*)—Dryopithecidae: Proconsulinae (*Proconsul*), Dryopithecinae (*Dryopithecus*)—Hylobatidae: Subfamily Indet. (*Krishnapithecus*), Hylobatinae (*Hylobates*)—Pongidae: Ramapithecinae (*Sivapithecus*), Ponginae (*Pongo*)—Panidae (*Pan* and *Gorilla*)—Hominidae (*Australopithecus* and *Homo*). Only one or two of the genera belonging to each group has been listed in this abbreviated classification.

Acknowledgments

I would like to thank Rob Corruccini for assistance and advice throughout the preparation of this chapter. For helpful background discussions on the topic of hominoid cladistics I thank Peter Andrews, P. K. Basu, Eric Delson, John Fleagle, Dave Gantt, and Lawrence Martin. I also acknowledge Joy Myers for providing background material and suggestions during the early development of this chapter. I am grateful to the UNCC Cartographic Laboratory Staff for drafting the figures, to Frankie Baucom for typing the text and tables, and to Wenda Trevathan for carefully proofreading the manuscript.

Notes

[1]   New excavations at Haritalyangar under the direction of faculty at Panjab University and with the proposed participation of this volume's editors may soon supplement the currently meager fossil record of *K. krishnaii*.

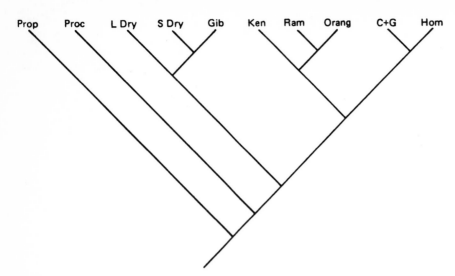

Figure 8 *Expanded cladogram of hominoid relationships supported specifically by the author.*

[2] Morphologically this specimen is somewhat different from both modern gorillas and chimpanzees, but in terms of size it is more like a chimpanzee. The structure of its enamel will soon be examined with a scanning electron microscope to establish its hominoid primate status (L. Martin, personal communication).

[3] Ward and Pilbeam (1983) and Ward and Kimbel (1983) suggest that the ancestor of the African ape–human lineage was characterized by the primitive "dryomorph" subnasal pattern. Yet, in many other respects this ancestor appears derived in the direction of the Ramapiths. I would like to suggest that the dryomorph–ramamorph maxillofacial dichotomy may not totally reflect the morphological variation in the Dryopith–Ramapith categories I have proposed. Suppose that the dryomorph–ramamorph differences are analogous to the sciuromorph–hystricomorph maxillofacial differences seen in living and fossil rodents (Wood, 1980; Lavocat, 1980). If so, the terms might be applied in a more descriptive sense. For instance, African Dryopiths might be termed "dryomorph Dryopiths," whereas Asian Ramapiths could be termed "ramamorph Ramapiths," while African Ramapiths would be termed "dryomorph Ramapiths." This scheme would preserve the nature of the shared derived morphology that unites Asian Ramapiths with the orangutan clade but leave open the possibility that the African ape–human clade was derived from a African Ramapith such as *Kenyapithecus*. The split between African and Asian Ramapiths could be as old as 15 m.y.a. (Bernor, 1983). Therefore this proposal would not alter any of Ward and Pilbeam's conclusions. This proposal also would be clearly testable with the recovery of an African Ramapith that has the maxillofacial region intact.

[4] In this discussion, one additional question should be posed: What factors can be *causally related* to the divergence of the African ape and hominid lineages? Wolpoff (1983) along with many other paleoanthropologists, considers the exploitation of open woodland/parkland and savanna habitats by the earliest hominids an important factor in the origin of the Homindae. What event in the 4–8 m.y.a. time range may have triggered the spread of these particular habitats in Africa? I suggest that the desiccation of the waters of the Mediterranean Basin and Black Sea together with the development of a cooling trend in the adjacent oceans (Hsü, 1972, 1978; Bensen, 1976, Hsü et al., 1978) is causally related to dramatic changes in climate on the African subcontinent. This drying out of the Mediterranean region, the "Messinian Event" as it is often termed, is dated to 6.5 m.y.a. Perhaps as huge deposits of salt accumulated in the Mediterranean Basin and desertic conditions prevailed in the region, this climatic shift could have triggered the development of more seasonal environments in parts of Sub-Saharan Africa resulting in the appearance of dry open-woodland and savanna habitats. Thus the Messinian Event may have been a determining factor in the divergence of the African ape and hominid lineages about 6.5 m.y.a.

Bibliography

Abel, O. 1902. Zwei neue Menschenaffen aus den Leithakalkbildungen des Weiner Beckens. *Sitz.—Berichte Kais Akad. Wiss. Wien, Math-Nat. Klasse,* 111(1):1171–1206.

Abel, O. 1931. *Die Stellung des Menschen in Rahmen der Wirbeltiere.* G. Fischer, Jena.

Aguirre, E. 1970. Identificacion de *"Paranthropus"* nen Makapansgat. *Cronica de XI Congreso Nacional de Arqueologia, Merida,* pp. 98–124.

Aiello, L. 1981. Locomotion in the Miocene Hominoidea, in: *Aspects of Human Evolution* (C. Stringer, ed.), pp. 63–97, Taylor and Francis, London.

Allen, L. L., Bridges, P. S., Evon, D. L., Rosenberg, K. R. Russell, M.D., Schepartz, L. A., Vitzthum, V. J., and Wolpoff, M. H. 1982. Demography and human origin. *American Anthropologist* 84:888–896.

Altner, G. 1971. Histologische und vergleichend-anatomische Untersuchungen zur Ontogenie und Phylogenie des Handskeletts von *Tupaia glis* (Diard 1820) und *Microcebus murinus* (J. F. Miller 1777). *Folia Primatol. Suppl.* 14:1–106.

Andrews, C. W. 1906 *A Descriptive Catalogue of the Tertiary Vertebrata of the Fayum, Egypt.* Brit. Mus. (Nat. Hist.), London.

Andrews, P. 1970. Two new fossil primates from the Lower Miocene of Kenya. *Nature (Lond.)* 288:537–540.

Andrews, P. 1973. *Miocene Primates (Pongidae, Hylobatidae) of East Africa.* Doctoral Thesis, St. Edmund's College, Cambridge, England.

Andrews, P. 1974. New species of *Dryopithecus* from Kenya. *Nature (Lond.)* 249:188–190.

Andrews, P. 1978. A revision of the Miocene Hominoidea of East Africa. *Bull. Brit. Mus. (Nat. Hist.) Geol.* 30(2):85–224.

Andrews, P. 1981a. Hominoid habitats of the Miocene. *Nature (Lond.)* 289:749.

Andrews, P. 1981b. Species diversity and diet in monkeys and apes during the Miocene, in: *Aspects in Human Evolution* (C. Stringer, ed.), pp. 25–62, Taylor and Francis, London. (See Chapter 26, this volume)

Andrews, P. 1982. Hominoid evolution. *Nature (Lond.)* 295:185–186.

Andrews, P. 1983. The natural history of *Sivapithecus,* in: *New Interpretations of Ape and Human Ancestry* (R. L. Ciochon and R. S. Corruccini, eds.), pp. 441–463, Plenum Press, New York.

Andrews, P., and Cronin, J. E. 1982. The relationship of *Sivapithecus* and *Ramapithecus* and the evolution of the orang-utan. *Nature (Lond.)* 297:541–546. (See Chapter 32, this volume)

Andrews, P., and Evans, E. M. N. 1979. The environment of *Ramapithecus* in Africa. *Paleobiology* 5:22–30.

Andrews, P., and Groves, C. P. 1975. Gibbons and Brachiation, in: *Gibbon and Siamang, Vol. 5* (D. Rumbaugh, ed.). pp. 167–218, Karger, Basel.

Andrews, P., and Tekkaya, I. 1976. *Ramapithecus* in Kenya and Turkey, in: *Les Plus Anciens Hominides* (P. V. Tobias and Y. Coppens, eds.), pp. 7–25, Union Internationale des Sciences Prehistoriques et Protohistoriques (IXth Congres).

Andrews, P., and Tekkaya, I. 1980. A revision of the Turkish Miocene hominoid *Sivapithecus meteai. Palaeontology* 23(1):85–95.

Andrews, P., and Tobien, H. 1977. A new Miocene locality in Turkey with evidence on the origin of *Ramapithecus* and *Sivapithecus. Nature (Lond.)* 268:699–701.

Andrews, P., and Van Couvering, J. H. 1975. Palaeoenvironments in the East African Miocene, in: *Approaches to Primate Paleobiology* (F. S. Szalay, ed.), pp. 62–103, Karger, Basel.

Andrews, P., and Walker, A. C. 1976. The primate and other fauna from Fort Ternan, Kenya, in: *Human Origins: Louis Leakey and the East African Evidence* (G. Ll. Isaac and E. R. McCown, eds.), pp. 270–304, W. A. Benjamin, Inc., Menlo Park.

Andrews, P., Hamilton, W. R., and Whybrow, P. J. 1978. Dryopithecines from the Miocene of Saudi Arabia. *Nature (Lond.)* 274:249–250.

Andrews, P., Lord, J. M., and Evans, E. M. N. 1979. Patterns of ecological diversity in fossil and modern mammalian faunas. *Bio. Jrnl. Linn. Soc.* 11:177–205.

Anfinsen, C. B. 1959. *The Molecular Basis of Evolution,* John Wiley and Sons, New York.

Ankel, F. 1965. Der Canalis sacralis als Indikator fur die Lange der Caudalregion der Primaten. *Folia Primatologica* 3:263–276.

Ansell, W. F. H. 1960. *Mammals of Northern Rhodesia,* Government Printer, Lusaka.

Antoniades, H. N. 1960. Circulating hormones, in: *The Plasma Proteins, Vol. I* (F. W. Putnam, ed.), pp. 105–125, Academic Press, New York.

Arambourg, C. 1959. Vertebres continentaux du Miocene superieur de l'Afrique du nord. *Publications du Service de la Carte Geologique de l'Algerie* 4:5–159.

Arambourg, C., and Coppens, Y. 1967. Sur la decouverte dans le Pleistocene inferieur de la vallee de l'Omo (Ethiopie) d'une mandibule d'Australopithecien. *C. R. Acad. Sci., Ser. D.,* 265:589–590.

Arambourg, C., and Coppens, Y. 1968. Decou-verte d'un Australopithecien nouveau des gisements de l'Omo (Ethiopie). *S. Afr. J. Sci.* 64:58–59.

Arambourg, C., Chevallon, J., and Coppens, Y. 1967. Premiers resultats de la nouvelle mission de l'Omo (1967). *C. R. Acad. Sci., Ser. D.,* 265:1891–1896.

Arambourg, C., Chavallon, J., and Coppens, Y. 1969. Resultats de la nouvelle mission de l'Omo (2e campagne 1968). *C. R. Acad. Sci., Ser. D,* 268:759–762.

Archibald, J. D. 1977. Ectotympanic bone and internal carotid circulation of eutherians in reference to anthropoid origins. *J. Hum. Evol.* 6:609–622.

Ardrey, R. 1961. *African Genesis,* Bell, New York.

Aronson, J. L., Schmitt, T. J., Walter, R. C., Taieb, M., Tiercelin, J. J., Johanson, D. C., Naeser, C. W., and Nairn, A. E. M. 1977. New geochronologic and paleomagnetic data for the hominid-bearing Hadar Formation, Ethiopia. *Nature (Lond.)* 267:323–327.

Ashley-Montagu, M. F. A. 1933. The anthropological significance of the pterion in the primates. *Am. J. Phys. Anthro.* 18:160–336.

Ashlock, P. D. 1974. The uses of cladistics. *Annu. Rev. Ecol. Syst.* 5:81–99.

Ashton, E. H., and Oxnard, C. E. 1963. The musculature of the primate shoulder. *Trans. Zool. Soc. Lond.* 29:553–650.

Ashton, E. H., and Oxnard, C. E. 1964. Locomotor patterns in Primates. *Proc. Zool. Soc. Lond.* 142:1–28.

Ashton, E. H., and Zuckerman, S. 1951. Some cranial indices of *Plesianthropus* and other primates. *Amer. J. Phys. Anthro.* 9:283–296.

Attenborough, D. 1961. *Zoo Quest to Madagascar,* Butterworth, London.

Avis, V. 1962. Brachiation: the crucial issue for man's ancestry. *Southwestern Journal of Anthropology* 18:119–148.

Axelrod, D. I. 1978. The roles of plate tectonics in angiosperm history, in: *Historical Biogeography, Plate Tectonics and the Changing Environment* (J. Gray and A. J. Boucot, eds.), pp. 435–447, Oregon State University Press, Corvallis.

Azzaroli, A. 1946. La scimmia fossile della Sardegna. *Riv. Sci. Presit.* 1:168–176.

Baba, M., Darga, L., and Goodman, M. 1980. Biochemical evidence on the phylogeny of Anthropoidea, in: *Evolutionary Biology of the New World Monkeys and Continental Drift* (R. L. Ciochon and A. B. Chiarelli, eds.), pp. 423–443, Plenum Press, New York.

Badrian, A., and Badrian, N. 1977. Pygmy chimpanzees. *Oryx* 13:464–468.

Bagenal, T. B. 1951. A note on the papers of Elton and Williams on the genetic relations of species in small ecological communities. *J. An. Ecol.* 20:242–245.

Baker, B. H., Williams, L. A. J., Miller, J. A., and Fitch, F. J. 1971. Sequence and geochronology of the Kenya Rift volcanoes. *Tectonophysics* 11:191–215.

Baker, P. E. 1973. Islands of the South Atlantic, in: *The Ocean Basins and Margins, Vol. I, The South Atlantic* (A. E. M. Nairn and F. G. Stehli, eds.). pp. 493–553, Plenum Press, New York.

Ba Maw, Ciochon, R. L., and Savage, D. E. 1979. Late Eocene of Burma yields earliest anthropoid primate, *Pondaungia cotteri*. *Nature (Lond.)* 282:65–67.

Banks, E. 1931. A popular account of the mammals of Borneo. *J. Malay. Br. Roy. Asiat. Soc.* 9:1–139.

Barbour, G. B. 1949. Ape or man? *Ohio Soc. J.* 49:4

Barnett, C. H., and Napier, J. R. 1952. The axis rotation at the ankle joint. *J. Anat.* 86:1–9.

Barnett, C. H., and Napier, J. R. 1953. The rotary mobility of the fibular in eutherian mammals. *J. Anat.* 87:11–21.

Barr, K. W. 1974. The Caribbean and plate tectonics—some aspects of the problem. *Verhandlungen NaturForschende Gesellschaft, Basel* 84:45–67.

Bartels, M., Jr., 1937. Zur Kenntnis der Verbreitung und der Lebenweise javanischer Saugetiere. *Treubia* 16:149–164.

Bartholomew, G. A., and Birdsell, J. A. 1953. Ecology and the protohominids. *Am. Anthropol.* 55:481–498.

Bartlett, A. D. 1863. Description of a new species of *Galago*. *Proc. Zool. Soc. Lond.* 1863: 231–233.

Baskin, J. A. 1980. Evolutionary reversal in *Mylogaulus* (Mammalia, Rodentia) from the Late Miocene of Florida. *Amer. Midl. Natural.* 104:155–162.

Bauchot, R., and Stephan, A. 1969. Encephalization et niveau evolutif chez les simiens. *Mammalia* 33:225–275.

Bauer, H. R. Chimpanzee bipedal locomotion in the Gombe National Park, East Africa. *Primates* 18:913–921.

Beach, F. A. 1947. Evolutionary changes in the physiological control of mating behavior in mammals. *Psychol. Rev.* 54:297–315.

Beach, F. A. 1978. Human sexuality and evolution, in: *Human Evolution: Biosocial Perspectives* (S. L. Washburn and E. R. McCown, eds.), pp. 123–154, Benjamin-Cummings, Menlo Park.

Beecher, R. M. 1977. Function and fusion at the mandibular symphysis. *Amer. J. Phys. Anthro.* 47:325–336.

Behrensmeyer, A. K. 1975. The taphonomy and paleoecology of Plio-Pleistocene vertebrate assemblages east of Lake Rudolf. *Bull. Mus. Comp. Zool., Harvard* 146:473–578.

Bennett, S. A., ed. 1958. *A Century of Darwin*, Heinemann, London.

Bensley, B. A. 1901a. On the question of an arboreal ancestry of the Marsupialia, and the interrelationships of the mammalian subclasses. *Amer. Natur.* 35:117–138.

Bensley, B. A. 1901b. A theory of the origin and evolution of the Australian Marsupialia. *Amer. Natur.* 35:245–269.

Benson, R. H. 1976. Testing the Messinian salinity crisis biodynamically: an introduction. *Palaeogeogr., Palaeoclimatol., Palaeoecol.* 20:3–11.

Benson, S. B., and Borell, A. E. 1931. Notes on the life history of the red tree mouse, *Phenacomys longicaudus*. *J. Mammal.* 12:226–233.

Benveniste, R. E., and Todaro, G. T. 1976. Evolution of type c viral genes: evidence for an Asian origin of man. *Nature (Lond.)* 261:101 108.

Berggren, W. A., and Hollister, C. D. 1974. Paleogeography, paleobiology, and the history of circulation in the Atlantic Ocean, in: *Studies in Paleo-oceanography* (W. W. Hay, ed.), pp. 126–186, Soc. Econ. Paleontol. Mineral. Spec. Publ. #20.

Berggren, W. A., and Van Couvering, J. A. 1974. The Late Neogene. *Palaeogeogr., Palaeoclimatol., Palaeoecol.* 16:1–216.

Berggren, W. A., McKenna, M. C., Hardenbol, J., and Obradovich, J. D. 1978. Revised Paleogene polarity time scale. *J. Geol.* 86:67–81.

Bergonnioux, F. M., and Crouzel, F. 1959. Les Pliopitheques de France. *Annales de la Paleontologie* 51:44–65.

Bernor, R. L. 1983. Geochronology and zoogeographic relationships of Miocene Hominoidea, in: *New Interpretations of Ape and Human Ancestry* (R. L. Ciochon and R. S. Corruccini, eds.), pp. 21–64, Plenum Press, New York.

Bernor, R., Andrews, P., Solounias, N., and Van Couvering, J. H. 1979. The evolution of 'Pontian' mammal faunas: some zoogeographic, paleoecologic and chronostratigraphic considerations. *Annals Geologique du Pays Hellenique,* Tome hors serie:81–89.

Biegert, J. 1963. The evaluation of characteristics of the skull, hands and feet for primate taxonomy, in: *Classification and Human Evolution* (S. L. Washburn, ed.), pp. 116–145, Aldine, Chicago.

Biegert, J., and Maurer, R. 1972. Rumpfskelettlange, Allometrien und Korperproportionen bie catarrhinen Primaten. *Folia Primatologica* 17:142–156.

Bielicki, T. 1966. On '*Homo habilis*'. *Curr. Anthro.* 7:576–578.

Bilsborough, A. 1971. Evolutionary change in the hominoid maxilla. *Man* 6:473–485.

Bingham, D. K., and Klootwijk, C. T. 1980. Paleomagnetic constraints on India's underthrusting of the Tibetan Plateau. *Nature (Lond.)* 284:336–338.

Bishop, A. 1964. Use of the hand in lower primates, in: *Evolutionary and Genetic Biology of Primates, Vol. II* (J. Buettner-Janusch, ed.), pp. 133–225, Academic Press, New York.

Bishop, W. W. 1958. Miocene mammalia from the Napak volcanics, Karamoja, Uganda. *Nature (Lond.)* 182:1480–1482.

Bishop, W. W. 1963. The later Tertiary and Pleistocene in Eastern Equatorial Africa, in: *African Ecology and Human Evolution* (F. C. Howell and F. Bourliere, eds.), pp. 246–275, Aldine, Chicago.

Bishop, W. W. 1967. The later Tertiary in East Africa-volcanics, sediments and faunal inventory, in: *Background to Evolution in Africa* (W. W. Bishop and J. D. Clark, eds.), pp. 31–56, University of Chicago Press, Chicago.

Bishop, W. W. 1968. The evolution of fossil environments in East Africa. *Trans. Leicester Lit. Philos. Soc.* 62:27–44.

Bishop, W. W. 1978. Geochronological framework for African Plio-Pleistocene Hominidae: as Cerberus sees it, in: *Early Hominids of Africa* (C. J. Jolly, ed.), pp. 255–265, St. Martin's, New York.

Bishop, W. W., and Chapman, G. R. 1970. Early Pliocene sediments and fossils from the northern Kenya rift valley. *Nature (Lond.)* 226: 914–918.

Bishop, W. W., Miller, J. A., and Fitch, F. J. 1969. New Potassium-argon age determinations relevant to the Miocene fossil mammal sequence in East Africa. *Amer. J. Sci.* 267:669–699.

Bluntschi, H. 1931. *Homunculus patagonius* und ihm zugereihten Fossil Funde aus den Santa-Cruz-Schichten Patagoniens. Eine morphologische Revison an Hand der Original Stucke in der Sammlung Ameghino zu La Plata. *Gegenbaurs Morph. Jb.* 67(2):811–892.

Boaz, N. T. 1983. Morphological trends and phylogenetic relationships from middle Miocene hominoids to late Pliocene hominids, in: *New Interpretations of Ape and Human Ancestry* (R. L. Ciochon and R. S. Corruccini, eds.), pp. 705–720, Plenum Press, New York.

Boaz, N. T., Gaziry, A. W., and Allei-Ei-Arnanti. 1979. New fossil finds from the Libyan Upper Neogene site of Sahabi. *Nature (Lond.)* 280:137–140.

Bock, W. J. 1969. Comparative morphology in systematics, in: *Systematic Biology,* pp. 411–447, National Academy of Sciences, Washington, D.C. (National Research Council, International Congress, Ann Arbor)

Bock, W. J. 1977a. Adaptation and the comparative method, in: *Major Patterns in Vertebrate Evolution* (M. K. Hecht, P. G. Goody and B. M. Hecht, eds.), pp. 57–82, Plenum Press, New York.

Bock, W. J. 1977b. Foundations and methods of evolutionary classification, in: *Major Patterns in Vertebrate Evolution* (M. K. Hecht, P. G. Goody, and B. M. Hecht, eds.), pp. 851–896, Plenum Press, New York.

Boekschoten, G. J., and Sondaar, P. Y. 1966. The Pleistocene of the Katharo Basin (Crete) and its *Hippopotamus*. *Bijdragen tot de Dierkunde* 36:17–44.

Boekschoten, G. J., and Sondaar, P. Y. 1972. On the fossil Mammalia of Cyprus, II. *Proc. Kon. Neder. Akad. Wetensch. Amsterdam, B.* 75: 326–338.

Bond, G. 1978. Evidence for Late Tertiary uplift of Africa relative to North America, South America, Australia and Europe. *J. Geol.* 86:47–65.

Booth, A. H. 1956. The Cercopithecidae of the Gold and Ivory Coasts: geographic and systematic observations. *Ann. Mag. Nat. Hist.* 9:476–480.

Booth, A. H. 1971. Observations on the teeth of the mountain gorilla *(Gorilla gorilla beringei)*. *Amer. J. Phys. Anthro.* 34:85–88.

Boskoff, K. J. 1977. Aspects of reproduction in ruffed lemurs *(Lemur variegatus)*. *Folia Primatologica* 28:241–250.

Boucot, A. J. 1976. Rates of size increase and phyletic evolution. *Nature (Lond.)* 26:694–696.

Boule, M. 1912. L'homme fossile de la Chapelle-aux-Saints. *Annales de Paleontologie* 7:85–192.

Boule, M. 1913. L'homme fossile de la Chapelle-aux-Saints. *Annales de Paleontologie* 8:1–67.

Bourliere, F. 1975. Mammals, small and large: the ecological implications of size, in: *Small Mammals: Their Productivity and Population Dynamics, IBP 5* (F. Golley, K. Petruse-wicz, and L. Ryszkowski, eds.), pp. 1–8, Cambridge University Press, Cambridge.

Bourne, G. H., and de Bourni, M. N. G. 1972. The histology and histochemistry of the chimpanzee tissues and organs, in: *The Chimpanzee, Vol. 5* (G. H. Bourne, ed.), pp. 1–76, Karger, Basel.

Bowen, B. E., and Vondra, C. F. 1973. Stratigraphical relationships of the Plio-Pleistocene deposits, East Rudolf, Kenya. *Nature (Lond.)* 242:391–393.

Bown, T. M. 1974. Notes on some early Eocene anaptomorphine primates. *Univ. Wyoming Contrib. Geol.* 13:19–26.

Bown, T. M. 1976. Affinities of *Teilhardina* (Primates, Omomyidae) with description of a new species from North America. *Folia Primatologica* 25:62–72.

Bown, T. M. 1979. New omomyid primates (Haplorhini, Tarsiiformes) from middle Eocene rocks of western central Hot Springs County, Wyoming. *Folia Primatologica* 31:48–73.

Bown, T. M., and Gingerich, P. D. 1973. The Paleocene primate *Plesiolestes* and the origin of Microsyopidae. *Folia Primatologica* 19:1–18.

Bown, T. M., and Rose, K. D. 1976. New early Tertiary primates and a reappraisal of some plesiadapiformes. *Folia Primatologica* 26:109–138.

Bown, T. M., Kraus, M. J., Wing, S. L., Fleagle, J. G., Tiffany, B. H., Simons, E. L., and Vondra, C. F. 1982. The Fayum primate forest revisited. *J. Hum. Evol.* 11(7):603–632.

Boyd, C. E., and Goodyear, C. P. 1971. Nutritive quality of food in ecological systems. *Arch. Hydrobiol.* 69:256–260.

Boyden, A. A. 1958. Comparative serology: aims, methods, and results, in: *Serological and Biochemical Comparisons of Proteins* (W. Cole, ed.), pp. 3–24, Rutgers University Press, New Brunswick.

Brace, C. L. 1962. Comment on 'Food transport and the origin of hominid bipedalism'. *Amer. Anthropol.* 64:606–607.

Brace, C. L. 1963a. Review of *Evolution and Hominisation* (G. Kurth, ed.) *Amer. J. Phys. Anthro.* 21:87–91.

Brace, C. L. 1963b. Review of *Ideas on Human Evolution* (W. W. Howells, ed.) *Hum. Biol.* 35:545–548.

Brace, C. L. 1963c. Structural reduction in evolution. *Amer. Natural.* 97:39–49.

Brace, C. L. 1967a. *The Stages of Human Evolution.* Prentice-Hall, Englewood Cliffs.

Brace, C. L. 1967b. Environment, tooth form and size in the Pleistocene. *J. Dent. Res.* 46:809–816.

Brace, C. L. 1973. Sexual dimorphism in human evolution. *Yrbk. Phys. Anthro.* 16:31–49.

Brace, C. L., and Montagu, A. 1977. *Human Evolution* (2nd ed.), Macmillan, New York.

Brace, C. L., Mahler, P. E., and Rosen, R. B. 1973. Tooth measurements and the rejection of the taxon *"Homo habilis"*. *Yrbk. Phys. Anthro.* 16:50–68.

Brain, C. K. 1958. The Transvaal ape-bearing cave deposits. *Transvaal Mus. Mem.,* No. 11.

Brain, C. K. 1967. The Transvaal Museum's fossil project at Swartkrans. *S. Afr. J. Sci.* 63:378–384.

Brain, C. K. 1970. New finds at the Swartkrans australopithecine site. *Nature (Lond.)* 225:1112–1119.

Bramblett, C. A. 1967. Pathology in the Darajani baboon. *Amer. J. Phys. Anthro.* 26:331–340.

Briggs, J. C. 1966. Zoogeography and evolution. *Evolution* 20:282–289.

Brock, A. J. P. v. d. 1938. Das Skelett einer weiblichen Efe-Pygmae. *Z. Morph. Anthro.* 40:121–169.

Broom, R. 1926. On the mammalian presphenoid and mesethmoid bones. *Proc. Zool. Soc. London* 1926:257–264.

Broom, R. 1927. Some further points in the structure of the mammalian basicranial axis. *Proc. Zool. Soc. London* 1927:233–244.

Broom, R. 1938. The Pleistocene anthropoid apes of South Africa. *Nature (Lond.)* 142:377–379.

Broom, R. 1950. The genera and species of the South African fossil ape-men. *Amer. J. Phys. Anthro.* 8:1–13.

Broom, R., and Robinson, J. T. 1949. A new type of fossil man. *Nature (Lond.)* 164:322.

Brown, J. H. 1975. Geographical ecology of desert rodents, in: *Ecology and Evolution of Communities* (M. L. Cody and J. M. Diamond, eds.), pp. 315–341, Harvard University Press, Cambridge.

Brundin, L. 1966. Transantarctic relationships and their significance, as evidenced by the chironomid midges, with a monograph of the subfamily Podonominae, Aphroteniinae and the austral Heptagyiae. *Kl. Sven. Ventenskaps akad. Handl., Ser. 4* 11:1–472.

Brundin, L. 1972. Phylogenetics and biogeography, a reply to Darlington's "practical criticism" of Hennig-Brundin. *Syst. Zool.* 21:69–79.

Buchardt, B. 1978. Oxygen isotope palaeotemperatures from the Tertiary period in the North Sea area. *Nature (Lond.)* 275:121–123.

Buettner-Janusch, J. 1966. *The Origins of Man,* Wiley, New York.

Buettner-Janusch, J. 1969. The nature and future of physical anthropology. *Trans. N. Y. Acad. Sci.* 31(2):128–138.

Buettner-Janusch, J. 1973. *Physical Anthropology: A Perspective,* Wiley, New York.

Bugge, J. 1974. The cephalic arterial system in insectivores, primates, rodents and lagomorphs, with special reference to the systematic classification. *Acta. Anat.* 82:1–160.

Bugge, J. 1980. Comparative anatomical study of the carotid circulation in New and Old World primates: Implications for their evolutionary history, in: *Evolutionary Biology of the New World Monkeys and Continental Drift* (R. L. Ciochon and A. B. Chiarelli, eds.), pp. 293–316, Plenum Press, New York.

Bush, M. E., Lovejoy, C. O., Johanson, D. C., and Coppens, Y. 1982. Hominid carpal, metacarpal and phalangeal bones recovered from the Hadar Formation: 1974–1977 collections. *Amer. J. Phys. Anthro.* 57:651–677.

Butler, P. M. 1956. The skull of Ictops and the classification of the Insectivora. *Proc. Zool. Soc., London* 126:453–481.

Butler, P. M. 1972. Some functional aspects of molar evolution. *Evolution* 26:474–483.

Butler, P. M. 1973. Molar wear facets of Tertiary North American primates. *Symp. IVth Ann. Congr. Primatol.* 3:1–37.

Butzer, K. W. 1974. Paleoecology of South African Australopithecines: Taung revisited. *Curr. Anthro.* 15:367–382.

Butzer, K. W. 1977. Environment, culture and human evolution. *Amer. Sci.* 65:572–584.

Cachel, S. 1979. A functional analysis of the primate masticatory system and the origin of the anthropoid post-orbital septum. *Amer. J. Phys. Anthro.* 50:1–18.

Cahen, L., and Lepersonne, J. 1952. Equivalence entre le Systeme du Kalahari du Congo Belge et les Kalahari Beds d'Afrique Australe. *Mem. Soc. Belge Geol., Sec. 8* 4:1–64.

Cain, A. J. 1953. Geography and coexistence in relation to the biological definition of the species. *Evolution* 7:76–83.

Cameron, T. W. M. 1960. Southern intercontinental connections and the origin of the southern mammals, in: *Evolution: Its Science and Doctrine* (Symposium Roy. Soc. Canada, 1959), pp. 79–89, University of Toronto Press, Toronto.

Campbell, B. G. 1966. *Human Evolution.* Aldine, Chicago.

Campbell, B. G. 1967. *Human Evolution,* Heinemann.

Campbell, B. G. 1968. Inspiration and Controversy: Motives for Research. *S. Afr. J. Sci.* 64:60–63.

Campbell, B. G. 1969. Early man in southern Africa. *S. Afr. Archeol. Bull.* 24:212.

Campbell, B. G. 1972. Man for all seasons, in: *Sexual Selection and the Descent of Man 1871–1971* (B. G. Campbell, ed.), pp. 40–58, Aldine, Chicago.

Campbell, C. B. G. 1966. The relationships of the tree shrews: the evidence of the nervous system. *Evolution* 20:276–281.

Cant, G. H. 1977. *Ecology, Locomotion, and Social Organization of Spider Monkeys (Ateles geoffroyi),* Ph.D. dissertation, Univ. Microfilms, Ann Arbor.

Carleton, A. 1936. The limb bones and vertebrae of the extinct lemurs of Madagascar. *Proc. Zool. Soc., London* 1936:231–233.

Carlsson, A. 1922. Uber die Tupaiidae und ihre Beziehungen zu den Insectivora und den Prosimiae. *Acta Zool. (Stockholm)* 3:227–270.

Carter, D. C. 1970. Chiropteran reproduction, in: *About Bats* (B. H. Slaughter and D. W. Walton, eds.), pp. 233–246, Southern Methodist Univ. Press, Dallas.

Cartmill, M. 1970. *The Orbits of Arboreal Mammals: a Reassessment of the Arboreal Theory of Primate Evolution,* Doctoral thesis, University of Chicago.

Cartmill, M. 1971. Ethmoid component in the orbit of primates. *Nature (Lond.)* 232:566–567.

Cartmill, M. 1972. Arboreal adaptations and the origin of the Order Primates, in: *The Functional and Evolutionary Biology of Primates* (R. Tuttle, ed.), pp. 97–122, Aldine, Chicago.

Cartmill, M. 1974a. Pads and claws in arboreal locomotion, in: *Primate Locomotion* (F. A. Jenkins, ed.), pp. 45–83, Academic Press, New York.

Cartmill, M. 1974b. *Daubentonia, Dactylopsila,* woodpeckers and klinorhynchy, in: *Prosimian Biology* (R. D. Martin, G. A. Doyle, and A. C. Walker, eds.), pp. 655–670, Duckworth, London.

Cartmill, M. 1974c. Rethinking primate origins. *Science* 184:436–443. (See Chapter 2, this volume)

Cartmill, M. 1975a. *Primate Origins,* Burgess, Minneapolis.

Cartmill, M. 1975b. Strepsirhine basicranial structures and the affinities of the Cheirogaleidae, in: *Phylogeny of the Primates: A Multidisciplinary Approach* (W. P. Luckett and F. S. Szalay, eds.), pp. 313–354, Plenum Press, New York.

Cartmill, M. 1978. The orbital mosaic in prosimians and the use of variable traits in systematics. *Folia Primatologica* 30:89–114.

Cartmill, M. 1980. Morphology, function and evolution of the anthropoid post orbital septum, in: *Evolutionary Biology of the New World Monkeys and Continental Drift* (R. L. Ciochon and A. B. Chiarelli, eds.), pp. 243–274, Plenum Press. New York.

Cartmill, M. 1982. Basic primatology and prosimian evolution, in: *Fifty Years of Physical Anthropology in North America* (F. Spencer, ed.), pp. 147–186, Academic Press, New York.

Cartmill, M., and Kay, R. F. 1978. Craniodental morphology, tarsier affinities, and primate suborders, in: *Recent Advances in Primatology, Vol. 3. Evolution* (D. J. Chivers and J. A. Joysey, eds.), pp. 205–214, Academic Press, London.

Cartmill, M., and Milton, K. 1974. The lorisiform wrist joint. *Amer. J. Phys. Anthro.* 41:471.

Cartmill, M., MacPhee, R. D. E., and Simons, E. L. 1981. Anatomy of the temporal bone in early anthropoids with remarks on the problem of anthropoid affinities. *Amer. J. Phys. Anthro.* 56:3–21.

Case, T. J. 1978. A general explanation for insular body size trends in terrestrial vertebrates. *Ecology* 59(1):1–18.

Cavagna, G. A., Saibene, F. P., and Margaria, R. 1964. Mechanical work in running. *J. Appl. Phys.* 19:249–256.

Cave, A. J. E. 1961. The frontal sinus of the gorilla. *Proc. Zool. Soc., London* 136:359–373.

Cave, A. J. E. 1967. Observations on the platyrrhine nasal fossa. *Amer. J. Phys. Anthro.* 26:277–288.

Cave, A. J. E. 1973. President's Address: The primate nasal fossa. *Biol. J. Linn. Soc.,* 5(4): 377–387.

Cave, A. J. E., and Haines, R. W. 1940. The paranasal sinuses of the anthropoid apes. *J. Anat.* 74(4):493–523.

Chace, F. A., Jr., and Manning, R. B. 1972. Two new caridean shrimps, one representing a new family, from marine pools on Acension Island (Crustacea: Decapoda: Natantia). *Smithson. Contrib. Zool.* 131:1–18.

Chance, M.R.A. 1955. The sociability of monkeys. *Man* 55:162–165.

Chance, M. R. A. 1962. Social behavior and primate evolution, in: *Culture and the Evolution of Man* (M. F. Ashley-Montagu, ed.), pp. 84–130, Oxford University Press, New York.

Chance, M. R. A. 1967. Attention structure as the basis of primate rank orders. *Man* (n.s.) 2:503–518.

Chantre, E., and Gaillard, C. 1897. Sur la faune du gisement siderolithique Eocene de Lissies (Rhone). *C. R. Acad. Sci. (Paris)* 125:986–987.

Chantrenne, H. 1961. *The Biosynthesis of Proteins,* Pergamon Press, Oxford.

Charles-Dominique, P. 1971. Eco-ethologie des Prosimiens du Gabon. *Biol. Gabon.* 7:121–228.

Charles-Dominique, P., and Martin, R. D. 1970. Evolution of lemurs and lorises. *Nature (Lond.)* 227:257–260. (See Chapter 9, this volume)

Chesters, K. I. M. 1957. The Miocene flora of Rusinga Island, Lake Victoria, Kenya. *Palaeontographica* 101(B):30–67.

Chiarelli, B. 1962. Comparative morphometric analysis of primate chromosomes. I. The chromosomes of anthropoid apes and of man. *Caryologia* 15:99–121.

Chiarelli, B. 1975. The study of primate chromosomes, in: *Primate Functional and Evolutionary Biology* (R. H. Tuttle, ed.), pp. 103–127, Mouton, The Hague.

Chiarelli, B. 1980. The karyology of South American primates and their relationship to African and Asian species, in: *Evolutionary Biology of the New World Monkeys and Continental Drift* (R. L. Ciochon and A. B. Chiarelli, eds.), pp. 387–398, Plenum Press, New York.

Chivers, D. S. 1972. The siamang and the gibbon in the Malay Peninsula, in: *Gibbon and Siamang Vol. 1.* (D. M. Rumbaugh, ed.), pp. 103–135, Karger, Basel.

Chivers, D. J. and Joysey, K. A. Eds. 1978. *Recent Advances in Primatology, Vol. 3: Evolution,* Academic Press, New York.

Chopra, S. R. K. 1962. The innominate bone of the Australopithecinae and the problem of erect posture. *Bibl. Primatol.* 1:93–102.

Chopra, S. R. K. 1983. The significance of recent hominoid discoveries from the Siwalik Hills of India, in: *New Interpretations of Ape and Human Ancestry* (R. L. Ciochon and R. S. Corruccini, eds.), pp. 539–557, Plenum Press, New York.

Chopra, S. R. K., and Kaul, S. 1979. A new species of *Pliopithecus* from the Indian Sivaliks. *J. Hum. Evol.* 8:475–477.

Ciochon, R. L. 1983. Hominoid cladistics and the ancestry of modern apes and humans: a summary statement, in: *New Interpretations of Ape and Human Ancestry* (R. L. Ciochon and R. S. Corruccini, eds.), pp. 783–843, Plenum Press, New York. (See Chapter 44, this volume)

Ciochon, R. L., and Chiarelli, A. B. 1980. Paleobiogeographic perspectives on the origin of the Platyrrhini, in: *Evolutionary Biology of the New World Monkeys and Continental Drift* (R. L. Ciochon and A. B. Chiarelli, eds.), pp. 459–493, Plenum Press, New York. (See Chapter 15, this volume)

Ciochon, R. L., and Chiarelli, A. B., Eds. 1980. *Evolutionary Biology of the New World Monkeys and Continental Drift,* Plenum Press, New York.

Ciochon, R. L., and Corruccini, R. S. 1975. Morphometric analysis of platyrrhine femora with taxonomic implications and notes on two fossil forms. *J. Hum. Evol.* 4:193–217.

Ciochon, R. L., and Corruccini, R. S. 1977a. The phenetic position of *Pliopithecus* and its phylogenetic relationship to the Hominoidea. *Syst. Zool.* 26:290–299.

Ciochon, R. L., and Corruccini, R. S. 1977b. The coraco-acromial ligament and projection index in man and other anthropoid primates. *J. Anat.* 124:627–632.

Ciochon, R. L., and Corruccini, R. S. 1982. Miocene hominoids and new interpretations of ape and human ancestry, in: *Advanced Views in Primate Biology* (A. B. Chiarelli and R. S. Corrucini, eds.), pp. 149–159, Springer-Verlag, Berlin.

Ciochon, R. L., and Corruccini, R. S., Eds. 1983. *New Interpretations of Ape and Human Ancestry,* Plenum Press, New York.

Ciochon, R. L., Savage, D. E., Thaw Tint, and Ba Maw. 1985. Anthropoid origins in Asia? New discovery of *Amphipithecus* from the Eocene of Burma. *Science* (in press).

Clark, J. D. 1941. An anaptomorphid primate from the Oligocene of Montana. *J. Paleontol.* 15:562–563.

Clark, J. D. 1970. *The Prehistory of Africa,* Praeger, New York.

Clarke, R. J. 1977. A juvenile cranium and some adult teeth of early *Homo* from Swartkrans, Transvaal. *South Afr. J. Sci.* 73:46–49.

Clarke, R. J. 1978. *The Cranium of the Swartkrans Hominid, SK 847 and its Relevance to Human Origins,* Doctoral thesis, University of Witwatersrand.

Clemens, W. A. 1974. *Purgatorius,* an early paromomyid primate (Mammalia). *Science* 184:903–905.

Clutton-Brock, T. H. 1977. Some aspects of intraspecific variation in feeding and ranging behavior in primates, in: *Primate Ecology* (T. H. Clutton-Bock, ed.), pp. 539–556, Academic Press, New York.

Clutton-Brock, T. H. 1979. Primate social organization and ecology, in: *Primate Ecology: Problem Oriented Field Studies* (R. W. Sussman, ed.), pp. 503–512, Wiley, New York.

Clutton-Brock, T. H., and Harvey, P. H. 1977. Primate ecology and social organization. *J. Zool., London* 183:1–39.

Coimbra-Filho, A. F., and Mittermeier, R. A. 1976. Exudate-eating and tree-gouging in marmosets. *Nature (Lond.)* 262:630.

Colbert, E. H. 1937. A new primate from the Upper Eocene Pondaung formations of Burma. *Amer. Mus. Novit.* No. 951:1–18.

Colbert, E. H. 1973. Continental drift and the distributions of fossil reptiles, in: *Implications of Continental Drift to the Earth Sciences, Vol. 1* (D. H. Tarling and S. K. Runcorn, eds.), pp. 395–412, Academic Press, London.

Collins, E. T. 1921. Changes in the visual organs associated with the adoption of arboreal life with the assumption of the erect posture. *Trans. Ophthal. Soc. U.K.* 41:10–90.

Comfort, A. 1956. *The Biology of Senescence,* Rinehart and Co., New York.

Connolly, C. J. 1950. *External Morphology of the Primate Brain,* Thomas, Springfield.

Conroy, G. C. 1972. Problems with the interpretation of *Ramapithecus:* with special reference to anterior tooth reduction. *Amer. J. Phys. Anthro.* 37:41–48.

Conroy, G. C. 1976a. Primate postcranial remains from the Oligocene of Egypt. *Contrib. Primatol.* 8:1–134.

Conroy, G. C. 1976b. Hallucial tarsometatarsal joint in an Oligocene anthropoid, *Aegyptopithecus zeuxis. Nature (Lond.)* 263:684–686.

Conroy, G. C. 1978. Candidates for anthropoid ancestry: some morphological and paleozoological considerations, in: *Recent Advances in Primatology, Vol. 3 Evolution* (D. J. Chivers and K. A. Joysey, eds.), pp. 27–41, Academic Press, London.

Conroy, G. C. 1980a. Ontogeny, auditory structures and primate evolution. *Amer. J. Phys. Anthro.* 52:443–451.

Conroy, G. C. 1980b. Evolutionary significance of cerebral venous patterns in paleoprimatology. *Z. Morphol. Anthropol.* 71:125–134.

Conroy, G. C., and Fleagle, J. G. 1972. Locomotor behavior in living and fossil pongids. *Nature (Lond.)* 237:103–104.

Conroy, G. C., and Pilbeam, D. 1975. *Ramapithecus:* a review of its hominid status, in: *Paleontology: Morphology and Paleoecology* (R. Tuttle, ed.), pp. 59–86, Mouton, The Hague.

Cook, N. 1939. Notes on a captive *Tarsius carbonarius. J. Mammal.* 20:173–178.

Cooke, H. B. S. 1972. The fossil mammal fauna of Africa, in: *Evolution, Mammals and Southern Continents* (A. Keast, F. C. Erk and B. Glass, eds.), pp. 89–139, S.U.N.Y. Press, Albany.

Coolidge, H. J. 1933. *Pan paniscus:* pygmy chimpanzee from south of the Congo River. *Amer. J. Phys. Anthro.* 18:1–59.

Coon, C. S. 1963. *The Origin of Races,* Alfred A. Knopf, New York.

Cope, E. D. 1872. On a new vertebrate genus from the northern part of the Tertiary basin of Green River. *Proc. Amer. Philos. Soc.* 12:554.

Cope, E.D. 1882. Contributions to the history of the Vertebrata of the lower Eocene of Wyoming and New Mexico made during 1881, I. The fauna of the Wasatch beds of the basin of the Big Horn River, II. The fauna of the Catathlaeus beds, or lowest Eocene, New Mexico. *Proc. Amer. Philos. Soc.* 20:139–197.

Cope, E. D. 1883. On the mutual relations of the bunotherian Mammalia. *Proc. Acad. Nat. Sci., Phila.* 1883:77–83.

Cope, E. D. 1885a. The Vertebrata of the Tertiary formations of the West. *Rep. U.S. Geol. Surv. Terr.* (F. V. Hayden, Geologist) *Washington* 3:1–1009.

Cope, E. D. 1885b. The Lemuroidea and the Insectivora of the Eocene period of North America. *Amer. Natural.* 19:457–471.

Coppens, Y. 1961. Decouverte d'un Australopithecine dans le Villafranchien du Tchad. *C. R. Acad. Sci.* 252:3851–3852.

Coppens, Y. 1970. Les restes d'hominides des series inferieures et moyennes des formations pliovillafrenchiennes de l'Omo en Ethiopie. *C. R. Acad. Sci., Ser. D* 271:2286–2289.

Coppens, Y. 1977. Evolution morphologique de la premiere premolaire inferieure chez certains primates superieurs. *C. R. Acad. Sci., Ser. D* 285:1299–1302.

Coppens, Y., Howell, F. C., Isaac, G. L., and Leakey, R. E. F., Eds. 1976. *Earliest Man and Environments in the Lake Rudolf Basin,* Univ. Chicago Press, Chicago.

Corruccini, R. S. 1975a. Morphometric affinities in the forelimb of anthropoid primates. *Zeit. Morpho. Anthro.* 67:19–31.

Corruccini, R. S. 1975b. Multivariate analysis in biological anthropology: some considerations. *J. Hum. Evol.* 4:1–19.

Corruccini, R. S. 1977. Crown component variation in hominoid lower third molars. *Zeit. Morpho. Anthro.* 68:14–25.

Corruccini, R. S. 1978. Comparative osteometrics of the hominoid wrist joint, with special reference to knuckle-walking. *J. Hum. Evol.* 7:307–321.

Corruccini, R. S., Baba, M., Goodman, M., Ciochon, R. L., and Cronin, J. E. 1980. Non-linear macro-molecular evolution and the molecular clock. *Evolution* 34:1216–1219.

Corruccini, R. S., and Ciochon, R. L. 1976. Morphometric affinities of the human shoulder. *Amer. J. Phys. Anthro.* 45:19–38.

Corruccini, R. S., and Ciochon, R. L. 1983. Overview of ape and human ancestry: phyletic relationships of Miocene and later Hominoidea, in: *New Interpretations of Ape and Human Ancestry* (R. L. Ciochon and R. S. Corruccini, eds.), pp. 1–19, Plenum Press, New York.

Corruccini, R. S., Ciochon, R. L., and McHenry, H. M. 1975. Osteometric shape relationships in the wrist joint of some anthropoids. *Folia Primatologica* 24:250–274.

Corruccini, R. S., Ciochon, R. L., and McHenry, H. M. 1976. The postcranium of Miocene hominoids: Were dryopithecines merely "dental apes"? *Primates* 17:205–223. (See Chapter 24, this volume)

Corruccini, R. S., and Henderson, A. M. 1978. Palatofacial comparison of *Dryopithecus (Proconsul)* with extant catarrhines. *Primates* 19:35–44.

Corruccini, R. S., and McHenry, H. M. 1980. Cladometric analysis of Pliocene hominoids. *J. Hum. Evol.* 9:209–221.

Coulson, C. 1966. The influence of the pair-bond and age on the breeding biology of the kittiwake gull *Rissa tridactyla. J. Anim. Ecol.* 35:269–279.

Coursey, D. G. 1973. Hominid evolution and hypogeous plant foods. *Man* 8:634–635.

Covert, H. H., and Kay, R. F. 1980. Dental microwear and diet-implications for early hominoid feeding behavior. *Amer. J. Phys. Anthro.* 52:216.

Cracraft, J. 1973. Continental drift, paleoclimatology, and the evolution and biogeography of birds. *J. Zool.* 169:455–545.

Cracraft, J. 1974a. Continental drift and vertebrate distribution. *Annu. Rev. Ecol. Syst.* 5:215–261.

Cracraft, J. 1974b. Phylogenetic models and classifications. *Syst. Zool.* 23:71–90.

Cracraft, J. 1975. Historical biogeography and earth history: perspectives for future synthesis. *Ann. Missouri Bot. Gard.* 62:494–521.

Cracraft, J. 1978. Science, philosophy and systematics. *Syst. Zool.* 27:213–216.

Cramer, D. L. 1977. *Craniofacial Morphology of Pan paniscus, Contrib. Primatol.* No. 10, Karger, Basel.

Cramer, D. L., and Zihlman, A. L. 1978. Sexual dimorphism in pygmy chimpanzees, *Pan paniscus,* in: *Recent Advances in Primatology, Vol. 3* (D. J. Chivers and K. A. Joysey, eds.), pp. 487–490, Academic Press, London.

Craw, R. C. 1979. Generalized tracks and dispersal in biogeography: A response to R. M. McDowall. *Syst. Zool.* 28:99–107.

Cray, P. D. 1973. Marsupialia, Insectivora, Primates, Credonta and Carnivora from the Headon Beds (Upper Eocene) of Southern England. *Bull. Brit. Mus. (Nat. Hist.) Geol.* 23:1–102.

Crick, F. H. C., Barnett, L., Brenner, S., and Watts-Tobin, R. J. 1961. General nature of the genetic code for proteins. *Nature (Lond.)* 192:1227–1232.

Croizat, L. 1958. *Panbiogeography, 3 vols.,* published by the author, Caracas.

Croizat, L. 1962. *Space, Time, Form: The Biological Synthesis,* published by the author, Caracas.

Croizat, L. 1971. De la "pseudovicariance" et de la "disjonction illusoire". *Anu. Soc. Broteriana* 37:113–140.

Croizat, L. 1979. Review of *Biogeographie: Fauna und Flora der Erde und ihre geschtliche Entwicklung* (P. Banaresu and N. Boscaiu) *Syst. Zool.* 28:250–252.

Croizat, L., Nelson, G., and Rosen, D. E. 1974. Centers of origin and related concepts. *Syst. Zool.* 23:265–287.

Crompton, A. W. 1971. The origin of the tribosphenic molar. *Zool. J. Linn. Soc., London* 50 (Supplement 1):65–87

Crompton, A. W., and Hiiemae, K. M. 1970. Molar occlusion and mandibular movements during mastication in the American opossum, *Didelphis marsupialis* Linn. *Zool. J. Linn. Soc., London* 49:21–47.

Crompton, A. W., and Jenkins, F. A., Jr. 1968. Molar occlusion in late Triassic mammals. *Biol. Rev. (Camb.)* 43:427–458.

Crompton, A. W., and Sita-Lumsden, A. 1970. Functional significance of the therian molar pattern. *Nature (Lond.)* 227:197–199.

Cronin, J. E. 1977. Pygmy chimpanzee *(Pan paniscus)* systematics. *Amer. J. Phys. Anthro.* 47:125.

Cronin, J. E. 1983. Apes, humans and molecular clocks: a reappraisal, in: *New Interpretations of Ape and Human Ancestry* (R. L. Ciochon and R. S. Corruccini, eds.), pp. 115–149, Plenum Press, New York.

Cronin, J. E., and Meikle, W. E. 1979. The phyletic position of *Theropithecus*: congruence among molecular, morphological and paleontological evidence. *Syst. Zool.* 28:259–269.

Cronin, J. E., and Meikle, W. E. 1982. Hominid and Gelada baboon evolution: agreement between molecular and fossil time scale. *Intl. J. Primatol.* 3(4):469–482.

Cronin, J. E., and Sarich, V. M. 1975. Molecular systematics of the New World monkeys. *J. Hum. Evol.* 4:357–375.

Cronin, J. E., and Sarich, V. M. 1978. Primate higher taxa: the molecular view, in: *Recent Advances in Primatology, Vol. 3* (D. J. Chivers and K. A. Joysey, eds.), pp. 287–289, Academic Press, London.

Crook, J. H. 1966. Gelada baboon herd structure and movement: a comparative report. *Symp. Zool. Soc., London* 18:237–258.

Crook, J. H. 1967. Evolutionary changes in primate societies. *Sci. J.* 3(6):66–72.

Crook, J. H. 1972. Sexual selection, dimorphism and social organization in the primates, in: *Sexual Selection and the Descent of Man* (B. G. Campbell, ed.), pp. 231–281, Aldine, Chicago.

Crook, J. H., and Aldrich-Blake, P. 1968. Ecological and behavioral contrasts between sympatric ground dwelling primates in Ethiopia. *Folia Primatologica* 8:192–227.

Crook, J. H., and Gartlan, J. S. 1966. Evolution of primate societies. *Nature (Lond.)* 210:1200–1203.

Crowson, R. A. 1970. *Classification and Biology*, Atherton Press, New York.

Crusafont-Pairo, M. 1967. Sur quelques prosimiens de l'Eocene de la zone preaxiale pyrenaique et un essai provisoire de reclassification. *C.N.R.S. (Paris)* 163:611–632.

Curtis, G. H., Drake, R. E., Cerling, T. E., and Hampel, J. H. 1975. Age of KBS Tuff in Koobi Fora Formation, East Rudolf, Kenya. *Nature (Lond.)* 258: 395–398.

Cutler, R. G. 1976. Evolution of longevity in primates. *J. Hum. Evol.* 5:169–202.

Darlington, C. D. 1959. *Darwin's Place in History*, Blackwell, Oxford.

Darlington, P. J., Jr. 1957. *Zoogeography: The Geographical Distribution of Animals*, Wiley, New York.

Darlington, P. J., Jr. 1959. Area, climate and evolution. *Evolution* 13:488–510.

Darlington, P. J., Jr. 1965. *Biogeography of the Southern End of the World*, Harvard University Press, Cambridge.

Darlington, P. J., Jr. 1970. A practical criticism of Hennig-Brundin "phylogenetic systematics" and Antarctic biogeography. *Syst. Zool.* 19:1–18.

Dart, R. A. 1925. *Australopithecus africanus*: The man-ape of South Africa. *Nature (Lond.)* 115:195–199.

Dart, R. A. 1949a. Innominate fragments of *Australopithecus prometheus. Amer. J. Phys. Anthro.* 7:301–334.

Dart, R. A. 1949b. The predatory implemental technique of *Australopithecus. Amer. J. Phys. Anthro.* 7:1–38.

Dart, R. A. 1955a. *Australopithecus prometheus* and *Telanthropus. Amer. J. Phys. Anthro.* 13:67–96.

Dart, R. A. 1955b. The first australopithecine fragment from the Makapansgat pebble culture stratum. *Nature (Lond.)* 176:170–171.

Dart, R. A. 1957a. The osteodontokeratic culture of *Australopithecus prometheus. Transvaal Mus. Mem.*, No. 10.

Dart, R. A. 1957b. The Makapansgat australopithecine osteodontokeratic culture. *Proc. 3rd. Pan-Afr. Cong. Prehist.* (Livingstone 1955): 161–171.

Dart, R. A. 1957c. The second adolescent (female) ilium of *Australopithecus prometheus. J. Palaeon. Soc. India* 3:73–82.

Dart, R. A. 1958a. A further adolescent ilium of Makapansgat. *Amer. J. Phys. Anthro.* 16:473–479.

Dart, R. A. 1958b. Bone tools and porcupine gnawing. *Amer. Anthropol.* 60:715–724 and 62:134–143.

Dart, R. A. 1960. The bone tool-manufacturing ability of *Australopithecus prometheus. Amer. Anthropol.* 63:134–142.

Dart, R. A. 1962a. The Makapansgat pink breccia australopithecine skull. *Amer. J. Phys. Anthro.* 20:119–126.

Dart, R. A. 1962b. Stalactites as a tool material for the australopithecines. *Ill. Lond. News* 241:1052–1055.

Dart, R. A. 1964. The ecology of the South African man-apes. *Monogr. Biol.* 14:49–69.

Darwin, C. 1859. *The Origin of Species by Means of Natural Selection or the Preservation of Favored Races in the Struggle for Life*, Murray, London.

Darwin, C. 1871. *The Descent of Man, and Selection in Relation to Sex*, Murray, London.

Dashzeveg, D. T., and McKenna, M. C. 1977. Tarsioid primate from the early Tertiary of the Mongolian People's Republic. *Acta Palaeontol. Polonica* 22(2):119–137.

Davis, D. D. 1962. Mammals of the lowland rain-forest of North Borneo. *Bull. Raffles Mus.* 31:1–129.

Davis, D. D. 1964. The giant panda: a morphological study of evolutionary mechanisms. *Fieldiana, Zool. Mem. #3*.

Davis, J. W. F. 1976. Breeding success and experience in the arctic skua, *Stercorarius parasiticus* (L.). *J. Anim. Ecol.* 45:531.

Davis, P. R., and Napier, J. 1963. A reconstruction of the skull of *Proconsul africanus* (R.S. 51). *Folia Primatologica* 1:20–28.

Davis, S. 1977. Size variation of the fox, *Vulpes vulpes*, in the palaearctic region today, and in Israel during the late Quaternary. *J. Zool., London* 182:343–351.

Day, M. H. 1969. Femoral fragment of a robust australopithecine from Olduvai Gorge, Tanzania. *Nature (Lond.)* 221:230–233.

Day, M. H., Ed. 1973. *Human Evolution*. Taylor and Francis, London.

Day, M. H., and Leakey, R. E. F. 1973. New evidence for the genus *Homo* from East Rudolf, Kenya (I). *Amer. J. Phys. Anthro.* 39:341–354.

Day, M. H., and Leakey, R. E. F. 1974. New evidence of the genus *Homo* from East Rudolf, Kenya (III). *Amer. J. Phys. Anthro.* 41(3):367–380.

Day, M. H., and Napier, J. R. 1964. Hominid fossils from Bed I, Olduvai Gorge: fossil foot bones. *Nature (Lond.)* 201:969–971.

Day, M. H., and Napier, J. R. 1966. A hominoid toe bone from Bed I, Olduvai Gorge, Tanzania. *Nature (Lond.)* 211:929–930.

Day, M. H., and Wood, B. A. 1968. Functional affinities of the Olduvai Hominid 8 talus. *Man* 3:440–445.

Day, M. H., and Wood, B. A. 1969. Hominoid tali from East Africa. *Nature (Lond.)* 222:591–592.

Day, M. H., Leakey, R. E. F., Walker, A. C., and Wood, B. A. 1974. New hominids from East Rudolf, Kenya, I. *Amer. J. Phys. Anthro.* 42:461–476.

Day, M. H., Leakey, R. E. F., Walker, A. C., and Wood, B. A. 1976. New hominids from East Turkana, Kenya. *Amer. J. Phys. Anthro.* 45:369–436.

De Bonis, L. G. 1978. Les primates hominoides du Miocene de Macedoine: Etude de la machiore superieur. *Ann. Palaeontol., Vert.* 64:185–202.

De Bonis, L. G. 1980. Nouvelles remarques sur l'anatomie d'un primate hominoide du Miocene, *Ouranopithecus macedoniensis*: Implications sur la phylogenie des Hominides. *C. R. Acad. Sci., Paris, Ser. D.* 290:755–758.

De Bonis, L. G. 1983. Phylogenetic relationships of Miocene hominoids and higher primate classification, in: *New Interpretations of Ape and Human Ancestry* (R. L. Ciochon and R. S. Corruccini, eds.), pp. 625–649, Plenum Press, New York.

De Bonis, L. G., and Melentis, J. 1977a. Les primates hominoides du Vallesien de Macedoine (Greece): Etude de la machiore inferieure. *Geobios* 10:849–885.

De Bonis, L. G., and Melentis, J. 1977b. Un nouveaus genre de primate hominoide dans le Vallesien (Miocene Superieur) de Macedonie. *C. R. Acad. Sci., Paris, Ser. D.* 284:1393–1395.

De Bonis, L. G., Bouvrain, G., Geraads, D., and Melentis, J. 1974. Premiere decouverte d'un primate hominoide dans le Miocene Superieur de Macedoine. *C. R. Acad. Sci., Paris, Ser. D.* 278:3063–3066.

De Bonis, L. G., Bouvrain, G., and Melentis, J. 1975. Nouveaux restes du Primates hominoides dans le Vallesien de Macedoine (Greece). *C. R. Acad. Sci., Paris, Ser. D.* 281:379–382.

Decker, R. L., and Szalay, F. S. 1974. Origins and functions of the pes in the Eocene Adapidae (Lemuriformes, Primates), in: *Primate Locomotion* (F. A. Jenkins, ed.), pp. 261–291, Academic Press, New York.

Dehm, R. 1983. Miocene hominoid primate dental remains from the Siwaliks of Pakistan, in: *New Interpretations of Ape and Human Ancestry* (R. L. Ciochon and R. S. Corruccini, eds.), pp. 527–537, Plenum Press, New York.

Delson, E. 1971. Estudo preliminar de unos restos de simios pliocenicos procedentes de Cova Bonica (Gava) (Prov. Barcelona). *Acta Geol. Hisp.* 6:54–57.

Delson, E. 1973. *Fossil Colobine Monkeys of the Circum-Mediterranean Region and the Evolutionary History of the Cercopithecidae (Primates, Mammalia).* Ph.D. thesis, Columbia University.

Delson, E. 1974a. Preliminary review of the cercopithecid distribution in the circum-Mediterranean region. *Bur. Rech. Geol. Min., Mem.* No. 78.

Delson, E. 1974b. The oldest fossil Cercopithecidae. *Amer. J. Phys. Anthro.* 41:474–475.

Delson, E. 1975a. Evolutionary history of the Cercopithecidae, in: *Approaches to Primate Paleobiology* (F. S. Szalay, ed.), *Contrib. Primatol.* 5:167–217, Karger, Basel.

Delson, E. 1975b. Toward the origin of the Old World monkeys, in: Evolution des Vertebres—Problemes Actuels de Paleontoloqie. *Colloq. Int. Cent. Nat. Rech. Sci.* 218:839–850. (See Chapter 23, this volume)

Delson, E. 1975c. Paleoecology and zoogeography of the Old World monkeys, in: *Primate Functional Morphology and Evolution* (R. Tuttle, ed.), pp. 37–64, Mouton, The Hague.

Delson, E. 1977a. Catarrhine phylogeny and classification: principles, methods and comments. *J. Hum. Evol.* 6:433–459.

Delson, E. 1977b. Vertebrate paleontology, especially of non-human primates in China, in: *Paleoanthropology in the People's Republic of China* (W. W. Howells and P. J. Tsuchitani, eds.), pp. 40–65, National Academy of Sciences, Washington.

Delson, E. 1979. *Prohylobates* (Primates) from the early Miocene of Libya: a new species and its implications for cercopithecid origins. *Geobios* 12:725–733.

Delson, E., and Andrews, P. 1975. Evolution and interrelationships of the catarrhine primates, in: *Phylogeny of the Primates* (W. P. Luckett and F. S. Szalay, eds.), pp. 405–446, Plenum Press, New York.

Delson, E., and Rosenberger, A. L. 1980. Phyletic perspectives on platyrrhine origins and anthropoid relationships, in: *Evolutionary Biology of the New World Monkeys and Continental Drift* (R. L. Ciochon and A. B. Chiarelli, eds.), pp. 445–458, Plenum Press, New York.

Delson, E., Eldredge, N., and Tattersall, I. 1977. Reconstruction of hominoid phylogeny: a testable framework based on cladistic analysis. *J. Hum. Evol.* 6:263–278.

De Lumley, H., and De Lumley, M. A. 1974. Pre-Neanderthal human remains from Arago cave in Southeast France. *Yrbk. Phys. Anthro.* 17:162–168.

Dene, H. T., Goodman, M., and Prychodko, W. 1976. Immunodiffusion evidence on the phylogeny of the primates, in: *Molecular Anthropology* (M. Goodman and R. E. Tashian, eds.), pp. 171–195, Plenum Press, New York.

DeVore, I. 1964. The evolution of social life, in: *Horizons in Anthropology,* First edition (S. Tax, ed.), pp. 25–36, Aldine, Chicago.

DeVore, I. 1965. Male dominance and mating behavior in baboons, in: *Sex and Behavior* (F. Beach, ed.), pp. 266–289, Wiley, New York.

DeVore, I., and Washburn, S. L. 1963. Baboon ecology and human evolution, in: *African Ecology and Human Evolution* (F. C. Howell and F. Bourliere, eds.), pp. 335–367, Aldine, New York.

Diamond, I. T., and Hall, W. C. 1969. Evolution of neocortex. *Science* 164:251.

Diard, M. 1820. Report of a meeting of the Asiatic Society of Bengal for March 10, 1820. *Asiat. J. Month. Reg.* 10:477–478.

Dietz, R. S., and Holden, J. C. 1970. The breakup of Pangaea. *Sci. Amer.* 223(4):30–41.

Dimpel, H., and Calaby, J. H. 1972. Further observations on the mountain pygmy possum *(Burramys parvus). Victorian Nat.* 89:101.

Dobzhansky, T. 1961. Man and natural selection. *Amer. Sci.* 49:285–299.

Dollo, L. 1899. Les ancetres des Marsupiaux etaient-ils arboricoles? *Trav. Stat. Zool. Wimereux* 7:188–203.

Dollo, L. 1900. Le pied de *Diprotodon* et l'origine arboricole des marsupiaux. *Bull. Sci. Fr. Belg.* 275–280.

Doolittle, R. F. 1981. Similar amino acid sequences: chance or common ancestry? *Science* 214:149–159.

Douglass, E. 1908. Vertebrate fossils from the Fort Union beds. *Ann. Carnegie Mus.* 5:11–26.

Doyle, G. A., and Martin, R. D., Eds. 1978. *The Study of Prosimian Behavior,* Academic Press, New York.

Dray, S., and Young, G. O. 1960. Genetic control of two gamma globulin isoantigenic sites in domestic rabbits. *Science* 131:738–739.

Drickamer, L. C. 1974. A ten year summary of population and reproductive data for the free ranging *Macaca mulatta* at La Parguera, Puerto Rico. *Folia Primatologica* 21:61–80.

D'Souza, F. 1974. A preliminary field report on the lesser tree shrew *(Tupaia minor),* in: *Prosimian Biology* (G. A. Doyle, R. D. Martin and A. C. Walker, eds.), pp. 167–182, Duckworth, London.

Ducker, G. 1957. Farb—und Helligkeitssehen und Instinkte bei Viverriden und Feliden. *Zool. Beit.,* n. f. 3:25–99.

Dunbar, R. I. M., and Dunbar, E. P. 1974. Ecological relations and niche separation between sympatric terrestrial primates in Ethiopia. *Folia Primatologica* 21:36–60.

Durette-Desset, M. C. 1971. (unpub.) *Essai de Classification des Nematodes Heligmosomes. Correlations avec la Paleobiogeographie des Hotes.* These Doctorat Sc. Nat., Univ. Paris (Orsay) Sud.

Earle, C. 1897. On the affinities of *Tarsius:* a contribution to the phylogeny of the primates. *Amer. Natural.* 31:569–575.

Eck, G., and Howell, F. C. 1973. New fossil *Cercopithecus* material from the lower Omo basin, Ethiopia. *Folia Primatologica* 18:325–355.

Edwards, W. E. 1967. The late Pleistocene extinction and diminution in size of many mammalian species, in: *Pleistocene Extinction* (P. Martin and H. Wright, Jr., eds.), pp. 141–154, Yale University Press, New Haven.

Eggeling, W. J. 1947. Observations on the ecology of the Budongo rain forest. *J. Ecol.* 34:20–87.

Eisenberg, J. F. 1966. The social organizations of mammals. *Handbk. Zool.* 10(7):1–92.

Eisenberg, J. F. 1978. The evolution of arboreal herbivores in the class Mammalia, in: *The Ecology of Arboreal Folivores* (G. G. Montgomery, ed.), pp. 135–152, Smithsonian Institute Press, Washington, D.C.

Eisenberg, J. F., and Leyhausen, P. 1972. The phylogeny of predatory behavior in mammals. *Z. Tierpsychol.* 30:59–93.

Eisenberg, J. F., and Thorington, R. W., Jr. 1973. A preliminary analysis of a Neotropical mammal fauna. *Biotropica* 5:150–161.

Eldredge, N., and Gould, S. J. 1972. Speciation and punctuated equilibria: an alternative to phyletic gradualism, in: *Models in Paleobiology* (T. J. M. Schopf, ed.), pp. 82–115, Freeman, San Francisco.

Eldredge, N., and Tattersall, I. 1975. Evolutionary models, phyletic reconstruction and another look at hominid phylogeny. *Contrib. Primatol.* 5:218–242.

Elftman, H., and Aikinson, W. B. 1950. The abdominal viscera of the gorilla, in: *The Anatomy of the Gorilla* (W. K. Gergory, ed.), pp. 197–204, Columbia University Press, New York.

Ellerman, J. R. 1940. *The Families and Genera of Living Rodents. Vol. 1. Rodents other than Muridae,* British Museum (Nat. Hist.), London.

Elliot-Smith, G. 1919. Discussion on the zoological position and affinities of *Tarsius. Proc. Zool. Soc. London* 1919:465–475.

Emlen, S. T., and Oring, L. W. 1977. Ecology, sexual selection and the evolution of mating systems. *Science* 197:215.

Emry, R. J. 1970. A North American Oligocene pangolin and other additions to the Pholidota. *Bull. Amer. Mus. Nat. Hist.* 142(6):455–510.

Enders, R. K. 1935. Mammalian life histories from Barro Colorado Island, Panama. *Bull. Mus. Comp. Zool.* 78:383.

Englemann, G. F., and Wiley, F. O. 1977. The place of ancestor-descendant relationships in phylogeny reconstruction. *Syst. Zool.* 26:1–11.

Erikson, G. E. 1963. Brachiation in New World Monkeys and Anthropoid Apes. *Symp. Zool. Soc. London* 10:135–164.

Erxleben, C. P. 1777. *Systema Regni Animalis,* Leipzig.

Every, R. G. 1974. Thegosis in prosimians, in: *Prosimian Biology* (R. D. Martin, G. A. Doyle, and A. C. Walker, eds.), pp. 579–619, Duckworth, London.

Ewer, R. F. 1973. *The Carnivores,* Cornell University Press, New York.

Falk, D. 1983. A reconsideration of the endocast of *Proconsul africanus:* implications for primate brain evolution, in: *New Interpretations of Ape and Human Ancestry* (R. L. Ciochon and R. S. Corruccini, eds.), pp. 239–248, Plenum Press, New York.

Feduccia, A. 1973. Dinosaurs as reptiles. *Evolution* 27:166–169.

Feldesman, M. R. 1982. Morphometric analysis of the distal humerus of some Cenozoic catarrhines. The late divergence hypothesis revisited. *Amer. J. Phys. Anthro.* 59:73–95.

Ferembach, D. 1958. Les limnopitheques du Kenya. *Ann. Paleontol., Paris* 44:149–249.

Ferris, V. R. 1980. A science in search of a paradigm?—Review of the symposium, "Vicariance Biogeography: A Critique". *Syst. Zool.* 29:67–76.

Ferris, S. D., Wilson, A. C., and Brown, W. M. 1981. Evolutionary trees of apes and humans based on cleavage maps of mitochondrial DNA. *Proc. Natl. Acad. Sci. USA* 78:2432–2436.

Filhol, H. 1873. Sur un nouveau genre de lemurien fossile, recemment decouvert dans les gisements de phosphirite de chaux du Quercy. *C. R. Acad. Sci., Paris* 77:111–1112.

Filhol, H. 1880. Note sur des mammiferes fossiles nouveaux provenant des phosphorites du Quercy. *Bull Soc. Philo. Paris, Ser. 7,* 4:120–125.

Filhol, H. 1887. *Recherches sur les Phosphorites du Quercy,* Masson, Paris.

Filhol, H. 1890. Description d'une nouvelle espece de lemurien fossile *(Necrolemur parvulus). Bull. Soc. Philo. Paris. Ser. 8,* 2:39–40.

Fisher, D. C. 1981. Crocodilian scatology, microvertebrate concentrations, and enamel-less teeth. *Paleobiology* 7(2):262–275.

Fisher, R. A. 1930. *Genetical Theory of Natural Selection,* Claredon, Oxford.

Fitch, F. J., and Miller, J. A. 1970. Radioisotopic age determination of Lake Rudolf artifact site. *Nature (Lond.)* 226:226–228.

Fitch, F. J., Findlater, I. C., Watkins, R. T., and Miller, J. A. 1974. Dating of the rock succession containing fossil hominids at Lake Rudolf, Kenya. *Nature (Lond.)* 251:213–215.

Fitch, F. J., Miller, J. A., Warrell, D. M., and Williams, S. C. 1974. Tectonic and radiometric age comparisons, in: *The Ocean Basins and Margins, Vol. 2. The North Atlantic* (A. E. M. Nairn and F. G. Stehli, eds.), pp. 485–538, Plenum Press, New York.

Fitch, W. W., and Langley, C. H. 1976. Evolutionary rates in proteins: Neutral mutations and the molecular clock, in: *Molecular Anthropology* (M. Goodman and R. E. Tashian, eds.), pp. 197–219, Plenum Press, New York.

Fleagle, J. G. 1978b. Size distributions of living and fossil primate faunas. *Paleobiology* 4:67–76.

Fleagle, J. G. 1979. Primate positional behavior and anatomy: naturalistic reconstruction and another look at hominid phylogeny. *Contrib. Primatol.* 5:218–242.

Fleagle, J. G. 1980. Locomotor behavior of the earliest apes: a review of the current evidence. *Z. Morphol. Anthro.* 71:149–156.

Fleagle, J. G. 1981. Book review of *Eternal Search. Evolution* 35:1029–1031.

Fleagle, J. G. 1983. Locomotor adaptations of Oligocene and Miocene hominoids and their phyletic implications, in: *New Interpretations of Ape and Human Ancestry* (R. L. Ciochon and R. S. Corruccini, ds.), pp. 301–324, Plenum Press, New York.

Fleagle, J. G. 1984. Are there any fossil gibbons in: *The Lesser Apes: Evolutionary and Behavioral Biology* (D. J. Chivers, H. Preuschoft, N. Creel, and W. Brockelman, eds.), Edinburgh University Press, Edinburgh.

Fleagle, J. G. and Bown, T. 1983. New primate fossils from late Oligocene (Colhuehuapian) localities of Chubut Province, Argentina. *Folia Primatol.* 41:240–266.

Fleagle, J. G., and Kay, R. F. 1983. New interpretations of the phyletic position of Oligocene hominoids, in: *New Interpretations of Ape and Human Ancestry* (R. L. Ciochon and R. S. Corruccini, eds.), pp. 181–210, Plenum Press, New York. (See Chapter 21, this volume)

Fleagle, J. G., and Rosenberger, A. L. 1983. Cranial morphology of the earliest anthropoids, in: *Morphologie, Evolutive, Morphogenese du Crane et Origine de l'Homme* (M. Sakka, ed.), pp. 141–153, CNRS, Paris.

Fleagle, J. G., and Simons, E. L. 1978a. *Micropithecus clarki,* a small ape from the Miocene of Uganda. *Amer. J. Phys. Anthro.* 49:427–440.

Fleagle, J. G., and Simons, E. L. 1978b. Humeral morphology of the earliest apes. *Nature (Lond.)* 273:705–707.

Fleagle, J. G., and Simons, E. L. 1979. Anatomy of the bony pelvis of parapithecid primates. *Folia Primatologica* 31:176–186.

Fleagle, J. G., and Simons, E. L. 1982. The humerus of *Aegyptopithecus zeuxis,* a primitive anthropoid. *Amer. J. Phys. Anthro.* 59:175–193.

Fleagle, J. G., Kay, R. F., and Simons, E. L. 1980. Sexual dimorphism in early anthropoids. *Nature (Lond.)* 287:328–330.

Fleagle, J. G., Simons, E. L., and Conroy, G. C. 1975. Ape limb bone from Oligocene of Egypt. *Science* 189:135–137.

Flynn, L. J., and Qi, G. 1982. Age of the Lufeng, China, hominoid locality. *Nature (Lond.)* 298:746–747.

Ford, S. 1980. Callithricids as phyletic dwarfs, and the place of Callithricidae in the Platyrrhini. *Primates* 21(1):31–43. (See Chapter 18, this volume)

Forster, R. 1978. Evidence for an open seaway between northern and southern proto-Atlantic in Albian times. *Nature (Lond.)* 272:158–159.

Foster, J. B. 1965. The evolution of the mammals on the Queen Charlotte Islands, British Columbia. *Brit. Columbia, Prov. Mus. Occas. Pap.* 14:1–130.

Fourtau, R. 1918. *Contribution a l'etude des Vertebres miocenes de l'Egypt,* Egyptian Survey Dept., Cairo.

Frakes, L. A., and Kemp, E. M. 1972. Influence of continental positions on Tertiary climates. *Nature (Lond.)* 240:97–100.

Frakes, L. A., and Kemp, E. M. 1973. Paleogene continental positions and evolution of climate, in: *Implications of Continental Drift to the Earth Sciences, Vol. 1* (D. H. Tarling and S. K. Runcorn, eds.), pp. 539–559, Academic Press, London.

Franzen, J. L. 1968. *Revision der Gattung* Palaeotherium *Cuvier 1804 (Palaeotheriidae, Perissodactyla, Mammalia).* Inaugural-Dissertation, Albert-Ludwigs-Universität zu Freiburg-im-Breisgau.

Frayer, D. W. 1973. *Gigantopithecus* and its relationship to *Australopithecus. Amer. J. Phys. Anthro.* 39:413–426.

Frayer, D. W. 1976. A reappraisal of *Rama-*

pithecus. Yrbk. Phys. Anthro. 18:19–30.

Frayer, D. W. 1978. The taxonomic status of *Ramapithecus,* in: *Krapinski pracoviek i evoluciaja hominida* (M. Malez, ed.), p. 255–268, Jugoslavenska Akademija Znanosti i Umjetnosti, Zagreb.

Freedman, L. 1957. The fossil Cercopithecoidea of South Africa. *Ann. Transvaal Mus.* 23:121–262.

Freedman, L. 1961. New cercopithecoid fossils, including a new species from Taung, Cape Province, South Africa. *Ann. S. Afr. Mus.* 46:1–14.

Freedman, L. 1965. Fossil and subfossil primates from the limestone deposits at Taung, Bolt's Farm and Witkrans, South Africa. *Palaeont. Afr.* 9:19–48.

Freedman, L., and Stenhouse, N. S. 1972. The *Parapapio* species of Sterkfontein, Transvaal, South Africa. *Palaeont. Afr.* 14:93–111.

Freeland, W. J., and Janzen, D. H. 1974. Strategies in herbivory by mammals: the role of plant secondary compounds. *Amer. Natural.* 108:269–289.

Frisch, J. E. 1973. The hylobatid dentition, in: *Gibbon and Siamang,* Vol. 2 (D. Rumbaugh, ed.), pp. 56–95, Karger, Basel.

Funnell, B. M., and Smith, A. G. 1968. Opening of the Atlantic Ocean. *Nature (Lond.)* 219:1328–1333.

Gabis, R. 1960. Les os des membres chez les singes cynomorphes. *Mammalia* 24:577–602.

Gadow, H. 1898. *A Classification of Vertebrata Recent and Extinct,* Adam and Charles Black, London.

Galdikas, B. M. F., and Teleki, G. 1981. Variations in subsistence activities of female and male pongids: New perspectives on the origins of hominid labor division. *Curr. Anthro.* 22:241–256.

Galton, P. 1977. The Upper Jurassic ornithopod dinosaur *Dryosaurus* and a Laurasia-Gondwanaland connection, in: *Paleontology and Plate Tectonics* (R. M. West, ed.), *Milwaukee Public Mus. Spec. Publ. Biol. Geol.* 2:41–54.

Gambarian, P. P., and Oganesian, R. O. 1970. Biomechanics of the gallop and of the primitive rebounding jump in small mammals. *Proc. Acad. Sci. USSR Biol.,* Ser. 3:441–447 (In Russian).

Gantt, D. G. 1977. *Enamel of Primate Teeth.* Ph.D. Dissertation, Washington University, St. Louis.

Gantt, D. G. 1979. Comparative enamel histology of primate teeth. *J. Dent. Res.* 58(B):1002–1003.

Gantt, D. G. 1980. Implications of enamel prism patterns for the origin of New World monkeys, in: *Evolutionary Biology of New World Monkeys and Continental Drift* (R. L. Ciochon and A. B. Chiarelli, eds.), pp. 201–217, Plenum Press, New York.

Gantt, D. G. 1983. The enamel of Neogene hominoids: structural and phyletic implications, in: *New Interpretations in Ape and Human Ancestry* (R. L. Ciochon and R. S. Corruccini, eds.), pp. 249–298, Plenum Press, New York.

Gantt, D. G., Pilbeam, D., and Steward, G.

1977. Hominoid enamel prism patterns. *Science* 189:135–137.

Gauss, G. F. 1934. *The Struggle for Existence,* Williams and Wilkins, Baltimore.

Gazin, C. L. 1952. The lower Eocene Knight Formation of western Wyoming and its mammalian faunas. *Smith. Misc. Coll.* 117:1–82.

Gazin, C. L. 1958. A review of the middle and upper Eocene primates of North America. *Smith. Misc. Coll.* 136:1–112.

Gazin, C. L. 1962. A further study of the lower Eocene mammalian fauna of southwestern Wyoming. *Smith. Misc. Coll.* 144:1–98.

Gazin, C. L. 1968. A new primate from the Torrejon middle Paleocene of the San Juan Basin, New Mexico. *Proc. Biol. Soc., Washington* 81:629–634.

Genet-Varcin, E. 1963. *Les Singes Actuels et Fossiles,* Boubee & Cie, Paris.

Genet-Varcin, E. 1969. *A la Recherche du Primate Ancetre de l'Homme,* Boubee, Paris.

Gentry, A. W. 1970. The Bovidae (Mammalia) of the Fort Ternan fossil fauna, in: *Fossil Vertebrates of Africa* (L. S. B. Leakey and R. J. G. Savage, eds.), pp. 243–324, Academic Press, London.

Gidley, J. W. 1919. Significance of divergence of the first digit in the primitive mammalian foot. *J. Wash. Acad. Sci* 9:273–280.

Gidley, J. W. 1923. Paleocene primates of the Fort Union, with discussion of relationships of Eocene primates. *Proc. U.S. Nat. Mus.* 63:1–38.

Gingerich, P. D. 1973a. Anatomy of the temporal bone in the Oligocene anthropoid *Apidium* and the origin of Anthropoidea. *Folia Primatologica* 19:329–337.

Gingerich, P. D. 1973b. First record of the Paleocene primate *Chiromyoides* from North America. *Nature (Lond.)* 244:517–518.

Gingerich, P. D. 1974a. *Cranial Anatomy and the Evolution of Early Tertiary Plesiadapidae (Mammalia, Primates).* Doctoral thesis, Yale University.

Gingerich, P. D. 1974b. Dental function in the Paleocene primate *Plesiadapis,* in: *Prosimian Biology* (R. D. Martin, G. A. Doyle and A. C. Walker, eds.), pp. 531–541, Duckworth, London.

Gingerich, P. D. 1974c. Function of pointed premolars in *Phenacolemur* and other mammals. *J. Dent. Res.* 53:497.

Gingerich, P. D. 1975a. Dentition of *Adapis parisiensis* and the evolution of lemuriform primates, in: *Lemur Biology* (I. Tattersall and R. W. Sussman, eds.), pp. 65–80, Plenum Press, New York.

Gingerich, P. D. 1975b. Systematic position of *Plesiadapis. Nature (Lond.)* 253:111–113.

Gingerich, P. D. 1975c. A new genus of Adapidae (Mammalia, Primates) from the late Eocene of Southern France, and its significance for the origin of higher primates. *Contrib. Mus. Paleontol., Univ. Michigan* 24:163–170.

Gingerich, P. D. 1975d. New North American Plesiadapidae (Mammalia, Primates) and a biostratigraphic zonation of the middle and upper Paleocene. *Contrib. Mus. Paleont. Univ. Michigan* 24:135–148.

Gingerich, P. D. 1976. Cranial anatomy and evolution of early Tertiary Plesiadapidae (Mammalia, Primates). *Univ. Michigan Papers Paleont.* 15:1–141.

Gingerich, P. D. 1977a. Correlation of tooth size and body size in living hominoid primates, with a note on relative brain size in *Aegyptopithecus* and *Proconsul. Amer. J. Phys. Anthro.* 47:395–398.

Gingerich, P. D. 1977b. Dental variation in early Eocene *Teilhardina belgica* with notes on the anterior dentition of some early Tarsiiformes. *Folia Primatologica* 28:144–153.

Gingerich, P. D. 1977c. Radiation of Eocene Adapidae in Europe. *Geobios, Mem. Spec.* 1:165–185.

Gingerich, P. D. 1978a. The Stuttgart collection of Oligocene primates from the Fayum Province of Egypt. *Paleont. Z.* 52:82–92.

Gingerich, P. D. 1978b. Phylogeny reconstruction and the phylogenetic position of *Tarsius,* in: *Recent Advances in Primatology* (D. J. Chivers and K. A. Joysey, eds.), pp. 249–255, Academic Press, New York.

Gingerich, P. D. 1979a. Phylogeny of Middle Eocene Adapidae (Mammalia, Primates) in North America: *Smilodectes* and *Notharctus. J. Paleont.* 53(1):153–163.

Gingerich, P. D. 1979b. Sexual dimorphism in Eocene Adapidae: implications for primate phylogeny and evolution. *Amer. J. Phys. Anthro.* 50:442.

Gingerich, P. D. 1979c. The stratophenetic approach to phylogeny reconstruction in vertebrate paleontology, in: *Phylogenetic Analysis and Paleontology* (J. Cracraft and N. Eldredge, eds.), pp. 41–77, Columbia University Press, New York.

Gingerich, P. D. 1980a. Eocene Adapidae, paleobiology, and the origin of South American Platyrrhini, in: *Evolutionary Biology of the New World Monkeys and Continental Drift* (R. L. Ciochon and A. B. Chiarelli, eds), pp. 123–138, Plenum Press, New York. (See Chapter 13, this volume)

Gingerich, P. D. 1980b. Evolutionary patterns in early Cenozoic mammals. *Ann. Rev. Earth. Planetary Sci.* 8:407–424.

Gingerich, P. D. 1980c. Dental and cranial adaptations in Eocene Adapidae. *Z. Morphol. Anthropol.* 71(2):135–142. (See Chapter 6, this volume)

Gingerich, P. D. 1981a. Cranial morphology and adaptation in Eocene Adapidae. 1. Sexual dimorphism in *Adapis magnus* and *Adapis parisiensis. Amer. J. Phys. Anthro.* 56:217–234.

Gingerich, P. D. 1981b. Early Cenozoic Omomyidae and the evolutionary history of Tarsiiform primates. *J. Hum. Evol.* 10:345–374.

Gingerich, P. D., and Martin, R. D. 1981. Cranial morphology and adaptation in Eocene Adapidae. 2. The Cambridge skull of *Adapis parisiensis. Amer. J. Phys. Anthro.* 56:235–257.

Gingerich, P. D., and Rose, K. D. 1977. Preliminary report on the American Clark Fork mammal fauna, and its correlation with similar faunas in Europe and Asia. *Geobios, Mem. Spec.* 1:39–45.

Gingerich, P. D., and Ryan, A. S. 1979. Dental and cranial variation in living Indriidae. *Primates* 20:141–159.

Gingerich, P. D., and Sahni, A. 1979. *Indraloris*

and *Sivaladapis:* Miocene adapid primates from the Siwaliks of India and Pakistan. *Nature (Lond.)* 279:415–416.

Gingerich, P. D., and Schoeninger, M. 1977. The fossil record and primate phylogeny. *J. Hum. Evol.* 6:483–505.

Gingerich, P. D., and Simons, E. L. 1977. Systematics, phylogeny and evolution of early Eocene Adapidae (Mammalia, Primates) in North America. *Contrib. Mus. Paleont., Univ. Michigan* 24:245–279.

Gingerich, P. D., Smith, B. H., and Rosenberger, K. 1980. Patterns of allometric scaling in the primate dentition and prediction of body size from tooth size (Abstract). *Amer. J. Phys. Anthro.* 52:231–232.

Gingerich, P. D., Smith, B. H., and Rosenberger, K. 1982. Allometric scaling in the dentition of primates and prediction of body weight from tooth-size in fossils. *Amer. J. Phys. Anthro.* 57:81–100.

Ginsburg, L. 1975. Le pliopitheque des faluns helvetiens de la Touraine et de l'Anjou. *Coll. Internat. C.N.R.S.* 218:877–886.

Ginsburg, L., and Mein, P. 1980. *Crouzelia rhodanica,* nouvelle espece de Primate catarhinien et essai sur la position systematique des Pliopithecidae. *Bull. Mus. Nat. Hist. Nat. Paris, Ser. 4* 2(C):57–85.

Glickstein, M. 1969. Organization of the visual pathways. *Science* 164:917.

Godinot, M. 1978. Un novel adapide (Primate) de l'Eocene inferieur de Provence. *C. R. Acad. Sci., Paris* 286:1869–1872.

Goffart, M. 1971. *Function and Form in the Sloth,* Pergamon Press, Oxford.

Goldstein, S., Post, D., and Melnick, D. 1978. An analysis of cercopithecoid odontometrics. I. The scaling of the maxillary dentition. *Amer. J. Phys. Anthro.* 49:517–532.

Good, R. A., and Gabrielson, A. E. 1968. The thymus and other lymphoid organs in the development of the immune system, in: *Human Transplantation* (F. T. Rapaport and J. Saussett, eds.), pp. 526–564, Grue and Stratton, New York.

Goodall, J. 1964. Tool-using and aimed throwing in a community of freeliving chimpanzees. *Nature (Lond.)* 201:1264–1266.

Goodall, J. 1965. Chimpanzees of the Gombe Stream Reserve, in: *Primate Behavior: Field Studies in Monkeys and Apes* (I. DeVore, ed.), pp. 425–473, Holt, Rinehart and Winston, New York.

Goodall, J. 1967. Mother-offspring relations in free-ranging chimpanzees, in: *Primate Ethology* (D. Morris, ed.), pp. 287–346, Aldine, Chicago.

Goodall, J. 1976. Continuities between chimpanzee and human behavior, in: *Human Origins: Louis Leakey and the East African Evidence* (G. L. Isaac and E. R. McCown, eds.), pp. 81–95, Benjamin Cummings, Menlo Park.

Goodman, M. 1960a. On the emergence of intraspecific differences in the protein antigens of human beings. *Amer. Nat.* 94:153–166.

Goodman, M. 1960b. The species specificity of proteins as observed in the Wilson comparative analyses plates. *Amer. Nat.* 94:184–186.

Goodman, M. 1961. The role of immunochemical differences in the phyletic development of human behavior. *Hum. Biol.* 33:131–162.

Goodman, M. 1962a. Evolution of the immunologic species specificity of human serum proteins. *Hum. Biol.* 34:104–150.

Goodman, M. 1962b. Immunochemistry of the Primates and primate evolution. *Ann. N.Y. Acad. Sci.* 102:219–234.

Goodman, M. 1962c. Problems in primate systematics attacked by the serological study of proteins, in: *Taxonomic Biochemistry and Serology* (C. Leone, ed.), pp. 467–486, Ronald Press, New York.

Goodman, M. 1963. Man's place in the phylogeny of the primates as reflected by serum proteins, in: *Classification and Human Evolution* (S. L. Washburn, ed.), pp. 204–235, Aldine, Chicago. (See Chapter 39, this volume)

Goodman, M. 1967. Deciphering primate phylogeny from macromolecular specifities. *Amer. J. Phys. Anthro.* 26:255–276.

Goodman, M. 1974. Biochemical evidence on hominid phylogeny. *Ann. Rev. Bio.* 3:203–288.

Goodman, M. 1975. Protein sequence and immunological specificity, in: *Phylogeny of the Primates, a Multidisciplinary Approach* (W. P. Luckett and F. S. Szalay, eds.), pp. 219–248, Plenum Press, New York.

Goodman, M. 1976a. Protein sequences in phylogeny, in: *Molecular Evolution* (F. J. Ayala, ed.), pp. 141–159, Sinauer, Sunderland.

Goodman, M. 1976b. Toward a genealogical description of the primates, in: *Molecular Anthropology* (M. Goodman and R. E. Tashian, eds.), pp. 321–353, Plenum Press, New York.

Goodman, M., and Cronin, J. E. 1982. Molecular anthropology: its development and current directions, in: *A History of Physical Anthropology 1930-1980* (F. Spencer, ed.), pp. 105–146, Academic Press, New York.

Goodman, M., and Lasker, G. W. 1975. Molecular evidence as to man's place in nature, in: *Primate Functional Morphology and Evolution* (R. H. Tuttle, ed.), pp. 71–101, Mouton, The Hague.

Goodman, M., and Poulik, E. 1961a. Effects of speciation in serum proteins in the genus *Macaca* with special reference to the polymorphic state of transferrin. *Nature (Lond.)* 190:171–172.

Goodman, M., and Poulik, E. 1961b. Serum tranferrins in the genus *Macaca*: species distribution of nineteen phenotypes. *Nature (Lond.)* 191:1407–1408.

Goodman, M., Baba, M. L., and Darga, L. L. 1983. The bearing of molecular data on the cladogenesis and times of divergence of hominoid lineages, in: *New Interpretation of Ape and Human Ancestry* (R. L. Ciochon and R. S. Corruccini, eds.), pp. 67–86, Plenum Press, New York.

Goodman, M., Hewett-Emmett, D., and Beard, J. M. 1978. Molecular evidence on the phylogenetic relationships of *Tarsius*, in: *Recent Advances in Primatology, Vol. 3* (D. J. Chivers and K. A. Joysey, eds.), pp. 215–224, Academic Press, Lodon.

Goodman, M., Moore, G. W., Farris, W., and Poulik, E. 1970. The evidence from genetically informative macromolecules on the phylogenetic relationships of the chimpanzees, in: *The Chimpanzee, Vol. 2* (G. H. Bourne, ed.), pp. 318–360, Karger, Basel.

Goodman, M., Poulik, E., and Poulik, M. D. 1960. Variations in the serum specificities of higher primates detected by two-dimensional starch-gel electrophoresis. *Nature (Lond.)* 188:78–79.

Goodman, M., Rosenblatt, M., Gottlieb, J. S., Miller, J., and Chen, C. H. 1963. Effect of sex, age, and schizophrenia on production of thyroid autoantibodies. *A. M. A. Arch. Gen. Psychiat.* 8:518–526.

Goudie, R. B., Anderson, J. R., and Gray, K. G. 1959. Complement-fixing anti-thyroid antibodies in hospital patients with asypmtomatic thyroid lesions. *J. Path. Bact.* 77:389–400.

Gould, S. J. 1966. Allometry and size in ontogeny and phylogeny. *Biol. Rev. Cambridge Phil. Soc.* 41:587–640.

Gould, S. J. 1975. On the scaling of tooth size in mammals. *Am. Zool.* 15:351–362.

Grand, T. 1968a. Functional anatomy of the upper limb. *Bibl. Primatol.* 7:104–125.

Grand, T. 1968b. The functional anatomy of the lower limb of the howler monkeys *(Alouatta caraya)*. *Amer. J. Phys. Anthro.* 28:163–182.

Granger, W. 1910. Tertiary faunal horizons in the Wind River Basin, Wyoming, with descriptions of new Eocene mammals. *Bull. Amer. Mus. Nat. Hist.* 28:235–251.

Grant, P. R. 1965. The adaptive significance of some size trends in island birds. *Evolution* 19:355–367.

Gray, J. E. 1870. *Catalogue of Monkeys, Lemurs and Fruit-eating Bats*, British Museum, London.

Greenfield, L. O. 1974. Taxonomic reassessment of two *Ramapithecus* specimens. *Folia Primatologica* 22:97–115.

Greenfield, L. O. 1975. A comment on relative molar breadth in *Ramapithecus. J. Hum. Evol.* 4:267–273.

Greenfield, L. O. 1977 *Ramapithecus and Early Hominid Origins*, Ph.D. Thesis, University Microfilms, University of Michigan.

Greenfield, L. O. 1978. On the dental arcade reconstructions of *Ramapithecus. J. Hum. Evol.* 7:345–359.

Greenfield, L. O. 1979. On the adaptive pattern of *"Ramapithecus". Amer. J. Phys. Anthro.* 50:527–548.

Greenfield, L. O. 1980. A late-divergence hypothesis. *Amer. J. Phys. Anthro.* 52:351–366. (See Chapter 30, this volume)

Greenfield, L. O. 1983. Toward the resolution of discrepancies between phenetic and paleontological data bearing on the question of human origins, in: *New Interpretations of Ape and Human Ancestry* (R. L. Ciochon and R. S. Corruccini, eds.), pp. 695–703, Plenum Press, New York.

Gregory, W. K. 1910. The order of mammals. *Bull. Amer. Mus. Nat. Hist.* 27:3–524.

Gregory, W. K. 1913. Relationship of the Tupaiidae and of Eocene lemurs, especially *Notharctus. Bull. Geol. Soc. Amer.* 24:247–252.

Gregory, W. K. 1916. Studies on the evolution of primates. *Bull. Amer. Mus. Nat. Hist.* 35:239–355.

Gregory, W. K. 1920a. Facts and theories of evolution, with special reference to the origin of man. *Dental Cosmos* 62:343–359.

Gregory, W. K. 1920b. Studies in comparative myology and osteology. IV. A review of the evolution of the lachrymal bone of vertebrates with special reference to that of mammals. *Bull. Amer. Mus. Nat. Hist.* 42:95–263.

Gregory, W. K. 1920c. On the structure and relations of *Notharctus*, an American Eocene primate. *Mem. Amer. Mus. Nat. Hist.* 3:45–243.

Gregory, W. K. 1922a. The origin and evolution of the human dentition. *J. Dent. Res. N.Y.* 2:89–175; 357–415; 607–717; 3:881.

Gregory, W. K. 1922b. *The Origin and Evolution of the Human Dentition*, Williams and Wilkins, Baltimore.

Gregory, W. K. 1927a. Two views on the origin of man. *Science* 65:601–605.

Gregory, W. K. 1927b. How near is the relationship of man to the chimpanzee-gorilla stock? *Quart. Rev. Bio.* 2:549–560.

Gregory, W. K. 1927c. The origin of man from the anthropoid stem—when and where? *Proc. Amer. Philos. Soc.* 66:439–463.

Gregory, W. K. 1927d. Dawn-man or ape? *Sci. Amer.* 137:230–232.

Gregory, W. K. 1927e. Did man originate in Central Asia? *Sci. Mnthly.* 24:385–401.

Gregory, W. K. 1928a. Were ancestors of man primitive brachiators? *Proc. Amer. Philos. Soc.* 67:129–150.

Gregory, W. K. 1928b. The upright posture of man: a review of its origin and evolution. *Proc. Amer. Philos. Soc.* 67:339–376.

Gregory, W. K. 1928c. Reply to Professor Wood-Jones's note "Man and the anthropoids". *Amer. J. Phys. Anthro.* 12:253–256.

Gregory, W. K. 1929. Is the pro-dawn man a myth? *Hum. Biol.* 1:153–165.

Gregory, W. K. 1930a. A critique of Professor Frederic Wood-Jones's paper "Some landmarks in the phylogeny of the primates". *Hum. Biol.* 2:99–108.

Gregory, W. K. 1930b. A critique of Professor Osborn's theory of human origin. *Amer. J. Phys. Anthro.* 14:133–161.

Gregory, W. K. 1930c. The origin of man from a brachiating anthropoid stock. *Science* 71:645–650.

Gregory, W. K. 1933. The new anthropogeny: twenty-five stages of vertebrate evolution from Silurian chordate to man. *Science* 77:29–40.

Gregory, W. K. 1934. *Man's Place Among the Anthropoids*, Claredon, Oxford.

Gregory, W. K. 1949. The bearing of the Australopithecinae upon the problem of man's place in nature. *Amer. J. Phys. Anthro.* 7:485–512.

Gregory, W. K., and Hellman, M. 1926. The dentition of *Dryopithecus* and the origin of man. *Anthro. Papers Amer. Mus. Nat. Hist.* 28:1–123.

Gregory, W. K., Hellman, M., and Lewis, G. E. 1938. Fossil anthropoids of the Yale-Cambridge India Expedition of 1935. *Carnegie Inst. Wash. Publ.* No. 495:1–27.

Groves, C. P. 1970. The forgotten leaf-eaters and the phylogeny of the Colobinae, in: *Old World Monkeys* (J. R. Napier and P. H. Napier, eds.), pp. 555–586, Academic Press, New York.

Groves, C. P. 1971. Distribution and place of origin of the gorilla. *Man* 6:44–51.

Groves, C. P. 1972a. Phylogeny and classification of primates, in: *Pathology of Simian Primates* (R. N. T.-W.-Fiennes, ed.), pp. 11–81, Karger, Basel.

Groves, C. P. 1972b. Systematics and phylogeny of gibbons, in: *Gibbon and Siamang*, Vol. 1 (D. M. Rumbaugh, ed.), pp. 1–80, Karger, Basel.

Grzimek, H. C. B., ed. 1975. *Animal Life Encyclopedia*, Van Nostrand Reinhold, New York.

Guilday, J. E. 1967. Differential extinction during the late Pleistocene and Recent times, in: *Pleistocene Extinctions* (P. Martin and H. Wright, eds.), pp. 121–140, Yale University Press, New Haven.

Guthrie, D. A. 1963. The carotid circulation in the Rodentia. *Bull. Mus. Comp. Zool.* 128(10):457–481.

Guthrie, D. A. 1969. The carotid circulation in the Aplodontia. *J. Mammal.* 50:1–7.

Haddow, A. J. 1952. Field and laboratory studies on an African monkey, *Cercopithecus ascanius schmidti* Matschie. *Proc. Zool. Soc. London* 122:297–394.

Haddow, A. J., Smithburn, K. C., Mahafey, A. F., and Bugher, J. C. 1947. Monkeys in relation to yellow fever in Bwanba County, Uganda. *Trans. R. Soc. Trop. Med. Hyg.* 40:677–700.

Haeckel, E. 1866. *Generelle Morphologie*, Reimer, Berlin.

Haeckel, E. 1868. *Naturliche Schopfungsgeschichte*, Reimer, Berlin.

Haeckel, E. 1874. *Anthropogenie oder Entwickelungsgeschichte des Menschen*, Englemann, Leipzig.

Heffer, J. 1969. Speciation in Amazonian forest birds. *Science* 165:131–137.

Haines, R. W. 1950. The interorbital septum in mammals. *J. Linn. Soc. London* 41:585–607.

Haines, R. W. 1955. The anatomy of the hand of certain insectivores. *Proc. Zool. Soc. London* 125:761–777.

Haines, R. W. 1958. Arboreal or terrestrial ancestry of placental mammals. *Quart. Rev. Biol* 33:1–23.

Hall, K. R. L. 1962. The sexual, agnostic and derived social behavior patterns of the wild chacma baboon, *Papio ursinus. Proc. Zool. Soc. London* 139:283–327.

Hall, K. R. L. 1965. Behavior and ecology of the wild Patas monkey, *Erythrocebus patas,* in Uganda. *J. Zool.* 148:15–87.

Hall, K. R. L., Boelkins, R. C., and Goswell, M. J. 1965. Behavior of patas monkeys *(Erythrocebus patas)* in captivity, with notes on the natural habitat. *Folia Primatologica* 3:22–49.

Hallam, A. 1974. Avian speciation in tropical South America. *Publ. Nuttall Orn. Club,* No. 14.

Hallam, A. 1975. Evolutionary size increase and longevity in Jurassic bivalves and ammonites. *Nature (Lond.)* 258:493–496.

Halstead, B. 1980. Popper: good philosophy, bad science? *New Sci.* 87:215–217.

Hamilton, W. J., Boyd, J. D., and Mossman, H

W. 1952. *Human Embryology,* Williams and Wilkins, Baltimore.

Hamilton, W. R. 1973. North African lower Miocene rhinoceroses. *Bull. Brit. Mus. Nat. Hist. Geol.* 24:349–395.

Hamilton, W. R. 1978. Fossil giraffes from the Miocene of Africa and a revision of the phylogeny of Giraffoidea. *Phil. Trans. Roy. Soc.* 283:165–229.

Hamilton, W. R., Whybrow, P. J., and McClure, H. A. 1978. Fauna of fossil mammals from the Miocene of Saudi Arabia. *Nature (Lond.)* 273:248–249.

Hamlett, G. W. D., and Wislocki, G. B. 1934. A proposed classification for types of twins in mammals. *Anat. Rec.* 61:81–96.

Hampton, S. H. 1974. Placental development in the marmoset, in: *Contemporary Primatology* (S. Kondo, M. Kawai and A. Ehara, eds.), pp. 106–114, Karger, Basel.

Harding, R. S., and Teleki, G., eds. 1981. *Omnivorous Primates: Gathering and Hunting in Human Evolution,* Columbia University Press, New York.

Harrington, J. E. 1978. Development of behavior in *Lemur macaco* in the first nineteen weeks. *Folia Primatologica* 29:107–128.

Harrison, J. L. 1951. Squirrels for birdwatchers. *Malay. Nat. J.* 5:134.

Harrison, T. 1982. *Small-Bodied Apes from the Miocene of East Africa,* Ph.D. Dissertation, University of London.

Harrisson, B. 1963. Trying to breed *Tarsius. Malay. Nat. J.* 17:218–231.

Hartenberger, J. L. 1973. Les rongeurs de l'Eocene d'Europe. Leur evolution dans leur cadre biogeographique. *Bull. Mus. Nat. Hist. Nat., Ser. 3* 132:49–70.

Hartenberger, J. L. 1975. Nouvelles decouvertes de rongeurs dans le Deseadien (Oligocene inferieur) de Salla Luribay (Bolivie). *C. R. Acad. Sci. Paris.. Ser. D* 280:427–430.

Harting, J. K., Glendenning, K. K., Diamond, I. T., and Hall, W. C. 1973. Evolution of the primate visual system: anterograde degeneration studies of the tecto-pulvinar system. *Amer. J. Phys. Anthro.* 38:383–392.

Harvey, P. H., and Clutton-Brock, T. H. 1978. Sexual dimorphism in primate teeth. *J. Zool. London* 186:475–485.

Hasek, M., Lengerova, A., and Hraba, T. 1961. Transplantation immunity and tolerance, in: *Advances in Immunology, Vol. 1.* (W. H. Taliaferro and J. H. Humphrey, eds.), pp. 1–66, Academic Press, New York.

Hatt, R. T. 1929. The red squirrel: Its life history and habits, with special reference to the Adirondacks of New York and the Harvard Forest. *Bull. N. Y. State Coll. For. (Syracuse)* 2:1–46.

Heaney, L. R. 1978. Island area and body size of insular mammals: evidence from the tricolor squirrel (*Callosciurus prevosti*) of Southeast Asia. *Evolution* 32(1):29–44.

Heaney, L. r. n.d. Correlates, predictors, and determinants of body size in North and Central American tree squirrels (*Sciurus* and *Tamiasciurus*). no reference cited.

Heiple, K. G., and Lovejoy, C. O. 1971. The distal femoral anatomy of *Australopithecus. Amer. J. Phys. Anthro.* 35:75–84.

Heller, F. 1930. Die Saugetierfauna der mitteleozanen Braunkohle des Geiseltales bei Halle a.S. *Jahrbk. des Hallescher Verbandes, Halle (Neue folge)* 9:13–41.

Hemmings, W. A., and Brambell, F. W. R. 1961. Protein transfer across foetal membranes. *Brit. Med. Bull.* 17:96–101.

Hennig, E. 1948. Quartarfaunen und Urgeschichte Ostafrikas. *Naturw. Rdsch. Jahrg.* 1 (Heft 5):212–217.

Hennig, W. 1950. *Grundzuge einer Theorie von phylogenetischen Systematik,* Deutscher Zentralverlag, Berlin.

Hennig, W. 1965. Phylogenetic systematics. *Annu. Rev. Entomol.* 10:97–116.

Hennig, W. 1966a. *Phylogenetic Systematics,* University of Illinois Press, Urbana.

Hennig, W. 1966b. the Diptera fauna of New Zealand as a problem in systematics and zoogeography. *Pac. Insects Mongr.* 9:1–81.

Hershkovitz, P. 1970a. Cerebral fissure patterns in platyrrhine monkeys. *Folia Primatologica* 13:213–240.

Hershkovitz, P. 1970b. Notes on Tertiary platyrrhine monkeys and description of a new genus from the Late Miocene of Columbia. *Folia Primatologica* 12:1–37.

Hershkovitz, P. 1972. The recent mammals of the Neotropical region: a zoographic and ecological review, in: *Evolution, Mammals, and Southern Continents* (A. Keast, F. C. Erk and B. Glass, eds.), pp. 311–431, State University of New York Press, Albany.

Hershkovitz, P. 1974a. A new genus of late Oligocene monkey (Cebidae, Platyrrhini) with notes on postorbital closure and platyrrhine evolution. *Folia Primatologica* 21:1–35.

Hershkovitz, P. 1974b. The ectotympanic bone and origin of higher primates. *Folia Primatologica* 22:237–242.

Hershkovitz, P. 1977. *Living New World Monkeys (Platyrrhini), with an Introduction to Primates, Vol. 1,* University of Chicago, Chicago.

Herz, N. 1977. Timing of spreading in the South Atlantic: Information from Brazilian alkalic rocks. *Geol. Soc. Am. Bull.* 88:101–112.

Hewes, G. W. 1961. Food transport and the origin of hominid bipedalism. *Am. Anthropol.* 63:687–710.

Hickman, V. V., and Hickman, J. L. 1960. Notes on the habits of the Tasmanian dormouse phalangers *Cercaertus nanus* (Desmarest) and *Eudromicia lepida* (Thomas). *Proc. Zool. Soc. London* 135:365–374.

Hiimae, K. M., and Kay, R. F. 1972. Trends in the evolution of primate mastication. *Nature (Lond.)* 240:486–487.

Hiimae, K. M., and Kay, R. F. 1973. Evolutionary trends in the dynamics of primate mastication. *Symp. 4th Int. Congr. Primatol.* 3:28–64.

Hildebrand, M. 1967. Symmetrical gaits of Primates. *Amer. J. Phys. Anthro.* 26:119–130.

Hildebrand, M. 1974. *Analysis of Vertebrate Structure,* John Wiley, New York.

Hill, C. A. 1973. The frequency of multiple births in the genus *Lemur. Mammalia* 37(1):101–104.

Hill, J. P. 1919. The affinities of *Tarsius* from the embryological aspect. *Proc. Zool. Soc. London* 1919:476–491.

Hill, J. P. 1932. The developmental history of the Primates. *Phil. Trans. Roy. Soc., Ser. B* 221:45–178.

Hill, W. C. O. 1936. The affinities of the lorisoids. *Ceylon J. Sci. (B)* 19:288–314.

Hill, W. C. O. 1949. The giant squirrels. *Zoo Life* 4:98–100.

Hill, W. C. O. 1953a. The blood-vascular system of *Tarsius*. *Proc. Zool. Soc. London* 123:655–692.

Hill, W. C. O. 1953b. *Primates: Comparative Anatomy and Taxonomy, Vol. 1. Strepsirhini,* University Press, Edinburgh.

Hill, W. C. O. 1955. *Primates: Comparative Anatomy and Taxonomy, Vol. 2. Haplorhini: Tarsoidea,* University Press Edinburgh.

Hill, W. C. O. 1957. *Primates: Comparative Anatomy and Taxonomy, Vol. 3. Hapalidae,* University Press, Edinburgh.

Hill, W. C. O. 1960. *Primates: Comparative Anatomy and Taxonomy, Vol. 4. Cebidae, Part A,* University Press, Edinburgh.

Hill, W. C. O. 1962. *Primates: Comparative Anatomy and Taxonomy, Vol. 5. Cebidae, Part B,* University Press, Edinburgh.

Hill, W. C. O. 1966. *Primates: Comparative Anatomy and Taxonomy, Vol. 6. Cercopithecinae,* University Press, Edinburg.

Hill, W. C. O. 1953–1966. *Primates: Comparative Anatomy and Taxonomy, Vol. 1–6,* University Press, Edinburgh.

Hill, W. C. O., Porter, A., and Southwick, M. D. 1952. The natural history, endoparasites and pseudoparasites of the tarsiers *(Tarsius cabonarius)* recently living in the Society's menagerie. *Proc. Zool. Soc. London* 122:79–119.

Himwich, H. E. 1951. *Brain Metabolism and Cerebral Disorders,* Williams and Wilkins, Baltimore.

Hladik, C. M. 1975. Ecology, diet and social patterning in Old and New World Primates, in: *Socioecology and Psychology of Primates* (R. Tuttle, ed.), pp. 3–35, Mouton, The Hague.

Hladik, C. M. 1977. Chimpanzees of Gabon and chimpanzees of Gombe: Some comparative data on the diet, in: *Primate Ecology* (T. H. Clutton-Brock, ed.), pp. 481–503, Academic Press, New York.

Hladik, C. M. 1978. Adaptive strategies of Primates in relation to leaf eating, in: *The Ecology of Arboreal Folivores* (G. G. Montgomery, ed.), pp. 373–395, Smithsonian Institute, Washington, D.C.

Hladik, C. M., Hladik, A., Bousset, J., Valdebouze, P., Viroben, G., and Delort-Laval, J. 1971. Le regime alimentaire des primates de l'ile de Barro-Colorado (Panama). Resultats des analyses quantitatives. *Folia Primatologica* 16:85–122.

Hockett, C. F., and Ascher, R. 1964. The human revolution. *Curr. Anthro.* 5:135–168.

Hofer, H. 1957. Uber das Spitzhornchen. *Natur. Volk.* 87:145–155.

Hofer, H. 1979. The external nose of *Tarsius bancanus borneanus* Horsfield, 1821 (Primates, Tarsiiformes). *Folia Primatologica* 32:180–192.

Hofer, H. O., and Wilson, J. A. 1967. An endocranial cast of an early Oligocene primate.

Folia Primatologica 5:148–152.

Hoffstetter, R. 1954. Les Mammiferes fossiles de l'Amerique du Sud et la biogeographie. *Rev. Gen. Sci.* 51(11–12):348–378.

Hoffstetter, R. 1968. Un gisement de Mammiferes deseadiens (Oligocene inferieur) en Bolivie. *C. R. Acad. Sci. (Paris) Ser. D.* 267:1095–1097.

Hoffstetter, R. 1969. Un primate de l'Oligocene inferieur sud-americain: *Branisella boliviana* gen. et sp. nov. *C. R. Acad. Sci. (Paris) Ser. D.* 269:434–437.

Hoffstetter, R. 1970a. Radiatin initiale des Mammiferes Placentaires et biogeographie. *C. R. Acad. Sci. (Paris) Ser. D.* 270:3027–3030.

Hoffstetter, R. 1970b. L'historire biogeographique des Masupiaux et la dichotomie Marsupiauz-Placentaires. *C. R. Acad. Sci. (Paris) Ser. D.* 271:388–391.

Hoffstetter, R. 1971. Le peuplement mammalien de l'Amerique du Sud. Role des continents austraux comme centres d'origine, de diversification et de dispersion pour certains groupes mammaliens. An. 1st Simp. Brasil. Paleont. (Rio de Janeiro, Sept. 1970) = *Suppl. An. Acad. Brasil. Ciencias.* 43:125–144.

Hoffstetter, R. 1972. Relationships, origins, and history of the ceboid monkeys and caviomorph rodents: a modern reinterpretation, in: *Evolutionary Biology, Vol. 6* (Th. Dobzhansky, M. K. Hecht, and W. C. Steere, eds.), pp. 323–347, Appleton-Century-Crofts, New York. (See Chapter 14, this volume)

Hoffstetter, R. 1974. Phylogeny and geographical deployment of the primates. *J. Hum. Evol.* 3:327–350.

Hoffstetter, R. 1977a. Origine et principales dichotonies des Primates Simiiformes (= Anthropoidea). *C. R. Acad. Sci. (Paris) Ser. D.* 284:2095–2098.

Hoffstetter, R. 1977b. Phylogenie de primates: Confrontation des resultats obtenus par les diverses voies d'approche du probleme. *Bull. M. Soc. Anthro. (Paris) Ser. 13,* 4:327–346.

Hoffstetter, R. 1980. Origin and deployment of the New World Monkeys emphasizing the southern continents route, in: *Evolutionary Biology of the New World Monkeys and Continental Drift* (R. L. Ciochon and A. B. Chiarelli, eds.), pp. 103–122, Plenum Press, New York.

Hoffstetter, R., and Lavocat, R. 1970. Decouverte dans le Desedien de Bolivie de genres pentalophodontes appuyabt les affinites africaines des Rongeurs Caviomorphes. *C. R. Acad. Sci. (Paris) Ser. D.* 271:172–175.

Holcombe, T. L., and Moore, W. S. 1977. Paleocurrents in the eastern Caribbean: geologic evidence and implications. *Mar. Geol.* 23:35–56.

Holloway, R. L. 1966a. Cranial capacity, neural reorganization and hominid evolution: a search for more suitable parameters. *Amer. Anthropol.* 68:103–121.

Holloway, R. L. 1966b. Structural reduction through the probable mutation effect: a critique with questions regarding human evolution. *Amer. J. Phys. Anthro.* 25:7–11.

Holloway, R. L. 1967. Tools and teeth: some speculations regarding canine reduction.

Amer. Anthropol. 69:63–67.

Holloway, R. L. 1969. Culture: a human domain. *Curr. Anthro.* 10:395–407.

Holloway, R. L. 1972. Australopithecine endocasts, brain evolution in the Hominoidea, and a model of hominid evolution, in: *Functional and Evolutionary Biology of Primates* (R. Tuttle, ed.), pp. 185–203, Aldine, Chicago.

Holloway, R. L. 1976. Paleoneurological evidence for language origins, in: *Origins and Evolution of Language and Speech* (S. R. Harnad, H. Steklis and J. Lancaster, eds.) *Ann. N. Y. Acad. Sci.* 280(7):330–348.

Holloway, R. L. 1980. Within-species brain-body weight variability: a reexamination of the Danish data and other primate species. *Amer. J. Phys. Anthro.* 53:109–121.

Hooijer, D. A. 1963. Miocene mammalia of Congo. *Ann. Mus. Roy. Afr. Cent., Ser. 8 vo., Sci. Geol.* 46:1–77.

Hooijer, D. A. 1967. Indo-Australian insular elephants. *Genetica* 38:143–162.

Hooijer, D. A. 1970. Miocene mammalia of Congo, a correction. *Ann. Mus. Roy. Afr. Cent., Ser. 8 vo., Sci. Geol.* 67:163–167.

Hooijer, D. A., and Colbert, E. H. 1951. A note on the Plio-Pleistocene boundary in the Siwalik Series of India and in Java. *Amer. J. Sci.* 249:533–538.

Hooton, E. A. 1946. *Up from the Ape,* Macmillan, New York.

Hopkins, B. 1965. *Forest and Savanna.* Heinemann, London.

Hopson, J. A. 1969. The origin and adaptive radiation of mammal-like reptiles and nontherian mammals. *Ann. N. Y. Acad. Sci.* 167:199–216.

Hopwood, A. T. 1933. Miocene primates from Kenya. *Linn. Soc. (Zool.)* 38:437–464.

Hopwood, A. T. 1947. The generic names of the mandrill and baboon, with notes on some of the genera of Brisson, 1762. *Proc. Zool. Soc. London* 11:533–536.

Horr, D. A. 1979. The Borneo Orang-utan, in: *Primate Ecology: Problem Oriented Field Studies* (R. W. Sussman, ed.), pp. 317–321, Wiley, New York.

Horsfield, T. 1851. *A Catalogue of the Mammalia in the Museum of the Hon. East-India Company,* J. and H. Cox, London.

Hospers, J. 1966. What is explanation?, in: *Essays in Conceptual Analysis* (A. Flew, ed.), pp. 94–119, Macmillan, London.

Howell, A. B. 1926. Voles of the genus *Phenacomys. N. Amer. Fauna* No. 48:1–66.

Howell, A. B. 1944. *Speed in Animals: Their Specialization for Running and Leaping,* University Press, Chicago.

Howell, F. C. 1959. The Villafranchian and human origins. *Science* 130:831–844.

Howell, F. C. 1967. Recent advances in human evolutionary studies. *Quart. Rev. Biol.* 42:471–513.

Howell, F. C. 1968a. Review of *Olduvai Gorge, Vol. 2. The Cranium and Maxillary Dentition of Australopithecus (Zinjanthropus) boisei* (P. V. Tobias) *Amer. Anthropol.* 70:1028–1030.

Howell, F. C. 1968b. Omo research expedition. *Nature (Lond.)* 219:576–582.

Howell, F. C. 1969. Remains of Hominidae from

Pliocene Pleistocene formations in the lower Omo Basin Ethiopia. *Nature (Lond.)* 223:1234–1239.

Howell, F. C. 1972. Recent advances in human evolutionary studies, in: *Perspectives in Human Evolution* (S. L. Washburn and P. Dolhinow, eds.), pp. 51–128, Holt, Rinehart and Winston, New York.

Howell, F. C. 1978. Hominidae, in: *Evolution of African Mammals* (V. J. Maglio and H. B. S. Cooke, eds.), pp. 154–248, Harvard University Press, Cambridge.

Howell, F. C., and Wood, B. A. 1974. Early hominid ulna from the Omo basin, Ethiopia. *Nature (Lond.)* 249:174–176.

Hrdlicka, A. 1935. The Yale fossils of anthropoid apes. *Amer. J. Sci.* 29:34–40.

Hrdy, S. B. 1976. The care and exploitation of non-human primate infants by conspecifics other than the mother, in: *Advances in the Study of Behavior* (J. S. Rosenblatt, R. A. Hinde, E. Shaw, and C. Beers, eds.), pp. 101–158, Academic Press, New York.

Hsu, C. -H., Wang, L. -H., and Han, K. -X. 1975. Australopithecine teeth associated with *Gigantopithecus. Vert. Pal-Asiat.* 13:81–88.

Hsu, K. J. 1972. When the Mediterranean dried up. *Sci. Amer.* 227(6):26–36.

Hsu, K. J. 1978. When the Black Sea was drained. *Sci. Amer.* 228(5):53–63.

Hsu, K. J., Mantadert, L., Bernoulli, D., Cita, M. B., Erickson, A., Garrison, R. E., Kidd, R. B., Melieres, F., Muller, C., and Wright, R. 1977. History of the Mediterranean salinity crisis. *Nature (Lond.)* 267:399–403.

Hubel, D. H., and Wiesel, T. N. 1970. Stereoscopic vision in macaque Monkey. *Nature (Lond.)* 225:41–42.

Hubrecht, A. A. W. 1896. Die Keimblase von *Tarsius. Fest. C. Gegenbaur* 2:149–178.

Hubrecht, A. A. W. 1908. Early ontogenetic phenomena in mammals and their bearing on our interpretation of the phylogeny of the vertebrates. *Quart. J. Micro. Soc.* 53:1–181.

Hudson, G. E. 1932. On the food habits of *Marmosa. J. Mammal.* 13:159.

Hürzeler, J. 1947. *Alsaticopithecus leemanni* nov. gen. nov. spec., ein neuer Primate aus dem unteren Lutetien von Buchweiler (Unterelsass). *Ecl. Geol. Helv.* 40:343–356.

Hürzeler, J. 1948. Zur Stammesgeschichte der Necrolemurien. *Schweizerischen Palaeont. Abhand.* 66(1):1–46.

Hürzeler, J. 1949. Neubeschreibung von *Oreopithecus bambolii* Gervais. *Schweizerischer Palaeont. Abhand.* 66(5):1–20.

Hürzeler, J. 1954a. Contribution à l'odontologie et à la phylogènese du genre *Pliopithecus* Gervais. *Ann. Paleont.* 40:1–63.

Hürzeler, J. 1954b. Zur systematischen Stellung von *Oreopithecus. Verh. Naturf. Ges. Basel* 65:88–95.

Hürzeler, J. 1958. *Oreopithecus bambolii* Gervais, a preliminary report. *Verh. Naturf. Ges. Basel* 69:1–47.

Hürzeler, J. 1968. Questions et reflexions sur l'histoire des Anthromorphes. *Ann. Paleont.* 54:195–233.

Hutchinson, G. E. 1965. *The Ecological Theater and the Evolutionary Play,* Yale University Press, New Haven.

Huxley, J. S., Dobzhansky, T., Niebuhr, R.,

Rieser,, O. L., and Nikhilananda, S. 1958. *A Book that Shook the World, Anniversary Essays on Charles Darwin's Origin of Species,* University of Pittsburgh Press, Pittsburgh.

Huxley, T. H. 1861. On the zoological relations of man with the lower animals. *Nat. Hist. Rev.* 1:67–84.

Huxley, T. H. 1863. *Evidence as to Man's Place in Nature,* Williams and Norgate, London.

Huxley, T. H. 1880. On the application of the laws of evolution to the arrangement of the Vertebrata, and more particularly of the Mammalia. *Proc. Zool. Soc. London* 1880:649–663.

Huxley, T. H. 1958. *The Living Thoughts of Darwin,* Cassell, London.

Hylander, W. L. 1975. Incisor size and diet in anthropoids with special reference to Cercopithecidae. *Science* 189:1095–1098.

Hylander, W. L. 1979. The functional significance of primate mandibular form. *J. Morphol.* 160:223–240.

Ingram, V. M. 1961. Gene evolution and hemoglobins. *Nature (Lond.)* 189:704–708.

Issac, G. 1965. The Peninj beds: an early Middle Pleistocene formation west of Lake Natron. *Quaternaria* 7:101–130.

Isaac, G. 1967. The stratigraphy of the Peninj group—Early Middle Pleistocene formations west of Lake Natron, Tanzania, in: *Background to Evolution in Africa* (W. W. Bishop and J. D. Clark, eds.), pp. 229–257, University Press, Chicago.

Isaac, G. 1978. Food sharing and human evolution *J. Anthro. Res.* 34:311–325.

Isaac. G., and Isaac, B. 1975. Africa, in: *Varieties of Culture in the Old World* (R. Stigler, ed.), pp. 8–48, St. Martins, New York.

Izawa, K. 1978. A field study of the ecology and behavior of the black-mantle tamarin *(Saguinus nigricollis). Primates* 19:241–274.

Izawa, K. 1979. Foods and feeding behavior of wild black-capped capuchin *(Cebus apella). Primates* 20:57–76.

Jacobs, L. L., and Pilbeam, D. R. 1980. Of mice and men: fossil-based divergence dates and molecular "clocks". *J. Hum. Evol.* 9:551–555.

Janzen, D. H. 1978. Complications in interpreting the chemical defences of trees against tropical arboreal plant-eating vertebrates, in: *The Ecology of Arboreal Folivores* (G. G. Montgomery, ed.), pp. 73–84, Smithsonian Institute, Washington, D.C.

Janzen, D. H., and Higgins, M. L. 1979. How hard are *Enterolobium cyclocarpum* (Leguminosae) seeds? *Brenesia* No. 16:61–68.

Jarman, P. J. 1974. The social organization of antelope in relation to their ecology. *Behav.* 48:215–266.

Jelliffe, D. B., and Jelliffe, E. F. P. 1978. *Human Milk in the Modern World,* Oxford University Press, Oxford.

Jenkins, F. A., Jr. 1970. Limb movements in a monotreme *(Tachyglossus aculiatus):* a cineradiographic analysis. *Science* 168:1473–1475.

Jenkins, F. A., Jr. 1971. Limb posture in the Virginia opossum *(Didelphis marsupialis)* and

non-cursorial mammals. *J. Zool. (London)* 165:303–315.

Jenkins, F. A., Jr. 1973. The functional anatomy and evolution of the mammalian humero-ulnar articulation. *Amer. J. Anat.* 137:281–298.

Jenkins, F. A., Jr. 1974. Tree shrew locomotion and the origins of primate arborealism, in: *Primate Locomotion* (F. A. Jenkins, Jr., ed.), pp. 85–115, Academic Press, New York. (See Chapter 1, this volume)

Jenkins, F. A., Jr. 1981. Wrist rotation in primates: a critical adaptation for brachiators. *Symp. Zool. Soc. London* 48:429–451.

Jenkins, F. A., Jr., and Fleagle, J. G. 1975. Knuckle walking and the functional anatomy of the wrist in living apes, in: *Primate Functional Morphology and Evolution* (R. H. Tuttle, ed.), pp. 213–227, Mouton, the Hague.

Jensen, Ad. S., Sparck, R., and Volsoe, H., eds. 1941. *Herluf Winge: The Interrelationships of the Mammalian Genera, Vol. 2 Rodentia, Carnivora, Primates,* C. A. Reitzels, Copenhagen.

Jepsen, G. L. 1930. New vertebrate fossils from the lower Eocene of the Bighorn Basin, Wyoming. *Proc. Amer. Philos. Soc.* 69:117–131.

Jepsen, G. L., and Woodburne, M. O. 1969. Paleocene hyracothere from Polecat Bench Formation, Wyoming. *Science* 164:543–547.

Jerison, H. J. 1973. *Evolution of the Brain and Intelligence,* Academic Press, New York.

Jerison, H. J. 1979. Brain, body and encephalization in early Primates. *J. Human. Evol.* 8:615–635.

Johanson, D. C. 1974a. *An Odontological Study of the Chimpanzee with some Implications for Hominoid Evolution.* Doctoral Dissertation, University of Chicago.

Johanson, D. C. 1974b. Some metric aspects of the permanent and deciduous dentition of the pygmy chimpanzee *(Pan paniscus). Amer. J. Phys. Anthro.* 41:39–48.

Johanson, D. C. 1976. Ethiopia yields first "family" of early man. *Natl. Geogr.* 150:790–811.

Johanson, D. C. 1980. Early African hominid phylogenesis: a reevaluation, in: *Current Arguement on Early Man* (L. K. Konigsson, ed.), pp. 31–69, Pergamon, Oxford.

Johanson, D. C., and Coppens, Y. 1976. A preliminary anatomical diagnosis of the first Plio-Pleistocene hominid discoveries in the central Afar, Ethiopia. *Amer. J. Phys. Anthro.* 45:217–234.

Johanson, D. C., and Taieb, M. 1976. Plio-Pleistocene hominid discoveries in Hadar, Ethiopia. *Nature (Lond.)* 260:293–297.

Johanson, D. C., and White, T. D. 1979a. A systematic assessment of early African hominids. *Science* 203:321–330. (See Chapter 37, this volume)

Johanson, D. C., and White, T. D. 1979b. On the status of *Australopithecus afarensis. Science* 297:1104–1105.

Johanson, D. C., White, T. D., and Coppens, Y. 1978. A new species of the genus *Australopithecus* (Primates: Hominidae) from the Pliocene of Eastern Africa. *Kirtlandia* 28:1–14.

Johanson, D. C., Lovejoy, C. O., Burstein, A. H., and Heiple, K. G. 1976. Functional implications of the Afar knee joint. *Amer. J. Phys. Anthro.* 44:188. (Abstract)

Johanson, D. C., Taieb, M., and Coppens, Y. 1978. Expedition internationale de l'Afar Ethiopie (4 eme et 5 eme campagne 1975--1977): Nouvelles decouvertes d'hominides et de couvertes d'industries lithiques pliocene a Hadar. *C. R. Acad. Sci., Paris, Ser. D.,* 287:237–240.

Johanson, D. C., Taieb, M., Gray, B. T., and Coppens, Y. 1978. Geological framework of the Pliocene Hadar Formation (Afar, Ethiopia) with notes on paleontology, including hominids, in: *Background to Fossil Man* (W. W. Bishop, ed.), pp. 549–564, Scottish Academic Press, Edinburgh.

Johnson, B. D., Powell, C. McA., and Veevers, J. J. 1976. Spreading history of the eastern Indian Ocean and Greater India's northward flight from Antarctica and Australia. *Geol. Soc. Am. Bull.* 87:1560–1566.

Johnson, G. L. 1901. Contributions to the comparative anatomy of the mammalian eye, chiefly based on opthalmoscopic examination. *Phil. Trans. Roy. Soc. London* 194:1–82.

Johnson, S. C. 1981. Bonobos: generalized hominid prototypes or specialized insular dwarfs? *Curr. Anthro.* 22:;363–375.

Jolly, A. 1966. *Lemur Behavior,* University of Chicago Press, Chicago.

Jolly, A. 1972. *The Evolution of Primate Behavior,* Macmillan, New York.

Jolly, C. J. 1963. A suggested case of evolution by sexual selection in primates. *Man* 63:178–179.

Jolly, C. J. 1965. *The Origins and Specializations of the Long-Faced Cercopithecoidea.* Doctoral Dissertation, University of London.

Jolly, C. J. 1966. Introduction to the Cercopithecoidea, with notes on their use as laboratory animals. *Symp. Zool. Soc. London* 17:427–457.

Jolly, C. J. 1967. The evolution of the baboons, in: *The Baboon in Medical Research, Vol. 2* (Vagtborg, ed.), pp. 427–457, University of Texas, Austin.

Jolly, C. J. 1970a. The large African monkeys as an adaptive array, in: *Old World Monkeys* (J. R. Napier and P. H. Napier, eds.), pp. 141–174, Academic Press, New York.

Jolly, C. J. 1970b. The seed-eaters: a new model of hominid differention based on a baboon analogy. *Man* 5:5–28. (See Chapter 41, this volume)

Jolly, C. J. 1970c. *Hadropithecus,* a lemuroid small object feeder. *Man* 5:525–529.

Jolly, C. J. 1972. The classification and natural history of *Theropithecus (Simopithecus)* (Andrews, 1916), baboons of the African Plio-Pleistocene. *Bull. Brit. Mus. Nat. Hist. Geol.* 22:1–122.

Jolly, C. J. 1973. Changing views of hominid origins. *Yrbk. Phys. Anthro.* 16:1–17.

Jolly, C. J., Ed. 1978. *Early Hominids of Africa,* St. Martin's, New York.

Jordan, C., and Jordan, H. 1977. Versuche zur Symbol-Ereignis-Verknupfung bei einem Zwergs-chimpansen (*Pan paniscus* Schwarz 1929). *Primates* 18:515–529.

Jukes, T. H. 1980. Silent nucleotide substitutions and the molecular evolutionary clocks. *Science* 210:973–978.

Jukes, T. H., and Holmquist, R. 1972. Evolutionary clock: Nonconsistence of rate in different species. *Science* 177:531–532.

Jungers, W. L. 1977. Hindlimb and pelvis adaptations to vertical climbing and clinging in *Megaladapis,* a giant subfossil prosimian from Madagascar. *Yrbk. Phys. Anthro.* 20:508–524.

Jungers, W. L. 1978a. The functional significance of skeletal allometry in *Megaladapis* in comparison to living prosimians. *Amer. J. Phys. Anthro.* 49:303–314.

Jungers, W. L. 1978b. On canine reduction in early hominids. *Curr. Anthro.* 19:155–156.

Jungers, W. L. 1979. Locomotion, limb proportions and skeletal allometry in lemurs. *Folia Primatologica* 32:8–28.

Jungers, W. L. 1980. Adaptive diversity in subfossil Malagasy prosimians. *Z. Morphol. Anthropol.* 71(2):177–186. (See Chapter 11, this volume)

Jungers, W. L. 1982. Lucy's limbs: skeletal allometry and locomotion in *Australopithecus afarensis. Nature (Lond.)* 297:676–678.

Kaas, J. H., Guillery, R. W., and Allman, J. M. 1972. Some principles of organization of the dorsal lateral geniculate nucleus. *Brain Behav. Evol.* 6:253–299.

Kälin, J. 1961. Sur les primates de l'Oligocene inferieur d'Egypte. *Ann. Paleontol.* 47:1–18.

Kälin, J. 1962. Uber *Moeripithecus markgrafi* Schlosser und die phyletischen Vorstufen der Bilophodontic der Cercopithecoidea. *Bibl. Primatol.* 1:32–42.

Kano, T. 1979. A pilot study on the ecology of pygmy chimpanzees, *Pan paniscus,* in: *The Great Apes: Perspectives on Human Evolution, Vol. 5* (D. A. Hamburg and F. R. McCown, eds.), pp. 123–135, Benjamin Cummings, Menlo Park.

Kaufmann, J. H. 1965a. A three year study of mating behavior in a free-ranging band of Rhesus monkeys. *Ecology* 46:500–512.

Kaufmann, J. H. 1965b. Studies on the behavior of captive tree shrews *(Tupaia glis). Folia Primatologica* 3:50–74.

Kay, R. F. 1973a. Humerus of robust *Australopithecus. Science* 182:396.

Kay, R. F. 1973b. *Mastication, Molar Teeth Structures, and Diet in Primates.* Doctoral Dissertation, Yale University.

Kay, R. F. 1975. The functional adaptations of primate molar teeth. *Amer. J. Phys. Anthro.* 43:195–215.

Kay, R. F. 1977a. The evolution of molar occlusion in the Cercopithecidae and early catarrhines. *Amer. J. Phys. Anthro.* 46:327–352.

Kay, R. F. 1977b. Diets of early Miocene hominoids. *Nature (Lond.)* 268:628–630.

Kay, R. F. 1977c. Post-Oligocene evolution of catarrhine diets. *Amer. J. Phys. Anthro.* 47:141–142.

Kay, R. F. 1978. Molar structure and diet in extant Cercopithecidae, in: *Development, Function and Evolution of Teeth* (P. Butler and K. A. Joysey, eds.), pp. 309–339, Academic Press, New York.

Kay, R. F. 1980. Platyrrhine origins: a reappraisal of the dental evidence, in: *Evolutionary Biology of the New World Monkeys and Continental Drift* (R. L. Ciochon and A. B. Chiarelli, eds.), pp. 159–187, Plenum Press, New York.

Kay, R. F. 1981. The nut-crackers: a new theory of the adaptations of the Ramapithecinae. *Amer. J. Phys. Anthro.* 55:141–151. (See Chapter 31, this volume)

Kay, R. F. 1982a. *Sivapithecus simonsi,* a new species of Miocene hominoid with comments on the phylogenetic status of the Ramapithecinae. *Int. J. Primatol.* 3:113–174.

Kay, R. F. 1982b. Sexual dimorphism in Ramapithecinae. *Proc. Natl. Acad. Sci. USA* 79:209–212.

Kay, R. F., and Cartmill, M. 1974. Skull of *Paleachthon nacimienti. Nature (Lond.)* 252:37–38.

Kay, R. F., and Cartmill, M. 1977. Cranial morphology and adaptation of *Paleachthon nacimienti* and other Paramomyidae (Plesiadapoidea, ?Primates), with a description of a new genus and species. *J. Hum. Evol.* 6:19–53. (See Chapter 4, this volume)

Kay, R. F., and Hiiemae, K. M. 1974a. Jaw movement and tooth use in recent and fossil primates. *Amer. J. Phys. Anthro.* 40:227–256.

Kay, R. F., and Hiiemae, K. M. 1974b. Mastication in *Galago crassicaudatus:* a cinefluorographic and occlusal study, in: *Prosimian Biology* (R. D. Martin, G. A. Doyle and A. C. Walker, eds.), pp. 501–530, Duckworth, London.

Kay, R. F., and Hylander, W. L. 1978. The dental structure of mammalian folivores with special reference to primates and Phalangeroidea (Marsupialia), in: *The Ecology of Arboreal Folivores* (G. G. Montgomery, ed.), pp. 173–191, Smithsonian Institute Press, Washington, D. C.

Kay, R. F., and Simons, E. L. 1980. The ecology of Oligocene African Anthropoidea. *Int. J. Primatol.* 1:22–37. (See Chapter 20, this volume)

Kay, R. F., and Simons, E. L. 1983. A reassessment of the relationship between late Miocene and subsequent Hominoidea, in: *New Interpretations of Ape and Human Ancestry* (R. L. Ciochon and R. S. Corruccini, eds.), pp. 577–624, Plenum Press, New York.

Kay, R. F., Fleagle, J. G., and Simons, E. L. 1981. A revision of the Oligocene apes from the Fayum Province, Egypt. *Amer. J. Phys. Anthro.* 55:293–322.

Kay, R. F., Sussman, R., and Tattersall, I. 1978. Dietary and dental variations in the genus *Lemur,* with comments concerning general dietary-dental correlations of Malagasy strepsirhines. *Amer. J. Phys. Anthro.* 49:119–128.

Keast, A. 1972. Continental drift and the evolution of the biota on southern continents, in: *Evolution, Mammals, and Southern Continents* (A. Keast, F. C. Erk, and B. Glass, eds.), pp. 23–87, State University of New York Press, Albany.

Keast, A. 1977. Zoogeography and phylogeny: the theoretical background and methodology to the analysis of mammal and bird fauna, in: *Major Patterns in Vertebrate Evolution* (P. C. Goody and B. M. Hecht, eds.), pp. 249–312, Plenum Press, New York.

Keast, A., Erk, F. C., and Glass, B., eds. 1972. *Evolution, Mammals and Southern Continents,* State University of New York Press, Albany.

Keith, A. 1903. The extent to which the posterior

segments of the body have been transmuted and suppressed in the evolution of man and allied primates. *J. Anat. Physio.* 37:18–40.

Keith, A. 1912. Certain phases in the evolution of man. *Brit. Med. J.* 1:734–737, 788–790.

Keith, A. 1914. *The Antiquity of Man,* Williams and Norgate, London.

Keith, A. 1923. Man's posture: its evolution and disorders. *Brit. Med. J.* 1:451–454, 499–502, 545–548, 587–590, 624–626, 669–672.

Keith, A. 1925. The fossil anthropoid ape from Taungs. *Nature (Lond.)* 115:234–235.

Keith, A. 1927. *Concerning Man's Origin,* Watts, London.

Keith, A. 1929. *The Antiquity of Man, Vol. 1 and 2,* Williams and Norgate, London.

Keith, A. 1934. *The Construction of Man's Family Tree,* Watts, London.

Keith, A. 1940. Fifty years ago. *Amer. J. Phys. Anthro.* 26:251–267.

Kellogg, D. E., and Hays, J. D. 1975. Microevolutionary patterns in late Cenozoic Radiolaria. *Paleobiol.* 1:150–160.

Kendrew, J. C., Dickerson, R. E., Strandberg, B. E., Hart, R. G., Davies, D. R., Phillips, D. C., and Shore, V. C. 1960. Structure of myoglobin, a three dimensional fourier synthesis at 2A resolution obtained by x-ray analysis. *Nature (Lond.)* 185:422–427.

Kennedy, G. E. 1978. Hominoid habitat shifts in the Miocene. *Nature (Lond.)* 271:11.

Kennedy, W. J., and Cooper, M. 1975. Cretaceous ammonite distributions and the opening of the South Atlantic. *Geol. Soc. London J.* 131:283–288.

Kent, P. E. 1972. Mesozoic history of the east coast of Africa. *Nature (Lond.)* 238:147–148.

Kern, H. M., and Straus, W. L. 1949. The femur of *Plesianthropus transvaalensis. Amer. J. Phys. Anthro.* 7:53–77.

Kiltie, R. 1979. Nutcrackers on the hoof: bite force and canine function in rainforest peccaries. Manuscript.

Kimura, T. 1968. Evolutionary rate at the molecular level. *Nature (Lond.)* 217:624–626.

King, J. L., and Jukes, T. H. 1969. Neo-Darwinian evolution. *Science* 164:788–798.

King, M. C., and Wilson, A. C. 1975. Evolution at two levels in humans and chimpanzees. *Science* 188:107–116.

Kingdon, J. 1971. *East African Mammals, Vol. 1,* Academic Press, New York.

Kinsey, A. C. 1936. *The Origin of Higher Categories in Cynips,* Indiana University Publ., Sci. Ser., No. 4.

Kinzey, W. G. 1971. Evolution of the human canine tooth. *Amer. Anthropol.* 73:680–694.

Kinzey, W. G. 1973. Reduction of the cingulum in Ceboidea, in: *Craniofacial Biology of Primates, Vol. 3* (M. A. Zingeser, ed.), pp. 101–127, Karger, Basel.

Kinzey, W. G. 1974. Ceboid models for the evolution of the hominoid dentition. *J. Hum. Evol.* 3:193–203.

Kinzey, W. G. 1978. Feeding behavior and molar features in two species of Titi monkey, in: *Recent Advances in Primatology, Vol. 1* (D. J. Chivers and J. Herbert, eds.), pp. 373–385, Academic Press, London.

Kinzey, W. G., Rosenberger, A. L., and Ramirez, M. 1975. Vertical clinging and leaping in a neotropical anthropoid. *Nature (Lond.)* 255:327–328.

Kirsch, J. A. W. 1968. Prodromus of the comparative serology of Marsupialia. *Nature (Lond.)* 217:418–470.

Kitts, D. B. 1977. Karl Popper, verifiability, and systematic zoology. *Syst. Zool.* 26:185–194.

Klaatsch, H. 1923. *The Evolution and Progress of Man* (A. Heilborn, ed.) T. Fischer Unwin, London.

Kloss, C. B. 1903. *In the Andamans and Nicobars,* John Murray, London.

Kloss, C. B. 1911. On a collection of mammals and other vertebrates from the Trengganu Archipelago. *J. Fed. Malay. States* 4:175–212.

Kluge, A. G. 1983. Cladistics and the classification of the Great Apes, in: *New Interpretations of Ape and Human Ancestry* (R. L. Ciochon and R. S. Corruccini, eds.), pp. 151–177, Plenum Press, New York.

Koford, C. B. 1963. Rank of mothers and sons in bands of Rhesus monkeys. *Science* 141:356. (Abstract)

Koford, C. B. 1965. Population dynamics of Rhesus monkeys on Cayo Santiago, in: *Primate Behavior: Field Studies of Monkeys and Apes* (I. DeVore, ed.), pp. 160–174, Holt, Rinehart and Winston, New York.

Kohne, D. E., Chiscon, J. A., and Hoyer, B. H. 1970. Nucleotide sequence change in nonrepeated DNA during evolution. *Carnegie Inst. Washington Yrbk.* 69:488–501.

Kohne, D. E., Chiscon, J. A., and Hoyer, B. H. 1972. Evolution of primate DNA: a summary, in: *Perspectives on Human Evolution,* Vol. 2 (S. L. Washburn and P. Dolhinow, eds.), pp. 166–168, Holt, Rinehart and Winston, New York.

Kolodny, R. C., Jacobs, L. S., and Daughaday, W. H. 1972. Mammary stimulation causes prolactin secretion in non-lactating women. *Nature (Lond.)* 238:284–285.

Kortlandt, A. 1962. Chimpanzees in the wild. *Sci. Amer.* 206:128–138.

Kortlandt, A. 1967. Experimentation with chimpanzees in the wild, in: *Progress in Primatology* (D. Starck, *et al.,* eds.), pp. 208–224, Gustav Fischer, Stuttgart.

Kortlandt, A. 1968. Handgebrauch bei ferilebenden Schimpansens, in: *Handgebrauch und Verstandigung bei Affen und Fruhmenschen* (B. Rensch, ed.), pp. 59–102, Hans Huber, Bern.

Kortlandt, A. 1972. *New Perspectives on Ape and Human Evolution,* University of Amsterdam, Amsterdam.

Kortlandt, A. 1980. The Fayum primate forest. Did it exist? *J. Hum. Evol.* 9:277–297.

Kortlandt, A. 1983. Facts and fallacies concerning Miocene ape habitats, in: *New Interpretations of Ape and Human Ancestry* (R. L. Ciochon and R. S. Corruccini, eds.), pp. 465–515, Plenum Press, New York.

Kraglievich, J. L. 1951. Contribuciones al conocimiento de los primates fosiles de la Patagonia. I. Diagnosis previa de un nuevo primate fosil del Ologiceno superior (Colhuehuapiano) de Gaiman, Chubut. *Com. Inst. Nac. Invest. Ciene. Nat. (Buenos Aires) Cienc. Zool.* 11(5):55–82.

Kramp, V. P. 1956. Serologische stammbaumforschung, in: *Primatologia. I. Systematik Phylogenie Ontogonie* (H. Hofer, A. H. Schultz and D. Starck, eds.), pp. 1015–1034, Karger, Basel.

Krantz, G. S. 1963. The functional significance of the mastoid process in man. *Amer. J. Phys. Anthro.* 21:591–593.

Kretzoi, M. 1975. New ramapithecines and *Pliopithecus* from the lower Pliocene of Rudabanya in north-eastern Hungary. *Nature (Lond.)* 257:578–581.

Kretzoi, M. 1976. Emberre Valas es az Australopithecinak. *Anthro. Kozlemenyek* 20:3–11.

Krishtalka, L. 1978a. Paleontology and geology of the Badwater Creek area, central Wyoming. Part 15. Review of the late Eocene primates from Wyoming and Utah, and the Plesitarsiiformes. *Ann. Carnegie Mus.* 47:335–360.

Krishtalka, L. 1978b. Review of *Systematics of the Omomyidae (Tarsiiformes, Primates) Taxonomy, Phylogeny and Adaptations. J. Mammal.* 59:901–903.

Krishtalka, L., and Schwartz, J. H. 1978. Phylogenetic relationships of Plesiadapiform-Tarsiiform Primates. *Ann. Carnegie Mus.* 47:515–540.

Kuhn, H. J. 1964. Zur Kenntnis von Bau und Funktion des Magens der Schlankaffen (Colobinae). *Folia Primatologica* 2:193–221.

Kuhn, T. S. 1970. *The Structure of Scientific Revolutions,* University of Chicago Press, Chicago.

Kumar, N., and Embley, R. W. 1977. Evolution and origin of Ceara Rise: an aseismic rise in the western equatorial Atlantic. *Geol. Soc. Am. Bull.* 88:683–694.

Kummer, H. 1967a. *Social Organization of Hamadryas Baboons, Biblioteca Primatol.* 6, Karger, Basel.

Kummer, H. 1967b. Tripartite relations in Hamadryas baboons, in: *Social Communication Among Primates* (S. Altman, ed.), pp. 63–72, University of Chicago Press, Chicago.

Kummer, H. 1971. *Primate Societies,* Aldine, Chicago.

Kurten, B. 1965. The Carnivora of the Palestine caves. *Acta Zool. Fennica* 107:1–74.

Kurten, B. 1966. Holarctic land connections in the early Tertiary. *Comment. Biol. Soc. Sci. Fenn.* 29(5):1–5.

Kurten, B. 1968. *Pleistocene Mammals of Europe,* Aldine, Chicago.

Kurten, B. 1972a. *The Age of Mammals,* Columbia University Press, New York.

Kurten, B. 1972b. *Not from the Apes,* Vantage Books, New York.

Lack, D. 1968. *Ecological Adaptations for Breeding in Birds,* Mentheun, London.

Ladd, J. W. 1976. Relative motion of South America with respect to North America and Caribbean tectonics. *Geol. Soc. Sm. Bull.* 87:969–976.

Ladd, J. W., Dickson, G. O., and Pittman, W. C., III. 1973. The age of the South Atlantic, in: *The Ocean Basins and Margins, Vol. 1 The South Atlantic* (A. E. M. Nairn and F. G. Stehli, eds.), pp. 555–573, Plenum Press, New York.

Lamberton, C. 1934. Contribution a la connaissance de la faune subfossile de Madagascar: Lemuriens et ratites. *Mem. Acad. Malagache* 17:19–22.

Lamberton, C. 1937. Contribution a la connaissance de la faune subfossile de Madagascar: Note 3. Les Hadropitheques. *Bull. Acad. Malgache* 20:1–44.

Lamberton, C. 1944–1945. Contribution a la connaissance de la faune subfossile de Madagascar. Bradytherium ou Paleopropitheque? *Bull. Acad. Malg.* 26:89–140.

Lamberton, C. 1946. Contribution a la connaissance de la faune subfossile de Madagascar. XX. Membre posterieur des Neopropitheques et des Mesopropitheques. *Bull. Acad. Malg.* 27:24–28.

Lamberton, C. 1956. Examen de quelques hypotheses de sera concernant les lemuriens fossils et actuels. *Bull. Acad. Malg.* 34:51–65.

Lancaster, J. B. 1972. Play-mothering: The relations between juvenile females and young infants among free ranging vervet monkeys, in: *Primate Socialization* (F. E. Poirier, ed.), pp. 83–104, Random House, New York.

Landry, S. O. 1957. The interrelationships of the New and Old World hystricomorph Rodents. *Univ. Calif. Publ. Zool.* 56:1–118.

Langdale-Brown, I., Osmasion, H. A., and Wilson, J. G. 1964. *The Vegetation of Uganda,* Entebbe, The Government Printer.

Langer, P. 1974. Stomach evolution in the Artiodactyla. *Mammalia* 38:295–314.

Larson, R. L., and Ladd, J. W. 1973. Evidence for the opening of the South Atlantic in the early Cretaceous. *Nature (Lond.)* 246: 209–212.

Latimer, B. N., White, T. D., Kimbel, W. H., Johanson, D. C., and Lovejoy, C. O. 1981. The pygmy chimpanzee is not a living missing link in human evolution. *J. Hum. Evol.* 10:475–488.

Lavocat, R. 1969. La systematique des rongeurs histricomorphes et la derive des continents. *C. R. Acad. Sci., Paris, Ser. D.,* 269:1496–1497.

Lavocat, R. 1971. Affinities systematiques des Caviomorphes et des Phiomorphes et origine africaine des Caviomorphes. An. I. Simposio Brasil. Paleont. (Rio de Janeiro, Sept. 1970)= *Suppl. An. Acad. Brasil. Ciencias.* 43:515–522.

Lavocat, R. 1973. Les Rongeurs du Miocene d'Afrique orientale. 1. Miocene inferieur. *Mem. Trav. E. P. H. E., Institut de Montpellier* 1:1–284.

Lavocat, R. 1974a. What is an hystricomorph?, in: *The Biology of Hystricomorph Rodents* (I. W. Rolands and B. J. Weir, eds.), pp. 7–20, Academic Press, London.

Lavocat, R. 1974b. The interrelationships between the African and South American rodents and their bearing on the problem of the origin of South American monkeys. *J. Hum. Evol.* 3:323–326.

Lavocat, R. 1977. Sur l'origine des faunes sudamericaines de Mammiferes du Mesozoique terminal et du Cenzoique ancien. *C. R. Acad. Sci. Paris, Ser. D.,* 285:1423–1426.

Lavocat, R. 1980. The implication of rodent paleontology and biogeography to the geographical sources and origin of platyrrhine primates, in: *Evolutionary Biology of the New World Monkeys and Continental Drift* (R. L. Ciochon and A. B. Chiarelli, eds.), pp. 93–102, Plenum Press, New York.

Layne, J. N., and Benton, A. H. 1954. Some speeds of small mammals. *J. Mammal.* 35: 103–104.

Leakey, L. S. B. 1959. A new fossil skull from Olduvai. *Nature (Lond.)* 184:491–493.

Leakey, L. S. B. 1960. *Adam's Ancestors,* Harper, New York.

Leakey, L. S. B. 1961a. New finds at Olduvai Gorge. *Nature (Lond.)* 189:649–650.

Leakey, L. S. B. 1961b. The juvenile mandible from Olduvai. *Nature (Lond.)* 191:417–418.

Leakey, L. S. B. 1962. A new lower Pliocene fossil primate from Kenya. *Ann. Mag. Nat. Hist.* 13:689–696.

Leakey, L. S. B. 1963. East African fossil Hominoidea and the classification within this superfamily, in: *Classification and Human Evolution* (S. L. Washburn, ed.), pp. 32–49, Aldine, Chicago.

Leakey, L. S. B. 1966. *Homo habilis, Homo erectus* and the australopithecines. *Nature (Lond.)* 209:1279–1281.

Leakey, L. S. B. 1967a. An early Miocene member of Hominidae. *Nature (Lond.)* 213: 155–163.

Leakey, L. S. B. 1967b. Notes on the mammalian faunas from the Miocene and Pleistocene of East Africa, in: *Background to Evolution in Africa* (W. W. Bishop and J. D. Clark, eds.), pp. 7–29, University Press, Chicago.

Leakey, L. S. B. 1968a. Bone smashing by late Miocene Hominidae. *Nature (Lond.)* 218: 528–530.

Leakey, L. S. B. 1968b. Lower dentition of *Kenyapithecus africanus. Nature (Lond.)* 217: 827–830.

Leakey, L. S. B. 1968c. Upper Miocene primates from Kenya. *Nature (Lond.)* 218:527–530.

Leakey, L. S. B., and Whitworth, T. 1958. Notes on the genus *Simopithecus,* with a description of a new species from Olduvai. *Coryndon Mem. Mus. Occ. Pap.* 6.

Leakey, L. S. B., Evernden, J. F., and Curtis, G. H. 1961. Age of Bed I, Olduvai Gorge, Tanganyika. *Nature (Lond.)* 191:478–479.

Leakey, L. S. B., Tobias, P. V., and Napier, J. R. 1964. A new species of the genus *Homo* from Olduvai Gorge. *Nature (Lond.)* 101:7–9.

Leakey, M. D. 1966. Review of the Oldowan culture from Olduvai Gorge. *Nature (Lond.)* 210:462–466.

Leakey, M. D. 1967. Preliminary survey of the cultural material from Beds I and II, Olduvai Gorge, Tanzania, in: *Background to Evolution in Africa* (W. W. Bishop and J. D. Clark, eds.), pp. 417–446, University Press, Chicago.

Leakey, M. D. 1970a. Stone artefacts from Swartkrans. *Nature (Lond.)* 225:1222–1225.

Leakey, M. D. 1970b. Early artefacts from the Koobi Fora area. *Nature (Lond.)* 226:228–230.

Leakey, M. D., Hay, R. L., Curtis, G. H., Drake, R. E., Jackes, M. K., and White, T. D. 1976. Fossil hominids from the Laetolil Beds, Tanzania. *Nature (Lond.)* 262:460–466.

Leakey, R. E. F. 1969. New Cercopithecidae from the Chemeron Beds of Lake Baringo, Kenya, in: *Fossil Vertebrates of Africa, Vol. 1* (L. S. B. Leakey, ed.), pp. 53–69, Academic Press, London.

Leakey, R. E. F. 1970a. Fauna and artifacts from a new Plio-Pleistocene locality near Lake Rudolf in Kenya. *Nature (Lond.)* 226: 223–224.

Leakey, R. E. F. 1970b. In search of man's past at Lake Rudolf. *Nat. Geogr.* 137:712–733.

Leakey, R. E. F. 1971. Further evidence of Lower Pleistocene hominids from East Rudolf, North Kenya. *Nature (Lond.)* 231: 241–245.

Leakey, R. E. F. 1972. Further evidence of Lower Pleistocene hominids from East Rudolf, North Kenya, 1971. *Nature (Lond.)* 237:264–269.

Leakey, R. E. F. 1973. Evidence for an advanced Plio-Pleistocene hominid from East Rudolf, Kenya. *Nature (Lond.)* 242:447–450.

Leakey, R. E. F. 1974. Further evidence of Lower Pleistocene hominids from East Rudolf, North Kenya 1973. *Nature (Lond.)* 248:653–656.

Leakey, R. E. F. 1976a. Hominids in Africa. *Amer. Sci.* 64:174–178.

Leakey, R. E. F. 1976b. New fossil hominids from the Koobi Fora Formation, North Kenya. *Nature (Lond.)* 261:574–576.

Leakey, R. E. F., and Lewin, R. 1977. *Origins,* Dutton, New York.

Leakey, R. E. F., and Lewin, R. 1978. *People of the Lake,* Doubleday, New York.

Leakey, R. E. F., and Walker, A. C. 1976. *Australopithecus, Homo erectus,* and the single species hypothesis. *Nature (Lond.)* 261:572–574. (See Chapter 36, this volume)

Leakey, R. E. F., and Wood, B. A. 1973. New evidence for the genus *Homo* from East Rudolf, Kenya (II). *Amer. J. Phys. Anthro.* 39:355–368.

Leakey, R. E. F., Mungai, J. M., and Walker, A. C. 1971. New australopithecines from East Rudolf, Kenya. *Amer. J. Phys. Anthro.* 35(2): 175–186.

Leakey, R. E. F., Mungai, J. M., and Walker, A. C. 1972. New australopithecines from East Rudolf, Kenya (II). *Amer. J. Phys. Anthro.* 32(2):235–252.

Lee, R. B. 1968. What hunters do for a living, or, how to make out on scarce resources, in: *Man the Hunter* (R. B. Lee and I. DeVore, eds.), pp. 30–48, Aldine, Chicago.

Le Gros Clark, W. E. 1924a. Notes on the living tarsier (*Tarsius spectrum*). *Proc. Zool. Soc., London* 1924:217–223.

Le Gros Clark, W. E. 1924b. The myology of the tree-shrew (*Tupaia minor*). *Proc. Zool. Soc., London* 1924:461–497.

Le Gros Clark, W. E. 1925. On the skull of *Tupaia. Proc. Zool. Soc., London* 1925:559–567.

Le Gros Clark, W. E. 1926. On the anatomy of the pen-tailed tree-shrew (*Ptilocercus lowii*). *Proc. Zool. Soc., London* 1926:1179–1309.

Le Gros Clark, W. E. 1927. Exhibition of photographs of the tree shrew (*Tupaia minor*). Remarks on the tree shrew, *Tupaia minor,* with photographs. *Proc. Zool. Soc., London* 1927:254–256.

Le Gros Clark, W. E. 1934. *Early Forerunners of Man,* Balliere, London.

Le Gros Clark, W. E. 1936. The problem of the claw in primates. *Proc. Zool. Soc., London* 1936:1–24.

Le Gros Clark, W. E. 1947. Observations on the anatomy of the fossil Australopithecinae. *J. Anat.* 81:300–333.

Le Gros Clark, W. E. 1950a. New paleontological evidence bearing on the evolution of the Hominoidea. *Quart. J. Geol. Soc. London* 105:225–264.

Le Gros Clark, W. E. 1950b. Hominid characteristics of the australopithecine dentition. *J. R. Anthro. Inst.* 80(1,2):37–54.

Le Gros Clark, W. E. 1952. Report on fossil hominoid material collected by the British Kenya Miocene expedition, 1949–1951. *Proc. Zool. Soc. London* 122(2):273–286.

Le Gros Clark, W. E. 1955. *The Fossil Evidence for Human Evolution,* University of Chicago, Chicago.

Le Gros Clark, W. E. 1959. *The Antecedents of Man,* Edinburgh University Press, Edinburgh.

Le Gros Clark, W. E. 1960. *The Antecedents of Man,* Quadrangle, Chicago.

Le Gros Clark, W. E. 1962. *The Antecedents of Man,* Edinburgh University Press, Edinburgh.

Le Gros Clark, W. E. 1963. *The Antecedents of Man,* Harper and Row, New York.

Le Gros Clark, W. E. 1964. *The Fossil Evidence for Human Evolution,* University of Chicago Press, Chicago, 2nd ed.

Le Gros Clark, W. E. 1967. *Man-apes or apemen?,* Holt, Rinehart and Winston, New York.

Le Gros Clark, W. E. 1970. *History of the Primates,* British Museum (Nat. Hist.), London.

Le Gros Clark, W. E., and Leakey, L. S. B. 1950. Diagnoses of East African Miocene Hominoidea. *Quart. J. Geol. Soc. London* 105:260–263.

Le Gros Clark, W. E., and Leakey, L. S. B. 1951. The Miocene Hominoidea of East Africa. *Fossil Mammals of Africa, No. 1.* British Museum (Nat. Hist.), London.

Le Gros Clark, W. E., and Thomas, D. P. 1951. Associated jaws and limb bones of *Limnopithecus macinnesi. Fossil Mammals of Africa. No. 3.* British Museum (Nat. Hist.), London.

Leidy, J. 1869. Notice of some extinct vertebrates from Wyoming and Dakota. *Proc. Acad. Nat. Sci., Phila.* 1869:63–67.

Leidy, J. 1873. Contributions to the extinct vertebrate fauna of the western territories. *Rep. U.S. Geol. Surv. Terr.* (F. V. Hayden, Geologist), *Washington* 1:14.

Lemoine, V. 1878. *Communication sur les Ossements Fossiles des Terrains Tertiares Inferieurs des Environs de Reims Faite a la Societe d'Histoire Naturelle de Reims.* Reims, 2 Parts: pp. 1–24; pp. 1–56.

Leopold, A. C., and Ardrey, R. 1972. Toxic substances in plants and the food habits of early man. *Science* 176:512–514.

Le Pichon, X. 1968. Sea-floor spreading and continental drift. *J. Geophys. Res.* 73(12):3661–3697.

Leutenegger, W. 1973. Maternal-fetal weight relationships in Primates. *Folia Primatologica* 20:280–293.

Lewin, R. 1981. Protohuman activity etched in fossil bones. *Science* 213:123–124.

Lewin, R. 1983. Is the organutan a living fossil? *Science* 222:1222–1223.

Lewis, G. E. 1934. Preliminary notice of new man-like apes from India. *Amer. J. Sci.* 22:161–181.

Lewis, G. E. 1937. Taxonomic syllabus of Siwalik fossil anthropoids. *Amer. J. Sci.* 234:139–147.

Lewis, O. J. 1964. The evolution of the long flexor muscles of the leg and foot. *Int. Rev. Gen. Zool.* 1:;165–185.

Lewis, O. J. 1969. The hominoid wrist joint. *Amer. J. Phys. Anthro.* 30:251–268.

Lewis, O. J. 1971a. Brachiation and the early evolution of the Hominoidea. *Nature (Lond.)* 230:577–579.

Lewis, O. J. 1971b. The contrasting morphology found in the wrist joints of semi-brachiating monkeys and brachiating apes. *Folia Primatologica* 15:248–256.

Lewis, O. J. 1972a. Evolution of the hominoid wrist, in: *The Functional and Evolutionary Biology of Primates* (R. H. Tuttle, ed.), pp. 207–222, Aldine, Chicago.

Lewis, O. J. 1972b. Osteological features characterizing the wrist of monkeys and apes, with a reconstruction of this region in *Dryopithecus (Proconsul) africanus. Amer. J. Phys. Anthro.* 36:45–58.

Lewis, O. J. 1972c. The evolution of the hallucial tarsometatarsal joint in the Anthropoidea. *Amer. J. Phys. Anthro.* 37:13–34.

Lewis, O. J. 1973. The hominoid os capitum, with special reference to the fossil bones from Sterkfontein and Olduvai Gorge. *J. Hum. Evol.* 2:1–11.

Lewis, O. J. 1974. The wrist articulation of the Anthropoidea, in: *Primate Locomotion* (F. A. Jenkins, ed.), pp. 143–169, Academic Press, New York.

Lewis, O. J. 1977. Joint remodeling and the evolution of the human hand. *J. Anat.* 123:157–201.

Lewis, O. J. 1980a. The joints of the evolving foot. Part I. The ankle joint. *J. Anat.* 130:527–543.

Lewis, O. J. 1980b. The joints of the evolving foot. Part II. The intrinsic joints. *J. Anat.* 130:833–857.

Lewis, O. J. 1980c. The joints of the evolving foot. Part III. The fossil evidence. *J. Anat.* 131:275–298.

Lewontin, R. C. 1969. The bases of conflict in biological explanation. *J. Hist. Bio.* 2:35–45.

Li, C.H. 1957. Properties of an structural investigations on growth hormones isolated from bovine, monkey and human pituitary glands. *Fed. Proc.* 16:775–783.

Li, C. 1978. A Miocene gibbon-like primate from Shihhung, Kiangsu Province. *Vertebr. Palasiat.* 16:187–192.

Liacopoulos, P., Halpern, B. N., and Perramant, F. 1962. Unresponsiveness to unrelated antigens induced by paralysing doses of bovine serum albumin. *Nature (Lond.)* 195:1112–1113.

Lillegraven, J. A. 1980. Primates from later Eocene rocks of southern California. *J. Mammal.* 61:181–204.

Lillegraven, J. A., Kraus, M. J., and Bown, T.

M. 1979. Paleogeography of the world of the Mesozoic, in: *Mesozoic Mammals* (J. A. Lillegraven, Z. Kielan-Jaworowska, and W. A. Clemens, eds.), pp. 277–308, University of California Press, Berkeley.

Lim, B. L. 1967. Note on the food habits of *Ptilocercus lowii* Gray (Pentail tree shrew) and *Echinosorex gymnurus* (Raffles) (Moonrat) in Malaya with remarks on "ecological labeling" by parasite patterns. *J. Zool.* 152:375–379.

Lim, B. L. 1969. Distribution of the primates of West Malaysia. *Proc. 2nd. Internat. Congr. Primatol.* 2:121–130.

Lind, E. M., and Morrison, M. E. S. 1974. *East African Vegetation,* Longman, London.

Lipson, S., and Pilbeam, D. R. 1982. *Ramapithecus* and hominoid evolution. *J. Hum. Evol.* 11:545–548.

Lisowski, F. P., Albrecht, G. H., and Oxnard, C. E. 1976. African fossil tali: further multivariate morphometric studies. *Amer. J. Phys. Anthro.* 45:5–18.

Livingstone, F. B. 1962. Reconstructing man's Pliocene pongid ancestor. *Amer. Anthropol.* 64:301–305.

Lloyd, J. L. 1963. Tectonic history of the south Central-American orogen, in: *Backbone of the Americas—A Symposium* (Childs and Beebe, eds.), pp. 88–100, Amer. Assoc. Petroleum Geol., Tulsa.

Loewenberg, J. B., ed. 1959. *Charles Darwin, Evolution and Natural Selection,* Beacon Press, Boston.

Loomis, F. B. 1906. Wasatch and Wind River primates. *Amer. J. Sci., Ser.f.* 21:277–284.

Lorenz von Liburnau, L. R. 1899. Einen fossilen anthropoiden von Madagaskar. *Anz. K. Acad. Wiss., Wein* 19:255–257.

Louis, P., and Sudre, J. 1975. Nouvelles donnces sur les primates de l'Eocene superieur europ- ean. *C.N.R.S., Paris* 218:805–823.

Lovejoy, C. O. 1974. The gait of australopithecines. *Yrbk. Phys. Anthro.* 17:147–161.

Lovejoy, C. O. 1975. Biomechanical perspectives on the lower limb of early hominids, in: *Primate Functional Morphology and Evolution* (R. H. Tuttle, ed.), pp. 291–326, Mouton, The Hague.

Lovejoy, C. O. 1978. A biomechanical view of the locomotor diversity of early hominids, in: *Early Hominids of Africa* (C. Jolly, ed.), pp. 403–429, St. Martin's, New York.

Lovejoy, C. O. 1979. A reconstruction of the pelvis of A. L. 288 (Hadar Formation, Ethiopia). *Amer. J. Phys. Anthro.* 50:460. (Abstract).

Lovejoy, C. O. 1981. The origin of man. *Science* 211:341–350. (See Chapter 38, this volume).

Lovejoy, C. O., and Heiple, K. G. 1970. A reconstruction of the femur of *Australopithecus africanus. Amer. J. Phys. Anthro.* 32:33–40.

Lovejoy, C. O., and Heiple, K. G. 1972. Proximal femoral anatomy of *Australopithecus. Nature (Lond.)* 235:175–176.

Lovejoy, C. O., and Meindl, R. S. 1972. Eukaryote mutation and the protein clock. *Yrbk. Phys. Anthro.* 16:18–30.

Lovejoy, C. O., Burstein, A. H., and Heiple, K. G. 1972. Primate phylogeny and immunological distance. *Science* 176:803–805.

Lovejoy, C. O., Heiple, K. G., and Burstein, A.

H. 1973. The gait of *Australopithecus*. *Amer. J. Phys. Anthro.* 38:757–780.

Lovejoy, C. O., Meindl, R. S., Pryzbeck, T. R., Barton, T. S., Heiple, K. G., and Kotting, D. 1977. Paleodemography of the Libben site, Ottawa County, Ohio. *Science* 198:291–293.

Lovtrup, S. 1973. Classification, convention and logic. *Zool. Scr.* 2:49–61.

Lowenstein, J. M. 1983. Fossil proteins and evolutionary time. *Pontificial Academy of Sciences, Scripta Varia* 50:151–162.

Lowenstein, J., Sarich, V. M., and Richardson, B. J. 1981. Albumin systematics of the extinct mammoth and Tasmanian wolf. *Nature (Lond.)* 291:409–411.

Luckett, W. P. 1969. Evidence for the phylogenetic relationship of tree shrews (family Tupaiidae) based osn the placenta and foetal membranes. *J. Reprod. Fert. Suppl.* 6:419–433.

Luckett, W. P. 1974. Comparative development and evolution of the placenta in primates. *Contrib. Primatol.* 3:142–234.

Luckett, W. P. 1975. Ontogeny of the fetal membranes and placenta: their bearing on primate phylogeny, in: *Phylogeny of the Primates: A Multidisciplinary Approach* (W. P. Luckett and F. S. Szalay, eds.), pp. 157–182, Plenum Press, New York.

Luckett, W. P. 1976. Cladistic relationships among primate higher categories: evidence of the fetal membranes and placenta. *Folia Primatologica* 25:245–276.

Luckett, W. P. 1980. Monophyletic or diphletic origins of Anthropoidea and Hystricognathi: evidence of the fetal membranes, in: *Evolutionary Biology of the New World Monkeys and Continental Drift* (R. L. Ciochon and A. B. Chiarelli, eds.), pp. 347–368, Plenum Press, New York.

Luckett, W. P., Ed. 1980. *Comparative Biology and Evolutionary Relationships of Tree Shrews*, Plenum Press, New York.

Luckett, W. P., and Szalay, F. S., Eds. 1975. *Phylogeny of the Primates: A Multidisciplinary Approach*, Plenum Press, New York.

Luckett, W. P., and Szalay, F. S. 1978. Clades versus grades in primate phylogeny, in: *Recent Advances in Primatology, Vol. 3. Evolution* (D. J. Chivers and K. A. Joysey, eds.), pp. 227–237, Academic Press, London.

Lyne, A. G. 1959. The systematic and adaptive significance of vibrissae in marsupials. *Proc. Zool. Soc. London* 133:79–133.

Lyon, M. W. 1913. Treeshrews: an account of the mammalian family Tupaiidae. *Proc. U.S. Nat. Mus.* 45:1–188.

McCann, C. 1928. Notes on the common Indian langur *(Pithecus entellus)*. *J. Bombay Nat. Hist. Soc.* 33:192–194.

McChance, R. A., Luff, M. C., and Widdowson, E. E. 1937. Physical and emotional periodicity in women. *J. Hygiene* 37:571–611.

McDowell, S. B., Jr. 1958. The Greater Antillean insectivores. *Bull. Amer. Mus. Nat. Hist.* 115:113–214.

McGrew, P.O., and Patterson, B. 1962. A picrodontid insectivore (?) from the Paleocene of Wyoming. *Breviora* 175:1–9.

McGrew, W. C., Tutin, C. E. A., and Baldwin, P. T. 1979. Chimpanzees, tools and termites: cross-cultural comparisons of Senegal, Tanzania and Rio Muni. *Man* 14:185–213.

McHenry, H. M. 1973a. Early hominid humerus from East Rudolf, Kenya. *Science* 180:739–741.

McHenry, H. M. 1973b. Humerus of robust *Australopithecus*. *Science* 182:396.

McHenry, H. M. 1975. Multivariate analysis of early hominid humeri, in: *Measures of Man* (E. Giles and J. S. Friedlander, eds.), pp. 338–371, Shenkman, Cambridge.

McHenry, H. M., and Corruccini, R. S. 1975. Distal humerus in hominoid evolution. *Folia Primatologica* 23:227–244.

McHenry, H. M., and Corruccini, R. S. 1976. The affinities of Tertiary hominoid femora. *Folia Primatologica* 26:139–150.

McHenry, H. M., and Corruccini, R. S. 1980. On the status of *Australopithecus afarensis*. *Science* 207:1103–1104.

McHenry, H. M., and Corruccini, R. S. 1981. *Pan paniscus* and human evolution. *Amer. J. Phys. Anthro.* 54:355–367.

McHenry, H. M., and Corruccini, R. S. 1983. The wrist of *Proconsul africanus* and the origin of hominoid postcranial adaptations, in: *New Interpretations of Ape and Human Ancestry* (R. L. Ciochon and R. S. Corruccini, eds.), pp. 353–367, Plenum Press, New York.

McHenry, H. M., Andrews, P., and Corruccini, R. S. 1980. Miocene hominoid palatofacial morphology. *Folia Primatologica* 33:241–252.

McKenna, M. C. 1960. Fossil Mammalia from the early Wasatchian Four Mile fauna, Eocene of Northwest Colorado. *Bull. Dep. Geol. Univ. Calif.* 37(1):1–30.

McKenna, M. C. 1963. The early Tertiary primates and their ancestors. *Proc. 16th Internat. Congr. Zool.* 4:69–74.

McKenna, M. C. 1966. Paleontology and the origin of primates. *Folia Primatologica* 4:1–25.

McKenna, M. C. 1967. Classification, range, deployment of the prosimian primates. *Collog. Int. C.N.R.S. (Paris)* 163:603–613.

McKenna, M. C. 1972a. Possible biological consequences of plate tectonics. *BioScience* 22:519–525.

McKenna, M. C. 1972b. Was Europe connected directly to North America prior to the middle Eocene?, in: *Evolutionary Biology, Vol. 6* (Th. Dobzhansky, M. K. Hecht, and W. C. Steere, eds.), pp. 179–189, Appleton-Century-Crofts, New York.

McKenna, M. C. 1973. Sweepstakes, filters, corridors, Noah's arks and beached Viking funeral ships in palaeogeography, in: *Implications of Continental Drift to the Earth Sciences, Vol. 1* (D. H. Tarling and S. K. Runcorn, eds.), pp. 295–308, Academic Press, London.

McKenna, M. C. 1975. Fossil mammals and early Eocene North Atlantic land continuity. *Ann. Missouri Bot. Gard.* 62:335–353.

McKenna, M. C. 1980. Early history and biogeography of South America's extinct land mammals, in: *Evolutionary Biology of the New World Monkeys and Continental Drift* (R. L. Ciochon and A. B. Chiarelli, eds.), pp. 43–77, Plenum Press, New York.

McKey, O. 1978. Soil vegetation and seed eating by black colobus monkeys, in: *The Ecology of Arboreal Folivores* (G. G. Montgomery, ed.), pp. 423–437, Washington, D.C., Smithsonian.

McKinley, K. R. 1971. Survivorship in gracile and robust australopithecines: a demographic comparison and a proposed birth model. *Amer. J. Phys. Anthro.* 34:417–426.

McMahon, T. A. 1973. Size and shape in biology. *Science* 197:1201–1204.

McNab, B. K. 1971. On the ecological significance of Bergmann's rule. *Ecology* 52(5):845–854.

Mabbutt, J. A. 1955. Erosion surfaces in Namaqualand and the ages of surface deposits in the South-western Kalahari. *Trans. Geol. Soc. S. Afr.* 58:13–30.

Mabbutt, J. A. 1957, Physiographic evidence for the age of the Kalahari sands of the southwestern Kalahari. *Proc. 3rd. Pan-Afr. Cong. Prehist. (Livingstone 1955):* 123–126.

MacArthur, R. H., and Wilson, E. O. 1967. *The Theory of Island Biogeography. Monogr. Pop. Biol. No. 1*, Princeton Univ. Press, Princeton.

MacDonald, J. R. 1963. The Miocene faunas from the Wounded Knee area of western South Dakota. *Bull. Amer. Mus. Nat. Hist.* 125:139–238.

MacInnes, D. G. 1943. Notes on the East African primates. *J. East Afr. & Uganda Nat. Hist. Soc.* 17:141–181.

MacIntyre, G. T. 1966. The Miacidae (Mammalia, Carnivora). I. The systematics of *Ictidopappus* and *Protictis*. *Bull. Amer. Nat. Hist.* 131:115–210.

MacPhee, R. D. E., and Cartmill, M. 1981. Further evidence for a tarsier–anthropoid clade within Haplorhini. *Amer. J. Phys. Anthro.* 54:248.

MacPhee, R. D. E., Cartmill, M., and Gingerich, P. D. 1983. New Paleogene primate basicrania and the definition of the order Primates. *Nature* 301:509–511.

Mackinnon, J. 1971. The orang-utan in Sabah today. *Oryx* 11:141–191.

Mackinnon, J. 1974. *In Search of the Red Ape*, Holt, Rinehart and Winston, New York.

Mackinnon, J. 1977. A comparative ecology of the Asian apes. *Primates* 18:747–772.

Maglio, V. 1973, Origin and evolution of the Elephantidae. *Trans. Amer. Phil. Soc.*, n.s. 63(3):5–149.

Mahe, J. 1965. *Les Subfossiles Malgache*, Imprimerie National, Tananarive.

Mahe, J., and Sourdat, M. 1972. Sur l'extinction des vertebres subfossiles et l'aridification du climat dans le sud-ouest de Madagascar. *Bull. Soc. Geol. France* 14:295–309.

Mahler, P. E. 1973. *Metric Variation in the Pongid Dentition*, Doctoral dissertation, University of Michigan.

Mai, L. L. 1983. A model of chromosome evolution in Primates and its bearing on cladogenesis in the Hominoidea, in: *New Interpretations of Ape and Human Ancestry* (R. L. Ciochon and R. S. Corruccini, eds.), pp. 87–114, Plenum Press, New York.

Maier, W. 1970. Neue Ergebnisse der Systematik und der Stammesgeschichte der Cercopithcoidea. *Z. Saugetierk.* 35:193–214.

Maier, W. 1971a. New fossil Cercopithecoidea from the Lower Pleistocene cave deposits of the Makapansgat Limeworks, South Africa. *Palaeontol. Afr.*13:69–108.

Maier, W. 1971b. Two new skulls of *Parapapio antiguus* from Taung and a suggested phylogenetic arrangement of the genus *Parapapio. Ann. S. Afr. Mus.* 59:1–16.

Maier, W. 1980. Nasal structures in Old World and New World primates, in: *Evolutionary Biology of the New World Monkeys and Continental Drift* (R. L. Ciochon and A. B. Chiarelli, eds.), pp. 219–241, Plenum Press, New York.

Maier, W., and Schneck, G. 1981. Konstructionsmorphologische Untersuchungen am Gebiss der hominoiden Primaten. *Z. Morphol. Anthro.* 72:127–169.

Major, C. I. F. 1894. On *Megaladapis madagascarensis*, an extinct gigantic lemuroid from Madagascar. *Phil. Trans. Roy. Soc. London, Ser. B* 185:15–38.

Major, C. I. F. 1901. On some characters of the skull in the lemurs and monkeys. *Proc. Zool. Soc. London* 1901:129–153.

Malbrant, R., and Maclatchy, A. 1949. *La Faune de l'Equateur Africain Francaise, Tome II*, P. Lechevelier, Paris.

Malfait, B. T., and Dinkelman, M. G. 1972. Circum-Caribbean tectonic and igneous activity and the evolution of the Caribbean plate. *Geol. Soc. Am. Bull.* 83:251–272.

Mann, A. E. 1968. *The Paleodemography of Australopithecus,* University Microfilms, Ann Arbor.

Mann, A. E. 1970. 'Telanthropus'and the single species hypothesis: a further comment. *Amer. Anthropol.* 72:607–609.

Mann, A. E. 1972. Hominid and cultural origins. *Man* 7:379–387.

Mann, A. E. 1975. Some paleodemographic aspects of the South African australopithecines. *Univ. Penn. Publ. Anthro.* 1.

Mann, A. E. 1981. The evolution of hominid dietary patterns, in: *Omnivorous Primates* (R. S. O. Harding and G. Teleki, eds.), pp. 10–36, Columbia Univ. Press, New York.

Marks, P. 1953. Preliminary note on the discovery of a new hominid from Java *Meganthropus* von Koenigswald in the Lower Middle Pleistocene of Sangiran, Central Java. *Indones. J. Nat. Sci.* 109:26–33.

Marsh, O. C. 1872. Preliminary description of new Tertiary mammals, II-IV. *Amer. J. Sci. & Arts., Ser. 3* 4:202–224.

Marshall, L. G., and Corruccini, R. S. 1978. Variability, evolutionary rates and allometry in dwarfing lineages. *Paleobiology* 4(2): 101–119.

Marshall, L. G., Pascual, R., Curtis, G. H., and Drake, R. E. 1977. South American geochronology: radiometric time scaling for middle to late Tertiary mammal-bearing horizons in Patagonia. *Science* 195:1325–1328.

Martin, D. E., and Gould, K. 1980. Comparative study of the sperm morphology of South American primates and those of the Old World, in: *Evolutionary Biology of the New World Monkeys and Continental Drift* (R. L.

Ciochon and A. B. Chiarelli, eds.), pp. 369–386, Plenum Press, New York.

Martin, L., and Andrews, P. J. 1982. New ideas on the relationships of the Miocene hominoids. *Primate Eye* 18:4–7.

Martin, R. D. 1966. Tree shrews: unique reproductive mechanism of systematic importance. *Science* 152:1402–1404.

Martin, R. D. 1968a. Reproduction and ontogeny in tree-shrews *(Tupaia belangeri),* with reference to their general behavior and taxonomic relationships. *Z. Tierpsychol.* 25:409–532.

Martin, R. D. 1968b. Towards a new definition of primates. *Man* 3:377–401.

Martin, R. D. 1972. Adaptive radiation and behavior of the Malagasy lemurs. *Phil. Trans. Roy. Soc. London* 264:294–352.

Martin, R. D. 1973. Comparative anatomy and primate systematics. *Symp. Zool. Soc. London* 33:301–337.

Martin, R. D. 1978. Major features of prosimian evolution: a discussion in the light of chromosomal evidence, in: *Recent Advances in Primatology, Vol. 3* (D. J. Chivers and K. A. Joysey, eds.), pp. 3–26, Academic Press, London.

Martin, R. D. Doyle, G. A., and Walker, A. C. Eds. 1974. *Prosimian Biology,* Duckworth, London.

Martyn, J., and Tobias, P. V. 1967. Pleistocene deposits and new fossil localities in Kenya. *Nature (Lond.)* 215:476–480.

Maser, R., and Maser, C. 1973. Notes on a captive long-tailed climbing mouse, *Vandeleuria oleracea* (Bennett, 1832) (Rodentia, Muridae). *Saugetierk. Mitt.* 21:336–340.

Mason, R. J. 1961. The earliest tool-makers in South Africa. *S. Afr. J. Sci.* 57:13–16.

Mason, R. J. 1962. *Prehistory of the Transvaal,* Univ. of the Witwatersrand, Johannesburg.

Matthew, W. D. 1904. The arboreal ancestry of the Mammalia. *Amer. Natur.* 38:811–818.

Matthew, W. D. 1909. The Carnivora and Insectivora of the Bridger Basin, Middle Eocene. *Mem. Amer. Mus. Nat. Hist.* 9:289–567.

Matthew, W. D. 1915. A revision of the lower Eocene Wasatch and Wind River faunas. Part IV. Entelonychia, Primates, Insectivora (part). *Bull. Amer. Mus. Nat. Hist.* 34: 429–483.

Matthew, W. D. 1917. The dentition of *Nothodectes. Bull. Amer. Mus. Nat. Hist.* 37:831–839.

Matthew, W. D. 1937. Paleocene faunas of San Juan Basin, New Mexico. *Trans. Amer. Phil. Soc.* 30:1–510.

Matthews, L. H. 1956. The sexual skin of the gelada. *Trans. Zool. Soc. London* 28:543–548.

Mayr, E. 1950 Taxonomic categories in fossil hominids. *Cold Spring Harbor Symp. Quant. Biol.* 15:108–118.

Mayr, E. 1963a. *Animal Species and Evolution,* Harvard Univ. Press, Cambridge.

Mayr, E. 1963b. The taxonomic evaluation of fossil hominids, in: *Classification and Human Evolution* (S. L. Washburn, ed.), pp. 332–346, Aldine, Chicago.

Mayr, E. 1965. What is a fauna? *Zool. Jahrbk. Syst. Geogr.* 92:473–486.

Mayr, E. 1969. *Principles of Systematic Zool-*

ogy, McGraw-Hill, New York.

Mayr. E. 1972. Sexual selection and natural selection, in: *Sexual Selection and the Decent of Man 1871–1971* (B. Campbell, ed.), pp. 87–104, Aldine, Chicago.

Mednick, L. W. 1955. The evolution of the human ilium. *Amer. J. Phys. Anthro.* 13: 203–216.

Medway. L. 1964. The marmoset rat, *Haplomys longicaudatus* Blyth. *Malay. Nat. J.* 18: 104–110.

Michael, R. P., and Zumpe, D. 1970. Rhythmic changes in the copulatory frequency of Rhesus monkeys *(Macaca mulatta)* in relation to the menstrual cycle and a comparison with the human cycle. *J. Reprod. Fert.* 21:199–201.

Michelmore, A. P. G. 1939. Observations on tropical African grasslands. *J. Ecol.* 27: 283–312.

Miller, D. A. 1977. Evolution of primate chromosomes. *Science* 198:1116–1124.

Miller, M. E., Christensen, G. C., and Evans, H. E. 1964. *Anatomy of the Dog,* W. B. Saunders, Philadelphia.

Mills, J. A. 1973. The influence of age and pairbond on the breeding biology of the Redbilled gull *Larus novaehollandiae scopulinus. J. Anim. Ecol.* 42:147–162.

Mills, J. R. E. 1963. Occlusion and malocclusion in primates, in: *Dental Anthropology* (D. R. Brothwell, ed.), pp. 29–51, Pergamon Press, Oxford.

Milner, A. R., and Panchen, A. L. 1973. Geographic variation in the tetrapod faunas of the Upper Carboniferous and Lower Permian, in: *Implications of Continental Drift to the Earth Sciences, Vol. 1* (D. H. Tarling and S. K. Runcorn, eds.), pp. 353–368, Academic Press, London.

Minoprio, J. D. L. 1945. Sobre el *Chlamyphorus truncatus* Harlan. *Acta Zool. Lilloana* 3:5–58.

Mitchell, G., and Brandt, E. M. 1972. Paternal behavior in primates, in: *Primate Socialization* (F. E. Poirier, ed.), pp. 173–206, Random House, New York.

Mittermeier, R. A., and Fleagle, J. G. 1977. The locomotor and postural repertoires of *Ateles geoffroyi* and *Colobus quereza* and a reevaluation of the locomotor category semibrachiation. *Amer. J. Phys. Anthro.* 45:235–256.

Mittermeier, R. A., and Van Roosmalen, M. G. M. 1981. Preliminary observations on habitat utilization and diet in eight Surinam monkeys. *Folia Primatologica* 36:1–39.

Mivart, St. G. 1873. On *Lepilemur* and *Cheirogaleus* and on the zoological rank of the Lemuroidea. *Proc. Zool. Soc. London* 1873: 484–510.

Mollison, I. D. 1910. Die Korperproportionen der Primaten. *Morph. Jrbk.* 42:79–299.

Molnar, S., and Gantt, D. G. 1977. Functional implications of primate enamel thickness. *Amer. J. Phys. Anthro.* 46:477–454.

Montagna, W. 1975. The skin of primates, in: *Biological Anthropology* (S. H. Katz, ed.), pp. 341–351, Freeman, San Francisco.

Moody, P. A., and Doniger, E. D. 1955. Serological light on porcupine relationships. *Evolution* 10(1):47–55.

Moore, J. C. 1959. Relationships among living squirrels of the Sciurinae. *Bull. Amer. Mus. Nat. Hist.* 118(4):159–206.

Morbeck, M. E. 1972. *A re-examination of the Miocene Hominoidea,* Ph.D. Dissertation, University of California, Berkeley.

Morbeck, M. E. 1975. *Dryopithecus africanus* forelimb. *J. Hum. Evol.* 4:39–46.

Morbeck, M. E. 1976. Problems in reconstruction of fossil anatomy and locomotor behavior: the *Dryopithecus* elbow complex. *J. Hum. Evol.* 5:223–233.

Morbeck, M. E. 1977. The use of casts and other problems in reconstructing the *Dryopithecus (Proconsul) africanus* wrist complex. *J. Hum. Evol.* 6:65–78.

Morbeck, M. E. 1979. Hominoidea postcranial remains from Rudabanya, Hungary. *Amer. J. Phys. Anthro.* 50:465–466.

Morbeck, M. E. 1983. Miocene hominoid discoveries from Rudabanya: implications from the postcranial skeleton, in: *New Interpretations of Ape and Human Ancestry* (R. L. Ciochon and R. S. Corruccini, eds.), pp. 369–404, Plenum Press, New York.

Moreau, R. E. 1951. Africa since the Mesozoic with particular reference to certain biological problems. *Proc. Zool. Soc. London* 121: 869–913.

Morris, D. 1967. *The Naked Ape,* Cape, London.

Morris, W. J. 1954. An Eocene fauna from the Cathedral Bluffs tongue of the Washakie Basin, Wyoming. *J. Paleontol.* 28:195–203.

Morse, J. C., and White, D. F., Jr. 1979. A technique for analysis of historical biogeography and other characters in comparative biology. *Syst. Zool.* 28:356–365.

Morton, D. J. 1922. Evolution of the human foot. I. *Amer. J. Phys. Anthro.* 5:305–336.

Morton, D. J. 1924a. Evolution of the human foot. II. *Amer. J. Phys. Anthro.* 7:1–52.

Morton, D. J. 1924b. Evolution of the longitudinal arch of the human foot. *J. Bone & Joint Surg.* 6:56–90.

Morton, D. J. 1926. Evolution of man's erect posture (Preliminary report). *J. Morpho. Physiol.* 43:147–179.

Morton, D. J. 1927. Human origin, correlation of previous studies on primate feet and posture with other morphological evidence. *Amer. J. Phys. Anthro.* 10:173–203.

Morton, D. J. 1935. *The Human Foot,* Columbia University Press, New York.

Morton, D. J., and Fuller, D. D. 1952. *Human Locomotion and Body Form,* Williams and Wilkins, Baltimore.

Mossman, H. W. 1937. Comparative morphogenesis of the fetal membranes and accessory uterine structures. *Contr. Embryol. Carnegie Inst.* 26:129–246.

Mossman, H. W., and Luckett, W. P. 1968. Phylogenetic relationship of the African mole rat, *Bathyergus janetta,* as indicated by the fetal membranes. *Amer. Zool.* 8:806.

Moynihan, M. 1976. *The New World Primates,* Princeton University Press, Princeton.

Muller, P. 1973. *The Dispersal Centres of Terrestrial Vertebrates in the Neotopical Realm,* Dr. W. Junk B.V., The Hague.

Murdock, G. P. 1949. *Social Structure,* Macmillan, New York.

Murray, P. 1975. The role of cheek pouches in cercopithecine monkey adaptive strategy, in: *Primate Functional Morphology and Evolu-*tion (R. H. Tuttle, ed.), pp. 151–194, Mouton, The Hague.

Musser, G. G. 1972. The species of *Haplomys* (Rodentia, Muridae). *Am. Mus. Nov.,* No. 2503.

Muul, I., and Lim, B. L. 1971. New locality records for some mammals of West Malaysia. *J. Mammal.* 52:430–437.

Napier, J. R. 1960. Studies of the hands of living primates. *Proc. Zool. Soc. London* 134: 647–657.

Napier, J. R. 1961. Prehensility and opposability in the hands of primates. *Symp. Zool. Soc. London* 5:115–132.

Napier, J. R. 1962. Fossil hand bones from Olduvai Gorge. *Nature (Lond.)* 196:409–411.

Napier, Jr. R. 1963. Brachiation and brachiators. *Symp. Zool. Soc. London* 10:183–195.

Napier, J. R. 1964. The evolution of bipedal walking in the hominids. *Arch. Biol. (Liege)* 75(suppl.):673–708.

Napier, J. R. 1967a. The antiquity of human walking. *Sci. Amer.* 216(4):56–66.

Napier, J. R. 1967b. Evolutionary aspects of primate locomotion. *Amer. J. Phys. Anthro.* 27:333–342.

Napier, J. R. 1970. Palecology and catarrhine evolution, in: *Old World Monkeys* (J. R. Napier and P. H. Napier, eds.), pp. 55–95, Academic Press, New York.

Napier, J. R. 1974. Book review: *The Functional and Evolutionary Biology of Primates* (R. H. Tuttle, ed.), Aldine-Atherton, Chicago. *Amer. J. Phys. Anthro.* 41:157–159.

Napier, J. R., and Davis, P. R. 1959. The forelimb skeleton and associated remains of *Proconsul africanus,* in: *Fossil Mammals of Africa No. 16.* British Museum (Nat. Hist.), London.

Napier, J. R., and Napier, P.H. 1967. *A Handbook of Living Primates,* Academic Press, London.

Napier, J. R. and Napier, P. H., Eds. 1970. *Old World Monkeys,* Academic Press, New York.

Napier, J. R., and Walker, A. C. 1967. Vertical clinging and leaping—a newly recognized category of locomotor behavior in primates. *Folia Primatologica* 6:204–219. (See Chapter 8, this volume)

Naylor, B. G. n.d. A paleontological view of cladistic analysis, ancestors and descendants, Manuscript.

Nelson, G. 1973. Comments on Leon Croizat's biogeography. *Syst. Zool.* 22:312–320.

Nelson, G. 1974. Historical biogeography: an alternative formalization. *Syst. Zool.* 23: 555–558.

Nelson, G. 1975. Biogeography, the vicariance paradigm, and continental drift. *Syst. Zool.* 24:490–504.

Nelson, G. 1978. From Candolle to Coizat: commentaries on the history of biogeography. *J. Hist. Biol.* 11:269–305.

Nelson, G., and Platnick, N. I. 1978. The perils of plesiomorphy: widespread taxa, dispersal and phenetic biogeography. *Syst. Zool.* 27: 474–477.

Nelson, G. J. 1973. Classification as an expression of phylogenetic relationships. *Syst. Zool.* 22:344–359.

Niemitz, C., Ed. 1984. *Biology of Tarsiers,* Gustav Fischer Verlag, Stuttgart.

Noback, C. R. 1975. The visual system of primates in phylogenetic studies, in: *Phylogeny of the Primates, a Multidisciplinary Approach* (W. P. Luckett and F. S. Szalay, eds.), pp. 199–218, Plenum Press, New York.

Novacek, M. J., and Marshall, L. G. 1976. Early biogeographic history of ostariophysan fishes. *Copeia* 1:1–12.

Nuttall, G. H. F. 1902. Progress report upon the biological test for blood as applied to over 500 bloods from various sources. *Brit. Med. J.* 1:825–827.

Nuttall, G. H. F. 1904. *Blood Immunity and Blood Relationship,* Cambridge University Press, Cambridge.

Oakley, K. 1959. Tools makyth man. *Smithson. Report* 1958:831–845.

O'Connor, B. L. 1974. *The Functional and Evolutionary Biology of the Cercopithecoid Wrist and Inferior Radioulnar Joints,* Ph.D. Thesis, University of California, Berkeley.

O'Connor, B. L. 1975. The functional morphology of the cercopithecoid wrist and inferior radioulnar joints and their bearing on some problems in the evolution of the Hominoidea. *Amer. J. Phys. Anthro.* 43:113–122.

O'Connor, B. L. 1976. *Dryopithecus (Proconsul) africanus:* quadruped or non-quadruped? *J. Hum. Evol.* 5:279–283.

O'Connor, B. L., and Rarey, K. E. 1979. Normal amplitudes of radioulnar pronation and supination in several genera of anthropoid primates. *Amer. J. Phys. Anthro.* 51:39–44.

Olson, T. R. 1981. Basicranial morphology of the extant hominoids and Pliocene hominids: The new material from the Hadar Formation and its significance in early human evolution and taxonomy, in: *Aspects of Human Evolution* (C. B. Stringer, ed.), pp. 99–128, Taylor and Francis, London.

Oparin, A. 1957. *Origin of Life on the Earth,* Academic Press, New York.

Opler, P. A. 1978. Interaction of plant life history components as related to arboreal herbivory, in: *The Ecology of Arboreal Folivores* (G. G. Montgomery, ed.) pp. 23–31, Smithsonian Institution, Washington, D.C.

Oppenheimer, A. M. 1964. Tool use and crowded teeth in Australopithecinae. *Curr. Anthro.* 5:419–421.

Orlosky, F. 1973. *Comparative Dental Morphology of Extant and Extinct Cebidae,* University Microfilms, Ann Arbor.

Orlosky, F. 1980. Dental evolutionary trends of relevance to the origin and dispersion of the platyrrhine monkeys, in: *Evolutionary Biology of the New World Monkeys and Continental Drift* (R. L. Ciochon and A. B. Chiarelli, eds.), pp. 189–200, Plenum Press, New York.

Orlosky, F., and Swindler, D. R. 1975. Origins of New World monkeys. *J. Hum. Evol.* 4:77–83.

Osborn, H. F. 1895. Fossil mammals of the Uinta Basin. *Bull. Amer. Mus. Nat. Hist.* 7:71–105.

Osborn, H. F. 1908. New fossil mammals from the Fayum Oligocene, Egypt. *Bull. Amer. Mus. Nat. Hist.* 24:265–272.

Osborn, H. F. 1909. New carnivorous mammals from the Fayum Oligocene, Egypt. *Bull. Amer. Mus. Nat. Hist.* 26:415–424.

Osborn, H. F. 1926a. The evolution of human races. *Nat. Hist.* 26:3–13.

Osborn, H. F. 1926b. Why Central Asia? *Nat. Hist.* 26:263–269.

Osborn, H. F. 1927a. Fundamental discoveries of the last decade in human evolution. *Bull. N. Y. Acad. Med.* 3:513–521.

Osborn, H. F. 1927b. Recent discoveries relating to the origin and antiquity of man. *Science* 65:481–488.

Osborn, H. F. 1927c. Recent discoveries in human evolution. *L. I. Med. J.* 21:563–567.

Osborn, H. F. 1928a. Recent discoveries relating to the origin and antiquity of man. *Paleobiology* 1:189–202.

Osborn, H. F. 1928b. The influence of habit in the evolution of man and the great apes. *Bull. N. Y. Acad. Med.* 4:216–249.

Osborn, H. F. 1928c. The influence of bodily locomotion in separating man from the monkeys and apes. *Sci. Monthly* 26:385–399.

Osborn, H. F. 1928d. The plateau habitat of the pro-dawn man. *Science* 65:570–571.

Osborn, H. F. 1928e. *Man Rises to Parnassus: Critical Epochs in the Prehistory of Man,* Princeton University Press, Princeton.

Osborn, H. F. 1928f. Present status of the problem of human ancestry. *Proc. Amer. Philos. Soc.* 67:151–155.

Osborn, H. F. 1929. Is the ape-man a myth? *Hum. Biol.* 1:4–9.

Osborn, H. F. 1930. The discovery of Tertiary Man. *Science* 71:1–7.

Oudin, J. 1960. L'allotype de certains antigens proteid du serum. Relations immunochimiques et genetiques entre six principaux allotypes observes dans le serum de lapin. *C. R. Acad. Sci., Paris* 250:770–772.

Oxnard, C. E. 1968a. A note on the fragmentary Sterkfontein scapula. *Amer. J. Phys. Anthro.* 28:213–217.

Oxnard, C. E. 1968b. A note on the Olduvai clavicular fragment. *Amer. J. Phys. Anthro.* 29:429–431.

Oxnard, C. E. 1968c. The architecture of the shoulder in some mammals. *J. Morph.* 126:249–290.

Oxnard, C. E. 1969. Mathematics, shape and function: a study in primate anatomy. *Amer. Sci.* 57:75–96.

Oxnard, C. E. 1972a. Functional morphology of primates: some mathematical and physical methods, in: *The Functional and Evolutionary Biology of Primates* (R. H. Tuttle, ed.), pp. 305–336, Aldine-Atherton, Chicago.

Oxnard, C. E. 1972b. Some African fossil foot bones: a note on the interpolation of fossils into a matrix of extant species. *Amer. J. Phys. Anthro.* 37:3–12.

Oxnard, C. E. 1973. Some locomotor adaptations among lower primates: implications for primate evolution. *Symp. Zool. Soc. London* 33:255–299.

Paluska, E., and Korinek, J. 1960. Studium der antigenen eiweibverwandtschaft zwischen menschen un einegen primaten mit hilfe neurer immunobiologischer methoden. *Z. Immun. Forschung und Exper. Therapie* 119:244–257.

Parra, R. 1978. Comparison of foregut and hindgut fermentation in herbivores, in: *Ecology of Arboreal Folivores* (G. G. Montgomery, ed.), pp. 205–230. Smithsonian Institute Press, Washington, D.C.

Partridge, T. C. 1973. Geomorphological dating of cave openings at Makanpansgat, Sterkfontein, Swartkrans and Taung. *Nature (Lond.)* 240:75–79.

Patterson, B. 1954. The geologic history of non-hominid primates in the Old World. *Hum. Biol.* 26(3):191–209.

Patterson, B., and Howells, W. W. 1967. Hominid humeral fragment from the early Pleistocene in Northwestern Kenya. *Science* 156:64–66.

Patterson, B., and Pascual, R. 1968. The fossil mammal fauna of South America. *Quart. Rev. Biol.* 43(4):409–451.

Patterson, B., and Pascual, R. 1972. The fossil mammals of South America, in: *Evolution, Mammals and Southern Continents* (A. Keast, F. Erk, and B. Glass, eds.), pp. 247–309, SUNY Press, Albany.

Paul, J. P. 1967 *Forces at the Human Hip Joint.* Ph.D thesis, University of Glasgow.

Pauling, L., and Corey, R. B. 1951. Configuration of polypeptide chains. *Nature (Lond.)* 168:550–551.

Perkins, E. M., and Meyer, W. C. 1980. The phylogenetic significance of the skin of primates: implications for the origin of New World monkeys, in: *Evolutionary Biology of the New World Monkeys and Continental Drift* (R. L. Ciochon and A. B. Chiarelli, eds.), pp. 331–346, Plenum Press, New York.

Perutz, M. F., Mossman, M. G., Cullis, A. F., Muirhead, H., Will. G., and North, A. C. T. 1960. Structure of hemoglobin. A three dimensional fourier synthesis at 5.5 A resolution obtained by x-ray analysis. *Nature (Lond.)* 185:416–422.

Peterka, H. E. 1937. A study of the myology and osteology of tree sciurids with regard to adaptation to arboreal, glissant and fossorial habaits. *Trans. Kans. Acad. Sci.* 39:313–332.

Peters, C. R. 1979. Toward an ecological model of African Plio-Pleistocene hominid adaptations. *Amer. Anthropol.* 81:361–278.

Petter, J. J. 1962. Recherches sur l'ecologie et l'ethologie des lemuriens Malagaches. *Mem. Mus. Nat. Hist. Nat. A* 27:1–146.

Petter, J. J., Albinac, R., and Rumpler, Y. 1977. Mammiferes Lemuriens (Primates Prosimiens). *Faune de Madagascar* 44:1–513.

Petter-Rousseaux, A. 1964. Reproductive physiology and behavior of the Lemuroidea, in: *Evolution and Genetic Biology of Primates,* Vol. 2 (J. Buettner-Janusch, ed.), pp. 91–132, Academic Press, New York.

Pfeffer, P. 1969. Considerations sur l'ecologie des forets claires du Cambodge oriental. *Terre Vie* 116:3–24.

Pfeiffer, J. E. 1969. *The Emergence of Man,* Harper and Row, New York.

Pianka, E. R. 1974. *Evolutionary Ecology,* Harper and Row, New York.

Pickford, M. 1975. Late Miocene sediments and fossils from the northern Kenya rift valley. *Nature (Lond.)* 256:279–284.

Pickford, M. 1978a. Paleoenvironments and vertebrate faunas of the mid-Miocene Ngorora Formation, Kenya, in: *Geological Background to Fossil Man* (W. W. Bishop, ed.), pp. 237–262, Scottish Academic Press, Edinburgh.

Pickford, M. 1978b. Stratigraphy and mammalian paleontology of the late Miocene Lukeino Formation, in: *Geological Background to Fossil Man* (W. W. Bishop, ed.), pp. 263–278, Scottish Academic Press, Edinburgh.

Pickford, M. 1982. New higher primate fossils from the middle Miocene deposits at Majiwa and Kaloma, western Kenya. *Amer. J. Phys. Anthro.* 58: 1-19.

Pickford, M. 1983. Sequence and environments of the lower and middle Miocene hominoids of western Kenya, in: *New Interpretations of Ape and Human Ancestry* (R. L. Ciochon and R. S. Corruccini, eds.) pp. 421-439, Plenum Press, New York.

Pilbeam, D. R. 1966. Notes of *Ramapithecus,* the earliest known hominid, and *Dryopithecus. Amer. J. Phys. Anthro.* 25: 1-6.

Pilbeam, D. R. 1967. Man's earliest ancestors. *Sci. J.* 3(2):47-53.

Pilbeam, D. R. 1968. The earliest hominids. *Nature (Lond.)* 219: 1335-1338. (See Chapter 28, this volume)

Pilbeam, D. R. 1969a. Tertiary Pongidae of East Africa: evolutionary relationships and taxonome. *Peabody Mus. Nat. Hist., Yale Univ. Bull.* 31: 1-185.

Pilbeam, D. R. 1969b. Possible identity of Miocene tali from Kenya. *Nature (Lond.)* 223:648.

Pilbeam, D. R. 1969c. Newly recognized mandible of *Ramapithecus. Nature (Lond.)* 222: 1093-1094.

Pilbeam, D. R. 1970. *Gigantopithecus* and the origins of the Hominidae. *Nature (Lond.)* 255: 516-519.

Pilbeam, D. R. 1972. *The Ascent of Man,* MacMillan, New York.

Pilbeam, D. R. 1976. Neogene hominids of South Asia and the origins of Hominidae, in: *Le Plus Anciens Hominides* (P. V. Tobias and Y. Coppens, eds.), pp. 39-59, C. N. R. S., Paris.

Pilbeam, D. R. 1978. Rethinking human origins. *Discovery* 13(1): 2-9. (See Chapter 29, this volume)

Pilbeam, D. R. 1979. Recent finds and interpretations of Miocene hominoids. *Ann. Rev. Anthro.* 8: 333-353.

Pilbeam, D. R. 1980. Major trends in human evolution, in: *Current Arguements on Early Man* (L. K. Konigsson, ed.), pp. 261-285, Pergamon, Oxford.

Pilbeam, D. R. 1982. New hominoid skull material from the Miocene of Pakistan. *Nature (Lond.)* 295: 232-234.

Pilbeam, D. R., and Simons, E. L. 1965. Some problems of hominid classification. *Amer. Sci.* 53: 237-259.

Pilbeam, D. R., and Simons, E. L. 1971. Humerus of *Dryopithecus* from Saint Gaudens, France. *Nature (Lond.)* 229: 408-409.

Pilbeam, D. R., and Walker, A. C. 1968. Fossil monkeys from the Miocene of Napak, Northeast Uganda. *Nature (Lond.)* 220: 657-660.

Pilbeam, D. R., Meyer, G. E., Badgley, C., Rose, M. D., Pickford, M. H. L., Behrensmeyer, A.

K., and Ibrahim Shah, S. M. 1977a. New hominoid primates from the Siwaliks of Pakistan and their bearing on hominoid evolution. *Nature (Lond.)* 270: 689-695.

Pilbeam, D. R., Berry, J., Meyer, G. E., Ibrahim Shah, S. M., Pickford, M. H. L., Bishop, W. W., Thomas, H., and Jacobs, L. L. 1977b. Geology and paleontology of Neogene strata of Pakistan. *Nature (Lond.)* 270: 684-689.

Pilbeam, D. R., Rose, M. D., Badgley, C., and Lipschutz, B. 1980. Miocene hominids from Pakistan. *Postilla* 181: 1-94.

Pilgrim, G. E. 1915. New Siwalik primates and their bearing on the question of the evolution of man and the Anthropoidea. *Rec. Geol. Surv. India* 45: 1-74.

Pilgrim, G. E. 1927. A new *Sivapithecus* palate, and other primate fossils from India. *Mem. Geol. Surv. India (Palaeontol. Ind.)* 14: 1-26.

Pitman, W. C., III, and Talwani, M. 1972. Sea-floor spreading in the North Atlantic *Geol. Soc. Am. Bull.* 83: 619-646.

Piveteau, J. 1948. Recherces anatomiques sue les lemuriens disparus: le genre *Archaeolemur. Ann. Paleontol.* 34: 127-171.

Piveteau, J. 1957. Primates, in: *Traite de Paleontologie, Vol 7* (J. Piveteau, ed.) Masson et Cie, Paris.

Platnick, N. I. 1976. Drifting spiders or continents?: vicariance biogeography of the spider family Maroniinae (Araneae: Gnaphosidae). *Syst. Zool.* 25: 101-109.

Platnick, N. I., and Nelson, G. 1978. A method of analysis for historical biogeography. *Syst. Zool.* 27:1-16.

Platt, J. R. 1962. A 'book model' of genetic information-transfer in cells and tissues, in: *Horizons in Biochemistry* (M. Kasha and B. Pullman, eds.) Academic Press, New York.

Pocock, R. I. 1918. On the external characters of the lemurs and of *Tarsius. Proc. Zool. Soc. London* 1918: 19-53.

Pocock, R. I. 1919. Discussion on the zoological position and affinities of *Tarsuis. Proc. Zool. Soc. London* 1919: 494-495.

Pocock, R. I. 1922 On external characteristics of the beaver (Castoridae) and of some squirrels (Sciuridae). *Proc. Zool. Soc. London* 1922: 1171-1212.

Pocock, R. I. 1925. External characters of the catarrhine monkeys and apes. *Proc. Zool. Soc. London* 1925: 1479-1579.

Poduschka, W. 1969. Erganzungen zum Wissen über *Erinaceus e. roumanicus* und kritische Überlegungen zur bisherigen Literatur uber europaische lgel. *Z. Tierpsychol.* 26: 761-804.

Polyak, S. 1957. *The Vertebrate Visual System,* University of Chicago Press, Chicago.

Popper, K. R. 1957. *The Poverty of Historicism,* Harper Torchbooks, New York

Popper, K. R. 1959. *The Logic of Scientific Discovery,* Harper and Row, New York.

Popper, K. R. 1963. *Conjectures and Refutations,* Harper and Row, New York.

Popper, K. R. 1980. Letter to the editor. *New Sci.* 87: 611.

Porter, R. R. 1953. The relation of chemical structure to the biological activity of the proteins, in: *The Proteins, Vol. I:B* (B. H. Neurath and K. Bailey, eds.), pp. 973-1015, Academic Press, New York.

Portman, O. W. 1970. Nutrition requirements (NRC) of nonhuman primates, in: *Feeding and Nutrition of Nonhuman Primates* (R. S. Harris, ed.), pp. 87-116. Academic Press, New York.

Poulik, M. D., and Smithies, O. 1958. Comparison and combination of the starch-gel and filter-paper electrophoretic methods applied to human sera: two dimensional electrophoresis. *Biochem. J.* 68: 636-643.

Powers, S. 1911. Floating islands. *Pop. Sci. Monthly* 79: 303-307.

Prasad, K. N. 1983. Historical notes on the geology, dating and systematics of the Miocene hominoids of India, in: *New Interpretations of Ape and Human Ancestry* (R. L. Ciochon and R. S. Corruccini, eds.), pp. 559-574, Plenum Press, New York.

Preuschoft, H. 1971. Body posture and mode of locomotion in early Pleistocene hominids. *Folia Primatologica* 14: 209-240.

Preuschoft, H. 1973. Body posture and locomotion in some East African Miocene Dryopithecinae, in: *Human Evolution* (M. Day, ed.) pp. 13-46, *Symp. Soc. Study Hum. Biol., Vol. 11,* Barnes and Nobles, New York.

Preuschoft, H. 1975. Body posture and mode of locomotion in fossil primates: Method and examples- *Aegyptopithecus zeuxis,* in: *Proc. 5th Symp. Congr. Internat. Primatol. Soc.* 1974: 345-359, Japan Science Press, Tokyo.

Preuss, T. M. 1982. The face of *Sivapithecus indicus:* description of a new, relatively complete specimen from the Siwaliks of Pakistan *Folia Primatologica* 38: 141-157.

Prince, J. H. 1953. Comparative anatomy of the orbit. *Brit. J. Physio. Optics* (n. s.) 10: 144-154.

Prost, J. H. 1980. Origin of bipedalism. *Amer. J. Phys. Anthro.* 52: 175-190.

Radinsky, L. B. 1967. The oldest primate endocast. *Amer. J. Phys. Anthro.* 27:385-388.

Radinsky, L. B. 1969. Outlines of Canid and Felid brain evolution. *Ann. N. Y. Acad. Sci.* 167: 277-288.

Radinsky, L. B. 1970. The fossil evidence of prosimian brain evolution, in: *The Primate Brain* (C. R. Noback and W. Montagna, eds.), pp. 209-224, Appleton-Century-Crofts, New York.

Radinsky, L. B. 1973. *Aegytopithecus* endocasts: oldest record of a pongid brain. *Amer. J. Phys. Anthro.* 39: 239-248.

Radinsky, L. B. 1974. The fossil evidence of anthropoid brain evolution. *Amer. J. Phys. Anthro.* 41: 15-28.

Radinsky, L. B. 1975. Primate brain evolution *Amer. Sci.* 63 (6): 656-663.

Radinsky, L. B. 1977. Early primate brains: facts and fiction. *J. Hum. Evol.* 6: 79-86.

Radinsky, L. B. 1978. Do albumin clocks run on time? *Science* 200: 1182-1183.

Radinsky, L. B. 1979. The fossil record of primate brain evolution. *James Arthur Lectures on the Evolution of the Human Brain* 49: 1-27, Amer. Mus. Nat. Hist., New York.

Raffles, T. S. 1821. Descriptive catalogue of a zoological collection, made on account of the Honourable East India Company, in the Island of Sumatra and its vicinity, under the direction of Sir Thomas Stamford Raffles, Lieutenant-Governor of Fort Marlborough; with additional notices illustrative of the natural history of those countries. *Trans. Linn. Soc. London* 13: 239-274.

Rahm, U. 1970. Ecology, zoogeography and systematics of some African forest monkeys, in: *Old World Monkeys* (J. R. Napier and P. H. Napier, eds.), pp. 589-626, Academic Press, New York.

Rand, A. L. 1935. On the habits of some Madagascar mammals. *J. Mammal.* 16: 89-104.

Rattray, J. M. 1960. *The Grass Cover of Africa,* F. A. O., New York.

Raven, H. C. 1950. *The Anatomy of the Gorilla* (W. K. Gregory, ed.), Columbia University Press, New York.

Read, D. W. 1975. Primate phylogeny, neutral mutations and "Molecular clocks". *Syst. Zool.* 24: 209-221.

Read, D. W., and Lestrel, P. 1970. Hominid phylogeny and immunology: a critical approach. *Science* 168: 578-580.

Read, D. W., and Lestrel, P. 1972. Phyletic divergence dates of hominoid primates. *Evolution* 26: 669-670.

Reig, O. A. 1962. Las Integraciones cenogeneticas en el desarrollo de la fauna de Vertebrados tetrapodos de America del Sur. *Ameghiniana* 2 (8): 131-140.

Reig, O. A., and Simpson, G. G. 1972. *Sparassocynus* (Marsupialia, Didelphidae), a peculiar mammal from the late Cenozoic of Argentina. *J. Zool.* 167: 511-539.

Remane, A. 1924. Einage Bemerkungen uber *Prohylobates tandyi* R. Fourtau und *Dryopithecus mogharensis* R. Fourtau. *Cbl. Min. Geol. Pal.* 14: 220-224.

Remane, A. 1951. Die Zähne des *Meganthropus africanus. Z. Morph. Anthro.* 42: 311-329.

Remane, A. 1959. Die primitivsten Menschenformen (Australopithecinae) und das Problem des tertiaren Menschen. *Naturwiss. Vereins fur Schlegswig-Holstein* 29: 310.

Remane, A. 1960. Zahne und Gebiss. *Primatologia* III (2) 637-846.

Remane, A. 1965. Die Geschichte der Menschenaffen, in: *Menscliche Abstammungslehre: Fortschritte der Anthropogenie 1863-1964.* (G. Heberer, ed.) pp. 249-309, Fischer, Stuttgart.

Rensberger, B. 1984. A new ape in our family tree. *Science 84* 5(1):16 (Jan./Feb).

Rensch, B. 1959. *Evolution Above the Species Level.* Menthuen, London.

Reyment, R. A., Benstson, P., and Tait, E. A., 1976. Cretaceous transgressions in Nigeria and Seripe-Alagoas (Brazil). *An. Acad. Brasil Cien.* 48: 253-264.

Reynolds, V. 1966. Open groups in hominoid evolution. *Man* 1: 441-452.

Reynolds, V. 1976. *The Biology of Human Action,* Freeman, San Francisco.

Reynolds, V., and Reynolds, F. 1965. Chimpanzees in the Budongo Forest, in: *Primate Behavior: Field Studies of Monkeys and Apes* (I. DeVore, ed.), pp. 368-424, Holt, Rinehart and Winston, New York.

Rich, P. V. 1978. Fossil birds of old Gondwanaland: A comment on drifting continents and their passengers, in: *Historical Biogeography, Plate Tectonics and the Changing Environment* (J. Gray and A. J. Boucot, eds.), pp.

321–332, Oregon State University Press, Corvallis.

Richard, A. F. 1978. *Behavioral Variation: Case Study of a Malagasy Lemur,* Bucknell University Press, Lewisburg.

Richards, P. W. 1952. *The Tropical Rain Forest.* Cambridge University Press, Cambridge.

Ridley, H. N. 1895. The mammals of the Malay Peninsula. *Nat. Sci.* 6:23–29.

Ridley, H. N. 1930. *The Dispersal of Plants Throughout the World,* L. Reeve, Ashford, Kent.

Rimoli, R. 1977. Una Nueva especie de monos (Cebidae: Saimirinae ?Saimiri) de la Hispaniola. *Cuad. del Cendia* (Univ. Autonoma de Santo Domingo) 242 (1): 1–16.

Ripley, S. 1967. The leaping langurs: a problem in the study of locomotor adaptation. *Amer. J. Phys. Anthro.* 26: 149–170.

Ripley, S. 1979. Environmental grain, niche diversification, and positional behavior in Neogene primates: an evolutionary hypothesis, in: *Environment, Behavior and Morphology: Dynamic Interactions in Primates* (M. E. Morbeck, H. Preuschoft and N. Gomberg, eds.), pp. 37–74, Gustay Fischer, Stuttgart.

Roberts, D., and Tattersall, I. 1974. Skull form and the mechanics of mandibular elevation in mammals. *Amer. Mus. Novit.* No. 2536: 1–9.

Robertson, M. 1979. Positional behavior in *Dryopithecus (Proconsul) africanus.* Paper presented at the 78th Annual Meeting of the Amer. Anthropol. Assoc.

Robinson, H. C., and Kloss, C. B. 1909. On mammals from the Rhio Archipelago and Malay Peninsula collected by Messers. H. C. Robinson, C. B. Kloss, and E. Seimund, and presented to the National Museum by the government of the Federated Malay States (Thomas, O., and Wroughton, R. C.). *J. Fed. Malay States* 4: 99–129.

Robinson, J. T. 1952. Some hominid features of the ape-man dentition. *Off. J. Dent. Assn. S. Afr.* 7:102–113.

Robinson, J. T. 1953. *Meganthropus,* Australopithecines and hominids. *Amer. J. Phys. Anthro.* 11:1–38.

Robinson, J. T. 1954a. The genera and species of the Australopithecinae. *Amer. J. Phys. Anthro.* 12:181–200.

Robinson, J. T. 1954b. Prehominid dentition and hominid evolution. *Evolution* 8:324–334.

Robinson, J. T. 1955. Further remarks on the relationship between *Meganthropus* and Australopithecines. *Amer. J. Phys. Anthro.* 13: 429–445.

Robinson, J. T. 1956. The dentition of the Australopithecinae. *Transvaal Mus. Mem.* No. 9.

Robinson, J. T. 1957. Occurrence of stone artifacts with *Australopithecus* at Sterkfontein. *Nature (Lond.)* 180:521–524.

Robinson, J. T. 1958. Cranial cresting patterns and their significance in the Hominoidea. *Amer. J. Phys. Anthro.* 16:397–428.

Robinson, J. T. 1959. A bone implement from Sterkfontein. *Nature (Lond.)* 184:583–585.

Robinson, J. T. 1960. The affinities of the new Olduvai Australopithecinae. *Nature (Lond.)* 186:458–468.

Robinson, J. T. 1961. The australopithecines and their bearing on the origin of man and of stone tool-making. *S. Afr. J. Sci.* 57:3–13.

Robinson, J. T. 1962a. Australopithecines and artefacts at Sterkfontein. *S. Afr. Archeol. Bull.* 17:87–126.

Robinson, J. T. 1962b. The origins and adaptive radiation of the Australopithecines, in: *Evolution and Hominization* (G. Kurth, ed.), pp. 120–140.

Robinson, J. T. 1963a. Australopithecines, culture and phylogeny. *Amer. J. Phys. Anthro.* 21:595–605.

Robinson, J. T. 1963b. Adaptive radiation in the Australopithecines and the origin of man, in: *African Ecology and Human Evolution (F. C. Howell and F. Bourliere, eds.),* pp. 385–416, Aldine, Chicago. *(See Chapter 34, this volume)*

Robinson, J. T. 1965. Homo 'habilis' and the Australopithecines. *Nature (Lond.)* 205: 121–124.

Robinson, J. T. 1966. The distinctiveness of *Homo habilis. Nature (Lond.)* 209:957–960.

Robinson, J. T. 1967. Variation and the taxonomy of early hominids, in: *Evolutionary Biology, Vol. 1* (T. Dobzhansky, M. K. Hecht, and W. C. Steere, eds.), pp. 69–100, Meredith, New York.

Robinson, J. T. 1970. Two new early hominid vertebrae from Swartkrans. *Nature (Lond.)* 225:1217–1219.

Robinson, J. T. 1972. *Early Hominid Posture and Locomotion,* University of Chicago Press, Chicago.

Robinson, J. T., and Mason, R. 1957. Occurrence of stone artefacts with *Australopithecus* at Sterkfontein. *Nature (Lond.)* 180:521–524.

Robinson, P. 1966. Fossil Mammalia of the Huerfano Formation, Eocene, of Colorado. *Bull. Peabody Mus. Nat. Hist., Yale Univ.* 21:1–85.

Robinson, P. 1967. The mandibular dentition of *? Tetonoides* (Primates, Anaptomorphidae). *Ann. Carnegie Mus.* 39:187–191.

Robinson, P. 1968. The paleontology and geology of the Badwater Creek area, central Wyoming. Part 4. Late Eocene primates from Badwater, Wyoming, with a discussion of material from Utah. *Ann. Carnegie Mus.* 39:307–326.

Rochi, J. 1971. Recherches mammalogiques en Guinee forestiere. *Bull. Mus. Nat. Hist. Naturelle Paris* 16:737–781.

Roitt, I. M., and Doniach, D. 1960. Thyroid auto-immunity. *Brit. Med. Bull.* 16:152–158.

Roitt, I. M., Campbell, P. N., and Doniach, D. 1958. The nature of the thyroid auto-antibodies in patients with Hashimoto's thyroiditis (lymphaodenoid goitre). *Biochem. J.* 69: 248–256.

Romer, A. S. 1945. *Vertebrate Paleontology,* University of Chicago Press, Chicago.

Romer, A. S. 1955. *Osteology of the Reptiles,* University of Chicago Press, Chicago.

Romer, A. S. 1966. *Vertebrate Paleontology,* University of Chicago Press, Chicago.

Romer, A. S. 1968. *Notes and Comments on Vertebrate Paleontology,* University of Chicago Press, Chicago.

Romero-Herrara, A. F., Lehmann, H., Castillo, O., Joysey, K. A., and Friday, A. E. 1976. Myoglobin of the orang-utan as a phylogenetic enigma. *Nature (Lond.)* 261:162–164.

Rose, K. D. 1975. The Carpolestidae, early Tertiary primates from North America. *Bull. Mus. Comp. Zool.* 147:1–74.

Rose, K. D. 1977. Evolution of Carpolestid primates and chronology of the North American middle and late Paleocene. *J. Paleont.* 51(3): 536–542.

Rose, K. D. 1978. A new Paleocene epoiditheriid (Mammalia), with comments on the Palaeanodonta. *J. Paleont.* 52:658–674.

Rose, K. D. 1980. Clarkforkian land-mammal age: revised definition, zonation and tentative intercontinental correlations. *Science* 208: 744–746.

Rose, K. D., and Fleagle, J. G. 1981. The fossil history of non-human primates in the Americas, in: *Ecology and Behavior of Neotropical Primates, Vol. 1* (A. F. Coimbra-Filho and R. A. Mittermeier, eds.), pp. 111–167, Academia Brasiliera de Ciencias, Rio de Janiero. (See Chapters 5, 7, 16, this volume)

Rose, K. D., and Gingerich, P. D. 1976. Partial skull of the plesiadapiform primate *Ignacius* from the early Eocene of Wyoming. *Contrib. Mus. Paleontol., Univ. Michigan* 24:181–189.

Rose, M. D. 1983. Miocene hominoid postcranial morphology: monkey-like, ape-like, neither, or both?, in: *New Interpretations of Ape and Human Ancestry* (R. L. Ciochon and R. S. Corruccini, eds.), pp. 405–417, Plenum Press, New York. (See Chapter 25, this volume)

Rose, K. D. 1984. Hominoid postcranial specimens from the middle Miocene Chinji Formation, Pakistan. *J. Hum. Evol.* 13:503–516.

Rosen, D. E., 1975. A vicariance model of Caribbean biogeography. *Syst. Zool.* 24:431-464.

Rosen, D. E. 1978. Vicariant patterns and historical explanation in biogeography. *Syst. Zool.* 27:159–188.

Rosenberger, A. L. 1977. *Xenothrix* and ceboid phylogeny. *J. Hum. Evol.* 6:461–481.

Rosenberger, A. L. 1979a. Cranial anatomy and implications of *Dolichocebus,* a late Oligocene ceboid primate. *Nature (Lond.)* 279:416-418.

Rosenberger, A. L. 1979b. *Phylogeny, Evolution and Classification of New World Monkeys (Platyrrhini, Primates)* Ph.D. Thesis, City Univ. of New York.

Rosenberger, A. L. 1980. Gradistic views and adaptive radiation of the platyrrhine primates. *Z. Morphol. Anthropol.* 71:157–163. (See Chapter 17, this volume)

Rosenberger, A. L., and Kinzey, W. G. 1976. Functional patterns of molar occlusion in Platyrrhine Primates. *Amer. J. Phys. Anthro.* 45:281–298.

Rosenberger, A. L., and Szalay, F. S. 1980. On the Tarsiiform origins of Anthropoidea, in: *Evolutionary Biology of the New World Monkeys and Continental Drift* (R. L. Ciochon and A. B. Chiarelli, eds.), pp. 139–157, Plenum Press, New York. (See Chapter 12, this volume)

Ross, H. H. 1974. *Biological Systematics,* Addison-Wesley, Massachusetts.

Rothe, H. 1972. Beobachtungen zum Bewegungsverhalten des Weissbuscheläffchens *Callithrix jacchus* Erxleben, 1777, mit besonderer Berücksichtingung der Hanfunktion. *Z. Morphol. Anthropol.* 64:90–101.

Rovner, I. 1971. Potential of opalphytoliths for use in paleoecological reconstruction. *Quaternary Res.* 1:343–359.

Rowell, T. E. 1967. Female reproductive cycles and the behavior of baboons and Rhesus macaques, in: *Social Communication Among the Primates* (S. Altman, ed.), pp. 15–32, University of Chicago Press, Chicago.

Rusconi, C. 1933. Nuevos restos de monos fosiles del terciario antiguo de la Patagonia. *Anal. Soc. Sci. Argent.* 116:286–289.

Rusconi, C. 1935. Las especies de primates del Oligoceno de Patagonia. *Rev. Argentina Paleont. Anthrop. Ameghino* 1:39–68, 71–100, 103–125.

Russell, D. 1964. Les mammiferes paleocenes d'Europe. *Mem. Mus. Nat. d'Hist. Natur., Ser C.* 13:1–324.

Russell, D. E., and Gingerich, P. D. 1980. Un nouveau primate omomyide dans l'eocene du Pakistan. *C. R. Acad. Sci., Ser D., Paris* 291:621–624.

Russell, D. E., Louis, P., and Savage, D. 1967. Primates of the French early Eocene. *Univ. Calif. Publ. Geol. Sci.* 73:1–46.

Saban, R. 1958. Insectivora, in: *Traite de Paleontologie, Tome VI-2* (J. Piveteau, ed.), pp. 822–909, Masson and Cie, Paris.

Saban, R. 1975. Structure of the ear region in living and subfossil lemurs, in: *Lemur Biology* (I. Tattersall and R. W. Susman, eds.), pp. 83–108, Plenum Press, New York.

Sade, D. S. 1967. Determinants of dominance in a group of free ranging Rhesus monkeys, in: *Social Communication Among the Primates* (S. Altman, ed.), pp. 99–115, University of Chicago Press, Chicago.

Sade, D. S., Cuching, K., Cuching, P., Dunaif, J., Figueroa, A., Kaplan, J. R., Lauer, C., Rhodes, D., and Schneider, J. 1976. Population dynamics in relation to social structure in Cayo Santiago. *Yrbk. Phys. Anthro.* 20:253–262.

Sahni, A., and Kumar, V. 1974. Paleogene paleobiogeography of the Indian subcontinent. *Palaeogeogr., Palaeoclimatol., Palaeoecol.* 15:209–226.

Said, R. 1962. *The Geology of Egypt,* Elsevier, Amsterdam.

Sanderson, I. T. 1940. Mammals of North Cameroons forest area. *Trans. Zool. Soc. London* 24:623–725.

Sarich, V. M. 1968a. The origin of the hominids: an immunological approach, in: *Perspectives on Human Evolution* (S. L. Washburn and P. C. Jay, eds.), pp. 94–121, Holt, Rinehart and Winston, New York.

Sarich, V. M. 1968b. Quantitative immunochemistry and the evolution of Old World Primates, in: *Taxonomy and Phylogeny of Old World Primates with Reference to the Origin of Man* (A. B. Chiarelli, ed.), pp. 139–140, Rosenberg and Sellier, Turino.

Sarich, V. M. 1970. Primate systematics with special reference to Old World monkeys, in: *Old World Monkeys* (J. R. Napier and P. H. Napier, eds.), pp. 175–226, Academic Press, New York.

Sarich, V. M. 1971. A molecular approach to the question of human origins, in: *Background for Man* (P. Dolhinow and V. M. Sarich, eds.), pp. 60–81, Little, Brown, Boston. (See Chapter 40, this volume.)

Sarich, V. M. 1973. Just how old is the hominid line? *Yrbk. Phys. Anthro.* 17:98–112.

Sarich, V. M., and Cronin, J. E. 1976. Molecular systematics of the primates, in: *Molecular Anthropology* (M. Goodman and R. E. Tashian, eds.), pp. 141–170, Plenum Press, New York.

Sarich, V. M., and Cronin, J. E. 1977. Generation length and rates of hominoid molecular evolution. *Nature (Lond.)* 269:354–355.

Sarich, V. M., and Cronin, J. E. 1980. South American mammal molecular systematics, evolutionary clocks and continental drift, in: *Evolutionary Biology of the New World Monkeys and Continental Drift* (R. L. Ciochon and A. B. Chiarelli, eds.), pp. 399–421, Plenum Press, New York.

Sarich, V. M., and Wilson, A. C. 1967a. Immunological time scale for hominid evolution. *Science* 158: 1200–1203.

Sarich, V. M., and Wilson, A. C. 1967b. Rates of albumin evolution in primates. *Proc. Nat. Acad. Sci.* 58: 142–148.

Sartono, S. 1971. Observations on a new skull of *Pithecanthropus erectus (Pithecanthropus* VIII) from Sangiran, Central Java. *Konikl. Ned. Akad. Wet.* 74: 185–194.

Savage, D. E., and Waters, B. T. 1978. A new omomyid primate from the Wasatch Formation of Southern Wyoming. *Folia Primatologica* 30: 1–29.

Savage, D. E., Russell, D. E., and Waters, B. T. 1977. Critique of certain early Eocene primate taxa. *Geobios, Mem. Spec.* 1: 159–164.

Savage, S., and Bakeman, R. 1978. Sexual morphology and behavior in *Pan paniscus,* in: *Recent Advances in Primatology* (D. J. Chivers and J. Herbert, eds.), pp. 613–616, Academic Press, New York.

Savage-Runbaugh, E. S., Wilkerson, B. J., and Bakeman, R. 1977. Spontaneous gestural communication among conspecifics in the pygmy chimpanzee *(Pan paniscus),* in: *Progress in Ape Research* (G. Bourne, ed.), pp. 97–116, Academic Press, New York.

Savin, S. M., Douglas, R. G., and Stehli, F. G. 1975. Tertiary marine paleotemperatures. *Bull. Geol. Soc. Am.* 86:1499–1510.

Schaeffer, B., Hecht, M. K., and Eldredge, N. 1972. Phylogeny and paleontology, in: *Evolutionary Biology, Vol. 6* (Th. Dobzhansky, M. K. Hecht, and W. C. Steere, Eds.), 31–46, Appleton–Century–Crofts, New York.

Schaffer, F. X. 1924. *Lehrbuch der Geologie. II. Teil. Grundzuge der historischen Geologie,* Franz Deuticke, Leipzig and Vienna.

Schaffer, W. M. 1968. Character displacement and the evolution of the Hominidae. *Amer. Natur.* 102:559–571.

Schaller, G. B. 1963. *The Mountain Gorilla,* University of Chicago Press, Chicago.

Schaller, G. B., and Emler, J. T. 1963. Observations on the ecology and social behavior of the mountain gorilla, in: *African Ecology and Human Evolution* (F. C. Howell and F. Bourliere, eds.), pp. 368–384, Aldine, Chicago.

Schaub, S. 1953. Remarks on the distribution and classification of the "Hystricomorpha". *Verh. Naturf. Ges. Basel* 64:389–400.

Schaub, S. 1958. Simplicidentata (=Rodentia), in: *Traite de Paleontologie,* Vol. 1–2 (J. Piveteau, ed.), pp. 669–818, Masson et Cie, Paris.

Scheltema, R. S. 1971. Larval dispersal as a means of genetic exchange between geographically separated populations of shallow-water benthic marine gastropods. *Biol. Bull.* 140:284–322.

Schlosser, M. 1888. Die Affen, Lemuren, Chiropteren, Insectivoren, Marsupialier, Creodonten und Carnivoren des europaischen Tertiars und deren Beziegungen zu inhren lebenden und fossilen aussereuropaischen Verwandten, II. *Beit. Palaeontol. Geol. Osterreich-Ungarns Orients, Vienna* 7:1–162.

Schlosser, M. 1907. Beitrag zur Osteologie und systematischen Stellung der Gattung *Necrolemur,* sowie zur Stammesgeschichte der Primaten uberhaupt. *Neues Jahrbh. Mineralogie, Geologie, Palaeontol. Festband* 1907: 197–226.

Schlosser, M. 1910. Uber einige fossile Saugetiere aus dem Ologacan von Aegypten. *Zool. Anz.* 35:500–508.

Schlosser, M. 1911. Beitrage zue Kenntnis der Oligozanen Lansaugetiere aus dem Fayum. *Beitrage Pal. Geol. Oesterr.—Ung.* 24:51–167.

Schlott, M. 1940. Beobachtungen an Tanas (*Tupaia tana* Raffl.). *Zool. Gart.* 12:153–157.

Schmid, P. 1979. Evidence of Microchoerine evolution from Dielsdorf (Zurich region, Switzerland)—a preliminary report. *Folia Primatologica* 31:301–311.

Schmid, P. 1981. Comparison of Eocene nonadapids and *Tarsius,* in: *Primate Evolutionary Biology* (A. B. Chiarelli and R. S. Corruccini, eds.), pp. 6–13, Springer, Berlin.

Schmid, P. 1983. Front dentition of the Omomyiformes (Primates). *Folia Primatologica* 40:1–10.

Schmidt, K. P. 1919. Contribution to the Herpetology of the Belgian Congo based on the collection of the American Museum Congo Expedition, 1909–1915. Part I. Turtles, crocodiles, lizards and chameleons. *Bull. Amer. Mus. Nat. Hist.* 39:385–624.

Schoener, T. W. 1974. Resource partitioning in ecological communities. *Science* 185:27–39.

Schon, M. A., and Ziemer, L. K. 1973. Wrist mechanisms and locomotor behavior of *Dryopithecus (Proconsul) africanus. Folia Primatologica* 20:1–11.

Schon-Ybarra, M., and Conroy, G. C. 1978. Nonmetric features in the ulna of *Aegyptopithecus, Alouatta, Lagothrix. Folia Primatologica* 29:178–195.

Schram, F. R. 1977. Paleozoogeography of late Paleozoic and Triassic Malacostraca. *Syst. Zool.* 26:367–379.

Schultz, A. H. 1926. Fetal growth of man and other primates. *Quart. Rev. Biol.* 1:465–521.

Schultz, A. H. 1927a. Observations on a gorilla fetus. *Eugenical News* 12:37–40.

Schultz, A. H. 1927b. Studies on the growth of a gorilla and of other higher primates with special reference to the fetus of the gorilla, preserved in the Carnegie Museum. *Mem. Carnegie Mus.* 11:1–87.

Schultz, A. H. 1930. The skeleton of the trunk and limbs of higher primates. *Hum. Biol.* 2:303–438.

Schultz, A. H. 1936. Characters common to

higher primates and characters specific for man. *Quart. Rev. Biol.* 11:259–283, 425–455.

Schultz, A. H. 1937. Proportions, variability, and asymmetrics of the long bones of the limbs and clavicles in man and apes. *Hum. Biol.* 9:281–328.

Schultz, A. H. 1940. The size of the orbit and of the eye in primates. *Amer. J. Phys. Anthro.* 25:398–408.

Schultz, A. H. 1948. The number of young at birth and the number of nipples in primates. *Amer. J. Phys. Anthro.* 6(1):1–23.

Schultz, A. H. 1950. The physical distinctions of man. *Proc. Phil. Soc.* 94:428–449.

Schultz, A. H. 1951. The specializations of man and his place among the catarrhine primates. *Cold Spring Harbor Symp. Quant. Biol.* 15:37–52.

Schultz, A. H. 1953a. Man's place among the primates. *Man* 53:7–9.

Schultz, A. H. 1953b. The place of the gibbon among the primates. *J. Roy. Anthro. Soc.* 53:3–12.

Schultz, A. H. 1953c. The relative thickness of the long bones and the vertebrae in primates. *Amer. J. Phys. Anthro.* 11:277–312.

Schultz, A. H. 1954. Bemerkungen zur Varibilitat und Systematik der Schimpanzen. *Saugertierkundl. Mitteil.* 2:159–163.

Schultz, A. H. 1956. Postembryonic age changes. *Primatologia* I:887–964.

Schultz, A. H. 1960. Einege Beobachtungen und Mass am Skelett von *Dreopithecus. Z. Morph. Anthro.* 50:136–149.

Schultz, A. H. 1961. Vertebral column and thorax. *Primatologia* I:887–964.

Schultz, A. H. 1963. Age changes, sex differences, and variability as factors in the classification of primates, in *Classification and Human Evolution* (S. L. Washburn, ed.), pp. 85–115, Aldine, Chicago.

Schultz, A. H. 1968. The recent hominoid primates, in: *Perspectives on Human Evolution, Vol. I* (S. L. Washburn and P. C. Jay, eds.), pp. 122–195, Holt, Rinehart and Winston, New York.

Schultz, A. H. 1969a. The skeleton of the Chimpanzee, in: *The Chimpanzee I. Anatomy, Behavior, and Diseases of Chimpanzees* (G. H. Bourne, ed.), pp. 51–103, University Park, Baltimore.

Schultz, A. H. 1969b. Observations on the acetabulum of primates. *Folia Primatologica* 11:181–199.

Schultz, A. H. 1969c. *The Life of Primates,* Universe Books, New York.

Schultz, A. H. 1978. Illustrations of the relation between primate ontogeny and phylogeny, in: *Human Evolution: Biosocial Perspectives* (S. L. Washburn and E. R. McCown, eds.), pp. 255–283, Benjamin Cummings, Menlo Park.

Schwalbe, G. 1899. Studien uber *Pithecanthropus erectus* Dubois. *Z. Morph. Anthrop.* 1:1–240.

Schwalbe, G. 1913. Kritische Besprechung von Boule's Werk: 'L' homme fossile de La Chapelle—aux—Saints' mit eigenes Untersuchungen. *Z. Morph. Anthrop.* 16:527–610.

Schwalbe, G. 1923. Die Abstammung des Menschen und die altesten Menschenformen, in: *Die Kultur der Gegenwart* (G. Schwalbe and

E. Fischer, eds.), pp. 223–338, Teubner, Leipzig.

Schwartz, J. H. 1978. If *Tarsius* is not a prosimian, is it a haplorhine?, in: *Recent Advances in Primatology, Vol. 3: Evolution* (D. J. Chivers and K. A. Joysey, eds.), pp. 195–202, Academic Press, New York.

Schwartz, J. H., Tattersall, I., and Eldredge, N. 1978. Phylogeny and classification of the Primates revisited. *Yrbk. Phys. Anthro.* 21:95–133.

Sclater, J. G., Hellinger, S., and Tapscott, C. 1977. The paleobathymetry of the Atlantic Ocean from the Jurassic to the present. *J. Geol.* 85:509–552.

Scott, W. B. 1928. Part IV: Primates of the Santa Cruz Beds, in: *Reports of the Princeton University Expedition to Patagonia 1896–1899, Vol. VI— Paleontology,* pp. 342–350.

Selander, R. K. 1966. Sexual dimorphism and differential niche utilization in birds. *Condor* 68:113–151.

Selander, R. K. 1972. Sexual selection and dimorphism in birds, in: *Sexual Selection and the Descent of Man* (B. G. Campbell, ed.), pp. 180–230, Aldine, Chicago.

Seligsohn, D. 1977. Analysis of species—specific molar adaptations in Strepsirhine Primates. *Contributions to Primatology, Vol. 11* pp. 1–116, S. Karger, Basel.

Senyurek, M. 1955. A note on the teeth of *Meganthropus africanus* from Tanganyika Territory. *Belleten (Ankara)* 19:1–55.

Seton, H. 1940. Two new primates from the lower Eocene of Wyoming. *Proc. New Eng. Zool. Club* 18:39–42.

Shaw, G. A. 1879. A few notes upon four species of Lemur, specimens which were brought back to England in 1878. *Proc. Zool. Soc. Lond.* 1879:132–136.

Sheine, W. S. 1971a. Digestibility of cellulose in prosimian primates. *Amer. J. Phys. Anthro.* 50:480–481.

Sheine, W. S. 1979b. *The Effect of Variations in Molar Morphology on Masticatory Effectiveness and Digestion of Cellulose in Prosimian Primates.* Ph.D. thesis, Duke University.

Sheine, W. S., and Kay, R. F. 1977. An analysis of chewed food particle size and its relationship to molar structure in the primates *Cheirogaleus medius* and *Galago senegalensis* and the insectivoran *Tupaia glis. Amer. J. Phys. Anthro.* 47:15–20.

Shorten, M. 1954. *Squirrels,* Collins, London.

Sibuet, J. C., and Mascle, J. 1978. Plate kinematic implications of Atlantic equatorial fracture zone trends. *J. Geophys. Res.* 83:3401–3421.

Sicher, H. 1944. Masticatory apparatus in the giant panda and the bears. *Field Mus. Nat. Hist., Zool. Series* 21:61–73.

Sickenberg, O. 1975. Die Gliederung des hoheren Jungtertiars und Altquartars in der Turkei nach Vertebraten und ihre Bedeutung fur die internationale Neogen-Stratigraphie. *Geol. Jahrbh., Hannover* 15:1–167.

Sillans, R. 1958. *Les Savanes de l'Afrique Centrale,* P. Lechevalier, Paris.

Simons, E. L. 1959. An anthropoid frontal bone from the Oligocene of Egypt: the oldest skull fragment of a higher primate. *Amer. Mus.*

Novit. No. 1976:1–16.

Simons, E. L. 1961a. Notes on Eocene tarsioids and a revision of some Necrolemurinae. *Bull. Brit. Mus (Nat. Hist.), Geology* 5:43–69.

Simons, E. L. 1961b. The dentition of *Ourayia:* its bearing on relationships of omomyid prosimians. *Postilla* 54: 1–20.

Simons, E. L. 1961c. The phyletic position of *Ramapithecus. Postilla* 57:1–9. (See Chapter 27, this volume)

Simons, E. L. 1962a. Two new primate species from the African Oligocene. *Postilla* 64:1–12.

Simons, E. L. 1962b. A new Eocene primate genus, *Cantius* and a revision of some allied European lemuroids. *Bull. Brit. Mus. (Nat. Hist.), Geology* 7:1–30.

Simons, E. L. 1963a. A critical reappraisal of Tertiary Primates, in: *Genetic and Evolutionary Biology of the Primates* (J. Buettner-Janusch, ed.), pp. 65–129, Academic Press, New York.

Simons, E. L. 1963b. Some fallacies in the study of hominid phylogeny. *Proc. 16th. Int. Congr. Zool.* 4:25–70.

Simons, E. L. 1964a. Old World higher primates: classification and taxonomy. *Sci. Rev.* 144:709–710.

Simons, E. L. 1964b. The early relatives of man. *Sci. Amer.* 211:50–62.

Simons, E. L. 1964c. On the mandible of *Ramapithecus. Proc. Natl. Acad. Sci., USA* 51: 528–535.

Simons, E. L. 1965a. New fossil apes from Egypt and the initial differentiation of the Hominoidea. *Nature (Lond.)* 205:135–139. (See Chapter 19, this volume)

Simons, E. L. 1965b. The hunt for Darwin's third ape. *Med. Opinion Rev.* 1965 (Nov.): 74–81.

Simons, E. L. 1967a. Review of the phyletic interrelationships of Oligocene Anthropoidea. *Coll. Int. Cent. Rech. Sci.* 163:597–602.

Simons, E. L. 1967b. The significance of primate paleontology for anthropological studies. *Amer. J. Phys. Anthro.* 27:307–332.

Simons, E. L. 1967c. New evidence on the anatomy of the earliest catarrhine primates, in: *Neue Ergebnisser der Primatologie* (D. Starck, R. Schneider, and H. -J. Kuhn, eds.), pp. 15–18, 1st Congr. Int. Primatol. Soc., Frankfurt, G. Fischer, Stuttgart.

Simons, E. L. 1967d. The earliest apes. *Sci. Amer.* 217(6):28–35.

Simons, E. L. 1967e. Fossil primates and the evolution of some primate locomotor systems. *Amer. J. Phys. Anthro.* 26:241–253.

Simons, E. L. 1968a. Assessment of a fossil hominid. *Science* 160:672–675.

Simons, E. L. 1968b. A source for dental comparison of *Ramapithecus* with *Australopithecus* and *Homo. S. Afr. J. Sci.* 64:92–112.

Simons, E. L. 1968c. Early Cenozoic mammalian faunas, Fayum Province, Egypt. I. African Oligocene mammals: introduction, history of study, and faunal succession. *Bull. Peabody Mus.* 28:1–21.

Simons, E. L. 1968d. Hunting the "Dawn apes" of Africa. *Discovery* 4:19–32.

Simons, E. L. 1969a. Miocene monkey *(Prohylobates)* from Northern Egypt. *Nature (Lond.)* 223:687–689.

Simons, E. L. 1969b. Origin and radiation of the primates. *Ann. N. Y. Acad. Sci.* 167:319–331.

Simons, E. L. 1969c. Recent advances in paleoanthropology. *Yrbk. Phys. Anthro.* 1967: 14–23.

Simons, E. L. 1970. The deployment and history of Old World monkeys (Cercopithecidae, Primates), in: *Old World Monkeys* (J. R. Napier and P. H. Napier, eds.), pp. 97–137, Academic Press, New York. (See Chapter 22, this volume)

Simons, E. L. 1971. Relationships of *Amphipithecus* and *Oligopithecus. Nature (Lond.)* 232:489–491.

Simons, E. L. 1972. *Primate Evolution, an Introduction to Man's Place in Nature (Lond.),* Macmillan, New York.

Simons, E. L. 1974a. The relationships of *Aegyptopithecus* to other primates. *Ann. Geol. Surv. Egypt* 4:149–156.

Simons, E. L. 1974b. *Parapithecus grangeri* (Parapithecidae, Old World Primates): new species from the Oligocene of Egypt and the initial differentiation of the Cercopithecidea. *Postilla* 166:1–12.

Simons, E. L. 1974c. Notes on early Tertiary prosimians, in: *Prosimian Biology* (R. D. Martin, G. A. Doyle and A. C. Walker, eds.), pp. 415–433, Duckworth, London.

Simons, E. L. 1976a. Relationship between *Dryopithecus, Sivapithecus,* and *Ramapithecus* and their bearing on hominid origins, in: *Les Plus Anciens Hominides* (P. V. Tobias and Y. Coppens, eds.), pp. 60–67, CNRS, Paris.

Simons, E. L. 1976b. The fossil record of primate phylogeny, in: *Molecular Anthropology* (M. Goodman, E. Tashian, and J. H. Tashian, eds.), pp. 35–62, Plenum Press, New York.

Simons, E. L. 1976c. The nature of the transition in the dental mechanism from pongids to hominids. *J. Hum. Evol.* 5:511–528.

Simons, E. L. 1977. *Ramapithecus. Sci. Amer.* 236:38–35.

Simons, E. L. 1978. Diversity among early hominids: a vertebrate paleontologist's viewpoint, in: *Early Hominids of Africa* (C. Jolly, ed.), pp. 543–566, Duckworth, London.

Simons, E. L. 1981. Mans immediate forerunners. *Phil. Trans. R. Soc.* 292:21–41.

Simons, E. L., and Chopra, S. R. K. 1969. *Gigantopithecus* (Pongidae, Hominoidea) a new species from North India. *Postilla* 138:1–18.

Simons, E. L., and Ettel, P. C. 1970. *Gigantopithecus. Sci. Amer.* 222(1):77–86.

Simons, E. L., and Fleagle, J. G. 1973. The history of the extinct gibbon-like primates, in: *Gibbon and Siamang, Vol. 2* (D. M. Rumbaugh, ed.), pp. 121–148, S. Karger, Basel.

Simons, E. L., and Pilbeam, D. 1965. Preliminary revisions of the Dryopithecinae (Pongidae, Anthropoidea). *Folia Primatologica* 3:81–152.

Simons, E. L., and Pilbeam, D. R. 1972. Hominoid paleoprimatology, in: *The Functional and Evolutionary Biology of Primates* (R. Tuttle, ed.), pp. 36–62, Aldine Atherton, Chicago.

Simons, E. L., and Pilbeam, D. R. 1978. *Ramapithecus* (Hominidae, Hominoidea), in: *Evolution of African Mammals* (V. J. Maglio and H. B. S. Cooke, eds), pp. 147–153, Harvard University Press, Cambridge.

Simons, E. L., and Russell, D. E. 1960. Notes on the cranial anatomy of *Necrolemur. Breviora* 127:1–14.

Simons, E. L., Andrews, P., and Pilbeam, D. R. 1978. Cenozoic apes, in: *Evolution of African Mammals* (V. J. Maglio and H. B. S. Cooke, eds.), pp. 120–146, Harvard University Press, Cambridge.

Simons, E. L., Kay, R. F., and Fleagle, J. G. 1981. A revision of the Oligocene apes of the Fayum Province, Egypt. *Amer. J. Phys. Anthro.* 55:293–322.

Simons, E. L., Pilbeam, D. R., and Ettel, P. C. 1969. Controversial taxonomy of fossil hominids. *Science* 166:258–259.

Simpson, G. G. 1928. *A Catalogue of the Mesozoic Mammalia,* Brit. Mus. (Nat. Hist.), London.

Simpson, G. G. 1931. A new insectivore from the Oligocene of Mongolia. *Amer. Mus. Novit.* No. 505:1–8.

Simpson, G. G. 1935. The Tiffany fauna. Upper Paleocene. 2. Structure of relationships of *Plesiadapis. Amer. Mus. Novit.* No. 816:1–30.

Simpson, G. G. 1937. The Fort Union of the Crazy Mountain Field, Montana and its mammalian faunas. *Bull. U.S. Nat. Mus.* 169: 1–287.

Simpson, G. G. 1940a. Mammals and land bridges. *J. Wash. Acad. Sci.* 30:137–163.

Simpson, G. G. 1940b. Studies on the earliest primates. *Bull. Amer. Mus. Nat. Hist.* 77: 185–212.

Simpson, G. G., 1945. The principles of classification and a classification of mammals. *Bull. Amer. Mus. Nat. Hist.* 85:1–350.

Simpson, G. G., 1947. Holarctic mammalian faunas and continental relationships during the Cenozoic. *Bull. Geol. Soc. Amer.* 58:613–688.

Simpson, G. G., 1949. *The Meaning of Evolution,* Yale University Press, New Haven.

Simpson, G. G., 1950. History of the fauna of Latin America, in: *Science in Progress* (G. Baiisell, ed.), pp. 369–408, Yale University Press, New Haven.

Simpson, G. G., 1952. Probabilities of dispersal in geologic time, in: *The Problem of Land Connections Across the South Atlantic, with Special Reference to the Mesozoic* (E. Mayr, ed.) *Bull. Amer. Mus. Nat. Hist.* 99:163–176.

Simpson, G. G., 1953a. *Evolution and Geography,* Condon Lectures, Eugene, Oregon.

Simpson, G. G., 1953b. *The Major Features of Evolution,* Columbia University Press, New York.

Simpson, G. G. 1955. The Phenacolemuridae, a new family of early primates. *Bull. Amer. Mus. Nat. Hist.* 105:415–441.

Simpson, G. G. 1956. Zoogeography of West Indian land mammals. *Amer. Mus. Novit.* No. 1759:1–28.

Simpson, G. G. 1961. *Principles of Animal Taxonomy,* Columbia University Press, New York.

Simpson, G. G. 1963. The meaning of taxonomic statements, in: *Classification and Human Evolution* (S. L. Washburn, ed.), pp. 1–31, Aldine, Chicago.

Simpson, G. G., 1964. *This View of Life,* Harcourt, Brace and World, New York.

Simpson, G. G., 1965a. Long-abandoned views. *Science* 147:1397. (Abstract)

Simpson, G. G., 1965b. *The Geography of Evolution,* Chilton Books, New York.

Simpson, G. G., 1966a. Mammalian evolution on the Southern continents. *N. Jahrb. Geol. Palaont. Abh.,* 125 (Festband Schindewolf): 1–18.

Simpson, G. G., 1966b. The biological nature of man. *Science* 152:472–478.

Simpson, G. G., 1967. The Tertiary Lorisiform primates of Africa. *Bull. Mus. Comp. Zool. Harvard* 136:39–61.

Simpson, G. G., 1969. South American Mammals, in: *Biogeography and Ecology in South America* (E. J. Fitkau, J. Illies, H. Klinge, G. H. Schwabe, and H. Siolo, eds.), 2:879–909, W. Junk, The Hague.

Simpson, G. G., 1975. Recent advances in methods of phylogenetic inference, in: *Phylogeny of the Primates: A Multidisciplinary Approach* (W. P. Luckett and F. S. Szalay, eds.), pp. 3–19, Plenum Press, New York.

Simpson, G. G., 1978. Early mammals in South America: fact, controversy and mystery. *Proc. Amer. Philos. Soc.* 122:318–328.

Simpson, G. G., Minoprio, J. L., and Patterson, B. 1962. The mammalian fauna of the Divisadero Largo Formation, Mendoza, Argentina. *Bull. Amer. Mus. Comp. Zool.* 127:237–293.

Skaryd, S. M. 1971. Trends in the evolution of the pongid dentition. *Amer. J. Phys. Anthro.* 35:223–239.

Smith, A. G., and Briden, J. C. 1977. *Mesozoic and Cenozoic Paleocontinental Maps,* Cambridge University Press, Cambridge.

Smith, E. L. 1957. Active site and structure of crystalline papain. *Fed. Proc.* 16:801–804.

Smith, G. E. 1912. The origin of man. *Smithsonian Inst. Ann. Rep. 1912:*553–572.

Smith, G. E. 1925. The fossil anthropoid ape from Taungs. *Nature (Lond.,)* 115:235.

Smith P. J. 1977. Origin of the Rio Grande rise. *Nature (Lond.)* 269:651–652.

Smith, R. J., and Pilbeam, D. R. 1980. Evolution of the orang-utan. *Nature (Lond.)* 284: 447–448.

Smith, R. T. 1961. Immunological tolerance of non-living antigens, in: *Advances in Immunology, Vol. 1* (W. H. Taliaferro and J. H. Humphrey, eds.), pp. 65–131, Academic Press, New York.

Sneath, P. H. A. 1962. Comparative biochemical genetics in bacterial taxonomy, in: *Taxonomic Biochemistry and Serology* (C. Leone, ed.), pp. 565–583, Ronald Press, New York.

Socha, W. W., and Moor-Jankowski, J. 1979. Blood groups of Anthropoid apes and their relationship to human blood groups. *J. Hum. Evol.* 8:453–465.

Sondaar, P. Y. 1977. Insularity and its effects on mammal evolution, in: *Major Patterns in Vertebrate Evolution* (M. Hecht, P. Goody, and B. Hecht, eds.), pp. 671–707, Plenum Press, New York.

Sondaar, P., and Boekschoten, G. J. 1967. Quaternary mammals in the South Aegean islands area, with notes on other fossil mammals from the coastal regions of the Mediterranean, I and II. *Proc. Kon. Neder. Akad. Wetensch. Amsterdam, B.* 70:556–576.

Sonntag, C. F. 1923. The anatomy, physiology and pathology of the chimpanzee. *Proc. Zool. Soc. London* 1923:323–429.

Sonntag, C. F. 1924. *The Morphology and Evolution of Apes and Man,* Bole and Dainelson, London.

Sorenson, M. W. 1970. Behavior of tree shrews, in: *Primate Behavior: Developments in Field and Laboratory Research, Vol. 1* (L. A. Rosenblum, ed.), pp. 141–193, Academic Press, New York.

Sorenson, M. W., and Conaway, C. H. 1964. Observations of tree shrews in captivity. *J. Sabah Soc.* 2:77–91.

Sorenson, M. W., and Conaway, C. H. 1966. Observations on the social behavior of tree shrews in captivity. *Folia Primatologica* 4:124–145.

Southwick, C. H., and Siddiqi, M. F. 1976. Demographic characteristics of semi-protected Rhesus groups in India. *Yrbk. Phys. Anthro.* 20:242–252.

Spatz, W. B. 1968. Die Bedeutung der Augen fur die Sagittale guestaltung des Schadels von *Tarsius* (Prosimiae, Tarsiiformes). *Folia Primatologica* 9:22–40.

Spatz, W. B. 1970. Binohulares Sehen ubd Kopfgestaltung: ein Beitrag zum Problem des Gestalt wandels des Schadels der Primaten, insbesondere der Lorisidae. *Acta Anat.* 75:489–520.

Sprankel, H. 1961. Uber Verhaltenweisen und Zucht von *Tupaia glis* (Diard 1820) in Gefangenschaft. *Z. Wiss. Zool.* 165:186–220.

Standing, H. 1908. On recently discovered subfossil primates from Madagascar. *Trans. Zool. Soc. London* 18:69–112.

Standing, H. 1909. Note sur less ossements subfossiles provenant des fouilles d'Ampasambazimba. Les Lemuriens. *Bull. Acad. Malg.* 1:61–64.

Stanley, S. M. 1979. *Macroevolution,* W. H. Freeman, San Francisco.

Starck, D. 1975. The development of the chondrocranium in primates, in: *Phylogeny of the Primates: A Multidisciplinary Approach* (W. P. Luckett and F. S. Szalay, eds.), pp. 127–155, Plenum Press, New York.

Stehlin, H. G. 1909. Remarques sur les faunules de mammiferes des couches Eocenes et Oligocenes du Bassin de Paris. *Bull. Soc. Geol. France, Ser. 4,* 9:488–520.

Stehlin, H. G. 1912. Die Saugetiere des schweizerischen Eocaens—*Adapis. Abh. Schweiz. Palaeont. Ges.* 38:1165–1298.

Stehlin, H. G. 1916. Die Saugetiere des schweizerischen Eocins. Eritischer Katalog der Materialien 7th Theil, Zweite Hälfte. *Abh. Schweiz. Pal. Ges.* 41:1297–1552.

Stehlin, H. G., and Schaub, S. 1951. Die Trigonodontie der simplicidentaten Nager. *Schweiz. Pal. Abh.* 67:1–385.

Steinbacher, G. 1940. Beobachtungen am Spitzhornchen und Panda. *Zool. Gart.* 12:48–53.

Stern, J. T., Jr. 1971. Functional myology of the hip and thigh of cebid monkeys and its implications for the evolution of erect posture. *Bib-*

lio. *Primatol.* 14:1–318.

Stern, J. T., Jr., and Oxnard, C. 1974. Primate locomotion: some links with evolution and morphology. *Bibl. Primatol.,* Vol. 4, S. Karger, Basel.

Stern, J. T., Jr., and Susman, R. L. 1981. Electromyography of the gluteal muscles in *Hylobates, Pongo,* and *Pan:* implications for the evolution of hominid bipedality. *Amer. J. Phys. Anthro.* 55:153–166.

Stern, J. T., Jr., and Susman, R. L. 1983. The locomotor anatomy of *Australopithecus afarensis.* Amer. J. Phys. Anthro. 60:279–317.

Stibbe, E. P. 1928. A comparative study of the nictitating membrane of birds and mammals. *J. Anat.* 62:159–176.

Stirton, R. A. 1951. Ceboid monkeys from the Miocene of Columbia. *Bull. Univ. Calif. Publ. Geol. Sci.* 28(11):315–356.

Stirton, R. A. 1953. Vertebrate paleontology and continental stratigraphy in Columbia. *Bull. Geol. Soc. Amer.* 64:603–622.

Stirton, R. A., and Savage, D. E. 1951. A new monkey from the La Venta Miocene of Columbia. Serv. Geol. Nac., Bogota 7:345–356.

Stock, C. 1933. An Eocene primate from California. *Proc. Natl. Acad. Sci., Washington,* 19:954–959.

Stock, C. 1934. A second Eocene primate from California. *Proc. Natl. Acad. Sci., Washington,* 20:150–153.

Stock, C. 1938. A tarsiid primate and a mixodectid from the Poway Eocene, California. *Proc. Natl. Acad. Sci., Washington,* 24:288–293.

Storr, G. C. C. 1780. *Prodromus Methodi Mammalian,* Wolffer, Tubingen.

Straus, W. L., Jr. 1940. The posture of the great ape hand in locomotion and its phylognentic implications. *Amer. J. Phys. Anthro.* 27:199–207.

Straus, W. L., Jr. 1941. The phylogeny of the human forearm extensors. *Hum. Biol.* 13:23–50, 203–238.

Straus, W. L., Jr. 1942a. Rudimentary digits in primates. *Quart. Rev. Biol.* 17:228–243.

Straus, W. L., Jr. 1942b. The homologies of the forearm flexors: Urodeles, lizards, mammals. *Amer. J. Anat.* 70:281–316.

Straus, W. L., Jr. 1949. The riddle of man's ancestry. *Quart. Rev. Biol.* 24:200–223.

Straus, W. L., Jr. 1953. Primate, in: *Anthropology Today* (A. L. Kroeber, ed.), pp. 77–92, Wenner-Gren Foundation, New York.

Straus, W. L., Jr. 1962. Fossil evidence for the evolution of the erect bipedal posture. *Clin. Orthopaed.* 25:9–19.

Straus, W. L., Jr. 1963. The classification of *Oreopithecus,* in: *Classification and Human Evolution* (S. L. Washburn, ed.), pp. 146–177, Aldine, Chicago.

Stromer, E. 1913. Mitteilungen uber die Wirbeltierreste aus dem Mittelpliocaen des Natrontales (Agypten). *A. Dtsch. Geol. Ges.* 65:350–372.

Struhsaker, T. T. 1967. Ecology of vervet monkeys *(Cercopithecus aethiops)* in the Masai, Amboseli Game Reserve, Kenya. *Ecology* 48:841–904.

Struhsaker, T. T. 1975. *The Red Colobus Monkey,* Univ. Chicago Press, Chicago.

Struhsaker, T. T. 1978. Interrelationships of red

colobus monkeys and rain forest trees in the Kibale Forest, Uganda, in: *The Ecology of Arboreal Folivores* (G. G. Montgomery, ed.), pp. 397–422, Smithsonian Institute, Washington, D.C.

Struhsaker, T. T., and Leland, L. 1977. Palm nut smashing by *Cebus a. apella* in Columbia. *Biotropica* 9:124–126.

Struhsaker, T. T., and Oates, J. F. 1979. Comparison of the behavior and ecology of Red Colobus and Black-and-White Colobus monkeys in Uganda: a summary, in: *Primate Ecology: Problem Oriented Field Studies* (R. W. Sussman, ed.), pp. 165–186, Wiley, New York.

Sudre, J. 1975. Un prosimien du Paleogene ancien du Sahara Nord-Occidental: Azibius trerki n.g.n.sp. *C. R. Acad. Sci. Paris, Ser. D.* 280:1539–1542.

Sugiyama, Y. 1968. Social organization of chimpanzees in the Budongo Forest, Uganda. *Primates* 9:225–258.

Sugiyama, Y. 1969. Social behavior of chimpanzees in the Budongo Forest, Uganda. *Primates* 10:197–225.

Susman, R. L. 1974. Facultative terrestrial hand posture in an orang-utan *(Pongo pygmaeus)* and pongid evolution. *Amer. J. Phys. Anthro.* 40:27–37.

Susman, R. L., Ed. 1984. *The Pygmy Chimpanzee: Evolutionary Biology and Behavior.* Plenum Press, New York.

Sussman, R. W., Ed. 1979. *Primate Ecology: Problem Oriented Field Studies,* Wiley, New York.

Swindler, D. R. 1976. *Dentition of Living Primates,* Academic Press, New York.

Swindler, D. R., McCoy, H. A., and Hornbeck, P. V. 1967. Dentition of the baboon *(Papio anubis),* in: *The Baboon in Medical Research, Vol. 2* (H. Vagtborg, ed.), pp. 133–150, Texas Univ. Press, Austin.

Szalay, F. S. 1968a. The beginnings of primates. *Evolution* 22:19–36.

Szalay, F. S. 1968b. The Picrodontidae, a family of early primates. *Amer. Mus. Novit.* No. 2329:1–55.

Szalay, F. S. 1969. Mixodectidae, Microsyopidae, and the insectivore-primate transition. *Bull. Amer. Mus. Nat. Hist.* 140:195–330.

Szalay, F. S. 1970. Late Eocene *Amphipithecus* and the origins of Catarrhine primates. *Nature (Lond.)* 227:355–357.

Szalay, F. S. 1972a. Cranial morphology of the early Tertiary *Phenacolemur* and its bearing on primate phylogeny. *Amer. J. Phys. Anthro.* 36:59–76.

Szalay, F. S. 1972b. Paleobiology of the earliest primates, in: *The Functional and Evolutionary Biology of Primates* (R. Tuttle, ed.), pp. 3–35, Aldine, Chicago.

Szalay, F. S. 1972c. *Amphipithecus* revisited. *Nature (Lond.)* 236:179–180.

Szalay, F. S. 1972d. Hunting-scavenging protohominids: a model for hominid origins. *Man* 10:420–429.

Szalay, F. S. 1973. New Paleocene primates and a diagnosis of the new suborder Paromomyiformes. *Folia Primatologica* 19:73–87.

Szalay, F. S. 1974a. A review of some recent advances in paleoprimatology. *Yrbk. Phys.*

Anthro. 17:39–64.

Szalay, F. S. 1974b. New genera of European adapid primates. *Folia Primatologica* 22: 116–133.

Szalay, F. S. 1975a. Where to draw the nonprimate-primate taxonomic boundary. *Folia Primatologica* 23:158–163. (See Chapter 3, this volume)

Szalay, F. S. 1975b. Phylogeny, adaptations, and dispersal of the tarsiiform primates, in: *Phylogeny of the Primates* (W. P. Luckett and F. S. Szalay, eds.), pp. 357–404, Plenum Press, New York.

Szalay, F. S. 1975c. Haplorhine phylogeny and the status of the Anthropoidea, in: *Primate Functional Morphology and Evolution* (R. Tuttle, ed.), pp. 3–22, Mouton, The Hague.

Szalay, F. S. 1975d. Phylogeny of higher primate taxa, in: *Phylogeny of the Primates* (W. P. Luckett and F. S. Szalay, eds.), pp. 91–125, Plenum Press, New York.

Szalay, F. S. 1976. Systematics of the Omomyidae (Tarsiiformes, Primates): Taxonomy, phylogeny and adaptations. *Bull. Amer. Mus. Nat. Hist.* 156:157–450.

Szalay, F. S. 1977a. Ancestors, descendants, sister groups and testing of phylognentic hypothesis. *Syst. Zool.* 26:12–18.

Szalay, F. S. 1977b. Constructing primate phylogenics: a search for testable hypotheses with maximum empirical content. *J. Hum. Evol.* 6:3–18.

Szalay, F. S. 1981. Phylogeny and the problem of adaptive significance: The case of the earliest primates. *Folia Primatologica* 36:157–182.

Szalay, F. S., and Dagosto, M. 1980. Locomotor adaptations as reflected in the humerus of Paleogene primates. *Folia Primatologica* 34:1–45.

Szalay, F. S., and Decker, R. L. 1974. Origins, evolution and function of the tarsus in late Cretaceous eutherians and Paleocene primates, in: *Primate Locomotion* (F. A. Jenkins, Jr., ed.), pp. 239–259, Academic Press, New York.

Szalay, F. S., and Delson, E. 1979. *Evolutionary History of the Primates,* Academic Press, New York.

Szalay, F. S., and Drawhorn, G. 1980. Evolution and diversification of the Archonta in an arboreal milieu, in: *Comparative Biology and Evolutionary Relationships of Tree Shrews* (W. P. Luckett, ed.), pp. 133–170, Plenum Press, New York.

Szalay, F. S., and Katz, C. C. 1973. Phylogeny of lemurs, galagos and lorises. *Folia Primatologica* 19:88–103.

Szalay, F. S., and Seligsohn, D. 1977. Why did the strepsirhine tooth comb evolve? *Folia Primatologica* 27:75–82.

Szalay, F. S., and Wilson, J. A. 1976. Basicranial morphology of the early Tertiary tarsiiform *Rooneyia* from Texas. *Folia Primatologica* 25:288–293.

Szalay, F. S., Tattersall, I., and Decker, R. L. 1975. Phylogenetic relationships of *Plesiadapis*—postcranial evidence. *Contrib. Primatol.* 5:136–166.

Taieb, M. 1975. La decouverte de restes d'homi-
nides vieux de plus trois millions d'annes en Ethiopie. *Bull. Mem. Soc. Anthropol., Paris, Ser. 2,* 13:87–89.

Taieb, M., Johanson, D. C., and Coppens, Y. 1975. Expedition internationale d l'Afar, Ethiopie (3e campagne 1974): Decouverte d'hominides plio-pleistocenes a Hadar. *C. R. Acad. Sci., Paris, Ser. D.* 281:1297–1300.

Taieb, M., Coppens, Y., Johanson, D. C., and Bonnefille, R. 1975. Hominides de l'Afar central, Ethiopie. *Bull. Mem. Soc. Anthropol., Paris, Ser. 2.* 13:117–124.

Taieb, M., Johanson, D. C., Coppens, Y., and Aronson, J. L. 1976. Geological and paleontological background of Hadar hominid site, Afar, Ethiopia. *Nature (Lond.)* 260:289–293.

Taieb, M., Johanson, D. C., Coppens, Y., and Tiercelin, J. J. 1978. Expedition internationale de l'Afar, Ethiopie (4 eme et 5 eme Campagne 1975–1977): Chronostratigraphie des gisements a hominides pliocene de l'Hadar et correlations avec les sites prehistoriques du Kada Gona. *C. R. Acad. Sci. Paris, Ser. D.* 287:459–461.

Taieb, M., Johanson, D. C., Coppens, Y., Bonnefille, R., and Kalb, J. 1974. Decouverte d'Hominides dans les series plio-pleistocenes d'Hadar (Bassin de l'Awash; Afar, Ethiopie). *C. R. Acad. Sci., Paris, Ser. D.* 279:735–738.

Tarling, D. H. 1979. Continental drift and the positioning of the circum-Atlantic continents throughout the last 100 million years of Earth history, in: *Abstracts of the VII Congress of the International Primatological Society* (M. Moudgal, ed.) Bangalore, India.

Tarling, D. H. 1980. The geologic evolution of South America with special reference to the last 200 million years, in: *Evolutionary Biology of the New World Monkeys and Continental Drift* (R. L. Ciochon and A. B. Chiarelli, eds.), pp. 1–41, Plenum Press, New York.

Tattersall, I. 1969a. Ecology of the north Indian *Ramapithecus. Nature (Lond.)* 221:451–452.

Tattersall, I. 1969b. More on the ecology of north Indian *Ramapithecus. Nature (Lond.)* 224:821–822.

Tattersall, I. 1971. Revision of the subfossil Indriinae. *Folia Primatologica* 16:257–269.

Tattersall, I. 1973a. Cranial anatomy of Archeolemurinae. *Anthropol. Pap. Amer. Mus. Nat. Hist.* 52(1):1–110.

Tattersall, I. 1973b. Subfossil lemuroids and the "adaptive radiation" of the Malagasy lemurs. *Trans. N.Y. Acad. Sci.* 35:314–324.

Tattersall, I. 1973c. A note on the age of the subfossil site of Ampasambazimba, Miarinarivo Province, Malagasy Republic. *Amer. Mus. Novit.* No. 2520:1–6.

Tattersall, I. 1974. Facial structure and mandibular mechanics in *Archaeolemur,* in: *Prosimian Biology* (R. A. Martin, G. A. Doyle, and A. C. Walker, eds.), pp. 563–577, Duckworth, London.

Tattersall, I. 1975a. *The Evolutionary Significance of Ramapithecus.* Burgess, Minneapolis.

Tattersall, I. 1975b. Notes on the cranial anatomy of the subfossil Malagasy lemurs, in: *Lemur Biology* (I. Tattersall, ed.), pp. 111–124, Plenum Press, New York. (See Chapter 10, this volume)

Tattersall, I. 1982. *The Primates of Madagascar,* Columbia University Press, New York.

Tattersall, I., and Eldredge, N. 1977. Fact, theory and fantasy in human paleontology. *Amer. Sci.* 65:204–211.

Tattersall, I., and Schwartz, J. H. 1974. Craniodental morphology and the systematics of the Malagasy lemurs (Prosimii, Primates). *Anthropol. Pap. Amer. Mus. Nat. Hist.* 52(3): 141–192.

Tattersall, I., and Sussmann, R. W., Eds. 1975. *Lemur Biology,* Plenum Press, New York.

Tattersall, I., and Sussman, R. W. 1975. Notes on topography, climate, and vegetarian of Madagascar, in: *Lemur Biology* (I. Tattersall, ed.), pp. 13–21, Plenum Press, New York.

Tax, S., Ed. 1960. *Evolution After Darwin,* 3 Vols., University of Chicago Press, Chicago.

Taylor, C. R., and Rowntree, V. J. 1973. Running on 2 or 4 legs: which consumes more energy? *Science* 179:186–187.

Tchernov, E. 1968. *Succession of Rodent Families During the Upper Pleistocene of Israel,* Verlag Paul Parey, Hamburg.

Tedford, R. H. 1974. Marsupials and the new paleogeography, in: *Paleogeographic Provinces and Provinciality* (C. A. Ross, ed.), *Soc. Econ. Palaeontol. Mineral. Spec. Publ.* 21: 109–126.

Teilhard De Chardin, P. 1921. Les mammiferes de l'Eocene inferieur francais et leurs gisements. *Ann. Palaeontol.* 10:171–176; 11:108.

Teilhard De Chardin, P. 1922. Les mammiferes de l'Eocene inferieur francais et leurs gisements. *Ann. Palaeontol.* 11:9–116.

Teilhard De Chardin, P. 1927. Les mammiferes de l'Eocene inferieur de la Belgique. *Mem. Mus. Roy. Hist. Nat. Belgique* 36:1–33.

Teleki, G. 1974. Chimpanzee subsistence technology: materials and skills. *J. Hum. Evol.* 3:575–594.

Teleki, G. 1976. Demographic observations (1963–1973) on the chimpanzees of Gombe National Park, Tanzania. *J. Hum. Evol.* 5: 559–598.

Thomas, O. 1913. On some rare amazonian mammals from the collection of the Para Museum. *Ann. Mag. Natur. Hist.* 11:130–136.

Thorington, R. W. 1976. The systematics of New World monkeys, in: *First Inter-American Conference on Conservation and Utilization of American Nonhuman Primates in Biomedical Research,* Pan African Health Organization, Sci. Publ. No. 317, pp. 8–18.

Thorington, R. W., and Groves, C. P. 1970. An annotated classification of the Cercopithecoidea, in: *Old World Monkeys* (J. R. Napier and P. H. Napier, eds.), pp. 631–647, Academic Press, New York.

Thorndike, E. E. 1968. A microscopic study of the marmoset claw and nail. *Amer. J. Phys. Anthro.* 28:247–262.

Tobias, P. V. 1965. Cranial capacity of *Zinjanthropus* and other Australopithecines. *Curr. Anthro.* 6(4):414–417.

Tobias, P. V. 1966. The distinctiveness of *Homo habilis. Nature (Lond.)* 209:953–957.

Tobias, P. V. 1967. *Olduvai Gorge, Vol. 2, The Cranium and Maxillary Dentition of Australopithecus (Zinianthropus) boisei,* Cambridge University Press, Cambridge.

Tobias, P. V. 1969a. Cranial capacity in fossil Hominidae. Lecture, Amer. Mus. Nat. Hist., May 1969.

Tobias, P. V. 1969b. The taxonomy and phylogeny of the Old World Primates with Reference to the Origin of Man (A. B. Chiarelli, ed.), pp. 277–315, Rosenberg and Sellier, Torino.

Tobias, P. V. 1971. *The Brain in Hominid Evolution,* Columbia University Press, New York.

Tobias, P. V. 1973a. Implications of the new age estimates of the early South African hominids. *Nature (Lond.)* 246:79–83.

Tobias, P. V. 1973b. New developments in hominid paleontology in South and East Africa. *Ann. Rev. Anthro.* 2:311–334.

Tobias, P. V. 1980a. A survey and synthesis of the African hominids of the late Tertiary and early Quaternary periods, in: *Current Argument on Early Man* (L. K. Konigsson, ed.) pp. 86–113, Pergamon, Oxford.

Tobias, P. V. 1980b. *"Australopithecus afarensis"* and *A. africanus:* Critique and alternative hypothesis. *Paleontol. Afr.* 23:1–17.

Tobias, P. V., and Hughes, A. R. 1969. The new Witwatersrand excavation at Sterkfontein. Progress report, some problems and first results. *S. Afr. Archeol. Bull.* 24:158–169.

Tobias, P. V., and Von Koenigswald, G. H. R. 1964. A comparison between the Olduvai hominines and those of Java and some implications for hominid phylogeny. *Nature (Lond.)* 204:515–518.

Tomasi, T. B., and Zibelbaum, S. 1963. The selective appearance of Y. A. globulins in certain body fluids. *J. Clin. Invest.* 42:1552.

Tomblin, J. F. 1975. The Lesser Antilles and Aves Ridge, in: *The Ocean Basins and Margins, Vol. 3, The Gulf of Mexico and the Caribbean* (A. E. M. Nairn and F. G. Stehli, eds.), pp. 467–500, Plenum Press, New York.

Tristram, G. R. 1953. The amino acid composition of proteins, in: *The Proteins, Vol. 1* (A. H. Neurath and K. Bailey, eds.), pp. 181–234, Academic Press, New York.

Trivers, R. 1972. Parental investment and sexual selection, in: *Sexual Selection and the Descent of Man: 1871–1971* (B. Campbell, ed), pp. 136–179, Aldine, Chicago.

Trouessart, E. L. 1879. Catalogue des mammiferes vivants et fossiles. *Rev. Mag. Zool., Paris, Ser. 3,* 7:219–285.

Tullberg, T. 1899. Ueber das System der Nagethiere, eine phylogenetische Studie. *Nov. Acta Reg. Soc. Sci. Upsal., Ser. 3,* 18:1–514, A1–A18.

Tuttle, R. H. 1965. *The Anatomy of the Chimpanzee Hand, with Comments on Hominoid Evolution,* University Microfilms, Ann Arbor.

Tuttle, R. H. 1967. Knuckle-walking and the evolution of hominoid hands. *Amer. J. Phys. Anthro.* 26:171–206.

Tuttle, R. H. 1969a. Knuckle-walking and the problem of human origins. *Science* 166:953–961.

Tuttle, R. H. 1969b. Quantitative and functional studies on the hands of the Anthropoidea. I. The Hominoidea. *J. Morphol.* 128:309–364.

Tuttle, R. H. 1969c. Terrestrial trends in the hands of the Anthropoidea: a preliminary report, in: *Proceedings of the 2nd International Congress of Primatology, Vol. 2,* 192–200, S. Karger, Basel.

Tuttle, R. H. 1970. Postural, propulsive and prehensile capabilities in the cheiridia of chimpanzees and other great apes, in: *The Chimpanzee, Vol. 2* (G. H. Bourne, ed.), pp. 167–253, S. Karger, Basel.

Tuttle, R. H. 1972a. Knuckle-walking hand postures in an orang-utan *(Pongo pygmaeus).* *Nature (Lond.)* 236:33–34.

Tuttle, R. H. 1972b. Functional and evolutionary biology of hylobatid hands and feet, in: *Gibbon and Siamang* (D. Rumbaugh, ed.), pp. 136–206, S. Karger, Basel.

Tuttle, R. H. 1972c. Relative mass of cheiridial muscles in catarrhine primates, in: *The Functional and Evolutionary Biology of Primates* R. H. Tuttle, ed.), pp. 262–291, Aldine, Chicago.

Tuttle, R. H. 1974. Darwin's apes, dental apes, and the descent of man: normal science in evolutionary anthropology. *Curr. Anthropol.* 15:389–398. (See Chapter 42, this volume)

Tuttle, R. H. 1975a. Knuckle-walking and knuckle-walkers: A commentary on some recent perspectives on hominoid evolution, in: *Primate Functional Morphology and Evolution* (R. H. Tuttle, ed.), pp. 203–212, Mouton, The Hague.

Tuttle, R. H. 1975b. Parallellism, brachiation and hominoid phylogeny, in: *Phylogeny of the Primates* (W. P. Luckett and F. S. Szalay, eds.), pp. 447–480, Plenum Press, New York.

Tuttle, R. H. 1981. Evolution of hominid bipedalism and prehensile capabilities. *Phil. Trans. R. Soc., London, Ser. B.* 292:89–94.

Tuttle, R. H., and Basmajian, J. V. 1974a. Electromyography of brachial muscles in *Pan gorilla* and hominoid evolution. *Amer. J. Phys. Anthro.* 41:71–90.

Tuttle, R. H., and Basmajian, J. V. 1974b. Electromyography of the manual long digital flexor muscles in gorilla, in: *Proceedings of the VIth International Congress of Physical Medicine* (F. Barndsell, ed.), pp. 311–315, Ministerio de Trabajo, Instuto Nacional de Prevision, Vol. 2.

Tuttle, R. H., and Basmajian, J. V. 1974c. Electromyography of forearm muscles in gorilla and problems related to knuckle-walking, in: *Primate Locomotion* (F. A. Jenkins, Jr., ed.), pp. 293–347, Academic Press, New York.

Tuttle, R. H., and Beck, B. B. 1972. Knuckle-walking hand postures in an orangutan *(Pongo pygmaeus). Nature (Lond.)* 236:33–34.

Tuttle, R. H., and Rogers, C. M. 1966. Genetic and selective factors in reduction of the hallux in *Pongo pygmaeus. Amer. J. Phys. Anthro.* 24:191–198.

Tuttle, R. H., Basmajian, J. V., and Ishida, H. 1975. Electromyography of the gluteus maximus muscle in *Gorilla* and the evolution of hominid bipedalism, in: *Primate Functional and Evolutionary Biology* (R. H. Tuttle, ed.), pp. 251–269, Mouton, The Hague.

Tuttle, R. H., Basmajian, J. V., Regenos, E., and Shine, G. 1972. Electromyography of knuc-

kle-walking: results of four experiments on the forearm of *Pan gorilla. Amer. J. Phys. Anthro.* 37:255–266.

Udry, J. R., and Morris, N. M. 1968. Distribution of coitus in the menstrual cycle. *Nature (Lond.)* 220:593–596.

Ulmer, F. A. 1963. Observations on the Tarsier in captivity. *Zool. Garten* 27:106–121.

Uzzell, T., and Pilbeam, D. R. 1971. Phyletic divergence dates of hominoid primates: a comparison of fossil and molecular data. *Evolution* 25:615–635.

Valentine, J. W. 1973. Plates and provinciality, a theoretical history of environmental discontinuities, in: *Organisms and Continents Through Time* (N.F. Hughes, ed.) *Spec. Pap. Palaeontol.* 12:79–92.

Valois, H. 1955. Ordre des Primates, in: *Traite de Zoologie* (P. P. Grasse, ed.), 17–2:1854–2206.

Van Andel, T. H., Theide, J., Sclater, J. G., and Hay, W. W. 1977. Depositional history of the South Atlantic Ocean during the last 125 million years. *J. Geol.* 85:651–698.

Van Couvering, J. A., and Miller, J. A. 1969. Miocene stratigraphy and age determinations, Rusinga Island, Kenya. *Nature (Lond.)* 221: 268–632.

Vandenbergh, J. G. 1963. Feeding, activity and social behavior of the tree shrew, *Tupaia glis,* in a large outdoor enclosure. *Folia Primatologica* 1:199–207.

Van Horn, R. N. 1972. Structural adaptations to climbing in the gibbon hand. *Amer. Anthropol.* 74:326–333.

Van Lawick-Goodall, J. 1967. Mother-offspring relationships in chimpanzees, in: *Primate Ethology* (D. Morris, ed.), pp. 287–346, Aldine, Chicago.

Van Lawick-Goodall, J. 1968. The behavior of free-living chimpanzees in the Gombe Stream area. *Anim. Behav. Monogr.* 1:161–311.

Van Lawick-Goodall, J. 1973. Cultural elements in a chimpanzee community, in: *Precultural Primate Behavior* (E. Menzel, ed.), pp. 144–185, S. Karger, Basel.

Van Valen, L. 1965. Treeshrews, primates and fossils. *Evolution* 19:137–151.

Van Valen, L. 1966. Deltatheridia, a new order of mammals. *Bull. Amer. Mus. Nat. Hist.* 132:1–126.

Van Valen, L. 1969. A classification of the primates. *Amer. J. Phys. Anthro.* 30:295–296.

Van Valen, L., and Sloan, R. E. 1965. The earliest primates. *Science* 150:743–745.

Verdcourt, B. 1963. The Miocene non-marine mollusca of Rusinga Island, Lake Victoria and other localities in Kenya. *Palaeontographia A* (Stuttgart) 121:1–37.

Verjyeyhen, W. N. 1962. Contribution a la craniologie comparee des primates. *Ann. Mus. Roy. Afr. Cent., Ser. 8vo, Sci. Zool.* 105: 1–247.

Verma, B. C. 1969. *Procynocephalus pinjori* sp.

nov.: a new fossil primate from the Pinjor Beds (Lower Pleistocene) east of Chandigarh. *J. Paleontol. Soc. India* 13:53–57.

Verma, K. 1965. Notes on the biology and anatomy of the Indian tree-shrew, *Anathana wroughtoni. Mammalia* 29:289–330.

Vinogradov, A. P., Ed. 1967. *Atlas of the Lithological-Paleogeographical Maps of the U.S.S.R., Vol. IV* (Paleogene, Neogene, and Quaternary) 1–55, Ministry of Geology of the U.S.S.R., Academy of Sciences of the U.S.S.R.

Vogel, C. 1966. Morphologische Studien am Gesichtsschadel catarrhiner Primaten. *Bibl. Primatol.* 4:1–226.

Vogel, C. 1975. Remarks on the reconstruction of the dental arcade of *Ramapithecus,* in: *Palaeoanthropology: Morphology and Paleoecology* (R. H. Tuttle, ed.), pp. 87–98, Mouton, The Hague.

Von Bartheld, F., Erdbrink, D. P., and Krommenhoek, W. 1970. A fossil incisor from Uganda and a method for its determination. *Proc. K. Ned. Akad. Wet. Amsterdam B.* 73:426–431.

Von Kampen, D. 1905. Die tympanalgegend des Saugetierschadde. *Gegenbaurs Morphol. Jahrb.* 34:321–722.

Von Koenigswald, G. H. R. 1957a. *Meganthropus* and the Australopithecinae. *Proc. 3rd. Pan-Afr. Congr. Prehist. (Livingstone 1955).* pp. 158–160.

Von Koenigswald, G. H. R. 1957b. Remarks on *Gigantopithecus* and other hominid remains from South China. *Prock. K. Ned. Akad. Wet. Amsterdam B.* 60:153–159.

Von Koenigswald, G. H. R. 1968. The phylogenetical position of the Hylobatidae, in: *Taxonomy and Phylogeny of Old World Primates with Reference to the Origin of Man* (A. B. Chiarelli, ed.), pp. 271–276, Rosenberg and Sellier, Torino.

Von Koenigswald, G. H. R. 1969. Miocene Cercopithecoidea and Oreopithecoidea from the Miocene of East Africa, in: *Fossil Vertebrates of Africa, Vol. 1* (L. S. B. Leakey, ed.), pp. 39–51. Academic Press, London.

Von Koenigswald, G. H. R. 1972a. Ein unterkiefer eines fossilen Hominoiden aus den Unterpliozan Griechenlands. *Proc. K. Ned. Akad. Wet. Amsterdam B,* 75:385–394.

Von Koenigswald, G. H. R. 1972b. Was ist *Ramapithecus? Natur und Museum* 102:173–183.

Von Koenigswald, G. H. R. 1973. The position of *Proconsul* among the Pongidae, in: *Symposia of the 4th International Congress of Primatology, Vol. 3, Craniofacial Biology of Primates,* pp. 148–153, S. Karger, Basel.

Von Koenigswald, G. H. R. 1981. A possible ancestral form of *Gigantopithecus* (Mammalia, Hominoidea) from the Chinji layers of Pakistan. *J. Hum. Evol.* 10:511–515.

Von Koenigswald, G. H. R. 1983. The significance of hitherto undescribed Miocene hominoids from the Siwaliks of Pakistan in the Senckenberg Museum, Frankfurt, in: *New Interpretations of Ape and Human Ancestry* (R. L. Ciochon and R. S. Corrucii, eds.), pp. 517–526, Plenum Press, New York.

Vorouison, H. H. 1962. The ways of food specialization and evolution of the alimentary system in Muroidea, in: *Symposium Theriologicund* (J. Kratochvil and J. Pelikan, eds.), pp. 360–377, Czech Academy of Sciences , Brno.

Voruz, C. 1970. Origine des dents bilophodontes des Cercopithecoidea. *Mammalia* 34:269–273.

Vrba, E. S. 1975. Some evidence of chronology and paleoecology of Sterkfontein, Swartkrans, and Kromdraai from the fossil Bovidae. *Nature (Lond.)* 254:301–304.

Walker, A. 1967a. Patterns of extinction among the subfossil Madagascan lemuroids, in: *Pleistocene Extinctions* (P. S. Martin and H. F. Wright, Jr., eds.), pp. 425–432, Yale University Press, New Haven.

Walker, A. 1967b. *Locomotor Adaptations in Recent and Subfossil Madagascan Lemurs.* Ph.D. Dissertation, University of London.

Walker, A. 1969. True affinities of *Propotto leakeyi* Simpson 1967. *Nature (Lond.)* 223: 647–648.

Walker, A. 1971. Late australopithecine from Baringo District, Kenya: partial australopithecine cranium. *Nature (Lond.)* 230:512–514.

Walker, A. 1972. The dissemination and segregation of early primates in relation to continental configuration, in: *Calibration of Hominoid Evolution* (W. W. Bishop and J. A. Miller, eds.), pp. 195–218, Scottish Academic Press, Glasgow.

Walker, A. 1973. New *Australopithecus* femora from East Rudolf, Kenya. *J. Hum. Evol.* 2:545–555.

Walker, A. 1974. Locomotor adaptations in the past and present prosimian primates, in: *Primate Locmotion* (F. A. Jenkins, Jr., ed.), pp. 349–382, Academic Press, New York.

Walker, A. 1976. Splitting times among hominoids deduced from the fossil record, in: *Molecular Anthropology* (M. Goodman and R. E. Tashian, ed.,), pp. 63–77, Plenum Press, New York.

Walker, A. 1979. S. E. M. analysis of microwear and it correlation with dietary patterns. *Amer. J. Phys. Anthro.* 50:489.

Walker, A., and Leakey, R. E. F. 1978. The hominids of East Turkana. *Sci. Amer.* 239(2): 54–66.

Walker, A., and Pickford, M. 1983. New postcranial fossils of *Proconsul africanus* and *Proconsul nyanzae,* in: *New Interpretations of Ape and Human Ancestry* (R. L. Ciochon and R. S. Corruccini, eds.), pp. 325–351. Plenum Press, New York.

Walker, A., and Rose, M. 1968. Fossil hominoid vertebra from the Miocene of Uganda. *Nature (Lond.)* 217:980–981.

Walker, A., Hoeck, H. N., and Perez, L. 1978. Microwear of mammalian teeth as an indicator of diet. *Science* 201:908–910.

Walker, E. P. 1948. *Ill. London News* 208: 738–739.

Walker, E. P. 1964. *Mammals of the World,* John Hopkins Press, Baltimore.

Walker, P., and Murray, P. 1975. An assessment

of masticatory efficiency in a series of anthropoid primates with special reference to the Colobinae and Cercopithecinae, in: *Primate Functional Morphology and Evolution* (R. H. Tuttle, ed.), pp. 203–212.

Wallace, J. A. 1975. Dietary adaptations of *Australopithecus* and early *Homo,* in: *Paleoanthropology, Morphology and Paleoecology* (R. H. Tuttle, ed.), pp. 203–223, Mouton, The Hague.

Walls, G. L. 1942. *The Vertebrate Eye and its Adaptive Radiation,* Hafner Publishing Co., New York.

Ward, S. C., and Pilbeam, D. R. 1983. Maxillofacial morphology of Miocene hominoids from Africa and Indo-Pakistan, in: *New Interpretations in Ape and Human Ancestry* (R. L. Ciochon and R. S. Corruccini, eds.), pp. 211–238, Plenum Press, New York.

Ward, S. C., and Kimbel, W. H. 1983. Subnasal alveolar morphology and the systematic position of *Sivapithecus. Amer. J. Phys. Anthro.* 61(2)157–171.

Warneke, R. M. 1967. *Aust. Mammal Soc. Bull.* 2:94–108.

Waser, P. 1977. Feeding, ranging and group size in the mangabey *Cercocebus albigena,* in: *Primate Ecology* (T. H. Clutton Brock, ed.), pp. 183–222, Academic Press, New York.

Washburn, S. L. 1950. The analysis of primate evolution with particular reference to the origin of man. *Cold Spring Harbor Symp. Quant. Bio.* 15:67–78.

Washburn, S. L. 1957. Australopithecines: the hunters or the hunted? *Amer. Anthropol.* 59:612–614.

Washburn, S. L. 1959. Speculations on the interrelationships of the history of tools and biological evolution. *Hum. Biol.* 31:21–31.

Washburn, S. L. 1960. Tools and human evolution. *Sci. Amer.* 203:63–75.

Washburn, S. L. 1963a. Behavior and human evolution, in: *Classification and Human Evolution* (S. L. Washburn, ed.), pp. 190–203, Aldine, Chicago.

Washburn, S. L., Ed. 1963b. *Classification and Human Evolution,* Aldine, Chicago.

Washburn, S. L. 1967. Behavior and the origin of man: The Huxley Memorial Lecure 1967. *Proc. Roy. Anthro. Inst. Great Britain and Ireland,* pp. 21–27.

Washburn, S. L. 1968a. Speculation on the problem of man's coming to the ground, in: *Changing Perspectives on Man* (B. Rothblatt, ed.), pp. 191–206, University of Chicago Press, Chicago.

Washburn, S. L. 1968b. *The Study of Human Evolution,* Condon Lecture Series, Oregon State System of Higher Education, Eugene.

Washburn, S. L. 1968c. On Holloway's 'Tools and teeth'. *Amer. Anthropol.* 70:97–101.

Washburn, S. L. 1971. The study of human evolution, in: *Background for Man* (P. Dolhinow and V. Sarich, eds.), pp. 82–121, Little, Brown and Co., Boston.

Washburn, S. L. 1972. Human evolution, in: *Evolutionary Biology, Vol. 6* (T. Dobzhansky, M. K. Hecht, and W. C. Steere, eds.), pp. 349–361, Appleton-Century-Crofts, New York.

Washburn, S. L. 1973. Primate studies, in: *Nonhuman Primates and Medical Research* (G. H. Bourne, ed.), pp. 476–485, Academic Press, New York.

Washburn, S. L. 1978. The evolution of man. *Sci. Amer.* 239:194–208.

Washburn, S. L., and Avis, V. 1958. Evolution and human behavior, in: *Behavior and Evolution* (G. G. Simpson and A. Roe, eds.), pp. 421–436, Yale University Press, New Haven.

Washburn, S. L., and Ciochon, R. L. 1974. Canine teeth: notes on controversies in the study of human evolution. *Amer. Anthropol.* 76:765–784.

Washburn, S. L., and Ciochon, R. L. 1976. The Single Species Hypothesis. *Amer. Anthropol.* 78:96–98.

Washburn, S. L., and Detweiler, S. R. 1943. An experiment bearing on the problems of physical anthropology. *Amer. J. Phys. Anthro.* 1:171–190.

Washburn, S. L., and Hamburg, D. A. 1965. The study of primate behavior, in: *Primate Behavior* (I. DeVore, ed.), pp. 1–13, Holt, Rinehart and Winston, New York.

Washburn, S. L., and Howell, F. C. 1960. Human evolution and culture, in: *Evolution After Darwin, Vol. 2,* University of Chicago Press, Chicago.

Washburn, S. L., and Lancaster, C. S. 1968. The evolution of hunting, in: *Man the Hunter* (R. B. Lee and I. DeVore, eds.), pp. 293–303, Aldine, Chicago.

Washburn, S. L., and Moore, R. 1974. *Ape into Man,* Little, Brown, and Co., Boston.

Washburn, S. L., and Moore, R. 1980. *Ape into Human: a Study of Human Evolution,* Little, Brown and Co., Boston.

Washburn, S. L., and Patterson, B. 1951. Evolutionary importance of the South African Man-apes. *Nature (Lond.)* 167:650–651.

Watson, J. M. 1951. The wild mammals of Teso and Karamoja—VII. *Uganda J.* 15:193–202.

Waterhouse, G. R. 1839. Observations on the Rodentia. *Mag. Nat. Hist.* 3:90–96., 184–188, 274–279, 598–600.

Waterhouse, G. R. 1848. *A Natural History of the Mammalia, Vol. 2, Rodentia,* H. Bailliere, London.

Webb, C. S. 1953. *A Wanderer in the Wind,* Hutchinson, London.

Webb, S. D. 1976. Mammalian faunal dynamics of the great American interchange. *Paleobiology* 2:220–234.

Weber, M. 1928. *Die Saugetiere, Bd. 2, Systematischer Teil,* Fischer Verlag, Stuttgart.

Wegner, R. N. 1964. Der Schadel des Beutelbaren *(Phascolarctos cinereus* Goldfuss 1819) und seine Umformung durch lufthaltige Nebenhohlen. *Abh. Deut. Akad. Wissenschaft. Berlin, Klasse Chem., Geol., Biol.* 4:1–86.

Weidenreich, F. 1943a. The 'Neanderthal Man' and the ancestors of '*Homo sapiens*'. *Amer. Anthropol.* 45:39–48.

Weidenreich, F. 1943b. The skull of *Sinanthropus pekinensis:* a comparative study on a primitive hominid skull. *Palaeontologica Sinica* 10:1–484.

Weidenreich, F. 1945. Giant early man from Java and South China. *Anthro. Pap. Amer. Mus. Nat. Hist.* 40:1–134.

Weigelt, J. 1933. Neue Primaten aus der mitteleozanen (oberlutetischen) Braunkohle des Geiseltals. *Nova Acta Leopoldina, Halle am Salle, Neue Folge* 1:97–156.

Weindert, H. 1951. *Stammesentwicklung der Menscheit,* Vieweg and Sohn, Braunschweig.

Weiner, J. S. 1972. *Man's Natural History.* Weidenfield and Nicolson, London.

Weinert, H. 1932. *Ursprung der Menscheit,* Enke, Stuttgart.

Weinert, H. 1944. *Ursprung der Menschenheit: Uber den engeren Anschluss des Menschengeschlechts an die Menschenaffen,* Ferdinand Enke, Stuttgart.

Weinert, H. 1950. Uber die Neuen vor- und Fruhmenschenfunde aus Africa, Java, China und Frankreich. *Z. Morph. Anthro.* 42:113–148.

Weinert, H. 1951. uber die Vielgestaltigkeit der Summoprimaten vor der Menschwerdung. *Z. Morph. Anthro.* 43:73–103.

Werth, E. 1928. *Der fossile Mensch,* Borntrager, Berlin.

Westcott, R. W. 1967. Hominid uprightness and primate display. *Amer. Anthropol.* 69:738.

Weyl, R. 1974. Die palaeogeographische Entwicklung Mittelamerikas. *Zentralbl. Geol. Palaeontol.* Teil 1, 5/6:432–466.

Wharton, C. H. 1950. Notes on the Philippine tree shrew, *Urogale everetti* Thomas. *J. Mammal.* 31:352–354.

White, R. S., Venkatamanan, S. V., and Dabadghao, P. M. 1954. The grassland of India. *Congr. Bot.* 8:46–53.

White, T. D. 1975. Geomorphology to paleoecology: *Gigantopithecus* reappriased. *J. Hum. Evol.* 4:219–233.

White, T. D. 1977a. New fossil hominids from Laetolil, Tanzania. *Amer. J. Phys. Anthro.* 46:197–230.

White, T. D. 1977b. *The Anterior Mandibular Corpus of Early African Hominidae: Functional Significance of Shape and Size,* Ph.D. thesis, University of Michigan.

White, T. D. 1981. Primitive hominid canine from Tanzania. *Science* 213:348.

White, T. D., and Harris, J. M. 1977. Suid evolution and correlation of African hominid localities. *Science* 198:13–21.

White, T. D., Johanson, D. C., and Kimbel, W. H. 1983. *Australopithecus africanus:* its phyletic position reconsidered, in: *New Interpretations of Ape and Human Ancestry* (R. L. Ciochon and R. S. Corruccini, eds.), pp. 721–780, Plenum Press, New York.

Wiener, A. S., Gordon, E. B., and Moor-Jankowski, J. 1964. Immunological relationship between serum globulins of man and of other primates, revealed by a serological inhibition test. *Transfusion* 4:347–350.

Wiley, E. O. 1975. Karl R. Popper, systematics and classification: a reply to Walter Bock andother evolutionary taxonomists. *Syst. Zool.* 24:233–243.

Williams, C. A., Jr. 1964. Immunochemical analysis of serum proteins of the primates a study in molecular evolution, in: *Evolutionary and Genetic Biology of Primates* (J. Buettner-Janusch, ed.), pp. 25–74, Academic Press, New York.

Williams, C. B. 1947. The generic relations of species in small ecological communities. *J. Anim. Ecol.* 16:11–18.

Williams, C. H., and Wemyss, C. T., Jr. 1961. Experimental and evolutionary significance of similarities among serum protein antigens of man and the lower primates. *Ann. N.Y. Acad. Sci.* 94:77–92.

Williams, E. E., and Koopman, K. 1952. West Indian fossil monkeys. *Amer. Mus. Novit.* No. 1546:1–16.

Wilson, A. C. 1965. Immunological time scale for hominid evolution. *Science* 158:1200–1203.

Wilson, A. C., and Sarich, V. M. 1969. A molecular time scale for human evolution. *Proc. Natl. Acad. Sci.* 63:1088–1093.

Wilson, A. C., Carlson, S. S., and White, T. J. 1977. Biochemical evolution. *Ann. Rev. Biochem.* 46:573–639.

Wilson, J. A. 1966. A new primate from the earliest Oligocene, West Texas, preliminary report. *Folia Primatologica* 4:227–248.

Wilson, J. A., and Szalay, F. S. 1976. New adapid primate of European affinities from Texas. *Folia Primatologica* 25:294–312.

Wilson, R. W. 1951. Preliminary survey of a Paleocene faunule from Angel's Peak area, New Mexico. *Univ. Kansas Publ., Mus. Nat. Hist.* 5:1–11.

Wilson, R. W., and Szalay, F. S. 1972. New paromomyid primate from the Middle Paleocene beds, Kutz Canyon area, San Juan Basin, New Mexico. *Amer. Mus. Novit.* No. 2499:1–18.

Winzler, R. H. 1960. Glycoproteins, in: The Plasma Proteins, Vol. 1 (W. Putnam, ed.), pp. 309–348, Academic Press, New York.

Wislocki, G. B. 1928. Observations on the gross and microscopic anatomy of the sloths *(Bradypus griseus griseus* Gray and *Choloepus hoffinani* Peters). *J. Morph. Physiol.* 46(2):317–377.

Wislocki, G. B. 1929, On the placentation of the Primates, with consideration of the phylogeny of the placenta. *Contr. Embryol. Carnegie Inst.* 20:51–80.

Wolfe, H. R. 1933. Factors which may modify precipitin tests in their applications to serology and medicine. *Physiol. Zoo.* 6:55–90.

Wolfe, H. R. 1939. Standardisation of the precipitin technique and its application to studies of relationships in mammals, birds, and reptiles. *Biol. Bull.* 76:108–120.

Wolfe, J. A. 1978. A paleobotanical interpretation of Tertiary climates in the northern hemisphere. *Amer. Sci.* 66:694–703.

Wolfe, J. A., and Hopkins, D. M. 1967. Climatic changes recorded by Tertiary land floras in northwestern North America. *Symp. Pacific Sci. Congr.* 25:67–76.

Wolpoff, M. H. 1968. '*Telanthropus*' and the single species hypothesis. *Amer. Anthropol.* 70:477–491.

Wolpoff, M. H. 1970. The evidence for multiple hominid taxa at Swartkrans. *Amer. Anthropol.* 72:576–607.

Wolpoff, M. H. 1971a. *Metric Trends in Hominid Dental Evolution,* Case Western Reserve Studies in Anthropology 2.

Wolpoff, M. H. 1971b. Interstitial wear. *Amer. J. Phys. Anthro.* 34:205–228.

Wolpoff, M. H. 1971c. Competitive exclusion among Lower Pleistocene hominids: the single species hypothesis. *Man* 6:601–614. (See Chapter 35, this volume)

Wolpoff, M. H. 1971d. Is the new composite

cranium from Swartkrans a small robust australopithecine? *Nature (Lond.)* 230:398–401.

Wolpoff, M. H. 1973. Posterior tooth size, body size, and diet in the South African gracile australopithecines. *Amer. J. Phys. Anthro.* 39:375–394.

Wolpoff, M. H. 1974. The evidence for two australopithecine lineages in South Africa. *Yrbk. Phys. Anthro.* 17:113–139.

Wolpoff, M. H. 1975. Comment on *Ramapithecus* as a hominid, in: *Paleoanthropology: Morphology and Paleoecology* (R. H. Tuttle, ed.), pp. 174–176, Mouton, The Hague.

Wolpoff, M. H. 1976a. Some aspects of the evolution of early hominid sexual dimorphism. *Curr. Anthro.* 7:579–606.

Wolpoff, M. H. 1976b. Data and theory in paleoanthropological controversies. *Amer. Anthropol.* 78:94–96.

Wolpoff, M. H. 1978. Analogies and interpretation in paleoanthropology, in: *Early Hominids of Africa* (C. Jolly, ed.) pp. 461–503, Duckworth, London.

Wolpoff, M. H. 1979a. Anterior dental cutting in the Laetolil hominids and the evolution of the bicuspid P3. *Amer. J. Phys. Anthro.* 51:233–234.

Wolpoff, M. H. 1979b. cited in "Form and Function; The Anatomists View." Mosaic 10(2): 23–29.

Wolpoff, M. H. 1980. *Paleoanthropology,* Knopf, New York.

Wolpoff, M. H. 1981. Comment on: Bonobos: Generalized hominid prototypes or specialized insular dwarfs? Curr. Anthro. 22: 370–371.

Wolpoff, M. H. 1982. *Ramapithecus* and hominid origins. *Curr. Anthro.* 23:501–522. (See Chapter 33, this volume)

Wolpoff, M. H. 1983. *Ramapithecus* and human origins: an anthropologist's perspective of changing interpretations, in: *New Interpretations of Ape and Human Ancestry* (R. L. Ciochon and R. S. Corruccini, eds.), pp. 651–676, Plenum Press, New York.

Wolpoff, M. H., and Russell, M. D. 1981. Anterior dental cutting at Laetolil. *Amer. J. Phys. Anthro.* 55:223–224.

Wood, A. E. 1949. A new Oligocene rodent genus from Patagonia. *Amer. Mus. Novit.* No. 1435:1–54.

Wood, A. E. 1950. Porcupines, paleogeography and parallelism. *Evolution* 4:87–98.

Wood, A. E. 1955. A revised classification of the rodents. *J. Mammal.* 36:165–187.

Wood, A. E. 1959. Eocene radiation and phylogeny of the rodents. *Evolution* 13(3):354–361.

Wood, A. E. 1962. The early Tertiary rodents of the family Paramyidae. *Trans. Amer. Philos. Soc.* 52(1):1–261.

Wood, A. E. 1965. Grades and clades among rodents. *Evolution* 19(1):115–130.

Wood, A. E. 1968. Early Cenozoic mammalian faunas, Fayum province, Egypt. The African Oligocene Rodentia. *Bull. Peabody Mus. Nat. Hist.* 28:23–105.

Wood, A. E. 1972. An Eocene hystricognathous rodent from Texas: its significance in interpretations of continental drift. *Science* 175:1250–1251.

Wood, A. E. 1974. The evolution of the Old World and New World hystricomorphs. *Symp. Zool. Soc. London* 34:21–60.

Wood, A. E. 1980. The origin of the caviomorph rodents from a source in Middle America: a clue to the area of the origin of the platyrrhine primates, in: *Evolutionary Biology of the New World Monkeys and Continental Drift* (R. L. Ciochon and A. B. Chiarelli, eds.), pp. 79–91, Plenum Press, New York.

Wood, A. E., and Patterson, B. 1959. The rodents of the Deseaden Oligocene of Patagonia and the beginnings of South American rodent evolution. *Bull. Mus. Comp. Zool.* 120(3):279–428.

Wood, A. E., and Patterson, B. 1970. Relationships among hystricognathous and hystricomorphous rodents. *Mammalia* 34:628–637.

Wood, B. A. 1974. Locomotor affinities of hominoid tali from Kenya. *Nature (Lond.)* 246:45–46.

Wood, B. A. 1979. Tooth and body size allometric trends in modern primates and fossil hominids. *Amer. J. Phys. Anthro.* 50:493.

Wood, P., Vaczek, L., Hamblin, D. J. and Leonard, J. N. 1972. *Life Before Man,* Time-Life Books, New York.

Wood, S. 1844. Record of the discovry of an alligator with several new Mammalia in the freshwater strata at Hordwell. *Ann. Mag. Nat. Hist.* 14:349–351.

Wood, S. 1846. On the discovery of an alligator and of several new Mammalia in the Hordwell cliff; with observations upon the geological phenomena of that locality. *London Geol. J.* 1:1–7, 117–122.

Wood-Jones, F. 1915. The influence of the arboreal habit in the evolution of the reproductive system. *Lancet* 50:1113–1124.

Wood-Jones, F. 1916. *Arboreal Man,* Edward Arnold, London.

Wood-Jones, F. 1918. *The Problem of Man's Ancestry,* Society for the Promoting of Christian Knowledge, London.

Wood-Jones, F. 1919a. On the zoological position and affinities of *Tarsius. Proc. Zool. Soc. London* 1919:491–494.

Wood-Jones, F. 1919b. The origin of man, in: *Animal Life and Human Progress* (A. Denby, ed.), pp. 101–131, Constable, London.

Wood-Jones, F. 1923. *The Ancestry of Man: Man's Place Among the Primates,* Gillies, Brisbane.

Wood-Jones, F. 1924. *The Mammals of South Australia, Part 1,* British Science Guild, Adelaide.

Wood-Jones, F. 1928. Man and the anthropoids. *Amer. J. Phys. Anthro.* 12:245–252.

Wood-Jones, F. 1929a. Some landmarks in the phylogeny of the primates. *Hum. Biol.* 1: 214–228.

Wood-Jones, F. 1929b. *Man's Place Among the Mammals,* Edward Arnold, London.

Wood-Jones, F. 1940. Attainment of the upright posture of man, *Nature (Lond.)* 146:26–27.

Wood-Jones, F. 1944. *Structure and Function as Seen in the Foot,* Bailliere, Tindall and Cox, London.

Wood-Jones, F. 1948a. *The Hallmarks of Mankind,* Balliere, Tindall and Cox, London.

Wood-Jones, F. 1948b. Present state of our knowledge of the anatomy of the primates. *Brit. Med. J.* 2:629–631.

Woollard, H. H. 1925. The anatomy of *Tupaia spectrum. Proc. Zool. Soc. London* 1925: 1071–1184.

Wortman, J. L. 1903a. Classification of the primates. *Amer. J. Sci., Ser. 4,* 15:399–414.

Wortman, J. L. 1903b. Studies of Eocene Mammalia in the Marsh collection, Peabody Museum, Part II. Primates. *Amer. J. Sci., Ser. 4,* 16:345–368.

Wortman, J. L. 1904. Studies of Eocene Mammalia in the Marsh collection, Peabody Museum, Part II. Primates. *Amer. J. Sci., Ser. 4,* 17:133–140.

Wrangham, R. W. 1977. Feeding behavior of chimpanzees in Gombe National Park, Tanzania, in: *Primate Ecology* (T. H. Clutton-Brock, ed.), pp. 504–538, Academic Press, New York.

Wrangham, R. W. 1980a. Bipedal locomotion as a feeding adaptation in gelada baboons and its implications for hominid evolution. *J. Hum. Evol.* 9:329–332.

Wrangham, R. W. 1980b. An ecological model of female-bonded primate groups. *Behaviour* 75:262–300.

Wright, R. V. S. 1978. Imitative learning of a flaked stone technology—the case of the orangutan, in: *Human Evolution: Biosocial Perspectives* (S. L. Washburn and E. R. McCown, eds.), pp. 215–238, Benjamin Cummings, Menlo Park.

Wu Rukang. (Woo Ju-Kang) 1962. The Mandibles and Dentition of *Gigantopithecus, Palaeontologica Sinica,* N.S., D., No. 11.

Wu Rukang. 1981. First skull of *Ramapithecus* found. *China Reconstructs* 30(4):68–69.

Wu Rukang, Han Defen, Xu Quinghua, Lu Qingwu, Pan Yuerong, Zhang Xingyong, Zheng Liang and Xiao Minghua. 1981. *Ramapithecus* skulls found first time in the world. *Kexue Tongbao* 26:1081–1021.

Xu Quinghua and Lu Qingwu. 1979. The mandibles of *Ramapithecus* and *Sivapithecus* from Lufeng, Yunnan. *Vert. PalAsiatica* 17:1–13.

Xu Quinghua and Lu Qingwu. 1980. The Lufeng skull and its significance. *China Reconstructs* 29(1):56–57.

Yamada, M. 1963. A study of blood-relationship in the natural society of the Japanese macaque. *Primates* 4:43–65.

Yulish, S. 1970. Anterior tooth reduction in *Ramapithecus. Primates* 11:255–270.

Yunis, J. J., and Prakash, O. 1982. The origin of man: a chromosomal pictorial legacy. *Science* 215:1525–1529.

Yunis, J. J., Sawyer, J. R., and Dunham, K. 1980. Striking resemblance of high-resolution G-banded chromosomes of man and chimpanzees. *Science* 208:1145–1148.

Zapfe, H. 1958. The skeleton of *Pliopithecus (Epipliopithecus vindobonensis)* Zapfe and Hurzeler. *Amer. J. Phys. Anthro.* 16:441–458.

Zapfe, H. 1960. Die Primatenfunde aus der Miozanen Spahenfullung van Neudorf an der March (Devinska Nova Ves), *Teschechoslowakei. Schweiz. Palaeontol. Abh.* 78:1–293.

Zihlman, A. L. 1969. *Human Locmotion: A*

Reappraisal of the Functional and Anatomical Evidence, Ph.D. Dissertation, University of California, Berkeley.

Zihlman, A. L. 1977. Implications for pygmy chimpanzee morphology for interpretation of early hominids. *Amer. J. Phys. Anthro.* 47:169. (Abstract)

Zihlman, A. L. 1979. Pygmy chimpanzee morphology and the interpretation of early hominids. *S. Afr. J. Sci.* 75:165–167.

Zihlman, A. L., and Cramer, D. L. 1978. Sexual differences between pygmy *(Pan paniscus)* and common chimpanzees *(Pan troglodytes)* *Folia Primatologica* 29:86–94.

Zihlman, A. L., and Lowenstein, J. M. 1983. *Ramapithecus* and *Pan paniscus:* significance for human origins, in: *New Interpretations of Ape and Human Ancestry* (R. L. Ciochon and R. S. Corruccini, eds.) pp. 677–694, Plenum Press, New York.

Zihlman, A. L., Cronin, J. B., Cramer, D. L., and Sarich, V. M. 1978. Pygmy chimpanzee as a possible prototype for the common ancestor of humans, chimpanzees and gorillas. *Nature (Lond.)* 275:744–746. (See Chapter 43, this volume)

Zuckerkandl, E., Jones, R. T., and Pauling, L. 1960. A comparison of animal hemoglobins by tryptic peptide pattern analysis. *Proc. Natl. Acad. Sci.* 46:1349–1360.

Zwell, M. and Conroy, G. C. 1973. Multivariate analysis of the *Dryopithecus africanus* forelimb. *Nature (Lond.)* 244:373–375.

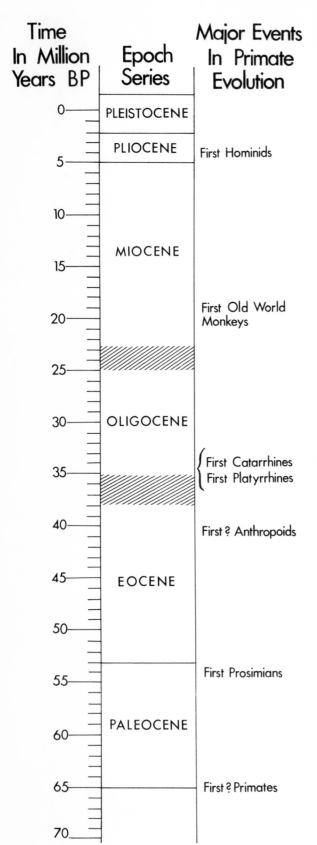

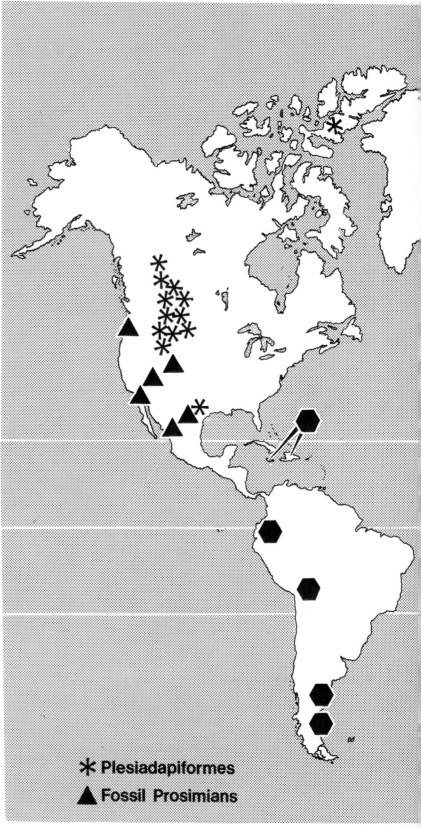

✳ Plesiadapiformes

▲ Fossil Prosimians

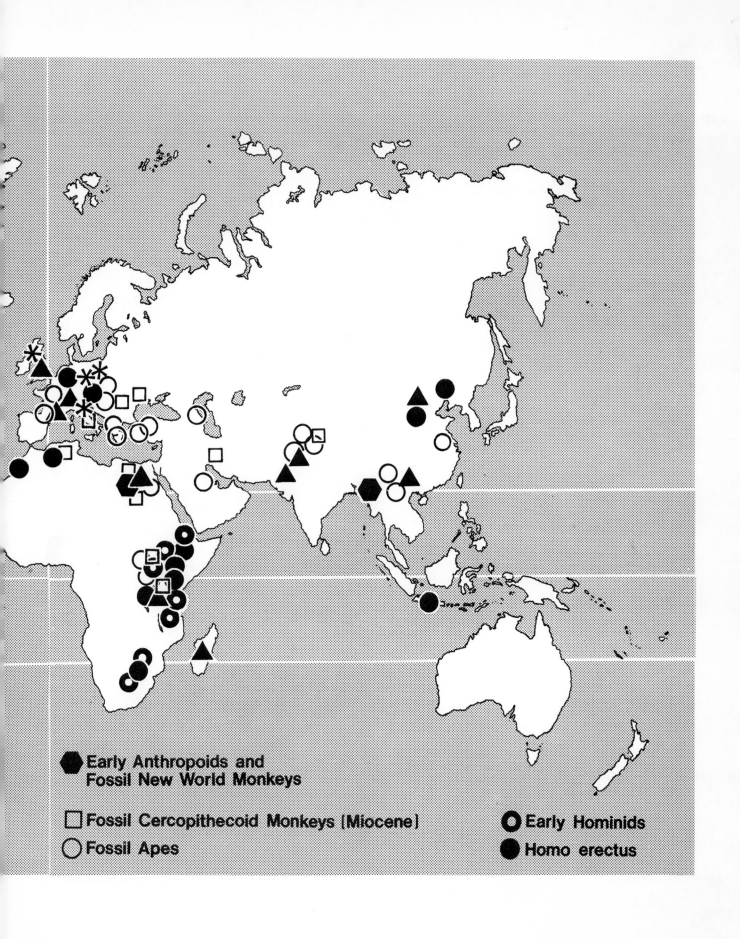

Early Anthropoids and Fossil New World Monkeys

☐ **Fossil Cercopithecoid Monkeys (Miocene)**

◯ **Fossil Apes**

◉ **Early Hominids**

● **Homo erectus**